T0207161

The lithosphere
An Interdisciplinary Approach

Modern Earth science suffers from fragmentation into a large number of sub-disciplines with limited dialog between them, and artificial distinctions between the results based on different approaches. This problem has been particularly acute in lithospheric research, where different geophysical techniques have given rise to a multitude of definitions of the lithosphere – seismic, thermal, electrical, mechanical, and petrological.

This book presents a coherent synthesis of the state-of-the art in lithosphere studies based on a full set of geophysical methods (seismic reflection, refraction, and receiver function methods; elastic and anelastic seismic tomography; electromagnetic, magnetotelluric, thermal and gravity methods; and rheological modeling), complemented by petrologic data on mantle xenoliths and laboratory data on rock properties. It also provides a critical discussion of the uncertainties, assumptions, and resolution issues that are inherent in the different methods and models. Most importantly, it discusses the relationships between methods and presents directions for their integration to achieve a better understanding of the processes that affect the lithosphere and thereby shape the Earth on which we live.

Multi-disciplinary in scope, global in geographical extent, and covering a wide variety of tectonic settings over 3.5 billion years of Earth history, this book presents a comprehensive overview of lithospheric structure and evolution. It is a core reference for researchers and advanced students in geophysics, geodynamics, tectonics, petrology, and geochemistry, and for petroleum and mining industry professionals.

Irina Artemieva is an Associate Professor at the University of Copenhagen, Denmark. She was awarded an M.S. in Physics from Lomonosov Moscow State University, a Ph.D. in Geophysics from the Institute of Physics of the Earth, Moscow and a Doctor of Science (dr. habil.) degree from the University of Copenhagen. Professor Artemieva has worked as a researcher at the Russian/USSR Academy of Sciences, Moscow, at the Universities of Uppsala (Sweden) and Strasbourg (France), and at the United States Geological Survey in California. She was Science Co-ordinator of the European Science Foundation/ International Lithosphere Programme EUROPROBE program 1999–2001. She is also a member of the Academia Europaea and a Fellow of the Royal Astronomical Society, London.

The lithosphere

An Interdisciplinary Approach

IRINA M. ARTEMIEVA

University of Copenhagen, Denmark
Also affiliated with the Institute of Physics of the Earth, Russian Academy of Sciences, Moscow, Russia

CAMBRIDGE
UNIVERSITY PRESS

CAMBRIDGE
UNIVERSITY PRESS

University Printing House, Cambridge CB2 8BS, United Kingdom

One Liberty Plaza, 20th Floor, New York, NY 10006, USA

477 Williamstown Road, Port Melbourne, VIC 3207, Australia

4843/24, 2nd Floor, Ansari Road, Daryaganj, Delhi - 110002, India

79 Anson Road, #06-04/06, Singapore 079906

Cambridge University Press is part of the University of Cambridge.

It furthers the University's mission by disseminating knowledge in the pursuit of education, learning and research at the highest international levels of excellence.

www.cambridge.org
Information on this title: www.cambridge.org/9781108448468

First published 2011
First paperback edition 2017

A catalogue record for this publication is available from the British Library

Library of Congress Cataloging in Publication data
Artemieva, I. M. (Irina Mikhailovna), 1961–
Lithosphere : an interdisciplinary approach / Irina M. Artemieva.
p. cm.
ISBN 978-0-521-84396-6 (hardback)
1. Lithosphere. 2. Earth – Crust. 3. Geodynamics. I. Title.
QE511.A78 2011
551–dc22
2011002467

ISBN 978-0-521-84396-6 Hardback
ISBN 978-1-108-44846-8 Paperback

To my teachers who became friends,
to the friends who became my teachers,
to all my family who have always been
my best friends and my best teachers.

Contents

The lithosphere is one of the most fundamental elements in solid Earth science in general, and in plate tectonics, in particular. As a result, solid Earth scientists from different backgrounds have studied the Earth's lithosphere intensively. In doing so, most studies have focused on particular areas and/or on the use of particular methodologies. Plate-tectonics, with its early breakthroughs in understanding first order patterns in sea-floor spreading controlling ocean floor bathymetry, age distribution, and horizontal motions, by its nature, has set the stage for a quantitative framework for the oceanic lithosphere.

The inferences from thermal modeling, seismological studies, and studies of the mechanical behaviour of the lithosphere, including results from marine geophysical studies of flexure of the lithosphere seaward of trenches and under seamounts, integrated with experimental studies of rock mechanics, had already led to an early realization of the need to reconcile different definitions of the lithosphere. This is even more the case for studies of the continental lithosphere, affected by a much longer geological evolution and characterized by significant heterogeneity in both its crustal and mantle components.

By now, the lithosphere is probably the best studied part of the plate-tectonics system. Seismic tomography has led to the realization that lithosphere slabs can interact with the mantle at depths much greater than had earlier been anticipated. At the same time growing awareness exists of the crucial role of the interaction of lithosphere and surface processes, including erosion and sedimentation, affected by climate.

Whereas large-scale lithosphere research programs, such as the International Lithosphere Programme, in their early studies mainly constrained crustal structure and lithosphere thickness and composition, these programs now make a major effort to integrate the study of the lithosphere in the context of connecting the deep Earth and the surface processes. The TOPO-EUROPE, EUROARRAY, EPOS, and the US EarthScope large-scale integrated research initiatives are exemplary for this. In the context of these recent developments, the interdisciplinary overview of lithosphere research developed by Irina Artemieva in this book is very timely. Of particular value here is her successful integration of an exceptionally broad suite of geophysical approaches with insights from petrology and geochemistry.

In doing so, the author, to the benefit of the readership also provides an in-depth coverage of the secular evolution of the lithosphere through all geological periods, much of which is based on her global research, including knowledge of the lithosphere in Eurasia. The book includes a comprehensive discussion of topics, including laboratory studies of physical properties of crustal and upper mantle rocks, methods in lithosphere research, and lithosphere structure in most oceanic and continental tectonic settings as constrained by a multiplicity of geophysical techniques.

As pointed out by the author, lithosphere research is at a very dynamic stage, with many fundamental questions to be further resolved. At the same time, it is to be expected that, with its very up-to-date coverage of recent progress on both oceanic and continental lithosphere and evolution, this book will serve as a reference for years to come.

Sierd Cloetingh
President, International Lithosphere Programme
Professor, VU University, Amsterdam

Preface

Recent progress in geophysical and geochemical studies has brought us far in the understanding of the structure, origin, and evolution of the lithosphere. The goal of this book is to summarize geophysical (and, to some extent, geochemical) data collected in the laboratory and in the field on the properties of the lithosphere. It reflects the state of the present understanding of the lithosphere structure and the processes that formed and shaped it. As any other book, it reflects the author's interpretations that may not necessarily be shared by other researchers. It also reflects the author's particular interests and, for this reason, the book has *a strong focus on the lithospheric mantle*, while the crustal structure is discussed in significantly less detail. The motivation for this discrimination is that, owing to historical reasons, the crustal structure is much better known and is much better understood than the structure of the lithospheric mantle. While it is universally understood that the crust is highly heterogeneous, many geophysical models still treat the lithospheric mantle as an almost homogeneous layer.

The suggestion to write a "Lithosphere" book came largely from Walter Mooney almost a decade ago. By that time I have already been much troubled by the question of how results from lithosphere studies using different geophysical techniques could be compared with each other, and how they could be compared with xenolith data. For example, if the seismic lithosphere is reported to be 220 km thick, does this estimate agree or disagree with a thermal estimate of 160 km, an MT estimate of 200 km, and a xenolith-based estimate of 180 km? Clearly, there are many assumptions behind each of the approaches, and there are theoretical and practical limitations and uncertainties associated with different techniques. It is not easy to find all of them; however, any cross-disciplinary comparison of the results would be meaningless without knowing model resolution and theoretical assumptions. Someone said that the writer is an unsatisfied reader. So my motivation in writing this book came from a wish to find the answers to questions like the one above. This is reflected in the structure of the book.

My general idea was to provide readers who specialize in a particular geophysical technique in lithospheric studies, with a simple reference book for comparison of their results with laboratory data and various geophysical–petrologic interpretations. For this reason, each chapter starts with a basic introduction for those unfamiliar with the subject. Advanced readers may skip these parts and move to an overview of laboratory data, modeling results, and interpretations. A summary of the resolution limitations and the uncertainties of different geophysical or petrological methods will facilitate understanding of how results can be compared across the various techniques used in lithosphere studies. It is not the aim of this book to teach *how to apply* these methods, but to provide information on *how to use the results* of these methods.

For this reason, the book intentionally omits mathematics and includes only very few equations; it will also facilitate reading by a large number of students. The corresponding theory can be found in other books, from general physics courses to specialized books in geodynamics, seismology, and potential fields. References to key publications are provided in the introduction to each chapter. The reference list includes the complete list of publications used in the preparation of the book. Its length reflects my wish to credit the authors of the original studies and those scientists who have pioneered different types of research on the upper 400 km of the Earth. It also gives the reader an opportunity to find the original publications and to form a personal judgment on various aspects of lithosphere research. The results of lithosphere studies are often controversial, and some of the topics are still the subject of heated debates. I have tried, as far as possible, to avoid making judgments when contradictions exist but rather to present the arguments of both sides.

The book reviews recent results in seismic, electromagnetic, gravity, thermal, rheological, and petrological studies of oceanic and continental lithosphere, followed by a chapter that focuses on the processes of lithosphere formation and evolution, and a chapter that summarizes the state of present understanding of lithosphere structure and methods used in its study. Owing to the very complex and controversial situation that exists with lithosphere definitions (part of which is due to semantics, the other part originating from the true multidisciplinarity of lithosphere research), Chapter 1 focuses on definitions of the lithosphere and related concepts such as thermal, mechanical, chemical, and rheological boundary layers, tectosphere, and perisphere.

Chapter 2 discusses applications of isotope geochronology to lithosphere studies. The structure of Chapters 3–8 is very similar. Each chapter begins with a review of laboratory data on the physical properties of the lithosphere that are the key parameters for the method discussed. This is followed by a brief summary of the method, with a major focus on basic assumptions, theoretical and practical limitations, and uncertainties. Then, field observations of lithosphere properties and their interpretation in terms of lithospheric structure and crustal and mantle processes are reviewed. Each chapter ends with a brief summary of the lithospheric structure of major tectonic provinces as constrained by the methods discussed. There is inevitably some redundancy between different chapters, since some lithospheric properties are studied using different techniques. Work on this book took several years; due to a continuous flow of new, and often contradictory results, some inconsistency may remain between the parts written at different times.

Many researchers have contributed to the writing through numerous formal and informal discussions, although most of them may have been unaware, particularly since work on the book took such a long time. I would like to acknowledge, in particular, my discussions with D. L. Anderson, N. Arndt, L. Ashwal, S. Bogdanova, G. Bokelmann, K. Burke, E. Burov, M. Cara, S. Cloetingh, K. Condie, E. Debayle, D. Eaton, Y. Elesin, K. Fuchs, A. Forte, D. Francis, A. O. Gliko, S. Grand, W. Griffin, H. Grütter, S. Haggerty, W. Hamilton, G. Houseman, C. Jaupart, A. Jones, T. Jordan, M. K. Kaban, S.-I. Karato, R. Keller, M. Kopylova, J. Korenaga, O. L. Kuskov, S. Lebedev, C.-T. Lee, A. Levander, D. Mainprice, J.-C. Mareschal, D. McKenzie, R. Meissner, W. D. Mooney, R. O'Connell, Y. Podladchikov, K. Priestly, G. Ranalli, T. Redfield, J. Ritsema, D. Scholl, N. Shapiro, A. Shulgin, F. Simons, S. V. Sobovev, W. Spakman, S. Stein, M. Talwani, H. Thybo,

T. Torsvik, R. van der Voo, L. P. Vinnik, R. Wortel, D. Yuen, V. N. Zharkov, M. Zoback. The book, however, reflects only the author's view on the subject which may differ from the opinions of other researchers, and all mistakes in the book are mine. My special thanks go to those of my colleagues who endured the pain of reading and commenting on the manuscript: W. Stratford, Y. Cherepanova, A. Frassetto, and above all H. Thybo. The friendly, professional atmosphere at USGS, Menlo Park, as well as its fantastic library were particularly important at the initial stages of work on the book. Financial support from Carlsbergfondet, Denmark, during the following years made it possible to complete the task. However, without the real long-term supporting patience of the editor, Susan Francis, the book would not have been possible. Sara Hoffritz has helped immensely with obtaining copyright permissions and I am happy to thank her. All of the figures have been prepared or redrawn by the author.

The book summarizes our present understanding of the lithosphere, its origin, structure, evolution, and impact on global mantle dynamics and plate tectonics. It shows that much remains to be understood, in particular on the path of merging geophysical and geochemical field and laboratory observations and their joint interpretations. I hope the book will be useful to those who are working in the exciting and still controversial field of multidisciplinary studies of the lithosphere.

Irina M. Artemieva
Copenhagen
July 2010

Acknowledgements

The publishers listed below are gratefully acknowledged for giving their permission to use original and redrawn figures based on illustrations in journals and books for which they hold the copyright. The original authors of the figures are cited in the figure captions, and I thank them for being able to use their figures. Every effort has been made to obtain permission to use copyrighted materials, and sincere apologies are rendered for any errors or omissions. The publishers would welcome these being brought to their attention.

Copyright owner	Journal	Figure number
American Geophysical Union	Journal of Geophysical Research	3.10, 3.11, 3.18, 3.20, 3.21, 3.22, 3.24, 3.45, 3.47, 3.49, 3.53, 3.54, 3.55, 3.56, 3.57, 3.58, 3.72, 3.77, 3.79, 3.80, 3.81, 3.84, 3.90, 3.93, 3.95, 3.97, 3.124, 3.127, 3.130, 4.11, 4.12, 4.28, 4.30, 4.41, 4.42, 4.43, 4.45, 5.19, 5.21, 6.4, 6.25, 7.5, 7.17, 7.34, 7.20, 8.32, 8.44, 8.45, 8.46, 8.55, 8.56
	Geophysical Research Letters	3.98, 3.111, 3.129, 5.20, 8.38, 8.53
	Geodynamics Series	3.14, 6.10
	Geophysical Monographs	7.20, 8.11, 8.17, 8.20, 8.23
	Geochemistry, Geophysics, and Geosystems	5.9, 6.26, 8.40
	Reviews of Geophysics	3.112, 3.114, 4.16, 4.22, 4.25, 4.33, 6.5, 8.53, 9.9
	Reviews of Geophysics and Space Physics	2.23
	AGU Reference Shelf	7.32
Geological Society of America	Geology	3.90
	GSA Today	9.2
	Tectonics	9.4
Elsevier	Earth-Science Reviews	2.2, 2.6
	Earth and Planetary Science Letters	3.5, 3.8, 3.43, 3.51, 3.105, 3.115, 3.125, 4.9, 5.10, 5.23, 6.8, 7.19, 8.21, 8.40, 8.41, 8.50, 8.61, 8.62
	Journal of Geodynamics	8.43
	Lithos	2.5, 2.7, 2.8, 2.13, 2.18, 2.20, 2.21, 3.78, 4.32, 6.15, 7.2

Copyright owner	Journal	Figure number
	Ore Geology Review	9.1
	Tectonophysics	3.59, 3.80, 3.102, 3.104, 3.119, 4.48, 4.49, 5.3, 5.14, 6.16, 6.17, 6.24, 8.12, 8.13, 8.15, 8.19, 8.26, 8.28, 8.33
	Physics of the Earth and Planetary Interior	3.7, 3.10, 3.13, 3.31, 3.62, 3.91, 6.14, 8.57
	Precambrian Research	6.8, 6.16, 6.23
	Quaternary Science Review	3.93
	Treatise on Geochemistry	2.6, 2.10
	Geochim. Cosmochimica Acta	2.11, 2.12
	Chemical Geology	2.11
Nature Publishing Group	Nature	2.9, 2.15, 2.18, 2.24, 2.25, 3.44, 3.63, 3.113, 3.116, 3.122, 9.20
The Center for American Progress (www. americanprogress. org)	Science progress	3.76
Oxford University Press	Journal of Petrology	6.6
National Research Council Canada	Canadian Journal of Earth Sciences	3.65, 3.67, 4.12
Wiley-Blackwell	Geophysical Journal International	3.83, 3.114, 4.40, 5.16, 7.5, 7.10, 8.39, 8.54, 8.56
American Association for the Advancement of Science	Science	2.4, 3.42, 3.106, 7.48, 9.23
Cambridge University Press		2.3, 3.17, 3.60, 3.61, 6.21, 7.19, 8.3, 8.16, 8.18, 8.20, 8.22, 8.31, 8.47, 8.48
Academic Press		3.66
Springer		8.6, 8.30
Princeton University Press		7.16
Geological Survey of Finland		9.3

What is the lithosphere?

"Unfortunately, the term lithosphere has recently been applied to many other concepts. The term is now in use with widely different meanings and implications."

D. L. Anderson (1995)

"How wonderful that we have met with a paradox. Now we have some hope of making progress."

Niels Bohr

The lithosphere forms the outer (typically, 50–300 km thick) rigid shell of the Earth. It includes the crust and, in general, some non-convecting part of the upper mantle called the lithospheric mantle (Fig. 1.1). Oceanic lithosphere is recycled into the mantle on a 200 Ma scale, whereas the study of the continental lithosphere is of particular importance since it offers the only possibility of unraveling the tectonic and geologic history of the Earth over the past *c.* 4 Ga. Knowledge of the structure, composition, and secular evolution of the lithosphere is crucial for the understanding of the geological evolution of the Earth since its accretion, including understanding the processes behind the formation of the early lithosphere, the processes behind plate tectonics, and lithosphere–mantle interaction. Many of these processes are closely linked to processes in the deep Earth and its secular cooling. Human society is strongly dependent on knowledge of the geodynamic processes in the lithosphere which manifest themselves as variations in topography and bathymetry, deposition of minerals many of which occur only in specific lithospheric settings, and high-impact geologic hazards. Understanding processes in the deep Earth is impossible without knowledge of lithosphere structure.

The specifics of the study of the Earth's deep interior, including the lithosphere, is that all of the parameters measured in direct and indirect geophysical and geochemical surveys are interrelated, being strongly dependent (among other factors) on pressure, temperature, composition, and the physical state of matter. This necessitates joint interpretation of the entire set of data provided using different techniques in the Earth sciences (such as seismic, gravity, thermal, electromagnetic, and petrological). Unfortunately, the true multidiscipli-narity of lithosphere research has led to a situation where numerous, and often significantly different, lithosphere definitions have emerged from different geotechniques (Fig. 1.2). This chapter makes an "inventory" and "systematizes" the definitions of the lithosphere and their relations between each other and with similar concepts.

1.1 Historical note

The early gravity studies of the eighteenth and nineteenth centuries gave birth to the first, empirically based concept of a solid, non-deformable outer layer of the Earth overlying a

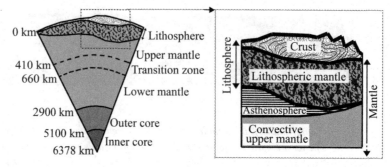

Fig. 1.1
Schematic Earth section showing various layers, not to scale (left). Zoom on the lithosphere (right).

Fig. 1.2   Confusion with various and numerous lithosphere definitions sometimes results in comparison of dissimilar parameters.

fluid, deformable interior. The thickness of this rigid layer, which supports sedimentary loads, was calculated from an analysis of gravity anomalies providing a value of about 100 km. Historically, the term "lithosphere" (from the Greek word meaning "rocky") appeared in geological literature probably only in the late nineteenth to early twentieth century. The first use of the term apparently dates back to the publication of J. Dana (1896), where in contrast to hydrosphere the solid Earth was described as "lithosphere". Soon afterwards, Barrel (1914) introduced the term "asthenosphere" to describe a fluid, deformable layer (estimated to be several hundred kilometers thick) below the "lithosphere".

The booming development of seismological methods in the first half of the twentieth century resulted in the discovery of the seismic low-velocity zone, the LVZ, in the Earth's mantle "centered at 100–150 km depth" (Gutenberg, 1954) with its base, the Lehmann discontinuity, determined to be at c. 220 km depth (Lehmann, 1961, 1962). [The existence of the Lehmann discontinuity had apparently been proposed almost a half-century earlier by the Russian Prince Boris Golitsyn (Galitzin, 1917), one of the founders of seismology and the inventor (1906) of the first electromagnetic seismograph.] Amazingly, the top of the seismic LVZ was at approximately the same depth as the transition between the solid outer layer of the Earth (the "lithosphere") and its low-viscous, deformable interior (the "asthenosphere"), as determined in early isostatic gravity studies. It was tempting to explain both gravity and seismic observations using the same physical mechanism, and the term "lithosphere" attained a new, seismic justification.

In the 1950s geothermics developed as an independent technique to assess the physical state of the deep Earth's interior. Although the first measurements of terrestrial heat flow were initiated by Everett (1883), the first reliable measurements only appeared in the late 1930s for the continents and a decade later for the oceans (Benfield, 1939; Bullard, 1939; Krige, 1939; Revelle and Maxwell, 1952). Data on surface heat flow allowed the calculation of crustal and upper mantle geotherms for a large range of continental and oceanic regions (e.g. Jaeger, 1965; McKenzie, 1967). These geotherms, combined with experimental and theoretical studies of melting conditions in the upper mantle (Uffen, 1952; Ito and Kennedy, 1967; Kushiro *et al.*, 1968), were used to explain the seismic LVZ in terms of peridotite partial melting. Thus, the term "lithosphere" obtained one more interpretation. It was an important step forward in upper mantle research because it linked observations, that came from independent fields of geophysics, into a joint picture of the physical state of the upper mantle.

Further development of geophysical techniques and a continuing accumulation of extensive geophysical (seismic, thermal, electromagnetic, gravity, and rheological) and geochemical data sets and models have led to much controversy in the use of the term "lithosphere", which had gradually become a convenient and widely used concept in geosciences. A drawback of the popularity of the term is that, depending on the geophysical techniques employed and physical properties measured, the lithosphere has different practical definitions. This has led to the situation where the "lithosphere" was stated to have "become an unnecessarily confusing concept" (Anderson, 1995) due to its excessive number of meanings. On the positive side, however, most lithosphere definitions are based on a pronounced change in the temperature-dependent physical properties of the upper mantle that occur at the transition from conductive (and rheologically strong) to convecting (and rheologically weak) upper mantle. They are supported by laboratory measurements of density, elastic moduli, and electrical conductivity of mantle rocks, parameters that have a strong temperature-dependence and a sharp change in properties at temperatures close to solidus (Murase and Fukuyama, 1980; Sato *et al.*, 1989; Constable *et al.*, 1992). A crucial dependence of most lithosphere definitions on the thermal regime of the upper mantle provides physical grounds for their direct comparison. Various lithosphere definitions are discussed in detail in the following sections.

1.2 Lithosphere definitions

1.2.1 Defining the lithospheric base

The lithosphere, as introduced at the beginning, is the outer layer of the Earth which includes the crust and, in general, the lithospheric mantle (Fig. 1.1). While the top of the lithosphere, obviously, coincides with the topographic surface, it is not an easy task to define the lithospheric base. The multiplicity of the existing practical definitions of the lithospheric base is related to:

- highly heterogeneous (both laterally and vertically) lithosphere structure;
- multiplicity of (geo)physical and (geo)chemical parameters by which it can be defined;
- multiplicity of methods that can measure these parameters;
- transitional (diffuse) nature of the lithospheric base;
- dualism in the lithosphere nature with respect to deep, global, and plate tectonic processes.

Changes in physical properties of mantle rocks measured in indirect geophysical surveys provide the basis for various definitions of the base of the lithosphere. The existing "lithosphere" definitions differ significantly, depending on the parameter under consideration. Furthermore, even the same parameter (for example, seismic velocity) will be measured with a significantly different resolution by different seismic techniques; physical assumptions and mathematical simplifications used in data interpretations often lead to significantly different practical "definitions" of the lithospheric base. For example, the thickness of seismic lithosphere constrained by seismic tomography can be biased by the choice of regularization procedure (smoothing of amplitudes versus gradients of velocity anomalies), the choice of the reference model, and by interpretations in terms of relative versus absolute velocity perturbations.

Most properties of the upper mantle change gradually with depth and do not exhibit sharp, "knife-cut", boundaries that could be uniquely associated with the lithospheric base. For this reason, the lithosphere–asthenosphere boundary (LAB) has a diffuse nature and, regardless of the definition employed, is always a transition zone over which a gradual change in physical and chemical characteristics occurs. As a result, the lithosphere thickness (the depth to the LAB) can be significantly different even for the same model of physical parameter variation with depth. The choice of a 1% or 2% velocity perturbation in a seismic model as the lithospheric base may lead to a ~50–100 km difference in the LAB depth (see Fig. 3.120). Similarly, defining the base of the mechanical lithosphere by the critical value of strength, strain rate, or viscosity leads to a large difference in the estimates of LAB depth (see Fig. 8.34).

The LAB is an important global boundary that has a dual nature, since it reflects the processes related to both global evolution (such as global mantle differentiation which led to crustal and lithosphere extraction; secular cooling of the Earth; styles and patterns of global mantle convection) and plate tectonics (such as lithosphere generation, recycling and modification by plate tectonics processes and secondary mantle convection) (Fig. 1.3). The duality in the nature of the lithosphere–asthenosphere boundary leads to significant differences in the LAB definitions from the "bottom" and from the "top". These definitions are discussed in detail in the corresponding chapters (see Sections 3.3.2; 3.6; 4.4.3; 5.1.2; 5.2.3–5.2.4; 6.1.4–6.1.5; 7.6; 8.1.3; 8.2.2); the correlations between some of them are shown in Fig. 1.4.

Four definitions of the lithosphere, *elastic, thermal, electrical, and seismic*, are widely used in geophysical studies. As is clear from their names, they are based on indirect

Dual nature of LAB
Lithosphere–asthenosphere boundary reflects:

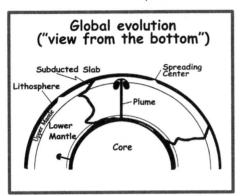

Global evolution ("view from the bottom")

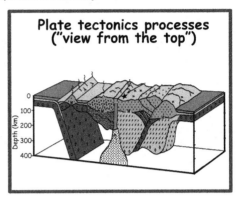

Plate tectonics processes ("view from the top")

Sketch illustrating the dual nature of the lithosphere–asthenosphere boundary (LAB). The LAB is an important global boundary that reflects the processes related to global evolution (mantle differentiation, secular cooling, global mantle convection, etc.) and to plate tectonics (the Wilson cycle, small-scale mantle convection, etc.). This duality causes difference in LAB definitions from the "top" and from the "bottom".

measurements of different properties of upper mantle rocks and thus they may refer to outer layers of the Earth of significantly different thickness. Petrologic studies of mantle xenoliths and interpretations of geophysical data in terms of chemical variations in the upper mantle provide additional definition of the *petrologic lithosphere*.

1.2.2 Elastic lithosphere

The concept of *elastic (flexural) lithosphere* (an elastic plate overlying viscous mantle) is close to the "classical" definition of the lithosphere as the upper rigid layer that moves mechanically coherently with plate motion (Barrell, 1914). The elastic thickness of the lithosphere is a measure of the mechanical strength of an elastic plate and it considers the lithosphere as a rheologically strong layer that provides the isostatic response of the plate to topographic and/or subsurface loads (i.e. due to variations in crustal thickness and density) and mechanically supports elastic stresses induced by lithospheric bending (flexure). In practice, the elastic thickness of the lithosphere is determined from the correlation between the topography and gravity anomalies (Fig. 1.5).

Although the concept of elastic lithosphere is related to the mechanical strength of the plate, the relationship between the two characteristics is not straightforward, and depends on lithosphere rheology, temperature, strain rate, and plate curvature. The mechanical thickness of the lithosphere (which can be defined as the depth where the yield stress becomes less than a particular limiting value) is, in general, greater than the thickness of the load-bearing lithospheric layer (the "elastic core") and the two parameters are only equal when the whole

Defining the base of the lithosphere and of the boundary layers

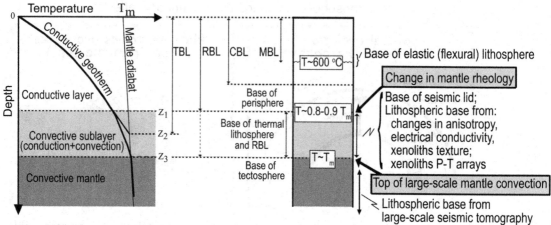

Fig. 1.4 Sketch illustrating relations between the conductive boundary layer and the convective mantle, on one side, and various approaches to define the base of the lithosphere, on the other side. The layer above depth Z_1 has a purely conductive heat transfer; in the transitional "convective boundary layer" between depths Z_1 and Z_3 the heat transfer mechanism gradually changes from convection to conduction. The base of the conductive boundary layer (or TBL) is between depths Z_1 and Z_3. Z_2 corresponds to the depth where a linear downward continuation of the geotherm intersects with mantle adiabat T_m that is representative of the convective mantle temperature profile. Thermal models commonly estimate Z_2, while large-scale seismic tomography images Z_3. The difference between Z_2 and Z_3 can be as large as 50 km, leading to a significant systematic difference in lithosphere thickness estimates based on seismic tomography and thermal data. Most practical definitions (except for chemical boundary layer and perisphere) are based on temperature-dependent physical properties of mantle rocks, and many lithosphere definitions correspond to the depth where a dramatic change in mantle rheology (viscosity) occurs. Layers RBL, TBL, CBL, and MBL are rheological, thermal, chemical, and mechanical boundary layers. Vertical dimensions are not to scale.

lithosphere thickness supports the load, such as in case of flexure of a single-layer elastic lithospheric plate (Sections 8.1.3 and 8.2.2).

The definition of an elastic (flexural) lithosphere implies that the lithosphere has elastic rheology and thus, from a rheological point of view, the base of the elastic lithosphere, in its simplest interpretation, corresponds to the elastic–plastic transition (Bodine *et al.*, 1981). The latter occurs at temperatures ranging from 250–450 °C for crustal rocks to 600–750 °C for upper mantle rocks (e.g. Meissner and Strehlau, 1982; Chen and Molnar, 1983). Thus the thickness of the elastic lithosphere is approximately one half those of the thermal, seismic, and electrical lithospheres, the bases of which are effectively controlled by lithospheric temperatures close to mantle adiabat (or to mantle solidus) (Fig. 1.4).[1]

---

[1] In a multicomponent geochemical system such as the Earth's mantle, the constituting minerals have different melting temperatures. Solidus temperature corresponds to the temperature when melting of the most fusible component starts, while at liquidus temperature all minerals, including the most refractory, are molten. Between the liquidus and solidus temperatures, the material consists of both solid and liquid phases.

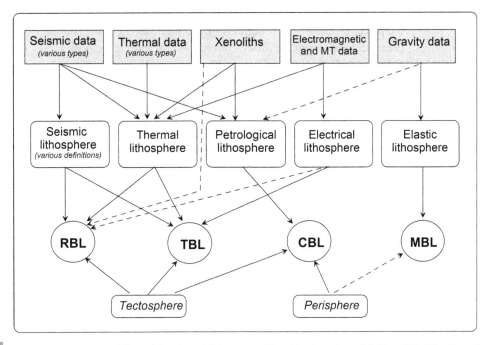

Fig. 1.5 Correspondence between different lithosphere definitions and links with other related definitions. RBL, TBL, CBL, and MBL are rheological, thermal, chemical, and mechanical boundary layers. See discussion in text.

1.2.3 Thermal lithosphere

Most practical geophysical definitions of the base of the lithosphere are based on temperature-dependent physical properties of mantle rocks measured indirectly in geophysical surveys (Fig. 1.5). From this point of view, the definition of the *thermal lithosphere (or TBL –* the layer with dominating conductive heat transfer above the convecting mantle) is the most straightforward and the least ambiguous. The base of the thermal lithosphere is commonly defined either by the depth to a constant isotherm (e.g. 1300 °C), or by the depth where a linear downward continuation of the geotherm reaches some pre-defined fraction of the ambient mantle temperature or mantle solidus (Fig. 1.4). The choice of ~0.8–0.9 times the mantle solidus (Pollack and Chapman, 1977) is supported by laboratory studies of physical properties of mantle rocks at high P–T conditions which indicate a sharp change in rheology and elastic properties of olivine-rich rocks at temperatures ~0.85–1.0 times that of solidus temperature (e.g. Sato and Sacks, 1989; Sacks *et al.*, 1989). The choice of mantle temperature 0.9–1.0 times that of the adiabatically upwelling mantle as the base of the thermal lithosphere is the most common; the potential temperature of the mantle is usually assumed to be in the range 1250–1350 °C since higher potential temperatures (1400–1450 °C) are inconsistent with laboratory data on P–T conditions at the 410 km depth (Katsura *et al.*, 2004).[2]

---

[2] The potential temperature is the temperature of the mantle if it were decompressed adiabatically to the surface.

In practice, lithospheric geotherms are constrained by surface heat flow, from pressure–temperature equilibrium conditions of mantle mineral phases constrained by xenolith data (xenolith geotherms), or from conversions of seismic velocities into temperatures (see Chapters 4–5). For example, in stable continental regions, where thermal equilibrium has been established since the last major thermo-tectonic event, lithospheric geotherms can be calculated from solution of the steady-state conductivity equation constrained by surface heat flow measurements assuming that the depth distribution of heat producing elements is known. However, the approach is invalid for tectonically active regions, where non-steady-state thermo-dynamical models should be used to calculate the lithospheric geotherms (strictly speaking, heat transfer in the cratonic lithosphere also is not steady state because the time-scale of radioactive decay for major heat producing elements is comparable with the time-scale of thermal diffusion through the thick continental lithosphere, Michaut and Jaupart, 2004).

Boyd (1973) proposed the use of mantle-derived xenoliths to constrain lithospheric geotherms (xenolith P–T arrays) and thus to estimate the lithospheric thermal thickness (Fig. 1.5). This approach is based on the calculation of equilibrium P–T conditions for different mineral systems. Owing to a significant discrepancy between mantle pressures and temperatures constrained by different sets of geothermobarometers (sometimes reaching ±10 kBar and ±100 °C for the same rock samples, Grutter and Moore, 2003), the thickness of the thermal lithosphere based on xenolith P–T arrays is constrained non-uniquely and its value should not necessarily agree with lithospheric thickness constrained by heat flow data.

A strong temperature-dependence of elastic and non-elastic behavior of mantle rocks at high P–T conditions (e.g. Duffy and Anderson, 1989; Jackson *et al.*, 1990; Karato, 1993; Karato and Spetzler, 1990) provides the basis for conversions of seismic velocities to upper mantle temperatures, and thus allows for independent estimates of the thickness of the thermal lithosphere (Fig. 1.5). Most of the recent models of this type take into account frequency-dependent anelastic effects (dissipation) (Karato, 1993; Faul and Jackson, 2005). However, large uncertainty remains with taking into account the effects of partial melt (which strongly depend on melt geometry and water presence), fluids, seismic anisotropy, and compositional variations on seismic velocities (e.g. Popp and Kern, 1993; Sobolev *et al.*, 1996; Deschamps *et al.*, 2002; McKenzie *et al.*, 2005). Furthermore, velocity–temperature conversions can lead to physically meaningless solutions with extremely short-wavelength contrasts in mantle temperatures: while seismic velocities that form the basis for velocity–temperature conversions can exhibit a sharp velocity contrast between two adjacent lithospheric blocks, horizontal heat transfer that has a smearing effect on lithospheric temperatures should reduce short-wavelength temperature contrasts. Despite model limitations, velocity–temperature conversions have been applied to high-resolution regional seismic tomography and seismic refraction models to estimate lithospheric temperatures in most continents (Chapter 5). These studies provide useful constraints on lithospheric geotherms and the depth to thermal LAB.

1.2.4 Seismic lithosphere

Changes in elastic and non-elastic properties of mantle rocks measured in indirect seismological studies provide the basis for definitions of the base of the seismic lithosphere.

"Seismic lithosphere" is one of the most diverse categories of lithosphere definitions (Chapter 3). According to the classical definition, the *seismic lithosphere*, or the lid, is a seismic high-velocity layer above the low-velocity zone (LVZ) or above a zone of high velocity gradient in the upper mantle, presumably caused by partial melting. Other mechanisms (high-temperature relaxation, contrast in volatile content or in grain size) can also explain the origin of the LVZ and imply that the base of the lid is diffuse and extends in depth over some tens of kilometers.

Interpretations of mantle layers with a negative seismic velocity contrast in terms of the lithospheric base require a word of caution. When seismic models are constrained in relative instead of absolute velocities (i.e. as a velocity perturbation with respect to a reference model), the seismic LVZ may be an artifact of the reference model. This is the case, in particular, when the PREM (preliminary reference Earth model) model (with a sharp velocity increase at 220 km depth, typical for the oceanic lithosphere, but not required by seismic data for the continental regions) is used as a reference model for stable continents (Kennett, 2006; Fig. 3.85). Another confusion arising from the definition of the seismic lid is related to the fact that LVZ can be produced without involving mantle melting: a velocity decrease with depth may happen when the seismic velocity increase due to pressure increase is less than the velocity decrease due to temperature increase with depth.

High-temperature laboratory measurements of density and elastic moduli of mantle rocks that indicate a sharp change in seismic velocities and seismic attenuation at temperatures close to mantle solidus provide the basis for interpretations of elastic and non-elastic mantle tomography in terms of lithospheric thickness. In such approaches, the seismic LAB is identified with a (to some extent, arbitrary) critical value of seismic velocity or attenuation. It is important to remember that the existence of the transitional layer in the mantle, where the mode of heat transfer gradually changes from conduction to convection, leads to a significant difference in lithospheric thickness constrained "from the top" and "from the bottom" (Fig. 1.4). Large-scale tomography models, which are sensitive to velocity anomalies associated with the convective mantle, treat the base of the lithosphere as the top of a large-scale mantle convection (Sleep, 2003). Thus, the lithospheric thickness estimated from large-scale seismic tomography models may extend some tens of kilometers deeper than the thermal lithosphere (or TBL).

If the seismic velocity decrease with depth is associated with a sharp decrease in upper mantle viscosity (e.g. due to the presence of partial melt), the lithospheric base should be a rheological boundary. Asthenospheric flow and shear in the rheologically weak layer below the lithosphere may produce alignment of minerals, including highly anisotropic olivine. As a result, the depth at which the axis of anisotropy changes orientation from fossil, frozen-in anisotropy typical of the lithospheric mantle to the anisotropy pattern related to the present plate motions and mantle flow typical for the asthenosphere (convective mantle) can also be considered as the base of the seismic lithosphere (Sections 3.1.5, 3.6.4). Similarly, a change in mantle anisotropy associated with the change in the mechanism of mantle deformation from dislocation creep to diffusion creep can be interpreted in terms of the seismic LAB, which in this case corresponds to the base of the rheological boundary layer.

1.2.5 Electrical lithosphere

The *electrical lithosphere* is usually defined as the highly resistive upper layer above the highly conducting asthenosphere. Its base corresponds to a sharp change in mantle conductivity (resistivity) and is commonly explained by the presence of 1–3% melt fraction (although the presence of a high-conducting phase such as graphite may produce a similar effect). In many regions the depths to the top of the asthenospheric conductor and to the top of the seismic LVZ are well correlated. Similarly to seismic velocity–temperature conversions, the strong temperature-dependence of the electrical conductivity of olivine can be used to estimate regional geotherms from magnetotelluric field observations (Ledo and Jones, 2005).

Anisotropic diffusivity of hydrogen in olivine crystals produces direction-dependent electrical conductivity. Thus, if olivine is partially aligned by asthenospheric flow, the depth where a change in mantle electrical anisotropy is observed can be interpreted as the electrical LAB. Since asthenospheric flow also produces lattice-preferred orientation of olivine crystals leading to seismic anisotropy, a joint analysis of seismic and electrical anisotropy data can provide important constraints on asthenospheric flow at the lithospheric base and the base of the rheological boundary layer.

1.2.6 Petrologic lithosphere

Clearly the most straightforward way of defining the base of the petrologic lithosphere is provided by mantle-derived xenoliths. For example, the base of the petrologic lithosphere can be determined by the change from depleted (lithospheric) to undepleted (asthenospheric) composition. Numerous petrologic studies of xenoliths from the Archean–Proterozoic cratons of Africa (Kaapvaal, Zimbabwe, and Tanzania cratons), Siberia, Europe (Baltic Shield), Canada (Superior and Slave cratons), and South America (Guyana craton) indicate that the depth of the transition from depleted to undepleted mantle composition often coincides with the depth to the isotherm of *c.* 1200–1300 °C and thus is close to the depth where a continental conductive geotherm intersects the mantle adiabat. Thus, in some cratonic regions, the petrologic and thermal lithospheres may have similar thicknesses (Chapter 5). However, this is not necessarily the general case, since in many cratons the transition from depleted to non-depleted composition is gradual, and in some cratons (e.g. the Slave craton in northern Canada) the lithosphere has compositional layering. In some petrologic definitions (such as those based on Y-depleted garnet concentrations (Griffin and Ryan, 1995)), the base of the lithosphere is a strongly non-isothermal boundary with temperatures at the lithospheric base (strictly speaking, at the base of the CBL) ranging from 950 °C to 1250 °C over lateral distances of 100–200 km (Griffin *et al.*, 1999a). These huge short-wavelength temperature variations at the base of the CBL preclude a physically meaningful comparison of the thus defined "petrological lithospheric thickness" with lithospheric thickness estimated by geophysical methods (e.g. seismic, magnetotelluric, Chapters 3, 7), in which the base of the lithosphere (or the LAB) is defined by a sharp change in temperature-dependent physical properties of the upper mantle. The base of the geophysically defined lithosphere should not necessarily be isothermal, since anomalies in mantle composition and the presence of fluids affect the physical properties of the lower-most lithospheric mantle measured in geophysical surveys. Stress and basal drag in the lower lithosphere can further enhance the non-isothermal structure (e.g. Artemieva and Mooney, 2002; Garfunkel, 2007).

Geophysical data can also be used to define the base of the petrologic lithosphere: variations of seismic velocity and density are caused not only by temperature variations in the upper mantle (and temperature-induced mantle melting), but also by compositional changes. Although the compositional effect on velocity variations in the lithospheric mantle is small, laboratory studies indicate that the Vp/Vs ratio is more sensitive to anomalies in iron-depletion than to temperature variations (Lee, 2003). Thus the depth in the upper mantle where a change in the Vp/Vs value occurs can be interpreted as the base of the *petrologic lithosphere* (Fig. 1.5). Similarly, gravity data interpreted in terms of density variations caused by compositional variations can be used to determine the petrologic lithosphere. However, in contrast to seismic data, gravity anomalies represent the integrated effect of density variations throughout the crust and the upper mantle and thus cannot be reliably associated with any particular depth.

1.3 Concepts related to the lithosphere

1.3.1 Boundary layers

Four boundary layers (i.e. the layers in the immediate vicinity of the upper bounding surface) are closely related to the concept of the lithosphere. They include the mechanical, rheological, thermal, and chemical boundary layers (MBL, RBL, TBL, and CBL, correspondingly) (Figs. 1.4, 1.5).

The classical definition of the "*lithosphere*" used in plate tectonics (as the upper rigid layer that moves mechanically coherently with plate motions) is close to the *mechanical boundary layer* (MBL). However, the elastic lithosphere is, in general, thinner than the MBL. The base of the MBL is associated with the transition (decoupling) zone from the upper layer dominated by brittle deformation to the underlying mantle dominated by plastic (viscous) deformation (Chapter 8). It approximately corresponds to the maximum depths of midplate and fracture zone earthquakes and should be close to the depth that the elastic–plastic transition occurs. The temperature range 550–750 °C, characteristic of the elastic–plastic transition in mantle rocks, also bounds isotherms corresponding to the maximum depth of seismicity. Because of the exponential decrease in mantle viscosity with increasing temperature, the mechanical boundary layer, which is effectively isolated from mantle convection, is thinner than the thermal boundary layer.

The *rheological boundary layer* (RBL) is another concept closely related to upper mantle rheology. Although MBL and RBL refer to significantly different depths in the upper mantle, they are sometimes confused since both depend on mantle rheology. In contrast to the MBL which extends down to the ~650 ± 100 °C isotherm, the base of the

RBL is commonly interpreted as being associated with mantle zones of reduced viscosity and asthenospheric flow, i.e. it corresponds to temperatures close to the mantle solidus of ~1100–1400 °C. Mantle zones of reduced viscosity cause alignment of minerals, such as highly anisotropic olivine, and they are thus associated with changes in seismic and electrical anisotropy (although, interpretations of the origin of electrical anisotropy are less unique due to the high sensitivity of mantle conductivity to the presence of high-conducting phases). Mantle flow also causes variation in the texture of xenoliths brought to the surface from different depths in the mantle, and in petrologic studies the base of the RBL can be constrained by the transition from non-sheared to sheared xenoliths (Nixon and Boyd, 1973). The thickness of the transition zone can reach 20–25 km (Garfunkel, 2007).

The *thermal boundary layer* (TBL) is the layer with dominating conductive heat transfer located above the convecting mantle. The thickness of the TBL should not necessarily coincide with the thickness of the thermal lithosphere: a gradual transition from convection to conduction mechanisms of heat transfer in the mantle leads to the existence of a depth interval where mantle temperature asymptotically approaches the mantle adiabat. In practice, a linear continuation of the geotherm in this depth interval to its intersection with the mantle adiabat is commonly used to estimate the lithospheric thermal thickness (Fig. 1.4). Since mantle viscosity is strongly temperature-dependent, the thickness of the RBL should be proportional to the thickness of the TBL, unless weak mantle rheology is caused by the presence of fluids. In contrast, the thickness of the MBL is only approximately one half of the TBL thickness.

Depth variations of mantle composition sampled by mantle-derived xenoliths provide information on the thickness of the *chemical boundary layer* (CBL), which is similar to the *petrologic lithosphere*. The concept of CBL is of particular importance in studies of stable continental (cratonic) regions (see the discussion on the tectosphere hypothesis below), although CBL plays an important role in stabilizing the thickness of the TBL both within the oceanic and the continental lithosphere (Lee *et al.*, 2005) (Chapter 6).

1.3.2 Perisphere and tectosphere

According to Anderson (1995), the lithosphere should be understood only in its classical sense as the mechanically strong layer that is approximately half the thickness of the TBL. Two more concepts, the tectosphere and the perisphere, have been introduced as alternatives to the classical concepts of the lithosphere and asthenosphere. Although they are somewhat close in definition to the CBL, they differ significantly from the petrologic lithosphere as defined by compositional variations in cratonic xenoliths.

The term *"perisphere"* is primarily a geochemical concept that has been introduced to describe *"a global, shallow, enriched reservoir or boundary layer"* (Anderson, 1995). The enriched mantle reservoir is likely to be shallow because melts from it "are most evident at new or slowly rifting regions, infant subduction zones, new backarc basins, slab windows, and midplate environments away from spreading induced upwelling". From the chemical point of view, "old lithosphere" is chemically "enriched" compared to the "depleted" upper mantle (the latter is also generally called the convecting upper mantle or asthenosphere). The "perisphere" is a hot, mechanically weak, but chemically distinct (enriched) sublithospheric layer (which may also include the lower part of the TBL since, according to Anderson (1995), the "lithospheric base" is defined by a $650 \pm 100\,°C$ isotherm); it is physically isolated from the depleted mantle and spreads across the top of the convecting mantle. Being a permanent part of the sublithospheric shallow mantle, the perisphere is an open system that is constantly refreshed by recycling; this makes a possible reservoir for continental flood basalts, while a delaminated continental lithosphere makes a source for ocean-island basalts and enriched mid-ocean ridge basalts.

A systematic difference between continents and oceans, persisting well below the seismic low-velocity zone down to at least 400–500 km depth, was found in the early 1970s from the analysis of surface-wave dispersion curves. This discovery led Jordan (1975a) to the "*tectosphere*" hypothesis. According to this hypothesis, the tectosphere is the outer layer, at least 400 km thick, (a) which moves mechanically coherently during horizontal plate motions, (b) is dominated by conductive heat transfer, and (c) is compositionally inhomogeneous so that cold cratonic regions should have a composition characterized by intrinsically lower densities to compensate for the gravitational effect of low temperatures (Jordan, 1978). Thus, the tectosphere concept equates the chemical boundary layer with the thermal boundary layer (the isopycnic, or equal-density, hypothesis). The existence of systematic variations in the composition of the lithospheric mantle has been confirmed by numerous petrologic studies of mantle xenoliths from different continental locations (e.g. Boyd and McCallister, 1976; Boyd, 1989; Jordan, 1988, 1997; Griffin *et al.*, 1998a; Boyd *et al.*, 1999; Lee and Rudnick, 1999) which indicate that depletion of the cratonic mantle in basaltic components (primarily Fe and Al) results in a systematic increase in the upper mantle density from the Archean cratons to Phanerozoic regions. However, the lithospheric mantle may not necessarily be perfectly isopycnic. In such a case, the bases of the CBL and the TBL need not coincide precisely.

1.4 An unnecessarily confusing concept?

The brief summary of lithosphere definitions presented in this chapter clearly shows the reasons why Anderson (1995) considered the term *lithosphere* "an unnecessarily confusing concept". A significant part of this confusion is caused not only by the existence of different lithosphere definitions, but also by the fact that the very same terms are used in approaches that utilize different techniques (Fig. 1.5). These techniques often assess different physical properties of mantle rocks and may therefore refer to different phenomena and to different depth intervals in the upper mantle. A good example is the "*thermal lithosphere*". Since continental geotherms constrained by surface heat flow measurements, models of mantle convection, xenolith P–T arrays, and seismic velocity-to-temperature conversions may disagree outside the uncertainty ranges for these methods, the thermal lithosphere defined by these approaches may differ significantly.

In contrast to Anderson (1995), one may argue that the concept of the lithosphere is "necessarily confusing" because of its complexity and its multi-disciplinary nature. This implies that any discussion of the lithospheric structure (including lithospheric thickness) is misleading without a proper specification, which should include a clear definition of the considered physical (or chemical) properties of the upper mantle, the techniques and approaches used to quantify them, and how the lithospheric base is defined. Throughout the book, the term "lithosphere" is used to define the outer, rheologically strong layer of the Earth that includes the crust and the uppermost (lithospheric) mantle and moves mechanically coherently with the lithospheric plates. The set of physical properties of the upper mantle, which defines the lithosphere, is specified for each interpretation technique discussed in the book. Since use of the concept of boundary layers is less ambiguous than using various lithosphere definitions, this is preferred where possible.

2

Age of the lithosphere

This chapter presents basic information on the age of the crust and the lithospheric mantle and provides a brief description of geochronological methods used in the geosciences. Full descriptions and technical details of these methods can be found in monographs by Faure (1986) and Dickin (1995) and in the recent reviews by Banner (2004) and Patchett and Samson (2004). Additional details can be found in Albarède, 2003, while a brief description of practical approaches is given in Fowler (2004).

2.1 Introduction to isotope geochronology

2.1.1 Geochemical classification of elements

Lithophile, siderophile, and chalcophile elements

The most widespread geochemical classification of elements has been proposed by Victor Moritz Goldschmidt (e.g. Kaufman, 1997). In Goldschmidt's classification, the chemical elements are grouped according to their preferred host phases into four categories:

- lithophile (silicate loving) elements,
- siderophile (iron loving) elements,
- chalcophile (sulfur or copper loving) elements, and
- atmophile (gas loving) elements.

Some elements have affinities to more than one phase and belong to several categories. *The lithophile elements* concentrate in rock-forming minerals of the lithosphere and the mantle. They include mainly the highly reactive metals (Na, K, Al, Mg, Ca), transition metals (Ti, Zr, V, Hf), and a small number of reactive nonmetals (Si, Cl, O, P). Large-ion lithophile elements which include large ions with weak charges (K, Rb, Cs, Ba) form a subclass of the lithophile elements. These elements cannot be effectively accommodated in the main mantle minerals (except for K-feldspar) and for this reason are concentrated in the continental crust. Most lithophile elements have a strong affinity for oxygen and associate very strongly with silica, forming relatively low-density silicate minerals. For this reason, lithophile elements are abundant in the crust, and concentrations of many of them are higher on Earth than in the solar system. However, the overall concentration of nonmetallic lithophile elements (P and the halogens) on Earth can be significantly reduced compared to their solar abundances.

Several transition metals (Cr, Mo, Fe, Mn) have both lithophile and *siderophile* characteristics. Similarly to lithophile elements, these metals form strong bonds with oxygen and do not exist in the Earth's crust in the free state. However, most of the other siderophile elements (including Re, Os, Au, Ni, Co) have almost no affinity for oxygen, but form relatively weak (as compared to the chalcophile elements) bonds with carbon or sulfur and very strong, metallic bonds with iron. As a result, the siderophile elements do not concentrate in the siliceous crust where they are rare. Metallic forms of these elements are expected to be present in the core, and due to high concentration of the siderophile elements in the mantle and core, their abundances on Earth are thought to be close to their solar abundances.

Chalcophile elements include metals (Cu, Ag, Zn, Pb, Hg) and heavier nonmetals (S) that prefer to bond with sulfur as highly insoluble sulfides but have a low affinity for oxygen. Zinc can form quite strong bonds with oxygen and is close to lithophile elements. Since sulfides are much denser than silicates, during early crustal crystallization chalcophile elements differentiated below the lithophile elements. As a result, the crust is depleted in chalcophile elements, but to a lesser degree than in siderophile elements. However, some of the chalcophile elements, that form volatile hydrides, are strongly depleted on Earth relative to cosmic abundances.

Atmophile elements (O, N, H, C, and the inert gases) exist chiefly or exclusively in the form of gases or form volatile hydrides. Because the atmophile elements were lost from the atmosphere during the formation of the Earth, they are strongly depleted on Earth relative to their solar abundances.

Melting, partitioning, and depletion

Geochemical classifications of the elements are based on their behavior during differentiation processes, such as partial melting or fractional crystallization of magma. Partition coefficients are used to describe the processes of substitution of minor or trace elements in the mineral lattice during differentiation processes for minerals that are in equilibrium with magmatic fluids or natural solutions. Partitioning behavior is characterized by the *compatibility* coefficient (mineral/melt partition coefficient) or its inverse, *incompatibility*, which is the element's tendency to go into silicate melt versus its affinity for crystalline silicate phases.

The most important parameters that determine the partitioning behavior of an element are its ionic size and charge. Elements that are unsuitable in size and/or charge to the cation sites of the rock-forming minerals are concentrated in the melt phase of magma (liquid phase) and are termed incompatible elements. For these, the partition coefficient between rock-forming minerals and melt is much smaller than one. There are two major groups of *incompatible elements*:

- *large-ion lithophile elements* (LILE) such as K, Rb, Cs, Sr, Ba, all of which have large ionic radius;
- elements of large ionic valences that can carry strong charges and develop strong electrostatic fields (*high field-strength elements*, HFSE), such as Zr, Nb, Hf, Th, U, Ta, and the rare-earth elements, REE.

The REEs include scandium, yttrium, and the fifteen lanthanides. They are further classified by mass: nine elements from Eu to Lu (with atomic numbers from 76 to 84) are named *heavy rare earth elements*, HREE, while the other eight elements (Sc, Y, La, Ce, Pr, Nd, Pm, Sm) are named *light rare earth elements*, LREE.

Mantle melting results in its chemical differentiation. Undifferentiated parts of the mantle are called *enriched or primitive*, whereas mantle reservoirs with a strong reduction in the concentration of lithophile elements are called *depleted*. In particular, rocks rich in LREE are referred to as *fertile*, in contrast to *depleted* rocks poor in LREE. A primitive composition is commonly observed in ocean-island basalts that are volcanic rocks from hotspot areas. A common assumption that many hotspots may originate from the deepest regions of the mantle (or at the core–mantle boundary in the so-called D″-layer) provides a basis for the geochemical assumption of the existence of a relatively closed reservoir of very primitive composition somewhere in the lower mantle. In contrast, depleted composition is typical of the continental lithospheric mantle, and in particular, of the oldest, cratonic, lithospheric keels.

Chondrites, Bulk Earth, and Bulk Silicate Earth

Modern geochemical hypotheses assume that all material of the solar system, including that forming the Earth, was derived from a nebula material. This nebula material had a characteristic chemical and isotopic composition that was essentially homogeneous. Most meteorites formed around 4.5 Ga ago and they are subdivided into approximately 12 major groups. The fundamental distinction is between meteorites that have undergone differentiation within parent bodies and those derived from undifferentiated parent bodies; the latter are termed *chondrites* and are the most abundant meteorites (~80%). They represent the oldest solid material in the solar system and provide cosmochemical constraints (the nearest approach) on the composition of the primordial solar nebula. This composition coined as a "*chondrite uniform reservoir*" (CHUR) (DePaolo and Wasserburg, 1976) is assumed to correspond to the composition of the *Bulk Earth* (BE). However, there are significant compositional differences between chondrites of different types. The major phases of chondrites are the Mg-silicates and Fe–Ni metal.

In contrast to geochemical classifications that are based on the ionic size and charge of the elements, the cosmochemical classifications are based on the volatility of the element in a gas of solar composition and on its affinity for metallic versus silicate (or oxide) phases. Since the condensation temperature is a convenient measure of an element's volatility, the cosmochemical classifications divide elements into refractory, common (Mg, Si, Cr, Fe, Ni, Co), volatile, and ice-forming (H, C, N, and the rare gases) elements. *Refractory elements* (with condensation temperatures above those of the major meteorite phases) include Al, Ca, Ti, Hf, Lu, Zr, Y, U, Th, Sr, Re, Os, and the REEs.

An important result of studies of chondritic meteorites is that all of them have identical (constant) ratios of refractory lithophile elements (e.g. Ca/Al, Ti/Al, Ti/Sc, Ti/REE) that hold to better than ±5% (O'Neill and Palme, 1998). This result makes the central postulate in studies of the Bulk Earth composition (and planetary composition, in general):

• relative abundances of refractory elements in BE are always the same as in meteorites.

In the ratio/ratio diagrams, that are common in geochemistry, mixtures generally form not straight lines but hyperbolas, because the weights assigned to the concentration balance

Table 2.1 Commonly used chondritic reference compositions

Isotope ratio	Value accepted for CHUR
$^{87}Sr/^{86}Sr$	0.6992
$^{187}Re/^{188}Os$	0.40186
$^{187}Os/^{188}Os$	0.1270
$^{147}Sm/^{144}Nd$	0.1966
$^{143}Nd/^{144}Nd$	0.512638
$^{176}Lu/^{177}Hf$	0.0332

CHUR = chondrite uniform reservoir (= Bulk Earth). References are given in the text.

Table 2.2 Models for the major-element composition of the primitive mantle (Bulk Silicate Earth)[1]

	Ringwood, 1975	Jagoutz et al., 1980	Palme and Nickel, 1985	Hart and Zindler, 1986	O'Neill and Palme, 1998
MgO	38.1	38.3	35.5	37.8	37.8
Al_2O_3	4.6	4.0	4.8	4.1	4.4
SiO_2	45.1	45.1	46.2	46	45
CaO	3.1	3.5	4.4	3.2	3.5
FeO	7.9	7.8	7.7	7.5	8.1
Total	98.8	98.7	98.6	98.6	98.8
mg # = Mg/[Mg+Fe]	0.896	0.897	0.891	0.899	0.893

[1] Bulk Silicate Earth is the primitive mantle reservoir formed as the residual after the Earth differentiated into the core and the mantle; the mantle has further differentiated into the continental and oceanic crusts, and the "depleted mantle".

depend on the denominator of the ratio (Albarède, 2003). Thus, the concavity of hyperbolas depend on the abundance ratios of the elements in the denominator in the components of the mixture. For this reason, the magnesium ratio, mg # = Mg/[Mg+Fe], which plays an important role in petrologic studies of the continental lithospheric mantle (see Chapter 6) requires special attention. The ratios of isotopes also play an important role in geochemistry because, in general, (i) they can be measured more precisely than isotope concentrations and (ii) they are less sensitive to secondary alterations. Some of the most important isotope ratios are listed in Table 2.1.

The concept of *Bulk Silicate Earth (BSE)*, introduced by Hart and Zindler (1986), is based on the fundamental difference between the outer silicate portion of the Earth (the mantle plus the crust) and the iron-rich core (Table 2.2). Most geochemical models assume that the core and the BSE are two chemically isolated reservoirs that have been separated for most of the Earth's history. This assumption is based on the observation that there is no notable change in the content of the siderophile elements (Ni, Co, Mo, Sn) in mantle-derived rocks over the past 4.0 Ga (the entire geological record) (O'Neill and Palme, 1998). Thus, the composition of the core plus the BSE should match the cosmochemical constraints on various element abundances in the solar nebula (i.e. CHUR), while the BSE is believed to have chondritic relative abundances only for the REEs, because they were not affected by core formation or modified during the Earth's accretion.

The BSE consists of several geochemical reservoirs (see also Section 9.1 and Fig. 9.10):

- the oceanic crust;
- the suboceanic lithosphere;
- the continental crust;
- the subcontinental lithospheric mantle;
- the depleted convecting upper mantle that is the source of mid-ocean ridge basalts (MORB);
- the enriched upper mantle that is the source of ocean-island basalts (OIB);
- (possibly) the primitive (undifferentiated) mantle (PUM) that, by definition, has BSE composition;
- hydrosphere and atmosphere.

The mass fraction of the continental crust in BSE is estimated as 0.0054, the mass fraction of the subcontinental lithospheric mantle is no more than 0.04, and the mass fraction of the depleted mantle is ~0.5 ± 0.15 (Taylor and McLennan, 1995; O'Neill and Palme, 1998).

2.1.2 Radioactive decay and the isochron equation

Knowledge of the age of the Earth and the timing of global-scale, regional-scale, and local tectonic and magmatic events which formed and modified the crust is critical to the understanding of the evolution of our planet. The dating of rocks and tectonic and magmatic events is based on the application of radioactive methods which employ natural radioactivity. An electrically neutral atom contains an equal number of protons (Z) and electrons (e^-). The number of protons determines the chemical element (the atomic number Z), whereas the number of neutrons (N) determines the isotope of the element. Most stable elements have approximately the same number of protons and neutrons. An atom with mass number M is denoted as $^{M}A_Z$ (here $M = Z + N$).

Radioactivity is caused by the radioactive decay of unstable nuclei which causes the nucleus to emit particles or electromagnetic radiation. The most common forms of radioactive decay are:

- α- (alpha-) decay which occurs when the nucleus emits an α-particle (i.e. a helium nucleus consisting of two protons and two neutrons). Alpha-decay is common at high mass and produces a new (daughter) element with a lower atomic number than the initial (parent) atom;
- β- (beta-) decay which is caused by a neutron–proton transformation in the nucleus and the emission of β-particles (electron or positron). The neutron-to-proton transformation is

accompanied by the emission of an electron β^- and an antineutrino $\bar{v}$, while proton-to-neutron transformation causes the emission of a positron β^+ and a neutrino v (the latter process takes place when $N < Z$, which is unusual for natural nuclides). As a result of β-decay, the atomic number of the nucleus increases or decreases by one;

• γ- (gamma-) decay results from the emission of electromagnetic radiation which can occur following the emission of α- or β-particles.

Processes of material removal from the mantle (due to melt extraction) or material addition to the mantle (due to crustal recycling) change the ratio between the parent and daughter isotopes. They cause upward transport of large-ion lithophile elements (K, Rb, Sr, U) and light rare-earth elements (Nd) and produce isotopic differences in the crust and mantle (Bennett, 2004). In a closed system, the total number of atoms remains constant. If for each parent atom P a stable daughter atom (or radioactive nuclide) D is created, then

$$P + D = P_o + D_o, \tag{2.1}$$

where the zero-index refers to time $t = 0$. Since $P = P_o e^{-\lambda t}$,

$$D = D_o + P(e^{\lambda t} - 1), \tag{2.2}$$

where λ is a decay constant (half-life time $T_{1/2} = \ln 2/\lambda$). Given that D_o (the number of daughter elements at time $t = 0$) is generally unknown, the equation cannot be used unless D_o is very small and can be neglected (as in K–Ar and U–Pb dating).

In geochronology, fractionation and instrumental biases are eliminated by considering the ratio of a number of parent atoms P of a radioactive nuclide to the number of atoms D_s of a stable isotope of the same element. This method is widely used for short-lived isotopes, for which the initial isotope ratio (at the time when the system was formed) can be determined. Following this approach, equation (2.2) can be normalized by the number of atoms D_s of a stable isotope of the same element as the radiogenic nuclide, so that it takes the form:

$$[D/D_s]_t = [D/D_s]_o + [P/D_s]_t(e^{\lambda t} - 1). \tag{2.3}$$

Equation (2.3) proposed by Nicolaysen (1961) is known as the *isochron equation*. It defines a straight line, the slope of which (defined by $(e^{\lambda t} - 1)$) gives the time t of rock (mineral) formation (modification) (Figure 2.1). The approach is valid under the following assumptions:

(a) the system is closed,
(b) all analyzed rock samples are of the same age, and
(c) the rock had initial isotopic homogeneity, i.e. all rock samples had the same initial (at time $t = 0$) isotope ratio $[D/D_s]_o$.

Isotope systems widely used in lithosphere dating because of their long radioactive half-life include K–Ar, U–Pb, Sm–Nd, Rb–Sr, Re–Os, and Lu–Hf (in geochronology, the parent–daughter isotope systems are denoted in the order P–D). Short-life isotopes such as those of carbon, beryllium, and thorium are used successfully in dating sedimentation and erosion

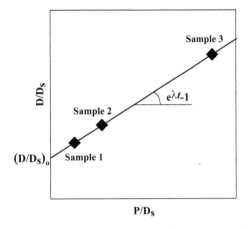

Fig. 2.1 An example of the isochron equation (2.3). The slope of the line defines the time t of rock formation; λ is a decay constant; P and D denote the parent–daughter isotope systems; D_s is the number of atoms of a stable isotope of the same element as the radiogenic nuclide; the zero subindex refers to time $t = 0$.

rates, cave and marine deposits. Technical details for methods can be found in specialized literature. Here, some basic information is provided for selected isotopic methods that are based on long-life isotopes (see also Table 2.3).

2.1.3 K/Ar

Potassium has three natural isotopes: ^{39}K (~93.26%), ^{40}K (~0.01%), ^{41}K (~6.73%). The radioactive ^{40}K decays into argon and calcium. The potassium–argon geochronological method is based on the reaction during which ^{40}Ar$_{18}$ is formed by capture of an electron by the ^{40}K$_{19}$ nucleus:

$$^{40}\text{K}_{19} + \text{e}^- \rightarrow {}^{40}\text{Ar}_{18}. \tag{2.4}$$

The decay rate of this process is $\lambda_{Ar} = 5.81 \times 10^{-11}$ y^{-1} (half-life ~11.93 Ga). Potassium-40 also decays into ^{40}Ca$_{20}$ by β-particle emission with decay rate $\lambda_{Ca} = 4.962 \times 10^{-10}$ y^{-1} (half-life ~1.4 Ga). The relative probability of ^{40}K$_{19}$ decaying to ^{40}Ar$_{18}$ is ~10.5%, and to ^{40}Ca$_{20}$, ~89.5%. Since the amount of radiogenic ^{40}Ca cannot be determined, only the decay to ^{40}Ar is used. Equation (2.2) then takes the form with the combined decay constant: $\lambda = \lambda_{Ca} + \lambda_{Ar} = 5.543 \times 10^{-10}$ y^{-1} (half-life ~1.25 Ga) in the exponent of eq. (2.2) and with the proportion of daughter nuclides that are ^{40}Ar atoms in the factor term, so that:

$$^{40}\text{Ar}_t = {}^{40}\text{Ar}_o + \lambda_{Ar}/[\lambda_{Ca} + \lambda_{Ar}]\ {}^{40}\text{K}\left(e^{\lambda t} - 1\right) = {}^{40}\text{Ar}_o + 0.1048\ {}^{40}\text{K}\left(e^{\lambda t} - 1\right). \tag{2.5}$$

A variant of the ^{40}K–^{40}Ar method is the ^{39}Ar–^{40}Ar method.

Application of equation (2.2) to the K–Ar isotope system assumes that the sample did not initially contain argon, i.e. that ^{40}Ar$_o$ can be neglected due to high parent/daughter ratios. Thus, application of the K–Ar method does not require the use of isochrons, and the age of the rock can be obtained from:

Table 2.3 Some isotopes systems used in geochronological studies of the lithosphere

Decay reaction (parent/ daughter)	Decay constant λ ($\times 10^{-11} y^{-1}$)	Half-life $T_{1/2} = \ln 2/\lambda$ (Ga)	Advantages	Disadvantages
$^{40}K/$ [$^{40}Ar+^{40}Ca$]	55.43	~1.25	• Does not require use of isochrons • Used for dating young basic rocks (paleomagnetism, exhumation rates, etc.)	• Possible contamination with atmospheric argon • Can be reset by metamorphism
$^{87}Rb/^{87}Sr$	1.42	48.9	• Used for dating igneous rocks (granites and basalts from both continental and oceanic settings)	• Can be reset by metamorphism
$^{238}U/^{206}Pb$ $^{235}U/^{207}Pb$	15.5 98.5	4.47 0.704	• Does not require use of isochrons • This pair of isotope systems allows exclusion of isotope ratios and use of concordia • Timing of tectonic or magmatic events can be determined from concordia • Used in zircon dating	• Can be affected by low-grade metamorphism, water circulation, and weathering • U–Th–Pb system may have been fractionated during core formation
$^{147}Sm/^{144}Nd$	0.654	106	• Strong fractionation between the crustal and mantle reservoirs • Generally immobile and are not altered by tectonism and metamorphism • Unaffected by core formation • Provide a well-defined model of the primitive mantle of the Earth	• A huge scatter in isochron ages is commonly observed, caused by tectono-magmatic resetting of the isotope system
$^{187}Re/^{187}Os$	1.61	43±5	• Extreme parent/daughter fractionation during mantle melting • Distinct Os isotopic composition of the oceanic and the continental lithospheric mantle • Altered by metamorphic processes less than Nd isotopes	• Very low decay energy which results in a large scatter of calculated half-life values
$^{176}Lu/^{176}Hf$	1.86	37.2	• Strong fractionation between the crustal and mantle reservoirs • Altered by tectonic and magmatic processes less than Nd isotopes • Has better resolution than Sm–Nd method due to shorter half-life • Negligible effect of *in situ* radiogenic growth in zircon crystals; Hf forms an integral part of the zircon lattice	

Decay constants and half-lives based on Dickin (1995), Van Schmus (1995).

$$e^{\lambda t} = 1 + 9.54 \; {}^{40}\text{Ar}/{}^{40}\text{K}. \qquad (2.6)$$

Measurement of the quantity of argon atoms constrains the time that has passed since a rock sample solidified. However, since early studies (Hart and Dodd, 1962), the presence of excess (inherited) argon has subsequently been found in samples from deep-seated (plutonic) environments and submarine lavas. Its presence was interpreted to be a product of the environment of crystallization or of incomplete degassing from lavas during eruption. Since argon is an inert gas, it does not bind with other atoms in a crystal lattice, and typically argon atoms formed during potassium–argon decay are trapped within the crystalline lattice because of their larger size. Rock melting creates conditions that are favorable for the diffusion of argon atoms (argon degassing). During crystallization, diffused argon that fails to degas from the magma, again becomes entrained in a lattice. Consequent ${}^{40}\text{K} - {}^{40}\text{Ar}$ decay will produce new argon that will accumulate along with the entrained (inherited) argon.

The K–Ar method is the only really good geochronometer for dating young basic rocks and it is widely used to calibrate magnetic reversals (York, 1978) and to quantify post-orogenic cooling rates and exhumation rates. Application of the method requires correction for contamination with atmospheric ${}^{36}\text{Ar}$ argon. Disadvantages include possible resetting of isotope ages by metamorphism because no mineral phase takes argon when it is lost from the host mineral.

2.1.4 Rb/Sr

Rubidium has two natural isotopes: ${}^{87}\text{Rb}$ (27.83%) and ${}^{85}\text{Rb}$ (72.17%). The ratio between the two isotopes (absolute isotopic abundance ratio) is a constant on the Earth, the Moon and in most meteorites, ${}^{85}\text{Rb}/{}^{87}\text{Rb} = 2.5932$, due to isotopic homogenization of the solar nebula (Catanzaro *et al.*, 1969). Isotope ${}^{87}\text{Rb}$ is radioactive (half-life 48.9 Ga) and decays to strontium. The decay process takes place through the emission of the electron and the antineutrino:

$$ {}^{87}\text{Rb} \rightarrow {}^{87}\text{Sr} + \beta^{-} + \bar{\nu}. \qquad (2.7)$$

Similarly to the K–Ar isotope system, melt extraction from the mantle and crustal recycling produce variations in the parent/daughter ratios in the rubidium–strontium system ${}^{87}\text{Rb}/{}^{87}\text{Sr}$. The equation (2.2) that allows calculation of the number of strontium atoms produced by rubidium decay since formation of the rock takes the form:

$$ {}^{87}\text{Sr}_t = {}^{87}\text{Sr}_o + {}^{87}\text{Rb}(e^{\lambda t} - 1). \qquad (2.8)$$

Since the absolute abundance of a nuclide is difficult to measure with high precision, relative abundances are commonly calculated by dividing the equation by the amount of isotope ${}^{86}\text{Sr}$. This isotope is not produced by radioactive decay, and its concentration remains constant through time. Thus, practical application of the Rb–Sr method is based on the isochron equation (2.3):

$$ \left[{}^{87}\text{Sr}/{}^{86}\text{Sr} \right]_t = \left[{}^{87}\text{Sr}/{}^{86}\text{Sr} \right]_o + \left[{}^{87}\text{Rb}/{}^{86}\text{Sr} \right]_t. \qquad (2.9)$$

The Rb–Sr method has been widely used for more than 40 years to date the emplacement of granites and other igneous rocks (such as basalts from both continental and oceanic settings). However, these isotope ages may be reset by metamorphism because both Rb and Sr are mobile in metamorphic and hydrothermal fluids.

Another problem arises with estimating the initial isotope ratio $[^{87}Sr/^{86}Sr]_o$ at the time when the system was formed (at time $t = 0$). Reliable constraints on Sr isotope evolution of the mantle are missing for much of the Earth's history (Bennett, 2004). For this reason, the $[^{87}Sr/^{86}Sr]_{BSE}$ composition of the Bulk Silicate Earth is not constrained directly, but is estimated from well-established correlations with the Nd isotope composition of modern oceanic basalts. Assuming that the strontium composition for the Bulk Silicate Earth corresponds to $\varepsilon_{Nd} = 0$,

$$[^{87}Sr/^{86}Sr]_{BSE} = 0.7045 \pm 0.0005 \tag{2.10}$$

(DePaolo and Wasserburg, 1976). (Epsilon Nd (ε_{Nd}) is introduced in equation (2.17) and it quantifies the deviation of the rock's Nd isotopic system from the chondritic, i.e. primitive, mantle that has BSE composition by definition). Estimates based on the composition of c. 4.5 Ga-old meteorites provide significantly lower values for the early Earth (Banner, 2004):

$$[^{87}Sr/^{86}Sr]_o \sim 0.699. \tag{2.11}$$

The Sr composition of the present-day upper mantle is defined from the analysis of normal mid-ocean ridge basalts (N-MORB):

$$[^{87}Sr/^{86}Sr]_t^{um} = 0.7025 \pm 0.0005. \tag{2.12}$$

The crust and the mantle, however, have significantly different ranges of Rb/Sr ratios due to their distinctly different mineral compositions. The Earth's crust has a large variety of Sr isotope ratios (see Fig. 2.7); its evolution and accretion included processes of mantle differentiation and crustal fractionation that produced a higher Rb/Sr ratio than in parent magmas. The large diversity of magmatic, sedimentary, and metamorphic processes that operate within the crust result in a wide range of Sr isotope ratios. Owing to the large uncertainty in the determination of ^{87}Sr composition in Precambrian rocks, the evolutionary trend of ^{87}Sr variation with lithospheric age cannot be reliably established. However, older crustal rocks apparently have higher present-day $[^{87}Sr/^{86}Sr]_t^{cc}$ values than younger rocks (Banner, 2004) (Fig. 2.2).

2.1.5 U/Pb

Uranium–lead is one of the oldest radiometric methods (Boltwood, 1907). The method is also one of the most refined and can provide an accuracy of dating of ~0.1% (Bowring *et al.*, 2007). The advantage of the uranium–lead method is that it has the pair of isotope systems: $^{238}U-^{206}Pb$ (^{238}U with half-life of 4.47 Ga, comparable to the age of the Earth) and $^{235}U-^{207}Pb$ (^{235}U with half-life of 704 Ma). The decays involve a series of α- and β-decays ($8\alpha + 6\beta$ for ^{238}U and $7\alpha + 4\beta$ for ^{235}U). However, all intermediate members in each series are short-lived and can be ignored for processes that occur on geological time-scales.

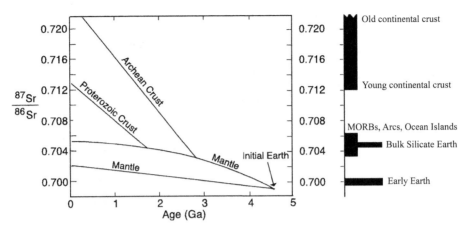

Sr isotope evolution of the Earth (after Faure, 1986). Two evolution lines shown for the continental crust represent only a small part of the range typical for different parts of the crust and constrained by broadly variable Rb–Sr ratios. Two mantle evolution lines cover most of the variability found in the Earth's mantle. Because of the low Rb–Sr ratio in the mantle and because the mantle's average Rb–Sr ratio decreases with time as the crust with relatively high Rb/Sr is progressively extracted from the mantle, the mantle evolution lines have a low slope and are curvilinear. The estimated initial ^{87}Sr/^{86}Sr value in the early Earth is based on analyses of well-preserved old basaltic achondrite meteorites (Papanastassiou and Wasserburg, 1969). The right insert shows typical Sr isotope ratios for different reservoirs. MORB = mid-ocean ridge basalts.

Since the U–Pb system has a high parent/daughter ratio, the term D_o in equation (2.2) may be neglected. The equations for each of the isotope systems then take the form:

$$^{206}\text{Pb}/^{238}\text{U} = \left(e^{\lambda[238]t} - 1\right); \tag{2.13a}$$

$$^{207}\text{Pb}/^{235}\text{U} = \left(e^{\lambda[235]t} - 1\right). \tag{2.13b}$$

Combining equations (2.13) for each of the U–Pb systems allows one to exclude the isotope ratios. An x–y plot of ^{207}Pb/^{235}U versus ^{206}Pb/^{238}U, made out of points where both ^{238}U–^{206}Pb and ^{235}U–^{207}Pb isotope systems give identical ages, is called *concordia* (Fig. 2.3). Some mineral grains may have discordant ages (i.e. not falling on the concordia diagram), but still form a linear array. The points of deflection and intersection of such arrays with the concordia line (labeled in ages) define the times of initial crystallization and later isotope system alteration by a tectonic or magmatic event.

A variant of the U–Pb method, the lead–lead method, utilizes the stable isotope ^{204}Pb (the only non-radiogenic isotope of lead) and two radiogenic isotopes ^{206}Pb and ^{207}Pb. The advantage of the Pb–Pb method is that it requires only a knowledge of isotope ratios of lead but not their concentrations. Patterson (1956) successfully used the U–Pb method to determine the age of the Earth and estimated it to be between 4.5 Ga and 4.6 Ga.

Possible contamination by lead at mineral surfaces or in grain fractures requires special attention in choosing for analysis the zones with (almost) no contamination. Due to the

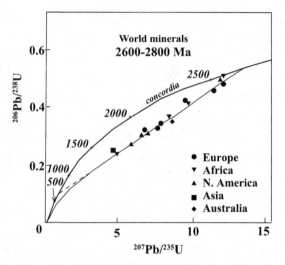

Fig. 2.3   Concordia diagram for Archean U-rich minerals from different continents, which intercepts the common concordia line at *c.* 600 Ma (after Dickin, 1995).

mobility of Pb and, in particular, U in the environments of low-grade metamorphism, water circulation, and weathering, the method cannot be used on silicate systems (Dickin, 1995). Furthermore, the system cannot be used to identify the isotopic composition of the Bulk Silicate Earth because the U–Th–Pb system may have been fractionated during core formation (e.g. due to extraction of Pb into the core as a sulfide phase).

Nowadays, the U–Pb method is widely applied to the dating of zircon grains ($ZrSiO_4$) in metamorphic and igneous rocks. Zircon provides unique possibilities for examining the evolution of the Earth, because (a) it has two radiogenic isotope systems (U–Pb–Th and Lu–Hf) widely used in geochronology and (b) the isotope ratios of these radiogenic systems are not reset during sedimentary and magmatic recycling and probably even during lower crustal delamination and sediment subduction (Gao *et al.*, 2004). The complex growth structure of crystals of zircons preserves unique information about the processes of crustal growth and magmatic events; the ages of discrete growth episodes of zircon crystals can be determined *in situ* by U–Pb isotope analysis (Figure 2.4).

2.1.6 Sm/Nd

Because of the very long half-life of ^{147}Sm (106 Ga), the samarium–neodymium system is used to study the evolution of the early Earth and the Solar System. Neodymium isotopes were first used in the late 1970s to calculate the age of the continents and to constrain models of the Earth's evolution (Jacobsen and Wasserburg, 1979; O'Nions *et al.*, 1979). The advantages of the Sm–Nd isotope system are:

● The Sm–Nd isotope system was not affected by core formation and it was not modified during Earth accretion; for this reason the Bulk Silicate Earth should have chondritic relative abundances of samarium and neodymium;

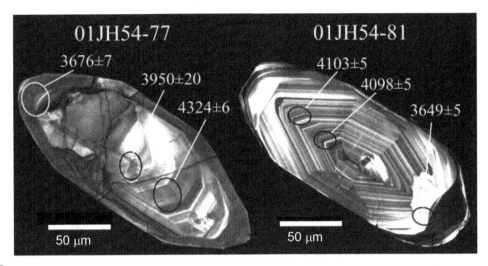

Fig. 2.4 Zircons from Jack Hills metaconglomerate (Australia) showing sites of U–Pb analyses with age (in Ma). Scale bars are 50 μm (from Valley *et al.*, 2006).

- Sm and Nd exhibit a pronounced fractionation between the crustal and mantle reservoirs;
- Sm and Nd are generally immobile and are not altered by secondary disturbances such as tectonism and metamorphism. However, some studies report alteration of Sm–Nd ages, such as their resetting by granulite-grade metamorphism in Antarctica (McCulloch and Black, 1984).

Samarium transforms to neodymium by α-decay:

$$^{147}\text{Sm} \rightarrow {}^{143}\text{Nd} + \alpha. \tag{2.14}$$

The stable reference isotope of neodymium is ^{144}Nd. Then the isochron equation (2.3) takes the form:

$$\left[{}^{143}\text{Nd}/{}^{144}\text{Nd}\right]_t = \left[{}^{143}\text{Nd}/{}^{144}\text{Nd}\right]_o + \left[{}^{147}\text{Sm}/{}^{144}\text{Nd}\right]_t \left(e^{\lambda t} - 1\right), \tag{2.15}$$

where $\lambda = 6.54 \times 10^{-12}$ y^{-1}. Isochron ages are commonly interpreted to reflect the time of major mineral equilibration during emplacement into the lithosphere or closure ages due to post-formation cooling. For example, Nd and Sr isotope studies of clastic metasediments from Isua, West Greenland indicate the presence of pre-3.8 Ga differentiated crustal components (Jacobsen and Dymek, 1988). However, because of the complex relationships between diffusion and radiogenic growth not each isochron age can be related to a specific tectonic-magmatic event (Zindler and Jagoutz, 1988). For example, some regions, like kimberlite fields in Siberia, demonstrate a huge scatter in ages constrained by Sm–Nd isochrons (Figure 2.5), which has been interpreted as resetting of the isotope system during kimberlite eruption or during lithospheric residence (Pearson, 1999).

Sm–Nd and Nd–Nd studies of chondritic meteorites provide a well-defined model of the primitive mantle of the Earth (Bennett, 2004):

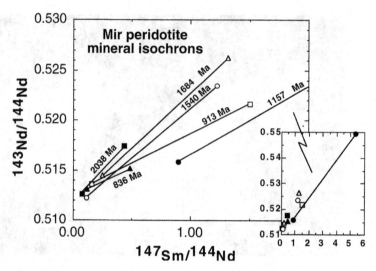

Fig. 2.5 Sm–Nd isochrons (equation 2.15) for peridotites from the Mir kimberlite pipe, Siberia (from Pearson, 1999).

$$\left[^{147}\mathrm{Sm}/^{144}\mathrm{Nd}\right]_{\mathrm{CHUR}} = 0.1966 \quad \text{and} \quad \left[^{143}\mathrm{Nd}/^{144}\mathrm{Nd}\right]_{\mathrm{CHUR}} = 0.512638, \qquad (2.16)$$

where sub-index CHUR refers to a chondrite uniform reservoir (i.e. the primitive mantle).

Chemical modification of the upper mantle due to melt extraction and crustal recycling produces variations in the parent/daughter ratios in the Sm–Nd and Nd–Nd systems. Owing to the large volume of the mantle and its large chemical inertia, it is important to distinguish small isotopic variations between different mantle reservoirs for isotope systems with a long half-life. For this reason, epsilon Nd ($\varepsilon_{\mathrm{Nd}}$) was introduced to quantify deviations of the Nd isotopic system in a rock from chondritic (i.e. primitive) mantle:

$$\varepsilon_{\mathrm{Nd}} = \left[\frac{^{143}\mathrm{Nd}/^{144}\mathrm{Nd}(t)}{[^{143}\mathrm{Nd}/^{144}\mathrm{Nd}]_{\mathrm{CHUR}}(t)} - 1\right] \times 10^4, \qquad (2.17)$$

where t is the crystallization age of the rock. Melt extraction from the primitive mantle causes light rare earth elements (LREE) such as Nd to partition into the melt. As a result, the LREE-enriched continental felsic crust has low $^{143}\mathrm{Nd}/^{144}\mathrm{Nd}$ ratio as compared to the Bulk Earth (BE), while the upper mantle has a radiogenic, high $^{143}\mathrm{Nd}/^{144}\mathrm{Nd}$ ratio. At any time, a large range of Nd isotope compositions characterizes the upper mantle. The highest positive value is assumed to characterize the depleted mantle composition (Figure 2.6). The convective upper mantle which is the source of mid-ocean ridge basalts (MORB) has some of the largest values of observed $\varepsilon_{\mathrm{Nd}}$ (+10±2).

Neodymium isotope ratios provide unique information on the chemical evolution of the mantle during the early evolution of the Earth, for which no rock record exists. Equation (2.17) implies that at the time of the Earth formation at 4.56 Ga the composition of the primitive mantle was $\varepsilon_{\mathrm{Nd}} = 0$. A well-established characteristic of the mantle is a (monotonous since at least 2.7 Ga) increase in $\varepsilon_{\mathrm{Nd}}$ from $\varepsilon_{\mathrm{Nd}} < 4$ in the Archean to $\varepsilon_{\mathrm{Nd}} > 6$ in the Phanerozoic (DePaolo et al., 1991) (Figure 2.6). Many Archean gneisses have a positive $\varepsilon_{\mathrm{Nd}}$

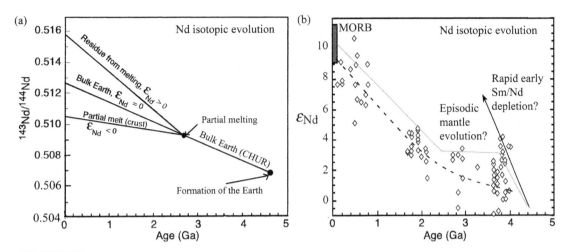

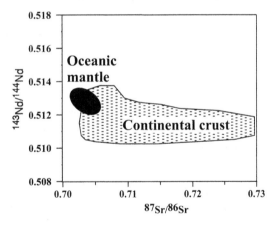

Fig. 2.6 (a) Nd isotope evolution of the bulk Earth, represented by a chondritic uniform reservoir (CHUR) (after Banner, 2004). The Nd isotope evolution of a reservoir is governed by its initial ^{143}Nd/^{144}Nd ratio and its ^{147}Sm/^{144}Nd ratio (equation 2.15). In contrast to the Rb–Sr system (Fig. 2.2), mantle sources are enriched in the parent–daughter ratio, relative to the continental crust. (b) Secular ε_{Nd} evolution of the upper mantle based on isotope studies of juvenile granites and Phanerozoic ophiolites (after Bennett, 2004). MORB = mid-ocean ridge basalts. Gray line and dashed line – possible interpretations of secular trends (with monotonous and episodic mantle differentiation).

Fig. 2.7 Sr–Nd isotope compositions for the continental crust and the oceanic mantle (modified after Pearson, 1999).

in the range from 0 to +4 with the prevailing value of $\varepsilon_{Nd} \geq +2$ at 3.8 Ga (Bennett, 2004). Unexpectedly, some of the oldest Archean rocks have an ε_{Nd} between −4 and +4 (the 4 Ga old Acasta gneisses in Canada, Bowring and Williams, 1999) and between − 3.7 and + 12.4 (the 3.81 Ga old Isua Supracrustal Belt in Western Greenland, Frei and Jensen, 2003). These positive values indicate a very early differentiation of the Earth with very early extraction of the continental crust (Armstrong, 1991).

The continental crust is characterized by large ε_{Nd} variations (Figure 2.7), sometimes even at rock scale. For example, garnet from the same Tanzanian eclogite xenolith shows 16 unites variations in ε_{Nd} (Jagoutz, 1988). The extreme variations in Nd isotope ratios in the Archean crust may reflect resetting of the Sm–Nd system during tectono-thermal (i.e. igneous, metamorphic, tectonic) processes which produce open-system behaviour (Moorbath *et al.*, 1997).

2.1.7 Re/Os

Similarly to the Sm–Nd system, the rhenium–osmium isotope system has applications in the geochronology of mantle evolution and magma genesis because Re and Os exhibit pronounced fractionation between the crustal and mantle reservoirs (Shirey and Walker, 1998). Like neodymium, rhenium partitions into melt, while Os, like Sm, resides in the mantle. Extreme parent/daughter fractionation during partial melting results in a large range in ^{187}Os isotopic variations: while the concentration of osmium in mantle rocks is *c.* 3 ppb, its concentration in crustal rocks is *c.* 0.03 ppb (Fig. 2.8). This huge difference between osmium concentrations in crustal and mantle rocks has been interpreted as an indicator that the crust is not the primary reservoir for mantle depletion and the presence of a reservoir of ancient basalt in the deep mantle (e.g. subducted oceanic crust) has been proposed to balance the mantle osmium depletion (Bennett, 2004). Because of the very low Os concentration in mantle rocks compared to that in silicate melts or other metasomatic agents, the Re–Os isotope system in mantle peridotites is considerably more stable to changes caused by metasomatism and contamination than are isotope systems based on incompatible elements (Sm–Nd and Rb–Sr).

Rhenium has a stable isotope ^{185}Re (37.4%) and an unstable isotope ^{187}Re (62.6%) with a very long half-life of 43 ± 5 Ga, and the ratio ^{185}Re/^{187}Re $= 0.5974$. Osmium has seven stable isotopes with masses 184, 186, 187, 188, 189, 190, and 192. The β-decay transforms ^{187}Re to ^{187}Os:

$$^{187}\text{Re} \rightarrow {}^{187}\text{Os} + \beta^- + \bar{\nu}. \tag{2.18}$$

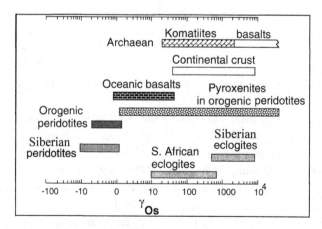

Fig. 2.8   Os isotope variations in different tectonic settings (from Pearson, 1999).

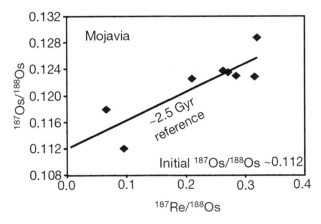

Fig. 2.9 Isochron plot for the Mojavia province in western USA (from Lee *et al.*, 2001). The slope of the isochron corresponds to an age of 2.5 ± 1.6 Ga with the initial ratio ^{187}Os/^{188}Os ≈ 0.112.

The available energy for this β-decay is one of the lowest (2.65 keV) known among all radionuclides, and the very low decay energy results in a large scatter of calculated half-life values. The reference isotope in the Re–Os system is ^{188}Os and the isochron equation (2.3) for the Re–Os isotope system takes the form:

$$\left[^{187}\mathrm{Os}/^{188}\mathrm{Os}\right]_t = \left[^{187}\mathrm{Os}/^{188}\mathrm{Os}\right]_0 + \left[^{187}\mathrm{Re}/^{188}\mathrm{Os}\right]_t \left(e^{\lambda t} - 1\right), \tag{2.19}$$

where $\lambda = 1.64 \times 10^{-11}$ y^{-1}. Because of extreme parent/daughter fractionation during mantle melting, the Re–Os method is best applied to direct samples of mantle material and is very successful in determining the age of the lithospheric mantle. For example, Re–Os studies of mantle-derived peridotites from tectonically active western USA indicate a Late Archean–Paleoproterozoic age for the lithospheric mantle beneath the Cenozoic Southern Basin and Range Province (Lee *et al.*, 2001) (Fig. 2.9). Furthermore, compared with Nd, Os is less prone to metamorphic disturbance (Pearson, 1999).

Similarly to $\varepsilon_{\mathrm{Nd}}$, γ_{Os} is introduced to characterize Os concentrations as deviations from a reference (chondritic) composition:

$$\gamma_{\mathrm{Os}} = \left[\frac{^{187}\mathrm{Os}/^{188}\mathrm{Os}(t)}{\left[^{187}\mathrm{Os}/^{188}\mathrm{Os}\right]_{\mathrm{CHUR}}(t)} - 1\right] \times 100. \tag{2.20}$$

The present-day average chondritic value is:

$$\left[^{187}\mathrm{Os}/^{188}\mathrm{Os}\right]_{\mathrm{CHUR}} = 0.1270 \tag{2.21}$$

with a slightly higher value estimated for the primitive upper mantle, PUM (Walker and Morgan, 1989; Martin, 1991):

$$\left[^{187}\mathrm{Os}/^{188}\mathrm{Os}\right]_{\mathrm{PUM}} = 0.1296 \pm 8. \tag{2.22}$$

This means that the present-day primitive upper mantle material has $\gamma_{\mathrm{Os}} = +2$. The oldest rocks studied for osmium composition, 3.81 Ga gneisses from the Itsaq complex in southwest

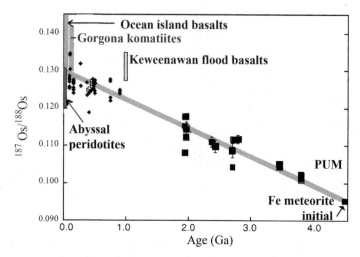

Fig. 2.10 Secular γ_{Os} evolution of the upper mantle (after Bennett, 2004). PUM = the primitive upper mantle.

Greenland, have a composition closest to chondritic $[^{187}Os/^{188}Os]_{CHUR}$ (Frei and Jensen, 2003). Similarly to ε_{Nd}, γ_{Os} increases from Precambrian ($\gamma_{Os} < 0.120$) to Phanerozoic, with an average value of 0.1247 reported for abyssal peridotites (MORB are characterized by a large range of osmium compositions) (Figs. 2.8, 2.10). Since the Os isotopic composition of the oceanic and the continental lithospheric mantles are clearly distinct, the Re–Os system is a powerful tool for the study of the evolution of the Earth's lithosphere.

2.1.8 Lu/Hf

The lutetium–hafnium isotope system is a relatively new tool in geochronology applied to studies of crystals of zircon, which are abundant in upper crustal rocks (Blichert-Toft and Albarède, 1997). Lutetium has two natural isotopes: ^{175}Lu (97.41%) and ^{176}Lu (2.59%), while Hf (named after the Latin name for "Copenhagen" where is was discovered in 1923) has six with masses of 174, 176, 177, 178, 179, 180. Like Os, Hf is formed by the β-decay process:

$$^{176}Lu \rightarrow {}^{176}Hf + \beta^- + \bar{\nu}. \tag{2.23}$$

The half-life of ^{176}Lu in terrestrial samples is 37.2 Ga. The isochron equation is obtained conventionally by dividing the decay equation by ^{177}Hf:

$$\left[^{176}Hf/^{177}Hf\right]_t = \left[^{176}Hf/^{177}Hf\right]_o + \left[^{176}Lu/^{177}Hf\right]_t\left(e^{\lambda t} - 1\right), \tag{2.24}$$

where $\lambda = 1.86 \times 10^{-11}$ y^{-1}.

Similarly to the Sm–Nd and Re–Os isotope systems, the Lu–Hf system fractionates during mantle melting producing a relatively non-radiogenic continental crust with low Lu/Hf and radiogenic $^{176}Hf/^{177}Hf$ ratio in a depleted mantle reservoir (Patchett et al., 1981). Thus, this isotope system provides important information on isotope evolution of the crust and the mantle (Fig. 2.11). The advantages of the Lu–Hf isotope system include:

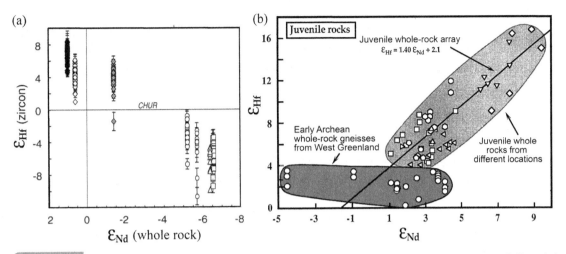

(a) Comparison of ε_{Hf} in zircons and the bulk rock ε_{Nd} in I-type granites from the Lachlan Orogenic Belt (Australia) (from Hawkesworth and Kemp, 2006). CHUR = chondrite uniform reservoir (primitive mantle). (b) Initial $\varepsilon_{Nd} - \varepsilon_{Hf}$ plot for juvenile rocks from the Canadian Cordillera, Arabian–Nubian Shield, Indian craton, Superior province, Labrador, Colorado Plateau, Baltic Shield, West Africa (light gray and different symbols). In these samples, Hf and Nd isotopic compositions are correlated and define the line. The exception are early Archean whole-rock gneisses from West Greenland (shown by dark shading and circles) that do not show a correlation between Nd and Hf compositions (from Vervoort and Blichert-Toft, 1999).

- better resolution than the Sm–Nd system because the half-life of ^{176}Lu is significantly shorter than that of ^{147}Sm;
- higher reliability for isotope studies because Hf isotopes are less altered by tectonic and magmatic processes than Nd isotopes;
- hafnium forms an integral part of the zircon lattice; the latter is very resistant to Hf mobility. Owing to a very low Lu/Hf ratio in zircons (because of a very high, $\sim 10^4$ ppm, Hf concentration), the effect of *in situ* radiogenic growth is negligible. Thus zircons preserve a ^{176}Hf/^{177}Hf ratio close to the initial, at the time of rock formation (Hawkesworth and Kemp, 2006). Metamorphic overprints can be revealed by Pb data.

Similarly to ε_{Nd}, ε_{Hf} is introduced in the Lu–Hf system as:

$$\varepsilon_{Hf} = \left[\frac{^{176}\text{Hf}/^{177}\text{Hf}(t)}{[^{176}\text{Hf}/^{177}\text{Hf}]_{\text{CHUR}}(t)} - 1 \right] \times 10^4 \qquad (2.25)$$

where t is the crystallization age of the sample. The commonly used reference chondritic composition value and the present-day isotope value are:

$$\left[^{176}\text{Lu}/^{177}\text{Hf}\right]_{\text{CHUR}} = 0.0332, \qquad (2.26)$$

$$^{176}\text{Hf}/^{177}\text{Hf} = 0.2828. \qquad (2.27)$$

For crustal rocks, zircons provide a measure of the crustal residence time (i.e. the time since the crustal rocks were extracted from the mantle). In particular, recent Lu/Hf isotope studies have confirmed the possible existence of the continental crust as early as 4.4 Ga (Harrison *et al.*, 2005); this result being the subject of debate (see discussion in Chapter 9).

2.1.9 Mantle evolution from Hf and Nd isotopes

The evolution of the depleted mantle as recorded by Hf and Nd isotope systems is similar (Figs. 2.11b, 2.12) (Vervoort and Blichert-Toft, 1999). The early Archean samples from West Greenland are the exception and do not follow the global trend. Neodymium and hafnium are light rare-earth elements and they fractionate during melt extraction: both Nd and Hf partition into melt and become part of a relatively non-radiogenic continental felsic crust, while the melting residue (the upper mantle) gains high-neodymium and high-hafnium radiogenic ratios ($> + 10$). The larger the volume of the produced continental crust, the more depleted is the mantle composition, and the higher the values of ε_{Nd} and ε_{Hf}. The global models for crustal evolution and mantle differentiation based on Hf and Nd isotope records show the following (Vervoort and Blichert-Toft, 1999):

- the oldest known rocks (~4.0 Ga, in West Greenland) have ε_{Hf} up to $+ 3.5$ and ε_{Nd} up to $+ 2.5$. These values require ~300–400 Ma of mantle differentiation, assuming mantle

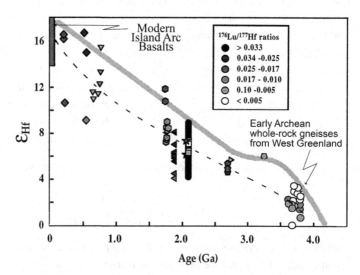

Fig. 2.12 Initial ε_{Hf} versus age of juvenile mantle-derived samples from the Canadian Cordillera, Arabian–Nubian Shield, Indian craton, Superior province, Labrador, Colorado Plateau, Baltic Shield, West Africa, and West Greenland (after Vervoort and Blichert-Toft, 1999). Gray line – approximate upper limit of the ranges; dashed line – alternative interpretation. The evolution of the Hf and Nd isotope systems is similar (compare with Fig. 2.6b).

fractionation factors similar to those in the modern MORB source regions. This implies significant mantle differentiation and crustal growth prior to ~4.0 Ga;

- ε_{Hf} increases with time from the near-zero values typical of the primitive upper mantle to $\varepsilon_{Hf} > +10$ in the Phanerozoic mantle; the highest ε_{Hf} values are characteristic of the modern island arc basalts ($\varepsilon_{Hf} > +14$) and normal mid-ocean ridge basalts N-MORB ($\varepsilon_{Hf} \sim +16 \pm 4$);
- the Lu/Hf and Sm/Nd ratios are generally well-correlated during mantle melting throughout at least the last 3 Ga, and initial isotopic compositions of samples of juvenile crust follow the relationship $\varepsilon_{Hf} = 1.4\,\varepsilon_{Nd} + 2.1$ ($r^2 = 0.7$).

The origin of high positive ε_{Nd} and ε_{Hf} values reported for the oldest Archean terranes (the Acasta gneiss complex of Canada, the Uivak gneisses of Labrador, and the Amitsoq gneisses of southwest Greenland) has been questioned (Moorbath *et al.*, 1997). A long history of metamorphic and tectonic evolution of these terranes suggests that highly positive ε_{Nd} and ε_{Hf} compositions can be artifacts of later alteration rather than original initial isotopic ratios.

2.1.10 Model ages

The isochron equation (2.3) can be considered as expressing a change in the $[D/D_s]$ ratio over time. For U–Pb, Nd–Sm, Rb–Sr, Re–Os, Lu–Hf isotope systems (in which the parent–daughter isotope ratios reflect mantle fractionation and crustal recycling) the equation rewritten once for a rock sample from the crust and once for a rock sample from the upper mantle allows one to eliminate $[D/D_s]_0$ and thus provides model ages for system fractionation (i.e. the time when the rock was separated from the mantle) (Albarède, 2003):

$$T_D = \frac{1}{\lambda_p}\frac{[D/D_s]_{rock} - [D/D_s]_{mantle}}{[P/D_s]_{rock} - [P/D_s]_{mantle}}. \tag{2.28}$$

For example, for the Sm–Nd system, D is ^{143}Nd, D_s is ^{144}Nd, and P is ^{147}Sm. Similarly to equation (2.3), equation (2.28) is valid only for a system that has remained closed since the time dated. Another limitation is the requirement for knowledge of the reservoir from which the rock has differentiated. Isotope measurements on meteorite samples give the isotope ratios at the time of Earth formation, which are:

$$^{143}\text{Nd}/^{144}\text{Nd} = 0.5067,\ ^{87}\text{Sr}/^{86}\text{Sr} = 0.6992,$$
$$^{206}\text{Pb}/^{204}\text{Pb} = 9.3,\ ^{187}\text{Os}/^{188}\text{Os} = 0.09531. \tag{2.29}$$

Two different model ages are used for the Re–Os isotope system because of its complex metasomatic alteration. The "conventional model age" T_{MA} is described by equation (2.30) (Shirey and Walker, 1998; Pearson, 1999):

$$T_{MA} = \frac{1}{\lambda_p}\ln\left[\frac{^{187}\text{Os}/^{188}\text{Os}_{BE}(t) - ^{187}\text{Os}/^{188}\text{Os}_{rock}(t)}{^{187}\text{Re}/^{188}\text{Os}_{BE}(t) - ^{187}\text{Re}/^{188}\text{Os}_{rock}(t)} + 1\right]. \tag{2.30}$$

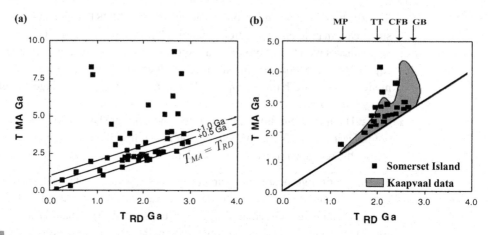

Fig. 2.13 Relationship between T_{MA} and T_{RD} for kimberlite-derived xenoliths. Both T_{MA} and T_{RD} are used to determine the time of melt depletion and the age of the lithospheric mantle. In general, T_{MA} provides the maximum bound and the T_{MA} ages may exceed the age of the Earth. The "Re depletion age" T_{RD} provides the minimum bound and is used when alteration of the isotope system due to metasomatic or melting events is suspected. (a) Data from Africa and Siberia (from Pearson, 1999). (b) Data for individual peridotite suites from the Somerset Island (northern Canada) and Kaapvaal (gray shading) (after Irvine et al., 2003). The line corresponds to $T_{MA} = T_{RD}$. Major tectono-magmatic events in North America are marked by arrows: GB = greenstone belt formation (2.74–2.70 Ga); CFB = Proterozoic basic magmatism (basic dykes and CFB, c. 2.2 Ga); TT = the Taltson–Thelon orogen (2.02–1.9 Ga); MP = Mackenzie plume (c. 1.27 Ga). CFB = continental flood basalts.

Here time t is the time of the rock differentiation from the mantle and BE denotes the Bulk Earth reservoir typically assumed to be chondritic and calculated for early solar system materials with the most primitive osmium isotope ratio. For chondrites, $^{187}Re/^{188}Os = 0.40186$. The conventional Re–Os model ages T_{MA} indicate the time when a rock sample had the same isotopic composition as the fertile mantle and provides the maximum bound on the time of melt depletion and the age of the lithospheric mantle. In case where a sample contains extra rhenium added during later metasomatic or magmatic events, T_{MA} will overestimate the true depletion time.

When alteration of the isotope system is suspected, the "Re depletion age" T_{RD} is used. This is calculated from equation (2.30) assuming that $^{187}Re/^{188}Os_{rock}(t) = 0$, i.e. that all ^{187}Re is consumed during rock (e.g. peridotite) formation. If a significant amount of Re is left in the residue (either because the degree of melting was low (<30%) and not all rhenium was exhausted or because rhenium was added later by metasomatism), the Re/Os isotope ratio is far from zero and T_{RD} will underestimate the true age of a rock (i.e. the time of its differentiation from the mantle). Thus, if some ^{187}Re is present in the rock, $T_{MA} > T_{RD}$ (Walker et al., 1989; Pearson et al., 1995). This situation is typical for most cratonic peridotites when T_{RD} gives the minimum estimate of rock age (Fig. 2.13, also Chapter 9.1.1 and Fig. 9.5). The accuracy of T_{RD} increases with an increase in degree of melting, because the Re/Os isotope ratio in the residue tends to zero.

2.2 Age of the crust and the lithospheric mantle

2.2.1 Continental crust

Tectono-thermal and geological ages

There is little geological record of the evolution of the planet during the first half-billion years, i.e. during most of the Hadean (the name originates from the Greek words "hades" which means "unseen" and it was first used in 1972 by P. Cloud referring to antique rocks). Heavy meteorite bombardment of the Earth from the moment of its formation at *c.* 4.56 Ga until *c.* 4 Ga has erased evidence of the early geological and tectonic evolution of the planet (Ryder, 1992). The oldest known well-preserved fragments of the oldest crust, *c.* 3.9–3.8 Ga old, are located in southern West Greenland (the Itsaq Gneiss Complex) (Nutman *et al.*, 1996), although the oldest known terrestrial rocks (the Acasta gneisses of the western Slave Province in Canada) are older, *c.* 3.96–4.03 Ga (Bowring and Williams, 1999) (Fig. 2.14). In the early 1980s isotope studies of detrital zircons from Western Australia had already demonstrated that minerals may be much older (4.2–4.1 Ga) than any known terrestrial rocks (Froude *et al.*, 1983). Recent progress in geochronology has shifted this time limit further backwards, to *c.* 4.4 Ga (Wilde *et al.*, 2001). Studies of Hf isotope ratios in 4.01–4.37 Ga old detrital zircons from the Jack Hills, Western Australia, have shifted the age of the

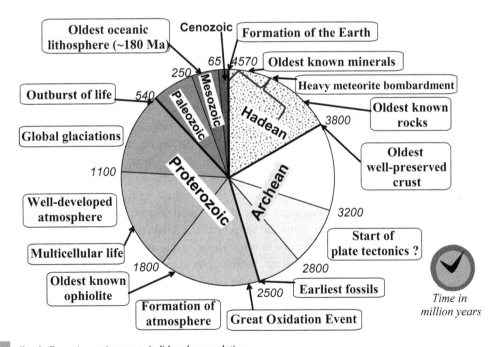

Fig. 2.14   Sketch illustrating major events in lithosphere evolution.

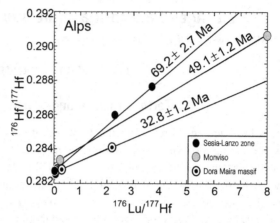

Fig. 2.15 Whole-rock Lu–Hf isochrons for eclogites from the Alps (after Duchène *et al.*, 1997).

earliest differentiated continental crust to 4.2–4.4 Ga (Figure 2.4) (Harrison *et al.*, 2005; Valley *et al.*, 2006; and discussion in Section 9.1).

Studies of post-3.5 Ga crust pose other questions about its age because several different age definitions are widely used in Earth sciences. The most common of these are:

- the geological age of the crust which is the age of crustal differentiation from the mantle,
- the tectono-thermal age which is the age of the last major tectono-thermal event.

In many continental regions, these two ages are essentially similar. This is true, for example, in many of the Precambrian cratons which have remained stable since the time of their differentiation. Similarly, concordant tectono-thermal and geological ages are typical for young tectonic provinces. For example, isotope studies of eclogites from the Alps show that their mineral ages are coeval with major tectonic episodes (Fig. 2.15). There are, however, many tectonic structures where the two ages are fundamentally different. Examples are discussed further in this section.

A global distribution of the age of the continental crust is presented in Fig. 2.16. Assuming that the continental crust and the lithospheric mantle were formed during the same mantle differentiation process, the map can be interpreted as reflecting the age of the subcontinental lithospheric mantle. This assumption is supported by isotope studies of crustal rocks and mantle-derived xenoliths from many of the cratonic regions. In some stable regions, in particular where the basement rocks are covered by thick sedimentary sequences, the geological ages are not well known. Spatial limitations in crustal- and mantle-derived xenoliths further restrict the database of crustal and lithospheric ages. For ice-covered Antarctica and Greenland, the crustal ages are essentially unconstrained, and the assumed ages are based on seismic data complemented by isotope dating of basement outcrops, chiefly along the ice-free coastlines.

The global distribution of continental crust of different ages is not monotonous (Fig. 2.17) with apparent peaks at *c.* 3.5–3.0 Ga and *c.* 2.0 Ga and a gradual increase in crustal areal extent and volume for the Neoproterozoic–Phanerozoic crust. It is not clear if this trend reflects periodic crustal growth or its selective survival (see discussion in Chapter 9). The

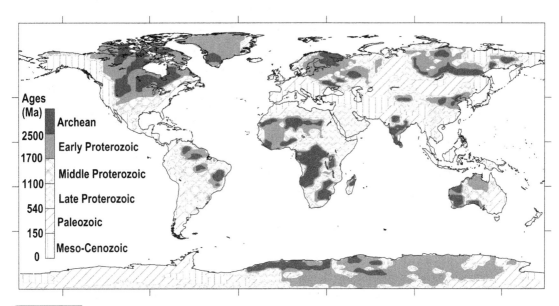

Fig. 2.16   Tectono-thermal age of the continental lithosphere (after Artemieva, 2006; www.lithosphere.info).

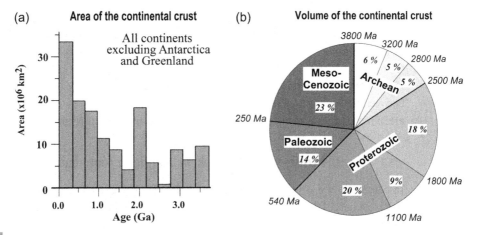

Fig. 2.17   Areal extent (a) and volume (b) of the continental crust of different ages (averaged for 0.3 Ga time bins). The crust of Antarctica and Greenland is excluded. Age data are based on the $1° × 1°$ TC1 global continental model (Artemieva, 2006), data on the crustal thickness are based on the $5° × 5°$ CRUST 5.1 global crustal model (Mooney *et al.*, 1998).

distribution of volume of continental crust of different ages is similar to their areal extent. The Archean crust covers approximately 16% of the continents, Proterozoic – *c.* 47%, Paleozoic – *c.* 14%, and Meso-Cenozoic – *c.* 23% (Figs. 2.16, 2.17). These numbers do not include the crust of Antarctica, where only *c.* 2% of the crust is exposed from the ice sheet.

Old lithosphere in tectonically young regions

In many continental regions the tectono-thermal age of the lithosphere may be significantly less than the differentiation ages of the crust and the lithospheric mantle. Isotope studies in tectonically active regions of western USA and the Canadian Cordillera with Cenozoic tectono-thermal ages provide impressive evidence for such a discrepancy. Rhenium–osmium isotope studies indicate that the Canadian Cordillera is underlain by a Proterozoic lithospheric mantle (Peslier *et al.*, 2000). Similarly, isotope studies of spinel peridotite xenoliths indicate the presence of Proterozoic lithospheric mantle beneath the Paleozoic crust in south-eastern Australia (Handler *et al.*, 1997). In the Vitim volcanic province in the Baikal rift region the Paleozoic-Cenozoic age of tectono-magmatic activity is in striking contrast both with the Proterozoic crustal basement ages (Kovalenko *et al.*, 1990) and with the Proterozoic ages of the lithospheric mantle (Pearson *et al.*, 1998).

Isotope studies indicate that the entire western USA, which is tectonically active in Meso-Cenozoic, may be underlain by ancient lithospheric mantle. Hafnium and neodymium isotope studies of basalts from the Rio Grande Rift indicate that their ratios differ significantly from the ocean-island Nd–Hf array. The low ^{176}Hf/^{177}Hf ratios in these basalts were interpreted as being derived from ancient sub-oceanic mantle (Johnson and Beard, 1993). Samarium–neodymium studies indicate late Archean–Paleoproterozoic ages for the crust of the southern Basin and Range province (2.0–2.6 Ga) and Mesoproterozoic ages for the Colorado Plateau (1.7–2.0 Ga) (Bennett and DePaolo, 1987; Ramo and Calzia, 1998; Wendlandt *et al.*, 1993). Since Re–Os studies of mantle-derived peridotites from the Mojavia province also indicate a *c.* 2.5 ± 1.6 Ga age for the lithospheric mantle beneath the southern Basin and Range province (Fig. 2.9), isotope data implies that the continental crust was coupled with the lithospheric mantle during their tectonic evolution (Lee *et al.*, 2001). This conclusion is supported by Re–Os isotope studies of mantle-derived xenoliths from southern Africa (Fig. 2.18): while there is little correlation between the age of the lithospheric mantle and depth, isotope ages of the lithospheric mantle are in remarkable agreement with the geological age of the crust (Pearson, 1999).

Old crust–young lithospheric mantle in continental settings

In contrast to the young crust–old mantle relations discussed above, some Archean cratons have lost part of their lithospheric mantle so that the geological age of the crust is older than the age of the underlying mantle. This situation is typical for several cratons (e.g. the Wyoming, North China, Tanzania, and Congo cratons) which have Archean differentiation ages but have been tectonically reworked in the Phanerozoic. Rhenium–osmium data for peridotite xenoliths from the eastern block of the North China craton indicate that the Archean (*c.* 2.7 Ga) crust is underlain by the lithospheric mantle that was formed during the major Proterozoic orogeny (*c.* 1.9 Ga ago) when the Archean lithospheric mantle was either delaminated or significantly reworked (Gao *et al.*, 2002). Furthermore, the Proterozoic lithospheric mantle was (at least, in part) further replaced by more fertile young lithospheric mantle during the Mesozoic–Cenozoic tectonic events, as evidenced by Paleozoic (457–500 Ma) kimberlites and Neogene (16–18 Ma) volcanic rocks from the North China craton (Griffin *et al.*, 1998b) (Fig. 2.19). Thus, isotope data from the North

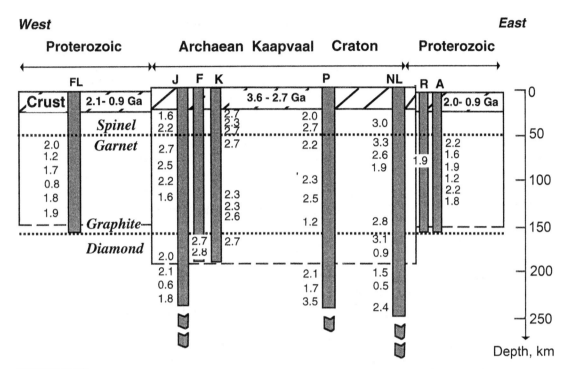

Fig. 2.18 Schematic cross-section through southern Africa showing variations of ages (in Ga) of the lithospheric mantle with depth based on Re–Os isotope data and geobarometry for xenoliths from different kimberlite pipes (shown by gray shading) (after Pearson, 1999).

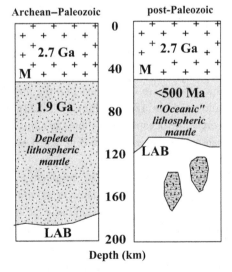

Fig. 2.19 A cartoon illustrating lithosphere evolution of the eastern North China craton (based on Re–Os, Nd–Sm, and Rb–Sr isotope studies by Griffin *et al.*, 1998b; Gao *et al.*, 2002, 2004). M = base of the crust, LAB = lithosphere-asthenosphere boundary. Numbers – age of the crust and the lithospheric mantle in Ga.

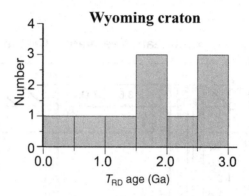

Fig. 2.20 Histogram of T_{RD} ages for peridotite xenoliths from the Wyoming craton (after Pearson, 1999).

China craton provide striking evidence that not only should geological and tectono-thermal ages not necessarily be concordant, but even geological ages of the crust and of the underlying lithospheric mantle may be significantly different.

Similarly, Re–Os isotope studies of peridotites from the Wyoming craton provide T_{RD} ages with the first peak, corresponding to the Archean differentiation of the Wyoming lithosphere, at $c.$ 2.5–3.0 Ga (Fig. 2.20). At the same time, U–Pb studies of Wyoming kimberlites yield Paleozoic ages (Fig. 2.21). As a result of Paleozoic tectono-magmatic processes (e.g. lithosphere delamination and metasomatism), ancient lithospheric mantle was replaced by fertile younger mantle. A similar situation is typical for the southern parts of the Archean–Paleoproterozoic East European craton. The Devonian plume caused significant modification and erosion of the cratonic lithospheric mantle along the Dnieper–Donets Rift, and possibly its complete loss at the intersection of two Devonian rifts in the central part of the PeriCaspian Basin, where the thickness of sediments reaches 20–22 km and formation of a new oceanic crust is suspected.

2.2.2 Oceanic crust

Since oceanic lithosphere is recycled into the mantle at subduction zones, the age of the oldest oceanic crust is $c.$ 180 Ma (Fig. 2.14, 2.22). In contrast to the age distribution of the continental crust, the oceanic crust shows an almost monotonous decrease of areal size with age for oceans older than $c.$ 20 Ma (Fig. 2.23), which reflects the survival of the oceanic lithosphere.

Oceanic lithosphere is formed by conductive cooling of the thermal boundary layer (Section 4.3), and the ages of oceanic crust and the underlying lithospheric mantle are expected to be the same. However, similarly to continental settings, recent isotope studies indicate that ancient (continental) lithosphere may be present in the suboceanic mantle.

Continental crust trapped in mid-ocean ridge

Zircons with ages of $c.$ 330 Ma and $c.$ 1.6 Ga have been found in gabbroic rocks from the central ($c.$ 23° N) Mid-Atlantic Ridge (MAR), 2000 km away from the continental margins

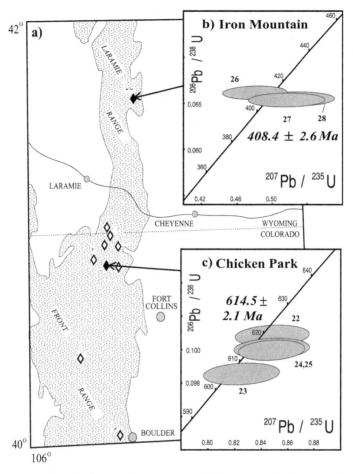

Fig. 2.21 U–Pb ages for kimberlites from the Archean Wyoming craton. (a) The location of kimberlites in Colorado and Wyoming. (b) and (c) $^{206}Pb/^{238}U$ age for two locations (from Heaman *et al.*, 2003). Most of the kimberlite magmatism in the Wyoming craton is Paleozoic or older (from 615 to 386 Ma).

(Pilot *et al.*, 1998). This unexpected finding has been interpreted as the presence of crustal material in the mantle beneath the MAR because if zircons were formed during magma generation along the ridge, their age would be *c.* 1 Ma. Two tectonic possibilities to explain the presence of Paleozoic and Proterozoic zircons in the mantle of the MAR include:

(i) delamination of the lithospheric mantle or continental crust during the opening of the Atlantic Ocean and its entrapment either into small-scale roll-like circulation cells along the ridge axis, or

(ii) entrapment of the continental material through a series of ridge jumps and transform migrations.

The Icelandic basalts further north along the Mid-Atlantic Ridge (*c.* 65° N) contain two major distinctive components: (a) a depleted component interpreted to be recycled oceanic crust (Chauvel and Hemond, 2002) and (b) an iron-enriched component derived either from

Age of ocean floor

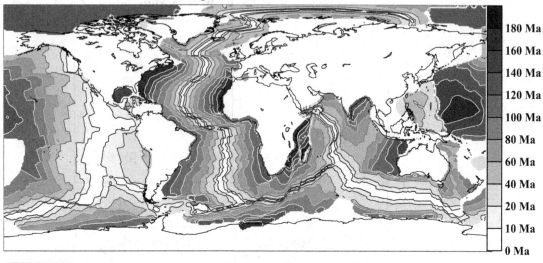

Fig. 2.22   Age of the ocean floor (based on digital data by Muller *et al.*, 2008).

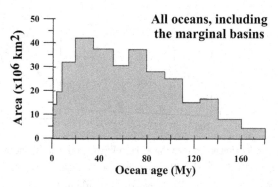

Fig. 2.23   Area as a function of age for the oceans (from Sclater *et al.*, 1980).

an ancient OIB seamount structure or from a predominantly eclogitic source (McKenzie *et al.*, 2004; Foulger *et al.*, 2005). Similarly to the Central Atlantic segment, isotope studies of the North Atlantic segment show that source rocks may also contain fragments of the Caledonian and older continental lithosphere, delaminated during the opening of the northern Atlantic Ocean (Korenaga and Kelemen, 2000).

Recycled Precambrian lithosphere in mantle plumes

Mantle plumes provide windows into the upper mantle. Another mantle plume that intersects the Mid-Atlantic Ridge at *c.* 39° N is located at the Azores with the islands scattered on both sides of the plume. This geometry of magmatism provides a unique possibility for examining variations in chemical and isotope composition of magmas across the mantle plume (Fig. 2.24).

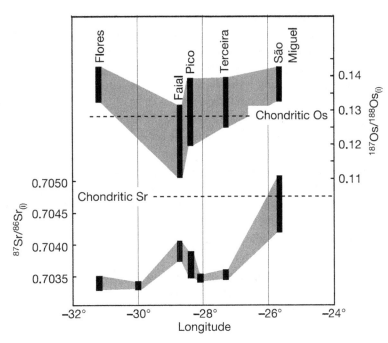

Fig. 2.24   NW–SE transect across the Azores archipelago showing the average $^{187}Os/^{188}Os$ and $^{87}Sr/^{86}Sr$ ratios for each island (from Schaefer *et al.*, 2002).

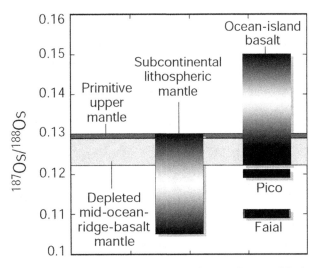

Fig. 2.25   Osmium isotope signatures in different mantle reservoirs (after Widom, 2002). Pico and Faial are two islands in the central part of the Azores archipelago.

While magmas in the Azores plume erupt through < 15 Ma old oceanic lithosphere, Re–Os isotope studies estimate the rhenium depletion age as 1.0–2.5 Ga in lavas from the islands of Pico and Faial. The value of T_{RD} provides the minimum estimate for ages, so the age of the depleted plume component should be at least late Archean (Schaefer *et al.*, 2002). Furthermore, many lavas show $^{187}Os/^{188}Os$ ratios lower than typical ocean-island basalts (Fig. 2.25). Lavas from three central islands of the Azores have subchondritic $^{187}Os/^{188}Os$ ratios (Fig. 2.24) which are typical for the Archean subcontinental lithospheric mantle. This unique osmium isotope signature of lavas requires their derivation from a depleted, harzburgitic mantle with more than 25% mantle melting (Schaefer *et al.*, 2002).

The proximity of the Azores archipelago to Iberia suggests that a fragment of the Iberian depleted lithospheric mantle could have been trapped during the opening of the Atlantic Ocean and could be the source of the mantle isotope anomaly (Widom and Shirey, 1996). However, the symmetric pattern of the Os anomaly with the lowest $^{187}Os/^{188}Os$ ratios observed in the central Azores islands as well as the Mesoproterozoic age of the Iberian lithospheric mantle question the validity of this hypothesis. Alternatively, the depleted mantle material may have originated from melting of the harzburgitic lithosphere from a subducted Archean oceanic plate (Schaefer *et al.*, 2002).

3

Seismic structure of the lithosphere

"One cannot embrace the unembraceable."

Kozma Prutkov

This chapter provides a short review of seismic methods used in lithosphere studies and their major results. It starts with an overview of laboratory measurements of seismic properties, which provide the basis for seismic interpretations, followed by a brief summary of seismic methods. For an introduction to seismic methods the reader is referred to the monograph of Shearer (1999). The advanced theory of seismic methods can be found in Aki and Richards (1980), Nolet (1987, 2008), Iyer and Hirahara (1993). Discussion of the major results of seismic studies of the lithosphere is structured by methods (converted waves, controlled source, and teleseismic tomography), which provide 1D, 2D, and 3D images of the lithosphere.

3.1 Laboratory studies of seismic properties of rocks

3.1.1 Introduction: Major elastic and anelastic parameters

An isotropic seismic model is defined by three parameters: density ρ of the material through which the wave is propagating, and the Lamé parameters λ and μ (see details in Section 8.1.1). The velocities of compressional (or P) waves Vp and shear (or S) waves Vs in an isotropic homogeneous medium are:

$$V_P = \sqrt{\frac{K + 4/3\mu}{\rho}},$$
$$V_S = \sqrt{\frac{\mu}{\rho}},$$

$$(3.1)$$

where K is the modulus of incompressibility (or the bulk modulus, see eq. 8.2) and μ (also denoted by G) is the shear modulus (a measure of material resistance to shearing). The bulk sound velocity is defined as:

$$V_B = \sqrt{\frac{K}{\rho}} = \sqrt{V_P^2 - 4/3 V_S^2}.$$

$$(3.2)$$

For an isotropic material, stress-induced deformation of the material in the direction of one axis will produce material deformation in three dimensions along the other axes (Hooke's

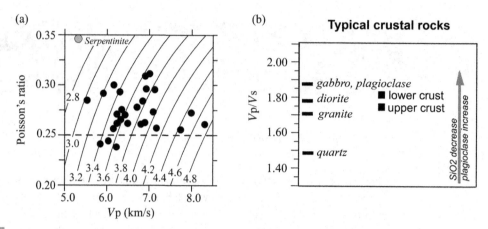

Fig. 3.1 (a) Relationship between Vp and Poisson's ratio based on laboratory measurements on crystalline (volcanic, igneous, and metamorphic) rocks (symbols – experimental data of Christensen, 1996). Thin dashed line corresponds to Poisson's ratio of 0.25 ($Vp/Vs = 1.73$). Thin lines – Vs (km/s) based on eq. 3.3. (b) Vp/Vs ratio for some crustal rocks.

law, eq. 8.1). The dimensionless Poisson's ratio v is the ratio of the relative transverse strain (normal to the applied load) to the relative axial strain (in the direction of the applied load); it quantifies the ratio of the lateral shortening to the longitudinal extension:

$$v = \frac{\lambda}{2(\lambda + \mu)} = 1 - \frac{V_P^2/2}{V_P^2 - V_S^2}. \tag{3.3}$$

The requirement for λ and μ to have positive values limits Poisson's ratio of a stable material to the range 0 to +0.5. (Some modern polymeric materials called auxetic may expand laterally when stretched and they have a negative Poisson's ratio (Lakes, 1987)). A perfectly incompressible material deformed elastically (e.g. liquids with $\mu = 0$) has a Poisson's ratio of exactly 0.5. For most crustal rocks, v varies between 0.25 and 0.30 (Fig. 3.1a). The mantle has $v \sim 0.25$ which is achieved when $\lambda = \mu$ (i.e. when $V_P = \sqrt{3}V_S$).

Since the Poisson's ratio concept is applicable only to an elastically deformed medium where Hooke's law is valid, the Vp/Vs ratio is widely used instead (Fig. 3.1b). The Vp/Vs ratio depends on many different factors such as composition, and there is no simple relation that exists between the Vp/Vs ratio and Vp or Vs, as indicated by studies of sedimentary (Castagna et al., 1985; Mavko et al., 1998) and igneous and metamorphic rocks (Christensen, 1996). Some other combinations of elastic moduli for isotropic materials that are in common use in seismology are:

$$K = (V_P^2 - 4/3V_S^2)$$
$$\lambda = \rho(V_P^2 - 2V_S^2) \tag{3.4}$$
$$\mu = \rho V_S^2.$$

Anelasticity leads to attenuation of seismic energy. Seismic attenuation is described by the "quality factor" Q which quantifies the amount of energy ΔE lost per cycle:

$$\frac{2\pi}{Q} = -\frac{\Delta E}{E}, \tag{3.5}$$

where E is the peak energy. Commonly, the inverse parameter Q^{-1} (seismic attenuation) is used in seismology. This includes two parts, intrinsic attenuation Q_i^{-1} (which is a material property and quantifies local conversion of elastic energy to heat) and scattering attenuation Q_s^{-1}:

$$Q^{-1} = Q_i^{-1} + Q_s^{-1}. \tag{3.6}$$

Attenuation results in dispersion, i.e. frequency dependence of the moduli. For a change in period from 1 sec to 1000 sec, seismic velocities change by $c.$ 1% for $Q_p \sim 200$ (typical for the upper mantle) and by $c.$ 4% for $Q_p \sim 60$ (typical for the low-velocity zone, LVZ) (Masters and Shearer, 1995).

3.1.2 Effects of pressure and temperature

Seismic velocities

Under the physical conditions of the Earth's mantle, the elastic constants of rocks vary significantly. With increasing depth, temperature and pressure increase rapidly and reach $c.$ 1400–1500 °C and $c.$ 13.5 GPa at the top of the transition zone (at $c.$ 410 km depth). While early experiments were made at ambient (standard pressure–temperature, STP) conditions (Birch, 1958), it was soon recognized that adequate comparison of observed seismic velocities with laboratory measurements on rock samples of different composition requires experiments at elevated (high) pressures and temperatures (Christensen, 1965; Kern, 1978). Experimental studies over the last half-century (with accuracy of individual measurements commonly better than 1%) have resulted in accumulation of an extensive database of seismic properties of rocks at crustal and upper mantle pressures and temperatures. Most of these measurements, in particular the early ones (Sato et al., 1989; Burlini and Kern, 1994), were made not at seismic, but at ultrasonic frequencies of 60–900 KHz (for comparison, the dominant period of the P-wave body waves is ~1 sec). As discussed later, laboratory experiments made at ultrasonic frequencies put certain limitations on the interpretation of seismic velocities in terms of temperature and composition, in particular at high temperatures (Fig. 3.2b).

The strong effect of temperature on elastic moduli (and hence on P- and S-wave velocities) has long been known from laboratory studies (e.g. Ide, 1937; Birch, 1943; Hughes and Cross, 1951; Christensen, 1965). Most measurements were made at temperatures lower than 600–700 °C (Kern, 1978; Christensen, 1979) and the amount of data for near-solidus temperatures ($T > 1000$ °C) remains very limited (Berckhemer et al., 1982; Murase and Kushiro, 1979; Murase and Fukuyama, 1980; Sato et al., 1989; Jackson, 1993). At low temperatures, the dependence of velocity on temperature is approximately linear (Sumino and Anderson, 1982):

$$\partial V_S / \partial T = -0.35 \text{ m/s/K}. \tag{3.7}$$

This relationship is based on laboratory experiments performed at high frequencies (~1 MHz, e.g. Kumazawa and Anderson, 1969) and accounts mostly for anharmonic effects

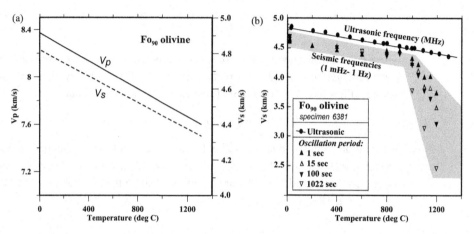

Fig. 3.2   Temperature-dependence of *Vp* (solid) and *Vs* (dashed) in fine-grained synthetic polycrystalline olivine samples (based on the results of Jackson *et al.*, 2005). (a) Results obtained at ultrasonic frequencies of 20–50 MHz. (b) Comparison of the results obtained at ultrasonic frequencies (circles and solid line based on theoretical predictions) with those at seismic frequencies (gray shading and other symbols). Note different vertical scales of the two plots.

(Karato, 1993). The latter are the deviation of mineral lattice vibrations from harmonic oscillations and result in pressure- and temperature-dependence of elastic constants. With temperature increase, at low homologous (near-solidus) temperatures *Tm/T* (where *Tm* is the solidus temperature below which a rock is completely solid), the temperature effect on elastic moduli becomes strong. It leads to viscous behavior in the material, a sharp decrease in *Vp* and *Vs*, and a sharp increase in Poisson's ratio and *Vp/Vs*. Importantly, a significant (*c.* 3%) reduction of seismic velocities occurs before any melting in the system starts (Fig. 3.3).

Pressure and temperature have opposite effects on seismic velocities: while pressure increase causes an increase in both *Vp* and *Vs* velocities, an increase in temperature leads to a decrease in both of the seismic velocities (Fig. 3.4). At relatively low temperatures typical for a shallow lithosphere, when no melt is present, pressure and temperature effects on elastic moduli approximately compensate each other, so that Poisson's ratio remains approximately constant and depends mostly on the composition. This fact is widely used in seismology for minera-logical interpretations of Poisson's ratio variations in the lithosphere (in particular, in the crust).

Seismic attenuation, anelasticity, and effect of frequency

The effect of temperature on seismic attenuation is much stronger than on seismic velocities (Fig. 3.5). Experimental measurements of seismic attenuation in upper mantle rocks carried out over a wide range of pressures and temperatures at seismic (0.01–1 Hz) frequencies (Berckhemer *et al.*, 1982; Jackson, 1993; Gribb and Cooper, 1998) show that at subsolidus temperatures the attenuation, Q^{-1}, in mantle rocks follows the Arrhenius law and exponen-tially increases with temperature *T*:

$$1/Q = A\tau^{\alpha} \exp\left(-\alpha E^{*}/RT\right), \tag{3.8}$$

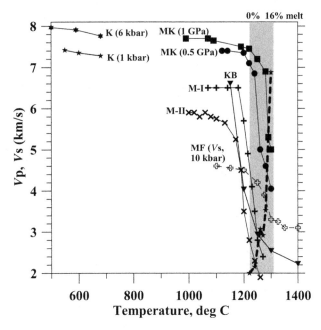

Fig. 3.3 Temperature-dependence of seismic velocities (based on Artemieva *et al.*, 2004). Solid lines – *Vp*, dashed lines – *Vs*. Thick dashed line and shaded area: changes in melt volume (Murase and Fukuyama, 1980). Sources: K = Kern (1978), *Vp* in peridotite at 1 and 6 Kbar; MK = Murase and Kushiro (1979), *Vp* in peridotite at 0.5 and 1.0 GPa; KB = Kampfmann and Berckhemer (1985), *Vp* in dunite (at 0.1 Hz); MF = Murase and Fukuyama (1980), *Vs* in peridotite at 10 Kbar; M-I and M-II = Murase *et al.* (1977), *Vp* in two peridotite samples.

where E^* is the activation energy, R is the gas constant, τ is the oscillation period, A is a scaling constant (commonly assumed to be 0.148, Sobolev *et al.*, 1996), and exponent α, determined from seismic studies and laboratory measurements on the upper mantle rocks, is almost temperature-independent and may range from 0.15 to 0.30 for different values of τ, T, and P (Jackson *et al.*, 1992).

Laboratory measurements of seismic attenuation in dry peridotite at high temperatures $0.95 < T_m/T < 1.17$ (where T_m is solidus temperature) and at pressures of 200–730 MPa demonstrate a linear dependence of the P-wave quality factor Q_p on the homologous temperature T_m/T (Sato and Sacks, 1989; Sato *et al.*, 1989). Laboratory results (performed at ultrasonic frequencies, 60–900 KHz) can be extrapolated to higher pressures through the pressure-dependent solidus temperature $T_m(P)$ (in Kelvin):

$$Q_p = Q_{pm}\exp[g(T_m(P)/T - a)], \tag{3.9}$$

where $Q_{pm} = 3.5 + P/0.073$ (here P is pressure in GPa), a and g are experimentally determined parameters (a is close to 1, while g is strongly T-dependent and increases from 6.75 to 13.3 with an increase in homologous temperature from 0.95 to 1.15).

Attenuation results in frequency-dependence of the elastic moduli which becomes particularly strong at near-solidus temperatures (Fig. 3.7). Jackson *et al.* (1992) measured seismic attenuation on dry dunite at seismic frequencies (0.01–1 Hz), at temperatures of 20°, 600°,

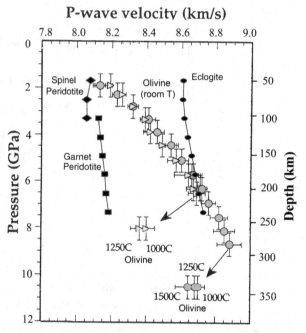

Fig. 3.4   Pressure-dependence of P-wave seismic velocity in mantle rocks. Gray symbols – San Carlos olivine at room temperature; the arrows show changes in velocity and pressure which result from heating (from Knoche *et al.*, 1998). Black symbols – calculated compressional velocities for hypothetical mantle eclogite and fertile peridotite compositions at *in situ* mantle temperatures (James *et al.*, 2004). Temperature is based on the cold Archean geotherm (Kaapvaal craton); pressure-to-depth conversion is calculated assuming densities of crust and mantle 2.7 and 3.3 g/cm^3, respectively.

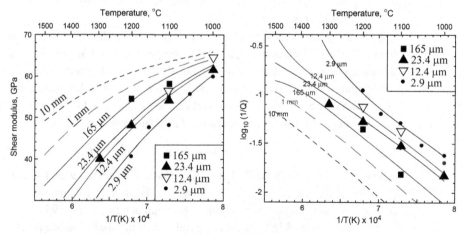

Fig. 3.5   Shear modulus μ and seismic attenuation as functions of temperature and grain size for a fixed oscillation period of 8.2 sec (after Faul and Jackson, 2005). Symbols show the experimental data, the lines are theoretical predictions.

800, and 1000 °C and at pressures up to 300 MPa. The experimental data allow for approximation of P-wave seismic attenuation as a function of temperature (T in °C) as:

$$1/Q_{p0} = \exp(0.00267^*T - 6.765),\eqno(3.10)$$

where $1/Q_{p0}$ is the value of $1/Q_p$ at 1 s period (Figs. 3.6a and 3.8b). The value of Q_p depends strongly on oscillation period, τ, and between 600 and 1000 °C increases approximately exponentially with increasing temperature T and the Arrhenius law (3.8) takes the form:

$$1/Q_p = 1/Q_{p0}\tau^{\alpha}.\eqno(3.8a)$$

At low frequencies (large oscillation period τ), attenuation is stronger and the shear modulus has smaller values than at high frequencies (Fig. 3.8).

With increasing temperature, anelasticity significantly affects seismic velocities, in particular for shear waves. In contrast to anharmonicity, anelasticity is accompanied by a loss of energy during wave propagation and thus anelastic effects are frequency-dependent. Their effect on seismic velocities becomes particularly important at high temperatures and low frequencies (at lower frequencies, seismic velocities are lower) and can be approximated as (Minster and Anderson 1981):

$$V_s(\text{anelastic}) = V_s(\text{elastic})[1 - C/Q(\tau, P, T)],\eqno(3.11)$$

where $C = 2/\tan(\pi\alpha/2)$ and $1/Q(\tau, P, T)$ is given by Eq. (3.8). Note that extrapolation of relationship (3.8) to seismic frequencies may not be valid, as mechanisms of seismic attenuation other than the grain boundary process (assumed by Sato and Sacks, 1989) such as dislocation and micro-creep processes can be equally important at seismic conditions (Karato, 1998; Romanowicz and Durek, 2000). In the upper mantle, the correction for anelasticity approximately doubles the temperature derivatives (eq. 3.7) caused by anharmonicity alone (Karato, 1993). However, it is impossible to measure anelasticity effects at seismic frequencies in laboratory experiments based on wave propagation because wavelengths of seismic waves are significantly larger than sample size.

3.1.3 Effect of grain size variations

Experimental studies performed on various ultrabasic and basic rocks and on polycrystalline olivine aggregates have demonstrated that shear modulus and attenuation depend not only on oscillation period but also on grain size (Berckhemer *et al.*, 1982; Kampfmann and Berckhemer, 1985). Recent experiments performed on melt-free polycrystalline aggregates of Fo90 olivine at fixed temperature and fixed frequency indicate that shear modulus increases and attenuation decreases with increase in grain size (Fig. 3.5). It is worth noting that the range of olivine grain sizes used in the experiments was significantly smaller than olivine grain size variations measured in upper mantle xenoliths from different continental locations (Fig. 3.9); a typical distribution of grain sizes in dunite samples that are widely used in laboratory studies is shown in Fig. 3.10.

Model calculations of grain size evolution in the oceanic upper mantle for a composite diffusion–dislocation creep rheology allow for estimation of grain size as a function of depth, seafloor age, and mantle water content (Behn *et al.*, 2009), assuming that grain size

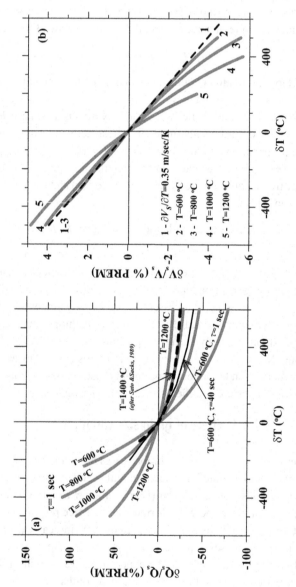

Relative anomalies of Q_s (a) and V_s (b) with respect to PREM as a function of temperature anomalies calculated for upper mantle temperatures of 600, 800, 1000, 1200, and 1400 °C (after Artemieva et al., 2004). (a) δQ_s anomalies calculated from the experimental data of Jackson et al. (1992), eqs. (3.10, 3.17), for $\tau = 1$ sec. For comparison, the results are also shown for $\tau = 40$ sec for $T = 600$ °C (thin black line). Dashed line – δQ_s calculated for high homologous mantle temperature T_m/T based on the experimental data of Sato and Sacks (1989), eq. (3.9). (b) δV_s anomalies caused by anelasticity (eqs. 3.8, 3.11), for $\tau = 40$ sec. For a comparison, dashed line: δV_s based on the linear relationship (3.7).

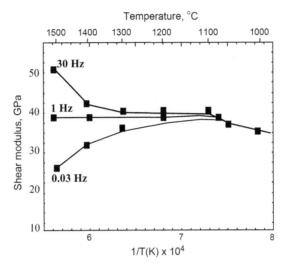

Fig. 3.7 Dispersion of the shear modulus of polycrystalline forsterite at high temperatures (after Berckhemer *et al.*, 1982).

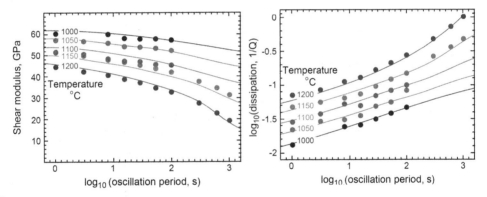

Fig. 3.8 Shear modulus μ and seismic attenuation as functions of temperature and period of oscillation for a sample with grain size 3 μm (from Faul and Jackson, 2005). The dots represent the experimental data, the lines are theoretical predictions.

evolution in the upper mantle is controlled by competition between dynamic recrystallization and grain growth (Austin and Evans, 2007). These theoretical calculations suggest that,
in a 60-Ma-old oceanic mantle, grain size has a minimal value of ~10 mm at *c.* 130 km depth
and increases to 20–30 mm above the transition zone in the case of a wet mantle (water
content of 1000 ppm), but it is only ~5 to ~10 mm at depths between 150 and 400 km in the
case of a dry mantle (water content 50 ppm).

The cratonic upper mantle, which is believed to be dry (Hirth *et al.*, 2000; for discussion
of this hypothesis see Sections 6.3 and 8.1.3), has similar values of olivine grain size
between 150 and 240 km depth as evidenced by mantle peridotite xenoliths from South
Africa (Fig. 3.9a). Since neither abrupt changes in stress regime nor phase changes are
expected at 140–150 km beneath the cratons, a pronounced drop in olivine grain size from

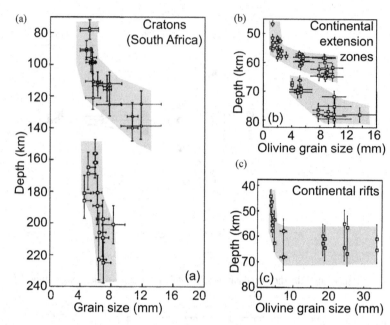

Fig. 3.9 Olivine grain-size distribution with depth for mantle peridotite xenoliths from (a) southern Africa kimberlites, (b) extensional Basin and Range Province, western USA, and (c) locations in France and Nevada typical of late stages of continental rifting (redrawn from Mercier, 1980).

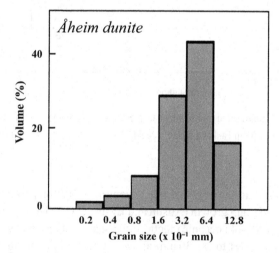

Fig. 3.10 Grain-size distribution in Åheim dunite samples that are widely used in laboratory studies (Berckhemer *et al.*, 1982).

c. 12 mm at 120–140 km depth to 4–8 mm below 150 km in the cratonic mantle has been attributed to a dramatic (temperature-dependent) change in the dominant recrystallization mechanism from subgrain rotation at low temperatures to grain boundary migration at high temperatures (Poirier and Nicolas, 1975; Poirier and Guillope, 1979; Mercier, 1980).

Data on depth variation of grain size allow for calculation of variations in shear wave velocity and seismic quality factor with depth. The strong dependence of shear modulus and attenuation on grain size allows for interpretation of the low-velocity zone beneath the ocean basins by variations in grain size without requiring the presence of melts and fluids (Faul and Jackson, 2005). However, this explanation only holds for wet mantle, while grain size variations in a dry upper mantle are insufficient to explain the seismic LVZ in the absence of water and/or melt.

3.1.4 Effect of mineralogy

Laboratory measurements of seismic velocities in rocks form the basis for petrologic interpretations of the seismic velocity structure of the crust and the upper mantle. These data are summarized in a review by Christensen and Mooney (1995) who discuss the effects of mineralogy, chemistry, and metamorphic grade on compressional wave seismic velocities in the crust (Fig. 3.11) and on their temperature-dependence (Fig. 3.12). Unfortunately, similar analysis for Vs seismic velocities has received only limited attention.

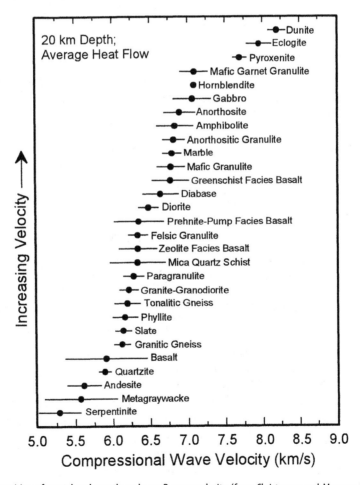

Fig. 3.11 Effect of composition of crustal and mantle rocks on P-wave velocity (from Christensen and Mooney, 1995).

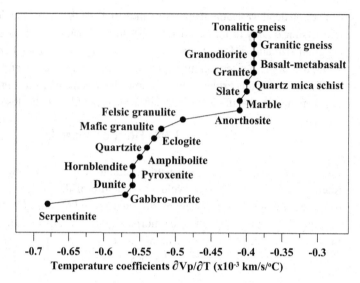

Fig. 3.12   Temperature coefficients for *Vp* seismic velocity in crystalline rocks (data of Christensen and Mooney, 1995).

An important conclusion of experimental seismic velocity measurements on crystalline rocks is the existence of a large variation in the range of elastic properties even for rocks with simple lithologies (such as granite and gabbro); for most crustal rocks velocity histograms have a normal distribution. Since variations in composition have different effects on shear and compressional wave velocities, the *Vp/Vs* ratio and Poisson's ratio provide better diagnostics of variations in the major element composition of crustal (in particular, lower crustal) and upper mantle rocks (e.g. Christensen and Fountain, 1975; Fig. 3.1). The effect of compositional variations in the upper mantle related to partial melting and melt extraction is discussed in Section 3.1.7.

3.1.5 Anisotropy

The lithosphere is heterogeneous. Wave propagation in heterogeneous media may result in seismic anisotropy, and seismic anisotropy is the rule rather than the exception (Montagner and Tanimoto, 1990; Maupin and Park, 2008). Heterogeneities that cause crustal and upper mantle seismic anisotropy can exist:

- *at the crystal scale*, caused by lattice-preferred orientation (LPO) due to alignment of anisotropic crystals such as olivine and orthopyroxene (Christensen and Lundquist, 1982; Nicolas and Christensen, 1987),
- *at the rock scale*, produced by rock fabric (Katayama *et al.*, 2004),
- *at the crustal scale*, caused by rock foliation and layering (Crampin, 1977),
- *at the rock scale or crustal scale*, caused by stress-aligned microcracks and pores (which can be filled with fluids and melts) (Nur, 1971; Crampin, 1987),
- *at large scale in the mantle*, caused by a temperature difference between upgoing and downgoing flow in mantle convective cells (Montagner and Guillot, 2002).

Pores and microcracks

Aligned microcracks, as well as shear fabric, rock foliation, and layered bedding in sedimentary formations are the major causes of seismic velocity anisotropy in the Earth's crust. Microcracks and pore-spaces that exist in most rocks are often preferentially aligned by the stress-field acting on the rock mass. Such aligned microcracks caused by regional stress-induced deformation are widely observed in nature, for example in metamorphic complexes (Etheridge, 1983). Crustal anisotropy resulting from crack alignment has received particular attention, since it can be used to determine the state of stress in the crust (Crampin, 1981). Laboratory experiments and field observations indicate that most microfractures are extensional features oriented perpendicular to the direction of the current minimum compressional stress (e.g. Tapponnier and Brace, 1976; Kranz, 1983). Stress changes associated with earthquakes modify crack geometry and cause changes in shear wave splitting that can be used to predict earthquakes with the same signatures (Crampin, 1987). The stress- and pressure-dependence of macroscopic seismic properties of rocks is also of a great importance for oil recovery, geothermal energy, and the disposal of radioactive wastes.

Geochemical studies show that a large amount of fluids exist in crustal rocks (e.g. Fyfe *et al.*, 1978) and greatly influence the macroscopic properties of rock masses. In crystalline rocks, fluids can exist in microinclusions (pores or partly healed microcracks). In the crust, the stress field and differential fluid pressure variations are necessarily dependent. The stress-induced variations of the internal crack geometry (due to opening, closure, subcritical and dynamic crack growth, and nucleations of cracks) promote changes in fluid pressure. The latter cause additional changes in macroscopic properties of fluid-saturated rock through their effects on both the elastic parameters and crack geometry. Furthermore, with changing pressure and temperature, pore fluids can change greatly their physical characteristics (density, conductivity, viscosity, acoustic velocities) which may result in a drastic change in the overall physical parameters of rock (Artemieva, 1996).

Pores and fluid-filled microcracks are essential only in the sedimentary and upper crustal rocks. It is widely accepted that fluid-filled cracks cannot exist deeper than 3 km because the confining pressure required to close elliptical cracks is proportional to the aspect ratio of the cracks and Young's modulus of the saturating fluid (e.g. Doyen, 1987). This means that for realistic values of crack aspect ratio and physical parameters of fluids, all cracks are expected to be closed at pressures ranging between 30–100 MPa. This conclusion is in agreement with *in situ* observations from the Kola Superdeep borehole which indicate that, due to the closure of microcracks under lithostatic pressure, their effect on seismic velocities becomes insignificant below *c.* 5 km depth (Kola Superdeep, 1984). In contrast, seismic studies suggest that fluid-filled microcracks are likely to exist in the continental crust down to a 10 to 20 km depth (Crampin, 1987). The existence of partially open cracks at these depths can be due to high pore pressures (Christensen, 1989) that can be produced by tectonic stress (subhorizontal tectonic compression, rapid tectonic loading due to sedimentation), chemical reactions (crack sealing by mineral deposition, metamorphic reactions involving pore-pressure buildup during dehydration), or phase changes in saturating fluid. According to numerical modeling of the closure of thermal microcracks (Zang, 1993), significant *in situ* porosity may be found at confining pressures up to 300 MPa (i.e. at a

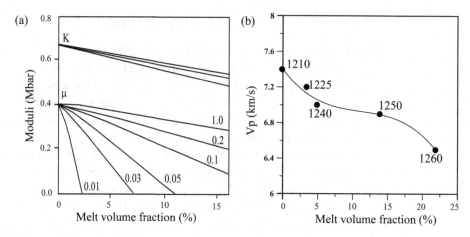

Fig. 3.13 (a) Elastic moduli for partially molten rock with ellipsoidal melt inclusions; numbers – aspect ratio of inclusions, i.e. the ratio of the short axis to the long axis (redrawn from Schmeling, 1985). (b) Changes in compressional velocity in spinel lherzolite (at 5 kbar) caused by changes in melt volume (based on laboratory data of Murase and Kushiro, 1979). Numbers – temperature in °C.

crustal depth of about 10 km), and the fraction of cracks that remain partially open under high pressures depends on the shape of the initial crack geometries that should be more realistic than elliptical.

Numerical modeling indicates that the amplitude of velocity reduction, as well as the amplitude of velocity anisotropy due to the presence of cracks in the upper crust or melt and fluid inclusions in the crustal and upper mantle rocks, depend critically on:

- the geometry of inclusions,
- the interconnectivity of the fluid phase (discussed in the next section).

For the same fluid (or melt) fraction, flattened elliptical inclusions with small aspect ratio (defined as the ratio between the short and the long axes) and thin films decrease shear modulus (and *Vs* seismic velocities) much more than inclusions with a more spherical geometry (Fig. 3.13a). Technical details on macroscopic elastic properties of cracked rocks can be found in numerous publications (e.g. Budiansky, 1965; Budiansky and O'Connell, 1976; Mavko, 1980; Schmeling, 1985a).

LPO anisotropy

The alignment of highly anisotropic olivine crystals (Fig. 3.14) by mantle flow and the associated deformation is believed to be the major cause of seismic anisotropy in the upper mantle (Babuska, 1984; Karato, 1992; Barruol and Kern, 1996; Ben Ismail and Mainprice, 1998). Two major deformation mechanisms play important roles under upper mantle conditions.

- *Dislocation creep* is caused by movement of crystalline dislocations within grains. It occurs at high stress, at large grain size, or at both, and leads to preferred orientation of

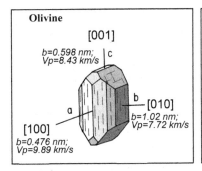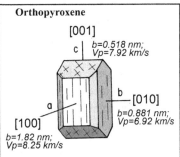

Fig. 3.14 Anisotropy of olivine and orthopyroxene crystals (based on Nicolas and Christensen, 1987). The length of the Burgers vector (*b*) and compressional wave velocity (*V_p*) are given for each slip direction. The favored slip direction is the one with the shortest Burgers vector.

minerals and to anisotropy. Additionally, dislocation-accommodated grain boundary sliding can play an important role in deformation, in particular of fine-grained samples (Hirth and Kohlstedt, 2003; Drury, 2005; Warren and Hirth, 2006).

- *Diffusion creep* is caused by solid-state diffusion between grain boundaries or across a crystal lattice, and occurs at relatively low stress, at small grain size, or at both. It does not lead to preferred orientation of minerals. It is commonly accepted that diffusion creep dominates in the upper mantle below a 250 km depth.

LPO (lattice preferred orientation) can develop not only by deformation due to dislocation slip, but by recrystallization as well. However, the former mechanism is dominant. Deformation by dislocation slip is simple shear and it is defined by the slip direction and the slip plane. The following notation (a three-value Miller index notation) is used in crystallography:

- planes are denoted by three indices in round brackets and each index denotes a plane orthogonal to a direction in the lattice; for example (100) refers to a plane normal to the *a*-direction; curly brackets like {100} denote the set of all planes that are equivalent by symmetry of the lattice (as in a cubic crystal);
- rectangular brackets denote directions in a crystal and the numbers are the smallest integers that correspond to the direction cosines; for example [100] is the direction parallel to the *a*-direction; triangular brackets as in <100> are used for equivalent directions.

The slip direction is characterized by the Burgers vector which specifies the magnitude and the direction of the lattice distortion in dislocations in a crystal lattice. The slip direction with the shortest Burgers vector is favored because of the lower strain energy associated with formation or motion of lattice dislocations. It is commonly assumed that olivine deformation is characterized by a single slip. Since Burgers vector along [100] is the smallest, the [100] slip is expected to dominate in olivine. However, the difference in the lengths of [100] and [001] vectors is small (Fig. 3.14) and under favorable physical/chemical conditions the dominant slip direction can change (Karato *et al.*, 2008). A significant LPO in olivine may develop in ~30 Ma at a typical geological strain rate of 10^{-15}/s (this value of strain rate corresponds to stress $\sigma = 10$ MPa and mantle viscosity $\eta = 10^{22}$ Pa s, eq. 8.11).

Experimental studies on the relationship between LPO and deformation in different minerals, including olivine, began in the 1970s (Carter and AveLallemant, 1970, Nicolas *et al.*, 1973). Until the last decade, all interpretations of seismic anisotropy had been based on these and similar laboratory results, many performed at upper mantle P–T conditions (e.g. Karato and Rubie, 1997; Cordier and Rubie, 2001). Studies of mantle xenoliths from kimberlites and Alpine massifs, as well as mantle rocks from ophiolites and oceanic islands demonstrate that the *a*-axis in olivine commonly orients parallel to the macroscopic flow direction, and the [100] slip has greatest influence on the seismic properties. Mainprice and Silver (1993) estimated that an upper mantle with 50% and 100% of olivine content can produce maximum anisotropies of 6% and 13.9% for P-waves and 7.1% and 9.5% for S-waves, respectively. These values are greater than the S-wave anisotropy of 3–7% reported for most types of mantle-derived xenoliths. However, mantle rocks such as peridotites are polymineralic, and deformation mechanisms can vary for different minerals. In particular, olivine and orthopyroxene have different slip systems (Fig. 3.14b), different elastic properties, and different strengths.

Laboratory measurements on mantle xenoliths demonstrate that at pressures up to 600 MPa and at temperatures up to *c*. 600 °C shear-wave anisotropy remains approximately the same, while higher pressures and temperatures affect anisotropy (Kern, 1993). A temperature of *c*. 900 °C is critical: it is difficult to reorient deformed olivine crystals below this temperature (Goetze and Kohlstedt, 1973). As a result, rocks deformed in a high temperature environment and then cooled below 900 °C will preserve a "frozen-in" anisotropy. Partial melts may cause grain size increase and thus increase anisotropy or extend the depth interval with dislocation creep (Nicolas, 1992).

To sum up, *the conventional point of view*, based on laboratory studies of deformation in olivine aggregates at upper mantle P–T conditions, is that:

- deformation of olivine in the upper mantle down to *c*. 250 km depth occurs essentially by dislocation creep;
- the dominant slip is [100];
- below a 250 km depth, diffusion creep becomes the dominant deformation mechanism, leading to a rapid decrease in anisotropy in the upper mantle.

Recent (post-2000) experimental studies of olivine deformation challenge this commonly accepted point of view, whereas the physical grounds for questioning the paradigm are based on the fact that several different LPOs can develop in olivine (reviewed by Katayama *et al.*, 2004; Mizukami *et al.*, 2004; Karato *et al.*, 2008). Different LPOs are produced by different kinematic processes, each with its own rate, and the physico-chemical conditions of deformation determine these rates and the type of LPO that is formed in different tectonic settings (Fig. 3.15c). In particular, olivine fabrics are sensitive not only to stress and temperature, but also to water content (Katayama *et al.*, 2004; Jung *et al.*, 2006). Under saturated conditions mantle olivine can undergo water-induced fabric transitions, when the orientation of the *a*-axis may become perpendicular to the mantle flow (Jung and Karato, 2001; Mizukami *et al.*, 2004). The results of some recent high-pressure, high-temperature simple-shear laboratory experiments on olivine aggregates indicate the following.

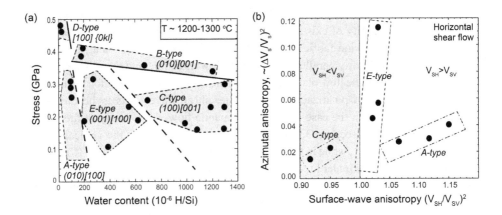

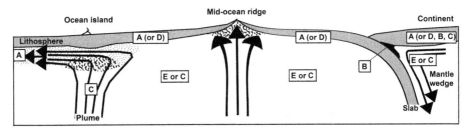

Fig. 3.15 (a) Deformation fabrics of olivine measured at high temperatures as a function of water content and stress. (b) The relation between the amplitudes of azimuthal and polarization anisotropy caused by horizontal shear flow for various types of olivine LPO. Symbols – laboratory data. (c) Diagram illustrating possible distribution of olivine fabrics (letters in boxes) in the upper mantle. (a – based on Katayama *et al.* 2004; b, c – based on Karato *et al.* 2008).

1) The olivine LPO at low pressure and high pressure is different, and the LPO may change with the pressure (depth) change (Couvy *et al.*, 2004). The exact values of stress at which different types of olivine fabric are formed are unknown because measurements were taken in stress–relaxation tests (Karato *et al.*, 2008). Similar conclusions were drawn by Mainprice *et al.* (2005) based on simple-shear laboratory experiments on olivine aggregates. They found that dislocation creep remains the dominant deformation mechanism at P–T conditions corresponding to *c.* 300 km depth, but with dominant activation of the [001] slip direction. These results were reported even for fine-grained, ~20–30 μm aggregates. A significant amount of water was detected in olivine samples in the experiments of both groups. Karato *et al.* (2008) classifies this as C-type olivine fabrics.

2) In the case of partial melting the olivine LPO is different from that in melt-free olivine (Holtzman *et al.*, 2003). Deformation experiments on olivine + basaltic melt at $P = 0.3$ GPa and $T = 1250$ °C indicate that olivine fabric is the same as in a melt-free olivine (A-type fabric) when a small amount of water is present. Reduction of melt permeability leads to its localization in narrow bands, and olivine deformation localizes in the melt-rich bands so that its [100] axes orient normal to the macroscopic flow direction and [001] axes align along the shear direction (a fabric similar to the olivine B-type fabric).

These results have important implications for interpretations of seismic anisotropy. The vast majority of existing interpretations are based on the widely accepted notion that the A-type olivine fabric dominates in the upper mantle, so that the olivine [100] axes are oriented nearly parallel to the mantle flow direction, and the olivine (010) planes are nearly parallel to the flow plane. This results in anisotropic propagation of surface waves so that the direction of polarization of the faster shear wave tends to become parallel to the flow direction. In the case of horizontal mantle flow, the velocity of the horizontally polarized wave V_{SH} (see Fig. 3.28) is greater than the velocity of the vertically polarized wave V_{SV}, $V_{SH}/V_{SV} > 1$, while in the case of vertical mantle flow, $V_{SH}/V_{SV} < 1$. The A-fabric is expected to dominate under lithospheric conditions which are characterized by low water content and moderate stress and temperature (Karato et $al.$, 2008). However, in case other olivine LPO fabrics can be formed under specific physico-chemical conditions, interpretations of seismic anisotropy can differ from the conventional. In particular, in the case of horizontal mantle flow, non-conventional $V_{SH}/V_{SV} < 1$ can be expected for C-type olivine fabrics (Fig. 3.15b).

Orthopyroxene which makes 20–30% of the upper mantle is also anisotropic (Fig. 3.14) and can also form LPO and contribute to seismic anisotropy (Skemer et $al.$, 2006). The number of experimental studies of orthopyroxene deformation is limited. Because the length of the [001] Burgers vector in orthopyroxene is much shorter than those of other vectors, deformation occurs on the [001] slip system (Mercier, 1985). However, the contribution of orthopyroxene to seismic anisotropy is thought to be small compared to olivine (Christensen and Lundquist, 1982).

3.1.6 Melt and fluid inclusions

The amount of water stored in the mantle is assumed to be small; there is, however, a significant variation in the reported values (see discussion in Section 8.1.3). Bulk water content in the upper mantle is estimated to be $c.$ 5–35 ppm in the oceanic lithosphere, 100–200 ppm in the MORB (mid-ocean ridge basalts) source, 200–500 ppm in the OIB (oceanic island basalts) source, and <200 ppm in the lower part of the upper mantle (Fig. 7.19). The continental lithospheric mantle of the Archean cratons, by comparison with "wet" oceanic lithosphere, is believed to be particularly "dry" (e.g. Hirth et $al.$, 2000). However, a significant amount of water can be stored in the mantle in such "anhydrous" minerals as olivine, pyroxenes, and garnets (e.g. Bell and Rossman, 1992). The presence of even a small amount of water or fluid in crustal and mantle rocks significantly reduces seismic velocities: for example, a 0.3% increase in H_2O content causes a 1% decrease in P-wave velocity in partially serpentinized ultramafic rocks.

Another important aspect of the presence of fluids in the upper mantle is their effect on melting temperature of mantle peridotite (Thompson, 1992; Wyllie, 1995). In particular, the solidus of "wet" peridotite is significantly lowered at depths below $c.$ 80 km (Fig. 3.16). This implies that at high temperatures, the presence of even small volumes of volatiles (water and carbon dioxide) in the upper mantle indirectly reduces seismic velocities by lowering mantle melting temperature.

Experimental studies of the properties of partially molten rocks demonstrate the strong role of the 3D geometry of a melt network and its interconnectivity on the seismic properties

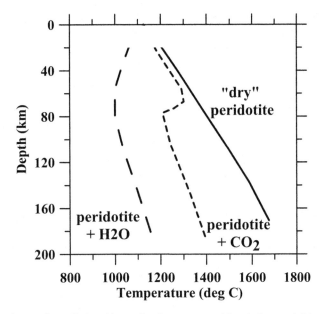

Fig. 3.16 P–T phase diagram for mantle peridotite with a major element composition similar to primitive mantle. Solid line – melting of a natural dry peridotite (Takahashi, 1986; Hirschmann, 2000). Dashed lines – melting of peridotite saturated with H_2O and CO_2 ($CO_2/(CO_2+H_2O)) = 0.8$ (compilation by Wyllie, 1995).

of crustal and mantle rocks. The interconnectivity of melt depends on the wetting (dihedral) angle between the solid and liquid phases (Fig. 7.15). Basaltic melt in the mantle (olivine-dominated) matrix is expected to have a wetting angle of 30–50° and to form a network of tubes along the grain boundaries (Fig. 7.14b). With such a geometry of melt distribution, partial melting should have a strong effect on the elastic properties of rocks; however the effect is significantly smaller than in the case of complete wetting which occurs when $\Theta = 0°$ (Stocker and Gordon, 1975). Recent results demonstrate that very small, unconnected grain boundary melt fractions can produce a significant shear wave velocity decrease such as observed in the LVZ (Faul and Jackson, 2007). Similarly, the presence of even a small amount of melt or fluids has a dramatic effect on seismic attenuation. However, the effect of melt or fluid inclusions on compressional wave velocities is small (Muellar and Massonne, 2001). In the case $\Theta > 60°$ (which may be typical of, at least, some upper mantle rocks), melt forms isolated pockets and its effect on the macroscopic rock properties depends on pocket geometry.

Up-to-date microstructural and experimental studies of melt distribution in mantle rocks, however, remain inconsistent. The reported grain-scale melt distributions include the following microstructures:

- melt concentration at two-grain boundaries at melt fractions above 0.05 (Hirth and Kohlstedt, 1995);
- melt concentration in a network of triple junction tubes with almost no melt along grain boundaries (for melt fraction c. 0.01) (Wark et al., 2003);

- domination of disc shaped inclusions or melt layers on two-grain boundaries with some melt distributed in a network of tubes along triple junctions (for melt fraction *c.* 0.1) (Faul, 1997[1]; ten Grotenhuis *et al.*, 2005);
- presence of transient melt features, corresponding to metastable structures in melt distribution, such as multigrain melt pools and melt lenses along grain boundaries (Walte *et al.*, 2003).

An increase in melt content apparently changes the geometry of a melt network which has a dramatic effect on seismic velocities (e.g. Fig. 3.13) (Anderson and Sammis, 1970; Shankland *et al.*, 1981). Based on this observation, seismic and electrical asthenospheres can be explained by the presence of partial melt in the mantle: while the observed anomaly in electrical conductivity requires the presence of several per cent of melt, the seismic velocity anomaly can be explained by a melt fraction of 0.1% in the case of very flattened melt inclusions (aspect ratio of 0.001) (Shankland *et al.*, 1981). Such interpretations have recently been challenged by Karato and Jung (1998), who argue that

> the common belief that the seismic low velocity and high attenuation zone (the astheno-sphere) is caused by the presence of a small amount of melt is not supported by recent mineral physics and seismological observations.

According to these authors, since water reduces seismic wave velocities, low percentage partial melting of the Earth's upper mantle will increase (not decrease, as common experience suggests) seismic wave velocities due to the removal of water from mantle olivine.

Figure 3.17 summarizes the effects of different parameters on variations of seismic velocities and attenuation.

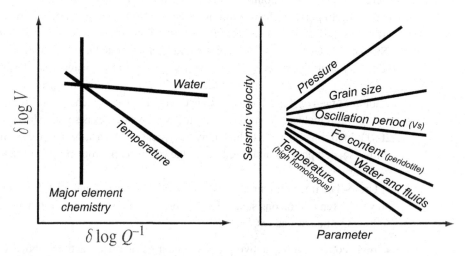

Fig. 3.17   Left: Relationship between attenuation and seismic velocity anomalies showing schematically the effects of water, temperature, and composition (redrawn from Karato, 2008; based on results of Shito and Shibutani, 2003). Right: Schematic diagram showing the effect of different parameters on seismic velocities.

3.1.7 Melt-depletion and mantle composition

Partial melting of the upper mantle and melt extraction cause its chemical differentiation with variations in major element chemistry that can be detected by seismological observations. Recent laboratory measurements combined with theoretical calculations for chemical and mineralogical compositions of natural peridotites indicate first-order differences between the composition of oceanic and continental lithospheric mantle (referred to as the oceanic and the continental trends). A detailed discussion of these differences is presented in Sections 6.1.2–6.1.4, while this section contains only general information useful in the interpretation of seismic velocity variations.

Oceanic melting trend

In the oceanic mantle, melting of mantle pyrolitic lherzolite produces the residual mantle rocks, such as lherzolite and harzburgite from mid-ocean ridges, that are formed after partial melts are derived from the uppermost mantle. The residue is depleted in (i.e. has a reduced concentration of) incompatible elements, such as SiO_2, Al_2O_3, CaO, with respect to compatible elements, such as MgO and Cr_2O_3. Atomic ratios $Mg\# = 100Mg/(Mg + Fe)$ of bulk rock and $Cr\# = 100\,Cr/(Cr + Al)$ of spinel (since melting commonly occurs within the spinel stability field at P <2.5 GPa) are indicators of the degree of mantle melt-depletion: both Mg# and Cr# increase with an increase in the degree of partial melting and melt extraction (Fig. 3.18). With increasing degree of partial melting, the olivine [$(Fe_xMg_{1-x})_2SiO_4$] content increases from ~55 to ~85 vol%, while the composition of the residue changes from lherzolite towards harzburgite. Oceanic mantle may have little variation in Mg# with depth (Fig. 3.19), at least in part due to homogenizing effect of the convective mixing in the sublithospheric mantle.

Major characteristics of the oceanic trend are as follows (Matsukage *et al.*, 2005).

- Compositional variations in peridotites are caused by one single melting event and a high Mg# corresponds to a low silica content.
- With Mg# increase, compressional and shear seismic velocities increase almost linearly, while density decreases.
- For P-wave velocity, the correlation of Vp with olivine content is significantly better than with Mg# (Fig. 3.20). However, the two parameters are not independent during partial melting in the shallow mantle, and seismic velocities in shallow oceanic peridotites are characterized by a single parameter such as Mg#.

Continental melting trend

In the subcontinental lithospheric mantle, melting may occur within both spinel and garnet stability fields. If melting occurs within the spinel stability field (as observed in many tectonically young provinces of the continents), chemical variations in spinel peridotites have trends similar to the oceanic settings. However, in cratonic settings, the composition of some

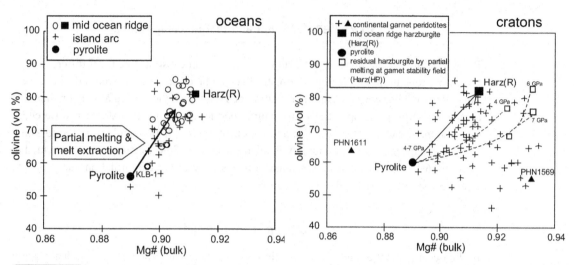

Fig. 3.18 Relationships between bulk Mg# = Mg/(Mg+Fe) atomic ratio and olivine content for mid-ocean ridge and island arc spinel peridotites (left) and for continental garnet peridotite xenoliths in kimberlite pipes (right) (from Matsukage *et al.*, 2005). The solid squares (labeled Harz(R)) show one of the most depleted mid-ocean ridge (MOR) harzburgites. Thick vectors indicate the residual trend of partial melting and melt extraction at the spinel stability field (left) and to the garnet stability field for subducted MOR peridotites originally formed at the spinel stability field (right). Dashed lines (right) show the depletion trends of partial melting of mantle pyrolite and melt extraction at the garnet stability field at pressures of 4, 6, and 7 GPa. Two solid triangles (right) indicate seismological end-members of continental garnet peridotites (Jordan, 1979).

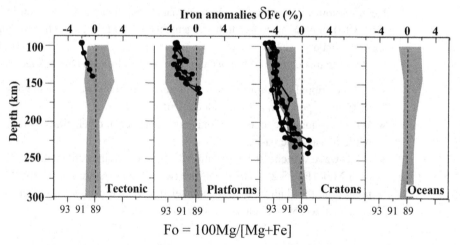

Fig. 3.19 Depth distribution of iron anomalies for several tectonic provinces delimited following the global tomography model 3SMAC (Nataf and Richard, 1996). Gray shading – iron anomalies calculated from a joint inversion of seismic tomographic model S16RLBM up to the spherical harmonic degree 16 (Woodhouse and Trampert, 1995) and the non-hydrostatic gravity anomalies derived from the geoid model EGM96 (Lemoine *et al.*, 1998). (Source: Deschamps *et al.*, 2002). Zero iron-anomaly corresponds to convective mantle with Fo = 89. Black lines – iron anomalies constrained by mantle peridotites (for details see Fig. 6.15).

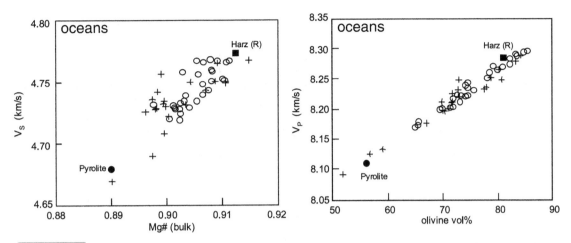

Fig. 3.20 Seismic velocities in mid-ocean ridge and island arc spinel peridotites at 1 GPa pressure as a function of bulk Mg/(Mg+Fe) atomic ratio and modal composition of olivine (from Matsukage *et al.*, 2005).

continental peridotites in the garnet stability field with very high bulk Mg# and high silica content cannot be explained by the partial melting of pyrolitic rocks alone and requires the presence (addition) of some orthopyroxene (Boyd, 1989; Walter, 1999). One proposed mechanism includes a two-stage process in which orthopyroxene can be formed at the expense of olivine by modal metasomatism in a reaction between melt-depleted peridotite and a silicic melt (Kelemen *et al.*, 1998). In the first stage, high Mg#, low modal orthopyroxene peridotites are created by large degrees of polybaric melting (at P < 30 kbar). At the second stage, interaction of these depleted residues with SiO_2-rich melts (generated mainly by partial melting of eclogitic basalt and sediment in a subduction zone) enriches them in orthopyroxene. As a result, the depletion trends for continental garnet peridotites differ significantly from the oceanic depletion trends produced by partial melting of pyrolite. In particular, no clear correlation between olivine fraction and Mg# exists for continental garnet peridotites (Fig. 3.18b).

Theoretical calculations of seismic velocities for natural peridotites and mineral assemblages typical of the continental lithospheric mantle indicate that shear-velocity is linearly correlated with Mg# and increases with iron-depletion (Fig. 3.21a; Fig. 6.12 and Section 6.1.4). This implies that, at the same temperatures, *V*s in the cratonic lithospheric mantle is faster down to *c*. 250–300 km depth than in lithospheric mantle of young continental and oceanic regions (Fig. 3.19). Unexpectedly, and in contrast to other continental locations, data for the Slave craton have been interpreted as showing no visible *V*s–Mg# correlation (Kopylova *et al.*, 2004). This conclusion is apparently biased by a very limited data set, since the Slave data clearly overlap with peridotite data from other locations (Fig. 3.21c). However, it cannot be excluded that the Slave trend is the true one due to a unique composition of its lithospheric mantle, while combining a large number of peridotite samples from diverse locations worldwide with a heterogeneous mantle composition (such as the dataset of Lee, 2003) obscures regional trends, which might also be present in datasets from individual cratons. This hypothesis is indirectly supported by the *V*p–Mg# trends for the Slave samples that are not evident in data from other continental locations (Fig. 3.21d).

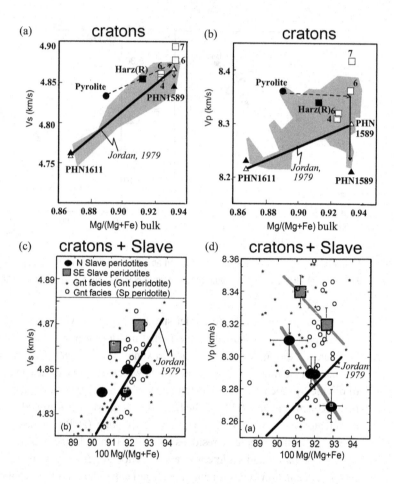

Fig. 3.21 Shear (a, c) and compressional (b, d) wave velocities in continental peridotites in garnet facies as a function of iron-depletion (expressed by Mg#). (a, b): Solid lines and open triangles – velocities calculated for two seismological end-members of garnet peridotites PHN1611 and PHN1589 (Jordan, 1979). Solid triangles – velocities calculated for residual harzburgites in garnet stability field at 1 atm and 300 K. Open squares and dashed arrows show the trends of the residue of partial melting at 4, 6, and 7 GPa. Thin arrows – the trends of orthopyroxene addition with constant bulk Mg# (Source: Matsukage *et al.*, 2005). (c, d): Comparison of velocities calculated for 133 natural peridotites from several continental locations (Lee, 2003) (shown by gray shading in (a, b) and small symbols in (c, d)) with velocities calculated for peridotites from the Slave craton at ambient conditions (large symbols) (Kopylova *et al.*, 2004). Gray lines – best fits for the Slave data; black lines as in (a, b). Gnt = garnet, Sp = spinel. Note different scales at (a, b) and (c, d).

For compressional wave velocity, the general Vp–Mg# trend has the same sign as the Vs–Mg# trend, but the correlation is poor due to a large scatter in values (Fig. 3.21b,d), and Vp velocity correlates better with olivine content (Fig. 3.20b). As a result of the strong Vs–Mg# and poor Vp–Mg# correlation, the Vp/Vs ratio exhibits a strong negative correlation with Mg# (Fig. 3.22).

Data for the Slave craton apparently display a clear trend of Vp decrease with an increase in iron-depletion, probably due to orthopyroxene enrichment (Kopylova *et al.*, 2004) given

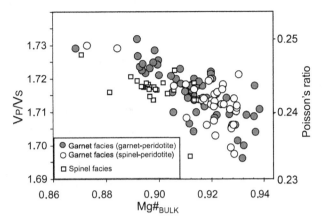

Fig. 3.22 The ratio Vp/Vs at room pressure–temperature conditions versus Mg# in mantle peridotites (based on Lee, 2003).

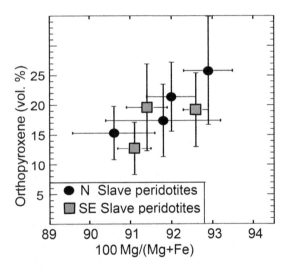

Fig. 3.23 Orthopyroxene enrichment in the Slave peridotites accompanies an increase in 100 Mg/(Mg+Fe) (based on Kopylova et al., 2004). The plot shows data averaged by rock type for 40 individual samples.

that orthopyroxene content in peridotites from the Slave mantle correlates with iron-depletion (Fig. 3.23), while orthopyroxene addition at constant bulk Mg# significantly reduces Vp wave speed (Fig. 3.21b). Note, however, that as for shear-velocity data, the strong negative Vp–Mg# correlation reported for the Slave cratonic mantle may be an artifact of a small data set (40 individual samples).

In summary, the continental trend has the following characteristics (Matsukage et al., 2005):

• The chemical composition of continental peridotites is controlled by at least two processes: partial melting and the addition of Si-rich materials (Matsukage et al., 2005). Both of these processes change Mg# and Opx#, but in different ways. For this reason, seismic

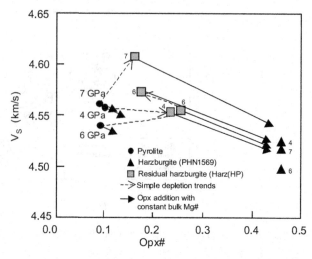

Fig. 3.24 Relationship between V_s seismic velocity and modal composition of orthopyroxene in pyrolite, in residual harzburgites in garnet stability field (Harz(HP)), and in continental garnet harzburgite with high orthopyroxene content (PHN1569) at pressures between 4 and 7 GPa. The dashed arrows show the trends of the residue of partial melting at 4, 6, and 7 GPa. The solid arrows show the trends of orthopyroxene addition with constant bulk Mg# (from Matsukage *et al.*, 2005).

velocities in deep continental peridotites are characterized by two parameters, Mg# and Opx# (volume fraction of orthopyroxene). The values of Mg# and Opx# in the continental mantle range from 0.89 to 0.94 and from 0.06 to 0.45, respectively.

- Seismic velocities in the garnet stability field (>70 km depth) increase with iron-depletion (i.e. with increasing Mg#) but decrease with increasing orthopyroxene content (Fig. 3.24). For the same Mg#, an increase in Opx# by 0.1 causes *c.* 0.5% decrease in V_s. A similar decrease in V_s can be produced by a decrease in Mg# by 0.01.
- In shallow upper mantle (70 to 190 km depth, or at pressures between 3 and 6 GPa) the effect of Mg# variations (due to depletion caused by partial melting and melt extraction) can be relatively small, so that variations in orthopyroxene content can become the major cause of velocity differences between pyrolite and continental harzburgite (Fig. 3.21) and may produce maximum $\delta \ln V_s$ and $\delta \ln V_p$ of −1.8% and −2.3% respectively (Matsukage *et al.*, 2005).
- Below a 190 km depth, the effect of iron-depletion on seismic velocities becomes particularly important and may produce maximum $\delta \ln V_s$ and $\delta \ln V_p$ of +1.1% and +1.0% respectively. Since seismic velocities in orthopyroxene increase rapidly with increasing pressure, the effects of iron-depletion and orthopyroxene-addition on seismic velocities can compensate each other with increasing depth.

3.1.8 Density–velocity relationship

In the strict sense, density is not a seismic parameter, but density and elastic moduli define seismic velocities (eq. 3.1). Birch's law (Birch, 1961) suggests a unique relationship

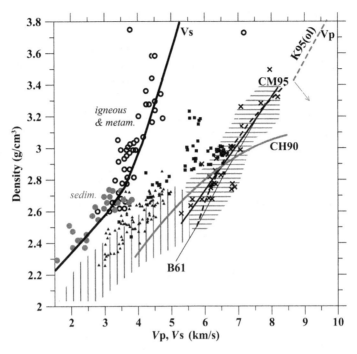

Fig. 3.25 Compressional and shear-wave velocity–density relationships for crustal and upper mantle rocks. *Vs* data: open black circles – data for igneous and metamorphic rocks; gray circles – data for sedimentary rocks. *Vp* data: thin black line (labeled B61) – Birch's relationship (Eq. 3.12) (after Birch, 1961). Hatching – laboratory measurements of P-wave velocity and density in sedimentary (vertical stripping) and metamorphic and igneous rocks (horizontal stripping) typical for the oceanic crust and upper mantle (after Ludwig *et al.*, 1970). Thick gray line (labeled CH90) – V_p–density relationship commonly used for the oceanic crust (after Carlson and Herrick, 1990). Data from the Mid-Atlantic Ridge: serpentinites (triangles) and gabbros (squares) (after Kelemen *et al.*, 2004). Crosses – experimental data (measured at pressure corresponding to a depth of 20 km and temperature of 309 °C) on average velocity versus average density for different rock types typical of the continental crust and mantle. Thick black line and dashed black line (both labeled CM95) – best linear fit to all data shown by crosses and best non-linear fit for mantle rocks only, respectively (after Christensen and Mooney, 1995). Gray dashed line labeled K95(ol) – calculated relationship for forsterite (mantle olivine) (after Karki *et al.*, 1995); thin arrow indicates the trend in velocity and density change in highly depleted cratonic mantle.

between compressional velocity V_p and density ρ for materials (isotropic polycrystals) with the same mean atomic weight M:

$$V_p = c_1(M) + c_2\rho. \qquad (3.12)$$

According to Birch, this relationship should be valid for velocity and density changes of any origin (i.e. caused by variations in temperature, pressure, or composition). Theoretical calculations indicate that although Birch's law provides a good approximation, it is not always valid: the violation is strongest for silicates and oxides, and several mantle minerals show a non-linear dependence of velocity on density (Karki *et al.*, 1995).

Despite these limitations, empirical relationships between compressional wave velocity V_p and density ρ, based on laboratory measurements and theoretical analyses, have been widely applied in seismology for more than 50 years (Nafe and Drake, 1957; Gardner *et al.*, 1974; Anderson and Bass, 1984). Correlation between compressional velocity and density is, in particular, important because it allows for gravity modeling constrained by seismic refraction data. However, even for rocks of similar composition, a wide range of densities (commonly with a scatter of up to $500\,kg/cm^3$ for crustal rocks) is possible for the same compressional wave velocities (Fig. 3.25). As a result of this scatter, any conversion of seismic velocities into densities, such as widely applied in gravity modeling, inevitably results in a significant uncertainty in gravity calculations, a fact almost never analyzed in the corresponding studies (an analysis of the problem can be found in Barton, 1986). In the worst case scenario, variations in observed gravity anomalies can be smaller than the uncertainty associated with the permissible range of densities as constrained by seismic data (Fig. 3.26), making gravity results meaningless.

Comparison of observed density-to-velocity heterogeneity ratios (i.e. the ratios of relative density variations to relative velocity variations):

$$R_{\rho/p} = \partial \ln \rho / \partial \ln V_p; \quad R_{\rho/s} = \zeta = \partial \ln \rho / \partial \ln V_s \quad (3.13)$$

with mineral physics predictions and laboratory measurements allows for speculation on the origin (e.g. thermal versus compositional) of seismic wave velocities and density variations. For example, $\zeta > 0$ for purely thermal anomalies because both density and seismic velocity increase (decrease) with temperature increase (decrease). Negative or low values of ζ cannot be explained by temperature variations alone and require the presence of compositional variations, e.g. in iron content (Fig. 3.19; Fig. 3.109). For the upper mantle, experimentally determined values of ζ are based on density and velocity measurements in olivine, and thus they refer to pure thermal effects, since elastic properties of other upper mantle minerals are not well documented. For olivine, which is the dominant mineral in the mantle, $\zeta = 0.35 - 0.45$, depending on mantle temperature (Isaac *et al.*, 1989). Anelasticity reduces ζ in the upper mantle to values between c. 0.1 and 0.2–0.3 (Karato, 1993), and in general agreement between ζ experimentally determined for olivine and ζ theoretically predicted for the upper mantle from geophysical data is poor (Deschamps *et al.*, 2001). Since elastic moduli depend on pressure, the density-to-velocity heterogeneity ratio can also be pressure-dependent.

3.2 Summary of seismic methods

3.2.1 Types of seismic waves

The main wave types traveling in solids and employed in seismic studies of the lithosphere are body waves (compressional P-waves, and shear S-waves) that travel through the volume of a solid material, and surface waves (Rayleigh and Love waves) that travel along free surfaces (interfaces). Surface waves propagate more slowly than body waves: in general, the velocities of Love and Rayleigh waves are c. 0.9 and 0.7 of the shear-wave velocity,

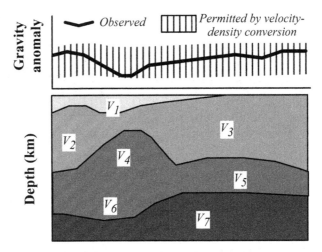

Fig. 3.26 Schematic illustration of the effect of a wide range of densities possible for rocks with the same seismic velocities on gravity modeling. In some cases, all variations of observed gravity anomalies along a seismic profile can be within the limits permitted by velocity–density conversion (calculated for the upper and lower bounds of permitted densities, Fig. 3.25). In such a case, a gravity model becomes meaningless.

respectively. However, the velocities of surface waves depend on the frequency of the propagating wave, and therefore on their depth of penetration (see below).

While Rayleigh waves exist near any free surface, Love waves require a velocity increase with depth. In Rayleigh waves particle motion is rotational (both horizontal and vertical components are present), while in Love waves particles have only transverse (horizontal) motion parallel to the surface but perpendicular to the direction of propagation (Fig. 3.27). Love waves propagate by multiple reflections as horizontally polarized S-waves in a lower-velocity near-surface layer (a waveguide), and are always dispersive (i.e. depend on wave frequency). Transverse waves are characterized by polarization, i.e. the orientation of the oscillations in the plane perpendicular to the direction of wave propagation. S-waves polarized in the horizontal plane are termed SH-waves and those polarized in the vertical plane are termed SV-waves (Fig. 3.28). In an anisotropic medium, transverse waves exhibit birefringence (shear-wave splitting).

3.2.2 Theoretical limits on seismic wave resolution

Seismic methods are classified into two broad categories: based on refracted (transmitted) and on reflected (scattered) waves. The former are used, for example, in seismic refraction profiling, travel time tomography, and surface wave dispersion analysis, while the latter are used in seismic reflection profiling. In practice, the resolution of seismic methods depends on data quality, which is commonly assumed to depend on the signal-to-noise ratio (i.e. the amplitude of the signal relative to noise). In general, it is accepted that the resolution increases when more seismic data are available.

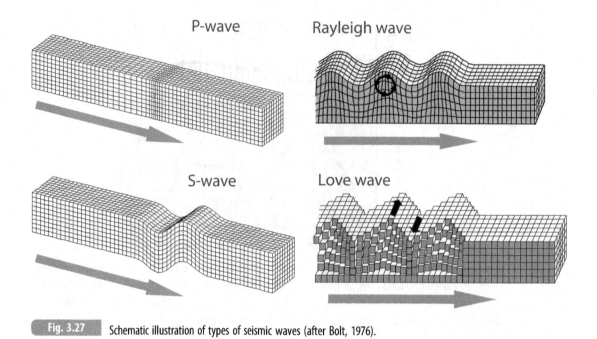

Schematic illustration of types of seismic waves (after Bolt, 1976).

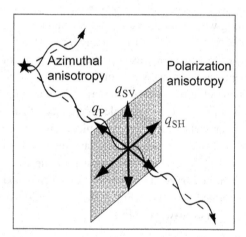

Sketch illustrating two types of seismic anisotropy. Azimuthal anisotropy is the variation of seismic wave speed
with the direction of wave propagation. Polarization anisotropy of surface waves is the variation of seismic wave
speed with the orientation of oscillations in the plane perpendicular to the direction of wave propagation. A
quasi-P wave (q_p) is polarized in the direction parallel to wave propagation, quasi-SH-waves are polarized
horizontally, while quasi-SV-waves are polarized vertically. See also Fig. 3.112.

The theoretical limit on the resolution of seismic methods based on transmitted
waves is determined by the Fresnel zone, which defines the volume in the radiation
pattern along the raypath between the source and the receiver. Its diameter d is
approximated as:

$$d = \sqrt{\lambda L / 2} = \sqrt{VL/2f}, \tag{3.14}$$

where L is the total length of the path traveled by the wave, λ is the dominant wavelength, V is the velocity, and f is the dominant frequency (e.g. Bostock, 1999). This means that small details in the lithospheric structure are best resolved by high frequency waves. For example, for crustal seismology with typical frequencies of 5–20 Hz the absolute accuracy of depth determination (by refracted Pn waves) is not better than $c.$ 1–2 km for the Moho depth.

For reflected waves, the lateral resolution at the discontinuity depends on the size of the Fresnel zone for the incident wave, while scattered waves are sensitive mainly to the structures of length scale comparable to the wavelength of the incident wave. For example, in a medium with vertical velocity gradient, waves with near-vertical incidence will produce reflected waves that will sample the zone with a width less than $\lambda/4$ (Richards, 1972).

Surface waves, since they are confined to the surface, decay more slowly with distance than body waves, which travel in three dimensions. The amplitude of surface waves decays as $1/\sqrt{L}$, where L is the distance the wave propagated from the source (e.g. earthquake), and it decreases exponentially with the depth. Dispersion is an important property of surface waves: lower frequency waves propagate faster and arrive ahead of higher frequency waves due to their deeper penetration into the Earth (because wave speed generally increases with increasing depth). For dispersed waves, the group velocity (i.e. the velocity with which the energy is propagated) is different from the phase velocity (i.e. the velocity with which the troughs and peaks propagate at a given frequency along the surface). Long-period seismology is an important practical application of the fact that waves with lower frequencies (longer wavelengths) penetrate to a greater depth. Seismic instruments used in such studies operate in the frequency range 0.05 to 0.01 Hz (period 20–100 s) which in the upper mantle corresponds to $\lambda \sim 160$–800 km and 90–500 km, for P- and S-waves correspondingly.

3.2.3 Methods of seismic data interpretation

One-dimensional techniques

The following discussion concentrates primarily on the methods used for studies of seismic structure of the subcrustal lithosphere. One-dimensional (1D) techniques used in seismic data interpretation assume a laterally uniform velocity structure. They are easy in use, efficient and fast in calculation. They include forward travel time and amplitude modeling, the Herglotz–Wiechert–Bateman inversion, and the τ_p inversion. As a result of one-dimensional velocity inversion, a 1D "average" velocity model can be constrained for a region from all available seismic data. The major limitation of these methods is that they often do not detect low-velocity zones (LVZ); if such zones are present in the real Earth, layers below LVZ will appear to be at shallower depths.

Ray theory

Ray theory (analogous to optical ray theory) is the primary method for the calculation of travel times in a velocity model: at each point the slope of a travel time curve determines a seismic velocity. The major limitations of ray theory are:

- usually it fails in predicting non-geometrical effects such as diffraction;
- it is a high-frequency approximation of the solution to the wave equation, which may be invalid at low frequencies (i.e. long periods);
- it may be invalid in structures with steep velocity gradients, where triplications in the travel time curve occur (caustics). The latter problem can be avoided by analyzing the delay time (τ_p) curve instead of the travel time curve, since the delay time curve is free of triplications and is single-valued;
- it fails to determine the velocity structure within a low velocity zone in the lithosphere: the LVZ refracts (or bends) the rays, so that no rays turn within the LVZ and a shadow zone is produced on the travel time or delay time curves.

The Herglotz–Wiechert–Bateman inversion

The Herglotz–Wiechert–Bateman (HWB) inversion method, invented about a century ago, is based on the assumption that the travel time curves are continuous and allows for calculation of a 1D velocity–depth model from a travel time curve. This model can be further modified and adjusted by iterative forward modeling to improve the overall travel time fit. In practice, travel time observations are scattered, and the HWB method is often used to find the best-fit travel time curve for the data. Since at each point the slope of the travel time curve defines a seismic velocity, seismic data are commonly inverted for a layered model with a small number of homogeneous layers. If only the first-arrival times are used, the solution is non-unique. A large number of possible models can effectively be discriminated if travel times of the secondary branches are available. The HWB method fails in the case of a LVZ since the solution becomes non-unique and both the velocity structure within the LVZ and the thickness of the LVZ cannot be uniquely determined (some limits can still be placed on the maximum thickness of the LVZ).

Non-uniqueness of 1D velocity inversions

The problem of non-uniqueness of seismic velocity inversions can be addressed by using regularization methods, which allow for finding the model in which a particular model parameter is maximized. Typical examples of regularization methods in seismology include finding solutions with smooth velocity–depth distribution (controlled by the first and the second velocity–depth derivatives). Since regularization tends to "smear" velocity anomalies to produce smooth models, "non-smooth" seismic structures constrained by such a model are considered to be real. If the fine-scale velocity structure cannot be reliably determined due to large scatter in travel time data, the upper and lower bounds on seismic velocity can be determined. Velocity bounds, in their turn, determine the bounds on the permitted depths thus permitting a large range of possible velocity–depth models (Fig. 3.29).

The velocity models based on travel time inversions are typically used only as a starting point for other seismic interpretations. More sophisticated applications of ray theory use data on amplitudes and phases. In particular, seismic amplitudes are very sensitive to velocity gradients and discontinuities. After the best fitting velocity–depth profile is obtained from

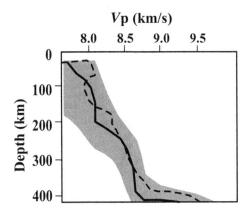

Fig. 3.29 The upper and lower bounds on V_p seismic velocity in the upper mantle calculated by τ_p travel time method
for earthquake data (based on Bessonova *et al.*, 1976). Gray shading – velocity–depth bounds for 98% confidence
level; solid and dashed lines – possible velocity–depth profiles.

travel times, velocity–depth models can be refined by trial-and-error amplitude fitting by
calculating synthetic seismograms (i.e. the seismic response of a medium to seismic wave
propagation which, ideally, should be calculated for the same frequency band as the observed
seismograms and should include converted and multiple phases). An important part of
amplitude and phase analysis is accounting for geometrical spreading effects such as reflection
and transmission at discontinuities and intrinsic attenuation (high frequencies are more
attenuated than low frequencies).

1D reference models of the Earth

Large-scale 1D reference models of the Earth were developed several decades ago and are
widely used in seismology. The PREM (preliminary reference Earth model) (Dziewonski
and Anderson, 1981) is the global spherically symmetric reference model which is based on
a large set of different observations and provides P- and S-velocities, density, and attenuation as functions of depth. The velocity structure of the upper mantle in the PREM model is
dominated by the oceanic mantle, given that more than 70% of the Earth is covered by the
oceans. In particular, in the depth range between 80 km and 220 km, the model includes the
low-velocity zone (LVZ); at the same depths the model is transversely isotropic (with a
vertical axis of symmetry) with the SH wave slightly faster than the SV wave. The depth
range between 220 km and 400 km is characterized by high velocity gradients (Fig. 3.30).
The presence of the LVZ in shallow mantle (80–220 km depth) is typical for the oceanic
upper mantle, but is not widely observed in the continental regions (see Section 3.3.2). As a
result, continent-scale seismic models which use the PREM as the reference model may
have an artifact velocity perturbation at 220 km depth, which can be interpreted erroneously
as the base of the seismic lithosphere (see Section 3.6.1).

In general, the PREM model should be used with caution as a reference model for regions
dominated by the continental lithosphere. The *iaspei* (or *iasp91*) and *ak135* reference models

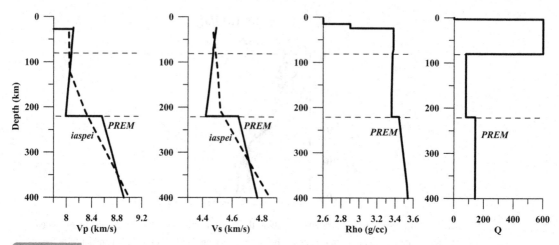

Fig. 3.30 Velocity, density, and Q structure of the upper mantle according to 1D reference models of the Earth based on the PREM (solid lines) and the *iaspei* (dashed lines) models (Dziewonski and Anderson, 1981; Kennett and Engdahl, 1991).

(Kennett and Engdahl, 1991; Kennett *et al.*, 1995) are widely used as reference models for the continental mantle since they do not include the low-velocity zone at depths between 80 km and 220 km (Fig. 3.30). However, it is desirable for different reference models to be used for different tectonic structures (Kennett, 2006).

Two-dimensional techniques

One-dimensional models assume a laterally homogeneous lithosphere structure and may provide misleading results in regions with significant lateral heterogeneity, such as produced by dipping layers, igneous intrusions, and faults. However, 1D models can serve as starting models for 2D seismic modeling.

Two-dimensional (2D) interpretations are based on seismic data from linear arrays and are widely used in wide-angle and refraction seismology. The first step is to calculate the travel times in the velocity model. This is commonly achieved:

1) by determining the average 1D model by assuming that the ray paths from source to receiver are along a series of nearly horizontal refractors with constant velocity and thus the method assumes lateral velocity homogeneity and the absence of velocity gradients within the refractors,
2) by calculating travel times along (infinite-frequency) rays in a medium with arbitrary vertical and lateral variations of velocity and inclining discontinuities.

In the next step, forward modeling of travel times is followed by calculation of synthetic seismograms, which are then compared to the observed data. The misfit between synthetic and observed data is used in an iterative procedure where adjustments to the velocity model lead to adjustments of the travel time model.

Three-dimensional techniques

Preamble

Seismic tomography represents a three-dimensional (3D) seismic technique. In regions with dense seismic data coverage, 1D "average" velocity models (see above) can be used as regional "reference" models. In large-scale (e.g. global) tomography studies, global reference models, such as the PREM model, are used. For all available data, the travel time predicted from the reference model is subtracted from the observed travel time and a 3D model of residual travel times is computed. If the average of residual travel times is non-zero, the reference model requires modification. The model of residual travel times is next inverted into velocity perturbations relative to the reference model used in the inversion, and thus a tomographic model is constrained in an iterative procedure. Many tomographic models do not proceed beyond this point and the final model is presented as relative velocity perturbations with respect to the reference model, rather than in absolute velocities. This may lead to significant misinterpretations of tomographic models (e.g. Fig. 3.85a,b; see Section 3.6 for a detailed discussion).

Regularization methods and damping

Models of three-dimensional velocity perturbations are parameterized either by continuous functions (commonly by spherical harmonics laterally and by polynomials vertically) or by dividing the model into blocks. In the latter case, the model is typically undefined in many blocks that are undersampled by seismic rays and regularization methods are used. The most common of these is the damped mean squares regularization in which the weighting parameter controls the difference between the misfit to the data and the variance of the model. This method minimizes the size of the model and permits rough models with large velocity perturbations. Solutions with small-scale velocity anomalies (compared to the size of the blocks) should be treated with caution. Further, if a strong contrast in the real seismic structure occurs such as at the ocean-shield transition, damping will not match either real structure, but will result in underestimation of the true velocity variations (Kennett, 2006).

The roughness of the model (i.e. the absence of strong contrasts in adjacent blocks) is controlled by introducing the average of the adjacent blocks (using the Laplacian operator). In such regularization (minimum roughness inversion), the weighting parameter controls the difference between the misfit to the data and the model roughness. The disadvantage of this approach is that blocks with no rays are interpolated between adjacent cells or extrapolated at the edges of the grid. As a result, cells with little data but with large-amplitude perturbations should be treated with caution.

Another complication arises from the fact that in teleseismic tomography seismic sources (earthquakes) are unevenly distributed and there can be directions with both very high and very low densities of seismic rays. Such problems cannot be "healed" by regularization methods, and the final inversion models may contain linear velocity perturbations along the most densely sampled rays.

Resolution tests and sensitivity analysis

In the case of small-scale regional tomography studies, when the number of parameters (blocks) is small, the reliability of the estimated model can be assessed by computing the full resolution matrix which contains complete information on the model. Mathematical analysis of teleseismic inversion methods examines the effects of errors in reference models and in the source parameters of the seismic events on the final model (Aki *et al.*, 1977). Resolution matrices for teleseismic S inversion indicate the parts of the block model where the number of rays is insufficient for reliable constraints of velocity anomalies.

In large-scale models with a very large number of unknowns, exact solutions are not possible, a formal resolution cannot be computed, and different tests are used. Sensitivity tests (e.g., Spakman and Nolet, 1988; van der Hilst and Engdahl, 1991; van der Hilst *et al.*, 1993) have become a standard way of assessing model resolution as they allow for relatively prompt computation of the results, even for a model with a large number of parameters, and for a comprehensive, visual inspection of spatial resolution of the model. The resolution is tested by the ability of the inversion scheme to retrieve a known input model for the same ray coverage as the one used for real data inversion. In a smearing test (which can be performed both for lateral and vertical resolution of the inversion scheme), synthetic travel times are computed for a model of velocity perturbations (such as an isolated velocity anomaly) for the same ray paths and with the same theoretical apparatus and simplifications as in the actual inversion of the real data. The synthetic data are then inverted to check how well the anomaly is recovered (both in size and amplitude) (see also Section 3.6). Another popular method is the checkerboard test (Lévêque *et al.*, 1993), in which, instead of an isolated anomaly, a model with a regular pattern of heterogeneities (velocity anomalies) of alternating sign (commonly with a ±2–5% amplitude perturbation) separated by an equal number of zero amplitude anomaly nodes is superimposed on the ambient velocity model (Fig. 3.31). Since seismic data are always noisy, in some checkerboard tests Gaussian random noise is added to the data (Piromallo and Morelli, 2003).

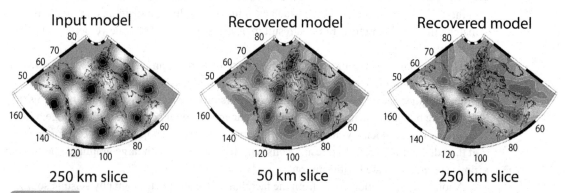

Fig. 3.31   Checkerboard resolution test for S-wave tomography of the Canadian Shield with velocity perturbations in the input model in the upper 500 km (after Frederiksen *et al.*, 2001). The amplitude of velocity perturbations at 250 km depth is ±550 m/s. Both the resolution of the shape of heterogeneities and the amplitude of S-velocity perturbations decrease with depth: the amplitude of velocity perturbations is *c.* ±350 m/s at 50 km depth and *c.* ±200 m/s at 250 km depth.

A few words of caution should be stated about sensitivity tests (Lévêque *et al.*, 1993; Shearer, 1999):

- the tests cannot check the uniqueness of the solution;
- they do not test the effect of errors in the theory because they are computed with the same theoretical model, assumptions, and simplifications that are used for the actual inversion;
- they are parameter-specific, i.e. in the case of an incorrect choice of parameters, the real Earth structure can be significantly different from the inverse "solution" even though sensitivity analysis will not necessarily reflect this;
- they only provide qualitative estimates of accuracy and resolution, but not necessarily a measure of the model true resolution as they only show a limited representation of the resolution matrix;
- they assume noise-free data and velocity perturbations with uniform amplitude thus effectively only controlling the higher amplitude anomalies;
- they check model resolution only for specific shapes of heterogeneities (e.g., as in checkerboard tests) that are very different from the real Earth structure.

Earthquake location problem

Seismology can be divided into two main categories by the sources of seismic waves: *passive-source seismology* which uses natural sources such as earthquakes, and *controlled-source seismology* in which the sources are man-made and which include primarily chemical explosions and seismic vibrators on land and airguns in marine seismic surveys. While the exact time and position of the seismic sources are known with high accuracy in controlled-source seismology, this is not the case for natural sources, where both the earthquake time and position are to be determined.

Major sources of error for earthquake location are assumptions of point source and lateral homogeneity of the medium. A common problem is mislocation of earthquakes that have occurred on a fault with different velocity structure on both sides of the fault. The problem of event location is relatively easily solved for 1D velocity models; however since travel times depend non-linearly on the earthquake location parameters, the location of seismic events for more sophisticated models remains a subject of seismic research. In particular, if earthquakes are located using a 1D reference model, their location can change in a 3D velocity model. In such cases, joint hypocenter and velocity inversions are used to simultaneously solve the problem for the velocity structure and the event locations.

3.3 Major seismic discontinuities in the lithosphere

3.3.1 The crust

Knowledge of the crustal structure is vital for the Earth sciences. On one side, the crust, which makes the upper (and thus best studied) part of the lithosphere, preserves a memory of tectonic evolution and planetary differentiation. On the other side, it is an "obstacle" in geophysical studies of the lithospheric and deep mantle: the crustal effect should be taken into account ("subtracted" from the mantle signal) in various types of geophysical studies such as seismic tomography, gravity, and thermal modeling. Detailed overviews of crustal structure can be found in Tanimoto (1995), Christensen and Mooney (1995), Mooney *et al.* (2002); an overview of crustal composition is presented by Rudnick and Fountain (1995), Rudnick and Gao (2003). These publications form the basis for the following brief summary of the crustal velocity structure. Crustal reflectivity and reflection Moho are discussed in Section 3.5.1.

Continental crust

Crustal layers

Four layers are traditionally recognized within the continental crust (Fig. 3.32). The top, sedimentary, layer is commonly defined by $Vp < 5.6$ km; however this boundary velocity can vary significantly between different regions depending on the composition and velocity structure of the consolidated sediments, and its regional value is constrained by drilling data from boreholes reaching the basement. The next three layers form the crystalline crust (the

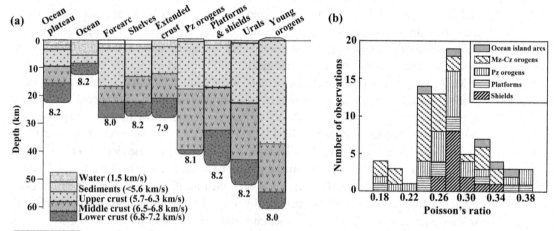

Fig. 3.32 (a) Structure of the primary types of continental and oceanic crust. The Paleozoic Urals orogen is shown separately because, in contrast to other Paleozoic orogens, it has a preserved thick crustal root. The upper crust is typically made of granitic–gneissic rocks with $5.7 < Vp < 6.4$ km/s; middle and lower crust with $6.4 < Vp < 6.8$ km/s and $6.8 < Vp < 7.2$ km/s, respectively, are composed of amphibolitic–granulitic–dioritic rocks; the lowermost crust with $7.2 < Vp < 7.8$ km/s is composed of mafic–granulitic rocks, while the upper mantle with $Vp > 7.6$–7.8 km/s has ultramafic composition. The Conrad discontinuity marks the transition between the upper and the middle-lower crust, while the Moho marks the boundary between the crust and the upper mantle. (b) Global variation of Poisson's ratio (after Zandt and Ammon, 1995). Note that regional crustal structure may differ significantly from any generalization (e.g. compare with Fig. 3.37).

basement). The upper crust with typical Vp velocities of $c.$ 5.7–6.3 km/s has a felsic (granites and low-grade gneisses) composition; the middle crust has typical Vp velocities of 6.5–6.8 km/s and is composed of rocks of intermediate composition; the lower crust with typical Vp velocities of 6.8–7.2 km/s has a mafic (basaltic) composition. Sometimes the transition between the upper and the middle crust is diffuse and they are considered together. In some continental settings, a layer with very high P-wave velocity of 7.2–7.6 km/s is observed in the lowermost crust.

Poisson's ratio depends on rock composition and increases with decreasing silica content (Fig. 3.1b). For this reason, Poisson's ratio usually increases with depth and is typically < 0.26 in felsic rocks, 0.26–0.28 in intermediate rocks, and > 0.28 in mafic rocks. Poisson's ratio in the continental crust varies from 0.18 to 0.38 with an average value of $c.$ 0.24–0.28 (Fig. 3.32b). Typical average crustal values are 0.29 ± 0.02 in shields, 0.27 ± 0.03 in platforms and Paleozoic orogenic belts, 0.25 ± 0.04 in young (Meso-Cenozoic) orogens, and 0.31 ± 0.05 in island arcs (Zandt and Ammon, 1995).

An example of the proportions of major rock units along the 3000 km long European Geotraverse which crosses the European continent from the Baltic Shield to the Mediterranean sea is presented in Fig. 3.33. This is based on seismic refraction data along the profile combined with mapping, petrological studies, and the computed abundance of major elements and thirty-two minor and trace elements. The composition of the lower crust is derived from worldwide data on felsic granulite terranes and mafic xenoliths. The crust along the profile ranges in age from Archean to Cenozoic (Fig. 3.34), and its thickness varies by more than 20 km. Because of the strong heterogeneity of the crustal structure, the average values along the Geotraverse should be treated with caution since they may not necessarily be typical of any particular crustal structure in Europe. The results indicate that the volume proportions of the upper crust to felsic lower crust to mafic lowermost crust are about 1:0.6:0.4 (in this balance, sedimentary rocks which make ~7 % of the total crustal volume are included in the upper crust) (Wedepohl, 1995).

The Conrad discontinuity is observed regionally in specific continental settings as a sharp increase in seismic velocities at a depth of 15 to 20 km. It is considered to mark the transition

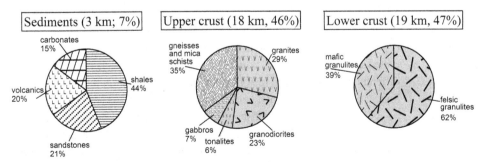

Fig. 3.33 Structure of the continental crust derived from the 3000 km long European Geotraverse (see Fig. 3.61 for location) generalized on the basis of mapping, petrological studies, and chemical balances (based on results of Wedepohl, 1995). Average crustal thickness along the profile is 40 km and it comprises 62% of Archean–Proterozoic crust (45.5 km thick) and 38% Proterozoic–Phanerozoic crust (30 km thick).

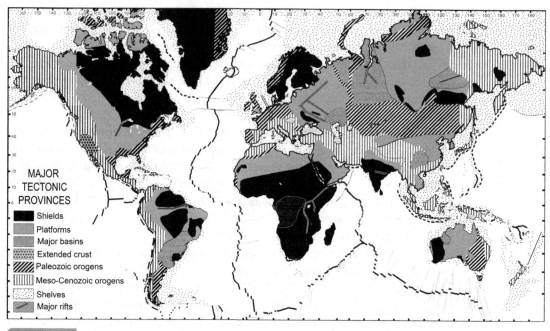

Fig. 3.34   Major tectonic provinces of the continents.

between the upper and the lower continental crust. Its origin remains unclear, although for several decades it has been interpreted as a compositional boundary between the sialic and the mafic crust.

The Moho and crustal thickness

The lowest P-wave velocity expected in the uppermost mantle is 7.6 km/s with the more common value of 7.8 km/s. Thus, conventionally a layer with $Vp > 7.6$–7.8 km/s is considered to be the mantle. The Mohorovičić discontinuity (commonly referred to as Moho or M-discontinuity) that marks the transition from the crust to the upper mantle was identified under the European continent from the analysis of P-waves in 1909 by Andrija Mohorovičić (1857–1936), a Croatian meteorologist and seismologist. The Moho is commonly interpreted as a compositional boundary, the origin of which is directly related to planetary differentiation. However, in some tectonic settings (e.g. in the rift zones) its formal seismological definition by P-wave velocity higher than 7.6 km/s leads to a situation where a part of a very high velocity lowermost crust can be formally interpreted as the mantle, or vice versa, when a part of a very low velocity uppermost mantle can be formally interpreted as the lower crust. The presence of a mixture of lower crustal and upper mantle material apparently observed in some tectonic provinces further complicates seismological constraints on the crustal thickness. In some tectonic settings, such as collisional (paleo)orogens with (paleo) subducting slabs, the double Moho is reported (e.g. Pechmann *et al.*, 1984; Ponziani *et al.*, 1995; Hansen and Dueker, 2009).

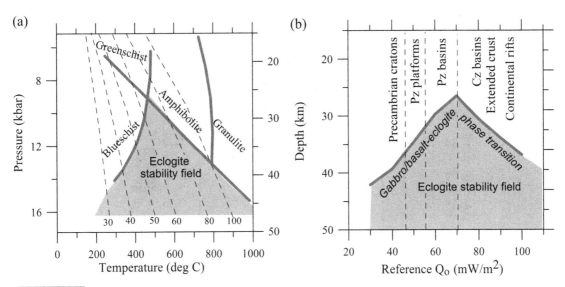

Fig. 3.35 Gabbro/basalt – eclogite phase transition in crustal rocks. Gray shading – eclogite stability field. Pressure–depth conversion is made assuming crustal density of 2.90 g/cm³. (a) Phase diagram (after Spear, 1993). Thin dashed lines – typical continental reference geotherms (Pollack and Chapman, 1977); numbers – surface heat flow in mW/m². (b) Depth to gabbro/basalt–eclogite phase transition (thick gray line) in different continental settings plotted versus continental reference geotherms labeled in heat flow values (Pollack and Chapman, 1977). Note that for the same lithospheric geotherms, measured surface heat flow values can significantly differ from the "reference" value, in particular due to large lateral variations in crustal heat production. To simplify interpretations, tectonic provinces are marked on the top based on the typical heat flow values observed there.

Gabbro/basalt–eclogite phase transition, which in a typical continental crust occurs at *c.* 40–50 km depth, is thought to control the maximal average thickness of the continental crust. Since the rate of gabbro/basalt–eclogite transformation is very slow and strongly temperature-dependent, all laboratory experiments have been made at $T > 800$–$1000\,°C$ and extrapolated to the lower temperatures (e.g. Ito and Kennedy, 1971). Importantly, the actual P–T stability field of the system at low temperatures is unknown (Fig. 3.35). At high crustal temperatures, typical for Phanerozoic regions, gabbro/basalt–eclogite phase transition requires time in the order of a million years (Artyushkov, 1993); the question of whether gabbro can remain metastable for a billion years remains unclear. Deep crustal roots extending down into the eclogite stability field (Mengel and Kern, 1992) are only known (a) in cold intracratonic environments (such as at the Archean–Svecofennian suture in the Baltic Shield) since the rate of metamorphic reactions depends strongly on temperature, and lower crustal mafic rocks can apparently remain metastable over a geologically long time, and (b) in young orogens such as the Alps as long as gabbro/basalt–eclogite phase transition is still in process.

The global average for thickness of the continental crust is *c.* 39–42 km; however large variations in crustal thickness (from less than 30 km to ~60 km) are observed between

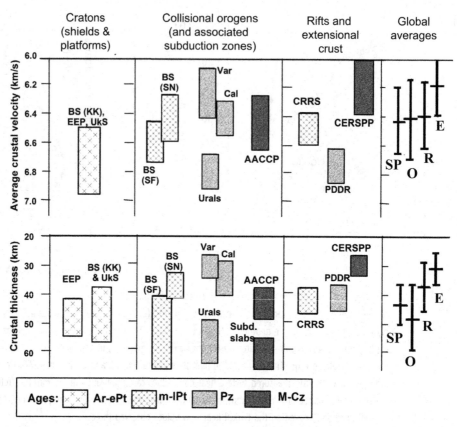

Fig. 3.36 Statistical properties of the on-shore European lithosphere for different tectonic structures (based on Artemieva *et al.*, 2006): top – average basement velocity; bottom – crustal thickness. Different shadings refer to tectono-thermal ages of different tectonic provinces. Abbreviations for tectonic provinces: *Archean–Paleoproterozoic (Ar-ePt)*: BS = Baltic Shield, KK = Kola-Karelia, EEP = East European (Russian) Platform, UkS = Ukrainian Shield; *Mesoproterozoic and Neoproterozoic (m-lPt)*: SF = SvecoFennian and SN = SvecoNorwegian provinces of the Baltic Shield, CRRS = Central Russia Rift System; *Paleozoic (Pz)*: Cal = Caledonides, Var = Variscides, Urals, PDDR = Pripyat–Dnieper–Donets Rift; *Meso-Cenozoic (M-Cz)*: AACCP = Alps–Andes–Carpathians–Caucasus–Pyrenees, CERSPP = Central European Rift System–Pannonian Basin–Po Basin. The last column – global statistical averages with standard deviations (Christensen and Mooney, 1995): SP – for shields and platforms, O – for orogens, R – for rifts, E – for extended crust.

continental structures of different ages and different tectonic evolution, and even within similar tectonic structures with similar ages (Figs. 3.36, 3.37). Additional complications arise due to a highly non-uniform coverage of the continents by seismic studies; large regions (most of Africa, large parts of South America, Greenland, and Antarctica) have no or very limited seismic data (Fig. 3.38). The "white spots" can be filled using statistical values based on the typical seismic structure of the crust of different ages and in different tectonic settings. This approach has been used to construct the global crustal

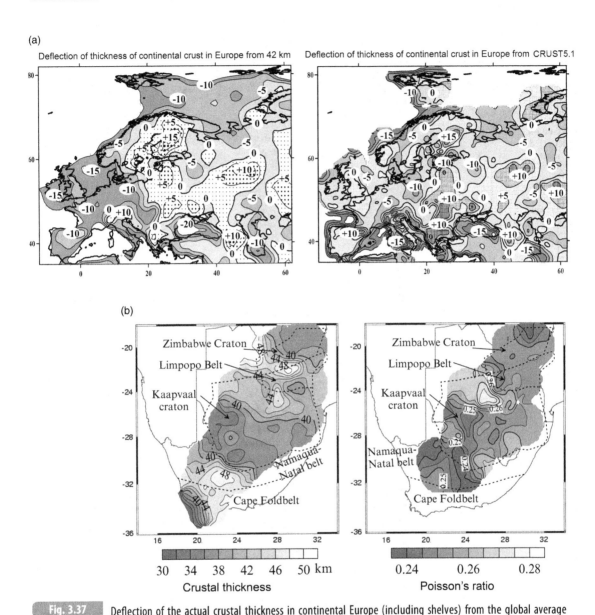

Fig. 3.37 Deflection of the actual crustal thickness in continental Europe (including shelves) from the global average for the continental crust (left) and global crustal model CRUST5.1 (right). Crustal thickness in Europe is smoothed with a low-pass filter for the seismic data compilation of Artemieva, 2007; Artemieva and Thybo, 2008, 2011. (b) Variations in crustal thickness and Poisson's ratio beneath South Africa based on receiver function analysis from 78 stations. Dashed lines – boundaries of major tectonic provinces (based on the results of Nair et al., 2006).

models CRUST 5.1 and CRUST 2.0 (Figs. 3.38, 3.39). However, any such generalization should be treated with caution because statistical crustal models, although useful in regions with no data, can be misleading and erroneous for individual tectonic structures. Thanks to an extremely high coverage by high quality seismic data (Fig. 3.38), Europe is

Crustal thickness

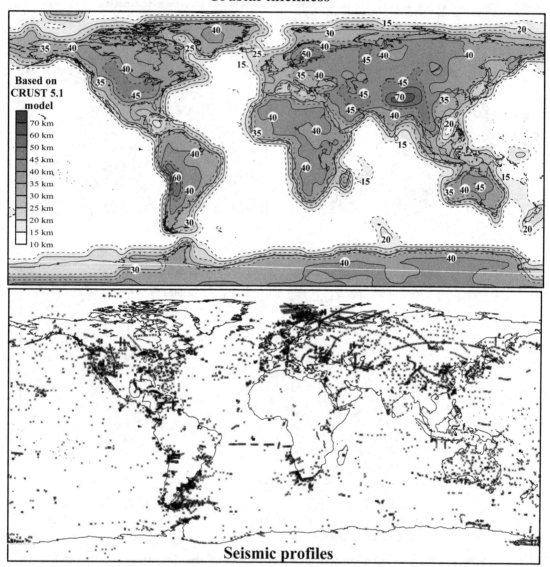

Fig. 3.38 Thickness on the crust (Source: CRUST 5.1 model, Mooney *et al.*, 1998). Because of a large averaging window on a regular (5° × 5°) grid, the individual structures are poorly resolved and the map provides a very generalized pattern similar to low-pass filtering. Bottom map shows coverage by seismic profiles in 2005.

the best region to compare statistical models for the continents with actual seismic crustal structure (Figs. 3.36, 3.37a). A comparison shows that very few of the tectonic structures in Europe have crustal thickness, thickness of sediments, or average crustal velocities that correspond to the global statistical averages for similar tectonic provinces of the world.

Thickness of sediments

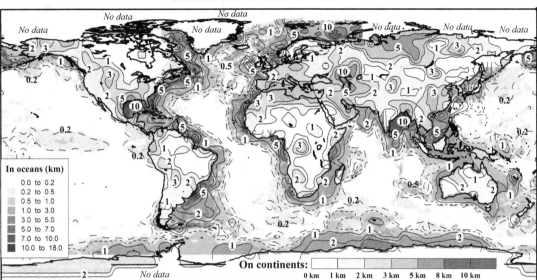

Fig. 3.39 Thickness of sediments in the continental regions (based on the CRUST 5.1 model, Mooney *et al.*, 1998) and in the oceans (based on a 5′ global compilation (Divins, 2008)). Continental shelves are excluded except for the Barents sea where high quality seismic data are available. Because of a large averaging window on a regular (5° × 5°) grid, the individual structures are poorly resolved and the map provides a very generalized pattern similar to low-pass filtering.

Similar discrepancies should be expected for all other continents, reflecting the long, complicated, and commonly asynchronous history of evolution of individual tectonic provinces worldwide. The discrepancy between actual seismic interpretations and the statistical model, which for crustal thickness reaches $\pm(5–10\,\mathrm{km})$ almost everywhere, stems from two major reasons:

- the existing statistical models are constrained by seismic reflection and refraction data that were available by 1995–1998 (CRUST 2.0 was additionally constrained by detailed borehole data on the thickness of the sedimentary cover); some of the regional interpretations used in statistical models were based not only on seismic data but on potential field data as well; in some regions recent seismic studies considerably revised old interpretations;
- CRUST 5.1 and CRUST 2.0 are constrained on 5° × 5° and 2° × 2° grids, respectively; as a result many short-wavelength variations in crustal structure, in particular in regions with complicated tectonics and highly heterogeneous crust, have been averaged (compare Figs. 3.39 and 3.40).

General trends in variations of crustal structure of the continents with age are exemplified by high-resolution seismic data from Europe (Fig. 3.41). The European continent provides

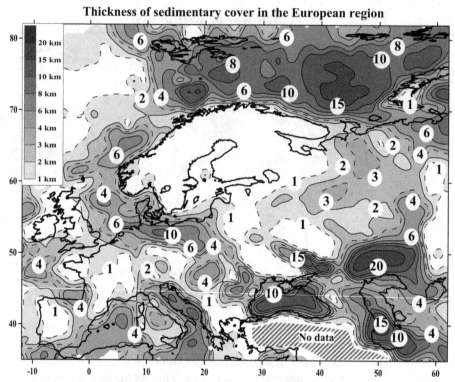

Thickness of sedimentary cover in the European region

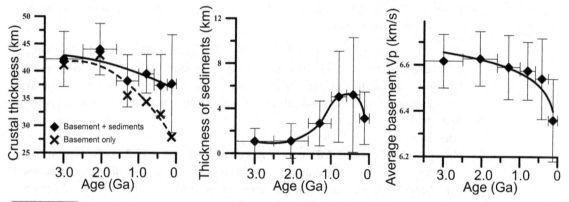

Fig. 3.40 Thickness of sediments in the European region constrained by the EXXON map (1985) complemented by numerous publications based on high-resolution regional seismic surveys and drilling (compilation of Artemieva and Thybo, 2011). The interpolation used to constrain the map filters out small-scale anomalies and thus the map show the minimum estimate of sedimentary thickness.

Fig. 3.41 Statistical properties of the continental crust in Europe averaged for six time intervals: Archean, Paleoproterozoic, Mesoproterozoic, Neoproterozoic, Paleozoic, and Meso-Cenozoic (after Artemieva, 2007). Horizontal bars – the time span used for averaging; vertical bars – standard deviation of the parameters.

a good test site owing to the diversity of its tectonic structures that cover *c.* 3 Ga of the continental evolution (Figs. 3.36, 3.60). The depth to the Moho generally increases with age; the pattern is even more pronounced for age variations of the thickness of the crystalline basement. This trend is, however, weak for the Precambrian crust, and although the Paleoproterozoic crust appears to have a slightly greater thickness than the Archean crust, the difference between them is within the typical uncertainty of seismic models. Furthermore, the basement age in most of the East European craton is Archean–Paleoproterozoic and the subdivision between the two ages is not straightforward, in particular, in the sediment-covered platform area. Typically, the thickness of sediments decreases with tectono-thermal age and is the largest in Paleozoic–Mesozoic basins (Fig. 3.40). A pronounced difference in average crustal velocity (the global average for the continental crust is 6.45 km/s) between the continental structures of different ages is the result of significant variations in thickness of crustal layers and average composition of the crust (Fig. 3.41).

Principal types of the continental crust

Stable parts of the continents

Cratons (from the Greek word meaning "strength") are tectonically stable parts of the continents that were formed in the Precambrian (>560 Ma) (Fig. 3.34; 2.16). Cratons with exposed Precambrian crystalline igneous and high-grade metamorphic rocks form Precambrian shields, the term introduced by German geologists in 1901. Platforms also form stable parts of the continents. Since the age of the platforms can be both Precambrian and Phanerozoic (most common Paleozoic), many of the platforms (in particular in Eurasia) were formed within the cratons and have remained their integral parts. For this reason, a subdivision of stable continents into terranes of Precambrian age, on one side, and platforms (with no age being specified), on the other side, while justified in some regional studies, is meaningless when employed in global crustal models. Contrary to shields, platforms are covered by a significant sequence of sedimentary rocks (commonly exceeding 2–3 km); in some basins the thickness of sediments can exceed 10 km (e.g. in the Peri-Caspian depression within the East European craton it exceeds 25 km, Fig. 3.39).

The crust in shields and platforms typically exceeds 40 km in thickness and commonly has a relatively thick lower crust (Fig. 3.32; Fig. 3.38), which results in a high average P-wave velocity in the crust (6.49 km/s for the shields). Poor coverage of stable continents by seismic data in the early 1990s, when data from cratonic South Africa and Australia dominated the global data set for the Archean terranes, led to the conclusion that the Archean crust is typically *c.* 35 km thick and thus is significantly thinner than the Paleoproterozoic crust that is typically *c.* 45 km thick (Durrheim and Mooney, 1991). New high-quality seismic data reported from different cratonic regions of the world do not support this conclusion (e.g. Fig. 3.37b). Since cratons preserve some of the oldest crust on the planet that has undergone a long tectonic evolution, the crustal structure of stable continents is highly heterogeneous both regarding thickness and the velocity structure. For example, within the Archean Kaapvaal craton, the crustal thickness is highly non-uniform over short distances with local variations between 50 km and 34 km (Nguuri *et al.*, 2001). A belt of thick crust of thickness of 51–53 km (locally reaching 60 km depth) has been reported for the

Archean Dharwar craton in India (Gupta *et al.*, 2003). Crustal thickness of 50 km and more is common in the Archean–Paleoproterozoic provinces of the East European Platform and the Ukrainian Shield (Trofimov, 2006; EUROBRIDGE, 2001). Values of *c.* 55 km (locally exceeding 60 km) are determined in the Baltic Shield (south-central Finland) at a paleo-collision zone between the Archean and Paleoproterozoic terranes (Hyvonen *et al.*, 2007). On the other hand, the Archean cratons of western Australia apparently have a relatively thin (less than 40 km) crust, while crustal thickness in the Proterozoic Central Australia locally exceeds 50–55 km (Clitheroe *et al.*, 2000).

Active parts of the continents

Young collisional orogens typically have thick crustal roots, although large variations in crustal structure exist between different Cenozoic orogens. The largest thickness of continental crust has been reported for the Andes and the Himalayas–Tibet region, where crustal roots extend down to *c.* 70 km. In other orogens crustal thickness commonly does not exceed 50–60 km. For example, in the Alps, crustal roots reach 50–55 km only in the western and central parts of the orogen and do not exceed 42–45 km in its eastern part (Grad *et al.*, 2009; Schreiber *et al.*, 2010). Seismic data on the crustal structure of other Cenozoic collisional orogens of Europe still remain limited; in particular little is known about the seismic structure of the crust beneath the Caucasus. Among Paleozoic orogens (the Appalachians, the Caledonides of Norway and Greenland, the Variscides, and the Urals), the Ural mountains are unique: this orogen which remained tectonically undisturbed within the continental interior since its formation has preserved thick crustal root that is *c.* 50 km deep and reaches 65 km in the central part and in the Polar Urals (Carbonell *et al.*, 1996; Avtoneyev *et al.*, 1992). In other Paleozoic orogens the crustal roots have been lost and crustal thickness is *c.* 40 km or less (Fig. 3.32).

The best known continental region of extended crust is the Cenozoic Basin and Range Province in western USA. The extended crust is typically only 30 km thick, and the lower crust is almost always missing. As a consequence, the Moho is marked by a sharp jump in seismic velocities. Crustal structure similar to the Basin and Range Province is observed in the Variscan belt of Europe, a 3000 km long Paleozoic orogen which extends from the North Sea to Iberia (Figs. 3.59, 3.60). Wide-spread anatectic granites indicate intensive melting of thickened continental crust at the late stages of the Variscan orogeny. Late Paleozoic large-scale normal faulting, crustal extension with possible rifting and delamination of the lower crust produced the modern crustal structure with a flat Moho at *c.* 30 km depth, analogous to the modern Basin and Range Province (Menard and Molnar, 1988). The Conrad discontinuity is often observed in the extended crust of the Variscides.

Various seismic studies indicate significant crustal thinning beneath some continental rift zones. For example, beneath the Kenya Rift the crust is thinned to *c.* 30 km and is underlain by a low-velocity mantle (with *Pn* velocity of 7.6 km/s). Supported by seismic studies in the East African Rift Zone, "classical" models for rifted terranes always have a thin crust. However, recent seismic studies indicate that continental rift zones can have normal continental crust with no crustal thinning. A seismic velocity model across the Baikal Rift Zone shows a gently deepening Moho from the Siberian Platform (41 km depth) into the Sayan–Baikal Fold Belt (46 km depth) with no Moho uplift beneath the rift axis (Thybo and Nielsen, 2009). Similarly, seismic reflection/refraction studies across the largest in Europe (*c*. 2000 km long, up to 170 km wide, and 22 km deep) Paleozoic intracratonic Dnieper–Donets rift (Ukraine) indicate that the crust is *c*. 40–42 km thick, and the rift basin is underlain by a high density and high velocity zone at middle to lower crustal levels (Lyngsie *et al*., 2007). Only a weak crustal thinning is observed beneath the Paleozoic Oslo graben, where the crust is *c*. 34–35 km thick as compared to 38–40 km in the adjacent regions (Stratford *et al*., 2009).

The continental shelves are the off-shore continuation of the closest onshore crustal terranes with a tendency towards some crustal thinning below the shelves. Many of the shelves, in particular in the Arctic, are the continuation of the cratonic crust and have crustal thickness of *c*. 30 km. Along the Arctic shelves (e.g. in the Barents Sea), where the crust has been subject neither to the effects of oceanic break-up nor to intensive rifting episodes with related crustal thinning as in the North Sea, the crust is 30–35 km thick. In island arcs (e.g. Japan) the crustal thickness is about 20–30 km and has a pronounced 3D structure. A low velocity mantle with *Pn* velocity of 7.5–7.8 km/s suggests high mantle temperatures.

Oceanic crust

General patterns

Compared to the continental crust, the oldest oceanic crust is Jurassic (~180 Ma). Classically the oceanic crust is divided into three layers (Fig. 3.32). Layer 1 is the sedimentary layer with a highly variable thickness. Layer 2 has P-wave velocities of 4.5–5.6 km/s, is *c*. 1.5–2.0 km thick, and is made of extrusive volcanic rocks (pillow basalts at shallow depths which grade downwards into sheeted dikes). Layer 3 (typical thickness 4.5–5.0 km) with P-wave velocities of 6.5–7.0 km/s has gabbro composition similar to the composition of ophiolite complexes. The top of Layer 3 is transitional with interfingering of sheeted dikes into the lower part of Layer 2. Further down, the upper part of Layer 3 consists of isotropic gabbro, underlain by layered gabbro and harzburgite. In mid-ocean ridges, where the new oceanic crust is formed, the base of the crust also marks the base of the lithosphere. The thickness of the oceanic crust (commonly defined as the thickness of Layers 2 and 3) is fairly uniform globally (5.6–7.1 km according to different authors) and does not show any age dependence (Tanimoto, 1995).

Anomalous oceanic crust

Some oceanic regions have crustal thickness significantly thicker than global observations. These regions are of a particular interest because, as a rule, bathymetry there does not follow the square root dependence on ocean floor age (see Chapter 4). This fact, together with the presence of anomalously thick oceanic crust, is attributed to mantle thermal anomalies with a high degree of melt generation. In ocean plateaux, crustal thickness reaches 20 km (and may be as thick as 35 km beneath the Ontong–Java plateau); the origin of this thick crust is commonly ascribed to a large amount of melt generated by a mantle plume. However, the seismic structure of many other ocean plateaux is still poorly known and some of them (such as the Falkland Plateau, Lord Howe Rise, the Kerguelen Ridge, the Seychelles Ridge, and the Arctic Ridge) may contain fragments of the continental crust (Mooney *et al.*, 1998 and references therein).

A thick (15–25 km) crust beneath several aseismic ocean ridges such as the Walvis and the Ninety-East Ridge in the Indian Ocean (Detrick and Watts, 1979) has been also explained by the presence of large volumes of melt associated with mantle upwellings (e.g. White *et al.*, 1992). Similarly, anomalously thick oceanic crust (30–35 km) underlies the Faeroe–Iceland–Greenland ridge that extends from Greenland to the British Isles across the Mid-Atlantic Ridge (Bott and Gunnarsson, 1980; White *et al.*, 2008). The presence of a thick oceanic crust can be interpreted in favor of the presence of a mantle plume with anomalously high melting temperature at Iceland, while the east–west symmetric crustal structure of the Faeroe–Iceland–Greenland ridge around Iceland is used as an argument against the plume origin of Iceland unless the plume location has been semi-stationary with respect to the plate boundary (Lundin and Dore, 2005).

Crustal thickness in Iceland is a topic of hot debate: two competing models, "thin crust" and "thick crust", have been proposed for Iceland (see review by Foulger *et al.*, 2005). Numerous seismic studies indicate that the oceanic Layer 3 (down to a depth of 10–20 km) has Vp velocity of 6.5–7.0 km/s. Below this depth, in Layer 4 (which may extend down to 60 km depth), the P-wave velocity gradually increases to 7.0–7.6 km/s. The principal difference between the two crustal models is the petrologic interpretation of Layer 4: in the "thin crust" model it is interpreted as anomalous peridotite mantle with *c.* 2% of melt, while in the "thick crust" model the same layer is interpreted as gabbroic "lower crust" with lenses of melt.

It should be noted that some oceanic regions have anomalously thin crust. They include regions with slow spreading rates (< 2 cm/y), non-volcanic rifted margins that have earlier undergone extreme extension, and fracture zones where Layer 3 is very thin or absent.

3.3.2 Seismic discontinuities in the upper mantle

LVZ, G-discontinuity and the base of seismic lithosphere

Seismic lithosphere (or the *lid*) is defined as the seismic high-velocity region on the top of the mantle. It generally overlies a low-velocity zone (LVZ) that was first recognized by Beno Gutenberg (1959). Thus, the top of the LVZ defines the bottom of the lid (Anderson, 1995). Since the LVZ is located between the thermal boundary layer (above) and the nearly adiabatic mantle (below), in the oceans the top of the LVZ and the top of the asthenosphere (a layer with lowered viscosity) are equivalent and are marked by the seismically sharp, Gutenberg (G-), discontinuity that is characterized by an abrupt seismic velocity decrease of roughly 9% (Bagley and Revenaugh, 2008).

Starting from the early works, a decrease in seismic velocities in the LVZ was attributed solely to partial melting (e.g. Lambert and Wyllie, 1970). The mechanism behind this melting is still the subject of discussion; volatile enrichment of the mantle is a possible cause. Alternatively, a sharp decrease of water solubility in mantle minerals at the depths that correspond to the LVZ may cause excess water to form a hydrous silicate melt (Fig. 3.42) (Mierdel *et al.*, 2007). If the LVZ below the seismic lithosphere results from

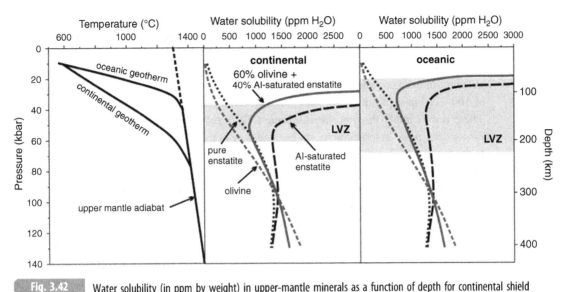

Fig. 3.42 Water solubility (in ppm by weight) in upper-mantle minerals as a function of depth for continental shield (middle panel) and oceanic mantle (right panel) calculated for typical oceanic and shield geotherms (left) (from Mierdel *et al.*, 2007). While water solubility sharply decreases with pressure (depth) in aluminous orthopyroxene (long dashes), it continuously increases with depth in olivine (short dashes). The low velocity zone (LVZ) coincides with the depth where the water solubility in mantle minerals is at a minimum. Gray shading – typical position of the LVZ below continental shields and oceans. Compare with Figs. 7.17 and 7.18.

partial melt, the lithospheric base should be a rheological boundary and it may be marked by a change in mantle anisotropy pattern from frozen-in anisotropy in the lithosphere to anisotropy in the asthenosphere due to mantle flow.

Recent studies favor explanations other than partial melting, such as high-temperature relaxation (Anderson, 1980), a contrast in volatile content from water-depleted lithosphere to water-rich asthenosphere (Karato and Jung, 1998), and grain size variations without requiring the presence of melts (Faul and Jackson, 2007). These mechanisms suggest that the base of a seismic lithosphere should be a diffuse boundary which extends over a certain depth interval. Seismic studies in the western and central Pacific based on multiple ScS reverberations put an upper limit of 30 km on the transition interval where most of the velocity decrease from the lid to the LVZ takes place (Bagley and Revenaugh, 2008), while a recent study based on converted P-to-S and S-to-P waves suggests that the velocity drop at the discontinuity occurs over depths of 11 km or less (Rychert and Shearer, 2007). The width of this transitional interval is more easily explained by the termination of the zone of partial melting rather than by a change in grain size. Furthermore, grain size variations with depth are insufficient to explain the seismic LVZ in the absence of water and/or melt and can only be applied to wet mantle.

The low-velocity zone, as its name indicates, is associated with negative velocity gradient: velocity in the upper mantle across the upper boundary of the LVZ decreases with increasing depth. Negative velocity gradients are a common feature in the upper mantle (e.g. Thybo, 2006; Fig. 3.70), in regions where a velocity increase due to compression becomes less than a velocity decrease due to heating (Anderson *et al.*, 1968). In particular, synthetic computation of the isotropic elastic wave velocities in a pyrolitic mantle indicates that a solid-state low-velocity zone can provide a satisfactory quantitative explanation of a drop in seismic velocities at 100–200 km depth in a 100 Ma Pacific Ocean (Stixrude and Lithgow-Bertelloni, 2005). These results imply that *not every seismic LVZ in the mantle should be interpreted as the lithospheric base*. In tomography studies, the base of the seismic lithosphere is closely linked to the top of a LVZ and in practice it is commonly defined either as the zone of high velocity gradient in the upper mantle or as the bottom of the layer with positive velocity anomalies, which should correspond to a transition from the high-velocity lithosphere to the LVZ. However, seismic tomography and seismic refraction models will not necessarily indicate the same depth to the lithospheric base.

Lehmann (L-) discontinuity

The Lehmann (or L-) discontinuity at a depth of *c.* 220 km associated with a *c.* 2% velocity increase is named after its discoverer, the Danish female seismologist Inge Lehmann (1888–1993). According to Francis Birch,

> Lehmann discontinuity was discovered through exacting scrutiny of seismic records by a master of a black art for which no amount of computerization is likely to be a complete substitute.

It was first observed in seismic refraction studies of earthquake data below Europe and in GNOME nuclear explosion data below the North American craton (Lehmann, 1959, 1961, 1962) (although the existence of a seismic discontinuity at around 200–220 km depth had been predicted much earlier by a Russian geophysicist Prince Golitzyn (Galitzin, 1917)).

For several years the Lehmann discontinuity was considered to be a global boundary at ~220 km depth (now extended to a depth range of 160–280 km) interpreted as the base of the low-velocity zone (Thybo and Perchuc, 1997a). Isotropic models of the Earth require the presence of a low-velocity zone in the upper mantle with a 5–10% shear-velocity decrease, implying significant melting of the upper mantle at this depth. Introduction of transverse isotropy can significantly reduce the magnitude of the velocity drop at the discontinuity. The PREM model (which includes transverse isotropy in the depth range between 24.4 km and 220 km), includes the 220 km depth as the first order discontinuity (Fig. 3.30). As a result, tomographic inversions in which PREM is used as a reference model tend to locate a velocity minimum at a 220 km depth (Su *et al.*, 1994). Since the velocity jump associated with the L-discontinuity is small (<5% in PREM), the uncertainties in travel time measurements are significantly larger than the uncertainties associated with the 410 km and 660 km discontinuities.

Global- or continent-scale seismic studies from the 1990s based on stacking of SH-components of surface waves showed little evidence for the L-discontinuity, suggesting that if it is a global feature, it should be extremely variable in depth. In fact, the depth to the

L-discontinuity is apparently correlated with tectonic setting, being shallow in tectonically active regions and deep in stable cratons (Revenaugh and Jordan, 1991). A recent targeted survey for the presence of the L-discontinuity based on long-period SS precursors indicated that it is a local feature with lateral velocity variations, regionally present under the continents and island arc regions (Vidale and Benz, 1992; Deuss and Woodhouse, 2004). Under the continental areas it is observed twice as often as under oceans, with the most robust observations under Eurasia and Africa (Fig. 3.43). The L-discontinuity is also detected under South America; weak L-reflections are detected under western Australia with no evidence for the L-discontinuity under eastern Australia. There is little evidence for the L-reflection under North America (e.g. Gu *et al.*, 2001). Furthermore, arrivals from the L-discontinuity differ significantly in phase and in amplitude from predicted arrivals and cannot be explained by a simple horizontal boundary associated with an increase in seismic velocity.

Interpretations of the nature of the Lehmann discontinuity indicate that its origin can be closely related to the lithospheric structure. For stable continental regions some interpretations of the origin of the L-discontinuity are closely related to the concept of the seismic lithosphere (the *lid*). Lehmann discontinuity is interpreted as:

- the base of the petrologically distinct chemical boundary layer under the continental cratons (Jordan, 1975a, 1978),
- the base of the zone of partial melting (Vidale and Benz, 1992) within a stable continental lithosphere (Lambert and Wyllie, 1970; Hales, 1991; Thybo and Perchuc, 1997a),
- the depth at which the grain size changes from millimeter scale in the shallow mantle to centimeter scale in the lower part of the upper mantle above the transition zone (Faul and Jackson, 2005),
- the depth where an abrupt change in upper mantle seismic anisotropy occurs (Montagner and Anderson, 1989; Gaherty and Jordan, 1995) (Fig. 3.44); this change may be caused by the change in the preferred orientation of olivine due to a change in deformation mechanism from dislocation creep in the shallow mantle (favorable for creating anisotropic structures) to diffusion creep in the deeper mantle (favorable for isotropic structures) (Karato, 1992).

The latter interpretation is a subject for debate. Deuss and Woodhouse (2004) have measured the seismological Clapeyron slopes by correlating discontinuity depths with local velocity perturbations from a tomographic model. Assuming that velocity perturbations are solely due to temperature variations, they found that (except for the Middle East region) the Lehmann discontinuity is globally characterized by a negative Clapeyron slope (Fig. 3.43c). These authors argue that since all known phase transitions in the upper mantle above the transition zone have positive Clapeyron slopes, the only hypothesis that can explain a negative Clapeyron slope is the transition from dislocation to diffusion creep. A positive Clapeyron slope that they observed in the Middle East could be attributed to a phase transition from coesite to stishovite. In contrast to these interpretations, an anisotropic origin of the Lehmann discontinuity has been recently questioned by seismic interpretations for North America (Vinnik *et al.*, 2005), the continent with little evidence for the L- reflection (Gu *et al.*, 2001; Deuss and Woodhouse, 2004).

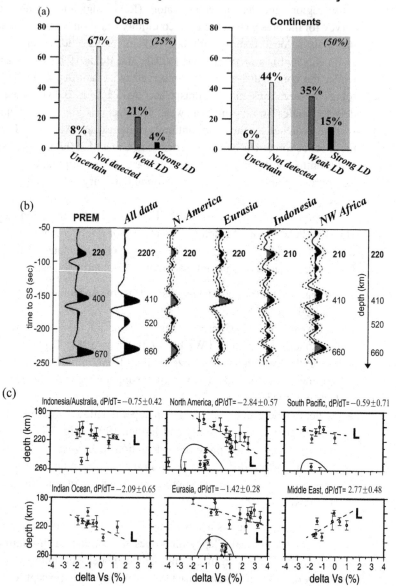

A global survey of the Lehmann discontinuity

Fig. 3.43 A global survey of the Lehmann discontinuity (a – after Gu *et al.*, 2001; b, c – after Deuss and Woodhouse, 2004). In the oceans, 75% of regions show no evidence for the L-discontinuity near the 220 km depth. (b) Stacking results for synthetic seismograms (PREM) and real seismic data (SS-precursor data set). The dashed lines show the 95% confidence levels for the regional stacks. The gray areas represent robust reflections with the lower confidence level above zero. (c) Depth of robust reflections versus velocity perturbations at the same depth. Dashed lines are approximations of Clapeyron slopes d*P*/d*T* for 180–240 km depth (the Lehmann discontinuity); clusters of reflections at 240–340 km depth are not a part of the L-populations. The values of d*P*/d*T* are in MPa/K; for West Africa (not shown) d*P*/d*T* = −1.33 ± 0.93 MPa/K.

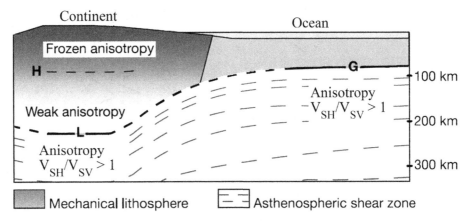

Fig. 3.44 Sketch illustrating an interpretation of the Lehmann (L) and Gutenberg (G) discontinuities in relation to seismic anisotropy; also shown is the Hales discontinuity (H) (from Gung *et al.*, 2003). Compare with Fig. 3.15 and 3.117.

3.4 Receiver function (converted waves) studies

3.4.1 The method

P- and S-converted waves

An incident seismic P-wave on an interface (crustal or mantle boundary with contrast seismic properties) continues as a direct (but deflected) P-wave and generates a converted S-wave (Figs. 3.45, 3.46). Similarly, an incident S-wave produces a converted P-wave. In the ideal situation of an isotropic horizontal-layered structure, P-to-S conversion results in SV phases so that all of the converted energy should be observed on the radial component and none on the transverse component, while the vertical component will be dominated by the high-amplitude initial P-wave (Vinnik, 1977). However, in teleseismic data P-to-SH conversions are commonly observed (Levin and Park, 1998).

The direct wave and a converted wave arrive at a receiver with a time difference t that depends on the depth H to the seismic interface, the velocities of the P- and S-wave in the layer above the interface, and the incidence angle of the impinging wave (i.e. the ray parameter = the slope of the travel time curve). In particular, for a P-to-S converted wave, the time separation between P_{direct} and Ps arrival is:

$$t = H \left(\sqrt{1/V_s^2 - p^2} - \sqrt{1/V_p^2 - p^2} \right), \tag{3.15}$$

where p is the ray parameter of the incident P-wave and V_s and V_p are average velocities in the layer between the velocity boundary and the receiver (Zhu and Kanamori, 2000).

If the velocity structure beneath the receiver is known, the converted phases may provide valuable information on the depth to seismic interfaces beneath the receiver. In

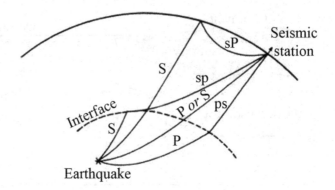

Fig. 3.45   Predicted ray paths for converted waves (from Sacks and Snoke, 1977).

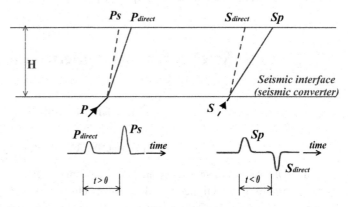

Fig. 3.46   Examples of P- and S-converted waves.

the real Earth, the incident wave produces not a single converted wave, but also multiples (Fig. 3.47). As for the primary P-to-S converted wave, their arrival times depend on the depth to the velocity contrast, the ray parameter of the incident P-waves, and the velocity structure of the layer between the seismic converter and the free surface. Since both the depth to the converter and the velocity structure above and below it are unknown, the ambiguity associated with the depth–velocity trade-off may be reduced by using the later arriving phases.

"Receiver functions"

To isolate the Earth structure near the receiver location, all other effects on the waveforms (such as distant structural variations and source effects) should be removed from the raw teleseismic data (Vinnik, 1977). "Receiver functions" are time-series that are computed from three-component seismograms after the response from the source and the vertically varying velocity structure beneath the receiver is isolated from the lateral heterogeneities.

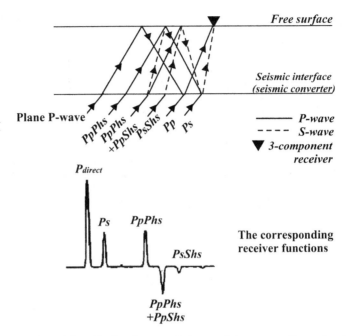

Fig. 3.47 A schematic ray diagram (upper) for major P-to-S converted phases and the corresponding receiver functions (lower) (redrawn after Ammon *et al.*, 1990). The following convention is used in denoting the seismic phases: except for the first arrivals, lowercase letters refer to the upgoing phases and uppercase letters to the downgoing phases; *h* denotes reflection from the interface.

This is done by deconvolution of the vertical component (which contains primarily compressional motion) from the horizontal components of ground (shear) motion after the recordings are rotated to a coordinate system defined by the ray direction before deconvolution. To isolate the source effect, a special technique that takes advantage of the redundant source information in three-component seismograms has been proposed (Vinnik, 1977; Langston, 1979); to isolate the structure response an inversion method of Ammon *et al.* (1990) is commonly applied. The resultant time-series are dominated by the effects of the local structure beneath the receiver. This method, known as the Receiver Function (RF) technique, is effectively a 1D seismic method: it provides a point measurement to the depth of the seismic converter because typically the conversion point is laterally close to the station (approximately 1/3 of the depth to the interface and often within 10 km for crustal studies) so that lateral velocity variations can be ignored.

The amplitudes of the converted waves are small (for example, the amplitudes of the phases converted from discontinuities in the mantle are a few percent in amplitude of the parent wave), and special signal processing procedures are required to extract them from noise. A combined analysis of recordings (stacking) of seismic waves arriving from the same source region and recorded at the same seismograph station is commonly used to improve the signal-to-noise ratio of weak secondary signals used in the receiver function approach. Assuming a similar level of noise in individual receiver functions (individual

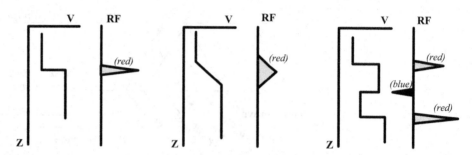

Fig. 3.48 Sketch illustrating receiver functions that correspond to three simplified velocity structures. Positive arrivals correspond to positive velocity contrast (conventionally shown in red); negative arrivals indicate the top of the low velocity zone (conventionally shown in blue).

waveforms radiated from different earthquakes), an improvement in the signal/noise ratio in the stack is proportional to the square root of the number of recordings. The amplitudes of the arrivals in RF depend on the incidence angle of the impinging wave and on the velocity contrast across the seismic converter. The latter property makes teleseismic RF (in which the arriving wave can be considered as a plane wave) a powerful tool for identifying the depth to major seismic interfaces in the lithosphere (primarily, depth to the Moho) and in the mantle (depth to the transition zone). To simplify visualization of RF interpretations, the positive polarity of the converted phase (which corresponds to a converter with a velocity increase with depth) is commonly shown in red and negative polarity (which corresponds to the top of a low velocity layer), in blue (Fig. 3.48).

Limitations of the RF method

The depth–velocity trade-off

RF inversions are non-unique and are primarily sensitive to the depth–velocity product (eq. 3.15) (Ammon *et al.*, 1990). Thus, to determine the depth to a seismic converter, *a-priori* information from other seismic experiments is necessary to reduce the velocity–depth ambiguity. A comparison of travel times in a seismic experiment with travel times for standard reference models may help to exclude extreme models and limit the range of possible solutions.

   An algorithm of stacking multiple converted phases (Zhu and Kanamori, 2000) is now in standard use to overcome the trade-off between the velocity structure and the depth to seismic converters. For these phases, the amplitudes of RF at all possible arrival times are summed for different thicknesses H of the layer (e.g. the crust) and different $k = Vp/Vs$. This procedure effectively transforms RF from the time domain into the depth (k–H) domain without the need to pick the arrival times of these phases. It provides the best estimations for both the depth to the converter and the Vp/Vs ratio (Fig. 3.49). An *average* crustal model can be obtained by stacking RF from different distances and different directions.

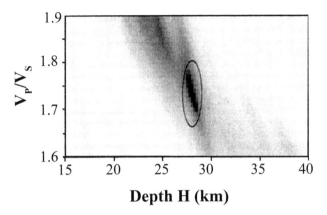

Depth H (km)

Fig. 3.49 Stacked receiver functions in the $Vp/Vs - H$ domain used to determine the crustal thickness, H, beneath the PAS station in Southern California. The best estimate is 28 km with a $Vp/Vs = 1.73$; the ellipse shows the 1σ uncertainty (Zhu and Kanamori, 2000).

Moho and intracrustal discontinuities

Since the amplitudes of the arrivals in a RF depend on the velocity contrast at the seismic converter, a boundary with the largest velocity contrast will produce the strongest converted waves. Therefore, in crustal studies the largest contrast need not necessarily be at the base of the crust but may be at an intracrustal boundary. For example, in the case of a high-velocity lowermost crust with $Vp \sim 7.5$ km/s, the velocity contrast at the Moho may be only $c.\ 0.5$ km/s, while the velocity contrast between the middle and the lowermost crust may be as large as 1.0 km/s. In such a case, the RF method will likely image the depth to the sharp intracrustal interface but not to the Moho. This may be the case in Greenland where conflicting depths to the Moho have been derived by RF, while refraction results agree with the deepest determination of the crust–mantle transition (Dahl-Jensen *et al.*, 2003; Kumar *et al.*, 2007). A similar example comes from Iceland, where the seismic signal from the Moho boundary varies from undetectable, to gradient zones and discrete discontinuities (Schlindwein, 2006). The complicated character of the signal is caused by the velocity structure of the crust, with the layer of rapid velocity increase at 8–14 km depth overlying the layer with almost constant velocities, which is masking seismic phases from the Moho.

Mantle discontinuities and the S-receiver functions

The RF method based on P-to-S converted waves has a disadvantage in studies of mantle seismic interfaces located at a depth of $c.\ 70–200$ km since in P receiver functions (PRF), the converted P phases from discontinuities in the mantle arrive in the time interval dominated by multiple reflections and scattering from crustal discontinuities (not necessarily from the Moho, but from the strongest crustal reflectors) (Fig. 3.50). To overcome this problem, the RF technique based on the S-to-P conversions has been proposed,

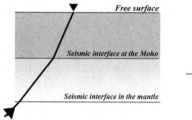 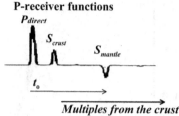 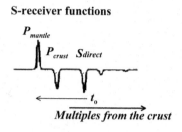

Fig. 3.50 A schematic diagram of P- and S-receiver functions for a seismic converter at 70–200 km depth in the mantle. In P-receiver functions phases from the mantle converter arrive within the crustal multiples and may not easily be identified. In contrast, in S-receiver functions it is the direct S-wave that arrives within the crustal multiples (due to high amplitude it can always be identified), while phases from the mantle converter arrive earlier and can also be easily identified.

in particular to study the depth to the lithosphere base (Farra and Vinnik, 2000). The S-receiver function technique (SRF) is based on the analysis of S-to-P converted phases (*Sp*) at seismic discontinuities in the upper mantle beneath stations. P-waves converted from a mantle velocity contrast arrive at the station much earlier than the direct S-waves, while all the multiple crustal reverberations arrive later than the S arrival (Fig. 3.50). Since high-amplitude direct S-waves can usually be easily identified even when they arrive within the crustal multiples, the S-receiver function analysis has advantages in studies of mantle discontinuities. Moreover, gradational boundaries in the mantle with a thickness of *c*. 30 km may be transparent at short periods (2–5 s) in PRF, but may be detectable at longer periods (10–15 s) in SRF.

A detailed discussion of the advantages and limitations of the S-receiver function technique is presented by Farra and Vinnik (2000) and is summarized here in brief:

- the time advance of the converted phase *Sp* relative to the direct S-wave depends on the depth of the discontinuity;
- the amplitude of *Sp* phase is nearly proportional to the *V*s velocity contrast at the discontinuity;
- the approximate position of the region sampled by the converted phase is at a roughly similar distance from the seismic station as the depth of the discontinuity;
- the greatest depth sampled by the SRF depends on epicentral distance;
- the method allows for detection of converted phases from the 410 km discontinuity implying that the SRF method is sensitive to detection of a *V*s velocity contrast of *c*. 0.2 km/s at a depth of about 400 km.

The apparent velocity of the *Sp* phase converted from a deep discontinuity may differ from the apparent velocity of the parent phase. For example, for the *Sp* phase from the 410 km discontinuity, the difference between the slowness of the *Sp* phase and its parent seismic phase (differential slowness) is around 0.6 s/deg for a standard Earth model. Because of noise or lateral heterogeneity of the Earth's structure, the observed slowness (the slowness of the trace with maximum amplitude of signal) may deviate from the theoretical predictions and for the phase converted at a dipping interface differential slowness may diverge from the standard value by about 0.2 s/deg for a tilt of 1 deg.

3.4.2 Examples of PRF and SRF studies of the crust and the upper mantle

Thickness of Precambrian crust in Greenland

Over the past two decades, receiver function analysis has become a routine technique for estimating the depth to the Moho. The method has been successfully used, in particular, in imaging the crust–mantle transition in complex tectonic settings such as subduction zones (e.g. Bostock *et al.*, 2002; Li *et al.*, 2003b).

One of the most spectacular examples is the RF application to the GLATIS (Greenland Lithosphere Analysed Teleseismically on the Ice Sheet) seismic data acquired in the interior parts of Greenland where the crustal thickness was previously completely unknown (Fig. 3.51) (Kumar *et al.*, 2007; Dahl-Jensen *et al.*, 2003). In spite of the excellent data quality, RF interpretations for four out of 16 temporally deployed broadband stations were complicated by strong converted phases generated at the base of the ice sheet (more than 3 km thick in some places). Since RF analysis is based on the assumption that the observed

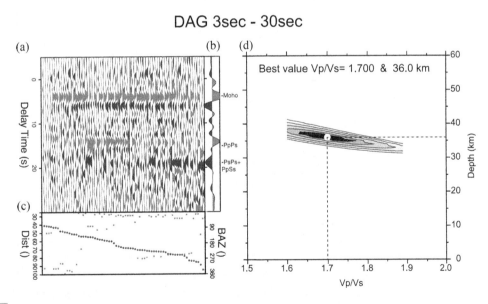

Fig. 3.51 Individual receiver functions (RFs) from all events used at the permanent seismic station DAG on the coast of northern East Greenland (a), the sum of the normal move-out-corrected individual RFs (a positive Ps has a peak to the right) (b), the distance in degrees (black symbols) and azimuth in degrees (gray symbols) for each event (c), and a plot of stack energy for various V_P/V_S ratios and the depth to Moho. V_P is assumed to be 6.5 km/s (from Dahl-Jensen *et al.*, 2003).

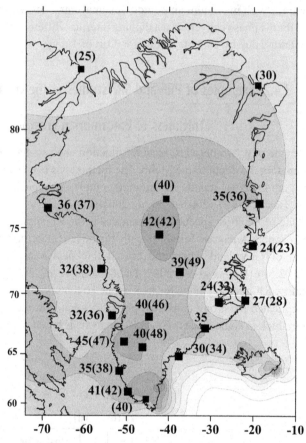

Fig. 3.52 Crustal thickness in Greenland (in km) based on receiver function analysis. Numbers – interpretations of Kumar *et al.* (2007) (no brackets) and Dahl-Jensen *et al.* (2003) (brackets). Seismic stations are shown as boxes.

wavefield is acquired at the free surface of an elastic half-space, seismic data collected on ice and on the ocean bottom have an added complication for RF interpretations as the time series constrained by standard methods may contain scattered energy from the ice or water column.

Receiver function studies in Greenland revealed significant lateral variations in the depth to the Moho. While the extended crust along the coast of East Greenland is less than 30 km thick, the Proterozoic part of central Greenland (north of the Archean core) has the largest values with an average depth to Moho close to 48 km (Fig. 3.52). Further to the north the Proterozoic crust is thinner, 37–42 km; similar values (around 40 km) were determined for the Archean crust in southern Greenland. Variations of 6–8 km in the thickness of the Proterozoic crust in Greenland may reflect the tectonic boundary between two ancient terranes (Dahl-Jensen *et al.*, 2003). However, later studies based on RF analysis of the same GLATIS data indicate significantly smaller values of crustal thickness in the Proterozoic terranes, 39–42 km, with little lateral variations in central

Greenland (Kumar *et al.*, 2007). A systematic difference in the values reported in the two studies for stations in central Greenland may suggest that the later study (with a systematically smaller crustal thickness) mapped, rather than the Moho, the top of the high-velocity lowermost crust, which is known to be present in regions with magmatic underplating in the crust. Alternatively, the difference between the two studies may be associated with strong converted phases from the ice column, since the largest discrepancy between the results is observed at the stations in inner Greenland where the ice thickness exceeds 2 km.

Lithosphere thickness in the collisional orogens of Central Asia

S-receiver functions (SRF) are a useful tool in mapping mantle discontinuities in the depth range obscured by crustal multiples present in P-receiver functions (PRF), such as the base of the lithosphere. Spectacular images of the upper mantle structure in Tibet, the Himalayas, and Central Asia constrained by RF for individual stations along a seismic profile provide 2D constraints on the dynamics of lithospheric plates in the zone of the largest present continent–continent collision and seismic images of Eurasian plate subduction (Kind *et al.*, 2002).

In the Tien Shan, RF studies of the lithosphere structure, initiated almost two decades ago (Kosarev *et al.*, 1993), clearly show converted phases from two interfaces: the upper with a positive velocity contrast is interpreted as the Moho. Different RF studies have revealed strong lateral variations in the crustal thickness of the Tien Shan and the Tarim Basin, with values ranging from 45 km to 65 km. The crustal thickness increases southwards to 80 km in northwest Tibet and exhibits a *c.* 20 km step in Moho depth across the Altyn Tagh Fault (Wittlinger *et al.*, 2004).

The lower interface has a negative velocity contrast and is interpreted as the top of a layer with reduced seismic velocities. Although this interface may not necessarily be the lithosphere–asthenosphere boundary (LAB), in most studies it is hypothesized to be the lithospheric base. Under this assumption, lithosphere thickness variations in the Tien Shan region as determined from S-receiver functions are moderate, 90 km to 110 km (Oreshin *et al.*, 2002). A later study, based on the same approach (Kumar *et al.*, 2005), however, reported significant variations in the lithosphere thickness of the region: arrival times for converted waves range from *c.* −10 sec to *c.* −30 sec, which for the *iasp91* model results in depths ranging from 90 km to almost 300 km (Fig. 3.53). The depth of the mantle interface with a negative velocity contrast is in fairly good agreement with a lower resolution, surface wave tomography image of the region, which shows a high velocity body at 80–180 km depth in the south (presumably the Indian plate) and a southwards dipping low-velocity body in the north, that was interpreted as evidence for subsidence of the rigid Tarim Basin block of the Asian lithosphere under the Pamir and Karakoram ranges (Kumar *et al.*, 2005). However, such an interpretation requires further support on the depth of the lithosphere–asthenosphere boundary since mantle discontinuity with a negative velocity contrast may not necessarily be the base of the lithosphere but, for example, a compositional intralithospheric boundary.

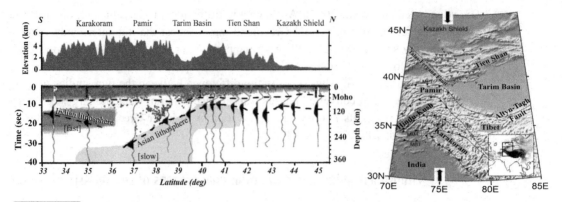

Fig. 3.53 Seismic section across the India–Asia collision zone (along 75 deg longitude) from S-receiver function analysis. Top panel – the topography along the profile. Bottom panel – depth to the Moho and to the upper mantle discontinuity with a negative velocity contrast. Zero time corresponds to the arrival time of the direct S-phase. Negative time-scale indicates times in front of the S arrival. Due to moveout correction the time axis is valid for a slowness of 6.4 s/deg. Gray shadings show velocity anomalies in the crust and the upper mantle based on surface wave tomography (dark shading – fast velocities, light shading – slow velocities). Dots are the earthquake hypocenters within a 100 km wide zone along the line. Right panel: location map with shaded topography (redrawn from Kumar *et al.*, 2005).

Compositional boundary within the cratonic lithospheric mantle?

The presence of a prominent interface with a 4.5% reduction in seismic velocity in a 10 km thick layer at approximately 150 km depth has been identified beneath the Kalahari craton by S-receiver functions (Savage and Silver, 2008). Since xenolith data as well as numerous seismic investigations based on various techniques consistently place the lithospheric base beneath the Kalahari craton at greater depth, 200–220 km, the converter at 150 km depth represents an intralithospheric boundary. This boundary crosses several Archean sutures and is unlikely to be related to the Archean tectonics. The base to the boundary corresponds approximately to the depth where olivine grain size decreases (Fig. 3.10a). A decrease in grain size should reduce seismic velocities (Figs. 3.6a, 3.14); however the effect is probably too weak to explain the seismic structure of the Kaapvaal mantle. Spatial correlation of the strongest velocity anomaly with intense Karoo volcanism suggests that the seismic discontinuity is a compositional boundary that marks the transition from a depleted (high-velocity) cratonic lithospheric mantle to the lower part of the lithosphere that experienced the influence of basaltic melts and metasomatic fluids, which have refertilized the cratonic lithosphere.

Similar conclusions based on buoyancy analysis have been made for the southern parts of the Archean–Paleoproterozoic East European (Russian) Platform that have been subsiding since the Paleozoic (Artemieva, 2003). Voluminous basaltic magmatism associated with the Devonian (possibly, plume-related) rifting in the Peri-Caspian Basin and in the Sarmatia subcraton (where it has led to formation of the huge, more than 20 km deep, Dnieper–Donets rift) could have refertilized the lower part of the cratonic lithosphere and therefore increased its density and created negative compositional buoyancy. The boundary between a highly depleted upper layer (with high seismic velocities) and a more fertile lower layer (with lower velocities) in the cratonic lithospheric mantle can produce a seismic discontinuity with negative velocity contrast similar to the one detected by SRF beneath the Kalahari craton (Savage and Silver, 2008).

Similarly, the presence of a seismic discontinuity with negative velocity contrast has been observed in other stable continental regions and can be a global feature for the Precambrian lithosphere; the depth to the top of this discontinuity varies from 80–100 km to c. 120–130 km (Thybo and Perchuc 1997a). A layer with reduced (c. 0 +1% as compared to the global continental model *iasp91*) seismic velocity has been identified within the high-velocity (c. +1 +3% as compared to the *iasp91* model) Proterozoic lithospheric mantle of the Baltic Shield along the FENNOLORA profile (Perchuc and Thybo, 1996), within the Archean Karelian mantle (Bruneton *et al.*, 2004), within the Precambrian lithospheric mantle of the northeastern East European Platform along the long-range profile Quartz (Ryberg *et al.*, 1996), within the lithospheric mantle of the West Siberian Basin and the Siberian craton (Nielsen *et al.*, 1999), within the Precambrian lithospheric mantle of the central US (Rodgers and Bhattacharyya, 2001), the Canadian Shield (LeFevre and Helmberger, 1989), and Greenland (Darbyshire, 2005). In some regions (the Baltic Shield, North America) the observed reduction in S-velocity may be stronger than the reduction in P-velocity. Possible explanations for a layer with reduced seismic velocity within a high-velocity lithospheric mantle of a cratonic root include the presence of fluids, small pockets of partial melts (or temperature close to the solidus), metasomatism, or compositional variations.

Mantle transition zone and thermal state of the upper mantle

The mantle transition zone is bounded by two principal discontinuities at approximately 410 and 660 km depths. Seismological observations and data from experimental mineralogy indicate that although both the 410 and 660 km discontinuities are sharp, they may spread over a depth interval of c. 4–10 km. For example, transformation of α-olivine to β-phase takes place over a pressure interval of c. 0.25 GPa where both phases coexist. This would result in a transitional discontinuity at 410 km depth spread over c. 7 km. The presence of even a small amount of water in the mantle stabilizes the β-phase over a wider range of pressure and temperature and further broadens the transition interval (Wood, 1995). Recent experimental high-pressure, high-temperature mineralogical studies indicate that water in the mantle has a significant effect on the topography of seismic discontinuities in the transition zone, in particular along the active margins. There is significant depression of the 660 km seismic discontinuity and an upward shift of the 410 km seismic discontinuity by 1–2 GPa which may be associated not only with cold subduction but also with wet subduction (Litasov *et al.*, 2005).

The RF technique allows for high-resolution determination of the transition zone thickness near a station (Fig. 3.54) (Vinnik, 1977). The P-to-S converted waves from the 410 and 660 km discontinuities travel practically identical paths through the upper mantle

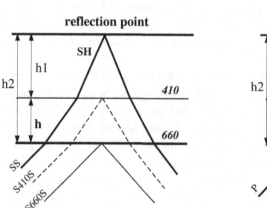
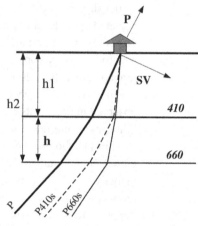

Fig. 3.54 Two approaches for calculation of the transition zone (TZ) thickness (from Gu and Dziewonski, 2002). Left: TZ thickness (*h*) is calculated from direct measurements of *S410S-S660S*, which is approximately twice the travel time of a near-vertical shear wave through the TZ. A schematic ray diagram shows SS and its precursors. Right: TZ thickness (*h*) is calculated from the differential times of *(P660s-P)-(P410s-P)* assuming that the ray parameter and both the P and the S velocities are known. A schematic ray diagram shows *Pds* phases (d denotes a TZ discontinuity).

above the 410 km phase boundary. Thus, the difference in their conversion times, $\Delta t = t_{660} - t_{410}$, characterizes the transition zone alone. The uncertainty in Δt estimates is typically less than 2 s (Gu and Dziewonski, 2002). The presence of velocity structure within the transition zone itself can produce another, up to 2 s, uncertainty in interpretations, but can successfully be overcome by computing the travel time residuals from two-way S travel time for a shear-wave that travels through the transition zone under each location (cap) (Fig. 3.55). Another complication in interpretations of the transition zone thickness arises from the fact that RF treats the arriving wave as a plane wave. Compared to a spherical wave, the plane wave approximation will lead to an overestimation of the time delay anomaly (up to 1.2 sec for an epicentral distance of 30–35°) and the transition zone thickness anomaly (up to 11 km for an epicentral distance of 30°) (Lawrence and Shearer, 2006). Due to velocity–depth trade-off in the RF analysis, larger Δt differences may indicate either greater thickness of the transition zone or lower seismic velocities above it. For the *iasp91* model, a change of 1 s in Δt corresponds approximately to a 10 km change in transition zone thickness.

An important property of the mantle transition zone is that its thickness provides important independent constraints on the thermal state of the overlying mantle. A decrease in mantle temperature causes elevation of the exothermic 410 km phase boundary (α-olivine → β-olivine phase transition) but depression of the endothermic 660 km phase boundary (γ-olivine → perovskite + magnesiowüstite phase transition) (see Fig. 9.12). An increase in mantle temperature has opposite effects on both of the discontinuities. Thus, the depths to the discontinuities are anticorrelated which results in a

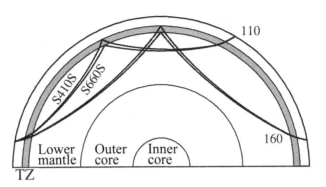

Fig. 3.55 A schematic diagram showing underside reflections *S410S* and *S660S* from the 410 km and 660 km discontinuities at distances of 110° and 160° (Gu and Dziewonski, 2002). The travel times *S410S-S660S* are primarily sensitive to transition zone (TZ) thickness near the reflection points.

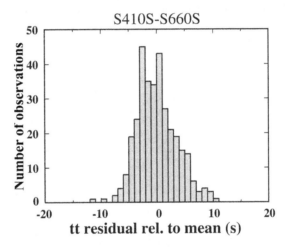

Fig. 3.56 Travel time residuals of the S410S-S660S measurements show an almost normal distribution with significant variations from the mean value of 242 ± 2 km (from Gu and Dziewonski, 2002).

decreased thickness of transition zone in regions with high mantle temperatures and increased thickness in regions where the ambient mantle is colder than average. The temperature effect on the conversion times difference Δt is commonly assumed to be $\partial(\Delta t)/\partial T = +0.009$ s/K. This means that a temperature variation in the mantle of *c.* 110 K will produce a Δt of *c.* 1 sec.

Global analyses of the travel time residuals of the S410S-S660S measurements show an almost normal distribution of transition zone thickness with a mean value of 242 ± 2 km and with significant (± 10 km) deviations from it (Fig. 3.56) (e.g. Flanagan and Shearer, 1998). Globally, the thickness of the mantle transition zone is correlated with surface tectonics and shows a significant difference between the oceanic and the continental mantle (Figs. 3.57, 3.58). Different interpretations are consistent in showing

Variations in the transition zone thickness

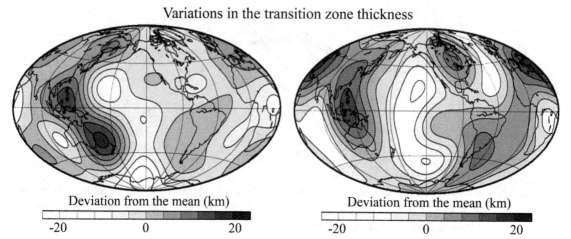

Fig. 3.57
Global variations in the transition zone thickness (left: interpretations of Flanagan and Shearer, 1998; right: interpretations of Gu and Dziewonski, 2002). Maps were filtered with a low-pass filter up to degree-6 spherical harmonics (from Gu and Dziewonski, 2002).

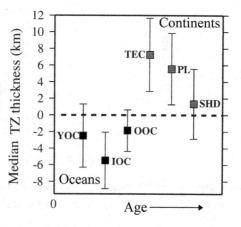

Fig. 3.58
Global variations in the transition zone thickness averaged for six tectonic provinces: YOC = young ocean; IOC = intermediate-age ocean; OOC = old ocean; TEC = tectonically active parts of the continents; PL = stable platforms; SHD = shields (from Gu and Dziewonski, 2002). The age-scale is non-linear.

significantly reduced thickness of the transition zone in oceanic regions, with the strongest anomaly in the Pacific Ocean. In continental regions the thickness of the transition zone is increased and the strongest anomaly is in Indochina. The thickness of the transition zone does not correlate with crustal age either in the oceans or on the continents (Fig. 3.58). The largest increase in transition zone thickness is observed in tectonically young continental structures and is apparently caused by cold subducting slabs in their mantle.

3.5 Controlled source methods: reflection/refraction studies of the upper mantle

3.5.1 Seismic reflection studies

The method and its resolution

Preamble

Application of the reflection method to geology was patented in 1916 in Germany by Ludger Mintrop and was first applied to hydrocarbon exploration in Oklahoma in 1920–1921. Since the mid 1970s the method has been widely used for crustal studies on the continents (e.g. profiles of COCORP in the USA, LITHOPROBE in Canada, ACORP in Australia, DEKORP in Germany, BIRPS in the U.K., ECORS in France, URSEIS in the Urals, FIRE in Finland, BABEL and MONA LISA profiles in the North and Baltic Sea) and, in particular, in the hydrocarbon exploration industry (for details see summary by Barazangi and Brown, 1986). Detailed reviews of reflection seismology can be found in Claerbout (1985) and Sheriff and Geldart (1982).

Seismic reflection data are typically recorded with dense regular shot and receiver spacing (commonly 50–500 m and 25–100 m, respectively) and to short offsets (<10 km). Controlled-source reflection seismology is based on the analysis of travel time (rather than amplitude) of seismic waves transmitted from the surface and reflected back to the surface from interfaces between geological layers with a contrast in the acoustic impedance (the product of seismic velocity and density). The amplitude of the reflected signal depends on the strength of the contrast in velocity or density. Such interfaces with a short wavelength change in elastic properties or density are loosely termed seismic discontinuities. Reflection seismology provides images of seismic discontinuities in the crust and the mantle and thus provides images of the structural geometry of the lithosphere, often with greater lateral and vertical resolution than refraction seismology.

Interpretation of seismic reflection data is based on application of:

(1) stacking of records to improve signal-to-noise ratio (stacking techniques use the normal moveout (NMO) correction for hyperbolic curvature of reflection travel times followed by the common midpoint (CMP) stacking),
(2) migration to compensate for energy scattering from dipping layers (effective for a dip less than 30°) and from lateral heterogeneities by summing along possible sources of scattered energy, and
(3) time shifts (static corrections) to compensate for variations in elevation and near-surface geology for on-shore profiles.

Resolution of reflection methods

The relative vertical resolution of seismic reflection studies is proportional to the wavelength of the seismic signal (Widess, 1982), i.e.

$$\delta \sim 0.25 V_{\mathrm{p}}/f, \qquad\qquad\qquad\qquad (3.16)$$

where $V\mathrm{p}$ is P-wave velocity (in km/s) and f is frequency (in Hz). This yields a vertical resolution of c. 60 m for $f = 25$ Hz in a continental crust with average velocity of c. 6.0 km/s. In the upper mantle, reflected P-waves are dominated by frequencies of c. 20 Hz which correspond to wavelengths of c. 400 m (Bostock, 1999). The resolution increases with frequency increase and in upper crustal studies may reach 10–20 m for high-frequency signals. The horizontal resolution of seismic reflection studies is poorer. It depends on the size of the Fresnel zone (see eq. 3.14) and decreases with depth. This means that reflectivity gaps with lateral dimensions less than c. 3 km cannot be resolved using reflection data.

The depth measure used in seismic reflection is in two-way travel time (TWT) for the vertical travel path below the shot point, and knowledge of the average velocity structure is required to convert it to depth. Typically, reflection data cannot provide background velocities. Thus, most seismic reflection profiles are plotted as time sections rather than depth sections, and reflection surveys are often complemented by refraction experiments, where errors in the calculated depth can be 5–6%. In cases where external velocity data are unavailable, velocity structure is estimated directly from reflection data.

Origin of crustal and mantle reflectivity

Crustal reflectivity

Seismic reflectors are caused by contrasts in seismic velocity and/or density. However, globally, lower crustal reflectivity does not correlate with any intracrustal velocity boundaries but is rather controlled by crustal viscosity (Mooney and Brocher, 1987). Several mechanisms (reviewed by Mooney and Meissner, 1992) have been proposed to explain crustal reflectivity, and one or more of these mechanisms can be responsible for seismic reflectivity in different tectonic settings:

- lithologic and metamorphic layering (i.e. presence of mylonite zones)
- compositional layering (e.g. due to igneous intrusives into the crust)
- layered crustal porosity and the presence of pore fluids (probably applicable only for the upper crust)
- lenses of partial melt
- reflections from fault zones
- mineral alignment caused by ductile flow
- velocity anisotropy introduced by the preferred orientation of olivine crystals.

Reflection Moho

As suggested by refraction seismic surveys, the Moho is usually a compositional boundary which marks the transition from the crust to the upper mantle. By comparison with refraction data in coincident reflection/refraction seismic experiments, the "reflection Moho" is commonly assigned to the base of a package of nearly continuous reflectors at

TWT 8–12 sec which is underlain by a nearly transparent zone with no (or very few) seismic reflectors. The mantle below the "reflection Moho" generally does not reflect seismic waves at near-vertical incidence with frequencies >5–8 Hz (except at local structures such as shear zones). This implies that the "reflection Moho" corresponds to an intralithospheric boundary with an abrupt change in the scale of scattering heterogeneities: lower crustal heterogeneities have much smaller dimensions than upper mantle heterogeneities (Fuchs and Wenzel, 1997).

There are no *a-priori* reasons to assume that the bottom of the lower crustal reflectivity package (which is a structural characteristic) should correlate with the depth with a sharp increase in seismic velocity (associated with a compositional boundary). The origin of the general agreement between the reflection and the refraction Moho as observed in many continental studies is the subject of some debate. It is proposed that the Moho is a compositional boundary formed during chemical differentiation within the lithosphere; the aspect ratios of velocity heterogeneities and density of their distribution may be directly linked to the process of lithosphere differentiation and to magmatic processes. Alternatively or additionally, relative motion between the crust and the mantle at detachment zones facilitates differential deformation between the crust and the mantle.

Seismic discontinuities in the upper mantle

Global seismic reflection observations indicate that the upper mantle is homogeneous at seismic reflection wavelength. However, some isolated mantle reflectors have been identified at depths of 70–100 km and more, for example in long-range PNE (Peaceful Nuclear Explosions) profiles recorded in the USSR between the 1970s and 1990s (see next section). By analogy with strong reflectivity of the lower crust, which is believed to be caused by ductile flow, transparency (weak reflectivity) of the subcrustal lithosphere may be controlled by mantle rheology. Thus, if reflectivity of the upper mantle is essentially controlled by its ability to form laminations and anisotropy through shear flow, the reflectivity pattern can be a measure of upper mantle (lithospheric) rheology. Rheological discontinuities within the lithosphere imaged by reflection seismology may also reflect changes in deformation mechanisms (Karato and Wu, 1993). The base of the lithosphere is generally accepted to be a rheological boundary separating rigid lithospheric plate from less viscous convective mantle, and thus it manifests itself as a seismic discontinuity. Deeper in the mantle, phase changes in the minerals, in particular in olivine, caused by pressure and temperature increase with depth, are responsible for seismic discontinuities at 410, 520, and 660 km depths.

Seismic reflection expression of tectonic styles

Most reflection profiles are limited to the crustal depths and only a few reach to 80–100 km depths. Since reflection seismology is a big topic, and in particular for crustal studies, the focus here will be solely on reflection expression of different tectonic styles with emphasis on reflection studies of the subcrustal lithosphere.

Extended continental crust

Reflectivity patterns observed in the lithosphere are well correlated with distinct tectonic settings. One of the most interesting results of seismic reflection surveys comes from reflection transects through regions with extended continental crust and passive margins. These regions have a highly reflective lower crust which is in sharp contrast with the almost transparent upper mantle and nearly flat reflection Moho (Fig. 3.59). Such a pattern is observed in the Cenozoic Basin and Range Province in western USA and the Paleozoic Variscides of western Europe which, despite the differences in ages of the crustal extensions, have similar crustal reflectivity: in both regions the reflection Moho is at a *c.* 30 km depth and ignores the diversity of tectonic regimes and structures at the surface (e.g. Allmendinger *et al.*, 1986; Meissner, 1999). The crust of the *c.* 3000 km long Variscan belt of Europe (Figs. 3.60; 3.61), which was formed by multiple Paleozoic terrane and continent–ocean collisional events, contains the Variscan collisional and subduction structures but has a nearly uniform crustal thickness. Flat Moho both in the Basin and Range Province and in the European Variscides is usually ascribed to extensional processes that have removed crustal roots below the pre-existing orogens (lower crustal eclogitization and delamination is an alternative explanation). For example, in western USA, crustal extension along a *c.* 323 km long profile varies from 300% in the western part of the profile to 15% in its central and eastern parts (Gans, 1987).

The lower crustal reflectivity in regions with extended crust is caused by compositional, metamorphic, and mineralogical layering (Mooney and Meissner, 1992). Its origin is attributed to mafic intrusions and ductile flow in the lower crust which make the reflectors

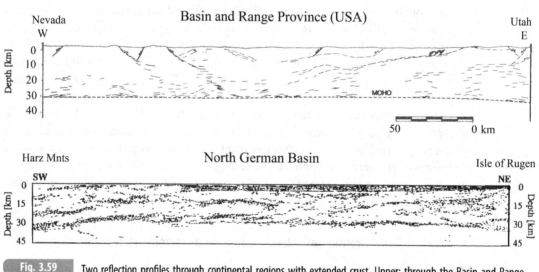

Fig. 3.59 Two reflection profiles through continental regions with extended crust. Upper: through the Basin and Range Province, western USA, along the COCORP 40° N seismic reflection transect (after Klemperer *et al.*, 1986). Lower: through the European Variscides (North German Basin) along the DEKORP profile (from Meissner, 1999). Note flat Moho and poor upper mantle reflectivity along both profiles.

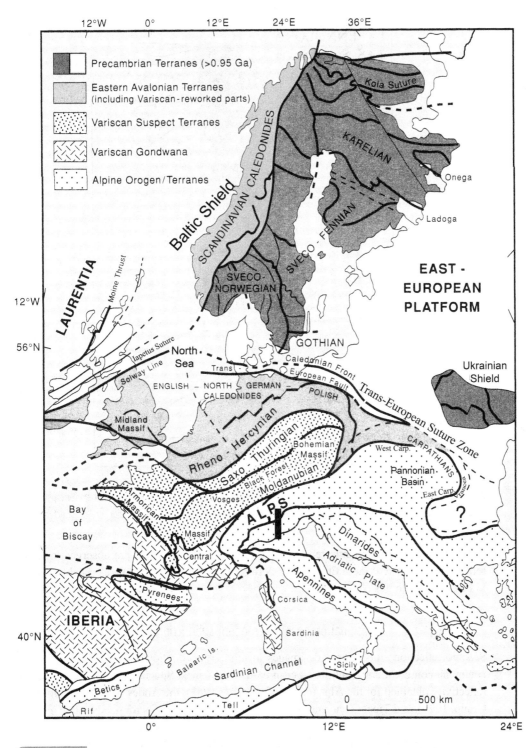

Fig. 3.60 Major tectonic units of western and northern Europe (after Berthelsen, 1992). Bold line – a part of the European Geotraverse (EGT) profile through the eastern Swiss Alps.

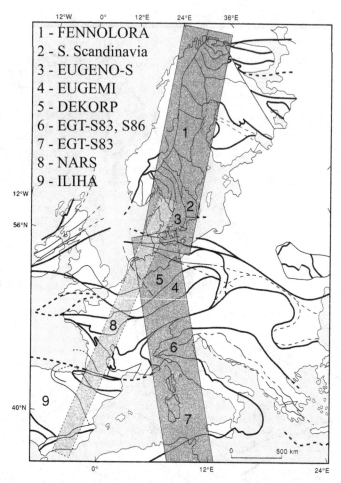

Fig. 3.61 European Geotraverse (EGT) seismic experiments in western Europe (from Blundell *et al.*, 1992). Numbers correspond to the individual projects within the joint program. Solid and dashed lines – boundaries of major tectonic structures.

near-horizontal. Upper mantle reflectors, both dipping and flat, that are observed in some reflection profiles in western Europe (e.g. between England and France) are probably related to Paleozoic tectonics (see below).

Collisional orogens and modern subduction zones

Seismic reflection studies provide images of modern subduction zones in the regions of continent–continent and continent–ocean collisional zones. Spectacular images of mantle reflectivity obtained for the Alps (e.g. Pfiffner *et al.*, 1988), the Andes (ANCORP Working Group, 1999), and Tibet (Zhao and Nelson, 1993) provide the basis for geodynamic models of evolution for these regions.

The Alpine deformation zone is formed by convergence of the Eurasian and African plates, which began at *c.* 120 Ma and resulted in plate collision and subduction at *c.* 65 Ma

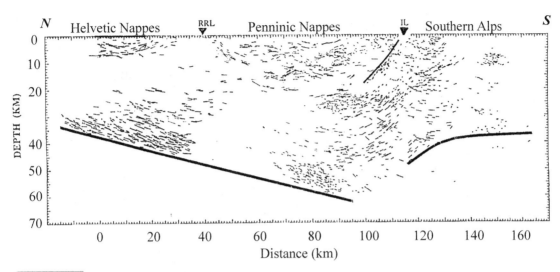

Fig. 3.62 Interpretation of migrated near-vertical reflection and wide-angle reflection and refraction data along the EGT profile through the eastern Swiss Alps (based on Kissling, 1993). Solid line – Moho (based mainly on wide-angle seismic data); thin solid line – Insubric Line (IL). For the location of the profile see Fig. 3.60 (black box) and Fig. 4.40a.

and uplift of the Alpine orogenic belt starting from *c.* 23 Ma. This convergence continues with a velocity of *c.* 9 mm/year. As a result, a highly complex and heterogeneous structure of the crust and upper mantle of the region has formed. High-resolution regional tomography models (although controversial, Lippitsch *et al.* 2003; Brückl *et al.*, 2010) indicate the presence of several subduction zones: beneath the western and central Alps a lithospheric plate (presumably continental European lower lithosphere) is steeply subducting south-eastwards, while in the eastern Alps the second plate (interpreted as the continental Adriatic lower lithosphere) may be subducting northeastward (possibly, almost vertically) beneath the European plate. In the subduction zone beneath the central Alps, the lower crust is characterized by strong dipping reflections continuing down to *c.* 60 km depth (Fig. 3.62); the deepest part of these reflectors is interpreted as the crustal root formed during the collision.

The ANCORP seismic experiment across the Central Andean forearc shows spectacular images of the Nazca plate subducting beneath the Andes (Fig. 3.63). Strong mantle reflectors observed in this and similar seismic profiles (e.g. across the Vancouver island in Canada) are interpreted as images of subducting oceanic slabs with strong reflectivity variations, possibly related to different rheology.

Mantle reflectors and paleosubduction zones

Some of the first detected mantle reflectors were observed in the North Sea in the Scottish Caledonides (the so-called W-reflectors, Fig. 3.64) (Smythe *et al.*, 1982). Originally they were interpreted as rift-related extensional shearing in the mantle. Later detailed waveform modeling indicated that they are produced by 6–8 km thick layers of high P-velocity

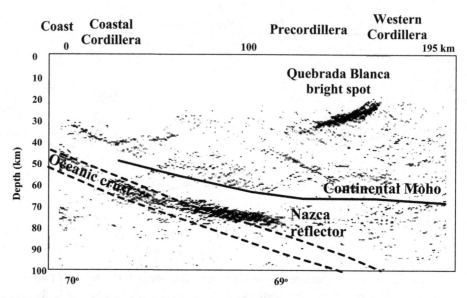

Fig. 3.63 Seismic reflection image of the Central Andean forearc and subduction zone from the ANCORP experiment (based on ANCORP WG, 1999). One of two strongest reflectors, the Nazca reflector, is interpreted as an image of the subducting oceanic crust.

(>8.4 km/s) and high density (>3.5 g/cc) and thus are likely to be remnants of fossil (Precambrian) subduction thrust faults. Consequent seismic reflection profiles in several parts of the Baltic region provided images of distinct mantle reflectors in the North and Baltic Seas (BABEL Working Group, 1990; 1993; MONA LISA Working Group, 1997). Similarity in the geometry of these reflectors with the reflectivity patterns observed in modern subduction zones and collisional sutures allows for their interpretation as traces of the ancient subduction of former oceanic crust. Their ages are constrained by those of known major tectonic events in the region (such as Paleoproterozoic plate convergence, subduction, and terrane accretion) and vary from c. 1.9 Ga for the northern part of the Bothnian Gulf to 1.6–1.4 Ga for the reflectors at the southern, Proterozoic, margins of the Baltic Shield, to c. 440 Ma for the reflector in the North Sea. The latter was probably formed during orogenic events related to the collision of Baltica, Laurentia, and Avalonia, closure of the Iapetus Ocean and Tornquist Sea, and subsequent amalgamation of a series of terranes at the edge of the East European craton.

Evidence for Paleozoic tectonic events has been found in seismic reflection studies in the Ural mountains. Mantle reflectors imaged at c. 175 km depth beneath the western margin of the Urals (Knapp et al., 1996) were interpreted as evidence for Paleozoic continent–ocean collision, followed by westward subduction and accretion of oceanic terranes along the eastern margin of the orogen. Alternatively, these reflections may mark the transition from lithosphere to ductile asthenosphere.

Images of dipping mantle reflectors have been produced in LITHOPROBE reflection profiles in the Canadian Shield in the Slave craton and in the Superior province (e.g., Calvert et al., 1995; Cook et al., 1999). Mantle reflectors extending from the lower crust down to

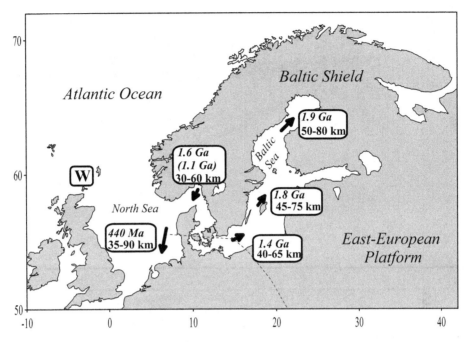

Fig. 3.64 Observations of subcrustal distinct dipping seismic reflectivity in the Baltic Shield (based on Balling, 2000). Arrows show dip direction and depth range of subcrustal reflectivity. These dipping seismic reflectors are interpreted as images of relict subduction zones and collisional sutures. Their ages range from *c*. 1.9 Ga for the northernmost reflector in the Bothnian Gulf to *c*. 440 Ma for the reflector in the southern part, in the North Sea, at the edge of the East European craton. W = W-reflectors of Smythe *et al.* (1982).

c. 100 km depth in the mantle were interpreted as Proterozoic subduction thrust below a 1.86–1.9 Ga volcanic arc at the western margin of the Slave craton. The most striking example of mantle reflectivity was imaged along LITHOPROBE Line 48 across the Archean Abitibi belt and the Paleoproterozoic Opatica plutonic belt in the Canadian Shield (Fig. 3.65). Dipping reflectors extending 30 km into the mantle were interpreted as images of a 2.69 Ga subduction thrust. These results provide strong evidence that some form of plate tectonics could have already been operating in the Archean.

Dipping tectonic boundaries in the lithosphere

Seismic reflection studies, in particular LITHOPROBE experiments in different parts of the Canadian Shield, provide evidence on the geometry of tectonic boundaries within the lithosphere. These results indicate that the boundaries between adjacent terranes are not necessarily vertical. A set of seismic transects across the Grenville Front, which separates the Archean lithosphere to the west from the Mesoproterozoic lithosphere to the east, clearly demonstrates that the deep expression of the edge of the Archean North American craton does not match its surface manifestation (Fig. 3.66). In particular, the Archean cratonic

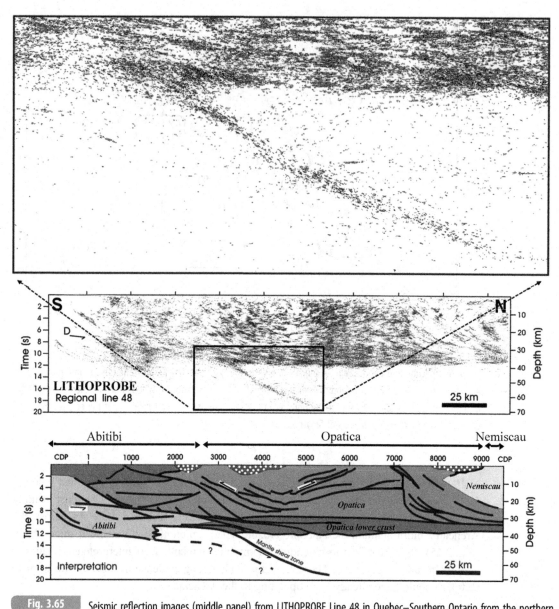

Fig. 3.65 Seismic reflection images (middle panel) from LITHOPROBE Line 48 in Quebec–Southern Ontario from the northern Abitibi belt to the Opatica plutonic belt (for location see Fig. 3.66). Upper part – zoomed central section of the reflection profile which shows a remarkable image of the fossil (*c.* 2.69 Ga) subduction zone and/or crustal delamination. Lower panel – tectonic interpretation based on geological, geochemical, and geophysical data (after Ludden and Hynes, 2000; www.lithoprobe.ca).

lithosphere of the eastern Canadian Shield underlies the Mesoproterozoic crust of the Grenville province over lateral distances of *c.* 200 km (Fig. 3.67). This implies that, within a few hundred kilometers of major tectonic boundaries, the age of the upper crust and the underlying lower crust and the upper mantle can differ dramatically.

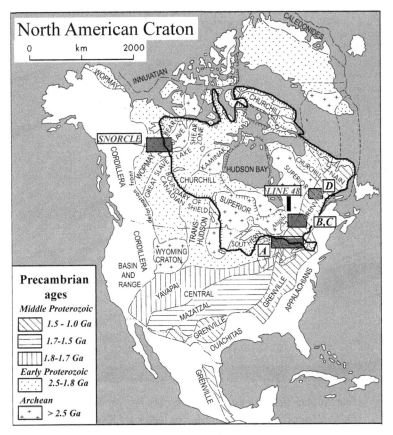

Fig. 3.66 Main geologic divisions of the North American craton and Greenland (after Goodwin, 1996 and Hoffman, 1989). Gray shadings show locations of the LITHOPROBE transects discussed in the text.

3.5.2 Seismic refraction and wide-angle reflection

The method

Seismic refraction studies are significantly younger than seismic reflection studies. The first cited paper on the application of this method to geology was only published in 1951, although the first refraction studies were initiated before the second world war. Since then, deep seismic sounding (DSS) profiles have provided important information on the velocity structure of the crust and the mantle. Compared to reflection seismology, a typical DSS profile is acquired on a linear array with long shot-to-receiver offsets (up to 500 km to allow for significant wave penetration below the crust), larger receiver spacing (typically 1–5 km, although some recent surveys deploy seismometers with smaller spacing), and larger shot spacing (20–100 km). Seismic waves are typically recorded at frequencies of 1–50 Hz which corresponds to wavelengths of 160 m–8 km for P-waves in

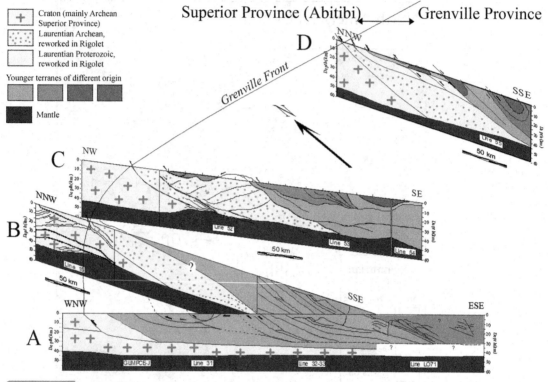

Fig. 3.67 Four composite sections of the Abitibi province and the Grenville province based on LITHOPROBE near-vertical-incidence seismic reflection profiles (based on Ludden and Hynes, 2000). For locations of the profiles see map in Fig. 3.66 (Boxes A, B, C, D). Note that the Archean terranes of the Canadian Shield underlay the Mesoproterozoic crust of the Grenville province.

the mantle. Thus, commonly, direct comparison of reflection data with refraction/wide-angle reflection is not possible due to dissimilar coverage by the two methods. The evolution of modern digital instruments and the availability of large pools of instruments have allowed seismologists to almost bridge the gap between the two techniques, although not yet fully.

Refraction profiles are usually reversed by positioning sources along the whole array (or at least at both ends) to allow for distinguishing lateral variations in the crust and mantle. It is common practice to combine refraction surveys with wide-angle reflection, since the latter provides important information on the structure of seismic discontinuities. The resolution of the refraction method is *c.* 5–10% for deeper crustal velocities and crustal thickness, and *c.* 3–5% for upper mantle velocities.

The number of seismic refraction profiles worldwide, including on-shore and marine seismic experiments, is enormous and they cannot be adequately covered in any review. Here the results of the method are illustrated by two unique seismic experiments:

(1) from the long-range profiles recorded between the 1960s and 1990s across Eurasia (northeastern Europe and Siberia) using nuclear explosions as the source, which allowed for determination of mantle structure down to *c.* 700 km depth and

(2) from one of the earliest (late 1960s) long-range refraction profiles in central North America (the Early Rise experiment) which provided important information on the structure of major seismic discontinuities in the upper mantle.

Heterogeneous upper mantle of Eurasia

Preamble

Impressive images of upper mantle seismic structure down to a depth of *c.* 700 km have been produced in unique and unprecedented seismic refraction and wide-angle reflection studies in Russia based on the PNE (Peaceful Nuclear Explosions) data acquired mostly in the 1980s (Fig. 3.68). These explosions were comparable in magnitude to 5–6 earthquakes and were complemented by chemical explosions. The profiles are 3000–4000 km long and reversed seismic data were recorded with 3-component seismographs located at *c.* 10 km intervals.

Interpretations of data along CRATON, QUARTZ, METEORITE, and RIFT (and partly HORIZONT) long-range seismic profiles allow the determination of seismic structure of the lithosphere and upper mantle in the Siberian Platform, West Siberian Basin, and the north-

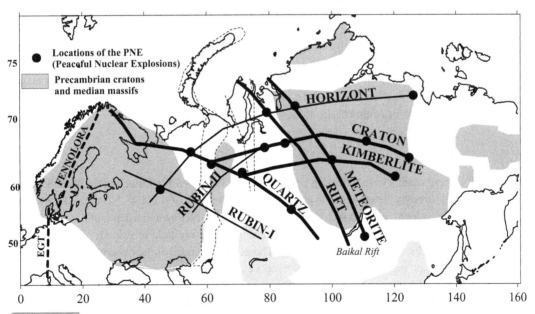

Fig. 3.68 Locations of the unique long-range PNE (Peaceful Nuclear Explosions) seismic profiles recorded in the USSR in the 1960s–1990s. Black circles – locations of nuclear explosions with energy release comparable to Mb = 5.0–6.0 earthquakes. Thin lines indicate profiles for which interpretations are limited or non-available. Also shown are the EGT (European Geotraverse) and the FENNOLORA seismic profiles. See Fig. 3.69 for the tectonic map.

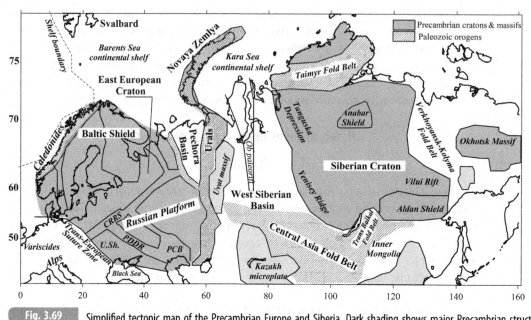

Fig. 3.69 Simplified tectonic map of the Precambrian Europe and Siberia. Dark shading shows major Precambrian structures (the East European craton, the Siberian craton and several median massifs). Dashed line – western margin of the Arctic Shelf. Abbreviations: CRRS = Central Russia Rift System; PCB = Peri-Caspian Basin; PDDD = Pripyat–Dnieper–Donets depression (rift); U.Sh. = Ukrainian Shield.

eastern parts of the East European (Russian) Platform and the Baltic Shield (Fig. 3.69). Early interpretations (Egorkin *et al.*, 1987; Burmakov *et al.*, 1987 and summaries in Fuchs, 1997; Mechie *et al.*, 1997) were followed by modeling based on modern seismic processing (e.g. Priestly *et al.*, 1994; Ryberg *et al.*, 1996; Mechie *et al.*, 1997; Morozova *et al.*, 1999). The major results of the studies show the following.

LVZs

The upper mantle of the Siberian craton, the East European Platform, and the West Siberian basin is strongly heterogeneous and incorporates a series of thin (30–50 km thick) alternating high-velocity and low-velocity layers from the Moho down to a depth of *c.* 150–220 km with significant lateral variations of depth range and amplitude of velocity variation in the layers (Fuchs, 1997) (Figs. 3.70, 3.71). The physical nature of the observed multi-LVZs in the Eurasian mantle is speculative since velocity anomalies in the upper mantle do not correlate with gravity anomalies. A simple explanation would be a solid-state low-velocity zone at depths where a velocity increase due to compression becomes less than a velocity decrease due to heating (Anderson *et al.*, 1968), an explanation that seems speculative for a cold cratonic mantle. However, the depths to the LVZs in the upper mantle of Eurasia are broadly consistent with global observations of a reduced velocity zone in the continental upper mantle, with the top at a *c.* 100 km depth, underlying a generally stratified uppermost mantle (Thybo and Perchuc, 1997a; Rodgers and Bhattacharyya, 2001). Importantly,

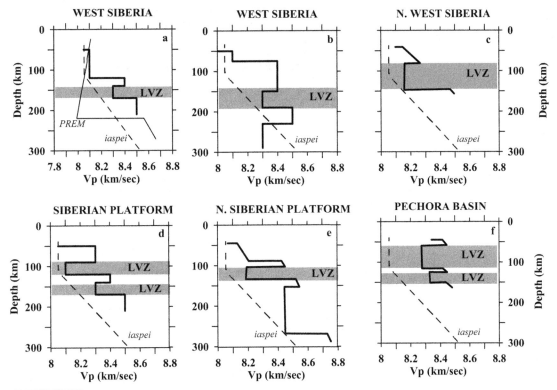

Fig. 3.70 1D velocity variations in the upper mantle of tectonic structures of Eurasia derived from 2D models based on long-range PNE profiles. (a) bold line – West Siberian Basin (WSB), profile CRATON (after Egorkin *et al.*, 1987), thin solid line – PREM model, thin dashed line – *iaspei* model; (b) WSB, profile QUARTZ (after Morozova *et al.*, 1999); (c) Ob paleorift, northern WSB, profile HORIZONT (after Burmakov *et al.*, 1987); (d) Tunguska Depression of the Siberian platform, profile CRATON (after Egorkin *et al.*, 1987); (e) northern Siberian Platform, profile HORIZONT (after Burmakov *et al.*, 1987). Note that seismic velocities in the LVZs are higher than in the global continental model *iaspei*.

seismic velocities in the LVZs are higher than the global continental average, similar to seismic observations in other Precambrian regions (see Section 3.4.2).

Morozova *et al.* (1999) interpreted a LVZ (with $Vp = 8.3$ km/s) at the QUARTZ profile in the depth range of *c.* 140 to 180 km beneath the West Siberian basin as being associated with partial melting and considered its top as the lithospheric base and its base as the Lehmann discontinuity. Similar interpretations of LVZs as the lithosphere–asthenosphere boundary were proposed for the RIFT profile (e.g. Priestly *et al.*, 1994). However, some authors (e.g. Mechie *et al.*, 1997) include upper mantle LVZ at depths of *c.* 130–160 km only for the QUARTZ and RUBIN-1 profiles, but not for other PNE profiles (CRATON, METEORITE, KIMBERLITE, RIFT), where interpretations allow for models without LVZ down to the 410 km discontinuity. Furthermore, since seismic data permit the existence of several LVZs in the upper 80–220 km of the mantle, it is unclear the top of which of the multi-LVZs should be interpreted as the lithospheric base if the formal definition of the seismic lid is

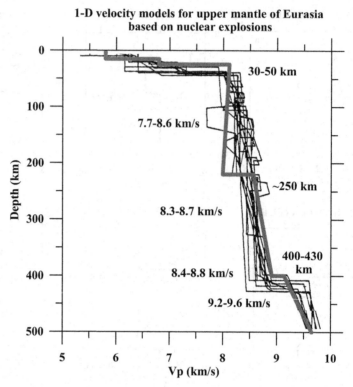

1-D velocity models for upper mantle of Eurasia based on nuclear explosions

30-50 km

7.7-8.6 km/s

~250 km

8.3-8.7 km/s

400-430 km

8.4-8.8 km/s

9.2-9.6 km/s

Fig. 3.71 1D velocity variations in the upper mantle of Eurasia based on long-range PNE profiles (based on Fuchs, 1997). Gray line – PREM model.

used for the lithosphere. It is more likely that the low-velocity zones are intralithospheric features which do not correspond to the lithospheric base. The presence of LVZs at a similar depth beneath all tectonic settings in Eurasia ranging from the Archean–Paleoproterozoic Siberian craton to Paleozoic rifts and basins (Fig. 3.70) suggests that they can be caused by changes in mantle mineralogy or olivine grain size with depth (compare with Fig. 3.9a).

The PNE seismic refraction and wide-angle reflection studies also detected the presence of seismic discontinuities at nominal 410 km, 520 km, and 660 km depth beneath Eurasia (Fig. 3.71). Along the QUARTZ profile, the 410 km discontinuity shallows northwards to *c.* 400 km depth beneath the Kola province of the Baltic Shield, while beneath the Altay mountains it is at 430 km depth (Morozova *et al.*, 1999).

Teleseismic Pn

Observation of seismic *Pn* and *Sn* phases with velocities of >8.0 and 4.7 km/s, respectively, are common in most tectonic settings and, together with seismic reflections from the Moho, are used for the determination of crustal thickness. They are interpreted as subhorizontally refracted phases (possibly also multiply reflected and scattered) which have been turning at the top of the mantle due to a positive vertical velocity gradient. Alternatively, existence of a sub-Moho waveguide has been proposed to explain *Pn* as waves reflected back towards the surface by scattering in a highly heterogeneous mantle (at a scale below the wavelength) (Fuchs, 1983). Such a waveguide can be generated by random successions of interchanging high- and low-velocity layers. Alternative explanations for the origin of *Pn* and *Sn* phases form the basis of recent debate when seismic data from the Soviet seismological program based on "Peaceful Nuclear Explosions" became available to several research groups in the west (Fuchs, 1997).

An intriguing feature observed in the PNE seismic data is the behaviour of a teleseismic *Pn* phase, a strong seismic phase with a long coda, that travels over unusually long (tele-seismic) distances of >3000 km with a group velocity of 8.0–8.1 km/s (Ryberg *et al.*, 1995). This phase, which is a secondary arrival observed behind the first arrivals and is relatively strong in the high-frequency band (>10 Hz), was termed a "high-frequency, teleseismic *Pn* phase". Modeling based on calculation of synthetic seismograms shows that the high-frequency component of this wave can be explained by forward and backward scattering in a laterally heterogeneous waveguide (with elongated anisotropic inhomogeneities with ±4% velocity fluctuations) between the Moho and a boundary at *c.* 100 km depth (Tittgemeyer *et al.*, 1996) (Fig. 3.72).

Alternatively, the observed amplitudes of the teleseismic *Pn* phase can be explained by a "whispering gallery" where phases travelling in the upper mantle experience multiple reflec-tions from the Moho (Morozov and Smithson, 2000). Scattering of "whispering gallery phases" from heterogeneities in the lower crust could explain the coda characteristics of the phase (Nielsen *et al.*, 2003). Furthermore, the teleseismic *Pn* phase has also been observed at low frequencies, but with different coda characteristics in different frequency bands (Nielsen and Thybo, 2003). While the mantle scattering model (Ryberg *et al.*, 1995) can only explain the high frequency part of the teleseismic *Pn* phase, and cannot explain the presence of the first arrivals (*Pn* phases), the "whispering gallery phase" can better explain all observations of the teleseismic *Pn* phase for realistic reflectivity properties of the lower crust, and it explains the observations of the first arrivals for a positive vertical velocity gradient in the upper mantle.

Seismic discontinuities in the upper mantle beneath North America

The Early Rise experiment was carried out in central North America in the mid 1960s (Iyer *et al.*, 1969). An underwater shot point with chemical explosives was located in Lake Superior; a series of linear arrays of seismometers at *c.* 30 km spacing radiated from the shot point over 2000–3000 km (Fig. 3.73). The disadvantage of the geometry of the project, as compared to the Soviet PNE profiles, was the absence of reversed and overlapping profiles, which are crucial for interpretation of heterogeneous mantle structure. In spite of these limitations, the project provided important pioneering data on seismic discontinuities in the upper mantle of North America.

Travel time interpretations along different profiles show the presence of seismic disconti-nuities at depths of 80–90 km and at *c.* 125 km depth (Green and Hales, 1968; Lewis and Meyer, 1968). Hales (1969) proposed that these boundaries have a common origin for all profiles and are caused by the spinel–garnet phase transition. Another boundary (the so-called 8° discontinuity) was identified at *c.* 700–900 km offsets (Fig. 3.74) and was

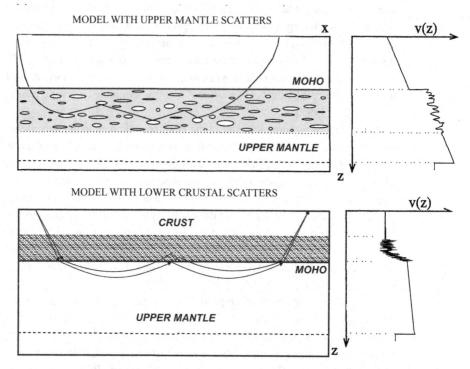

Fig. 3.72 Alternative velocity models for the origin of teleseismic *Pn* waves observed at the long-range PNE seismic profiles in Eurasia. One-dimensional velocity profiles (right) are extracted from the 2D velocity models for a 2000 km long QUARTZ profile. Upper: a sub-Moho waveguide model with transparent crust and Gaussian velocity fluctuations in the upper mantle which are generated by scatterers with anisotropic structure; their horizontal and vertical correlation lengths are 20 and 0.5 km, respectively (based on Ryberg *et al.*, 1995). Lower: a "whispering gallery" model with lower crustal velocity fluctuations (at 15 to 40 km depth) with horizontal and vertical dimensions of 2.4 and 0.6 km, respectively (redrawn from Nielsen and Thybo, 2003).

interpreted as the top of the low-velocity zone beginning at near a depth of *c*. 100 km (Thybo and Perchuc, 1997a,b). The refracted branch from the 410 km discontinuity was identified on several profiles.

Reinterpretations of travel time data from the Early Rise profiles provided new information on the seismic and thermal structure of the upper mantle of North America. While all profiles show similar seismic structure for offsets less than 800 km, at greater offsets there is a significant difference between profiles 1 and 2 located entirely within the Canadian Shield and other profiles, in particular profiles 6, 8, and 9 trending west and southwest (Fig. 3.74). This difference can be explained in terms of cold-to-hot transition in the upper mantle, which corresponds to the transition from the cold cratonic lithosphere of the Canadian Shield to the hot upper mantle of Phanerozoic regions, in particular in western USA (Thybo *et al.*, 2000). This transition in the mantle determined from the refraction data is in striking agreement with the margin of the North American craton as determined from surface wave tomography (Fig. 3.73). It also corresponds to the transition from almost aseismic to seismic regions.

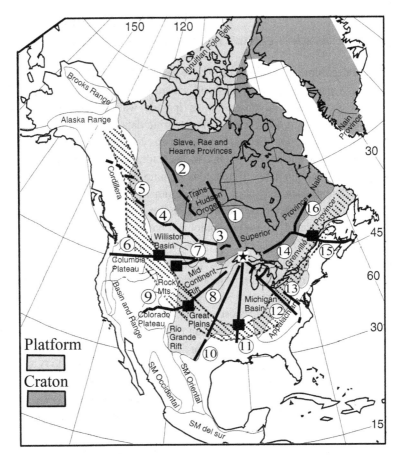

Fig. 3.73 Location map for the Early Rise long-range seismic profiles (numbered 1 to 16) radiating from the shot point in lake Superior (shown by a star). The transition from cold-to-hot regions is marked by black boxes (based on Thybo and Perchuc, 1997b); hatching – boundary of the North American craton as identified from surface wave tomography (van der Lee and Nolet, 1997).

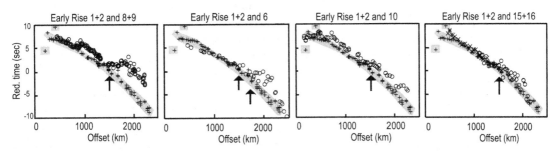

Fig. 3.74 Travel time plots of first arrivals for the Early Rise profiles (based on Thybo and Perchuc, 1997b). Numbers refer to profile numbers (for locations see Fig. 3.73). Profiles 1–2 which traverse only the cold cratonic lithosphere of the Canadian Shield are shown on each plot by crosses. Travel times for other profiles are shown by circles. The characteristic delays at c. 1300–1600 km offsets (arrows) are interpreted as the transition from cold-to-hot regions.

3.6 Teleseismic seismology

Teleseismic seismology uses large earthquakes (typically with magnitude greater than 4.5) as seismic sources. Broad-band waveforms from these earthquakes (teleseisms) recorded at global seismic observatories include waves radiating at different angles from the hypocenters, typically providing seismic data with good ray path coverage and in a wide frequency band. This section discusses teleseismic elastic tomography based on body-waves and surface-waves and anelastic tomography with the goal of recognizing the robust features common to different tomographic models. To facilitate a comparison of results from different research groups, a brief summary of the methods and their resolution is provided prior to discussion of global and regional seismic tomography results. Selected technical details related to tomographic inversion have been discussed earlier in this chapter (Section 3.2).

In the ocean of publications on methodological aspects of teleseismic seismology and the results of tomographic modeling, even a simple listing of the research groups and major large-scale and regional-scale tomographic models is not possible. The author apologizes to those researchers whose studies are not mentioned here. Overviews of elastic seismic tomography methods can be found in Romanowicz (1991), Iyer and Hirahara (1993), Nolet *et al.* (1994), Masters and Shearer (1995), Ritzwoller and Lavely (1995), Bijwaard *et al.* (1998), Dahlen and Tromp (1998), Boschi and Ekström (2002), Vasco *et al.* (2003), Ritsema *et al.* (2004), Kennett (2006), Nolet (2008). Some useful reference information can be found at: http://cfauvcs5.harvard.edu/lana/rem/index.htm, where several widely used global elastic tomography models are compared. Anelastic tomography is reviewed by Mitchell (1995), Bhattacharyya *et al.* (1996), Romanowicz and Durek (2000), while Jackson (2000) provides an overview of recent progress in laboratory measurements of seismic wave dispersion and attenuation.

3.6.1 Elastic tomography: methods and uncertainties

This section aims to provide some basic information on the advantages and limitations of body-wave and surface-wave tomography models. Shear-wave velocities in the upper mantle can be constrained with surface-wave phase- and group-velocities, body-wave travel times, and free-oscillation spectra, while P-wave velocity structure is constrained by much fewer data types. This section focuses primarily on surface-wave and body-wave tomography, and models constrained by free oscillations are omitted from the discussion. Because of differences in mantle sampling by the various body-wave and surface-wave types (see also Section 3.2), resolution of the deep structure can vary significantly between tomographic models constrained by different data, so that their comparison is, in general, like comparing apples and pears. In addition, long-period seismic waves have limited sensitivity to P-velocity structure.

A significant diversity exists even between tomography models based on the same seismic data, further influenced by model parameterization and the regularization used in the inversion to stabilize the inverse problem (in particular, in poorly constrained regions). As discussed in Section 3.2.3, when damping is used, cells with small-scale velocity perturbations should be treated with caution. Roughness control (i.e. the absence of strong contrasts in adjacent blocks) results in models with significant but smoothly varying heterogeneity, and cells with few data but with large-amplitude perturbations should be treated with caution. Furthermore, regularization can dominate an inversion result in poorly resolved cells (Vasco et al., 2003). If different tomography models use the same type of regularization for poorly resolved regions, a strong correlation between model results does not exclude the possibility of other patterns of heterogeneity (in shape and amplitude) that are consistent with observations.

With these words of caution in mind, a comparison of large-scale tomographic models allows for recognition of robust, model- and inversion-independent, features of the upper mantle. In comparing different models, it is important to recognize reliably determined features which supply important information on the physical properties and structure of the upper mantle. Due to the limitations of the approach, only well-resolved structures should be compared between different tomography models.

Body-wave seismic tomography: uncertainty and resolution

Resolution of a tomography model is determined by seismic wavelength, which in turn is proportional to the seismic wave period and inversely proportional to frequency. Generally, the higher the frequency, the higher the model resolution, and regional body-wave tomography typically utilizes seismic waves with a frequency of 0.1–10 Hz (e.g. Sipkin and Jordan, 1975; Inoue et al. 1990; Grand, 1994; Su et al., 1994; Masters et al. 1996; Trampert and Woodhouse 1996). Recent global travel time P-wave tomography models (e.g. Spakman et al. 1993; Zhou, 1996; Bijwaard et al., 1998) can image the seismic velocity structure of particular regions of the upper mantle with lateral resolution as detailed as that in regional tomographic studies (50–100 km) allowing us to distinguish even localized anomalies such as slabs in the upper mantle of the present subduction zones (e.g. van der Hilst et al., 1997; Grand, 2002) and lower mantle slabs associated with ancient subduction zones (e.g. van der Voo et al., 1999b; Fukao et al., 2001). However, lateral resolution of teleseismic body-wave tomography based on direct arrivals is significantly non-uniform and is limited by ray path coverage which is good only in tectonically active regions and areas with dense distribution of seismometers (direct arrivals are assumed to be sensitive mainly to the upper mantle structure in a narrow zone beneath sources and receivers). To overcome this problem, teleseismic body-waves multiply reflected from the surface and turned in the upper mantle have been used in some regional studies (e.g. Grand and Helmberger, 1985).

Most teleseismic phases arrive at the receiver along near-vertical ray paths, in particular phases from the core which are widely used in body-wave tomography studies (Figs. 3.75, 3.76). The near-vertical propagation in the upper mantle puts strong limitations on vertical resolution which at present does not exceed 50–100 km. One way to improve the vertical resolution is to also use travel times of waves from local earthquakes observed at local

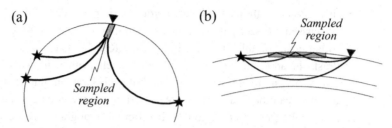

Fig. 3.75 Sketch showing ray paths in body-wave (a) and surface-wave (b) tomography. Gray shading – region sampled by seismic waves. Due to near-vertical propagation of body-waves, they provide low vertical resolution. In contrast, surface waves smear lateral velocity structure.

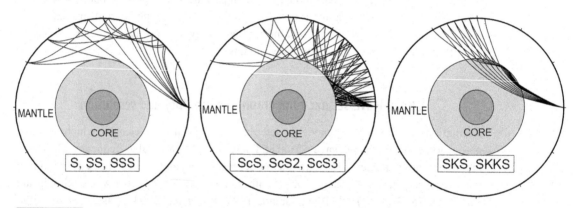

Fig. 3.76 Geometric ray paths for the shear wave phases (i.e. body-wave phases that propagate through the mantle with shear wave speed). These phases (i) propagate through the lower mantle (S) or diffract along the core–mantle boundary, (ii) reflect once (SS), twice (SSS), or three times (SSSS) off the Earth's surface, (iii) reflect once (ScS), twice (ScS2), or three times (ScS3) off the core–mantle boundary, or (iv) propagate as compressional waves through the core (SKS and SKKS) (from Ritsema and and van Heijst, 2000b).

stations, as these waves tend to travel sub-horizontally in the upper mantle. However, this approach is limited by local seismicity and also by the tendency of earthquakes to occur only along distinct fault zones.

Another approach is used in body-wave tomographic models of S. Grand (e.g. Grand, 1994, 2002; Grand *et al.*, 1997), where all velocity anomalies are put ad hoc into the upper 200 km. The approach is validated by the spectrum of mantle velocity anomalies as revealed by both body-wave and surface-wave global whole mantle tomography: velocity perturbations with amplitude exceeding 2% are restricted to the shallow mantle and the core–mantle boundary, while the spectrum of middle-mantle anomalies is "white" (e.g. Ritsema *et al.*, 2004).

Computations of the spatial resolution of body-wave tomography based on several traveltime phases have demonstrated the following (Vasco *et al.*, 2003):

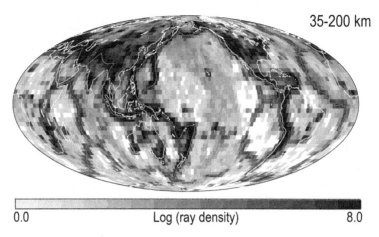

Fig. 3.77 Ray path sampling by compressional waves in the upper mantle. The logarithmic scale shows the number of rays intersecting each $3° \times 3°$ cell (unsampled blocks are shown in white) (from Vasco *et al.*, 2003).

- The errors associated with different phases differ significantly and core phases can lead to greater errors than first arriving P phases. As a result, inclusion of various phases into the inversion does not guarantee an improvement in model resolution.
- Model parameter resolution is correlated with ray density which is controlled by distribution of seismic events and seismic stations. This leads to highly variable resolution of tomographic images.
- The best resolved cells lie beneath the southern parts of Eurasia and North America, and a narrow zone of subduction zones encircling the Pacific Ocean (Fig. 3.77). Localized high velocity anomalies interpreted as slabs often coincide with narrow zones of high resolution in the mantle along event-station corridors and are a consequence of ray density–model resolution correlation.
- The mantle structure is moderately resolved beneath mid-ocean ridges and in the northern parts of Eurasia and North America. The worst resolved cells include the Pacific, Atlantic, and Indian ocean basins, the Arctic, Siberia, and most of the southern hemisphere except for Australia, southern Africa, and the Andes. In poorly constrained cells, the errors can exceed several percent.
- Mantle anisotropy affects ray density–model resolution correlation, in particular beneath the central Pacific: cells with a high ray density may still be poorly resolved.
- Existing crustal models allow for calculation of local velocity perturbations due to heterogeneities in crustal thickness and average crustal velocities. Although such corrections are of particular importance in surface-wave modeling (see below in this section), teleseismic frequency-dependent site effects (as compared to frequency-independent station corrections commonly used to correct for site effects) may have a significant effect on teleseismic P-wave amplitudes with the major contribution of amplitude losses in the sedimentary layer (Zhou *et al.*, 2003).
- Depth "leakage" due to smoothing in tomographic inversions reduces vertical resolution of the models (see also next section).

Surface-wave tomography: uncertainty and resolution

Dispersion, vertical resolution, and depth leakage

The advent of surface-wave tomography methods (Toksöz and Anderson, 1966; Cara, 1979; Woodhouse and Dziewonski, 1984; Nolet, 1990; Kennett, 1995; Trampert and Woodhouse, 1995, 1996) provided some of the best constraints on the structure of the Earth's upper mantle (e.g. Zhang and Tanimoto, 1993; Zielhuis and Nolet, 1994; Laske and Masters, 1996; Ekström *et al.*, 1997; van der Lee and Nolet, 1997a; Shapiro and Ritzwoller, 2002; Ritsema *et al.*, 2004; Debayle *et al.*, 2005; Panning and Romanowicz, 2006; Kustowski *et al.*, 2008a). Surface waves are commonly the strongest arrivals recorded at teleseismic distances. Seismic surface waves that propagate through the upper mantle include Rayleigh waves and Love waves. The latter are essentially horizontally polarized shear waves (SH waves) and their sensitivity to compressional velocities is exactly zero (Fig. 3.27).

The amplitudes of surface waves generally decrease exponentially with the depth of the source. Simple rules-of-thumb state that for fundamental modes, the depth best resolved with phase velocity is *c.* 0.4 times the wavelength of Rayleigh waves and *c.* 0.25 times the wavelength of Love waves (Love waves are particularly sensitive to the velocity structure in the upper 40 km of the Earth). Because of the frequency dependence of surface wave velocities (Fig. 3.78), their waveforms are strongly dispersed. As a result, surface waves, Rayleigh waves in particular, are sensitive to seismic velocity structure in the upper mantle over a relatively broad depth range. Backus–Gilbert averaging kernels (Backus and Gilbert, 1968) for Rayleigh waves illustrate three important features of surface wave tomography:

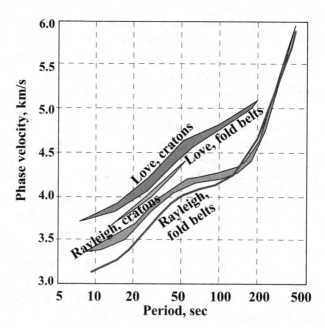

Fig. 3.78 An example of frequency-dependent phase velocity data for Rayleigh waves and Love waves from the cratonic regions and Proterozoic fold belts (after Lebedev *et al.*, 2008).

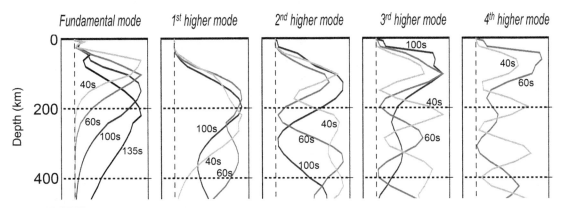

Fig. 3.79 Sensitivity kernels that relate phase velocities of the fundamental and the 1st, 2nd, 3rd, and 4th overtone Rayleigh waves to shear velocity in the PREM model (after Ritsema *et al.*, 2004). They illustrate that Rayleigh waves are sensitive to seismic velocity structure in the upper mantle over a relatively broad depth range. Numbers near the curves – seismic wave period in sec. Fundamental mode at periods of 40 sec or less samples only the upper 200 km; higher modes allow for sampling of deeper mantle. Long-period Rayleigh waves sample deeper layers in the upper mantle and propagate faster than shorter period Rayleigh waves because mantle structure at greater depth has higher wave speeds.

- The depth of wave penetration is very different for different modes (Fig. 3.79). Fundamental mode samples only the upper layers of the mantle. The higher the mode, the deeper the Rayleigh waves penetrate. Although the upper 300 km of the Earth are best sampled by surface waves, higher-mode surface-wave dispersion allows for constraining shear velocity variations down to at least 1000 km depth (Ritsema *et al.*, 2004).
- The depth of penetration depends on wave period. Long-period Rayleigh waves sample deeper layers in the upper mantle. They also propagate faster than shorter period Rayleigh waves because mantle structure at greater depth has higher wave speeds.
- Sensitivity kernels illustrate effective "depth resolution" and lateral resolution of Rayleigh wave tomography models (Fig. 3.79). To achieve the best vertical resolution, sensitivity kernels should be close to the delta-function. In reality, in particular for fundamental mode models, depth resolution is low and depends on wave period.

The last point is of particular importance for meaningful interpretation and comparison of surface-wave tomography models, especially fundamental mode models, but is often forgotten. Backus–Gilbert averaging kernels illustrate that fundamental modes of surface waves with period less than 40 sec do not resolve mantle structure below a *c.* 200 km depth. For this reason, Rayleigh waves with periods of 60–80 sec are widely used in tomography studies of the lithosphere–asthenosphere boundary. However, models based on 60 sec and 100 sec fundamental mode Rayleigh waves integrate velocity structure over the top *c.* 200 km and the top *c.* 400 km of the mantle, respectively (Fig. 3.79). For surface-wave models that are constrained not only by fundamental modes, vertical resolution in the uppermost 200 km of the mantle is of the order of 30–50 km. This means that in the uppermost mantle seismic velocity variations within a depth range of 30–50 km are

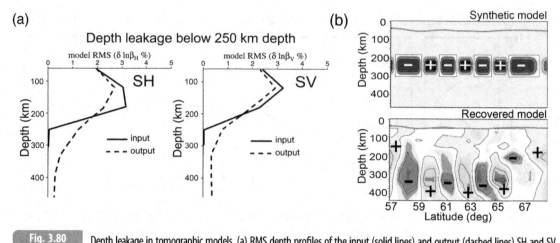

Fig. 3.80 Depth leakage in tomographic models. (a) RMS depth profiles of the input (solid lines) and output (dashed lines) SH and SV shear velocity models calculated for a global surface-wave tomographic model (Zhou *et al.*, 2006). The velocity anomaly in the input model was truncated below 250 km depth. However, in the output model some velocity anomaly persists down to the transition zone. Depth leakage (i.e. depth smoothing) is inevitable in tomographic inversions due to limited path coverage and random noise in synthetic data. (b) Synthetic test for body-wave tomography (after Eken *et al.*, 2008). Top: synthetic model which includes relative negative and positive *V*s anomalies (±3%) at depths of 180–300 km. Bottom: recovered synthetic model. Pluses and minuses indicate the sign of the recovered anomalies. The amplitude of the recovered anomalies is significantly weaker than that of the anomalies in the synthetic model.

indistinguishable and the depth extent of mid-ocean ridges (~200 km) and of the cratons (~250 km) cannot be better resolved than ~50 km (Ritsema *et al.*, 2004). The worst resolution is for the depth range *c.* 300–400 km, the depth to which the deepest roots of the cratonic lithosphere (or tectosphere) may extend (Jordan, 1975a).

Smoothing used in tomographic inversions leads to another limitation on vertical resolution of tomographic models: depth leakage, i.e. velocity anomalies that in the real Earth terminate at a certain depth in the upper mantle, "leak" down in tomographic inversions, and produce a tail of weak anomalies extending well below the bottom of the real anomaly (Fig. 3.80). Such depth smoothing (leakage) is inevitable because of limited data coverage and noise in the data, and weak anomalies should be interpreted with great caution (Zhou *et al.*, 2006).

Ray path coverage and lateral resolution

In the horizontal direction, the amplitudes of surface waves decay as the square root of the distance the wave has traveled from the source. Thus they decay more slowly with distance than do body waves and can orbit the Earth several times along great circle paths. An uneven distribution of seismic sources (since most earthquakes occur at plate boundaries) and receivers (most of which are located on land) results in incomplete data coverage of the mantle by surface waves. As for teleseismic body-wave tomography, this puts additional limitations on the effective resolution of surface-wave tomography, although surface waves provide more uniform

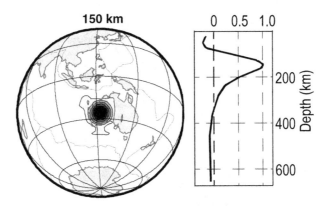

Fig. 3.81   Backus–Gilbert resolution kernels for a point beneath Australia, depth 150 km (left) and the radial dependence of the kernel (on the right) for the tomography model S20RTS which is a degree 20 shear velocity model based on fundamental and higher-mode Rayleigh wave dispersion, teleseismic body-wave travel times, and normal-mode splitting data (after Ritsema *et al.*, 2004). The resolution kernels illustrate lateral (left) and vertical (right) averaging in the final tomography model.

data coverage than body waves. As a result of uneven ray path coverage, the resolution of any tomographic model is heterogeneous and high resolution is geographically restricted.

The minimum lateral resolution of a tomography model is proportional to the wavelength $\lambda = V_s T_o$ and is at least *c.* 400 km (here V_s is the average seismic speed in the mantle and T_o is the period). However, the real lateral resolution is much poorer since surface waves orbit several times around the Earth before they are recorded at a seismic station and average the Earth structure along their path (Fig. 3.75). For example, at 150 km depth beneath Australia lateral resolution of the S20RTS surface-wave model (which is a degree 20 shear velocity model based on fundamental and higher-mode Rayleigh wave dispersion, teleseismic body-wave travel times, and normal-mode splitting data, Ritsema *et al.*, 1999) is almost half of the continent (~1000 km) (Fig. 3.81). The diameter of the resolution kernel indicates the size of the area over which the velocity anomaly at a given location is averaged.

An uneven ray path coverage may create additional problems in interpretation of regional velocity perturbations in tomographic models as further illustrated by an example from Australia. Because of poor ray path coverage in western Australia, tomography models based on the SKIPPY seismic data could not reliably resolve the upper mantle structure under the western part of the continent (van der Hilst *et al.*, 1994). Since in Australia the Archean cratons are located exactly in a poorly resolved domain (Fig. 3.82), some regional tomographic models have been interpreted as evidence for, contrary to globally observed patterns, thinner lithosphere beneath the Archean cratons than beneath the Proterozoic parts of the continent (e.g. Simons *et al.*, 1999). Relatively thin lithosphere (~200 km) beneath the Archean cratons of Australia is apparently an artifact of poor model resolution (Fig. 3.83) as evidenced by recent regional tomographic models with an improved ray path coverage for the western domain (B. Kennett, personal communication).

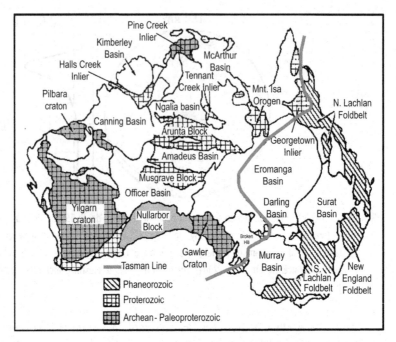

Fig. 3.82 Tectonic map of Australia (based on Goodwin, 1996). The Tasman line marks the transition from the Precambrian western-central part of the continent to the Phanerozoic eastern part of the continent. Western Australia includes two Archean cratons: Yilgarn and Pilbara; central Australia is formed by Mesoproterozoic–Neoproterozoic orogenic belts. Large inner parts of the continent are covered by a thick sedimentary cover and their detailed geology is poorly known.

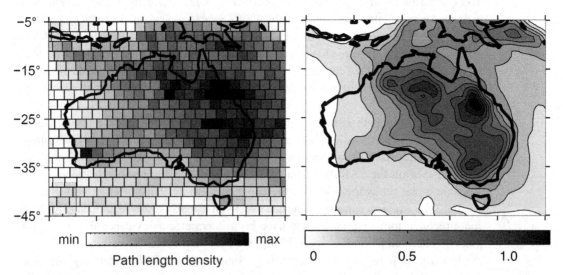

Fig. 3.83 Ray path coverage for Australia based on SKIPPY data (left) and the corresponding resolution test for isotropic surface-wave model (right, darker shades refer to higher resolution) (from Simons *et al.*, 2002). Except for western Australia, the tomography model gives a fair representation of the upper mantle shear-velocity structure under Australia.

Crustal corrections

The 21.4 km-thick globally averaged crust incorporated in PREM differs significantly in thickness from both the oceanic and the continental crust. However, propagation of surface waves in the upper mantle is significantly affected by the laterally variable properties of the Earth's crust, and tomographic inversion can erroneously propagate crustal structure down into the mantle. To compensate for this effect, a crustal correction is applied to seismic data before inverting for the mantle structure. Crustal corrections require independent knowledge of the crustal structure since global-scale and regional-scale tomographic models cannot resolve this. Surface waves are sensitive to Vp and Vs seismic velocities and density, and thus crustal models should include all of these parameters.

One of the early crustal models used in tomographic modeling was the model of global variations in crustal thickness of Soller et al. (1982) complemented by estimated average seismic velocities and densities for oceans and continents (Smith, 1989). A different crustal model was provided by the 3SMAC global tomographic model (Nataf and Richard, 1996). Global crustal models of the next generation are CRUST 5.1 and CRUST 2.0 (Mooney et al., 1998; Bassin et al., 2000). As compared to earlier crustal models, they are based on an extensive set of seismic reflection/refraction data available at that time for continents and oceans, complemented by high-resolution data on thickness of sedimentary cover (e.g. Laske and Masters, 1997). These models are parameterized on 5 deg x 5 deg and 2 deg x 2 deg grids, respectively, and include average values of thickness, density, Vp and Vs velocities for seven crustal layers; cells with no seismic data are filled with statistical averages for a crust of the same type (tectonic setting and age) (see also Section 3.3.1 and Figs. 3.38 and 3.39).

The effect of crustal structure on the phase velocities of surface waves has been analyzed in detail by Mooney et al. (1998), Boschi and Ekstrom (2002) and Ritsema et al. (2004) for the CRUST 5.1 model and by Bassin et al. (2000) for the CRUST 2.0 model. Additional discussions can be found in Laske and Masters (1997) and the effect of ice is discussed by Ritzwoller et al. (2001). The following summary is based on the results of these studies.

The effect of crustal heterogeneity is different for fundamental and higher-mode surface waves:

- For fundamental mode Rayleigh waves and the highest overtones, the crustal effect on velocity perturbations in the model is of critical importance because phase velocity variations caused by crustal heterogeneity and derived from the seismic data have similar amplitudes (Fig. 3.84).
- For the first and second Rayleigh overtones, the effect of crustal corrections is small as compared to the observed seismic data. Although these overtones provide constraints on mantle structure that are largely independent of the crust, they average the entire upper mantle and have low sensitivity to the lithospheric structure (Fig. 3.79).

The effect of crustal heterogeneity on velocity perturbations in the model depends on the period of surface waves:

Long-period surface waves:

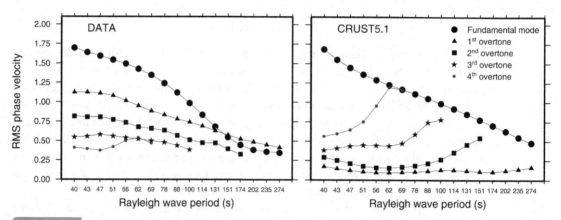

Fig. 3.84 Root-mean-square (RMS) phase velocity perturbation with respect to PREM (in %) of the Rayleigh wave phase velocity fundamental-mode and overtones in the data (left) and produced by crustal correction (for the CRUST 5.1 model, Mooney *et al.*, 1998) due to thickness and velocity variations in the crust (right) (from Ritsema *et al.*, 2004). For the fundamental-mode Rayleigh wave and the highest overtones, the RMS of the phase velocity distribution caused by variations in the crustal structure and the RMS derived from the data have similar amplitudes; for these modes the crustal effect on velocity perturbations in the model is of a critical importance. For the first and second Rayleigh overtones, the RMS of the crustal corrections is small compared to the observed RMS; this means that these overtones provide constraints on mantle structure that are largely independent of the crust. However, these overtones average the entire upper mantle and have low sensitivity to the lithospheric structure (Fig. 3.79).

- Variations in crustal structure have little effect on lateral variations in phase velocities.
- Crustal corrections, however, increase the variance of observed phase velocities in the mantle since seismic signals from the crust and from the upper mantle are often anti-correlated. For example, the peak-to-peak amplitude of 167 s Rayleigh waves increases by a factor of 1.3 after the crustal correction is applied.
- For Rayleigh waves at a period of 150 s the crustal contribution to phase velocity variations can be as large as 50% of the total signal and with sign opposite to that of the mantle contribution.

Short-period surface waves:

- Phase velocity variations caused by crustal heterogeneities are very strong. If crustal correction is applied, the peak-to-peak amplitude of phase velocity anomalies for 40 s Rayleigh waves is twice as large as for 167 s Rayleigh waves.
- Crustal corrections to Rayleigh waves significantly redistribute phase velocity anomalies and in many cases make them more localized.
- Rayleigh waves are sensitive to Vp velocity heterogeneity in the crust, in particular in the near-surface layers, and the presence of large sedimentary basins may have a significant effect on phase velocities. For example, the analysis performed for Rayleigh waves at 35 s indicates that underestimation of sedimentary thickness by 1 km produces a +1% erroneous increase in phase velocities, while a 3 km overestimation of sediment thickness in large parts on the continents has the opposite effect and produces a velocity error of

c. +2.5% (the consequences of different crustal models on tomographic results can be assessed by comparing Figs. 3.39 and 3.40 for Eurasia).

- Similarly, the presence of a thick ice sheet such as in Antarctica and Greenland, strongly affects surface wave velocities at short and intermediate periods: because of the very low density of ice at short periods the ice reduces the velocities similar to a sedimentary basin, and its presence on top of the crust also reduces the surface-wave velocities at intermediate periods.
- Since the sensitivity of Love waves is greatest in the upper 40 km of the Earth, these waves mainly sample the crust and are highly sensitive to variations in crustal structure. Note that Love waves are affected only by variations in Vs and density and are insensitive to Vp variations. As a result, they sample the crust and the mantle in different ways than Rayleigh waves at the same period. The effect of crustal correction on Love waves is twice as large as for Rayleigh waves at the same period. Strong dependence of very short wavelength Love waves on crustal structure can be used to refine crustal models in regions with poor coverage by seismic refraction data (Ekström *et al.*, 1997).

Note that CRUST 5.1 and CRUST 2.0 models average crustal structure over large areas (*c.* 550 km x 550 km and 200 km x 200 km, respectively). Use of even these advanced models for crustal correction can lead to significant artifacts at tectonic boundaries that separate crustal types with significantly different thickness, velocity, and density structure. This effect may be particularly important at the continent–ocean transition or at the transition from the Precambrian to Phanerozoic crust with significantly different average crustal structure (e.g. Fig. 3.41). Because of its higher lateral resolution the CRUST 2.0 model should provide a better correction for crustal heterogeneity.

Absolute versus relative velocities

As discussed earlier, a simple comparison between different tomographic models is hindered by differences in:

- data sets used to constrain them,
- parameterizations and regularization schemes,
- crustal models used for crustal corrections.

Most of the tomographic models work not with real velocities but with velocity perturbation with respect to some reference velocity model (for details see the overview by Kennett, 2006). Use of different reference models, when the results are presented as velocity perturbation with respect to them, but not as absolute velocities, further complicates comparison between different tomographic results.

The most commonly used reference models are *iaspei* (or *ak135*) and PREM. The latter is dominated by the oceanic velocity structure (as oceans comprise *c.* 70% of the Earth) and thus has the step at 220 km depth not required by the continental seismic data. As a result, all tomographic models that use PREM as a reference model inherit the velocity drop at 220 km depth (e.g. Su *et al.*, 1994). Tuning of the reference model in tomographic models for the continents to a "zero-level" model smooths the step but does not exclude it entirely (Fig. 3.86). This leads to

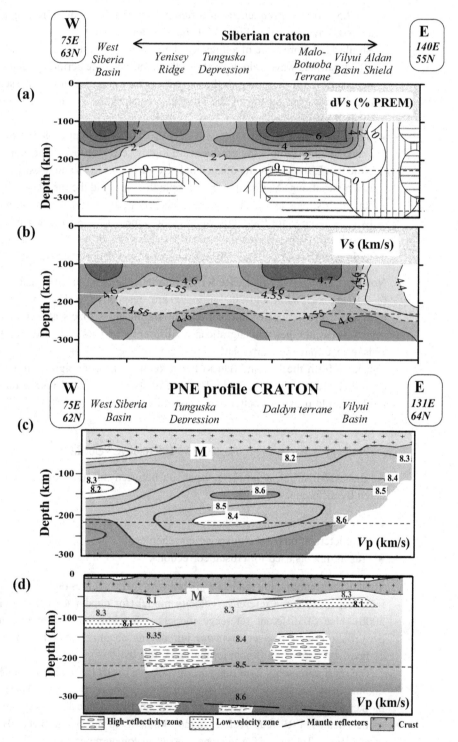

Fig. 3.85 Velocity structure of the upper mantle beneath Siberia. (a) Relative *V*s velocity perturbations with respect to PREM based on Rayleigh wave tomography (results of Priestley and Debayle, 2003); (b) the same velocity model recalculated to absolute velocities; (c) and (d) *V*p velocity structure along the part of the long-range PNE profile CRATON that crosses similar tectonic

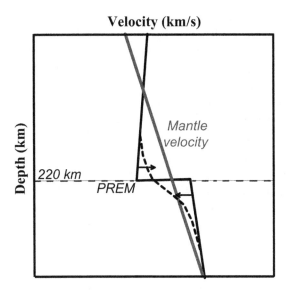

Fig. 3.86 Sketch illustrating that using PREM as a reference model in tomographic inversions for continental regions may lead to erroneous interpretations of the thickness of the seismic lithosphere. Inversion tends to 'smooth' the step at 220 km depth, which is not required for the continental mantle, by increasing velocities above and decreasing velocities below 220 km depth (dashed line), thus artificially enhancing the velocity gradient across the 220 km depth.

artificially perturbed seismic velocities in a *c.* ±70 km thick corridor above and below 220 km depth (at *c.* 150–300 km depth). Since seismic velocities beneath many stable continental regions still show positive velocity anomalies below a 300 km depth, a velocity decrease around 220 km depth and a strong velocity gradient zone commonly observed at *c.* 220 km depth beneath stable continents may be an artifact of using PREM in inversion.

Unfortunately, 220 km is also the depth where the deepest kimberlite magmas are generated and from which deep mantle xenoliths are brought to the surface. As xenolith data seem to indicate lithospheric thickness in cratons to be 200–220 km (perhaps only because greater depths are not sampled by kimberlite magmatism, see detailed discussion in Chapter 5), it is not surprising that many seismologists are tempted to interpret a sharp change in relative velocities at the same depth as the base of the seismic lithosphere. However, a high velocity gradient zone may not even exist if the absolute velocities are considered.

A dramatic effect of the reference model on tectonic interpretations is illustrated by Fig. 3.85 (see Fig. 3.68 for the location map). A tomographic model for Siberia was

Caption for figure 3.85 (cont.)

structures to the tomographic model (based on results of (c) Burmakov *et al.* (1987) and (d) Pavlenkova *et al.*, 1996). Apparent location of the base of seismic lithosphere at *c.* 200 km depth as suggested by a sharp change in the sign of relative *Vs* velocity perturbations in (a) is inherited from a velocity step at 220 km depth (marked by dashed lines) in PREM and is also not supported by the absolute *Vs* velocities which show only a moderate velocity decrease at around 200 km depth (b) or by *Vp* velocities along the subparallel CRATON profile (c, d). See Fig. 3.68 for the location map. A reduced-velocity zone at a depth of 150–200 km is observed along the PNE profiles beneath many terranes of Siberia (Fig. 3.70).

calculated from fundamental and higher mode *S*V Rayleigh waves as relative perturbations with respect to PREM (Priestley and Debayle, 2003). As a result, the final model inherited a velocity discontinuity at around 220 km depth. A sharp change in the sign of relative *V*s velocity perturbations at *c.* 200 km depth was interpreted by the authors as the base of the seismic lithosphere beneath the Siberian craton. However, absolute velocities show only a small decrease (by 0.05 km/s) in seismic velocities at *c.* 170–220 km depth with higher velocities below 220 km. Although this reduced-velocity zone may be an artifact associated with PREM, similar low-velocity zones at a depth of 150–200 km are observed along PNE profiles beneath many terranes of Siberia (Fig. 3.70). The *V*p velocities along the long-range PNE profile CRATON that crosses the Siberian craton further north through the same tectonic structures show some lower velocity anomaly only beneath the Tunguska Basin (Fig. 3.85c), while other interpretations along the same profile demonstrate a gradual increase in *V*p with depth down to the transition zone (Pavlenkova *et al.*, 1996; Nielsen *et al.*, 1999).

3.6.2 Elastic tomography models of the upper mantle

Global patterns

Teleseismic tomography images calculated by different research groups over the past 25 years provide a coherent qualitative portrait of the 3D velocity structure of the upper (and lower) mantle. Recent S-velocity tomographic models are well correlated at the long wavelengths (spherical harmonics < degree 12), and P- and S-velocity variations in the upper mantle are well correlated also (e.g. Ritzwoller and Lavely, 1995; Ritsema and van Heijst, 2002). Since the heterogeneity spectrum of the Earth is distinctly "red" (i.e. dominated by large-wavelength components (Su and Dziewonski, 1991)), the large-scale velocity anomalies are easier to recover and they are the most reliable in tomographic modeling. Furthermore, Dziewonski (2003) argues that

> the observed anomalies in seismic data result from integration of perturbation in structure either along the ray path, for travel times, or volume for splitting of low order normal modes. Perturbation is a smoothing operation and, assuming constant amplitude, the effect of the component of the structure with a wavelength λ_1 compared to that of λ_2 is equal to the ratio λ_1/λ_2. Thus long-wavelength components of the structure are much easier to recover than short-wavelength ones.

Significant discrepancies in the amplitudes and in the radial distributions of velocity anomalies, in particular in the uppermost mantle, result, in part, from incomplete crustal corrections, unaccounted variations in topography of first-order seismic discontinuities in the mantle, and from incorrect assumptions on the relations between *V*p and *V*s heterogeneity.

The most important robust result of global tomography is the observation of a strong correlation of velocity structure of the upper 200–300 km with surface tectonics observed in all tomographic models (Fig. 3.87). Strong (~15%) velocity heterogeneity in the upper mantle (down to *c.* 300 km depth) is related to plate tectonics and the ocean–continent variation (Nataf and Ricard, 1996). The ocean–continent variation is the first-order feature revealed by all types of seismic data. In the upper 200 km, P-velocity heterogeneity in body-wave tomography models is well correlated with major surface tectonics elements such as cratons and subduction zones, where peak anomalies exceeding 2% are observed (Vasco *et al.*, 2003). Low-velocity anomalies associated with back arc basins can be traced down to 200–400 km depth, while some cratons are seen as high-velocity anomalies (although of a smaller lateral extent) down to 400 km depth.

Similarly, surface-wave tomography indicates the existence of a strong correlation between surface tectonics and the patterns of velocity heterogeneity in the upper mantle: fast regions (with positive velocity anomalies with respect to PREM) are observed beneath the cratons, while slow regions are observed beneath the oceans. Except for tectonically active regions such as western USA, regional continental models have high shear velocities throughout the entire upper mantle (higher than 4.5–4.6 km/s) and, in a good agreement with PREM, flat-gradient velocity profiles between 200 km and 410 km depth (Fig. 3.88). Linear belts of high mantle velocities clearly correlate with major continent–continent and continent–ocean collisional zones (e.g. the Andes, the Himalayas–Tibet, the Altaids of Central Asia, and Java–Sumatra) and are associated with subducting slabs. Tectonically active regions, although not so well resolved due to their smaller size, show slow velocities in the upper mantle. Similarly, the oceanic upper mantle has slow velocities, in particular at shallow depth along mid-ocean ridges (Fig. 3.87).

In the cratonic regions high upper mantle velocities may persist down to 200–300 km depth (Fig. 3.89). As discussed earlier, the apparent termination of high-velocity anomalies at these depths can be a systematic artifact due to:

(1) use of PREM in tomographic inversions for the continents (Fig. 3.86);
(2) use of fundamental modes that (for periods <40 sec) cannot resolve upper mantle deeper than ~200 km (Fig. 3.79);
(3) a general decrease in resolution of tomographic models at 300–400 km depth.

Global observations of the velocity structure of the upper mantle are supported by recent high-resolution continent-scale and regional tomographic models. Seismic velocity structure of the continents and oceans is discussed in more detail below. First, an overview of continent-scale and ocean-scale patterns is provided, followed by the characteristic velocity structures for various continental and oceanic tectonic settings.

Continents

A detailed description of regional studies is outside the scope of this book which aims to discuss general patterns in lithospheric structure related to the tectonic evolution of the Earth. A brief (and inevitably incomplete due to the huge number of regional publications) overview of continent-scale tomographic models is presented below. The limitations on vertical, lateral, and amplitude resolution of tomographic models are discussed in detail in the previous section.

North America and Greenland

Since the early tomographic models (Grand and Helmberger, 1984), it has been recognized that the velocity structure of the upper mantle in North America is well correlated with

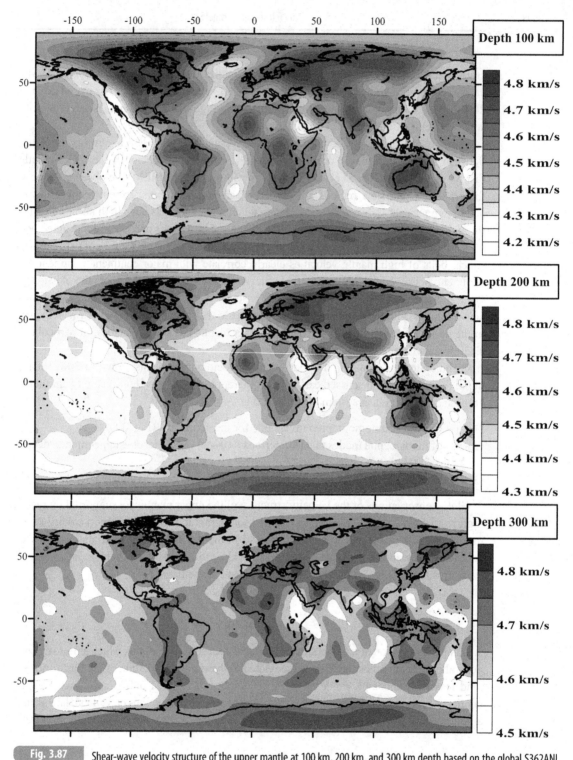

Fig. 3.87 Shear-wave velocity structure of the upper mantle at 100 km, 200 km, and 300 km depth based on the global S362ANI tomography model of Kustowski *et al.* (2008). In this model, radial anisotropy is confined to the uppermost mantle (that is it becomes very small below a depth of 250 km and disappears at 410 km). Including anisotropy in the uppermost mantle significantly improves the fit of the surface-wave data. The inversion is based on a new spherically symmetric reference model REF.

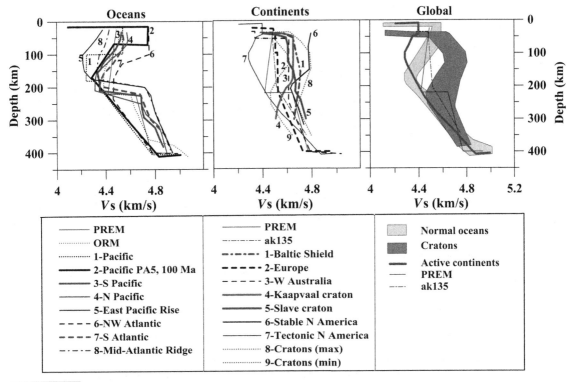

Fig. 3.88 Mean isotropic shear velocity profiles for regional and global oceanic (left) and continental (middle) tomographic models. Regional models commonly provide a better fit to observed seismograms from these regions than the existing global tomographic models. Right panel: typical shear velocity profiles for normal oceans, cratons, and active continents. Reference models are shown as a comparison: global model *PREM* (Dziewonski and Anderson, 1981), global continental model *ak135* (Kennett *et al.*, 1995), and global reference models for normal oceans (ORM) and continents (CRM) based on S20RTS model (Ritsema *et al.*, 2004). References for regional models: Grand (1994, 1997); Lerner-Lam and Jordan (1987); Kennett *et al.* (1994); Gaherty *et al.* (1999); Simmons *et al.* (2002), Pedersen *et al.* (2009), Lebedev *et al.* (2009). Compare with Fig. 3.101a.

surface geology (e.g. Grand, 1994; Humphreys and Dueker, 1994; van der Lee and Nolet, 1997a; Nettles and Dziewonski, 2008):

(1) In the Archean–Paleoproterozoic provinces of the Canadian Shield, the strongest shear-velocity anomaly is observed in the upper 200 km of the mantle with a weaker velocity anomaly extending down to *c.* 300 km depth (Fig. 3.90).

(2) A recent high-resolution seismic tomography model indicates the presence of two distinct lithospheric domains in the Canadian Shield with the boundary at approximately 86–87° W (the Wabigoon-Wawa/Quetico boundary) (Frederiksen *et al.*, 2007). To the west of this boundary, the lithospheric mantle has high velocities, while to the east the average shear-velocity anomaly is *c.* 2.5% weaker (relative to *iasp91*).

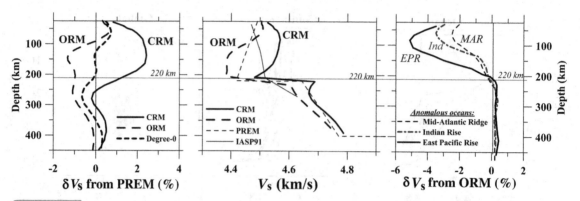

Fig. 3.89 Depth profiles of average shear velocity anomalies in the upper mantle based on the S20RTS global tomographic model of Ritsema *et al.* (2004). Left: average seismic velocity profiles with respect to PREM beneath continents (the Continental Reference Model, CRM) and beneath those oceanic regions where water depth is 5500–6500 m and lithosphere is 40–80 My old (the Oceanic Reference Model, ORM; for these ocean regions, cooling plate models explain the bathymetry and the surface heat flow well); dashed line labeled "Degree–0" – average anomaly of the entire S20RTS model (Ritsema *et al.*, 2004). Middle: the same reference models converted to absolute velocities; global models PREM and *iasp91* are shown for comparison. CRM and ORM inherit the LVZ at around 220 km depth when converted to absolute velocities. Right: Depth profiles of average S20RTS shear velocity heterogeneity in the upper mantle of anomalous oceanic structures (the East Pacific Rise, Indian Rise, and the Mid-Atlantic Ridge) with respect to ORM (after Ritsema *et al.*, 2004). Below 220 km depth and down to the transition zone, both oceanic and continental velocity profiles based on the S20RTS model have systematically higher values (by 0.1–0.2 km/s) than global models.

(3) At present, tomographic models of the Archean Slave craton are largely restricted to models based on the POLARIS seismic array in the central part of the craton. They indicate that high shear velocities in the upper mantle persist down to at least a 300 km depth (Pedersen *et al.*, 2009).

(4) In the terranes of Mesoproterozoic age in south-central North America (Fig. 3.66), a positive velocity anomaly in the upper mantle is more shallow than beneath the Archean–Paleoproterozoic terranes and terminates at around 200–250 km depth.

(5) Western North America, as other tectonically active continental regions, is, in general, characterized by slow upper mantle velocities which already start at *c.* 50–70 km depth and have a minimum at *c.* 100–120 km depth (Figs. 3.88, 3.90b).

However, recent high-resolution regional tomographic models for northwestern USA image a very complex mosaic of velocity anomalies in the upper mantle related to post-Laramide tectono-magmatic events and the tectonic evolution of the Cascadia Subduction Zone (e.g. Roth *et al.*, 2008). The major features of the upper mantle P-wave velocity structure are well-resolved fast velocities associated with the Juan de Fuca–Gorda slab (which is traced down to at least 500 km), beneath the Idaho batholith and beneath Nevada where a slab-like

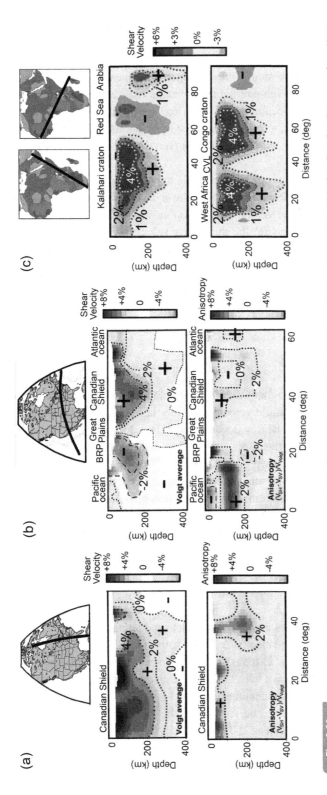

Fig. 3.90 Shear velocity perturbations in the upper mantle of (a, b) North America (based on Nettles and Dziewonski, 2008) and (c) Africa (based on Ritsema and van Heijst, 2000a). (a, b) Top – locations of two profiles: (a) N–S from the Hudson Bay to the Gulf of Mexico and (b) W–E from the Pacific to the Atlantic Ocean across North America. Middle – Voigt average shear-wave velocity along the profiles, plotted with respect to the global average at each depth. Bottom – anisotropy along the profiles (no average has been removed). Dotted lines outline contours of 0%, +2%, and +4% velocity anomalies; dashed lines – contours of negative anomalies. BRP = Basin and Range Province. (c) Top – locations of two profiles: SE–NW from the Cape Fold belt to the Persian Gulf and NE–SW from West Africa to Madagascar. Middle and bottom – shear wave velocity perturbations along the profiles, plotted with respect to the PREM model at each depth. Dotted lines outline contours of +1%, +2%, and +4% velocity anomalies. CVL = Cameroon Volcanic Line.

eastward-dipping feature of increased P-wave velocities is observed down to at least 300 km depth (and perhaps as deep as 700 km). Slow upper mantle velocities are observed beneath the Yellowstone–Snake River plain, the north-central Oregon, and along the Pacific coast. Several tomographic studies have imaged fast linear velocity anomalies below a 200–300 km depth that are sub-parallel to the Pacific coast and extend down to *c.* 1500 km depth beneath the western-central North American continent. They are interpreted as fragments of the Farallon plate, a huge oceanic plate that has subducted beneath the American plates during the Mesozoic and Cenozoic eras (van der Lee and Nolet, 1997b; Bunge and Grand, 2000).

For Greenland, the number of regional-scale tomographic models is so far very limited and restricted to the GLATIS project (Darbyshire *et al.*, 2004). The analysis of *fundamental-mode* Rayleigh wave phase velocity dispersion curves (over the period range 25–160 s) suggests that in the "lithosphere" the shear wave velocity is 4–12% above global reference models (for resolution kernels, see Fig. 3.79). The thickness of the lithosphere is estimated to be *c.* 100 km along the southeast coast of Greenland increasing to *c.* 180 km beneath central-southwestern Greenland where the highest velocities are reported. A body-wave tomographic model (e.g. Grand, 2002) indicates that the region of high shear velocities (2% above PREM) may extend down to *c.* 300 km.

South America

Until recently, regional tomographic models of South America suffered from insufficient ray coverage, and the upper mantle velocity structure beneath most of the stable part of the continent to date remained largely unconstrained (Fig. 3.91). Recent tomographic models provide reliable images for the upper 100–200 km of the mantle, but their resolution is still limited due to the poorly known crustal structure of the continent. The best known is the velocity structure of the upper mantle beneath the Andes (both for compressional and for shear waves) which has been resolved in great detail by numerous studies. They indicate a complex short-wavelength pattern of fast and slow velocities which suggests that lateral variations in composition, melting, and water content play an important role in velocity variations in the mantle wedge.

The major large-scale features of the South American upper mantle revealed by recent tomographic studies include:

(1) slow subhorizontal shear-velocity anomalies in the upper mantle down to *c.* 150 km depth beneath the Andes associated with flat subducting slab;
(2) Eastward-dipping slow-velocity anomaly in the mantle continues beneath the Chaco Basin and extends as far east as the Paraná Basin where it is imaged at a 200–300 km depth; low velocities down to *c.* 200 km beneath the Chaco and the Pantanal Basins can be associated with the subducting slab;
(3) high Vs velocities beneath the Paraná and the Parnaiba Basins down to 100–150 km depth and perhaps as deep as 250 km (Lebedev *et al.*, 2009);
(4) fast-velocity anomaly down to *c.* 200 km depth beneath the Archean Amazonian craton as consistently imaged by different tomographic models; some regional tomographic

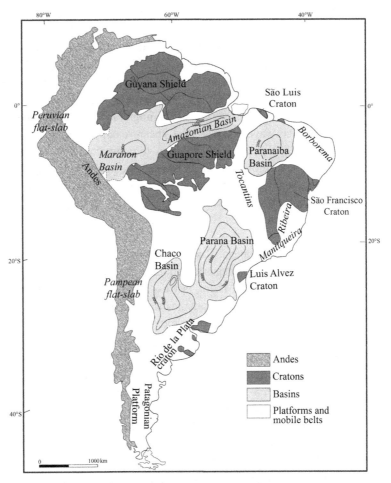

Fig. 3.91 Major tectonic provinces of South America (based on Heintz *et al.*, 2005).

studies image a reduced-velocity anomaly beneath the Amazon Basin (a paleorift?) which cuts the oldest part of the continent into two subcratons (the Guayana and the Guapore Shields);

(5) high-velocity region in the upper mantle (down to 150–200 km depth) beneath the Archean Saō Francisco craton.

Most tomographic results for South America are based on fundamental mode surface waves (e.g. Feng *et al.*, 2007) and thus cannot reliably resolve the upper mantle structure below 200 km depth (Fig. 3.79).

Europe

Tomographic models for Eurasia, one of the best studied continents, are numerous, both on the continent-scale and at regional scales (e.g. Panza *et al.*, 1980; Spakman, 1991; Zielhuis and Nolet, 1994; Villaseñor *et al.*, 2001; Piromallo and Morelli, 2003; Kustowski *et al.*,

2008b). The major large-scale patterns are as follows (for details and references see Artemieva *et al.*, 2006).

High-velocity compressional and shear-velocity anomalies are imaged in the upper 200–250 km beneath the cratonic regions, with the strongest anomalies beneath south-central Finland in the Baltic Shield where they extend down to at least 250 km. Below this depth, most published tomography models lose resolution since they are based on fundamental-mode surface waves. Tomographic models based on body waves and higher-mode surface waves display fast shear-velocity anomalies ($Vs > 4.7$ km/s) down to at least 300 km depth. High-resolution tomographic shear-velocity models of the Baltic Shield based on the SVEKALAPKO seismic tomography experiment image a surprisingly complex upper mantle structure with large velocity and anisotropy variations at relatively small lateral and vertical scales. While a certain correlation is recognized between shallow upper mantle anomalies and surface geology, deeper anomalies are difficult to correlate with tectonics.

For the East European (Russian) Platform which lacks seismic events, high resolution regional tomographic images are limited mostly to the southern and western parts, where ray path coverage is better. There, fast Vs- and Vp- velocities are observed down to *c.* 150 km depth with smaller-scale anomalies extending down to at least 250 km. Recent analysis based on broad-band surface wave dispersion indicates that the high-velocity "lid" of the platform can be 250–300 km thick (Lebedev *et al.*, 2009). The fastest upper mantle velocities ($Vs > 4.7$ km/s at 250–320 km depth) are reported for the Archean Volga–Uralia province in the eastern part of the platform and for the Archean Ukrainian Shield–Vitebsk block in the western part of the East European craton (Fig. 3.69). Surprisingly, the velocity structure of the upper mantle beneath the huge Paleozoic Dnieper–Donets rift in the southern part of the platform does not differ from the surroundings, probably due to lack of lateral resolution. A strong positive-velocity anomaly is found beneath the Caspian basin at 100 km depth.

The western edge of the East European craton is marked by a remarkably sharp change in the upper mantle velocity structure at the transition from the cratonic to Phanerozoic crust of western and central Europe (the Trans-European suture zone, the major tectonic boundary in Europe). While in cratonic Europe fast velocities extend down to *c.* 200 km, they typically terminate at *c.* 70–100 km depth in most of Phanerozoic Europe. The maximum velocity contrast across the cratonic margin is between the Pannonian Basin and the East European Platform (12% at 80 km depth) (Zielhuis and Nolet, 1994). A zone of low shear velocity at 300–400 km depth beneath the cratonic margin has been interpreted as a relic from the paleosubduction zone (Nolet and Zielhuis, 1994).

The recent seismic tomography experiment (TOR) across the Trans-European suture zone (TESZ), from the stable lithosphere of the Baltic Shield to the Phanerozoic lithosphere of western Europe, has demonstrated a contrasting upper mantle seismic structure between lithospheric terranes of different ages. The interpretation of the lithospheric base, constrained by a transition from positive to negative velocity anomalies with respect to the reference model, suggests that over the lateral distance of 200–250 km the lithospheric thickness decreases step-wise from *c.* 220 km beneath the Baltic Shield to *c.* 50 km beneath Phanerozoic Europe (Fig. 3.92). Since the thermal regime of the lithosphere along the TOR

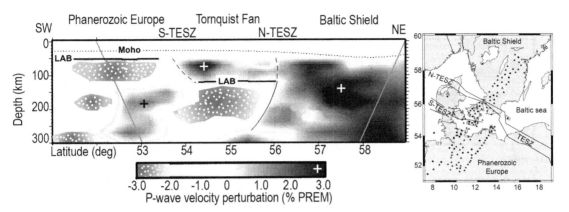

Fig. 3.92 P-wave tomography model across the margin of the Baltic Shield (after Arlitt, 1999). LAB = Lithosphere–asthenosphere boundary; TESZ = Trans-European Suture Zone which marks the transition from the East European craton to Phanerozoic Europe; it appears as two major trans-lithospheric sutures in the Danish area, where the TESZ splits into two branches bounding the Tornquist Fan.

profile is approximately the same on the cratonic and on the Phanerozoic sides, the observed sharp change in velocity anomalies in the upper mantle is likely to be of a compositional origin (Artemieva *et al*., 2006). In this case, the transition from positive to negative velocity anomalies seen in TOR tomography should be interpreted as the base of the *chemical* lithosphere. Such an explanation has petrological grounds: iron-depleted cratonic lithosphere should have higher seismic velocities than the fertile Phanerozoic lithosphere of western Europe and the Proterozoic lithosphere of the Tornquist Fan (the area located between two split branches on the TESZ in the north, Fig. 3.92) which has been subject to wide spread Paleozoic magmatism and thus could have been metasomatised by basaltic melts.

The velocity structure of the upper mantle of the tectonically young structures of Europe is highly heterogeneous at small scales. Low-velocity anomalies at *c*. 70 km depth are imaged in the upper mantle beneath the Pannonian Basin and some parts of the French Massif Central. Seismic tomography images a mosaic of high- and low-velocity bodies in the upper mantle of southern Europe and the Mediterranean (e.g. Bijwaard *et al*., 1998; Piromallo and Morelli, 2003). High-velocity anomalies are commonly interpreted as images of subducting slabs pushed into the mantle by complex tectonic stresses associated with the collision of the European and African plates.

High-resolution regional tomography across the Alpine orogen has revealed several high-velocity bodies in the upper mantle that have been interpreted as subducting slabs. One of them dips southeastwards beneath the Adriatic microplate in the western and central Alps and is interpreted as subducted European lower lithosphere. Another dips northeastwards in the eastern Alps and is interpreted as Adriatic lower lithosphere subducting beneath the European plate (Fig. 3.93, compare with Fig. 3.62). Similarly, a near-vertical localized high-velocity body is imaged in the upper mantle beneath the Vrancea zone in the Carpathians. Tectonic interpretations of its origin remain controversial. Since the velocity anomaly

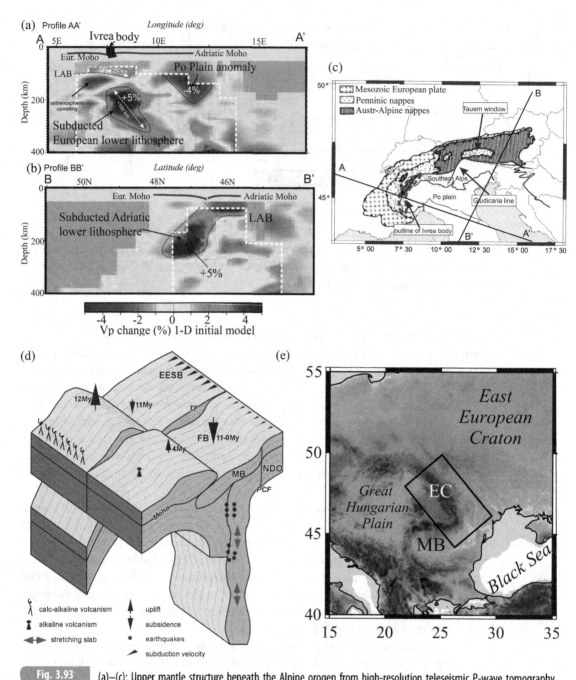

Fig. 3.93 (a)–(c): Upper mantle structure beneath the Alpine orogen from high-resolution teleseismic P-wave tomography (based on Lippitsch *et al.*, 2003). Regions with high resolution are outlined by white dashed lines. (c) Tectonic map of the Alps with the location of the profiles across the western and eastern Alps. (d): 3D cartoon of the subduction pattern in the East Carpathians (the Vrancea zone) (from Cloetingh *et al.*, 2005). (e) Shaded topography of the region; the box shows the area illustrated in (d). EC – East Carpathians; EESB – East-European/Scythian block; MB – Moesian block; NDO – North Dobrogea orogen; TF – Trotus Fault; PCF – Peceneaga-Camena Fault.

correlates with the zone of high seismicity, most interpretations invoke the presence of a subducting slab or a delaminated lithosphere fragment in the upper mantle of the region (Section 9.3 and Fig. 9.25).

Siberia and Kazakhstan

Tomographic models for the Paleozoic West Siberian Basin (which contains buried terranes of Archean ages) and for the largely Archean–Paleoproterozoic Siberian craton (the southeastern parts of which were rifted in the Paleozoic) remain controversial, in part due to significant differences in the crustal models used in tomographic inversions (in particular the CRUST5.1 model significantly, >5 km, underestimates the crustal thickness in some parts of the Siberian Platform). Fundamental-mode surface-wave tomography models image fast shear velocities down to at least 200–250 km depth beneath the Siberian craton (Villaseñor *et al.*, 2001), with some of them displaying fast upper mantle velocities (2% higher than in the *iaspei* reference model) down to *c.* 250–300 km depth, below which they lose resolution (Shapiro and Ritzwoller, 2002). However, a body-wave model (Grand, 2002) displays fast *V*s velocities (+2+4% with respect to *ak135*) beneath all of the Siberian Platform at 175–250 km depth and beneath major kimberlite fields at 250–325 km depth. At 325–400 km depth the Siberian Platform has near-zero velocity anomalies, while the Volga–Uralia province of the East European craton displays +1+2% *V*s anomalies even at these depths.

The West Siberian Basin (considered by Sengör *et al.* (1993) to be part of the Paleozoic Altaids orogen) was assembled in the Paleozoic as a result of progressive convergence between the Siberian and the North China cratons. It has a very complex tectonic structure which includes fragments of ancient (Archean or Paleoproterozoic) massifs, Meso-Neoproterozoic oceanic complexes, and island arcs. Similarly to the Siberian craton, fast velocities in the upper mantle of the basin persist down to *c.* 200 km; however at 150 km depth shear velocity beneath the basin is *c.* 0.2 km/s slower than beneath the Siberian craton. The Kazakh Shield to the south from the West Siberian Basin apparently may have fast upper mantle velocities (above the *ak135* reference model) down to the transition zone, however the deflection from the reference model is significant only in the upper 200 km (Lebedev *et al.*, 2009).

Tomographic models of the Baikal Rift zone at the southern margin of the Siberian craton are controversial. Most tomographic models, in general, do not display slow velocities in the upper mantle beneath the rift zone and interpret it as a "passive rift", the geodynamic origin of which is in one way or another related to convective (or asthenospheric) flow at the edge of the Siberian craton (Petit *et al.*, 1998; Achauer and Masson, 2002; Lebedev *et al.*, 2006; Artemieva, 2009; Thybo and Nielsen, 2009). An exception is one teleseismic P-wave tomography model that displays a localized low-velocity anomaly beneath the rift axis at depths 50–200 km interpreted as the signature of a mantle plume (Zhao *et al.*, 2006). A low-velocity anomaly to the southwest from the rift zone is imaged in all tomographic models. High-velocity anomalies in the lower mantle west of Lake Baikal seen in P-wave tomographic models are interpreted as remnants of slabs subducted in the Jurassic when the Mongol–Okhotsk and Kular–Nera oceans closed between the Siberian craton and the Mongolia–North China block (van der Voo *et al.*, 1999a).

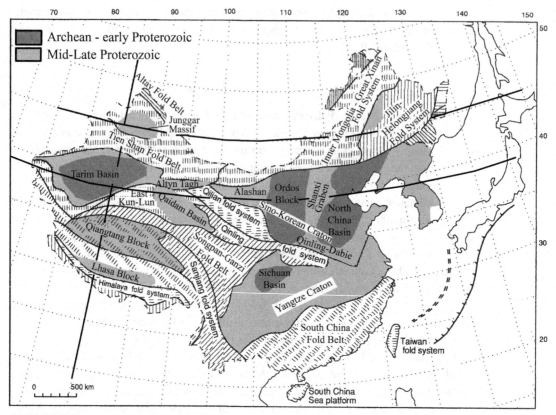

Fig. 3.94 Tectonic map of China and adjacent regions (based on Goodwin, 1996; Ren *et al.*, 1981). Two-dimensional profiles along bold lines are shown in Fig. 3.95.

China, Mongolia, and India

Several existing high-resolution tomographic models of China (Fig. 3.94) image a highly heterogeneous structure for the upper mantle that is well correlated with surface tectonics. A tomographic model based on group velocity dispersions of fundamental Rayleigh waves (Huang *et al.*, 2003) images the following upper mantle shear-velocity anomalies (see the previous section for a discussion on the resolution) (Fig. 3.95):

(1) fast shear velocities in the upper mantle down to 150–200 km depth beneath the cratonic central-northern India;

(2) a thin high-velocity "lid" (lithospheric keel) beneath the North China craton that terminates at *c.* 100–120 km depth (compare with Fig. 2.20); a localized (*c.* 500 km across) high-velocity block extending down to 200–250 km depth beneath the Archean Ordos block of the North China craton is, however, imaged in a regional P-velocity tomography model (Tian *et al.*, 2009);

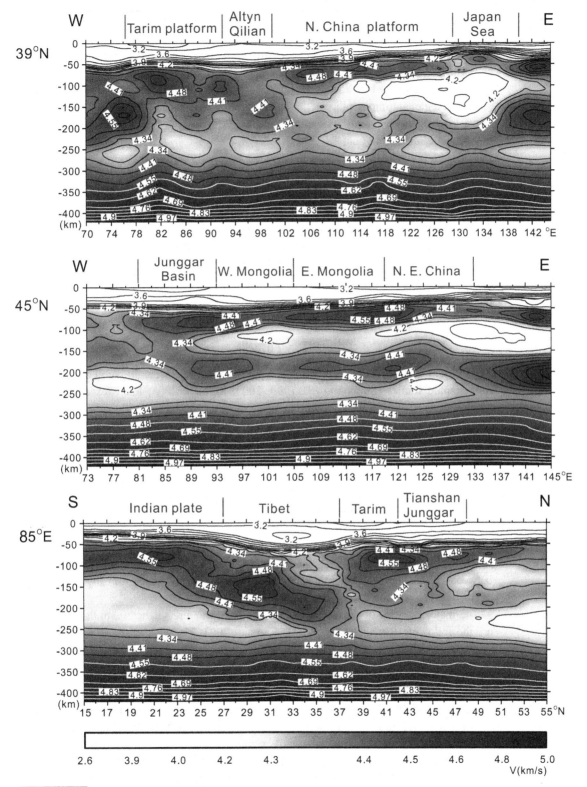

Fig. 3.95 Shear-velocity in the upper mantle of China constrained by surface wave tomography (from Huang *et al.*, 2003). See Fig. 3.94 for profiles location (bold lines).

(3) high-shear velocities down to *c*. 200 km depth beneath the Yangtze craton (Lebedev *et al.*, 2009), the Tarim Basin, and the Jungar Basin; a strong high-velocity shear-velocity anomaly beneath the Sichuan Basin at 150–250 km depth is also imaged in other regional tomographic models (Lebedev and Nolet, 2003);

(4) a low-velocity anomaly extending from 50–70 km depth down to *c*. 300 km depth beneath Inner Mongolia;

(5) beneath southern Mongolia, a low-velocity anomaly at 100–150 km depth which is underlain by a zone of fast velocities at 150–200 km depth;

(6) low upper mantle velocities in coastal areas and in the continental shelves of China and Indochina;

(7) a northwards-dipping high-velocity zone that resembles a subducting plate and reaches a 200 km depth is imaged beneath the Tibetan Plateau up to the Kunlun Mountains in the north; the tomographic models do not support the presence of any low-velocity anomalies beneath Tibet that may indicate an uplift of hot asthenospheric material or delamination of the lower lithosphere (Griot *et al.*, 1998).

Australia

The velocity structure of the Australian upper mantle has been the subject of several tomographic studies (Simons *et al.*, 1999, 2002; Kennett, 2003; Fishwick *et al.*, 2005; Pedersen *et al.*, 2009). Tomographic models based on SKIPPY data have had problems resolving the structure of western Australia, where the oldest Archean cratons are located (Fig. 3.83). Furthermore, since these cratons are close to the continent–ocean transition with a sharp contrast in velocity structure (Fig. 3.82), damping used in tomographic inversion will result in an underestimate of the true velocity variations (Kennett, 2006). With this limitation in mind, tomographic models that show a thin (down to *c*. 175–250 km depth) high-velocity lid (lithospheric keels) beneath the Archean Pilbara and Yilgarn cratons in western Australia as compared to thick (down to *c*. 250–300 km depth) high-velocity lid beneath the Proterozoic Central Australia should be interpreted with caution (Simons *et al.*, 1999). The conclusion based on such tomographic models that no general relationship exists between seismic signature and lithospheric age (not supported by data from any other stable continental region) could be questioned.

The major features of the upper mantle velocity structure of the Australian continent are the following (Simons *et al.*, 2002; Kennett, 2003; Fishwick *et al.*, 2005; Pedersen *et al.*, 2009):

(1) the high-velocity lid beneath the Archean cratons of Western Australia extends, at least, down to *c*. 200–250 km depth. Small-scale velocity variations in the upper mantle of the cratons correlate with terrane boundaries. A recent comparison of shear-velocity structure beneath four Archean cratons indicates that, at depths between 50 km and 170 km, the lithosphere of the Yilgarn craton has significantly faster shear velocities (by *c*. 0.08–0.15 km/s) than the lithospheres of the Kaapvaal, the Slave craton, and the Baltic Shield;

(2) the transition from the cratonic lithosphere in western and northern Australia to Phanerozoic upper mantle in eastern Australia is marked by a velocity contrast down

to *c*. 200 km depth; the eastern margin of the high-velocity region in the upper mantle coincides with the Tasman Line which marks the eastern limit of outcrops of the Proterozoic basement; at 200–260 km depth a high-velocity region beneath the northern part of Australia extend eastwards beyond the Tasman Line;

(3) a fast wave-speed anomaly continues down to at least 250 km depth beneath the North Australian Craton, suggesting that the Archaean lithosphere may extend underneath areas with Proterozoic crust. High upper mantle velocities down to *c*. 200 km depth are observed northwards from Australia beneath the Indian Ocean and Papua New Guinea and may indicate an off-shore extension of the North Australia craton;

(4) beneath Proterozoic Central Australia a region of low seismic velocities is observed at *c*. 75 km depth in the uppermost mantle. Below, down to 200–250 km depth, the upper mantle has fast shear velocities typical of cratonic lithosphere;

(5) in eastern Australia the region of high upper mantle velocities terminates at *c*. 100 km depth and is underlain by a slow mantle with a velocity minimum at *c*. 150 km depth.

Africa

Continent-scale tomographic models are not available for Africa. Knowledge on upper mantle velocity structure is based either on regional tomographic studies available for southern Africa (e.g. SASE, the Southern Africa Seismic Experiment (Carlson *et al*. 1996 and several papers published in *Geophys. Res. Lett.*, 2001, v. 28)), for the Tanzanian craton (Weeraratne *et al.*, 2003), for the Kenya Rift (KRISP, Kenya Rift International Seismic Project, Prodehl *et al.*, 1997), and for the Ethiopian Plateau–Afar region (EAGLE), or on high-resolution global wavespeed models that provide information on the upper mantle structure of other tectonic provinces of Africa, including the West African craton, the Congo craton, and the East Saharan craton (Fig. 3.96). In particular, high-resolution multi-mode global surface wave tomography model S20RTS (Ritsema *et al.*, 2004) shows fast mantle (with +4% velocity with respect to PREM) extending to at least 200 km depth beneath the cratons of Africa (Fig. 3.90). In West Africa the depth extent of the fast anomaly (seismic lithosphere) may be significantly deeper than beneath the cratons of central and southern Africa.

Regional surface-wave tomographic studies of the Tanzanian craton and the adjacent segments of the East African Rift indicate that upper mantle beneath the craton is faster than the global average down to a depth of 150 ± 20 km (Fig. 3.97). Below this depth, shear velocity rapidly decreases from *c*. 4.65 km/s at 140 km depth to 4.20 ± 0.05 km/s at depths of 200–250 km, where it has a minimum. At the depth interval between 200 km and 400 km, shear velocities beneath the Tanzanian craton are slower than beneath the adjacent branches of the East African Rift. This significant drop in shear velocities can be attributed to high mantle temperatures and possible presence of melts (Weeraratne *et al.*, 2003). The same surface-wave tomographic model displays a sharp minimum in upper mantle shear velocities (3.7–4.0 km/s) at a depth of 120–170 km beneath the East African Rift. Tomographic models of the Kenya Rift, developed as a result of the KRISP broadband seismic experiment, also display a cylindrical low-velocity (–0.5–1.5%) anomaly below the rift extending to about 150 km depth with steep, near-vertical boundaries

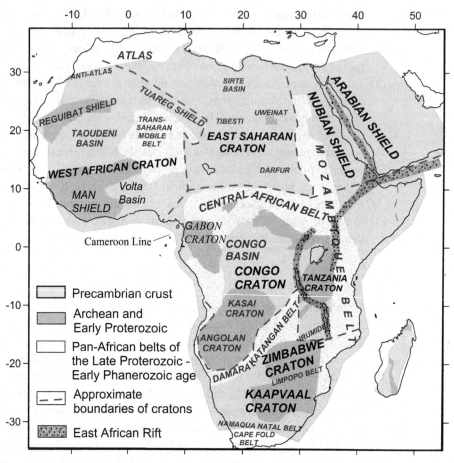

Fig. 3.96   Tectonic map of Africa.

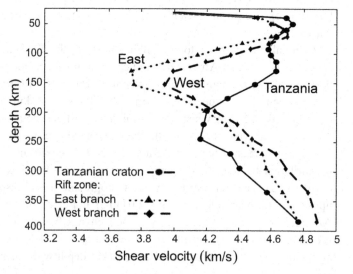

Fig. 3.97   Shear-velocity structure of the upper mantle beneath the Tanzanian craton and the eastern and western branches of the East African Rift from Rayleigh wave tomography (based on Weeraratne et al., 2003).

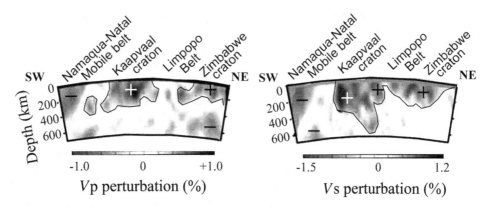

Fig. 3.98 Vp and Vs velocity structure of the upper mantle beneath South Africa from body-wave tomography (based on James *et al.*, 2001). The profile goes from approximately 20E, 35S to 30E, 20S. Plus and minus signs indicate the sign of velocity perturbations (for an unspecified reference model). The resolution of the S-velocity model is weaker due to fewer observations and greater uncertainties in relative time delays.

located under the rift border faults (e.g. Achauer and Masson, 2002). Recent models suggest that a low wave-speed anomaly along much of the East African rift (observed in the upper mantle of Kenya, Tanzania, and Ethiopia) may be dipping westwards below depths of ~150–200 km.

Regional high-resolution tomographic studies of southern Africa provide a wealth of data on the upper mantle velocity structure beneath the Kalahari craton. However, up-to-date seismic interpretations are highly controversial, in particular on the depth extent of the high-velocity lid (seismic lithosphere or tectosphere) and on regional seismic anisotropy. The upper mantle of southern Africa cratons is one of the best sampled by mantle-derived xenoliths from more than 1000 kimberlite pipes (see Chapter 5). However, despite several attempts to correlate velocity anomalies in the upper mantle with petrological data and diamond occurrences, no convincing conclusions have been made.

Recently body-wave tomography has imaged high velocity upper mantle beneath southern Africa down to 300–400 km depth. In particular, a one-dimensional model for P-wave velocities displays a high-velocity anomaly (*c.* 3% higher than global averages) down to 300 km depth (Zhao *et al.*, 1999). The results for the linear travel-time inversion for P- and S-waves support this finding (Fig. 3.98). Although the resolution of the S-wave model is weaker, both models display similar patterns: high-velocity mantle anomalies coincide with the tectonic boundaries of the Archean Kaapvaal and Zimbabwe cratons and extend down to at least a 300 km depth, while the upper mantle Proterozoic Namaqua–Natal mobile belt and the Archean Limpopo belt (disrupted by the Bushveld complex emplaced at *c.* 2.05 Ga) has significantly lower velocities (James *et al.*, 2001). Vp and Vs seismic velocities beneath the Bushveld (although loosely constrained) are, respectively, ~0.5% and ~0.8% lower than beneath the craton; and the mantle zone of reduced velocities associated with the Bushveld extends from beneath the intrusion itself well to the west into Botswana.

The low mantle velocities beneath the Bushveld complex may be interpreted as indicating chemical modification of the mantle during a Proterozoic magmatic event (James *et al.*, 2001). This hypothesis is consistent with Re–Os results which indicate Proterozoic resetting of the age of mantle nodules from the Bushveld region (Carlson *et al.*, 2000).

In contrast to body-wave studies, a recent multi-mode teleseismic surface wave tomographic model displays a high-velocity anomaly beneath the cratonic region of southern Africa only down to 175 ± 25 km depth (Priestley *et al.*, 2006). The authors argue that because of the technique employed in the inversion, the final velocity structure is weakly dependent on the reference model (smoothed PREM) and thus should not inherit a step at *c.* 220 km depth. Unfortunately, according to the presented checkerboard resolution test, the model cannot reliably resolve the upper mantle velocity structure at 250 km depth and exhibits a strong vertical and lateral smearing already at 150 km depth. Another tomographic study based on *fundamental-mode* Rayleigh and Love waves (with limited resolution below *c.* 200 km depth) supports the conclusion that mantle beneath the Kaapvaal craton is characterized by high seismic velocities in the upper 200 km with no evidence for a substantial LVZ down to this depth (Freybourger *et al.*, 2001). In a recent study (Pedersen *et al.*, 2009), a pronounced decrease in shear velocity (from *c.* 4.65–4.70 km/s at 200 km depth to *c.* 4.50 km/s at 300 km depth) was resolved beneath the Kaapvaal craton, while such a velocity decrease below 200–250 km depth has not been recognized in the same study beneath the Slave, Yilgarn, and South-Central Finland cratons. Since the latter study is based on phase velocities of *fundamental-mode* Rayleigh waves, these conclusions are speculative.

Antarctica

Tomographic modeling of Antarctica is challenging due to the unknown crustal structure and the presence of a thick ice cap. Regional models are hampered by the limited number of stations and seismic events on the continent (Kobayashi and Zhao, 2004; Morelli and Danesi, 2004; Kuge and Fukao, 2005). Recent surface wave tomographic inversions for the Antarctic region display high upper mantle velocities in the cratonic East Antarctica at periods of 90–150 s, suggesting a very thick seismic lithosphere of East Antarctica underneath the Dronning Maud Land, Enderby Land, Gamburtsev Mountains, and Wilkes Land. In contrast, mantle velocities in tectonically young West Antarctica are low at all periods between 20 s and 150 s, suggesting on-going tectonic activity in the West Antarctic Rift System. In the Ross Embayment low seismic velocities are imaged down to 250 km. The transition from fast to slow upper mantle correlates with the Transantarctic Mountains.

Oceans

For the "normal" oceans, the bathymetry follows the square root of age pattern (Schroeder, 1984) and the cooling half-space model provides a good approximation for the thickness of the thermal boundary layer (lithosphere) (see Chapter 4). The amplitude of velocity

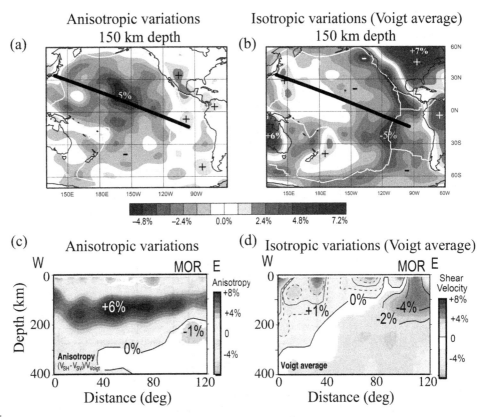

Fig. 3.99 Shear-wave velocity variations beneath the Pacific Plate. (a) Difference between the SV and SH velocities at 150 km depth; (b) the Voigt averaged isotropic variation in S-wave velocity at 150 km depth. The isotropic (thermal) S-wave variations in general correlate with the age of the ocean floor (the Pacific Superswell is a clear exception); anisotropic velocity variations are as large as isotropic variations. (c) Shear velocity anisotropy along the profiles shown in (a, b). (d) The Voigt average shear-wave velocity along the profile, plotted with respect to the global average at each depth. Note a strong negative anomaly beneath the mid-ocean ridge (MOR) and a gradual deepening of a zero-velocity anomaly with ocean floor age. (a) and (b) based on Ekström and Dziewonski (1998); (c) and (d) based on Nettles and Dziewonski (2008).

perturbations in the normal oceanic upper mantle is age dependent: the strongest low-velocity anomalies are clearly resolved along mid-ocean ridges (near-zero age), in particular at 100 km depth (Figs. 3.87; 3.99; 3.100cd; 3.125). In general, low shear velocities beneath mid-ocean ridges are restricted to the upper 200 km of the mantle. In old (>80 Ma) oceans, shear velocities in the upper mantle become similar so that the difference between velocity anomalies in 80 Ma and 140 Ma old oceans becomes statistically insignificant (Fig. 3.100c).

Tomographic models for normal oceans have a high-velocity lid in the upper 80–100 km, particularly pronounced beneath the Pacific Ocean and south-west Atlantic Ocean (Fig. 3.88). This high-velocity lid is absent in anomalous ocean regions such as the East Pacific Rise, the Indian Rise, and the Mid-Atlantic Ridge. At c. 100–200 km depth, the oceanic upper mantle

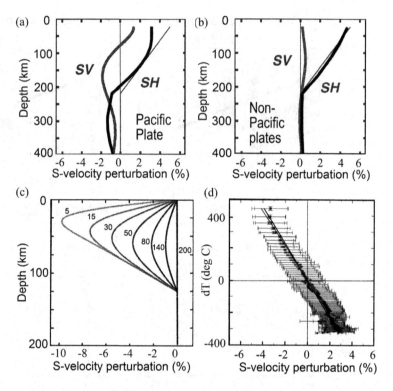

Fig. 3.100 (a–b) Average profiles of S velocity perturbations in the Pacific plate for different age of plate (based on Ekström and Dziewonski, 1998). The profiles are referenced to the oldest profile. (c–d) Correlation between seismic and thermal structure of oceanic upper mantle. (c): The predicted evolution of S velocity with depth as a function of ocean age. The profile is referenced to a profile for 200 Ma age. Numbers on the curves: ocean floor age in Ma (from Ekström and Dziewonski, 1997). (d) Comparison of seismic shear velocity and temperature anomalies in the oceanic lithosphere (based on Forte, 2007). Seismic shear velocity anomalies are based on tomography models SAW24 (Megnin and Romanowicz, 2000) and TX2002 (Grand, 2002). The age-dependent temperature variations in the oceanic lithosphere are based on a cooling half-space model with thermal parameters as for the GDH1 plate model (Stein and Stein, 1992). Both shear-velocity and temperature anomalies are vertically averaged in the depth interval 0–100 km and are spatially averaged into 2 Ma age bins, and are shown with respect to the global oceanic mean value.

has a distinct LVZ with $Vs \sim 4.3$–4.4 km/s. The strongest negative Vs anomalies with respect to global or oceanic reference models are observed at 100–200 km depth beneath the East Pacific Rise (Fig. 3.89c), while the anomalies beneath the Mid-Atlantic Ridge are significantly weaker, and the mantle beneath the Indian Rise has intermediate velocity anomalies (Ritsema and Allen, 2003). In the Pacific Ocean the upper mantle is on average 1% slower with respect to PREM down to at least 400 km depth (Ekström and Dziewonski, 1998). In contrast to a flat-gradient region below 220 km depth in the PREM model, a high gradient zone is observed between 200 km and 410 km depth in the oceanic mantle (Fig. 3.88a,c).

Surface-wave tomographic models require that age-dependent variations in oceanic upper mantle extend down to at least 200 km depth, while the cooling half-space model predicts lithosphere thickness of less than 100 km (Fig. 3.100c). For example, at 150 km depth

SH-velocity anomalies in the mantle below the Pacific Ocean vary from -4% at the East Pacific Rise to $+3\%$ in the Western Pacific (Dziewonski, 2003). Decomposition of the shear-wave velocity model into isotropic (Voigt average) and anisotropic parts helps us to understand the age dependence of mantle velocity structure below the base of the thermal boundary layer. For the Pacific Ocean, the isotropic part of velocity perturbations becomes more consistent with the cooling half-space model (below 175 km depth no age-dependent differences are observed for isotropic shear velocities, Fig. 3.99d), except for anomalous ocean between the East Pacific Rise and the Pacific "Superswell" which has low shear velocities at 150 km depth. The anisotropic part of the model has the peak of the anomaly at 100–200 km depth and a rapid decrease in anomalies at 200–250 km depth (Figs. 3.90b, 3.99c). Thus the greater depth extent of the SH anomalies than would be required by lithospheric cooling can be interpreted as a change in polarization anisotropy with age (see discussion in Section 3.6.4).

Synopsis for some tectonic settings

Stable continents

Global and regional tomography models provide strong support for the tectosphere hypothesis and the existence of thick lithospheric roots beneath the Precambrian cratons with high seismic velocities. High velocities can be due to low temperatures and/or petrologically distinct composition (FeO-poor, olivine-rich, and garnet-poor). Fast velocity anomalies beneath the ancient parts of all of the continents are clearly seen in all tomographic models down to at least 200–250 km depth (Fig. 3.87). In most of the cratons, fast velocities persist down to $c.$ 300 km depth (below this depth many tomographic models lose resolution), with peak seismic velocities at $c.$ 100–200 km depth (Fig. 3.88). Lebedev $et\ al.$ (2008) report isotropic-average Vs velocities in the range 4.58–4.94 km/s at depths between the Moho and ~250 km beneath the Archean Baltic Shield, East European (Russian) Platform, Guyana Shield, Parana Basin, Congo Craton, and the Archean Western Australia; beneath the Proterozoic mobile belts of Kazakhstan and China (Yangtze craton) Vs velocities range from 4.49 to 4.75 km/s (for a comparison, at depths 40–200 km the reference, continental average Vs velocity in the $ak135$ model is 4.5 km/s).

There are, however, significant differences in the velocity structure of the upper mantle of different cratons, even when constrained by the same data set and the same inversion, and when xenoliths suggest similar geotherms. Recent surface-wave tomography based on analysis of dispersion curves for four Archean cratons (Kaapvaal (South Africa), Slave (north-central Canada), Yilgarn (western Australia), and Baltic Shield (central Finland)) indicate significantly different velocities in the lithospheric mantle (Pedersen $et\ al.$, 2009). The highest velocities (0.08–0.12 km/s above Kaapvaal velocities) were determined in the Yilgarn lithosphere with half of that velocity anomaly determined for the Baltic Shield. This implies that seismic velocity models for the upper mantle do not necessarily support a simple relationship between crustal age and upper mantle composition.

Polet and Anderson (1995) have analyzed the shear-velocity structure of the cratonic upper mantle constrained by body-wave and surface-wave global tomographic models. Their analysis (still statistically valid despite significant progress in tomographic studies over the past

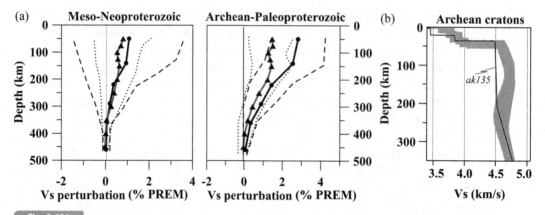

Fig. 3.101 (a) Depth extent of shear-velocity perturbations (with respect to PREM) in the continental upper mantle for cratons of two ages (based on the analysis of Polet and Anderson, 1995). Shear velocities are derived from global tomography models based on body waves (Grand, 1997) (shown by black lines and circles, standard deviation indicating the 66% confidence level is outlined by dashed lines) and on surface waves (Zhang and Tanimoto, 1993) (shown by gray lines and triangles, standard deviation is outlined by dotted lines). (b) Summary profile of isotropic-average shear-wave velocity (gray shading) beneath the cratons of the Baltic Shield, East European Platform, Guyana Shield, Parana Basin, Congo Craton, and Yilgarn–Pilbara cratons (after Lebedev et al., 2008). Solid line – global continental reference model ak135 (Kenneth et al., 1994). Compare with Fig. 3.88.

years) is based on numerical cross-correlations for two age groups (Archean–Paleoproterozoic cratons with ages greater than 1.7 Ga and Mesoproterozoic–Neoproterozoic cratons with ages between 0.8 Ga and 1.7 Ga). It is performed on a 1 deg x 1 deg spatial grid which is significantly finer than the lateral resolution of the surface-wave model used in the analysis (constrained on a 5 deg x 5 deg spatial grid). The major conclusions of the analysis are the following (Fig. 3.101):

- High velocity anomalies are restricted to the oldest parts of the continents.
- The Archean–Paleoproterozoic cratons have fast Vs velocities down to $c.$ 500 km, although not statistically significant below $c.$ 250 km depth.
- For the Mesoproterozoic–Neoproterozoic regions, the velocity anomalies are not statistically significant at any depth (even in the shallow mantle).
- Cratons of the same age have a different velocity structure of the upper mantle.
- Beneath the West African craton, the Canadian Shield, and the cratonic part of Europe significant high-velocity anomalies are observed down to 300–450 km depth.
- At depths between 100 and 250 km, the strongest high-velocity anomalies are observed beneath the Canadian, Baltic, Siberian, Australian, and the West African cratons, but not beneath southern Africa.
- In the Archean–Paleoproterozoic cratons of South America, Central Africa, India, and China, high velocity anomalies are relatively shallow (down to 150–250 km depth). The Neoproterozoic Arabian Shield has low seismic velocities (atypical for other shields) in the upper mantle.

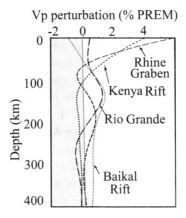

Fig. 3.102 Compressional velocity versus depth in the upper mantle of four continental rifts (based on Achauer and Masson, 2002). P-velocity deviations from PREM are estimated from S-velocity perturbations.

Continental rifts

The tectonic evolution of rifts is governed by complex regional and far-field stresses produced by plate motion and by the rheological strength of the lithosphere, so that continental rifts tend to follow the pre-existing zones of rheological and tectonic weakness. A comparative analysis of absolute upper mantle velocities derived from recent global 3D surface-wave tomography of four Cenozoic continental rifts (the Kenya, Baikal, and Rio Grande rifts, and the Rhine Graben) indicates that the mantle structure beneath the rifts differs substantially down to depths of ~300 km (Fig. 3.102). While the Kenya and Rio Grande rifts may be considered as active rifts where large mantle upwellings control their tectonic evolution, the southern Rhine Graben and the Baikal Rift zone are more likely to be passive rifts (Achauer and Masson, 2002): at depths between 50 km and 300 km the upper mantle beneath the latter two rifts is faster than a global reference model. Long-offset seismic refraction and teleseismic studies of the East African Rift and the Rio Grande Rift provide compelling evidence for hot mantle down to a depth of at least 100–200 km, and support the hypothesis of their deep mantle origin (Braile *et al.*, 1995; Baldridge *et al.*, 1995). In contrast, recent tomographic models show fast-velocity anomalies beneath the Baikal Rift zone at depths of 100–500 km and thus support the conclusion of its passive origin. Similarly, paleorifts within stable continents such as the Proterozoic Mid-Continent Rift in North America, the Paleozoic Dnieper–Donets Rift in southern Russia, the Paleozoic Oslo Graben in Baltica, and the Paleozoic Viluy Rift in the eastern part of the Siberian craton do not display low-velocity anomalies in the upper mantle with respect to global averages, implying that the present thermal regime of the mantle may not differ significantly from the surrounding (adjacent) cratonic mantle.

Past and present subduction zones

Tomographic images only provide a snapshot of present-day mantle velocity anomalies, and any "temporal interpretations" of them, e.g. in terms of images of subducting slabs or plumes, should be made with caution. Elongated high-velocity anomalies in the upper and lower mantle seen in tomographic models are often interpreted as evidence for past and present subduction processes. Together with data on regional seismicity, the presence of subducting slabs in the upper mantle beneath most of the presently active zones of continent–continent and continent–ocean collisions is imaged by global and regional tomographic models. This includes the flat-subducting slab beneath the Andes along the Pacific coast of South America, subduction of the Indo-Australian plate beneath the Himalayas and Tibet, and numerous subduction zones imaged in the upper mantle of the Mediterranean and Asia Minor, the origins of which are related to the plate collision between Africa and Europe and the closure of the Paleo–Tethys Ocean. Low-velocity anomalies in the Trans-Pacific mantle can be explained by temperature anomalies of 200–300 K associated with subducted slabs (Fukao *et al.*, 2004).

Ancient subduction zones, many of which presently located far from the continental margins, have been imaged by tomographic studies beneath several continents. A pronounced high-velocity anomaly clearly seen in a regional surface-wave tomographic model at a distance of ~2000 km from the present Pacific margin has been interpreted as the fragments of the subducted Farallon slab beneath North America (van der Lee and Nolet, 1997b). A linear shear-wave anomaly along the western margin of the East European craton is interpreted as the signature of a paleosubduction along the cratonic margin with an increased fluid content in the upper mantle (Nolet and Zielhuis 1994). High-velocity P-wave anomalies in the lower mantle beneath Siberia, west of Lake Baikal, are interpreted as Jurassic lithospheric slabs associated with the closure of a paleoocean (van der Voo *et al.*, 1999a). One should remember, as noted above, that in some tomographic inversions localized high-velocity anomalies can be a consequence of ray density – model resolution correlation (Vasco *et al.*, 2003).

High-resolution P-wave and S-wave tomography studies of presently active subduction zones indicate that while most of the subducted slabs are deflected near the 660 km discontinuity, some fast seismic velocity anomalies that may relate to subducting slabs reach into the lower mantle (van der Hilst *et al.*, 1997; Bijwaard *et al.*, 1998; Fukao *et al.*, 2001) (Fig. 3.103). This result has an important bearing on models of mantle convection (whole-mantle, layered, or hybrid) and provides indirect support for the hypothesis of episodic catastrophic overturns in mantle associated with sinking of the slabs to a compositionally heterogeneous layer at 1700–2300 km depth (van der Hilst and Karason, 1999) or to the core–mantle boundary (Chapter 9).

Mantle plumes and hotspots

Hotspots, the areas that have experienced long-term volcanism, exist both at intraplate settings (like Hawaii, Yellowstone, and Afar) and on mid-ocean ridge axes (like Iceland and Azores). Different authors recognize different numbers of hotspots; a widely accepted compilation lists 37 (Sleep, 1990). Mantle plumes are a generally accepted hypothesis for their formation (Morgan, 1972) as well as a petrologically based estimate that they are associated with a 200–250 °C access temperature in the mantle (Sleep, 1990).

The very existence of mantle plumes and their number, if they exist, is a subject for hot debate (e.g. see http://www.mantleplumes.org; Morelli *et al.*, 2004). This paragraph overviews the competing opinions in the most general way, without any attempt to discriminate between them. Five criteria have been proposed by Courtillot *et al.* (2003) as specific for plumes:

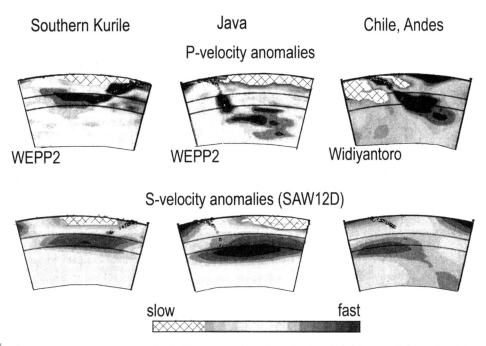

Southern Kurile — **Java** — **Chile, Andes**

P-velocity anomalies

WEPP2 — WEPP2 — Widiyantoro

S-velocity anomalies (SAW12D)

slow — fast

Fig. 3.103 Vertical cross-sections across the mantle of subduction zones beneath the Southern Kuriles, Java, and Chile (based on Fukao *et al.*, 2001). P-velocity models (WEPP2 of Obayashi *et al.*, 1997 and Widiyantoro, 1997) show velocity perturbations relative to the spherical average of the final aspherical model; S-velocity model (SAW12D of Li and Romanowicz, 1996) shows velocity perturbations relative to PREM. Thin lines show the locations of 410 km and 660 km discontinuities.

(1) hotspot track (volcanic chain),
(2) large igneous province at one of the ends,
(3) high buoyancy flux,
(4) high ^{3}He/^{4}He ratio, and
(5) low seismic velocities down to a 500 km depth in the mantle.

These "plume criteria" have been expanded to twelve to include the parameters that characterize the seismic structure of the entire mantle (i.e. the presence of low-velocity anomalies at several depths in the upper and lower mantle) and its thermal state (Anderson, 2005). Global analysis of proposed mantle plumes indicates that many of them exhibit equally many of the characteristics of both a "plume" and of a "plate tectonics" model.

Seismic tomography provides some of the most important arguments for the on-going debate on the existence of mantle plumes (Nataf, 2000; Sleep, 2006). However, since seismic velocity anomalies provide only a snapshot of present-day mantle structure, their extrapolations in time are uncertain (and may even be misleading). In particular, a new insight into the debate on the existence and temporal evolution of mantle plumes is provided by laboratory modeling of thermo-chemical convection. Simulations of plume generation, evolution, and death indicate that in a dying plume negative compositional buoyancy becomes uncompensated by positive thermal buoyancy. This causes downward material flow along a still hotter-than-normal "plume channel" so that *dying plumes start disappearing from the bottom up, sometimes even before reaching the upper boundary, ... they finally fade away by thermal diffusion. This sequence of events shows that time-dependence is a key-factor when interpreting present-day tomographic images of mantle upwellings. In particular, it could be erroneous to identify the depth of a present-day slow seismic anomaly with the depth of its origin, or to interpret the absence of a long tail as the absence of a plume"* (Davaille and Vatteville, 2005).

The global multi-mode surface-wave tomographic model RTS20S indicates large shear-velocity variations in the upper mantle beneath different hotspots (Ritsema and Allen, 2003): beneath 9 out of 37 hotspots (including the Afar, Hawaii, and Iceland hotspots) the upper mantle is 1% slower than the Oceanic Reference Model (ORM) (Fig. 3.89); beneath 14 hotspots that are located on or near mid-ocean ridges (including Tristan, Azores, and Galapagos) low velocity anomalies are observed only in the upper 200 km of the mantle; 15 hotspots (including Reunion and Yellowstone) do not display a low shear-wave velocity anomaly anywhere in the upper mantle.

Seismic evidence for the existence of a mantle plume beneath Iceland remains controversial. Body-wave tomographic models display a low-velocity anomaly below Iceland that persists below the transition zone and thus argue in favor of a narrow whole mantle plume (Bijwaard and Spakman, 1999; Morelli *et al.* 2004). Regional surface-wave tomography models image a significant low-velocity anomaly around Iceland with a diameter of *c.* 1000 km extending down to a depth of at least 600 km (e.g. Ritsema and Allen, 2003). The observed amplitude (*c.* 5–10%) of the S- and P-wave velocity anomaly in the shallow mantle can be explained by a temperature anomaly of *c.* 50–100 °C with less than 1% of melt and perhaps as little as <0.1% (Foulger *et al.*, 2005 and references therein). In contrast with tomographic models, receiver function analysis of seismic data from Iceland (Vinnik *et al.*, 2005) indicates that a low-velocity zone is restricted to the shallow mantle only (with the velocity minimum centered at a depth of 100 km) and that the transition zone has a normal thickness; a weak depression of the 410 km discontinuity can be explained by a *c.* 50 °C thermal anomaly in the mantle and does not support the presence of a plume beneath Iceland.

In contrast to Iceland, the presence of a mantle plume beneath the Hawaiian hotspot, as supported by geophysical and geochemical data, is generally accepted. Joint interpretation of global seismic tomography and semi-global electromagnetic tomography for the depth range 350–850 km in the mantle shows consistently high temperature anomalies (200–300 °C) in the mantle transition region beneath Hawaii (Fukao *et al.*, 2004).

3.6.3 Origin of seismic velocity anomalies in the upper mantle

The major mechanisms responsible for seismic velocity anomalies in the upper mantle are temperature variations, compositional variations, presence of melts and fluids, and anisotropy. Other parameters such as grain size variations may also contribute to seismic velocity perturbations. The effect of upper mantle anisotropy is addressed in detail in the next section, while Section 3.1 provides background information on the effect of temperature and compositional variations on seismic velocities in the mantle. Since interpretation of seismic velocity variations is impossible without discussing compositional effects, the cross-references between this section and Chapter 6, where the upper mantle composition is addressed in detail, were difficult to avoid (readers unfamiliar with the topic may find it useful to first read Section 6.1).

Correlations with the thermal regime

Velocity variations in the upper mantle are strongly correlated with surface tectonics and tectono-thermal ages; surface heat flow and thermal state of the continental and of the oceanic upper mantle also depend on tectono-thermal ages (see Chapter 4). This leads to the logical conclusion, supported by laboratory data on a strong temperature dependence of seismic velocities, that a significant part of seismic velocity variations in the mantle can be attributed to temperature variations. This conclusion is further supported by a strong correlation between seismic shear velocity and temperature anomalies in the oceanic lithosphere (Fig. 3.100d), except for regions with a strong seismic velocity anisotropy in the upper mantle (e.g. Fig. 3.99c). These observations provide a basis for estimation of mantle temperatures from seismic velocity models (this approach is discussed in detail in Chapter 5). Importantly, the effect of lateral temperature variations on velocities is strong in the uppermost mantle, where lateral temperature variations are the strongest, but decreases with depth whilst the lithospheric geotherm approaches the mantle adiabat. However, at near-solidus temperatures other processes become important in reducing seismic velocities.

A significant (up to several percent) drop in seismic velocities occurs at near-solidus temperatures before any melting in the mantle starts (Fig. 3.3), while the presence of even small amounts of melt leads to a dramatic drop in seismic velocities (Fig. 3.13). Near-solidus temperatures and partial melting are widely used to explain low-velocity anomalies in the upper mantle of tectonically active regions where high mantle temperatures may be expected, such as below the continental rift zones and regions of high lithosphere extension. Similarly, melting beneath the mid-ocean ridges is the major mechanism responsible for low mantle velocities at oceanic spreading centers.

Correlation between the upper mantle velocity structure and the thickness of (depth to) the transition zone provides further support to the hypothesis that many of the velocity variations in the upper mantle are of thermal origin, since both velocity anomalies in the upper mantle and the thickness of the mantle transition zone show a significant difference between the oceanic and the continental mantle (Fig. 3.57). However, this correlation holds only for large-scale tectonic structures. Furthermore, the thickness of the transition zone poorly correlates with the crustal ages of the continents and thus mechanisms other than thermal should be invoked to explain seismic velocity anomalies in the subcontinental mantle.

Effect of water

A significant amount of water is present in the upper mantle as solid solution in major minerals, rather than in hydrous phases. Experimental studies show that, under upper mantle conditions, up to 0.1 wt% of water can be dissolved in mantle olivine (Bai and Kohlstedt,

1992; Kohlstedt *et al.*, 1995). A particularly large amount of water is expected in the mantle wedge above subducting oceanic lithosphere (Fig. 7.19). The presence of even a small amount of water or fluids in mantle rocks significantly (by a few percent) reduces seismic velocities. This fact has allowed for interpretation of a strong low-velocity shear-wave velocity anomaly at 300–500 km depth along the western boundary of the East European Platform by water injection into the mantle during Paleozoic subduction associated with the closure of the Tornquist Ocean (Nolet and Zuilhuis, 1994).

Based on mineral physics observations, Karato (1995) analyzed the possible effects of water on seismic wave velocities in the upper mantle. He examined three mechanisms:

(1) the direct effect of water through the change in bond strength,
(2) an indirect effect due to enhanced anelastic relaxation caused by the presence of water, and
(3) indirect effects due to the possible change in preferred orientation of olivine leading to changes in seismic anisotropy.

The analysis by Karato indicates that, for a reasonable range of water content in the upper mantle, the first mechanism yields negligibly small effects on seismic velocities. However, the two indirect effects, which involve the motion of crystalline defects, can be significant in reducing seismic velocities. The effect of anisotropy is considered in the section to follow. Here it is worth mentioning that the effect of water on creep in olivine is anisotropic: large for [001] but small for [100], and thus deformation of olivine under high water fugacities may result in a change of the dominant slip systems.

Enhanced anelastic relaxation may reduce the seismic wave velocities by a few percent. Under dry conditions, anelasticity Q reduces seismic wave velocities and the amplitude of the effect depends on temperature (eq. 3.8). The presence of water will produce a change in Q which will result in a significant change in seismic wave velocities (Fig. 5.17). The effect of water on enhanced anelastic relaxation is particularly strong for low Q values (~50) under dry pre-conditions. For $\alpha = 0.1$–0.3 and Q between 50 and 100, seismic velocity reduces by 1–6% (eq. 3.8, 3.11). Since water solubility increases with pressure, the effect of water on the reduction of seismic velocities should be stronger at greater depth within the upper mantle.

Effect of grain size

Another candidate is a change in grain size with depth, since grain size in the upper mantle is controlled by competing stress-dependent processes of dynamic recrystallization and grain growth. It is intriguing that in the cratonic mantle of the Kaapvaal craton a change in grain size occurs at *c.* 140–150 km depth (Fig. 3.104), the depth where a significant change in mantle composition is documented by petrological studies of mantle-derived xenoliths from different cratonic settings worldwide (Gaul *et al.*, 2000). However, this depth is far too shallow to correspond to the base of the seismic lithosphere, since high-velocity anomalies beneath the Kaapvaal are observed down to 250–300 km depth (Fig. 3.98). Numerous studies of mantle-derived peridotites from cratonic settings indicate that iron depletion decreases with depth (Figs. 3.104a, 6.15b). Group I kimberlites from the Kaapvaal mantle erupted at ~90 Ma indicate fertile mantle composition at depths below 150–180 km. If this picture is representative of the Kaapvaal mantle, then high seismic velocities in the depth

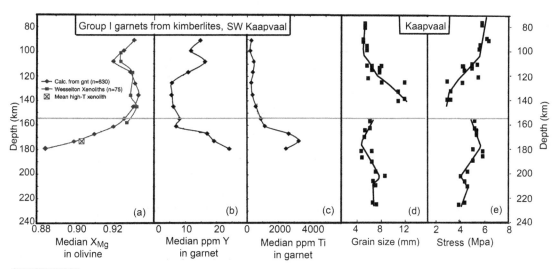

Fig. 3.104 (a–c) Variations in composition with depth in group I garnets from kimberlites of SW Kaapvaal (based on O'Reilly and Griffin, 2006); (d) olivine grain size variations in xenoliths from Kaapvaal (best fit to data of Mercier, 1980, Fig. 3.9). Note a synchronous change in composition and grain size at a depth of 150–160 km (marked by a thin line). The change in composition is related to metasomatism of the lower lithosphere by small-scale asthenospheric melts, while the discontinuity in olivine grain size is explained by a change in the recrystallization mechanism, from subgrain rotation at shallow depths to grain boundary migration below 150 km. Since recrystallization processes are stress dependent, grain size variations with depth allow for calculating deviatoric stresses in the lithosphere (e) (based on Mercier, 1980). Grain size deformation has been calibrated by olivine aggregates, and the results may not be entirely representative of the upper mantle.

range 150–250 km are unrelated to iron depletion and should have a primarily thermal origin. In contrast, older kimberlites suggest significant depletion of the Kaapvaal upper mantle down to 200–220 km depth (Fig. 6.15a).

Compositional anomalies

Compositional heterogeneity is another strong candidate for velocity variations in the upper mantle. While high temperatures, fluids, and partial melts are successful in explaining low-velocity anomalies, changes in composition may be important in explaining some high-velocity anomalies. Griffin *et al.* (1998a) argue that compositional anomalies can account for at least 50% of the seismic velocity anomalies observed in tomographic models for the continents, while Deen *et al.* (2006) reduce this value to ~25%, suggesting that temperature variations account for ~75% of seismic velocity variations in the upper mantle worldwide. Artemieva *et al.* (2004) argue that in 50% of continents, half of the amplitude of velocity anomalies in the lithospheric mantle cannot be explained by temperature variations alone.

The presence of compositional variations in the upper mantle is supported by geochemical data. Fundamental differences in composition of the subcontinental and suboceanic

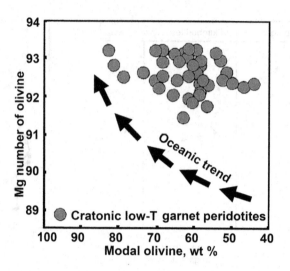

A plot of Mg number versus modal proportion of olivine for various tectonic settings (redrawn from Boyd, 1989). Mg number is 100 Mg/(Mg + Fe) of olivine for ophiolite and abyssal peridotites and a bulk value for compositional models. Oceanic trend – a trend line joining the compositions of residues and model sources that include residues of oceanic volcanism together with model compositions for fertile, parental, upper-mantle peridotite. Cratonic low-T garnet peridotites are coarse, low-temperature, garnet peridotite xenoliths from the kimberlites of the Kaapvaal craton, southern Africa. See Fig. 6.8 and related discussion.

mantles have been recognized from analyses of the major-element compositions of oceanic peridotites and mantle-derived xenoliths from different tectonic settings (Fig. 3.105 and Chapter 6). This conclusion is supported by geophysical analysis based on gravity and seismic data that shows that the ocean–continent difference in seismic tomography is not purely thermal in origin and internally stabilized continent–ocean chemical differences are present in the upper mantle (Forte et al., 1995).

On the continents, compositional variations in lithospheric mantle (primarily in iron-content) are well correlated with the age of the overlying crust (Fig. 6.11 and Chapter 6). This correlation is interpreted as a secular, irreversible trend from highly depleted (due to loss of iron during melt extraction) Archean continental lithospheric mantle to fertile Phanerozoic continental and suboceanic mantle. It forms the basis for geophysical interpretations, most of which consider iron depletion as the only source of compositional velocity and density anomalies in the upper mantle.

For example, joint inversions of a large set of geodynamic data related to mantle convection, including global tomography models, have been used for mapping depth variations in iron content in the subcontinental upper mantle (Forte and Perry, 2000). The results suggest, in general agreement with petrological data, that iron depletion is restricted to the upper 200–300 km. Depending on modeling assumptions, the depletion anomaly either has the peak at c. 150 km depth, or gradually decreases in amplitude from the Moho down to the transition zone (Fig. 3.106).

Regional geophysical studies support the presence of significant compositional differences in the continental upper mantle, particularly between the cratonic and Phanerozoic

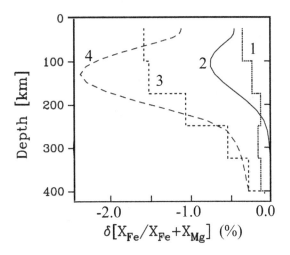

Fig. 3.106 Iron depletion anomalies in the subcratonic upper mantle (from Forte and Perry, 2000; compare with Fig. 3.19). The calculations are based on the assumption that perturbations in iron and garnet content in the tectosphere are correlated as a result of basalt depletion. Curves 1 and 2: the internally consistent estimate of iron depletion in the tectosphere based on $\partial \ln \rho / \partial \ln V_s$ scaling factor and calculated for two tomographic models. Since the approach is invalid when partial melting or thermally induced V_s attenuation are present in the thermal mantle, independent mineral physics estimates of average $\partial \ln \rho / \partial \ln V_s = 4$ in the upper mantle (Karato, 1993) are used to calculate curves 3 and 4. Tomographic models used in geodynamic calculations: curves 1 and 3 – Grand et al., 1997; curves 2 and 4 – Ekström and Dziewonski, 1998.

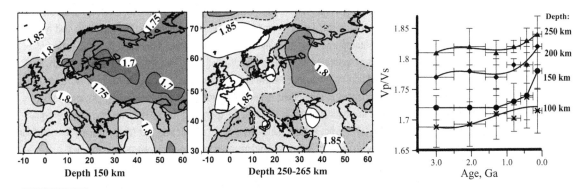

Fig. 3.107 Vp/Vs variations in the upper mantle of Europe at depths of 150 km and 250–265 km (smoothed by Gaussian filtering), based on the tomography models of Bijwaard and Spakman (2000), Piromallo and Morelli (2003), Shapiro and Ritzwoller (2002). The Vp/Vs ratio is more sensitive to compositional than thermal anomalies in the upper mantle. Note a sharp expression of the cratonic margin at depth 150 km. Right-hand plot shows variations of the Vp/Vs ratio at four depths as a function of the tectono-thermal age of the lithosphere of Europe (based on Artemieva, 2007).

mantle. For example, the Vp/Vs ratio, which is more sensitive to chemical than to temperature variations, displays a sharp change across Europe that correlates remarkably well with the Trans-European suture zone that marks the western margin of the East European Craton (Fig. 3.107). It is tempting to attribute the change in the Vp/Vs ratio across the cratonic margin to changes in mantle Mg# (Fig. 3.22).

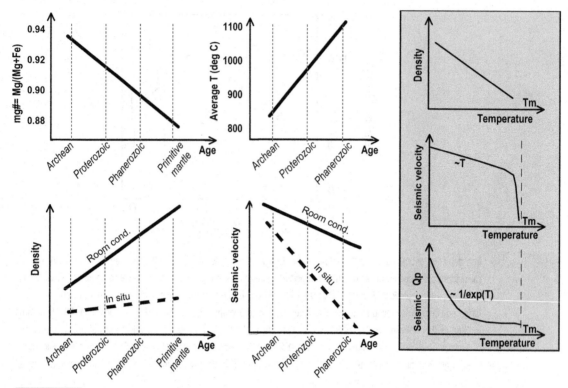

Fig. 3.108 Sketch illustrating secular variations in iron content (expressed by Mg#), density, and seismic velocity in mantle rocks as indicated by laboratory measurements (at room conditions) on mantle-derived xenolith peridotites and typical average temperatures in the lithospheric mantle of different tectonic provinces. Gray insert shows the effect of temperatures on density, seismic velocities, and inverse seismic attenuation. Dashed lines for densities and seismic velocities refer to *in situ* conditions in the mantle, where high temperatures cause density and velocity to decrease. Systematic age-dependent variations in lithospheric temperatures from hot tectonically young provinces to old and cold cratonic lithosphere mask variations in *in situ* densities (providing the basis for the isopycnic (i.e. equal density) hypothesis of Jordan (1978)), but enhance variations in *in situ* seismic velocities (thus providing "grounds" for interpretation of seismic "blue-versus-red" velocity perturbations in terms of "cold-versus-hot" mantle).

Large-scale compositional heterogeneity of the upper mantle should be reflected in seismic tomography and gravity models. However, compositional and temperature effects are difficult to separate due to strong lateral and vertical temperature variations in the upper mantle (see Chapter 4) and the strong effect of temperature variations on physical properties of mantle rocks (Section 3.1). In the subcontinental upper mantle both iron depletion and mantle temperatures (at the same depth within the lithospheric mantle) systematically decrease with age (Figs. 3.108, 4.33, 6.11, 6.15). High upper mantle temperatures produce a density decrease in Phanerozoic regions due to thermal expansion, thus reducing *in situ* density contrast between the low-dense depleted cratonic and young, iron-rich, fertile and heavy lithospheric mantle. The existence of such a thermally balanced *in situ* density equilibrium between depleted and fertile lithospheric mantle (the isopycnic, or equal-density hypothesis) was proposed by Jordan (1978, 1988) based on the analysis of geo-physical (including gravimetric and geothermic) data (see discussion in Chapter 6). It explains the absence of significant geoid anomalies in the cratons. A global analysis of gravity data, corrected for temperature-induced density variations in the upper mantle, indicates a strong contrast in composition between the oceanic and continental upper mantle, and shows that a strong compositional heterogeneity exists within subcontinental mantle even for regions of the same age (Kaban *et al.*, 2003; Fig. 6.32).

In contrast with density, temperature increase causes reduction in seismic velocity; as a result a difference in mantle temperatures between the Archean and Phanerozoic regions enhances the seismic velocity contrast between depleted and fertile mantles thus supporting a common interpretation of seismic velocity anomalies in terms of "cold" versus "hot" regions (Fig. 3.108). Although variations in the volumic fraction of olivine and in the iron content certainly have an effect on seismic velocities and densities, it is not so strong as the effect of variations in mantle temperature (Karato and Karki, 2001). An analysis of the effects of temperature and iron content on seismic velocities (Deschamps *et al.*, 2002) shows that a 1% velocity anomaly can be produced by an iron anomaly (i.e. the volumic fraction of iron Fe/[Fe+Mg]) of 4% or by a temperature anomaly of 50–100 °C (which is about the accuracy of thermal constraints at any lithospheric depth) (Fig. 6.14a).

In contrast with this prediction, a recent geophysical analysis has indicated the presence of strong non-thermal velocity variations in the upper mantle of the continents if the effect of temperature variations is removed (Artemieva, 2009). The results are constrained by shear-wave tomographic velocity models for the upper mantle that were corrected for lateral temperature variations (Figs. 3.109 and 6.35a). On all continents, the surface transition from the cratonic to the Phanerozoic lithosphere is marked by a sharp decrease in non-thermal seismic velocity anomalies in the lithospheric mantle, which is correlated with a sharp change in the non-thermal part of mantle gravity anomalies (i.e. increase in mantle density) (Kaban *et al.*, 2003). Lateral, non-thermal in origin, variations in seismic velocity and density structure may be attributed, in part, to a change in olivine content at the transition from the cratonic to non-cratonic lithospheric mantle, although a significant part of these variations may be caused by other compositional and non-compositional effects.

Interpretations of seismic velocity anomalies in the cratonic lithospheric mantle in terms of iron depletion are challenged by recent laboratory measurements and theoretical calculations for chemical and mineralogical compositions of natural peridotites (Kopylova *et al.*, 2004; Matsukage *et al.*, 2005). These indicate that seismic velocities in the continental lithosphere are characterized by two parameters, Mg# and Opx# (volume fraction of orthopyroxene) (Figs. 3.21, 3.24). While compressional and shear seismic velocities increase almost linearly with increasing Mg#, addition of low-velocity orthopyroxene decreases seismic velocities. Maximum velocity anomalies produced by variations in Opx# (-1.8% and -2.3% for Vs and Vp, respectively) are larger than Vs and Vp anomalies caused by iron depletion ($+1.1\%$ and $+1.0\%$, respectively). Matsukage *et al.* (2005) argue that variations in orthopyroxene content can be the main cause of seismic velocity variations at garnet stability field (depths between 70 to 190 km). Thus, at present the conclusions of laboratory and theoretical studies on the effects of compositional variations in the subcontinental lithospheric mantle remain incomplete.

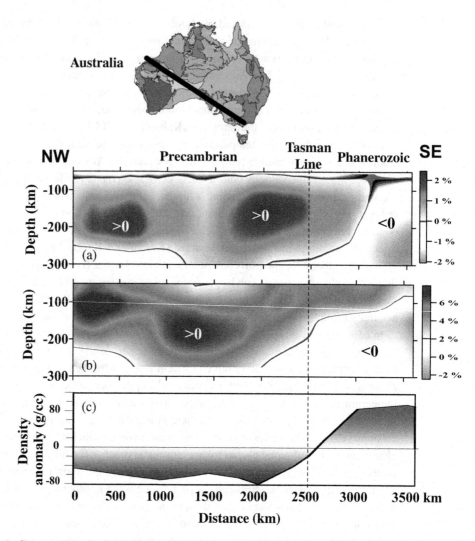

Fig. 3.109 Profile across Australia showing compositional heterogeneity in the upper mantle (based on Artemieva, 2009). (a–b): *V*s seismic velocities corrected for lateral and vertical temperature variations based on (a) surface-wave model (Shapiro and Ritzwoller, 2002) and (b) body-wave tomographic model (Grand, 2002). (c) Density variations in the upper mantle of Australia constrained by temperature-corrected gravity anomalies (Kaban *et al.*, 2003). The depth distribution of these anomalies is unconstrained and the values reflect density anomalies integrated over the entire lithosphere.

Density-to-shear-wave velocity scaling factor

Owing to strong temperature-dependence, seismic velocity anomalies alone are insufficient to distinguish chemical variations in the mantle. An independent set of data, such as density anomalies, is needed to separate the compositional and thermal effects on seismic wave speeds and to infer chemical variations in the mantle. A relative density-to-shear-wave velocity scaling factor $\zeta = \partial \ln \rho / \partial \ln V_s$ has been proposed to discriminate between thermal and chemical heterogeneity (see eq. 3.13 and Section 3.1.8). It provides a crude diagnostic of the thermal versus compositional origin of density and seismic velocity anomalies in the mantle. For purely thermal anomalies, when temperature variations produce density and velocity variations of the same sign, $\zeta > 0$. Unlike temperature variations, the effect of chemical variations is more complicated since compositional changes may have a different effect on the scaling factor and both cases, $\zeta > 0$ or $\zeta < 0$, are possible depending on composition. For example, two end-member upper mantle compositions have opposite signs of synthetic ζ values (positive for olivine–garnet but negative for olivine–diopside mineralogies) since an increase in the volume fraction of diopside leads to an increase in both density and velocity, whereas an increase in the volume fraction of garnet produces a density increase, but smaller velocities (Deschamps *et al.*, 2001). Tables 3.1 and 3.2 summarize information on the effects of various factors on mantle density and seismic velocities.

On the whole, for the real upper mantle the values of ζ are not well known and remain controversial (Fig. 3.110). Joint inversions of gravity and tomographic models for the whole mantle commonly use low-degree spherical harmonics which are sensitive to the structure of the deep mantle. The reported ζ values range from $-0.3 < \zeta < 0.3$ with the change of the sign at $c.$ 300 km depth for inversions for degrees $l = 2$–8 of spherical harmonics ζ (Forte *et al.*, 1994). Inversions for $l = 11 - 16$ based on global tomographic model S16RLBM (Trampert and Woodhouse, 1995) and on the EGM96 geoid model (Lemoine *et al.* 1998) predict $-0.05 < \zeta < 0.05$ for the upper 400 km (Deschamps *et al.*, 2001). The results of the latter inversion suggest that in the subcontinental upper mantle the sign of ζ changes from positive to negative at $c.$ 200 km depth, while beneath the oceans $\zeta < 0$ at between the depths of $c.$ 120 km and 220–240 km (Fig. 3.110b). The negative sign of the scaling factor in the lower parts of the subcontinental lithosphere (deeper than 200 km), where temperatures approach mantle adiabat and thus the thermal effect on velocity variations decreases, indicates the important role of chemical anomalies on seismic velocities and mantle density. Compositional anomalies with intermediate length scale are expected to be present in the uppermost part of the subcontinental mantle, where the shapes of depth variations of ζ and relative V_s anomalies are significantly different (Deschamps *et al.*, 2001).

The results of Deschamps *et al.* (2001) are in contrast, both for continents and oceans, with similar studies performed by other groups. Kaban and Schwintzer (2001) use the isotropic part of shear-wave tomography model S20 (Ekstrom and Dziewonski, 1998) and the EGM96 spherical harmonic expansion of the gravitational geopotential (Lemoine *et al.* 1998) to derive mean gravity values for $1° \times 1°$ blocks. They further correct gravity data for the effect of crustal density heterogeneity using the CRUST 5.1 model (Mooney *et al.*, 1998) and calculate the density-to-shear velocity scaling factor from the thus derived "crust-free" gravity model, performing the inversion for the whole mantle (for spherical harmonic degrees $l = 3 - 20$). Their analysis indicates the presence of compositional layering in the oceanic upper mantle with a transition between the two layers at 70 km depth (although lateral temperature variations produce most of the density variations in the oceanic upper mantle). Above a 75 km depth and below a 225 km depth, the velocity-to-density scaling factor is close to predictions based on mineral physics analysis and is consistent with the composition of dry partially depleted peridotite (Fig. 3.110d). The oceanic layer between a

Table 3.1 Non-thermal effects on variations of upper mantle density and Vs seismic velocity

Parameter	Affected depth	$\delta\rho$	δVs	Reference
Iron content increase	CLM	+ 1.4%(for $X_{Fe} = +4\%$)	−1.0% (for $X_{Fe} = +4\%$)	D02, L03
Garnet content increase	CLM	> 0	> 0	D02
Substitution of pyrore-rich garnet by grossular garnet	CLM	< 0, Weak effect	> 0	D02
Clinopyroxene content increase	CLM	< 0	−0.4 % (for $X_{di} = +6\%$)	NZ94
Orthopyroxene content increase	CLM (70 to 190 km depth)	No effect	Up to +1.8%	M05
Fluids	CLM	Indirectly, through the effect on solidus	Indirectly, through the effect on anelastic relaxation −1 to − 6%	T81
Anelasticity at high homologous temperature $T/Tm > 0.95$	The lower part of the TBL	No effect		S89
Melts	Chiefly the lower part of the TBL	< 0	− 10% for ~ 5% melt (strongly depends on melt geometry)	MK79
Anisotropy	CLM	No effect	Either sign	BIM98

(Source: Artemieva, 2009).

Comments: CLM = continental lithospheric mantle. Where absolute values are unavailable, only the sign of δVs and δp variation is given. Positive values (>0) indicate a velocity or density increase, while negative values (<0) refer to a decrease in the corresponding parameter.

References: BIM98 – Ben Ismail and Mainprice, 1998; D02 – Deschamps et al., 2002; L03 – Lee, 2003; MK79 – Murase and Kushiro, 1979; M05 – Matsukage et al., 2005; NZ94 – Nolet and Zielhuis, 1994; S89 – Sato et al., 1989; T81 – Tozer, 1981.

Table 3.2 Factors affecting Density and velocity anomalies in the upper mantle of different tectonic provinces

Tectonic setting	Major factors affecting ρ and Vs	$\delta\rho/\rho_0$ (%)*	$\delta Vs/V_0$ (%)*
Archean cratonic mantle	• Low Fe (Fo = 93)	− 1.8%	+ 1.3%
Proterozoic cratonic mantle	• Intermediate Fo = 92	− 0.6 –1.2%	+ 0.4%
Phanerozoic or metasomatised Fe-enriched cratonic mantle (Fo = 90)	• In the case of high garnet content • In the case of high diopside content	> 0 < 0	> 0 − 0.4%
Orogens	• Metamorphism (in a 20 km thick layer) • Fluids • Melts?	+ 1.0% (overall effect for 80 km thick mantle) No effect < 0	No effect < 0 − 10% for ~5% melt
Active (plume-related) continental regions	• Melts • Subsolidus T	< 0 No effect	− 10% for ~5% melt − 6% (in a 10 km thick basal layer)
Subducted Precambrian slab (made of 40 km thick crust + 110 km thick low-Fe (Fo92) mantle)	• Composition of crust and mantle • Metamorphism (in a 20 km thick layer) • Fluids	− 6.2 % (overall effect) + 0.4% (overall effect) No effect	< 0 (crust), + 1% (mantle) No effect < 0
Subducted Phanerozoic slab (made of 40 km thick crust + 80 km thick harzburgite mantle (high ol, low gnt, no cpx))	• Composition of crust and mantle • Metamorphism (in a 20 km thick layer) • Fluids	− 6.7 % (overall effect) + 0.4% (overall effect) No effect	< 0 (crust), No effect (mantle) No effect < 0

(Source: Artemieva, 2009).

* Relative density and velocity anomalies are given at SPT (room conditions, thermal effect excluded) with respect to "normal" (Fo90) stable continental lithosphere (such as is typical of Paleozoic platforms).

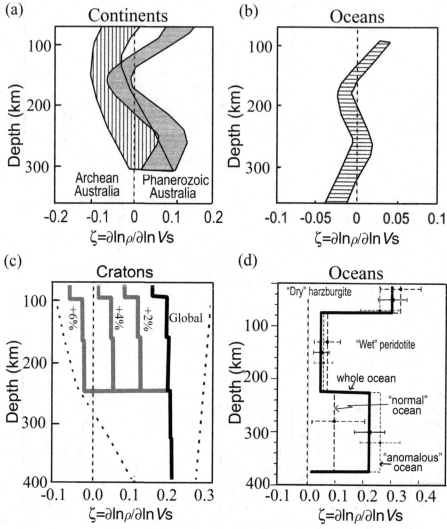

Fig. 3.110 A relative density-to-shear-wave velocity scaling factor ζ in the subcontinental and suboceanic upper mantle (eq. 3.13) based on different inversions for the whole mantle. Negative values of the scaling factor indicate the presence of compositional variations, whereas positive values may be caused either by temperature or by chemical variations. Note different horizontal scales. (a) ζ calculated for the Archean and Phanerozoic regions of Australia for the SKIPPY-based isotropic surface-wave tomography model AUS-04VS (results of van Gerven *et al.*, 2004); (b) ζ calculated in a joint inversion of the global tomographic model S16RLBM and the EGM96 geoid model (results of Deschamps *et al.*, 2001). Shaded areas in (a–b) indicate both the mean values and the error bars estimated by introducing randomly generated errors in the seismic model. (c) Optimal global 1D scaling based on simultaneous inversion of shear-wave travel-time data and convection-related geodynamic constraints. Solid gray lines – profiles corrected for velocity perturbations of $dVs = 2\%, 4\%, 6\%$ within the cratonic roots; dotted lines – range of 3D variations of the scaling factor (results of Simmons *et al.*, 2009). (d) Scaling factor for oceanic mantle calculated from the isotropic part of shear-wave tomography model S20 and global gravity based on the EGM96 geoid model and corrected for the effect of crustal density heterogeneity (results of Kaban and Schwintzer (2001)).

70 km and 225 km depth has a significantly lower density/velocity ratio and may have the composition of wet fertile peridotite.

Simmons *et al.* (2009) simultaneously inverted shear-wave travel-time data and a set of geodynamic convection-related constraints consisting of the global free-air gravity field, tectonic plate divergences, dynamic surface topography, and the excess ellipticity of the core–mantle boundary, represented by surface spherical harmonic functions up to degree 16. The temperature-dependence of the relative density-to-shear-wave velocity scaling factor $\zeta = \partial \ln \rho / \partial \ln V_{\mathrm{s}}$ in the upper mantle is significant due to the temperature-dependence of seismic attenuation. The inversion provides an optimum 1D thermal profile of ζ that is generally compatible with all considered data except for the cratonic regions where dynamic topography produced by density anomalies in shallow mantle is expected (Fig. 3.110c). The results, based on calculation of the relative magnitudes of the thermal and compositional (non-thermal) contributions to mantle density anomalies, indicate that within the non-cratonic mantle thermal effects dominate shear-wave and density heterogeneity. However, within the cratonic mantle, strong negative correction factors to ζ are required due to the significant compositional buoyancy of cratonic roots. Similar results have been obtained in a recent regional inversion for central Australia (van Gerven *et al.*, 2004). This study reports $\zeta < 0$ at 75–150 km depths and near-zero ζ values down to *c.* 300 km depth which are interpreted as indicating strong compositional heterogeneity of the upper part of the cratonic upper mantle (Fig. 3.110a).

3.6.4 Seismic anisotropy in the upper mantle

The following paragraph largely follows an excellent review of seismic anisotropy studies of the upper mantle published by Savage (1999). It has been updated to include major results from the past decade. A discussion of crustal anisotropy (the major mechanisms for which are anisotropic rock fabric, alignments of cracks and fractures within the crust, and alternate layers of fast and slow material) is omitted here; a review can be found in Crampin (1977). Since the total measured anisotropy results from a combination of crustal and mantle components, which may have different orientations (e.g. Christensen *et al.*, 2001), the effect of crustal anisotropy must be separated out in studies of mantle anisotropy.

Types of anisotropy

Theoretical studies of the propagation of elastic waves in anisotropic media were started by Christoffel (1877) and were further developed by Lord Kelvin (1904), Love (1944) and Musgrave (1959). Experimental studies of seismic anisotropy began largely in the early 1960s with the works of Birch (1961), simultaneously with field observations of seismic anisotropy in the uppermost mantle of the oceans (e.g., Hess, 1964; Francis, 1969).

Three major types of anisotropy include intrinsic, azimuthal, and polarization anisotropy.
- *Intrinsic anisotropy* refers to the material itself and is defined by the difference between the maximum and minimum velocities in a medium (Birch, 1961).
- *Azimuthal anisotropy* is the variation of wavespeed for a certain type of wave as a function of the azimuth of the propagation direction (Figs. 3.28, 3.111) and in early studies it was constrained by *P*n velocity variations based on refraction data (e.g., Hess, 1964).

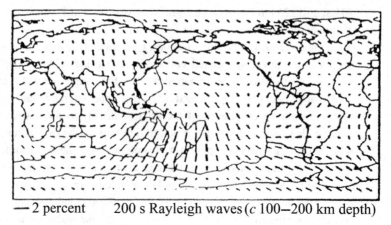

— 2 percent   200 s Rayleigh waves (c 100–200 km depth)

Fig. 3.111   Azimuthal anisotropy of Rayleigh waves of 200 s period. Lines show the propagation direction of the fast waves (after Tanimoto and Anderson, 1984).

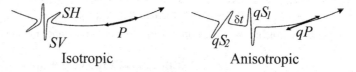

Fig. 3.112   Sketch of isotropic and anisotropic wave propagation (redrawn from Savage, 1999). In the isotropic case, the P-wave has linear particle motion in the propagation direction and the S-wave has a particle motion perpendicular to propagation; two S-wave components are arbitrarily defined as oriented horizontally (SH) and vertically (SV). In the anisotropic case, linear particle motion in a quasi-P wave is not quite parallel to the propagation direction, while two quasi-S waves (qS_1 and qS_2) with polarizations parallel and perpendicular to the fast direction for the propagation direction are separated by time δt.

- *Polarization anisotropy* is the variation of wavespeed of phases with different polarization that travel along the same direction; it is less sensitive to heterogeneous isotropic structure than azimuthal anisotropy. Examples of polarization anisotropy include:
  (1) Love and Rayleigh surface waves (Fig. 3.27) since horizontally polarized Love waves travel faster than vertically polarized Rayleigh waves (e.g., Anderson, 1961);
  (2) SH (fast) and SV (slow) phases of body waves (e.g., Shearer, 1991);
  (3) Shear-wave splitting (e.g. of body waves) that has the same physical meaning as optical birefringence (e.g. Crampin, 1977) (Fig. 3.112).

Laboratory measurements of anisotropy on mantle-derived samples indicate that most of them have either orthorhombic or hexagonal symmetry (Fig. 3.14). For horizontal alignment of the *a*-axis and near-vertical incidence angle, the anisotropy patterns produced by these symmetry systems do not differ significantly. Elastic anisotropy in the mantle is commonly attributed to the preferred orientation of olivine crystals which results from deformation. Large-strain deformation aligns the olivine [100] axes along the flow direction or parallel to the direction of extension. For horizontal flow it leads to $V_{SH} > V_{SV}$ and for vertical flow to

$V_{SH} < V_{SV}$. These interpretations, commonly accepted in most seismic studies, have been challenged by recent laboratory experiments on deformation of olivine aggregates (see Section 3.1.5). In particular, in the case of partial melting such as in the mantle wedge, the [001] olivine axes may align along the shear direction with its [100] axes oriented normal to the macroscopic flow direction (Fig. 3.15c).

Origin of mantle anisotropy

Tectonic processes that produce deformation of mantle rocks and are thought to be responsible for mantle anisotropy include the following.

(1) *Fluid-filled cracks deformed by local stress*. This mechanism produces fast polarizations aligned parallel to the maximum compressional stress and is likely to be the major cause of anisotropy beneath oceanic spreading centers where the fast axis of anisotropy is parallel to the spreading direction (Vinnik *et al.*, 1989; Blackman and Kendall, 1997; Wolfe and Silver, 1998), continental rift zones (Sandvol *et al.*, 1992; Vinnik *et al.*, 1992), and the East Pacific Rise (Forsyth, 1975; Forsyth *et al.*, 1998).

(2) *Lithosphere deformation by strain*. In the case of collisional events caused by uniaxial compression, *b* is the axis of symmetry (the slow direction) and it is parallel to the direction of shortening; fast polarization is parallel to the strike of the mountain belts, e.g. due to "escape" tectonics (Christensen and Crosson, 1968; Milev and Vinnik, 1991; Silver, 1996; Meissner *et al.*, 2002).

(3) *Frozen-in anisotropy*. In many regions anisotropy in the crust, in the shallow and in the deep lithosphere reflects different tectonic episodes and preserves frozen-in anisotropy from past tectonic events (i.e. deformation at temperatures above 900 °C followed by cooling, so that temperature never rose again to the critical value of *c*. 900 °C). In the continents the crust and the lithospheric mantle may experience coherent deformation through geologic history. In such a case, fast polarization in the mantle should correlate with crustal tectonics (Silver and Chan, 1988, 1991; Vinnik *et al.*, 1992; Silver, 1996). Anisotropy at deeper, hotter depths may be different and caused by recent deformations. Similarly, in the Pacific Ocean shallow (<100 km) frozen-in anisotropy is related to fossil spreading whereas deeper mantle anisotropy is related to the present-day flow direction (Nishimura and Forsyth, 1988; Montagner and Tanimoto, 1991).

(4) *Large-scale asthenospheric flow*. Asthenospheric flow produces the preferred orientation of olivine and mantle anisotropy (Vinnik *et al.*, 1992). In the case where flow occurs by simple shear, fast polarization tends to be aligned with the flow direction (within 30° of the direction of flow for large strain and at about 45° for small strain). Examples of this are the Pacific and the Indio-Australian plates (Montagner, 1994). However, anisotropy in the North Atlantic is inconsistent either with the fossil seafloor spreading direction or with present-day plate motion (Montagner and Tanimoto, 1991). Asthenospheric flow plays an important role in producing azimuthal anisotropy (Vinnik *et al.*, 1992).

(5) *Relative motion of lithospheric plates and convective mantle*. Similarly to asthenospheric flow, fast polarization tends to be parallel to the direction of relative plate motion

(spreading direction) (Ribe, 1989). In subduction zones slabs may channel astheno-spheric flow parallel to the slab dip and the anisotropy axis is expected to be parallel to the direction of subduction. In some regions (e.g. the North Atlantic) plate motion may be decoupled from convective flow deeper in the mantle (Bjarnason *et al.*, 1996).

(6) *Small-scale asthenospheric flow.* It may be important in some tectonic settings. Corner flow at mid-ocean ridges may produce horizontal stretching lineation and the horizontal foliation plane (Hess, 1964; Silver, 1996). In some cases, due to small-scale flow, fast polarizations may be non-related to surface tectonics (Makeyeva *et al.*, 1992).

(7) *Temperature difference* in mantle convective cells (Montagner and Guillot, 2002).

Body-wave studies

Preamble

Body-wave studies of seismic anisotropy have the same limitations regarding vertical resolution as body-wave tomography: while shear-wave splitting provides lateral resolution as good as <50 km, the depth extent of the observed anisotropy cannot be reliably con-strained by seismic data. The most popular body-wave studies are SKS and similar phases. They begin propagation as S-waves, convert to P-waves at the liquid outer core, and are then converted back to S-waves when they exit from the core. The advantage of these phases is based on the fact that any splitting from the source side of the path is destroyed during their propagation in the outer core and any anisotropy produced at the source side is lost. The polarization of the S-wave when it enters the mantle is known (because it is generated by a P-wave traveling in the core, the emerging wave has SV polarization) and is independent of the earthquake focal mechanism, while S-wave splitting occurs on the path between the core–mantle boundary and the receiver. Near-vertical S-wave propagation in the upper mantle provides high lateral resolution (within the first Fresnel zone).

Two parameters characterize shear-wave splitting (Fig. 3.112): polarization of the first (fast) shear wave and a time delay between the two quasi-S-waves. The latter depends on the wavespeed difference between them and on the path length in anisotropic medium. However, for most symmetry systems both fast polarization direction and time delay depend on the back azimuth and incidence angle. The simplest interpretation of shear-wave splitting assumes transverse anisotropy with a horizontal symmetry axis. Most of the reported values of time delays range between 1.0 s and 2.0 s. If shear-wave splitting is distributed over a 400 km thick layer, minimum values for average anisotropy of 1.0 s and 2.0 s are 1.1% and 2.2%, respectively; in the case of a 250 km thick layer these values are 1.8% and 3.6%. Laboratory measurements of mantle xenoliths that came from a 40–170 km depth indicate comparable values of shear-wave splitting anisotropy, between 3.5% and 7% (Mainprice and Silver, 1993; Kern, 1993).

Frozen versus asthenospheric anisotropy

The observation of shear-wave splitting on broadband seismic waveforms has led to two competing models for anisotropy within the upper mantle (Fig. 3.113). According to one group of authors, anisotropy is primarily within the asthenosphere as indicated by agreement

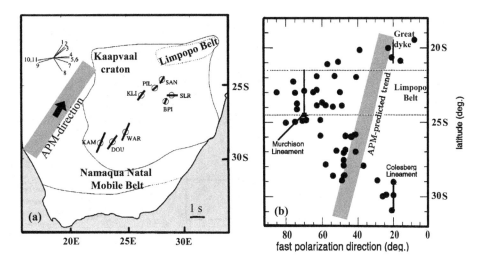

Fig. 3.113 Two alternative views on the origin of seismic anisotropy beneath southern Africa based on shear-wave splitting analyses of SKS and similar phases (a – after Vinnik *et al.*, 1995; b – based on Silver *et al.*, 2001). (a) The direction of polarization of the fast split wave at different stations is shown by straight lines, the lengths of which are proportional to the time lag of the slow wave (see scale bar in the lower right corner). The upper left corner shows azimuths of the recorded events. The direction of the Absolute Plate Motion (APM) since the end of the Jurassic is shown by gray shading with an arrow. The results for all stations (except SLR) indicate that the direction of mantle flow beneath southern Africa is close to the APM and suggest that mantle anisotropy is related chiefly to shearing of the sublithospheric mantle at depths between 150 km and 400 km by the plate above. (b) Direction of fast shear-wave polarization in the mantle plotted versus station latitude. The crustal component of shear-wave splitting has been isolated in the analysis. Fast polarization directions predicted for the present-day APM model for the African plate (shown by gray shading) are in sharp contrast to the data. The anisotropy pattern seems to be better correlated with geological features.

between present-day plate motion directions and the orientation of the fast axis of wave-speed (Vinnik *et al.*, 1995, 1998). Such correlations are found in several continental regions (South Africa, North America) and in most oceanic regions at 100–200 km depth (e.g., Montagner, 1994; Vinnik *et al.*, 1999; Lévêque *et al.*, 1998; Bokelmann, 2002). Although the depth distribution of anisotropy is unknown, some indirect assessments can be made from mantle temperatures since olivine mobility below *c.* 1000 °C is low. Beneath the South African craton this temperature is reached at *c.* 150 km depth and it has been proposed that the deep part of the continental root (between 150 km and 400 km) can experience deformation associated with plate motion and mantle flow. This hypothesis is supported by the distribution of olivine grain-sizes and deviatoric stress in the lithosphere beneath the Kaapvaal craton (Fig. 3.104). An estimated lateral displacement of the deep portion of the root at 400 km depth with respect to a 150 km depth is a few hundred kilometers as compared to a *c.* 3000 km shift of the lithospheric plate since the Jurassic (Vinnik *et al.*, 1995). In cratonic North America, where a large-scale coherent SKS anisotropy is present, while large-scale azimuthal anisotropy is absent in long-range seismic refraction data for the upper 100 km of the mantle, shear-wave splitting seems to be generated below a

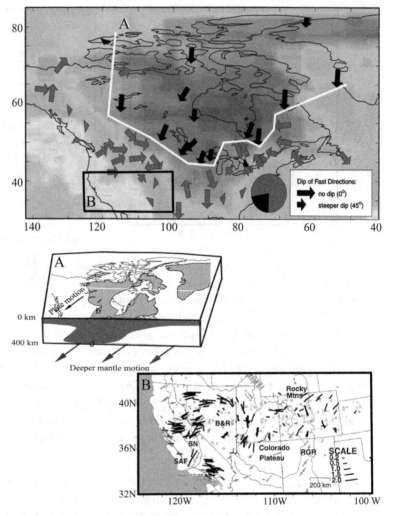

Fig. 3.114 Seismic anisotropy in North America (after Bokelmann, 2002). Orientation of fast axes for stations in North America are shown by arrows (black for azimuths between 180° and 260°). Shaded background – vertically integrated travel-time residuals in the upper mantle in the S-wave tomographic model (Grand, 1994). Insert A: proposed interpretation of anisotropy origin beneath the cratonic part of North America (schematically outlined by a white line in the map) caused by simple shear due to the relative motion of thick cratonic root over deeper mantle (after Bokelmann, 2002). Insert B: Shear-wave splitting measurements in western USA plotted at 220 km depth projection of the ray. Bars are oriented parallel to the fast polarization direction, their length is proportional to the delay time (after Savage, 1999).

100 km depth. Thus it is likely to be caused by present-day flow in the mantle rather than reflect past tectonic processes (Vinnik *et al.*, 1992; Bokelmann, 2002) (Fig. 3.114).

The other competing point of view on the origin of mantle anisotropy is that anisotropy originates primarily within the lithosphere ("frozen-in anisotropy") as indicated by the agreement between the geometry of surface tectonics (continental accretion) and the

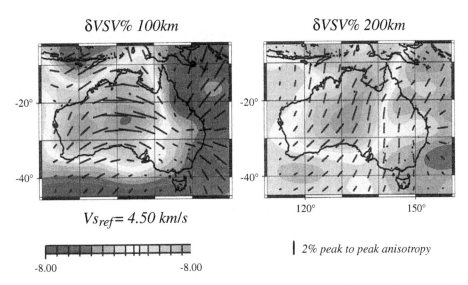

δVSV% 100km   δVSV% 200km

Vs$_{ref}$= 4.50 km/s

-8.00   -8.00

| 2% peak to peak anisotropy

Fig. 3.115 SV wave heterogeneities and azimuthal anisotropy (from Debayle and Kennett, 2000). Gray shading shows velocity anomalies and bars show azimuthal anisotropy (the length indicates the strength of anisotropy).

orientation of the fast axis in North America and South Africa (Silver and Chan, 1988, 1991; Silver *et al.*, 2001). This means that mantle deformation can have been preserved in the lithosphere structure since the Precambrian, implying coherent deformation history for the crust and the lithospheric mantle. Similarly to continents, anisotropy in the suboceanic mantle can also be frozen-in and associated with the spreading direction (e.g. Nishimura and Forsyth, 1989).

These two points of view need not be incompatible since seismic observations are best modeled by a combination of lithospheric and asthenospheric anisotropy (Vinnik *et al.*, 1992). Differences in the depth distribution of radial and azimuthal surface-wave anisotropies provide strong support for this conclusion (Lévêque *et al.*, 1998). In particular, below the Australian continent the drastic change in the direction of anisotropy occurs at a depth of 150–200 km, while a positive velocity anomaly is clearly seen down to ~300 km depth. Observation of polarization anisotropy down to at least 200 km beneath the Australian continent supports a two-layered anisotropic pattern as constrained by the azimuthal anisotropy of SV waves. In the upper layer down to *c.* 150 km depth, anisotropy has roughly a sublongitudinal pattern which does not correspond to the present plate motion; it is interpreted as a frozen-in anisotropy related to past deformation. In the lower layer, below *c.* 150 km depth anisotropy has a sublatitudinal pattern that may reflect present deformation due to plate motion (Debayle and Kennett, 2000) (Fig. 3.115).

Similarly, stratified anisotropy in the upper mantle has been reported recently beneath the North American craton (Deschamps *et al.*, 2008). SKS splitting and long-period teleseismic waveform analysis indicate stratification of the lithospheric mantle in North America with transition between the two layers at *c.* 150 km depth (Yuan and Romanowicz, 2010). Other studies propose that the source of shear-wave splitting beneath the North American craton is deeper than 200 km (Li *et al.*, 2003a). A recent study for North America suggests the presence of two anisotropy layers with different orientations of the fast axis in the upper mantle of the craton: at shallow depth (80–200 km) the fast axis is oriented roughly north–south, while deeper (200–400 km), the fast axis is subparallel to the absolute plate motion (Marone and Romanowicz, 2007).

Stratified anisotropy was long ago reported for the Pacific Ocean (Nishimura and Forsyth, 1988; Montagner and Tanimoto, 1991): while the shallow (<100 km) upper mantle anisotropy of the Pacific Ocean is related to fossil spreading, mantle anisotropy below a 100 km depth is related to the present-day flow direction.

Global patterns

On the global scale, studies of azimuthal anisotropy display the following patterns.

- In the regions of the present plate convergence, the fast direction of anisotropy is usually parallel to the plate boundary suggesting similar stress orientation in the crust and the upper mantle (Christensen and Crosson, 1968; Milev and Vinnik, 1991; Silver, 1996). For example, studies of azimuthal anisotropy in Tibet indicate similar directions for shear-wave splitting and crustal strain and suggest a coherent deformation of the entire lithospheric plate. However, SKS anisotropy in mountain regions around Tibet shows a very complicated pattern.
- In many strike-slip regions the fast direction of anisotropy is parallel to the strike-slip motion. The results from the San Andreas Fault are controversial, probably due to the presence of several anisotropic layers (Vinnik *et al.*, 1989, 1992).
- Other tectonically active regions such as western North America display a very complex anisotropy pattern which is apparently not well related to surface tectonics but may be controlled by flow deeper in the mantle (e.g. Fig. 3.114).
- In some continental rifts and oceanic spreading centers (the Baikal Rift, the Red Sea, the southern Mid-Atlantic Ridge, the East Pacific Rise, Iceland) the fast direction of anisotropy (and inferred mantle flow) is often parallel to the direction of extension in the crust (Sandvol *et al.*, 1992; Vinnik *et al.*, 1992; Forsyth *et al.*, 1998). In other rifts and spreading centers (the Rio Grande Rift, the East African Rift, and the East Pacific Rise according to several studies) the fast direction is parallel or subparallel to the tectonic structure, suggesting either olivine alignment in ridge-parallel flow or anisotropy produced by fluid-filled cracks (Gao *et al.*, 1997).
- The anisotropy pattern in subduction zones is very complicated (see review by Savage, 1999). The mantle above and below the slabs has anisotropy between 0.5% and 2.0%, while anisotropy within the slab may reach 5%. Fast ScS polarizations perpendicular to the slab are observed in Japan and western South America. However, SKS data for stations above subduction zones often indicate fast polarizations parallel to the plate boundary (e.g. Shih *et al.*, 1991).
- The anisotropy of oceanic mantle is fossil and was formed at the spreading ridge with the fast axis parallel to the spreading direction as indicated by seismic data and by ophiolite sequences. However, in the Pacific mantle anisotropy is two-layered with frozen-in anisotropy in the shallow layer (down to 100 km) related to the fossil spreading direction and parallel to the present day plate motion at greater depths (Montagner and Tanimoto, 1991). The pattern for the Atlantic Ocean is controversial (Bjarnason *et al.*, 1996).

- Stable continents have a combination of frozen anisotropy in the lithospheric mantle and recently formed anisotropy in the asthenosphere and, globally, display a positive correlation between the directions of absolute plate motion and the fast direction of anisotropy (Vinnik *et al.*, 1992; Bokelmann, 2002). However, there are exceptions and controversies. In the cratons of Australia and North America anisotropy is two-layered: the shallow layer preserves fossil anisotropy, but anisotropy in the deeper lithosphere (below 150 km) correlates with the present plate motion (Yuan and Romanowicz, 2010). Surface-wave studies (see next section) indicate that the upper mantle beneath the Baltic Shield may also have a two-layer anisotropy, but the direction of azimuthal anisotropy in the deep layer differs significantly from the present direction of absolute plate motion (Bruneton *et al.*, 2004).

Surface-wave anisotropy

Preamble

Studies of polarization anisotropy of surface waves are based on comparisons between propagation of Love and Rayleigh waves. Rayleigh waves are produced as interference waves by the interaction of incident P and SV waves with the free surface and are mostly sensitive to SV velocity, whereas Love waves are produced by interference of reflections of SH waves generated at the free surface and trapped in a near-surface waveguide (e.g. Lay and Wallace, 1995). Therefore, separate inversions of Love waves and Rayleigh waves provide insight into the radial anisotropy of the upper mantle.

Simple symmetry systems are commonly used to interpret seismic anisotropy, in particular because fundamental modes are sensitive to a limited number of elastic constants (Montagner and Tanimoto, 1991). Most techniques can distinguish velocity anisotropy only in two orthogonal directions, e.g. vertical and horizontal, which is characterized by V_{SH}/V_{SV} (see Section 3.1.5 and Fig. 3.15). Radial anisotropy (hexagonal, or cylindrical, anisotropy with a vertical symmetry axis and phase velocity independent of the propagation azimuth) explains the polarization anisotropy of Love and Rayleigh waves and the differences between SH and SV arrivals (Anderson, 1961). This type of anisotropy with a vertical symmetry axis is best explained by horizontal shearing (flow).

Radial shear-wave anisotropy is present in the anisotropic version of the PREM model (Dziewonski and Anderson, 1981), where it is confined to the upper mantle down to 220 km depth with maximum anisotropy just below the Moho and its monotonic decrease with depth (Fig. 3.116). Other 1D reference models (such as *iasp91*, *ak135*) include similar anisotropy for P- and S-waves in the upper mantle with isotropic mantle beneath. Three-dimensional tomographic models calculated as a perturbation with respect to PREM display globally averaged anisotropy in the upper mantle that is similar to the reference model. However, tomographic inversions based on isotropic starting models without the 220 km discontinuity display an anisotropic pattern of shear velocities in the upper mantle that differs significantly from PREM (e.g. Nettles and Dziewonski, 2008).

A brief remark on comparison of body-wave and surface-wave anisotropy is appropriate (Montagner *et al.*, 2000). In the simplest case of a horizontal fast symmetry axis, upper mantle anisotropy constrained by surface-wave dispersion curves and body-wave SKS data can be explained by the same three anisotropic parameters. Although it allows for direct comparison

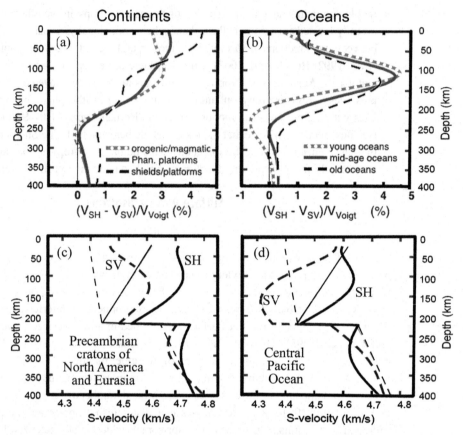

Fig. 3.116 Depth profiles showing average shear-velocity anisotropy in the upper mantle of continents and oceans constrained by surface-wave tomography (a and b – based on Nettles and Dziewonski, 2008; c and d – based on Ekström and Dziewonski, 1998). (a) and (b): Average anisotropy profiles calculated for the continental and oceanic regions (for tectonic regionalization of Jordan, 1981). (c) and (d): average velocity profiles of shear-wave velocities SH (solid lines) and SV (dashed lines) beneath the Precambrian cratons of North America and Eurasia and beneath the central part of the Pacific plate. In all other oceanic plates except the Pacific plate, the average SH and SV velocities in the upper mantle are close to that of the PREM model. Thin solid and dashed lines – SH and SV velocities in the anisotropic PREM model. The step in seismic SH and SV around 220 km depth is a consequence of the PREM reference model.

of the SKS delay time and surface-wave anisotropy tomography, large-scale comparison between the two approaches happens to be disappointing due to the limited areal coverage for both data types. In particular, surface-wave tomography is commonly used for studies of suboceanic mantle, while most SKS splitting data are available only for subcontinental mantle. However, good agreement between surface-wave and body-wave upper mantle anisotropy was found in tectonically active regions, such as the western United States and in Central Asia. Comparison of surface-wave tomography and body-wave SKS data is further complicated by different vertical and lateral resolutions of the two different data types. In a way, the two approaches are complementary since body-wave SKS data provides high lateral

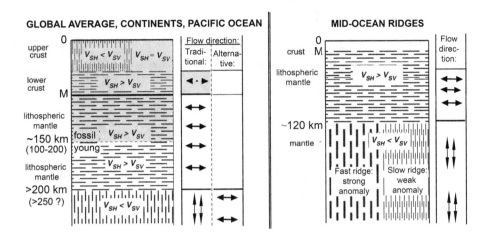

Fig. 3.117 Schematic summary of a typical depth distribution of anisotropy styles in the upper mantle. Left panel: the global average and anisotropy beneath stable continents and the Pacific Ocean. Right panel: anisotropy patterns beneath fast- and slow-spreading mid-ocean ridges.

resolution (but cannot determine the depth distribution of anisotropy), while surface-wave tomography provides constraints on the depth of anisotropy (with certain limitations, such as for example depth leakage associated with tomographic inversion, Fig. 3.80). The major results from surface-wave anisotropy studies of the upper mantle are summarized below.

Continents

Major results for shear-wave anisotropy in stable continents (Precambrian shields and platforms) include observations that polarization anisotropy extends down to at least 200 km depth. The depth extent of seismic anisotropy in the upper mantle is the subject of strong debate (e.g. Gung *et al.*, 2003; Mainprice *et al.*, 2005). It has some bearing on the controversy of whether mantle anisotropy is due primarily to strain at the base of the plate in the asthenosphere (Vinnik *et al.*, 1995), or rather is due to remnant, orogenic events that have been frozen into the lithosphere (Silver and Chan, 1991).

Figures 3.116 and 3.117 summarize major results of surface-wave anisotropic tomography. Strong shear-wave anisotropy reported for the upper mantle down to 200–250 km depth is a robust feature revealed by all tomographic studies. Even at 250–350 km depth, surface-wave anisotropy is significantly stronger beneath the cratons than in any other regions (Fig. 3.116). Maximum anisotropy is observed in the shallow mantle and gradually decreases with depth. For shear waves horizontally propagating in this anisotropic upper mantle layer, horizontally polarized waves, SH, have faster velocities than vertically polarized waves, SV. Radial anisotropy with a vertical symmetry axis produced by horizontal shearing is a widely accepted model for these observations. Studies of mantle-xenoliths from different cratonic settings indicate that both in spinel-peridotites and garnet-peridotites deformation in olivine occurs by dislocation creep with dominant [100] slip down to a depth of at least 150 km.

Below 220–250 km the upper mantle becomes weakly anisotropic with a strong decrease in anisotropy with depth and with a change of seismic anisotropy pattern to $V_{SH} < V_{SV}$ (in particular

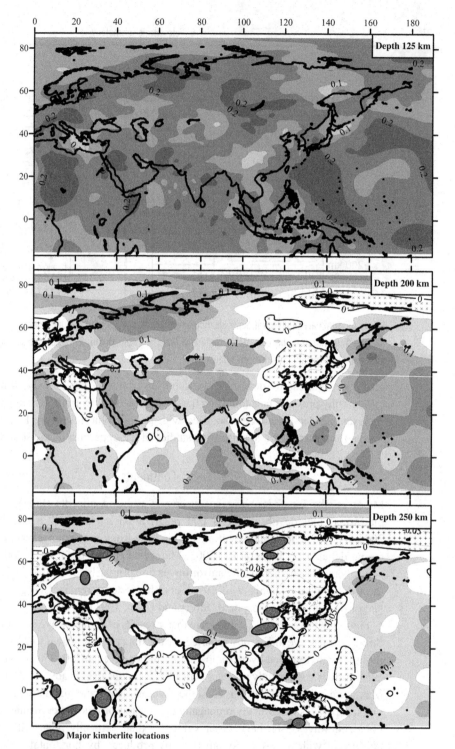

Fig. 3.118 The difference between the SH and SV velocity perturbations at 125 km, 200 km, and 250 km depth beneath Eurasia, northern Africa, and the western Pacific Ocean (based on the surface-wave tomography model of Kustowski *et al.*, 2008). Gray shading – regions where $V_{SH} > V_{SV}$; hatching – regions where $V_{SH} < V_{SV}$. Black shading on the lower map – locations of major kimberlite fields (based on the database of Faure, 2006).

beneath the Precambrian cratons and the Central Pacific Ocean). However, beneath the oceanic basins of the Pacific and Indian plates, anisotropy with $V_{SH} > V_{SV}$ may persist down to 250–300 km depth (Lévêque *et al.*, 1998).

Global observations, that at *c.* 220–250 km depth the anisotropy sharply decreases and its pattern changes from $V_{SH} > V_{SV}$ to $V_{SH} < V_{SV}$, are commonly interpreted by a transition in deformation mechanism in mantle olivine from dislocation creep with dominant [100] slip to diffusion creep, and making the basis for interpretations of the origin of the Lehmann discontinuity (Fig. 3.44). Then 220–250 km depth would also correspond to the zone of mechanical decoupling between the lithosphere and the underlying mantle and can be interpreted as the base of the rheological lithosphere (Gung *et al.*, 2003).

This point of view is challenged by two arguments.

(1) If the continental lithosphere were indeed confined to the upper mantle layer with $V_{SH} > V_{SV}$, the results of recent surface wave tomography would indicate a *c.* 200 km thick lithosphere beneath the major kimberlite fields of the Siberian craton (Fig. 3.118). In conflict with this conclusion, the deepest xenoliths from the lithospheric mantle beneath Siberia come from a 240-250 km depth (Griffin *et al.*, 1996; Aschepkov *et al.*, 2003).

(2) Recent synthetic calculations and simple-shear laboratory measurements examine deformation of olivine aggregates at high pressure (11.8 GPa) and high temperature (1380 °C) which correspond to a depth of 355 km in the mantle (Mainprice *et al.*, 2005). The results indicate that dislocation creep is still possible under high P–T conditions, but with a different dominant slip: both [100] slip and [010] slip become normal to the shear plane, while [001] concentrates parallel to the shear plane and thus the [001] slip system becomes dominant. As a result, the fastest P-wave velocity is no longer parallel to the shear direction as is observed for the dominant [100] slip in the upper mantle above 220 km depth, but propagates at a high angle to the shear plane. If deformation of the upper mantle below 220 km depth still occurs by horizontal shearing, almost no azimuthal variation of P-wave velocity is expected and a vertically propagating P-wave is only slightly faster than a horizontally propagating P-wave. Similarly, the fastest shear waves are the waves polarized at a high angle to the shear plane and the largest delay times are expected for shear waves propagating at a high angle to the shear direction in the shear plane. If shear waves propagate in the shear plane, the fastest shear waves are polarized at a high angle to the shear plane. This implies that for horizontal shearing, $V_{SH} < V_{SV}$. The results also indicate that the transition from dislocation creep with the dominant [100] slip to dislocation creep with the dominant [001] slip results in a significant decrease in anisotropy to *c.* 0.9% and 1.9% for compressional and shear waves, respectively.

These results have important implications for the thickness of the rheologically strong upper mantle layer (the lithosphere):

- the upper mantle even below 220–250 km depth can be deformed by dislocation creep;
- the transition from strong to weak anisotropy at 220–250 km depth may not reflect the extent of mechanical coupling between the lithosphere and the asthenosphere, but may reflect a change in slip direction in dislocation creep;

- the upper parts of the upper mantle can be strongly coupled with the deeper mantle down to the transition zone.

In general agreement with SKS studies, shear-wave polarization anisotropy in the subcontinental upper mantle provides evidence for the existence of two (or perhaps, several) anisotropic layers with different orientations of the fast polarization axes. A surprisingly complex structure with considerable variations in radial and azimuthal anisotropy has been revealed by analysis of fundamental-mode Rayleigh and Love waves and P-waves under the SVEKALAPKO array in the Baltic Shield (e.g. Plomerová et al., 2005; Bruneton et al., 2004). Although azimuthal anisotropy down to c. 200–250 km depth is negligible, data on P-wave residuals indicate a complex anisotropy pattern in the area. The apparent absence of azimuthal anisotropy can result from different orientations of the olivine fast axis beneath the array. Significant azimuthal anisotropy appears below 200–250 km depth, but its direction (roughly 0–40°N) differs significantly from the present direction of absolute plate motion (55–62°N). A possible explanation of the discrepancy between the directions of the plate motion and anisotropy in the lower parts of the lithosphere is that it reflects fossil flow direction, while the short time elapsed since the change in the flow direction was not sufficient to change the olivine alignment in the case of simple shear (Kaminski, 2002).

Seismic anisotropy preserves information on multi-episodic tectonic activities in the subcontinental lithosphere, such as cratonic amalgamation and later tectonic deformations and rejuvenation. For example, a recent high-resolution seismic tomography and shear-wave splitting study of the Precambrian Canadian Shield revealed the presence of two distinct lithospheric domains with a boundary at approximately 86–87°W which corresponds to the tectonic boundary between the Wabigoon and Wawa/Quetico provinces (Frederiksen et al., 2007). Strong anisotropy in the lithospheric mantle is observed only to the west of this boundary, while to the east the average split times are half as long and the average velocity anomaly is smaller by 2.5%.

Regional tomographic studies in western Europe reveal the presence of paleosubduction zones in the European lithosphere associated with continental growth during the Paleozoic orogeny (Babuska et al., 1998). The lithosphere beneath the Armorican Massif in NW France consists of two seismologically distinct domains that have different seismic velocities and different anisotropy down to ~150 km depth. These lateral variations have been interpreted as resulting from collision-related subduction-type processes that predated the Variscan orogeny (Judenherc et al., 2002). In young orogens the largest difference between the SH and SV waves is observed in the upper 150 km with a maximum at c. 100–150 km depth (Fig. 3.116). However, since lateral resolution of teleseismic surface-wave tomographic models is low, tectonically young structures with small lateral dimensions are better resolved by body-wave SKS data.

Oceans

The oceanic upper mantle is highly anisotropic with the strongest SH–SV anisotropy at around 100–120 km depth (Fig. 3.116). The dependence of anisotropy on the age of the ocean floor is insignificant down to c. 150 km depth, below which the difference between SH and

SV velocities becomes age dependent with higher values for old oceans and $V_{SH} > V_{SV}$ in young oceans below 200 km depth. Below 300 km depth suboceanic upper mantle, in general, becomes isotropic. However, significant $V_{SH} < V_{SV}$ anomalies are observed beneath the Mid-Atlantic Ridge and the Red Sea down to at least 400 km depth (Zhou *et al.*, 2006).

Unique upper mantle anisotropy is observed beneath the Pacific plate (Ekström and Dziewonski, 1998). In contrast to PREM and to other oceanic plates that display SH and SV close to PREM down to 400 km, the Pacific Ocean displays anomalous anisotropy in the upper mantle. Above the 220 km discontinuity it does not show a smooth decrease with depth: polarization anisotropy is small in the shallow mantle and increases with depth down to 150–180 km where maximum anisotropy is confined to a relatively narrow layer (Fig. 3.99). Significant anisotropy is also observed at 250–350 km depth.

The most anomalous mantle is beneath the Central Pacific plate around the East Pacific Rise, where strong anomalous (as compared to PREM) anisotropy extends from *c.* 50 km (where SH and SV velocities are very close) to 220 km depth. The maximum difference between SH and SV velocities (*c.* +5%) is in the upper mantle southwest of Hawaii. Large geographical variations in upper mantle anisotropy beneath the Pacific plate are not correlated with ocean floor age; however the isotropic part calculated from SH and SV velocities displays a clear correlation with age (except for very low velocities beneath the Pacific Superswell). A possible explanation for the unique anisotropy of the Pacific mantle includes strong olivine alignment by small-scale convection flow (Ekström and Dziewonski, 1998).

A recent finite-frequency surface-wave tomography study of global shear-wave velocity structure and radial anisotropy in the upper mantle with a high resolution of small-scale mantle heterogeneities reveals a peculiar pattern of anisotropy beneath the spreading centers (Zhou *et al.*, 2006). Significant $V_{SH} < V_{SV}$ anomalies are observed beneath mid-ocean ridges at depths greater than 120 km with distinctly different patterns beneath slow- and fast-spreading ridges (Fig. 3.117). Polarization anisotropy beneath fast-spreading ridges is stronger, but rapidly decreases below 250 km depth, while beneath slow-spreading centers (the northern Mid-Atlantic Ridge, the Red Sea), weak anisotropy extends down to at least the top of the transition zone. This observation implies that the primary driving force of seafloor spreading may be different beneath fast- and slow-spreading ridges.

3.6.5 Lithosphere thickness from elastic tomography

Definitions and uncertainties

The base of the lithosphere is a diffuse boundary that stretches over a particular depth interval and thus any practical definition of its base is somewhat arbitrary (see also Section 3.3.2). The seismic lithosphere (or the lid) is defined as the layer above a rheologically weak zone caused by high temperatures and/or partial melting and in tomographic studies its base is defined as:

(1) the bottom of the seismic high-velocity region on the top of the mantle that generally overlies a low-velocity zone;

(2) a depth where the mantle anisotropy pattern changes from frozen-in anisotropy in the lithosphere to anisotropy in the asthenosphere caused by plate motion and mantle flow.

The first definition of the lithosphere as the lid above a LVZ is based on the common assumption that the LVZ corresponds to the asthenosphere where partial melting occurs. However, a LVZ can be solid state associated with mineralogy, and thus the top of a LVZ may not necessarily be associated with the lithosphere–asthenosphere boundary (see discussion for Fig. 3.130).

In practice, the first approach is used when the velocity structure of the upper mantle is known, so that the base of the lithosphere is defined either by the zone of high velocity gradient in the upper mantle or by the bottom of the layer with positive velocity anomalies that should correspond to the transition from a high-velocity lithosphere to LVZ. Use of velocity gradients in the mantle for defining the lithospheric base is hampered by a wide use of reference models in tomographic studies; in particular continental models based on PREM inherit a velocity gradient at around 220 km depth, not necessarily required by the data. Furthermore, absolute velocities may not always preserve a gradient zone in the upper mantle seen in relative velocity perturbations (e.g. Fig. 3.85).

The base of the lithosphere is not marked by a sharp change in temperature or in composition, and thus it is not marked by a sharp change in seismic velocities. Since the lithospheric base has a diffuse character with a transition zone where seismic velocities gradually decrease from lithospheric to asthenospheric values, the lithospheric thickness depends significantly on the amount of velocity perturbation chosen to define the lithospheric base. Commonly the base of the seismic lithosphere is determined as the bottom of the layer where positive velocity anomalies drop below +1% or +2% with respect to a reference model. The difference in lithospheric thickness estimates based on a +1% or a +2% velocity anomaly may be significant. For example, the tomographic cross-section for Africa (Fig. 3.90) illustrates that the lithosphere thickness beneath West Africa is *c*. 200 km if defined by a +2% velocity anomaly, but may extend down to *c*. 400 km depth if defined by a +1% velocity anomaly.

Further problems arise when comparing results of different studies, since there may be a significant difference between reference models, so that the same relative velocity perturbation corresponds to different absolute velocities. While PREM is a widely used reference model for global studies and for the oceanic regions, *ak135* or similar reference models (e.g. *iasp91*) are more suitable for studies of the continents.

Another problem is related to the amplitude of velocity perturbations (used to define the lithospheric base) which depends on the regularization used in tomographic inversion to stabilize the inverse problem (in particular, in poorly constrained regions) (see Section 3.2.3). In particular, damping permits rough models with large velocity perturbations, while roughness control (i.e. the absence of strong contrasts in adjacent blocks) results in models with smoothly varying heterogeneity. Combined with an arbitrary chosen (say, +1% or +2%) velocity perturbation defining the lithospheric base, ambiguity in lithosphere thickness determinations can hardly be assessed.

Depth resolution of tomographic models puts additional limitations on thickness constraints, and in many inversions the details of the velocity structure cannot reliably be resolved (Fig. 3.119). The near-vertical propagation of body waves in the upper mantle limits their vertical resolution obtained in tomographic studies to 50–100 km. Resolution of surface-wave tomographic models that use only fundamental modes rapidly decreases with

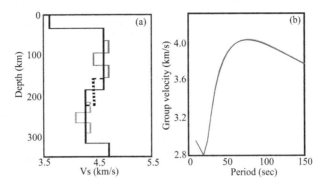

Fig. 3.119 Inversion tests for surface-wave tomography showing: (a) velocity profiles and (b) dispersion curves for tests of velocity variability with depth (gray lines) and of LAB discontinuity sharpness (dotted line). Note that neither velocity variations in the upper mantle nor variations in lithosphere thickness can be distinguished in dispersion curves from the initial velocity profile (black solid line in (a)) (after Pasyanos, 2009).

depth, and models based on Rayleigh waves at 40–60 s cannot resolve mantle structure below *c*. 200 km depth (Fig. 3.79). Furthermore, resolution kernels indicate that tomographic inversions provide velocity structure averaged over a significant depth interval in the upper mantle (Fig. 3.81). The worst resolved is the depth between 300 and 400 km. Depth leakage of velocity anomalies, an inevitable consequence of inversion, further complicates estimates of lithosphere thickness (Fig. 3.80).

Finally, uneven ray path coverage may further bias results of tomographic inversions (Fig. 3.83). This problem is particularly important at the transitions between structures with a high contrast in mantle velocities and lithospheric thickness (such as the ocean–continent transition).

Global patterns

With all of the uncertainties in defining the thickness of the seismic lithosphere, any reported absolute numbers should be treated with caution, while a comparative analysis of different tectonic regions is more helpful in recognizing robust global patterns. Figure 3.120 shows the thickness of seismic lithosphere defined by a +1% and +2% shear-wave velocity perturbations with respect to the reference models derived from a global body-wave tomography model (Grand, 1997, 2002) and a global surface-wave tomography model (Shapiro and Ritzwoller, 2002). Since the maps are smoothed by interpolation, the lithosphere thickness appears underestimated for deeply extending continental roots and overestimated for regions with thin lithosphere. In spite of different vertical and lateral resolution of the two tomographic models, they are consistent in depicting general patterns in lithosphere thickness variations.

In the oceanic regions seismic lithosphere is typically thinner than 100 km. According to surface-wave models, the lithospheric thickness shows a pronounced correlation with the age of the ocean floor and increases from near-zero values in spreading centers to *c*. 80 km in old oceans (Fig. 3.121). The pattern is not observed in body-wave models, perhaps due to lack of vertical resolution and geographical limitations.

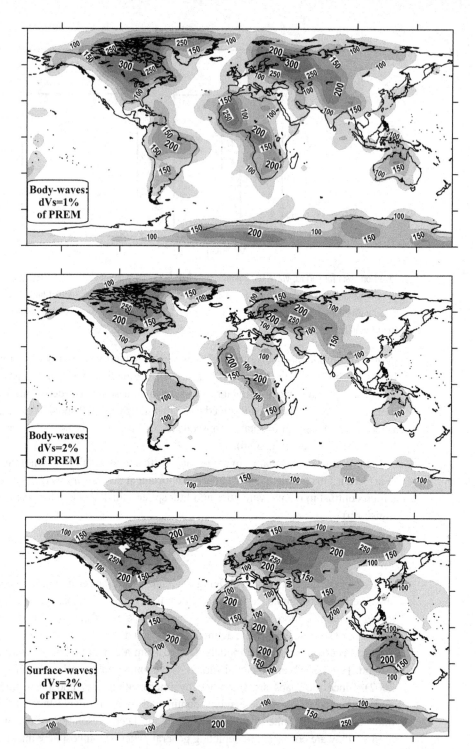

Fig. 3.120 Thickness of seismic lithosphere based on a global body-wave tomography model (Grand, 1997, 2002) and a global surface-wave tomography model (Shapiro and Ritzwoller, 2002). The base of the lithosphere is defined by a 1% (upper map) and 2% (middle and lower maps) shear-wave velocity perturbation with respect to the reference model (PREM for the upper and middle maps and *ak135* for the lower map). The maps are smoothed by a Gaussian filter. As a result, the thickness is underestimated for deeply extending continental roots and overestimated for regions with thin lithosphere. In most of the oceans, seismic lithosphere is thinner than 100 km. In stable continents, lithosphere thickness may exceed 300 km.

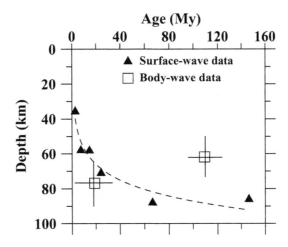

Fig. 3.121 Thickness of the oceanic lithosphere versus age of the ocean floor (data of Forsyth, 1975; Revenaugh and Jordan, 1991; Gaherty *et al.*, 1996) (based on Karato, 2003b).

Below the continents, lithosphere thickness is correlated with surface tectonics and crustal ages (compare with Fig. 2.16). However because of significant variations in velocity structure in tectonic provinces of any age, the age dependence is much less pronounced than for the suboceanic lithosphere. In general, thick lithosphere is observed in old continents and thin lithosphere in tectonically young structures (except for zones of continent–continent or continent–ocean collision where subducting lithosphere slabs can be traced deep into the mantle). The deepest lithospheric roots are clearly associated with the cratons.

The depth extent of the cratons as constrained by tomographic models is a subject of ongoing debate (Fig. 3.122). An analysis by Polet and Anderson (1995) shows that the depth extent of continental lithospheric roots is strongly age dependent. If the lithospheric base is defined by a +1% Vs anomaly, the Archean and Paleoproterozoic cratons typically have lithospheric roots that are 240–280 km thick, whereas the lithosphere beneath Meso-Neoproterozoic terranes is only about 150 km thick (for a +0.5% Vs anomaly the lithosphere is about 60 km thicker). Furthermore, cratons of the same age have a different depth extent of the lithospheric roots. In particular, seismic tomography models indicate that Archean–Paleoproterozoic cratons that have undergone Phanerozoic reworking (e.g. the Sino-Korean craton) have lost deep lithospheric roots.

Depth variations in seismic anisotropy have been proposed to define the lithospheric thickness (Babuska *et al.*, 1998). In this approach the lithosphere–asthenosphere boundary (LAB) is considered to be the transition from 'frozen-in' anisotropy in the lithosphere to mantle flow-related anisotropy in the asthenosphere. Since these two anisotropic zones may have different alignment of mantle olivine, the depth where a change in the direction of anisotropy occurs can be interpreted as the LAB. Based on this definition, the lithospheric thickness is estimated to be *c.* 350 km beneath the Superior province in the Canadian Shield and the Siberian craton, and *c.* 220–250 km beneath the Baltic Shield and the Hudson Bay in Canada. This approach is complicated by recent studies that indicate that many cratons may have (at least) a two-layer anisotropy with significantly different orientations of the fast axis.

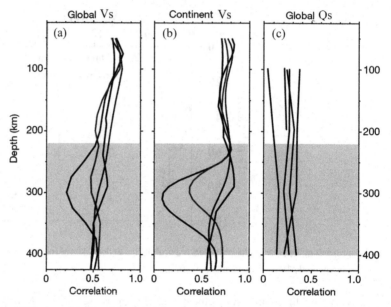

Fig. 3.122 Correlation coefficient as a function of depth between five global tomographic shear velocity (a and b) and shear attenuation (c) models (after Gung *et al.*, 2003; Dalton *et al.*, 2008). For S-velocity models correlation is computed between an SH model SAW24B16 and models S20A_SH9 (an SH model), SB4L186 (a hybrid model), S20A_SV9 and S20RTS7 (both are SV models) (a) over the whole globe and (b) over the continental areas only (defined as all areas of elevation >−500 m to include the shelves). Note that in the continental regions correlation between SH/hybrid models and SV models is strongly reduced below 250 km depth. For shear attenuation models correlation (c) is computed between model QRLW8 and models WS02, MQCOMB, SW02, QR19, and QRFSI12. Correlation between all global shear attenuation models is weak.

Surface-wave tomographic models are being widely used to assess depth variations in SV and SH waves (e.g. Gung *et al.*, 2003). As discussed in the previous section, the depth extent of the cratonic roots as constrained by polarization anisotropy is strongly debated. Most tomographic models observe strong anisotropy only down to a 200–250 km depth and the change from $V_{SH} > V_{SV}$ to $V_{SH} < V_{SV}$ at the same depth. This change in anisotropy pattern is commonly interpreted as the transition from dislocation to diffusion creep. If this is correct, the upper part of the mantle becomes mechanically decoupled below this depth from the deeper mantle, and the depth of 200–250 km corresponds to the base of the lithosphere. As discussed in detail in Section 3.6.4, this conclusion is non-unique and is questioned by laboratory studies on olivine deformation at P–T conditions which correspond to a 300–350 km depth (Mainprice *et al.*, 2005). Experimental results suggest that the change from $V_{SH} > V_{SV}$ to $V_{SH} < V_{SV}$ does not necessarily imply mechanical decoupling of the layers, which can remain coupled down to the transition zone. Thus, the question of whether the change in anisotropy at 200–250 depth beneath the continents may be interpreted as the lithospheric base, remains open.

3.6.6 Anelastic tomography

General remarks

Studies of seismic attenuation in the Earth's mantle began in the 1960s (Anderson and Archambeau, 1964; Ben-Menahem, 1965; Anderson et al., 1965); but global tomographic models of seismic attenuation have only been calculated recently (Romanowicz, 1994, 1995; Bhattacharyya et al., 1996; Durek and Ekström, 1996; Reid et al., 2001; Selby and Woodhouse, 2002; Dalton et al., 2008). A significant number of papers have focused on regional studies, in particular for the oceans. These papers as well as the review papers of Mitchell (1995) and Romanowicz and Durek (2000) provide a comprehensive overview of attenuation studies of the crust and the upper mantle.

Seismic waves attenuate with travelled distance faster than predicted by geometrical spreading of wave fronts. This energy loss is caused by anelastic processes (internal friction, or intrinsic attenuation). Anelasticity is generally quantified in seismology by the quality factors, Q_p and Q_s, for compressional and shear waves, respectively. The loss of seismic energy during wave propagation is quantified by the inverse quality factor, or internal friction. Scattering attenuation (caused by scattering off small-scale heterogeneities, focusing/defocusing, refraction and reflection) also reduces the amplitude of seismic waves, but does not change the integrated energy in the total wavefield.

Laboratory studies indicate that seismic attenuation is sensitive to seismic frequency, temperature, and grain size (see Section 3.1). The strong temperature dependence of attenuation (eq. 3.9–3.10) is of fundamental interest for geophysics since it provides higher sensitivity of anelastic tomography to elevated temperatures and melts than elastic tomography. Tomographic inversions based on surface waves provide values of seismic Q_s that can be converted into Q_p (Anderson et al., 1965):

$$Q_p^{-1} = 4/3\, V_s^2 / V_p^2 Q_s^{-1}. \qquad (3.17)$$

For the upper mantle velocities and at near-solidus temperatures $Q_p = 2.25\, Q_s$. In shallow low-temperature mantle, however, thermo-elastic mechanisms of attenuation lead to $Q_p \sim Q_s$ (Frankel, 1982). In the globally spherical PREM model the radial average Q_s has a value of 600 at depths between 40 km and 80 km and of 80 between 80 km and 220 km depth.

Anelastic tomography suffers from the same limitations and uncertainties as elastic tomography (see Section 3.6.1). Additional uncertainties are caused by scattering attenuation, since commonly the effects of intrinsic and scattering attenuation are difficult to separate. On the whole, the amplitudes of surface wave attenuation depend on:

(1) the intrinsic attenuation along the ray path,
(2) elastic focusing/defocusing effects along the ray path (Selby and Woodhouse, 2002),
(3) uncertainties in the strength of excitation at the source,
(4) uncertainties in the response at the station (e.g. Dalton et al., 2008), and
(5) the damping used in the inversion.

Simultaneous inversion of amplitude data for models of attenuation and seismic velocities facilitates separation of focusing effects on amplitude anomalies and thus improves amplitude resolution of anelastic models. Similarly to elastic tomography, crustal corrections have a strong effect on the amplitude of attenuation, in particular in shallow mantle.

Correlations with surface tectonics

Global and regional models of attenuation in the crust and upper mantle differ significantly, in particular in amplitude (Fig. 3.122). However, a large span of calculated attenuation values (>2 orders of magnitude) allows for meaningful comparison of different models and recognition of global and regional trends despite large uncertainties in attenuation amplitudes.

Similarly to elastic tomography, upper mantle seismic attenuation exhibits a strong correlation with surface tectonics (Fig. 3.123). Stable continents have high Q values (low attenuation) down to at least around 150 km depth. Apparent exceptions include cratons of South Africa, South America, India, and China. However, attenuation anomalies beneath these cratons may not be resolved due to their small lateral sizes: lateral resolution of global tomographic models is limited by spherical harmonic representation; most of the models are constrained in degrees up to 8–12 which corresponds to c. 3000–5000 km and, at present, is lower than the lateral resolution of modern elastic tomography models. A recent attenuation model QRFSI12 constrained up to degree 12 by fundamental mode Rayleigh waves at periods between 50 and 250 s (Dalton *et al.*, 2008) shows a similar pattern and displays low attenuation beneath the stable continental interiors of Europe, Siberia, North America, Greenland, Australia, west-central Africa, and Antarctica down to at least 200–250 km depth with a possible weaker continuation of high-Q anomalies down to 300–400 km depths. Since some smearing of the signal may occur below 200–250 km depth (e.g. Fig. 3.79), the model cannot resolve if the correlation between attenuation and surface tectonics is restricted to the upper 200–250 km (Fig. 3.124).

Low attenuation is found beneath Tibet–Himalayas, in particular at c. 100–150 km depth (Bhattacharyya *et al.*, 1996). Phanerozoic platforms such as western Europe and West Siberia and tectonically active continents such as western North America have a higher upper mantle attenuation than cratonic upper mantle. A recent study of the coda of Lg waves in Eurasia (Mitchell *et al.*, 2008) further supports the conclusion that upper mantle attenuation structure is strongly correlated with surface tectonics and that the Q value is proportional to the tectono-thermal age (i.e. the time since the most recent episode of major tectonic or orogenic activity).

The strongest attenuation is observed beneath young oceans that have high attenuation through the entire period range (Mitchell, 1995). "Normal" oceanic regions display a correlation between the age of the ocean floor and attenuation in the upper mantle. Strong attenuation anomalies are associated with the East Pacific Rise, Mid-Atlantic Ridge, and other ridge systems down to 150–200 km and perhaps deeper. For example, a regional shear-wave study displays high attenuation in the Pacific Ocean down to 100–150 km depth

50 km depth

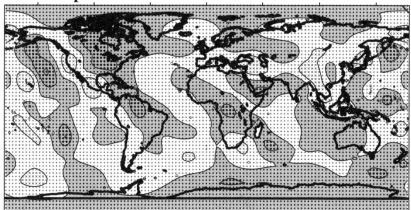

100 km depth

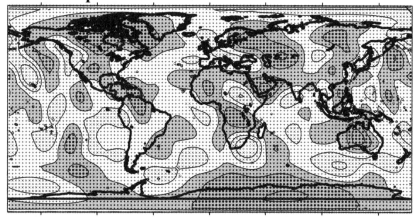

150 km depth

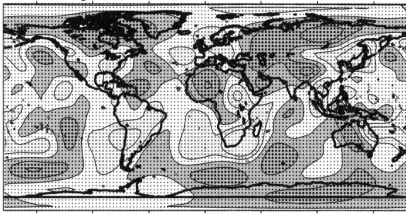

Fig. 3.123 Inverse shear attenuation in the upper mantle (as percent of the PREM value) (based on the tomography model of Billien *et al.*, 2000). The anelastic tomography model is calculated from fundamental-mode Rayleigh waves at periods between 40 and 150 sec and includes both anelastic and scattering effects (the effect of focusing due to velocity heterogeneities has been accounted for). The model is represented by spherical harmonics up to degree 12. Gray shading – positive anomalies.

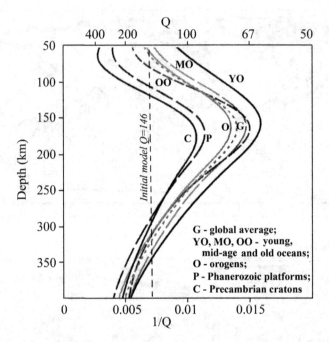

Fig. 3.124 Regionally averaged shear attenuation profiles for six tectonic provinces and the global average (short dashes) based on surface-wave global tomographic modeling (the reference model has constant $Q_s = 146$ throughout the mantle) (based on Dalton *et al.*, 2008).

(Fig. 3.125). However, Q anomalies beneath the Pacific Ocean display a highly asymmetric pattern which becomes more pronounced at 300–450 km depth (Fig. 3.126). Joint interpretations of attenuation and shear velocity models for the Pacific Ocean indicate that melt may be present at 25–100 km depth beneath the MELT (the Mantle Electromagnetic and Tomography) seismic array (age of ocean floor *c.* 2 My) and at 40–100 km depth beneath the GLIMPSE area about 400 km away from the East Pacific Rise.

Correlations with temperature and surface heat flow

Strong, Arrhenius-type, exponential dependence of attenuation on temperature as indicated by laboratory measurements (e.g. Berckhemer *et al.*, 1982; Kampfmann and Berckhemer, 1985; Sato *et al.*, 1989; Gribb and Cooper, 1998; Jackson, 2002) implies that regions of high attenuation in the mantle should be correlated with high mantle temperatures. The temperature dependence of seismic attenuation has been used in a number of studies to calculate mantle geotherms from seismic attenuation (e.g., Kampfmann and Berckhemer, 1985; Sato *et al.*, 1988; Sato and Sacks, 1989).

A global analysis indicates that in the upper 250 km, attenuation anomalies calculated from the amplitudes of low-frequency Rayleigh waves on a $10° \times 10°$ grid (the QR19 model) are, to some extent, correlated with surface heat flow (the correlation coefficient r is 0.20–0.35 with the correlation peak at $z \sim 200$ km). Deeper in the mantle (below the

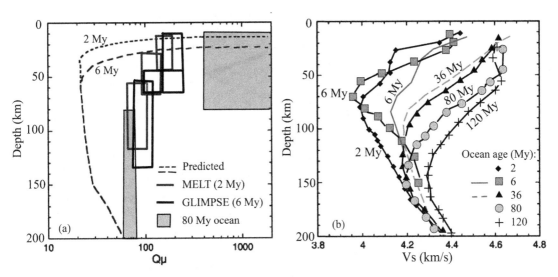

Fig. 3.125 Quality factor (a) and shear velocity (b) as a function of depth and age of seafloor in the Pacific Ocean (based on Yang *et al.*, 2007). (a) Quality factor: empty boxes show range of acceptable quality factor based on MELT and GLIMPSE seismic arrays which have resolution down to 100 km and 150 km depth, respectively. Gray boxes – average oceanic values; dashed lines – theoretical predictions for a simple half-space cooling model based on the data of Faul and Jackson (2005). (b) SV component of shear velocities: different symbols refer to ocean floor with different ages (tomographic model of Nishimura and Forsyth, 1989). Gray lines – theoretical predictions for a simple half-space cooling model. The difference between observed and predicted values at 50–80 km depth beneath a young ocean is attributed to the presence of melts.

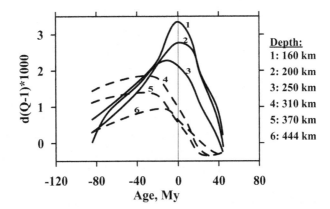

Fig. 3.126 Dependence of the *Q* anomaly (degree-6 model QR19) on seafloor age in the southern Pacific Ocean (based on Romanowicz, 1998) at six different depths: 160–250 km (solid lines) and 310–444 km (dashed lines). Note that the symmetry around the ridge is broken at any depth, and in particular in the deeper depth range.

transition zone) *Q*-anomalies are better correlated with hotspot distribution ($r \sim 0.35$–0.38) (Fig. 3.127). This conclusion is supported by global anelastic models which display low seismic attenuation in the upper mantle beneath stable continents where mantle temperatures are low (e.g. Romanowicz, 1994, 1995; Bhattacharyya *et al.*, 1996). Regional seismic

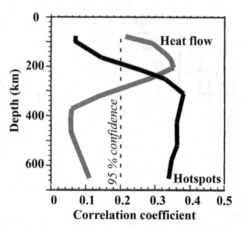

Fig. 3.127   Global correlations of shear attenuation model QR19 (up to degree 6 spherical harmonics) with surface heat flow and hotspot distribution (after Romanowicz, 1995).

studies of shear-wave attenuation also reveal a qualitative correlation between regional attenuation anomalies in the crust and upper mantle and tectonic provinces, thus supporting the idea of the thermal origin of most of the upper mantle Q_s anomalies (e.g., Canas and Mitchell, 1978; Durek and Ekström, 1996; Sarker and Abers, 1998, Selby and Woodhouse, 2002; Mitchell *et al.*, 2008).

Note that relatively high global correlations are based mostly on high attenuation anomalies documented for oceanic regions such as mid-ocean ridges (Fig. 3.127). Although an essential part of attenuation anomalies in the oceanic upper mantle can be explained by temperature variations, seismic studies of the North Atlantic and western Pacific subduction zones (Nakanishi, 1978; Sheehan and Solomon, 1992; Tsumura *et al.*, 2000) suggest that partial melting is important in producing high attenuation anomalies.

Statistical analysis shows that, globally on the continents, correlation between quality factor Q_s and upper mantle temperature is significantly weaker than one might expect from laboratory experiments (Fig. 3.128). Although a weak correlation between quality factor and temperature can, in part, result from different lateral and vertical resolutions of seismic and thermal models, non-thermal effects, such as compositional variations and the presence of fluids and melts, play an important role in producing attenuation anomalies in the upper mantle (Artemieva *et al.*, 2004). Studies of the Lg coda wave attenuation for Eurasia confirm this conclusion: while most attenuation anomalies are apparently related to the tectonic history of the Eurasian crust, attenuation in Eurasia is most easily explained by fluid flow or by scattering from fluid-enhanced zones that may be associated with past subduction zones (Mitchell *et al.*, 2008).

Correlations with seismic velocities

Correlation between seismic elastic and anelastic properties has been examined in a number of seismic and laboratory studies. Both global and regional anelastic tomography models for the oceans indicate that anomalies of seismic velocity and attenuation are correlated in the

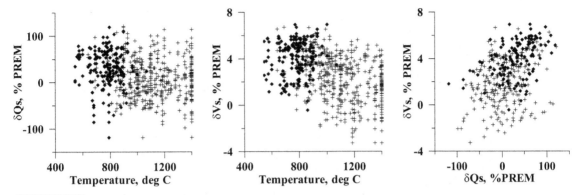

Correlation between inverse shear attenuation, shear velocity, and temperature at 100 km depth beneath continents (from Artemieva *et al.*, 2004). Note that attenuation and velocity were constrained by the same tomographic inversion (Billien *et al.*, 2000). Crosses — all continents; dots — cratonic regions only.

upper mantle, at least qualitatively (Romanowicz, 1995; Romanowicz & Durek, 2000; Sheehan & Solomon, 1992; Roth *et al*, 2000; Tsumura *et al*, 2000; Dalton *et al.*, 2008). A quantitative correlation is significantly hampered:

(1) by significantly different sensitivities of seismic attenuation and velocity to temperature (exponential and linear, respectively, Fig. 3.3), and
(2) by large uncertainties in amplitudes of attenuation anomalies resulting in a low global correlation (Fig. 3.128).

An attempt to quantitatively relate attenuation and velocity anomalies has been made for the Fiji–Tonga region (Roth *et al.*, 2000). The assumption behind the study was that both anomalies have a similar thermal origin. Theoretical calculations aimed to match seismic observations indicate that, at depths greater than 100 km, velocity and attenuation anomalies are strongly correlated and can be fit by an exponential equation, while at shallower depths seismic data show a high degree of scattering (Fig. 3.129). Analyses of the correlation between attenuation and shear-wave velocity anomalies globally and for the continents further indicate that the strong correlation between Qs and Vs observed in the Fiji–Tonga subduction zone is more a regional phenomenon than a general rule (Roth *et al.*, 2000). Comparison of seismic velocities and seismic attenuation with theoretical predictions based on laboratory measurements of the temperature sensitivity of attenuation and velocity (Faul and Jackson, 2005) indicate that temperature variations alone can only explain about half of the observed global variations in velocity and attenuation in the continental regions (Artemieva *et al.*, 2004). Similar global analysis suggests that the observed velocity–attenuation relationship agrees with the predicted values (calculated from temperatures) in the upper 100–150 km of the oceanic regions, while beneath the cratonic regions seismic data deviate from theoretical predictions down to 250 km depth (recall limitations on resolution) that may represent the base of a chemical boundary layer (Dalton *et al.*, 2008).

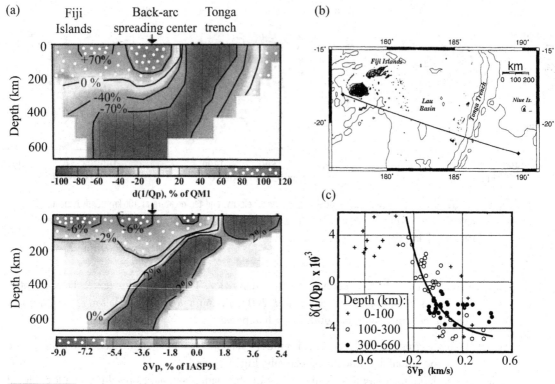

Fig. 3.129 Seismic structure through the Tonga–Fiji subduction zone (from Roth *et al.*, 2000). (a) 2D cross-sections of P-wave attenuation (upper) and velocity (lower) anomalies as percent difference between the tomographic and reference models. (b) Location of the profile; thin lines – bathymetry contours. (c) Correlation between attenuation and velocity anomalies at different depths. Solid line – the best exponential fit through all data.

3.7 Seismic lithosphere: summary

Factors affecting seismic velocities and attenuation

Seismic velocities and attenuation in the lithosphere are primarily controlled by temperature variations and the presence of (even small quantities of) melt or fluids (Section 3.1). Variations in pressure, composition, and grain size have further effects on both velocity and attenuation. Porosity may have a strong effect on seismic velocities in the shallow crust, while anisotropy is important through the entire lithosphere.

Temperature effects on velocities and attenuation are different and complicate a direct comparison of elastic and anelastic seismic tomography models, in particular at high temperatures: while velocities display a linear dependence on temperature, attenuation follows an exponential law (Fig. 3.3).

Seismic velocities and attenuation depend on frequency and thus laboratory studies performed at ultrasonic frequencies should not necessarily be applicable to data obtained at seismic frequencies (Section 3.1).

Resolution problems

Resolution of seismic studies has both theoretical and practical limitations. The size of the Fresnel zone (eq. 3.14) puts theoretical limits on the resolution of seismic methods based on transmitted waves (refraction seismology, travel time tomography, surface-wave dispersion analysis), so that higher frequency waves provide higher resolution. However, lower frequency waves penetrate deeper into the Earth. Practical limitations on resolution also depend on the method employed. In the receiver-function method there is an inevitable trade-off in resolving both the velocity structure and depths to seismic converters (Section 3.4.1; Fig. 3.49).

Surface-wave tomographic models have better vertical resolution, but poorer lateral resolution than teleseismic body-wave tomographic models (Section 3.6.1). However, surface-wave tomography based on fundamental modes commonly cannot resolve structure deeper than 200–300 km where it loses resolution (Fig. 3.79). Sensitivity kernels provide a measure for both lateral and vertical resolution of tomographic models (Fig. 3.81). In large-scale tomographic models, vertical resolution is not better than 50 km (i.e. results displayed for, for example, 150 km depth refer to the depth layer between 125 km and 175 km); moreover, any tomographic inversion inevitably results in depth leakage (Fig. 3.80).

The amplitudes of velocity anomalies depend on the regularization method used in tomographic modeling (Section 3.2.3), which results in a relatively low quantitative correlation between different models (Fig. 3.122). Crustal corrections (in particular corrections for the effects of sediments and ice) have an important effect on the amplitudes of both surface-wave and body-wave tomographic models (Section 3.6.1).

Reference models used in seismic inversion have a strong effect on seismic models. In particular, continental models based on PREM commonly inherit a velocity step at around 220 km depth, not necessarily required by the seismic data. A seismic model presented in absolute and in relative velocities may display significantly different qualitative features (Section 3.6.1).

Major results and global trends

In spite of certain limitations of seismic modeling, all seismic models display a strong correlation between seismic velocity and attenuation structure of the lithosphere and surface tectonics (i.e. lithospheric ages). Since crustal structure is highly heterogeneous even as short scales, this correlation is especially evident for the upper mantle (Sections 3.6.2, 3.6.3).

In many regions there is a clear correlation between the velocity and attenuation structure of the upper mantle and surface heat flow (Sections 3.6.2, 3.6.3). In particular, in normal oceans (i.e. oceans where the bathymetry can be explained by a cooling half-space model), seismic structure correlates with the ocean floor age. Beneath the continents, faster velocities and lower attenuation in the upper mantle are observed in the oldest regions with (commonly) low surface heat flow. The thickness of the mantle transition zone is also well correlated with surface tectonics (Section 3.4.2).

The unquestionable global correlation between seismic (both velocity and attenuation) and thermal structure of the upper mantle indicates that a significant part of seismic anomalies is produced by temperature variations. However, chemical variations in the upper mantle have a significant effect on the seismic structure of the mantle (Sections 3.4.2; 3.6.2).

Thickness of seismic lithosphere: results and uncertainties

The seismic lithosphere (the lid) is commonly defined as the high-velocity layer above the low-velocity zone (LVZ) in the upper mantle (Section 3.3.2). The LVZ is traditionally attributed to the top of the zone with partial melting, although other mechanisms such as high-temperature relaxation, a contrast in volatile content, or a drastic change in grain size are suggested by recent studies. All of these mechanisms imply that the base of the lid has a transitional character and extends over some tens of kilometers in depth.

Based on the definition of the seismic lid, the lithospheric base in a receiver-function (or converted waves) method is defined as a mantle interface with negative velocity contrast (Section 3.4.2). However, negative velocity gradients characteristic for the LVZ (i.e. velocity decreasing with increasing depth) can be explained without melting. A solid-state LVZ is possible when the seismic velocity increase due to compression is less than the velocity decrease due to heating. For example, theoretical calculations for a pyrolitic upper mantle show that, except for very young oceans (< 5 Ma), seismic observations in a 100 Ma Pacific Ocean can be quantitatively explained by a solid-state low-velocity zone (Fig. 3.130). The possibility of the existence of a solid-state low-velocity zone due to compositional variations in the upper mantle has been demonstrated by velocity–temperature inversions for the Kaapvaal craton (Kuskov *et al.*, 2006) (Fig. 5.23).

Interpretations of the base of the seismic lithosphere based on seismic tomography velocity models are subject to large uncertainty:

- reference models may introduce artifact low-velocity anomalies, such as at around 220 km depth beneath the continents when PREM is used as the reference model; use of velocity perturbations relative to a reference model may produce similar artifacts;
- choice of a 2% (or any other) velocity perturbation with respect to a reference model as the lithospheric base is arbitrary and may result in significant differences in lithosphere thickness estimates (Fig. 3.120); significant differences in the amplitudes of velocity perturbations in different tomographic models further complicate the matter;
- significant smearing may occur at the boundaries between tectonic structures with significantly different lithospheric thickness (such as ocean–craton transition);
- body-wave tomographic models do not have strong control on the vertical distribution of seismic velocities, while surface-wave tomographic models either integrate velocity structure over a large depth interval (which weakens the depth resolution) or (in the case of fundamental modes) lose resolution below 200–300 km depth (Figs. 3.79, 3.81, 3.119); depth leakage further reduces depth resolution of tomographic inversions (Fig. 3.120).
- In case where the origin of the LVZ below the seismic lithosphere is associated with the presence of partial melt, the lithospheric base is likely to be a rheological boundary. A change in mantle rheology may lead to a change in the mechanism of mantle deformation from dislocation creep to diffusion creep and, as a result, to a change in mantle anisotropy

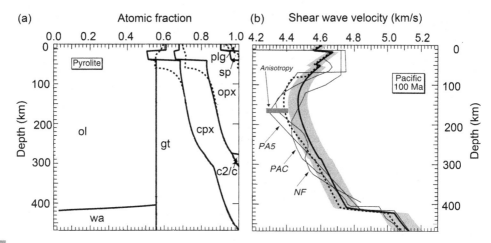

Fig. 3.130 Explanation of LVZ in the oceans by means of a solid-state low-velocity zone (after Stixrude and Lithgow-Bertelloni, 2005). (a) Phase proportions as atomic fraction computed along the 1600 K mantle adiabat (dashed) and along a geotherm for 100 Ma oceanic lithosphere (solid). (b) Shear-wave velocity calculated for pyrolite composition along the 100 Ma oceanic geotherm (bold line). Shading represents the uncertainty in the calculated velocity. The bold dashed line – velocity corrected for attenuation model QR19 for 100 Ma Pacific. Thin lines – seismological models for Pacific ocean. Gray bar – the approximate magnitude of SH–SV anisotropy. For younger lithospheric age the low velocity zone will shift to shallower depths and the value of the minimum velocity will decrease.

pattern. However, the change in the pattern of SV and SH velocities in the upper mantle may not necessarily indicate the transition from dislocation creep to diffusion creep in the mantle (Sections 3.1.5, 3.6.2).

In spite of significant controversy and ambiguity in interpretations of the base of the seismic lithosphere by different methods and research groups, the results obtained within each approach consistently indicate that the thickness of the lid increases with the age of the lithosphere both in the oceans (except for anomalous regions) and in the continents. In the oceans, the thickness of the seismic lithosphere increases from 30–60 km in young oceans to *c*. 80 km in oceans older than 80 Ma. However, in case of solid-state LVZ which is also in quantitative agreement with seismic observations from the Pacific Ocean (Fig. 3.130), lithospere of old oceans can be much thicker than 80 km.

In the continents, the thickness of the seismic lithosphere clearly correlates with lithospheric age. In Phanerozoic regions the base of the lid is typically at 50 to 100 km depth and deepens to more than 150 km in stable continents. In the collisional orogens, lithospheric slabs extend down to 150–250 km. Although the values of *c*. 200–220 km are often reported as thickness of seismic lithosphere beneath the Precambrian terranes, they do not provide compelling evidence that the tectosphere cannot extend down to the transition zone. Some high-resolution tomographic models indicate high mantle velocities and no LVZ down to a 300–400 km depth beneath the cratonic regions of West and Central Africa, Europe, Siberia, and Canada. More details on the upper mantle structure of different tectonic provinces are provided in Table 3.3.

Table 3.3 Summary of seismic structure of the mantle for some tectonic provinces

Tectonic structure	Summary of seismic structure of the mantle
Normal oceans	• Age-dependent thickness of the high-velocity lid: increases with age from mid-ocean ridges to $c.$ 80–100 km in old oceans. Velocity structure of the upper mantle beneath oceans >80 Ma is similar. • LVZ with $Vs \sim 4.3$–4.4 km/s at 100–200 km depth. • High gradient in seismic velocity at 200–410 km depths. • Strong anisotropy down to 150–200 km depth with a peak at $c.$ 100 km depth. Depth distribution shows only a weak dependence on age. • Decreased thickness of the mantle transition zone.
Pacific Ocean, East Pacific Rise	• Low Vs at 100–200 km depth (the strongest anomaly for the oceanic regions). • Strong anisotropy $V_{SH} > V_{SV}$ at depths between $c.$ 100 km and 200 km. • Strong attenuation anomaly, in particular in the upper 250 km, with a pronounced a-symmetric pattern with respect to the spreading center.
Mid-ocean ridges (MOR)	• Strong low-velocity anomalies down to 150–200 km depth. • Strong anisotropy $V_{SH} > V_{SV}$ in the upper 120 km associated with mantle flow in the spreading zone. • $V_{SH} < V_{SV}$ below 120 km depth, strong anomalies beneath fast ridges, weak but deep anomalies beneath slow ridges. • Strong attenuation anomalies along MOR, at least down to 200–250 km.
Collisional structures	• Slab-like high-velocity anomalies in the upper mantle down to 150–250 km. • Beneath some collisional orogens (e.g. the Caucasus) subducting slabs in the upper mantle have not been reliably resolved. • Presence of paleo-slabs (the oldest – Archean in age) in the upper mantle of several cratonic regions as indicated by seismic reflection studies.
Cratons	• High seismic velocities down to 200–250 km depth as reported by all studies (with smaller values for some cratons, such as China and the Arabian Shield), possibly down to 300–400 km in some cratons. • Fastest and deepest high-velocity lids beneath the Archean–Paleoproterozoic cratons; in some interpretations down to the transition zone (in particular, the Canadian, Baltic, Siberian, and West African cratons). Velocities peak at $c.$ 100–150 km depth ($Vs \sim 4.6$–4.8 km/s), below their amplitude decreases with depth and becomes uncertain at 300–400 km depth due to resolution problems. • Strong anisotropy in the upper mantle, in particular at shallow depths. • Two-layered anisotropy in the upper mantle of several cratons: frozen-in in the upper lithospheric layer and related to present-day plate motion in the lower (lithospheric or asthenospheric) layer. • Low attenuation in the upper mantle, down to at least 200–250 km. • Increased thickness of the mantle transition zone.
Phanerozoic platforms	• Seismic velocities close to global continental averages through the entire upper mantle. • Strong anisotropy in the shallow mantle. • Increased thickness of the mantle transition zone.
Rifts	• Velocity structure significantly different for rifts of the same age. • In East African Rift low Vp velocities down to $c.$ 150 km depth ($Vp \sim 7.7$ km/s; $Vs \sim 3.7$–4.0 km/s). • In Baikal (and Rio Grande) rifts, no low-velocity anomaly in the upper mantle.

Table 3.3 (cont.)	
Tectonic structure	Summary of seismic structure of the mantle
Young continents (except for orogens)	• Low seismic velocities from 50–100 km down to $c.$ 200 km depth. • Strong anisotropy in the shallow mantle.
Data on the crustal structure are summarized in Section 3.3.1.	

4

Thermal regime of the lithosphere from heat flow data

This chapter discusses the thermal regime of the lithosphere constrained by surface heat flow measurements. Other methods used to assess thermal structure of the lithosphere, such as xenolith geotherms and conversions of seismic velocities and seismic attenuation to mantle temperatures are discussed in Chapter 5. The goal of such a subdivision of constraints on crustal and upper mantle temperatures is to separate thermal models related to the global heat flow balance of the Earth from thermal models coming from, largely, non-potential field data.

The number of broad overviews of geothermal studies is limited (e.g. Sclater *et al.*, 1980; Stein, 1995). Pollack *et al.* (1993) provide an overview of global patterns of surface heat flow based on their compilation of a significant portion of borehole measurements available at that time. Overviews of the thermal regime of the lithosphere are given by Stein and Stein (1992) for the oceans and by Artemieva and Mooney (2001) and Artemieva (2006) for the continents. Beardsmore and Cull (2001) review practical applications of heat flow measurements, largely in relation to geothermal exploration. Jaupart and Mareschal (2004) discuss data on heat production and its contribution to surface heat flow, and a new monograph addressing various aspects of geothermal research is being published (Jaupart and Mareschal, 2010).

4.1 Field observations and laboratory data

4.1.1 Heat flow measurements

The thermal regime of the Earth provides critical information on its evolution through geological history by putting energy constraints on the processes of planetary accretion and differentiation. Although the energy that the Earth receives from the Sun is about two orders of magnitude greater than the global heat loss from the planet's interior, most of the solar energy is reradiated and the temperature in the Earth is controlled by internal heat: at present about 80% of heat flow from the planet's deep interior comes from radiogenic heat production, while the remaining 20% is the result of secular cooling of the Earth which includes latent heat from solidification of the core (Malamud and Turcotte, 1999).

Knowledge of the present thermal state of the Earth is crucial for models of crustal and mantle evolution, mantle dynamics and processes in the deep interior, which are reflected in geophysical observations such as seismic elastic and anelastic models, electromagnetic and gravity models, and geoid observations. Physical properties of crustal and mantle rocks determined by remote geophysical sensing are temperature dependent, and interpretations of

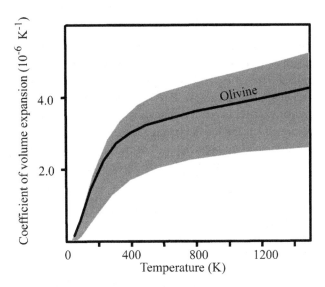

Fig. 4.1 Range of thermal expansion coefficients for common mantle minerals (source: Duffy and Anderson, 1989).

geophysical data sets are impossible without information on the thermal regime of the crust and the mantle. For example, numerous laboratory measurements indicate that seismic velocities and seismic attenuation are strongly temperature dependent (in particular at sub-solidus temperatures), implying that the temperature signal is an important part of any seismic model (Chapter 3). Mantle viscosity, which defines the vigor of mantle convection, depends exponentially on temperature. Similarly, electrical conductivity of mantle and crustal rocks also shows exponential dependence on temperature (Chapter 6), and the temperature effect on density through thermal expansion manifests itself in gravity and geoid anomalies (Fig. 4.1).

Mechanisms of heat transfer and dimensionless numbers

Three mechanisms transfer heat to the surface of the Earth: conduction, convection, and advection (the role of the fourth heat transfer mechanism, radiation, is minor). Most heat loss through the Earth's surface is by conduction, and thus it is the *conductive heat flow* that is measured in geophysics and commonly quoted as terrestrial heat flow. However, locally near-surface convective or advective heat flow components can be significant, such as forced (tectonically driven) convection in volcanic regions and at mid-ocean ridges and free convection caused by water circulation through the oceanic crust. In the deep interior, convection in the mantle and in the liquid outer core plays an important role in the thermal regime of the planet on geological times. Advection of heat driven by free or forced convection plays an important role in near-surface heat transfer in regions with high hydraulic crustal permeability, such as sedimentary basins.

The Péclet number Pe is a dimensionless number that relates the rate of advection due to material flow to the rate of thermal diffusion:

$$\text{Pe} = [\text{advection rate}]/[\text{thermal diffusion rate}] = VL/\chi \qquad (4.1)$$

where V is the velocity of material flow, L is a characteristic length scale of the process and χ is thermal diffusivity. Advective heat transfer dominates when $\text{Pe} \gg 1$, whereas conduction dominates when $\text{Pe} \ll 1$. For $\chi = 10^{-6}$ m²/s and $V = 1$ cm/year (typical for magma uplift beneath spreading ridges), $\text{Pe} \sim 3 \times 10^{-4} L$ (in m). This means that the thickness of the conductive layer beneath a spreading ridge is less than 3 km (i.e. less than thickness of the oceanic crust), and the thermal regime beneath spreading ridges is controlled by advection. This has important implications for interpretation of heat flow measurements in the oceans.

The Nusselt number Nu is the dimensionless ratio of convective to conductive heat transfer:

$$\text{Nu} = [\text{convective heat}]/[\text{conductive heat}] = hL/\lambda \qquad (4.2)$$

where h is the convective transfer coefficient (proportional to heat flow change and inversely proportional to the heat transfer surface area, temperature difference between the layers, and time), L is a characteristic length scale of the process and λ is thermal conductivity. When the Nusselt number is close to unity the flow is laminar; at $\text{Nu} \sim 100\text{--}1000$ the convection pattern becomes turbulent.

The Prandtl number Pr is the dimensionless ratio of kinematic viscosity v $(v = \mu/\rho)$ to thermal diffusivity χ:

$$\text{Pr} = [\text{viscous diffusion rate}]/[\text{thermal diffusion rate}] = v/\chi = C_p\mu/\lambda \qquad (4.3)$$

where C_p is specific heat, ρ is density, and μ is dynamic viscosity. The Prandtl number does not depend on a length scale, but depends only on material properties. For the mantle, $\text{Pr} \sim 10^{25}$. In heat transfer, the Prandtl number controls the relative thickness of the thermal boundary layer. For example, when Pr is large, the heat diffuses very slowly and the thermal boundary layer is thick.

The Rayleigh number Ra is a dimensionless number associated with buoyancy-driven flow (i.e. free convection). When the Rayleigh number is below the critical value, heat transfer is primarily by conduction; when Ra exceeds the critical value, heat transfer is primarily by convection. For the Earth's mantle, the Rayleigh number indicates the vigor of convection and (in the case of internal heating) is:

$$\text{Ra} = \frac{gp^2\alpha AD^5}{v\chi\lambda} \qquad (4.4)$$

where g is gravity acceleration, ρ is density, α is the thermal expansion coefficient, χ is thermal diffusivity, λ is thermal conductivity, v is kinematic viscosity, A is the rate of radiogenic heat production, and D is the thickness of the convective layer.

Accuracy of heat flow measurements

Data on surface heat flow together with geological and tectonic information are interpreted in terms of the thermal structure of the deep interior. Heat conduction in a medium is described by Fourier's law:

$$Q_i = \lambda_{ij} \frac{\partial T}{\partial z_j} \tag{4.5}$$

where Q_i is heat flow (the vector of specific energy flow rate), λ_{ij} is the thermal conductivity tensor, and the last term is the vector of temperature gradient. This equation provides the basis for near-surface heat flow measurements: the temperature gradient along vertical profiles in near-surface boreholes is the primarily directly measurable parameter, which, for known thermal conductivity of near-surface rocks, is converted into surface heat flow values.

The typical accuracy of thermal conductivity measurements on borehole core samples is 7–10% (largely because laboratory measurements are not done at *in situ* conditions, while *in situ* variations in porosity and fluid saturation of shallow crustal rocks have a significant effect on thermal conductivity; commonly laboratory measurements on low-porosity crystalline rocks provide conductivity values lower than at *in situ* conditions). The accuracy of modern measurements of temperature gradient is 5% or less; however it is important that measurements are made when the temperature distribution in a borehole (disturbed by drilling) equilibrates to steady state. Unfortunately, the latter requirement is not always met, in particular when industrial boreholes are used for geothermal research. The resulting uncertainty in surface heat flow measurements (even assuming temperature equilibrium was reached in a borehole) can amount to *c*. 15% as indicated by a comparison of heat flow data from nearby sites (Jaupart *et al.*, 1982). The major source of noise (in addition to measurement accuracy) is shallow water circulation.

Since in most cases only the vertical component of the temperature gradient is measured, the strong anisotropy of thermal conductivity of most sedimentary and metamorphic rocks may cause lateral flow of heat. Thus, in most cases data on near-surface heat flow require corrections for factors, such as sedimentation (e.g. von Herzen and Uyeda, 1963), surface topography variations, presence of rocks with anomalous thermal conductivity (such as salt domes) (for details see Beardsmore and Cull, 2001). Other factors, such as groundwater circulation, especially in shallow boreholes, are difficult to account for. While the effects of sedimentation are important only in a limited number of locations where sedimentation rates exceed 10^{-3}–10^{-4} m/year and a thick sedimentary cover is present, fluctuations of near-surface temperatures associated with paleoclimate changes may lead to variations of deep temperatures and temperature gradients in the shallow crust and thus have an important effect on heat flow measurements in most continental locations. In particular, the most pronounced climatic variations in the northern parts of Europe and North America are associated with glaciations.

Paleoclimatic corrections

Fluctuations of near-surface temperatures, such as caused by glaciation, can lead to variations in deep temperatures and temperature gradients in the shallow crust (Fig. 4.2). Variations in temperature gradient $\Delta\gamma$ can be found from solution of the (1D) thermal conduction boundary problem:

$$\frac{\partial T}{\partial t} = \chi \frac{\partial^2 T}{\partial z^2} \tag{4.6}$$

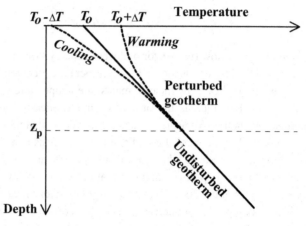

Fig. 4.2   Relation of borehole temperatures to climate changes.

$$T(0,t) = \begin{cases} 0,\, t < \tau_1 \\ \Delta T,\, \tau_1 < t < \tau_2 \end{cases}$$

$$T|_{z \to \infty} \to T_0 + \gamma z$$

where γ is the equilibrium temperature gradient, ΔT is the difference in surface temperature during and after glaciation, τ_1 and τ_2 are the times elapsed from the end and from the beginning of glaciation, respectively. Solution of the boundary problem provides the value of the unperturbed temperature gradient in the crust (Beck, 1977):

$$\Delta \gamma = \Delta T \left\{ \frac{1}{\sqrt{\pi \chi \tau_1}} \exp\left[-\left(\frac{z}{2\sqrt{\chi \tau_1}}\right)^2\right] - \frac{1}{\sqrt{\pi \chi \tau_2}} \exp\left[-\left(\frac{z}{2\sqrt{\chi \tau_2}}\right)^2\right] \right\} \qquad (4.7)$$

When τ_2 is unknown, paleoclimatic disturbances of the temperature gradient can be found from the first term of this equation. The value of $\Delta \gamma$ decreases with depth and with increasing τ_1 (Fig. 4.3). Typically, at depths $z > 1.5$ km the effect of glaciation on the surface heat flow is less than 2 mW/m^2 and is often neglected. However, in certain locations where near-surface temperature changes during a recent glaciation were significant, present-day temperatures can be perturbed even in deep boreholes. In particular, in some parts of the Baltic Shield perturbation of the present surface heat flow can be as large as 15–20 mW/m^2 (Balling, 1995).

4.1.2 Thermal conductivity

For a known thermal conductivity of near-surface rocks, the temperature gradient measured in heat flow studies is converted into surface heat flow values (Eq. 4.5). Although radiative heat transfer may play some role at high temperatures (higher than 600 °C, and in particular at temperatures above 1200 °C due to the strong temperature dependence of the radiative conductivity of polycrystalline rocks, e.g. Hofmeister, 1999), its role in the thermal regime

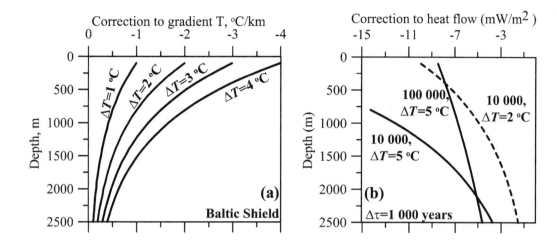

Fig. 4.3 Corrections to temperature gradient (a) and surface heat flow (b) due to recent glaciations. (a) Paleoclimate corrections for the Baltic Shield for different values of near-surface temperature changes with lower ΔT values for higher latitudes. (b) An example illustrating the effects of the age of glaciation (numbers on the curves) and variations in near-surface temperature ΔT on surface heat flow, assuming that glaciation lasted for 1000 years and the thermal conductivity of shallow crust is 3.0 W/m/K.

of the Earth remains unknown, and it is usually quantified by introducing effective (conductive plus radiative) thermal conductivity. High-temperature measurements of thermal conductivity always include a radiative component.

Laboratory measurements of thermal conductivity of sedimentary and crystalline rocks together with seismic models of crustal structure provide the basis for models of depth and lateral variations of thermal conductivity (Horai, 1971). An overview of the subject can be found in Clauser and Huenges (1995). Laboratory measurements of thermal conductivity of rocks form the basis for several extended collections. Although these compilations are commonly dominated by rocks from certain tectonic provinces, some general conclusions are summarized in the following.

Mineral composition

Mineral composition is the major factor controlling the thermal conductivity of crustal and mantle rocks (Fig. 4.4).

- For any given rock type, thermal conductivity may vary by as much as a factor of two to three.
- Thermal conductivity in plutonic and metamorphic rocks depends significantly on feldspar and quartz content, respectively, while porosity variations have a strong effect in sedimentary and volcanic rocks.
- Thermal conductivity of the lithospheric mantle is controlled by the relative amount of olivine and clinopyroxene with ~30% difference in their conductivities (Schatz and Simmons, 1972). However, since changes in olivine and clinopyroxene content between

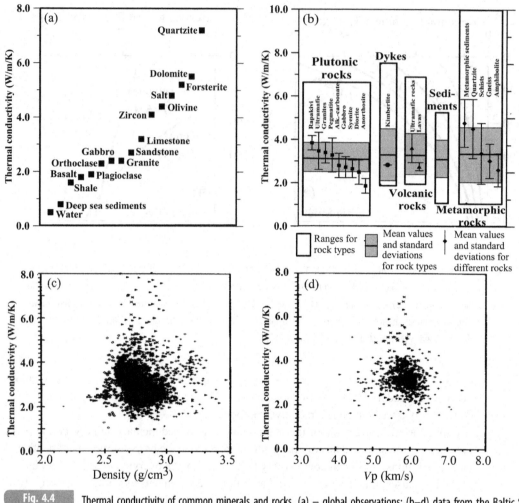

Fig. 4.4   Thermal conductivity of common minerals and rocks, (a) – global observations; (b–d) data from the Baltic Shield (source: Kukkonen and Peltoniemi, 1998).

Archean and post-Archean provinces are complimentary to each other, the net effect of compositional variations on the bulk mantle conductivity is ~0.1 W/m/K, and no significant variations in thermal conductivity of the lithospheric mantle of different ages are expected (Levy *et al.*, 2010).

- There is no straight-forward correlation between thermal conductivity and seismic velocities or between thermal conductivity and density (Fig. 4.4).

Porosity and fluid saturation

Thermal conductivity depends significantly on porosity: according to some studies, a 5% porosity increase can produce a 25% drop in thermal conductivity (Jessop, 1990). However,

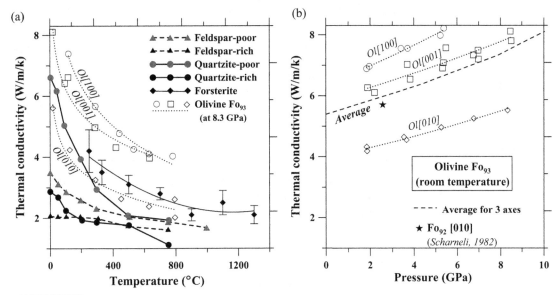

Fig. 4.5 (a) Temperature dependence of the thermal conductivity of plutonic rocks (dashed lines), metamorphic rocks (thick solid lines), and mantle forsterite (diamonds with uncertainties and thin solid line) (data of Clauser and Huenges, 1995; Schatz and Simmons, 1972). Plutonic rocks include feldspar-rich (syenite, anorthosite, hypersthenite) and feldspar-poor (granite, alkali granite, diorite, quartz diorite, monzonite, tonalite, gabbro, hornblende gabbro, peridotite, lherzolite, bronzitite, dunite, olivinite, granodiorite). Metamorphic rocks include quartzites and quartz-poor rocks (marble, serpentinite, eclogite, albitite, slate, amphibolite). The thermal conductivity in olivine (a and b) is highly anisotropic and is shown along three crystallographic axes (dotted lines with open symbols that refer to different measurements) (data of Osako et al., 2004).

the effect depends critically on the shape of the pore space (Horai, 1971; Rocha and Acrivos, 1973) and on the composition of the saturating fluid (Artemieva and Chesnokov, 1991).

Effect of temperature

Thermal conductivity of most rocks decreases with temperature.

- In metamorphic and plutonic rocks, the temperature-induced conductivity decrease depends significantly on the content of the dominant mineral phase (quartz and feldspar, respectively); the temperature effect is stronger for feldspar-poor and quartz-rich rocks (Fig. 4.5). A number of approaches based on laboratory data have been proposed to extrapolate thermal conductivity to high temperatures. The following temperature-dependent function is used for rock types representative of the lithosphere (limestones, metamorphic, felsic, and mafic igneous rocks) (for details on individual rock types see overview by Zoth and Haenel, 1988):

$$\lambda[\mathrm{W/m/K}] = 0.86 + 662/(350 + T[^\circ\mathrm{C}]). \tag{4.8}$$

- Another relationship between thermal conductivity λ and temperature T often used for the upper crust is:

$$\lambda[W/m/K] = \lambda_0/(1 + cT[°C]), \tag{4.9}$$

where λ_0 is the thermal conductivity at $T = 0$ °C and at near-surface pressure conditions and c is a material constant (in the range 0–0.003 °C^{-1}) empirically determined from experimental studies on different rock samples (e.g. $c = 0.001$ for the upper, "granitic", crust) (Cermak and Rybach, 1982).

- At temperatures typical for the lower crust (200–600 °C), the thermal conductivity of basalts is almost constant (≈ 2.0 W/m/K) and does not depend on temperature.

- The temperature dependence of the thermal conductivity of peridotite is not well known and commonly a constant value of 4.0 W/m K is assumed for the upper mantle (Schatz and Simmons, 1972). Recently Xu *et al.* (2004) have proposed the following expression for olivine:

$$\lambda[W/m/K] = \lambda_0(298/T[K])^n \tag{4.10}$$

where $\lambda_0 = 4.08$ W/m/K and $n = 0.406$.

For mantle olivine, another expression based on experimental studies of the temperature dependence of thermal conductivity along three crystallographic axes is (Osako *et al.*, 2004):

$$\lambda[W/m/K] = \lambda_0 + cT[K] \tag{4.11}$$

where $\lambda_0 = 1.91, 0.84$, and 2.08 W/m/K along the [100], [010], and [001] axes, respectively, with the corresponding values of constant $c = 2088, 1377$, and 1731 W/m (Fig. 4.5).

- In contrast with lattice (or phonon) conductivity, the "radiative thermal conductivity" follows a T^3-law and becomes significant at $T > 1000$–1200 °C (Clauser, 1988). For this reason, thermal conductivity typically decreases with temperature until around 1000–1200 °C, when the radiative component balances this decrease. Analytical expressions for the temperature dependence of radiative and phonon thermal conductivity in the mantle based on a review of experimental and theoretical estimates are proposed by Hofmeister (1999).

Effect of pressure

Conductivity increases slightly with pressure and typically linear relations are applied for its pressure dependence (e.g. Horai and Susaki, 1989; Seipold, 1992). In granites and metamorphic rocks, thermal conductivity increases by $c.$ 10% at pressures 0–500 MPa, with the stronger increase over the first 50 MPa due to porosity decrease. The pressure dependence of thermal conductivity in mantle olivine has been parameterized as (Osako *et al.*, 2004):

$$\lambda[W/m/K] = \lambda_0 \exp(b\, P[GPa]), \tag{4.12}$$

where $\lambda_0 = 6.61, 3.98$, and 5.91 W/m/K along the [100], [010], and [001] axes, respectively, with the corresponding values of constant $b = 0.038, 0.042$, and 0.034 [GPa]$^{-1}$ (Fig. 4.5).

The first-order effect of both temperature and pressure on thermal conductivity can be approximated as (Ketcham, 1996):

$$\lambda[W/m/K] = 0.86 + 662/(350 + T[°C]) + 0.041\, P\,[Kb]. \tag{4.13}$$

Anisotropy

Most volcanic and plutonic rocks are, to first order, isotropic, while the thermal conductivity of many sedimentary and metamorphic rocks is strongly anisotropic. For example, in sheet silicates thermal conductivity in the directions parallel and perpendicular to the main cleavage plane may vary by a factor of 6–8.

In olivine, the anisotropy of thermal conductivity and thermal diffusivity is observed for all ranges of upper mantle temperatures and pressures (Fig. 4.5). The corresponding expressions for the temperature and pressure dependence of thermal conductivity along three crystallographic axes are given by Eq. 4.11–4.12 (Osako *et al.*, 2004).

4.1.3 Thermal expansion, thermal diffusivity, and specific heat

Other thermal parameters that are widely used in thermal studies include the thermal expansion coefficient α, thermal diffusivity χ, and specific heat capacity at constant pressure C_p. The latter parameter is not very well constrained and is typically considered to be $C_p = 1000$–1200 J/kg/K in most rocks (Oxburgh, 1980). Later experimental studies indicate that specific heat is temperature dependent and can be parameterized as (Berman and Aranovich, 1996):

$$C_p(T)[\text{J/kg/K}] = a_1 + a_2/T^{1/2} + a_3/T^3, \qquad (4.14)$$

where temperature T is in K and coefficients $a_1 = 233.18$, $a_2 = -1801.6$, $a_3 = -29.794 \times 10^7$ for forsterite and $a_1 = 252$, $a_2 = -2013.7$, $a_3 = -6.219 \times 10^7$ for fayalite.

The thermal expansion coefficient α is typically assumed to be in the range $(3.0$–$3.2) \times 10^{-5}$ K^{-1}. However, an experimental study indicates that in the temperature range 325 °C to the melting point the thermal expansion of forsterite increases smoothly from 2.8 to 4.5 K^{-1} (Fig. 4.1) and can be approximated by a linear dependence on temperature as (Bouhifd *et al.*, 1996):

$$\alpha(T) = \alpha_o + cT[\text{K}], \qquad (4.15)$$

where $\alpha_o = 2.832 \times 10^{-5}$ K^{-1} and $c = 3.79 \times 10^{-8}$ K^{-2}.

Thermal diffusivity can be expressed through other parameters as

$$\chi = \lambda/\rho C_p, \qquad (4.16)$$

where λ is the thermal conductivity and ρ is the density. In crustal rocks, thermal conductivity λ can range within broad limits leading to large variations in thermal diffusivity values. For example, in the lower crust with typical values of $\rho = 2900$ kg/m^3, $C_p = 1000$ J/kg/K and $\lambda = 2.0$ W/m/K, $\chi = (0.6$–$0.7) \times 10^{-6}$ m^2/s, which is 50–70% of its upper mantle value. For the upper mantle rocks, $\lambda = 4.0$ W/m/K (Schatz and Simmons, 1972), $\rho = 3350$ kg/m^3, and eq. (4.16) yields a thermal diffusivity value $\chi \sim (1.0$–$1.2) \times 10^{-6}$ m^2/s.

Experimental studies of thermal diffusivity in mantle olivine indicate that its strongly anisotropic temperature dependence can be approximated by a linear dependence, similar to eq. (4.11), with the coefficients of $\chi_o = -0.06$, -0.13, and -0.03 ($\times 10^{-6}$ m^2/s) along the

Table 4.1 Half-lives of ^{232}Th, ^{238}U, ^{40}K, and ^{235}U isotopes

Parent isotope	Present-day relative concentration*	Daughter isotope	Half-life, by mass
^{238}U	0.9927	^{206}Pb	4.47 Ga
^{235}U	0.0072	^{207}Pb	0.704 Ga
^{232}Th	4.0	^{208}Pb	14.01 Ga
^{40}K	1.6256	^{40}Ca (~89.5%)	1.25 Ga
^{40}K	1.6256	^{40}Ar (~10.5%)	11.93 Ga

(after Dickin, 1995; Van Schmus, 1995).
* Relative concentration normalized by the abundance of total U for K:U:Th $\approx$ 12700:1:4.

[100], [010] and [001] axes, respectively, with the corresponding values of constant $c = 938$, 626, and 832 ($\times 10^{-6}$ m^2/s/K). The pressure dependence of thermal diffusivity has the form of eq. (4.12) with coefficients $\lambda_o = 2.50$, 1.53, and 2.16 ($\times 10^{-6}$ m^2/s) along the [100], [010], and [001] axes, respectively, with the corresponding values of $b = 0.033$, 0.040, and 0.035 [GPa]$^{-1}$ (Osako *et al.*, 2004).

4.1.4 Heat production

Major heat-producing isotopes

Radioactive decay of long-lived isotopes (^{232}Th, ^{238}U, ^{40}K, and ^{235}U, listed in order of decreasing importance for the present-day thermal balance of the Earth) provides internal thermal energy and largely defines the temperatures in the planetary interior. Present-day relative concentrations and half-lives of ^{232}Th, ^{238}U, ^{40}K, and ^{235}U isotopes are listed in Table 4.1. Potassium is one of the so-called large-ion lithophile elements (LIL); it is an important component of several rock-forming minerals (e.g. orthoclase, biotite, leucite) and concentrates in evolved crustal rocks such as granites and shales. Uranium and thorium are also LIL elements, but in contrast to potassium they are found only as trace minerals in common rocks and their abundances can vary over several orders of magnitude. Because U, Th, and K are all LIL elements, they display a similar presence in the Earth when bulk concentrations are considered. The K/U ratio is $(1.0–1.3) \times 10^4$ for the bulk Earth, primitive mantle, and continental crust and the Th/U ratio is 3.7–4.0 (Van Schmus, 1995), and relative abundances of radioactive isotopes in the Earth are near:

$$K : U : Th \approx 10,000 : 1 : 3.7. \qquad (4.17)$$

Data, although not well constrained, on radiogenic heat production in the bulk Earth, primitive mantle, and the crust, together with data on abundances of Th, U, and K isotopes allow for modeling of the thermal history of the Earth (Fig. 4.6). For a thick (~ 250 km) continental lithosphere, the time-scale of thermal diffusion:

$$\tau \sim L^2/\chi \qquad (4.18)$$

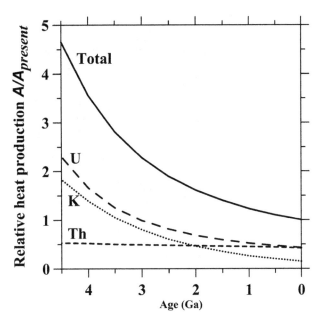

Fig. 4.6 Heat production in the Earth relative to the present total (source: van Schmus, 1995). Assumed present abundances are: K = 200 ppm, Th = 74 ppb, U = 20 ppb (eq. 4.17).

(where L is lithosphere thickness and χ is thermal diffusivity, typically 1 mm^2/s) is comparable to the half-lives of radioactive isotopes and thus the steady-state regime is not achieved (Michaut and Jaupart, 2004). In particular, long half-lives of radioactive isotopes imply that, even at constant thermal boundary conditions at the lithospheric base, it undergoes secular cooling due to radioactive heat loss. This heat loss is estimated to be ~50–150 °C/Ga. In a Proterozoic (~1.5 Ga old) region, secular cooling contributes ~3 mW/m^2 to the surface heat flow; although this value is smaller than the accuracy of heat flow measurements, it is comparable to the radiogenic contribution of the lithospheric mantle to the total surface heat flow.

The bulk amount and the depth distribution of radiogenic elements in the crust and in the mantle are still largely unknown. However, information on radioactive heat production in the crust is critical for thermal modeling, since the contribution of crustal heat production to surface heat flow can vary from <20% to >80% (Fig. 4.7). Several complementary approaches are used to address these questions, since no single technique for estimating the crustal radiogenic heat production $A(z)$ has proved to be consistently accurate. Up to the present, in most thermal studies the distribution of crustal heat production is an assumption rather than a constraint.

Laboratory measurements

Near-surface samples

Laboratory measurements of heat production in near-surface rocks form the basis for models of the heat production in sedimentary cover and the upper crust. In the laboratory, radiogenic

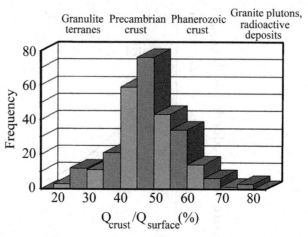

Fig. 4.7 Contribution of radiogenic heat flow generated in the continental crust to surface heat flow (after Artemieva and Mooney, 2001).

heat production is usually determined by measuring the radioactive isotope abundances in ~1 kg samples through their natural decay. The bulk heat production H (in W/kg) of a rock sample is the sum of the contributions from ^{232}Th, ^{238}U, ^{40}K, and ^{235}U isotopes, and the heat production per unit volume A is calculated from H and the rock density ρ as: $A = \rho H$. For mantle heat production the expression has the form (Ashwal *et al.*, 1987):

$$A = \rho(0.097\text{U} + 0.026\text{Th} + 0.036\text{K}), \tag{4.19}$$

where the concentrations of isotopes of U and Th are in ppm (parts per million), the concentration of K is in wt. %, density ρ is in g/cm^3, and A is in μW/m^3. The concentrations of Th and K in the Earth are well constrained and the average ratios Th/U and K/U between isotope concentrations are known (eq. 4.17). Since most crustal rocks contain less than 20% of potassium, the intrinsic radiogenic heat production of Earth materials is largely defined by the amount of uranium and thorium present (Jaupart and Mareschal, 2004).

Systematic regional sampling of near-crustal rocks either on a regular grid or by calculating a weighted average of values for various lithologies present in a geological province provides unbiased distributions of heat production over large areas. Such detailed studies have been performed in various continental locations (Western Australia, Canada, Baltic Shield, and China) and indicate the presence of a significant scatter of heat production at all scales: the values may be quite variable laterally and vertically within a single pluton and within a geological province as well as in rocks of the same bulk composition from different locations (Fig. 4.8). For example, heat production in gneisses from the Yilgarn craton (Western Australia, Fig. 3.82) varies from near-zero values to >6 μW/m^3 with an average value of 2.6 ± 1.9 μW/m^3, while in gneisses from the Gawler craton (Central Australia) the average value is 3.7 ± 3.6 μW/m^3; variations of heat production in all near-surface rocks of the Yilgarn craton and the Baltic Shield span from near-zero values to >12 μW/m^3. The Archean granulites from the Jequie-Bahia terrane (Brazil) is another example: in contrast to

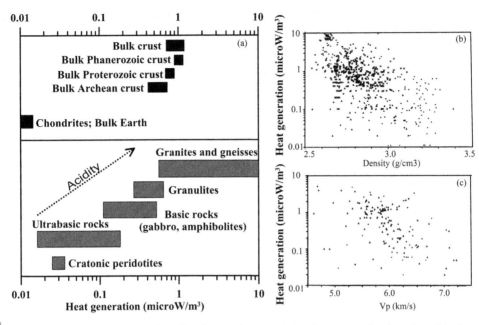

Fig. 4.8 Heat production in rocks. (a) General trends in heat production variations in common minerals, rocks, and in the crust. Sources: Nicolaysen *et al.*, 1981; Weaver and Tarney, 1984; Taylor and McLennan, 1985; Fountain, 1986; Shaw *et al.*, 1986; Ashwal *et al.*, 1987; Fountain *et al.*, 1987; Pinet and Jaupart, 1987; Rudnick and Presper, 1990; Ketcham, 1996; McLennan and Taylor, 1996; Gao *et al.*, 1998; Rudnick *et al.*, 1998; Rudnick and Nyblade, 1999; Beardsmore and Cull, 2001. (b–c) Heat production versus rock density and Vp velocity for the Baltic Shield (source: Kukkonen and Peltoniemi, 1998).

global observations ($A < 1\ \mu W/m^3$), these granulites have high values of heat production, up to $>10\ \mu W/m^3$ with an average value $>2.5\ \mu W/m^3$ (Iyer *et al.*, 1984).

To complicate matters, shallow rocks may not be representative of deeper rocks from the same massif: for example, near-surface water circulation leads to uranium depletion in granites and may thus lead to an erroneous estimate of uranium concentration in the upper crust (Jaupart and Mareschal, 2004). Another factor modifying the concentration of heat-producing elements in near-surface rocks is leaching of radioactive elements by weathering. As a result, the error in heat production estimates in near-surface samples is *c.* 20% (Jaupart *et al.*, 1982), although analytical errors do not exceed 5% (with the largest errors for low-radioactivity samples).

Correlation between seismic velocity and heat production

Experimental measurements on near-surface rocks indicate that heat production may not be predicted reliably from seismic velocity (e.g., Fountain, 1986; Kern and Siegesmund, 1989) (Fig. 4.8), in contrast to earlier laboratory measurements on a small data set from a few European locations. Data from the KTB and the Kola Superdeep boreholes also do not confirm that seismic velocities and heat production are correlated. These results are of little

Table 4.2 Exposed cross-sections of the continental crust

Structure	Exposed crust (km)	Total crustal thickness (km)	Reference
Vredefort structure, South Africa	20	36	Nicolaysen *et al.*, 1981
Lewisian fragment, Scotland	20	40	Weaver and Tarney, 1984
Wawa–Kapuskasing transect, Canada	25	43	Ashwal *et al.*, 1987
Pikwitonei structure, Canada	30	40	Fountain *et al.*, 1987
Southern Norway	?	28	Pinet and Jaupart, 1987

surprise: the isotopes of Th, U and K are incompatible elements, i.e. they are not readily incorporated into mineral lattices. Radioactive isotopes tend to concentrate in silicic igneous rocks and at grain boundaries, and thus

(i) their abundances do not closely follow major element abundances and
(ii) heat production is not directly related to bulk rock chemistry.

Although a very general trend in the increase of heat production from ultrabasic to basic and acid rocks can be recognized (Fig. 4.8), a large scatter in heat production values is documented for rocks of the same lithology.

Exposed cross-sections of the deep crust

While laboratory measurements on rocks from ophiolite complexes provide data on heat production in the deep parts of the oceanic crust, several continental locations provide a unique opportunity to measure directly heat production in middle crustal rocks and to determine directly an average vertical profile of heat production in the crust (Fountain and Salisbury, 1981; Nicolaysen *et al.*, 1981; Weaver and Tarney, 1984; Ketcham, 1996; Ashwal *et al.*, 1987; Fountain *et al.*, 1987; Pinet and Jaupart, 1987). The upper 15–30 km of the crust, which is equivalent to 50% to 75% of the total crustal thickness, is available for direct studies in the exposed (overturned) sections of several Precambrian regions: the Vredefort structure (South Africa), the Wawa–Kapuskasing transect and the Pikwitonei structure (both in the Superior Province, Canada), the Lewisian terrane (Scotland), the Sveconorwegian–Svecofennian province (southern Norway), and metamorphic core complexes from Arizona (USA) (Table 4.2). This means that in all exposed sections, the lower parts of the crust (as much as 15–21 km as in the Vredefort, Wawa–Kapuskasing, and Pikwitonei sections) are unavailable for direct study. As a result, the heat production of the lower crust is poorly constrained. Additionally, it is not known how representative the exposed cross-sections of the crust in general are.

Laboratory measurements on rocks sampled from exposed crustal cross-sections indicate that in the middle crustal rocks heat production typically ranges from 0.2 to 0.4 $\mu W/m^3$. However, significantly higher heat production values, 0.4 to 0.5 $\mu W/m^3$, have been determined for Precambrian granulite terranes of the Canadian Shield along the 100 km long Wawa–Kapuskasing transect which has been interpreted as a cross-section through the upper 25 km of the crust (Fig. 4.9). These granulites were interpreted as mid-crustal rocks

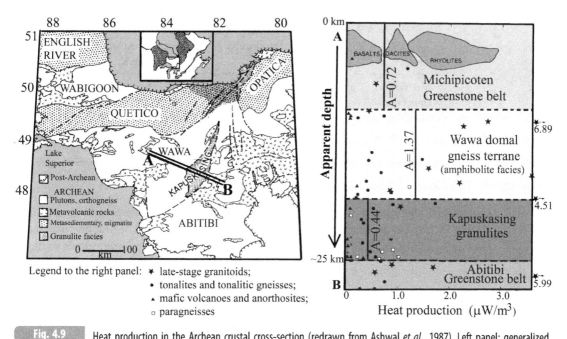

Fig. 4.9   Heat production in the Archean crustal cross-section (redrawn from Ashwal *et al.*, 1987). Left panel: generalized geological map of the Superior Province, Canadian Shield, showing the location of the transect A–B. Right panel: heat production in individual rock samples (symbols) and average heat production (vertical lines) in different terranes along the transect A–B. The boundaries between the terranes are shown by dashed lines. The transect A–B is interpreted as an oblique cross-section of the upper ~25 km of the Archean crust, which was uplifted along a major thrust fault.

located at apparent depths of ~18–25 km. Some authors speculate that these high heat production values may continue into the lower crust, below 25–30 km depth (Ashwal *et al.*, 1987; Rudnick and Presper, 1990).

   In the Sveconorwegian–Svecofennian province of the Baltic Shield (southern Norway) two outcrops of granulite facies rocks are interpreted to be parts of the same lower crustal horizon; they are overlain by amphibolite facies terranes interpreted as representative of metamorphic conditions at middle crustal levels. The average heat production is 1.6 $\mu W/m^3$ in the Norwegian amphibolites and 0.4 $\mu W/m^3$ in granulites. Heat production in mafic granulites varies globally from 0.06 to 0.4 $\mu W/m^3$, which provides a range of possible variations in the lowermost crust (e.g., Pinet and Jaupart, 1987; Rudnick and Fountain, 1995), and values of 0.1 $\mu W/m^3$ or 0.4 $\mu W/m^3$ are commonly used in thermal modeling (Table 4.3). It is worth noting that for a typical 13-to-16-km-thick continental lower crust (Rudnick and Fountain, 1995), a 0.3$\mu W/m^3$ difference in assumed values of lower crustal heat production results in a 4–5 mW/m^2 difference in estimates of mantle heat flow and in a 20–70 km difference in estimates of lithospheric thermal thickness (Artemieva and Mooney, 2001). Moreover, thermal models with high values of heat production (0.4–0.5 $\mu W/m^3$) throughout the entire lower crust would predict zero (or even negative) mantle heat flow in

Table 4.3 Typical thermal properties of the continental lithosphere

Layer	Heat production, $\mu W/m^3$	Thermal conductivity, W/m/K
Sediments	0.7–1.3	0.5–4.0
Upper crust	0.4–7.0	2.5–3.0
Middle crust	0.2–0.5	2.0–2.5
Lower crust*	0.1–0.4	2.0
Lithospheric mantle*	0–0.01	4.0

* Inferred values

continental regions with low surface heat flow (such as parts of the Siberian craton, the Baltic Shield, and the Canadian Shield), implying that lower values of heat production (as low as $0.1~\mu W/m^3$) may be characteristic of the lower crust, at least in some tectonic settings.

Bulk heat production of the crust

Chemical constraints

In spite of the small volume of continental crust on the planetary scale (it constitutes only ~0.53% of the mass of silicate Earth (mantle left after core formation but prior to continental crust extraction)), it contains 20% to 70% of incompatible elements, which include heat-producing radioactive isotopes of Th, U, and K (Rudnick *et al.*, 1998 and references therein). This fact defines the important role the continental crust plays in any chemical mass-balance models and in thermal modeling. Calculations of global crustal/mantle chemical budgets are based on redistribution of heat-producing elements in the Bulk Silicate Earth between the continental crust and mantle reservoirs. In spite of a large set of assumptions (e.g. on the composition of the Bulk Silicate Earth and mantle reservoirs, and on the structure of mantle convection), they provide a narrow range of estimates of bulk heat production in the continental crust, 0.74–$0.86~\mu W/m^3$ (O'Nions *et al.*, 1979; Allègre *et al.*, 1988). For a typical crustal thickness of ~40 km on the continents, these values yield ~30–35 mW/m^2 of crustal contribution to surface heat flow. This implies that in the Precambrian terranes where surface heat flow is, on average, 41–49 mW/m^2 (Nyblade and Pollack, 1993), mantle heat flow should be 6–19 mW/m^2. However, in the cratonic terranes with very low surface heat flow (<30 mW/m^2 as reported for some terranes of the Superior, Siberian, and West African cratons (Sass and Behrendt, 1980; Duchkov *et al.*, 1987; Levy *et al.*, 2010)) these estimates of bulk crustal heat production are, clearly, too high.

Petrologic studies provide independent constraints on average concentrations of radioactive isotopes in crust and mantle. These estimates are based on studies of mantle and crustal xenoliths, chemical composition of granulite facies terranes, and models of crustal growth complemented by heat flow constraints (e.g., Weaver and Tarney, 1984; Taylor and McLennan, 1985; Shaw *et al.*, 1986; Rudnick and Fountain, 1995; McLennan and Taylor, 1996; Gao *et al.*, 1998; Rudnick *et al.*, 1998; Rudnick and Nyblade, 1999). However, because of the large variability in heat production at all scales, extrapolations from rock samples to the

Table 4.4 Average bulk heat production of the continental crust*		
	Heat production ($\mu W/m^3$)	Reference
Bulk continental crust	0.58–0.70	Taylor and McLennan, 1985; McLennan and Taylor, 1996
	0.74–0.86	O'Nions *et al.*, 1979; Allègre *et al.*, 1988
	0.92–0.93	Weaver and Tarney, 1984; Rudnick and Fountain, 1995
	1.00	Gao *et al.*, 1998
	1.12	Christensen and Mooney, 1995
	1.25	Wedepohl, 1995
	1.31	Shaw *et al.*, 1986
Archean crust	0.35–0.55**	Artemieva and Mooney, 2001
	0.4–0.5	Rudnick *et al.*, 1998
	0.48	Taylor and McLennan, 1985
	0.50	Rudnick and Fountain, 1995
	0.61	Weaver and Tarney, 1984
	0.64	McLennan and Taylor, 1996
	0.93 (North China)	Gao *et al.*, 1998
Proterozoic crust	0.7–0.9**	Artemieva and Mooney, 2001

* Inferred values
** Typical values, the ranges estimated for different cratons are given in Table 4.5.

whole crust are subject to many uncertainties. As a result, published estimates of bulk heat production in the crust vary by a factor of two. Typically, estimates range from 0.48 to 0.61 $\mu W/m^3$ in the Archean crust (with some regional values as low as 0.35 $\mu W/m^3$ and as high as 0.93 $\mu W/m^3$) to 0.58–0.93 $\mu W/m^3$ in the Proterozoic crust, and to 0.96–1.38 $\mu W/m^3$ in the Phanerozoic crust (Table 4.4). Furthermore, a poorly known value for heat production in the lower crust (rarely exposed for direct studies below 25–30 km depth) results in significant uncertainty in petrologic models of bulk crustal heat production. Apparently systematic variations in bulk crustal heat production with age reflect the chemical heterogeneity of the continental crust: the composition of the Archean crust is dominated by Na-granitoids, while younger crust is dominated by K-rich granitoids (Martin, 1993).

Morgan (1985) has argued that the global trend of a decrease in surface heat flow with age from Phanerozoic to Archean can be explained by a decrease in crustal radioactivity. Although both petrologic data and thermal constraints indicate that, in general, bulk crustal heat production is lowest in the Archean crust and largest in the Phanerozoic crust (Sclater *et al.*, 1980; Vitorello and Pollack, 1980; Morgan, 1985; Taylor and McLennan, 1985; Rudnick *et al.*, 1998; Artemieva and Mooney, 2001; Jaupart and Mareschal, 2004), this trend only holds globally. Because of large regional variability, a particular geological province should not necessarily fit global values for its age group. Furthermore, a contrast in crustal heat production between Precambrian terranes of different ages is not necessarily required by heat flow data from South Africa (Ballard *et al.*, 1987; Artemieva and Mooney, 2001) (Table 4.5).

Table 4.5 Average bulk heat production (in $\mu W/m^3$) in the Precambrian crust*

Archean and Paleoproterozoic crust	A	Meso-Neoproterozoic crust	A
South Africa	0.49–0.78	South Africa	0.61–0.77
Yilgarn	0.39–0.52	Northern Australia	0.78–1.40
Pilbara	0.56	Southern Australia	1.03–1.55
Sino-Korean Craton	0.57–1.00	Yangtze Craton	0.58–1.24
Dharwar Craton	0.36–0.50	Gondwana basins (Indian Shield)	0.86–1.18
Baltic Shield (Kola-Karelia) and Ukrainian Shield	0.36–0.49	Atlantic Shield (South America)	0.62–0.83
Superior Province	0.29–0.47	Grenville	0.39–0.95
Siberian craton (Anabar Shield and Yenisey Ridge)	0.17–0.38		
Siberian Platform	0.37–0.49		
Aldan Shield	0.61		

* Inferred values
(Source: Artemieva and Mooney, 2001)

Estimates from surface heat flow

Surface heat flow observations provide additional constraints on bulk crustal heat production (e.g. Artemieva and Mooney, 2001; Table 4.5). The simplest requirement is that mantle heat flow should not exceed the lowest measured surface heat flow in a region. Another constraint comes from the fact that surface heat flow measurements are effectively 1D (in the vertical direction) and ignore the horizontal component of heat conduction. The wavelength of surface heat flow variations provides important information on the depth distribution of heat-producing elements: the contribution of deep layers of high heat production to surface heat flow is wiped out by lateral heat conduction caused by lateral heat production variations in the crust, so that only the near-surface heat-producing bodies affect surface heat flow variations (Jaupart, 1983). This observation effectively allows for separation of surface heat flow anomalies produced by lateral and vertical variations in crustal heat production from anomalies of subcrustal (mantle) origin: the latter will produce detectable surface heat flow perturbations with wavelength of several hundred kilometers (Fig. 4.10). The effect of lateral variations in heat production on surface heat flow is discussed in more detail later.

A combination of (i) heat flow observations with (ii) lower crustal and mantle xenolith data, (iii) models of crustal structure and evolution, and (iv) laboratory data on heat production in rocks has been used to estimate the composition of the continental crust and its bulk heat production (Taylor and McLennan, 1985; McLennan and Taylor, 1996; Rudnick *et al.*, 1998; Rudnick and Nyblade, 1999). These and other similar studies indicate that in Precambrian terranes crustal contribution to surface heat flow is between 20 and 44 mW/m^2 and typically makes >50% of observed surface heat flow. However, at present any tight constraints on the relative contributions of the crust and mantle to surface heat flow are not possible. For example, a comparison of conductive geotherms based on model concentrations of K, Th, and U isotopes in the Archean continental crust with xenolith geotherms

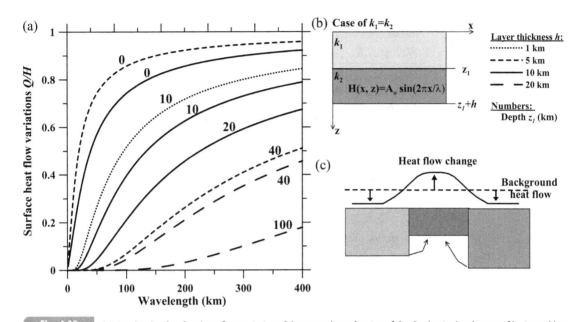

Fig. 4.10 (a) Amplitude of surface heat flow variations Q (expressed as a fraction of the Q value in the absence of horizontal heat conduction) produced by a horizontal layer with laterally sinusoidal (with wavelength λ) and vertically uniform distribution of heat production over thickness h (schematically shown in insert b). The layer is located at depth z_1 (the values of z_1 are indicated by numbers on the curves) and generates additional heat flow H. Surface heat flow is poorly sensitive to radioactivity contrasts located at large depth, and short wavelength radioactivity variations are not reflected in the surface heat flow. A heat flow anomaly (of any origin) in the mantle will produce a surface heat flow anomaly of wavelength of several hundred kilometers. (c) Horizontal heat transfer smoothes lateral variations in radioactivity: heat flow in the thin block is enhanced by contributions from the deep parts of the neighboring blocks where, in turn, heat flow becomes reduced. (After Jaupart, 1983.)

(see Chapter 5) indicates that either the crust should have very low heat production or that xenolith geotherms reflect frozen-in mineral equilibria and correspond to a transient thermal regime associated with magmatism (Rudnick *et al.*, 1998). In the latter case, crustal heat production can be relatively high.

Heat production in the mantle

Little is known about heat production in the lithospheric mantle. The only direct information is based either on studies of mantle xenoliths or massif peridotites (fragments of lithospheric mantle in Phanerozoic fold belts). There is a large variation in heat-producing elements in peridotites from different settings (Table 4.6). Mean values of heat production calculated for massif peridotites from Phanerozoic fold belts and for lithospheric peridotites carried in alkali basalts in Proterozoic continental regions are 0.018 $\mu W/m^3$ and 0.033 $\mu W/m^3$, respectively. In cratonic settings, mean values of heat production range from 0.028 $\mu W/m^3$ in non-

Table 4.6 Heat production in mantle peridotites (in $\mu W/m^3$)		
Settings	Mean	Median
Off-craton, massif peridotites	0.018	0.006
Off-craton, spinel peridotites	0.033	0.013
On-craton, all	0.093	0.044
On-craton, kimberlite-hosted	0.104	0.050
On-craton, non-kimberlite-hosted	0.028	0.019

(Source: Rudnick *et al.*, 1998)

kimberlite hosted peridotites to $0.104\,\mu W/m^3$ in peridotites contaminated by kimberlite magmas, i.e. *mantle metasomatism increases the content of heat-producing elements* (Rudnick *et al.*, 1998). In contrast, peridotites from the Phanerozoic Oman ophiolite have extremely low heat production ($0.002\,\mu W/m^3$) that is comparable to heat production in the N-MORB mantle source (Jochum *et al.*, 1983).

High heat production values in the lithospheric mantle may be difficult to reconcile with xenolith geotherms from the Archean cratons and surface heat flow, since for Kaapvaal data the best agreement is achieved for mantle heat production values between 0 to $0.03\,\mu W/m^3$ (Rudnick and Nyblade, 1999). This would imply that the concentration of heat-producing elements in cratonic peridotites may be non-representative of the Archean lithospheric mantle (Rudnick *et al.*, 1998). A possible explanation for radioactive enrichment of continental lithosphere is that radioactive elements can accumulate there during metasomatic processes (Petitjean *et al.*, 2006). However, for the Kaapvaal craton geotherms calculated for mantle peridotite values as high as 0.07–$0.08\,\mu W/m^3$ still fall within the 95% confidence limits for the P–T xenolith data (Rudnick and Nyblade, 1999).

Geochemical studies indicate that the amount of apatite, which is widespread in Phanerozoic lithospheric mantle, may be greatly underestimated (O'Reilly and Griffin, 2000). However, apatite has high concentrations of U and Th (60 ppm and 200 ppm, respectively), and its abundance may have a dramatic effect on the heat budget of the mantle and lithospheric temperatures. Estimates show that in the case where the apatite content in the lithospheric mantle is ~0.5%, heat production of the lithospheric mantle due to apatite alone would be ~$0.2\,\mu W/m^3$, and apatite abundance in a 70 km-thick Phanerozoic lithospheric mantle would contribute *c.* 12 mW/m^2 to mantle heat flow.

A commonly assumed value of $0.01\,\mu W/m^3$ for heat production of the lithospheric mantle (Table 4.3) can be justified by a simple analysis of the heat balance of the Earth (Petitjean *et al.*, 2006). If heat production in the primitive mantle is 5 pW/kg (Jochum *et al.*, 1983; McDonough and Sun, 1995) and bulk heat production in the crust is known (e.g. the values reported by Rudnick and Fountain, 1995), then the radiogenic heat generated in the present-day mantle is 3.15 pW/kg and, assuming a mantle density of 3300 kg/m^3, its heat production is $0.01\,\mu W/m^3$ (see p. 232).

Vertical and lateral variations in heat-producing elements

Correlation between heat flow and heat production

Early analyses of heat flow and heat production measurements in plutonic rocks demonstrated that these two parameters are related linearly (Birch *et al.*, 1968; Roy *et al.*, 1968) (Fig. 4.11a). This empirical observation led to the concept of a "heat flow province", i.e. a region in which surface heat flow Q is related to surface radioactivity A_0 as:

$$Q = Q_r + D\, A_0 \qquad (4.20)$$

where Q_r is termed the "reduced heat flow" and the slope parameter D with units of length is termed the "characteristic depth". The physical meaning of the constants Q_r and D permits several interpretations. Constant Q_r is commonly interpreted as the heat flow below the uppermost crustal layer of thickness D (typically $D \sim 10\,\text{km}$), where most of the heat-producing elements are concentrated (i.e. granitic upper crust).

Two observations are important in relation to equation (4.20):

(1) The heat flow province concept is a thermal, but not a petrological concept, and thus a single tectonic structure may comprise several "heat flow provinces".
(2) In general, the correlation between heat flow and near-surface heat production weakens when data from both plutonic and metamorphic sites are included, and the smallest scatter of data around the best-fit line is observed when all data are from plutonic sites.

Statistical correlation between heat flow and near-surface heat production has been observed in most of the continental regions (see Artemieva and Mooney, 2001 for an overview). It has been argued, however, that in some regions relationship (4.20) does not hold, in particular when data from sites of diverse tectonic origin are forced into a single "heat flow province". The Paleoproterozoic Trans-Hudson Orogen (the central Canadian Shield) is an example. The orogen is formed by several distinct belts with different origins and lithologies: the Flin Flon Belt is made up of arc volcanic and plutonic rocks; the Lynn Lake Belt is formed by island arc volcanic rocks; the Thompson Nickel Belt contains metasedimentary rocks; other domains are made up of Archean gneisses. While a correlation between surface heat flow and surface radioactivity hardly exists for the entire orogen, the two parameters are correlated statistically within the individual orogen-forming belts, each of which apparently makes an individual "heat flow province" (Fig. 4.11b).

Since equation (4.20) is effectively 1D and it ignores lateral heat transfer, the scatter around the linear fit is, in part, caused by 2D and 3D heat flow effects which smooth lateral variations in heat production. The effect of horizontal variations in heat production in the crust on surface heat flow and thus on the $Q - A_0$ relationship (4.20) has been examined for harmonic (Jaupart, 1983) and random (Vasseur and Singh, 1986) lateral distributions of heat sources in the crust with a constant or exponential vertical distribution. The analytical expressions for a cylindrical pluton and for a 2-layer structure with variable thermal conductivities in the layers and for different cases of uniform and exponential depth distribution of heat production (with laterally uniform or sinusoidal distribution within the layers) can be found in Jaupart (1983). The results of these studies indicate the following:

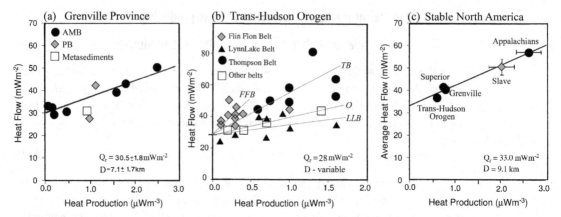

Fig. 4.11 Plots of heat flow as a function of radiogenic heat production in different tectonic provinces of stable North America (after Pinet *et al.*, 1991; Mareschal *et al.*, 1999; Perry *et al.*, 2006). (a) Data from the crystalline terranes of the Allochtonous Monocyclic Belt (AMB) and the Parauchtonous Belt (PB) of the Grenville province (the eastern Canadian Shield). Most of the measurements were made on low and intermediate grade metamorphic rocks (charnockites, granites, gneisses) and anorthosites. The parameters Q_r and D are for the best-fit linear relation through these data (Eq. 4.20). (b) Data from the Paleoproterozoic Trans-Hudson Orogen (the central Canadian Shield). The orogen is formed by several distinct belts with different origins and different lithologies. The correlation between heat flow and heat production is marginal for the entire orogen due to fundamental geological and geochemical contrasts between the component belts. However, data for the individual belts will fit linear relations with the same values of Q_r for all belts but with different D. (c) Average values for the Superior provinces and the Slave craton (for the latter the amount of data remains limited) of the Canadian Shield, and data for the Trans-Hudson Orogen (excluding the Thompson Belt), the Grenville Province, and the Appalachians.

- Lateral heat flow smoothes variations in heat production so that only shallow anomalies are well reflected in surface heat flow (Fig. 4.10).
- The slope D (depth-scale), due to the averaging effects of lateral heat transfer, is an average of various thicknesses of radioactive layer in the region of study (Fig. 4.11b).
- The wavelength of lateral variations in radioactivity is important: heat flow smoothing by lateral heat transfer in the crust is efficient over lateral distances of ~350 km (Jaupart, 1983). For example, for $D = 10$ km (a typical estimate of the D-layer thickness in many continental regions) all heat production anomalies of wavelength less than ~100 km will be reduced to less than 80% of their amplitude in surface heat flow (Fig. 4.10).
- In almost all cases, lateral heat flow results in an apparently lower D-value as obtained from the heat flow–heat production plot than the actual thickness of the radioactive layer. The amplitude of this reduction depends on the radial distance over which the (randomly distributed) heat production is correlated. If the thickness of the radioactive D-layer is much smaller than the scale of horizontal fluctuations in radioactivity, vertical heat flow prevails and the effect of lateral heat production heterogeneity on Q and on linear correlation (4.20) is insignificant. However, if the scale of horizontal fluctuations in radioactivity is small compared to D, the contribution of lateral heat flow to surface heat flow variations becomes important (Vasseur and Singh, 1986).

- The fact that the apparent depth-scale D is more sensitive to horizontal variations in radioactivity than to the actual thickness of the D-layer questions the meaning of a single linear heat flow–heat production correlation. To resolve this contradiction, Jaupart (1983) proposed that relationship (4.20) should have the form:

$$Q = Q_r + D_U A_U + D_{Th} A_{Th} + D_K A_K, \qquad (4.21)$$

where A_U, A_{Th}, A_K are heat production and D_U, D_{Th}, D_K are the corresponding depth-scales for uranium, thorium, and potassium, respectively, and in the simplest interpretation, it is D_{Th} that characterizes the depth scale of radioactive enrichment.
- Since horizontal heat flow caused by lateral variations in radioactivity effectively averages radiogenic heat production in the crust (over lateral distances of ~350 km), Q_r corresponds to a large-scale average of surface heat flow from which radioactivity contrasts in the shallow crust are subtracted and thus, it allows for calculation of lithospheric geotherms (Jaupart, 1983).

Depth variations of heat production

Theoretical constraints

A compilation of published D-values together with a global analysis of heat flow and heat production data for the regions, where linear relationship (4.20) is statistically valid, demonstrate significant regional variations of D with values ranging from 4 to 16 km. These variations do not seem to correlate with any physical or geological phenomena, which prevents straightforward interpretation of D in physical, petrologic, and tectonic terms. In particular, there is no global correlation between the D parameter and the age of crustal provinces, although within the same continent the D value commonly increases from the Archean to the Proterozoic terranes (Artemieva and Mooney, 2001).

Equation (4.20) implies that (in the case of 1D heat conduction) the vertical distribution of heat-producing elements within a heat flow province can be described by a single function. Several functions have been proposed for this purpose, the most common are linear (or step-like) and exponential (Fig. 4.12ab). The exponential function, which assumes that crustal radioactivity decreases exponentially with depth as $A(z) = A_o \exp(-z/D)$ has received special attention (Lachenbruch, 1970) since it is the only function that takes into account differential erosion at the surface. In the case of an exponential function, Q_r approximately corresponds to the base of the crust. The assumption of an exponential decrease of A with depth has however been challenged by numerous studies. Another way to meet the erosion argument is to question the assumption that the D-value remains constant through geological history (Jaupart, 1983).

A broad correlation between heat production and rock acidity (Fig. 4.8) together with lithological stratification of the crust (where acidity decreases with depth while mafic assemblages increase with depth) suggests that there should be a general trend of heat production decrease with depth. This conclusion is, in general, supported by regional studies on exposed cross-sections of the continental crust (e.g. Swanberg et al., 1974; Nicolaysen et al., 1981; Weaver and Tarney, 1984; Ashwal et al., 1987; Pinet and Jaupart, 1987; Ketcham, 1996); but, only a few reported vertical cross-sections show a simple pattern of

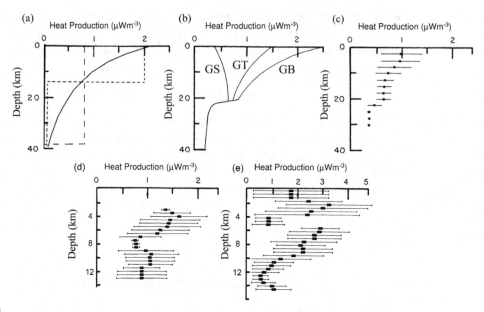

Fig. 4.12   Depth distribution of heat production in continental crust (a–c from Fountain *et al.*, 1987; d–e from Ketcham, 1996). (a) Theoretical models of exponential, linear, and constant heat production distribution. The bulk (integral) heat production of the crust is approximately the same for all three models shown. (b) Theoretical models of Allis (1979) for greenstone terrane (GS), gneiss terrane (GT), and granitoid batholith (GB). (c) Weighted average of heat production in the crust along the cross-section of exposed crust in the Pikwitonei structure (the Superior Province, Canada). (d–e) Depth distribution and standard deviations in two Arizona metamorphic core complexes obtained by a moving average. Note different depth scales in (d–e) and different horizontal scale in (e).

heat production that resembles a linear or an exponential function (Fig. 4.12c–e). In most cases, no systematic trend is observed for the vertical distribution of heat production.

Data from the Kola Superdeep Borehole

The most striking evidence for non-systematic variation of heat production with depth is provided by data from deep drillholes, such as the SG-3 in the Kola peninsula (Russia) and the KTB in Germany. Unfortunately, the number of publications in international journals describing the unique data from the Kola borehole is limited. Several monographs published in Russian summarize major scientific results (e.g. Kola Superdeep, 1984; 1998).

Scientific drilling in the Kola peninsula started in 1970 and a depth of 7263 m was reached in April 1975. This marked the end of the first stage of drilling, after which no further drilling took place until October 1976, when the second drilling stage, based on different technology and equipment, started. At the second stage, several drillholes were drilled below a 7 km depth mark and in 1984 a depth of 12 000 m was reached (Fig. 4.13). The temporary termination of drilling for 1.5 year in 1975–1976 is important from a scientific point of view: thermal equilibrium in the borehole was re-established during this time and, as a result, high quality thermal measurements were made in 1976 down to ~7 km depth. Thermal parameters measured during the next decade down to ~9 km depth and deeper are less reliable since they were measured under transient thermal conditions.

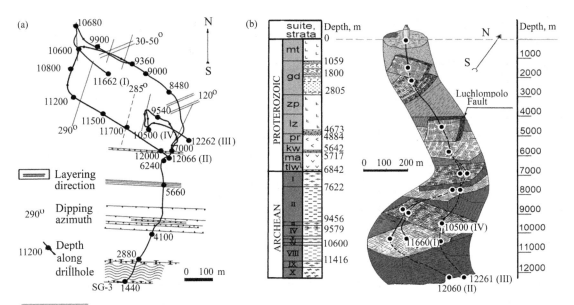

Fig. 4.13 Kola Superdeep Borehole SG-3 (after Gorbatsevich and Smirnov, 1998). Structural and anisotropic properties of the upper-middle crust estimated from the deviation of the drillholes from the vertical (left and right panels) shown together with (a) horizontal projections of four drillholes (marked I to IV) of SG-3, (b) lithological and structural cross-section along the borehole. A simplified list of rock lithology along the boreholes includes: metapelitic tuff of ultrabasic composition and metadiabases at 0–1 km depth; metaphyllites, metatuffs, metadiabases, metasandstones at 1–2.8 km depth; schistozed metadiabases, amphibolitized metagabbro-diabases at 2.8–4.7 km depth; dolomites, sandstone, and dolomitized sandstone at 4.7–4.9 km depth; schists at 4.9–5.6 km depth; quartzitic sandstones, carbonate-micas, and quartz-micas schists at 5.6–5.7 km depth; amphibolites and schists at 5.7–6.8 km depth; pegmatites, gneisses, biotite-amphibole-plagioclase paraschists at 6.8–7.6 km depth; gneiss-schist, amphibolites, migmatites down to a 10 km depth (data source: Atlas of physical rock properties from the cross-section of Kola Superdeep Borehole; Moscow, 1982, Ministry of Geology, USSR. Unpublished Report).

The upper 6842 m of the superdeep borehole SG-3 cut Paleoproterozoic plutonic rocks of the Karelian basement; the Archean granites and metamorphic rocks extend from 6842 m to 12 262 m (Fig. 4.14). The age of the Archean basement is 2.93–2.76 Ga (the U–Pb method), with the oldest rocks being ~3.15–2.89 Ga old (the Sm–Nd method). The major Luchlompolo fault crosses the borehole at a depth of ~4.8 km.

Heat production in the Proterozoic formations varies from ~0.4 $\mu W/m^3$ in plutonic rocks to ~1.3 $\mu W/m^3$ (with maximum values of ~2.5 $\mu W/m^3$) in metasediments (Fig. 4.14). The peak in heat production at *c.* 4.8 km depth correlates with the major fracture zone, and high heat production is explained by fluid migration and redistribution of radioactive elements. Heat production in the Archean gneisses is between 0.8 and 1.2–1.7 $\mu W/m^3$. On the whole, no decrease in heat production is observed down to ~12 km depth. Laboratory measurements on core samples from the Kola drillhole indicate that heat production is controlled primarily by U and Th content; the content of Th increases through the entire sampled depth, both in the Proterozoic and in the Archean rocks (Kola Superdeep, 1984; 1998).

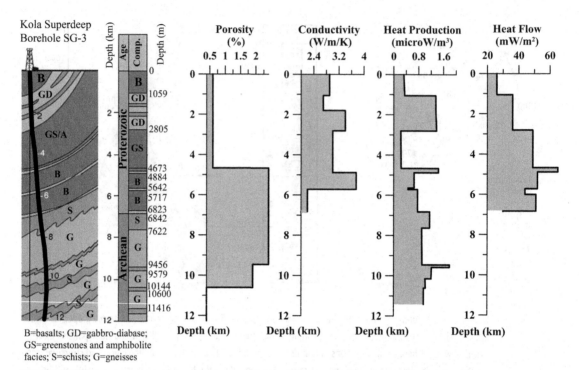

B=basalts; GD=gabbro-diabase;
GS=greenstones and amphibolite
facies; S=schists; G=gneisses

Fig. 4.14 Schematic cross-section of the Kola Superdeep Borehole SG-3 (Kola Peninsula, Baltic Shield, Russia) and depth distribution of several parameters (porosity, thermal conductivity, and heat production averaged for each lithological unit, and heat flow). Note that most variations in all parameters correlate with changes in rock lithology. Peaks in heat production and heat flow at c. 4.8 km and 9.5 km depths correlate with major fracture zones. Measurements of thermal parameters are highly reliable down to ~7 km depth (the borehole was sealed for 1.5 years prior to measurements to re-establish thermal equilibrium); thermal measurements down to ~9 km depth were made at transient thermal conditions (not shown). Thermal gradient is c. 16 °C/km at 2.8–4.3 km, 16.5–20 °C/km at 4.3–5.0 km, 22 °C/km at 5.0–6.6 km, 25 °C/km down to 6.8 km, 15 °C/km at 6.8–7.0 km, and 23 °C/km at 7.0–7.2 km depth. Near-surface heat flow (in the upper 1.0 km) is 26 ± 2 mW/m^2 and increases to 36 ± 4 mW/m^2 at 1.0–2.8 km depth. Heat flow at depths of 2.8–4.3 km (49 ± 1 mW/m^2), where rock composition is relatively homogeneous and porosity is low, is considered to be representative of a deep thermal regime. A localized narrow zone of anomalously high heat flow (65 ± 7 mW/m^2) at depth 4.3–4.9 km is probably caused by fluid circulation in the Luchlompolo fault zone. Data sources: http://superdeep.pechenga.ru/; Kola Superdeep, 1984; 1998.

One of the big surprises from the Kola borehole is data on surface heat flow. As mentioned earlier, high quality measurements of thermal gradient are available down to ~7 km depth. The average value of the thermal gradient increases with depth from 10 °C/km at near-surface to ~18 °C/km in the lower levels of the borehole (~7 km). Changes in thermal gradient approximately correlate with boundaries between different formations (suites) (Fig. 4.14). The temperature at 12 km depth is 212 °C, in contrast to the expected 120 °C. High crustal temperatures are likely to be caused by high radioactivity in the Archean basement. Based on borehole data, the temperature at 42 km depth (Moho?) is expected to be as high as 580 °C.

Thermal conductivity along the entire depth of the borehole varies from 2.5 to 3.6 W/m/K with typical values ~2.8–2.9 W/m/K. Data on thermal conductivity and thermal gradient allow for calculation of heat flow, which surprisingly is highly variable along the borehole depth and, in contrast to expectations, increases in the upper 5 km of the cross-section. Heat flow in the upper 1.0 km is 26 ± 2 mW/m^2 and reaches 49 ± 1 mW/m^2 at depths of 2.8–4.3 km. It is heat flow at this depth interval, where the rock composition is relatively homogeneous and porosity is low, that is considered by some authors to be representative of a deep thermal regime of the region. Low values of near-surface heat flow are attributed to the effects of water circulation, recent glaciation, and denudation. A localized narrow zone of anomalously high heat flow (65 ± 7 mW/m^2) at depth 4.3–4.9 km is probably caused by fluid circulation in the Luchlompolo fault zone. There is a sharp decrease in heat flow below the fault, to 48–56 mW/m^2 with a further decrease below 6.8 km depth (since measurements below this depth were made at transient thermal conditions, they should be treated with caution).

To conclude, the patterns of depth distribution of porosity, thermal gradient, heat production, heat flow (and many other parameters) measured in the upper 12 km of the crust of the Baltic Shield are in strong contrast with expectations based on theoretical models. Geothermal studies from several other deep drillholes worldwide show similar complicated patterns of depth distribution of thermal parameters and heat flow. It is still debated which heat flow values should be considered as "surface heat flow" in each of the cases.

Reduced heat flow

In relation (4.20), Q_r is the difference between the surface heat flow and the product DA_o, which is the heat generated within the upper crustal layer enriched in heat-producing elements. Thus the value of Q_r gives a rough approximation of the non-radiogenic component of heat flow. Sometimes Q_r is equated with heat flow at the Moho (mantle heat flow). However, this is only correct for some specific assumptions on depth variations of heat production (Fig. 4.12a). In most cases, the term Q_r does not incorporate heat generated by radioactive isotopes in the middle and lower crusts. The latter can amount to 10–15 mW/m^2.

In spite of these limitations, Sclater *et al.* (1980) suggested that

> Q_r *must have a general geophysical significance for old continental crust.*

A global analysis of the heat flow–heat production relationship does indeed show a clear trend with an increase in Q_r value with age from Archean through Proterozoic to Paleozoic regions (Fig. 4.15). The mean values range from ~20 mW/m^2 in the Archean regions of West Africa and Eurasia to ~30 mW/m^2 in the Archean regions of South Africa, Canada, South America, and Australia, and to ~30–35 mW/m^2 in the Proterozoic regions with a scatter in Q_r values of ~20 mW/m^2 between different cratons of a similar age (Artemieva and Mooney, 2001). A decrease Q_r with age is likely to imply a decrease in mantle heat flow beneath the ancient continental regions. However, one should not forget that

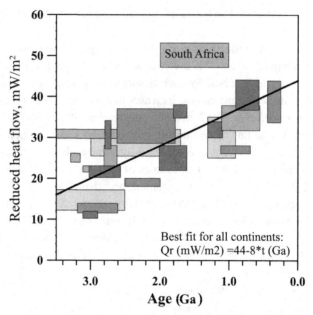

Fig. 4.15 Reduced heat flow versus age (after Artemieva and Mooney, 2001). Boxes correspond to individual continental terranes; the horizontal size shows age ranges, the vertical size reflects variations in estimated values of the reduced heat flow obtained by different authors. South Africa remains apart from the general trend.

estimates of Q_r are biased by both the limitations associated with Eq. 4.20 (in particular, a quasi-1D assumption) and the accuracy of heat flow and heat production measurements.

Values of Q_r for the Archean cratons are close to the estimates of mantle heat flow in the best-studied Archean cratons of Canada and South Africa (see Table 4.11). Based on the analysis of surface heat flow and crustal structure, heat flow through the Moho is estimated to be ~10–15 mW/m^2 in the Canadian shield and ~17 mW/m^2 in the Kaapvaal craton (Jaupart and Mareschal, 1999). Xenolith geotherms for the Kalahari craton (South Africa) are fit best by thermal models which imply ~18–19 mW/m^2 for heat flow across the Moho (Rudnick and Nyblade, 1999). A comparison of Q_r with mantle heat flow in these cratons indicates that only about 10–15 mW/m^2 should be generated in the crustal radiogenic layer (see Section 4.4.2 for details). If this layer is 10 km thick, its mean heat production should amount to 1.0–1.5 μW/m^3.

A decrease in Q_r with age correlates with a similar trend for surface heat flow on the continents, and worldwide on the continents the difference between surface and reduced heat flow appears to be almost independent of crustal age, ~17 mW/m^2 for best-fit linear approximations, with slightly lower values for the Archean cratons and slightly higher values for the Phanerozoic regions (Artemieva and Mooney, 2001). Although this observation apparently argues in favor of a nearly uniform contribution by crustal radioactivity DA_0 into variations in surface heat flow, one should not forget that this global generalization does not necessarily hold in individual geological provinces.

4.2 Heat flow data

This section provides an overview of heat flow observations worldwide. Because of the principal differences in the thermal structure of the oceanic and the continental lithosphere, they will be discussed in detail in Sections 4.3 and 4.4, respectively.

4.2.1 Global compilations of surface heat flow

Measurements of temperature gradient in near-surface boreholes together with data on thermal conductivity of near-surface rocks allow calculation of surface heat flow (Eq. 4.5). The uncertainty (accuracy) of high quality surface heat flow measurements is ~15% and is associated with:

(1) the accuracy of analytic measurements of thermal conductivity (7–10%) and temperature gradient (<5%);
(2) natural factors, some of which cannot be reliably assessed (groundwater circulation, denudation history, changes in paleoclimate); the effect of natural factors on the quality of heat flow data is minimized by eliminating measurements from shallow (<100–300 m) boreholes;
(3) the conditions for borehole measurements (in particular, if steady-state thermal regime was re-established in the borehole after drilling);
(4) the assumption on quasi-vertical heat flow (Eq. 4.5); this assumption fails in the case of lateral heat transfer due to high contrasts in thermal properties of the crust; in such cases borehole measurements may significantly underestimate (overestimate) the actual heat flow from the deep interior;
(5) the assumption on conductive origin of surface heat flow (Eq. 4.5); this assumption fails in regions where the convective (and advective) heat flow component is expected to be significant and thus borehole measurements can be non-representative of the deep thermal regime. Such regions include (a) tectonically active regions (in particular, continental regions of magmatism and mid-ocean ridges), and (b) ocean floor where water circulation through the oceanic crust may lead to an underestimate of heat flow (see Section 4.3).

Heat flow measurements were initiated in the UK at the end of the nineteenth century (Everett, 1883). However, the first modern measurements of terrestrial heat flow began only in the late 1930s on the continents (Benfield, 1939; Bullard, 1939; Krige, 1939) and the 1950s in the oceans (Revelle and Maxwell, 1952). The first compilation of available heat flow measurements into a database is almost half a century old (Lee and Uyeda, 1965) and it included about 1000 measurements, most of them off-shore. Progress in thermal studies has resulted in routine measurements of heat flow and a fast accumulation of new data, which by 1976 included ~1700 on-shore and ~3700 off-shore measurements (Jessop et al., 1976). This compilation formed the basis for the first global analysis of heat flow data and global heat loss which was performed in two classical studies of Chapman and Pollack (1975) and Sclater et al. (1980). The major results of these studies remain undisputed, although the number of heat flow measurements has increased dramatically over the past decades.

The heat flow compilation by Jessop *et al.* (1976) made the basis of the most recently published global heat flow database which includes a total of 24 774 heat flow measurements (Pollack *et al.*, 1993). The database includes all available measurements but with a large diversity in data quality and reliability. Low quality heat flow measurements include primarily industrial measurements in shallow boreholes affected by groundwater circulation, in boreholes where a steady-state thermal regime was not achieved at the time of measurement, and measurements using non-conventional techniques (in particular, in South America). The database can be downloaded at: http://www.heatflow.und.edu; interpolated values of heat flow based solely on high-quality measurements can be downloaded at: www.lithosphere.info (Fig. 4.16). Up to date, the compilation

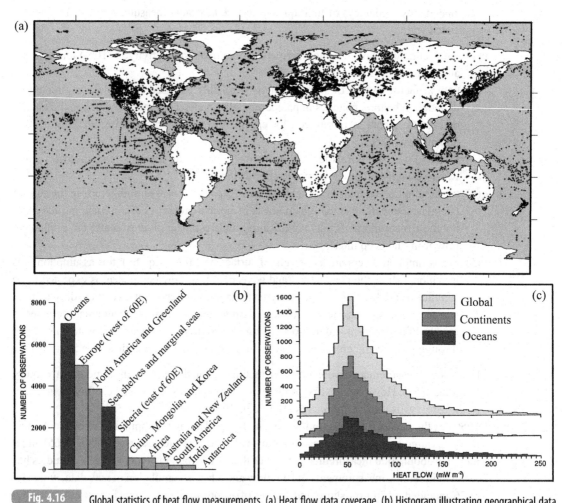

Fig. 4.16   Global statistics of heat flow measurements. (a) Heat flow data coverage. (b) Histogram illustrating geographical data coverage (based on an updated compilation of Pollack *et al.*, 1993). (c) Statistics of measured heat flow values (redrawn from Pollack *et al.*, 1993). Since 1993, the number of borehole measurements has substantially increased for North America, India, China, South America, and the oceans. However, new data do not affect the global statistics.

of Pollack and co-authors remains the most complete published database of surface heat flow. Newer versions updated for continuing borehole measurements are becoming available at the websites of the IHFC (the International Heat Flow Commission) and NOAA (National Oceanic and Atmospheric Administration).

4.2.2 Global trends in surface heat flow

Fig. 4.16a,b displays global coverage by heat flow measurements. Although at present the number of heat flow measurements is twice the size it was in 1993 (in particular, a significant number of new data became available in Canada, India, Norway and in the oceans), the general pattern of data coverage has not changed significantly: heat flow measurements are still available for only about 50% of the globe. Most measurement campaigns have focused either on easily accessible areas or on tectonically important regions. No, or very few, measurements have been taken in large areas of South America, Arctic North America, Africa, Arabia, Greenland, and Antarctica. In the oceans, most of the measurements have been made in the tropical and middle latitudes, with very few data in high latitudes. Even the oceans with a sufficient number of measurements suffer from patchy data coverage. In spite of the existence of "white spots", uneven coverage, and geographical biases, which result in significant spatial limitations for thermal models, several robust features can be clearly recognized in the available data. They allow statistical interpretation of data and meaningful conclusions on the thermal regime of the Earth's interior.

- Continental and oceanic distributions of heat flow appear very similar (Fig. 4.16c). However, this similarity is misleading because of a strong hydrothermal effect on measurements in the oceans (see discussion in Section 4.3.3).
- The mean heat flux on continents is lower than in the oceans (Figs. 4.16c; 4.17), 65 mW/m^2 and 101 mW/m^2, respectively, with a global mean of 87 mW/m^2.
- Surface heat flow is highly variable (Fig. 4.18); strong lateral variability is observed even at small distances. The reported values range from near-zero to several hundred mW/m^2, with the most common values 0–125 mW/m^2 and a range of vertical temperature gradient of 10–80 °C/km.
- In general, higher heat flow is observed in young structures both on continents and in the oceans (Fig. 4.19). In oceans the highest values typically correlate with mid-ocean ridges.

4.3 Thermal regime of oceanic lithosphere

4.3.1 Age dependence of seafloor topography and heat flow

According to plate tectonics, oceanic lithosphere is created at mid-ocean ridges by passive upwelling of hot mantle material (magma). Magma cools to form oceanic lithosphere which behaves like a plate and, as more magma is intruded at the spreading center, it moves away from the ridge. The combined effect of ocean spreading and lithosphere consumption at subduction zones results, in general, in a decrease in ocean area with age (Fig. 4.20) (Sclater

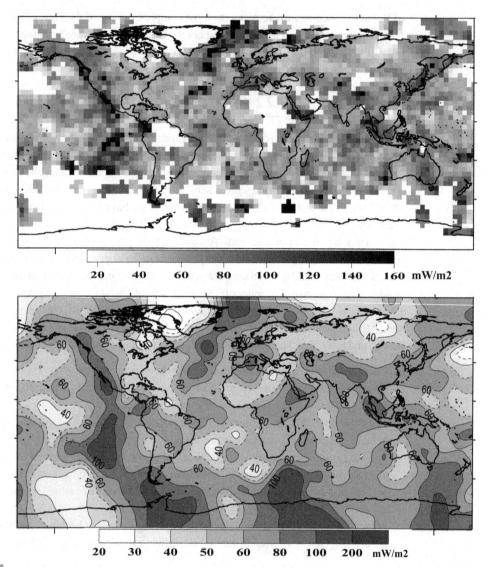

Fig. 4.17   Global surface heat flow (based on an updated compilation of Pollack *et al.*, 1993). Top: averaged on a 5 deg x 5 deg grid; bottom: global interpolation with a low-pass filter. Solid lines – heat flow contours with a 20 mW/m^2 step; dashed lines – contours of 30 and 50 mW/m^2. Small-scale anomalies are smeared both laterally and in amplitude.

et al., 1980). Most of the young oceanic lithosphere is generated in the fast spreading South Pacific Ocean (the spreading rate along the East Pacific Rise is up to 130 mm/yr). Oceans with slow spreading rates, such as the Atlantic and Indian Oceans, where the spreading rate is *c*. 4 times slower than in the South Pacific, have area–age distributions similar to the North Pacific Ocean, which has only the western branch.

The general age dependence of ocean floor topography and oceanic heat flow has been recognized since the 1960s (e.g. von Herzen and Uyeda, 1963; Langseth *et al.*, 1966; Sclater

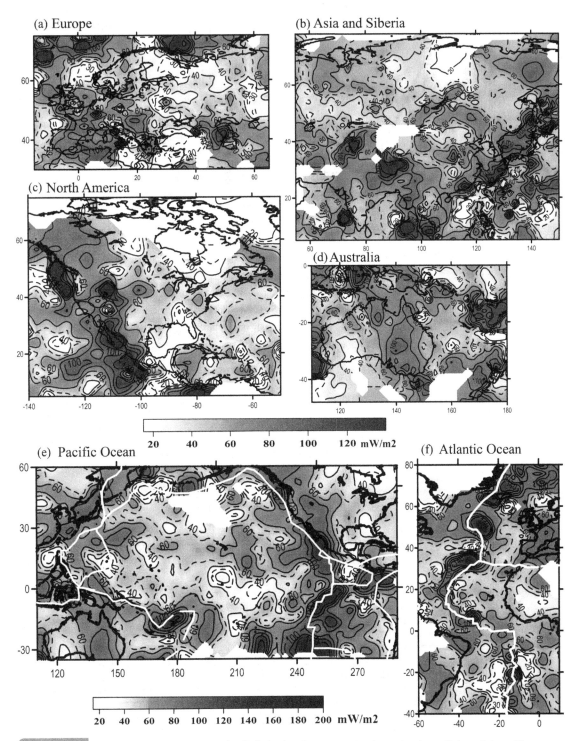

Fig. 4.18 Surface heat flow in regions with a high density of measurements, interpolated on a 5 deg x 5 deg grid:
(a) Europe; (b) Asia and Siberia; (c) North America; (d) Australia and New Zealand; (e) Pacific Ocean; (f) Atlantic Ocean.
White – areas with no heat flow data; white lines in (e–f) – plate boundaries.

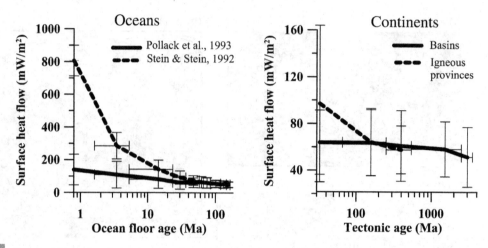

Fig. 4.19 Mean heat flow values for continental and oceanic regions versus age (Source: Pollack *et al.*, 1993; Stein and Stein, 1992). Bars – standard deviations. The large discrepancy in two sets of estimates for young (<20–30 Ma) oceans is caused by the strong effect of sea water circulation through the oceanic crust that is not covered by deep-sea low-permeability sediments (see discussion in Section 4.5). Conventional heat flow measurements only determine the conductive component of heat flow and thus significantly underestimate the heat loss in young oceans. On the continents, a significant difference exists between stable and young (Meso-Cenozoic) tectonic regions.

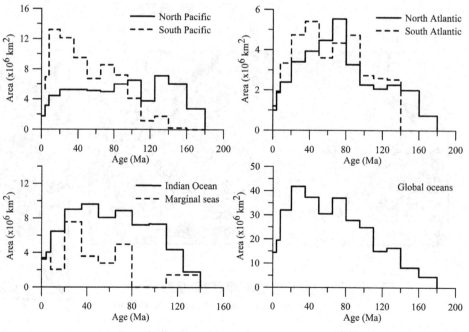

Fig. 4.20 Ocean floor area as a function of ocean age (after Sclater *et al.*, 1980). Note different vertical scales.

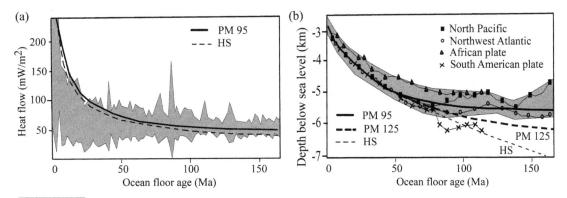

Fig. 4.21   Heat flow (a) and depth below sea level (b) in the oceans as a function of ocean age (after Stein and Stein, 1992). Gray shading – upper and lower bounds of observations. Thick solid and dashed lines in (a,b) – theoretical predictions for plate models (PM) with 125-km- and 95-km-thick plates (Parsons and Sclater, 1977; Stein and Stein, 1992). Thin dashed line labeled HS – theoretical predictions based on instantaneous boundary layer cooling of an infinite half-space for the same parameters as the plate models. Symbols and thin lines in (b): depth data averaged for 5 Ma bins for four oceans. Significant differences between plates hamper any global reference model for ocean bathymetry.

and Francheteau, 1970). The highest heat flow and the shallowest topography are observed at mid-ocean ridges; heat flow decreases from more than 200 mW/m^2 in the youngest oceanic lithosphere near mid-ocean ridges to 50–60 mW/m^2 in the old (>70 Ma) ocean floor, while ocean depth, in general, increases from ~2.5–3 km in young oceans to ~5–6 km in old oceans (Fig. 4.21; compare with Fig. 3.121). Age variations in seafloor bathymetry and heat flow can be approximated by the square root of ocean age: $1/\sqrt{\text{age}}$ for depth variations and $\sqrt{\text{age}}$ for heat flow variations (Davis and Lister, 1974). There is, however, a large scatter in topography, especially for the old oceanic lithosphere, where most of the seamounts, swells, and oceanic plateaus are located. Because of hydrothermal circulation, for heat flow the square root of age law only holds approximately for crust older than ~2–6 Ma, covered by a thick layer of impermeable sediments (Fig. 4.22); old oceans (>80 Ma), however, also may deviate from a simple square root of age relationship.

4.3.2 Normal oceans

Cooling, contraction, and sedimentation of oceanic lithosphere explain qualitatively heat flow decrease and ocean depth increase with ocean floor age. Cooling of lithospheric plates is described by the heat conduction equation (e.g. Turcotte and Schubert, 1982). The boundary condition at the base of the plate can be specified either by heat flux (with variable plate thickness) or by temperature (with fixed plate thickness). Shear heating associated with plate motion (Schubert et al., 1976) or a uniform distribution of radioactive elements in the upper 300 km of the mantle (Forsyth, 1977) have been proposed as maintaining constant heat flux at the lithospheric base. In contrast, small-scale convection in the upper mantle has been proposed as a mechanism for maintaining constant temperature at the lithospheric base (McKenzie, 1967; Richter, 1973). The choice of boundary condition implies different

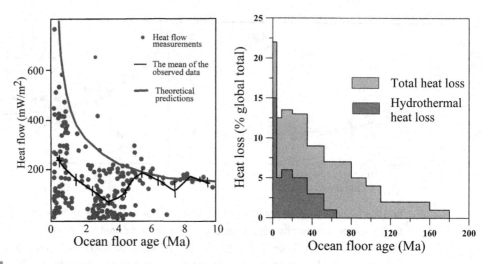

Fig. 4.22 Heat flow near ocean spreading centers. Left: Heat flow at the Galapagos spreading center (Source: Anderson and Hobart, 1976; modified after Sclater *et al.*, 1980). Right: Estimated heat loss through oceanic provinces (Source: Sclater *et al.*, 1980).

assumptions about the nature of the thermal boundary layer and results in two different types of models of thermal evolution of the oceanic lithosphere: cooling half-space and plate model. In oceanic thermal models, radiogenic heat production is commonly ignored because radiogenic elements (which are present in the mantle in very low concentrations, see Section 4.1.4) are further removed from the mantle during melt generation.

Cooling half-space model

Model predictions and empirical relationships

Conductive cooling of a half-space (an oceanic plate as it moves away from the mid-ocean ridge axis where it has been created) explains the observed $1/\sqrt{\text{age}}$ variations in seafloor depth and $\sqrt{\text{age}}$ heat flow variations for the young oceans (Davis and Lister, 1974; Turcotte and Schubert, 1982; Carlson and Johnson, 1994). Temperature variation $T(\text{age}, z)$ with depth z and ocean floor age is given by:

$$T(\text{age}, z) = T_{\mathrm{m}} \, \mathrm{erf}(z\sqrt{4\chi \; \text{age}}) \qquad (4.22)$$

where T_{m} is the initial temperature of the half-space (mantle temperature), χ is the thermal diffusivity of the mantle,

$$\mathrm{erf}(x) = \frac{2}{\sqrt{\pi}} \int_0^x \mathrm{e}^{-y^2} \mathrm{d}y \qquad (4.23)$$

is the error function, and the thermal boundary layer thickens with time t as:

$$\delta \sim \pi\sqrt{\chi t}. \qquad (4.24)$$

In the real Earth, the geotherm approaches mantle adiabat T_m asymptotically. The constant on the right-hand side of eq. (4.24) depends on the choice of temperature T which is assumed to be close enough to T_m to be considered as the basal temperature of the thermal boundary layer. The thickness of oceanic lithosphere L (i.e. the thickness of the thermal boundary layer over the cooling half-space in the case of no radioactive sources) is (Schubert *et al.*, 2001, eq. 4.22):

$$L[\text{km}] = 2 \; \text{erfc}^{-1}(\Delta T)\sqrt{\chi t} \sim c_2\sqrt{\text{age}} \tag{4.25}$$

where $\text{erfc}(\Delta T) = 1 - \text{erf}(\Delta T)$ is the complementary error function of (ΔT) and age is in Ma. For a thermal diffusivity value $\chi = 10^{-6} \, \text{m}^2/\text{s} \sim 31.5 \, \text{km}^2/\text{Ma}$, and assuming that the base of TBL is defined by a 10% temperature anomaly with respect to T_m ($\Delta T = 0.1$), the constant $c_2 \sim 12.6 \, \text{km/Ma}^{1/2}$. For 5% and 20% temperature anomalies at the base of TBL, constant c_2 has similar values, $12 \, \text{km/Ma}^{1/2}$ and $14.5 \, \text{km/Ma}^{1/2}$, correspondingly.

For the oldest oceanic lithosphere of ~180 Ma which is present in the western Pacific Ocean (Fig. 2.22), formal solution of eq. (4.25) gives a lithospheric thickness of 150–175 km. These values are about twice the traditional geophysical estimates for old oceans (Fig 3.121). However, a recent tomographic model apparently indicates that a +2% Vs anomaly beneath the western Pacific persists down to ~120 km depth, and a +1% Vs anomaly down to 150–170 km (Fig. 3.99d).

Aging oceanic lithosphere, which thickens as $\sqrt{\text{age}}$, becomes heavier and its surface subsides. The depth to the unsedimented basement D (assuming that the cooling thermal boundary layer subsides by the Airy isostatic compensation) is:

$$D(\text{age}) = D_r + c_1\sqrt{\text{age}}, \tag{4.26a}$$

$$\text{where } c_1 = \alpha T_m\sqrt{4\chi/\pi} \; \rho_m/(\rho_m - \rho_w), \tag{4.26b}$$

D_r is the depth to the ridge crest, ρ_m and ρ_w are mantle and seawater densities, respectively, and α is the volume thermal expansion coefficient for mantle. A simple empirical relationship based on bathymetry and age data from the North Pacific and North Atlantic Oceans younger than ~80 Ma yields:

$$D[\text{km}] = 2.5 + 0.35 \; \sqrt{\text{age}}, \tag{4.26c}$$

where age is in Ma. In fact, the subsidence rate c_1 is commonly estimated to be between 336 and 344 m/Ma$^{1/2}$, but a commonly used value is 350 m/Ma$^{1/2}$. There are, however, large regional variations in parameters. For example, a number of systematic studies along the East Pacific Rise show variations in ridge depth from 2000 to 3200 m, and variation in the associated subsidence rate between 50 and 450 m/Ma$^{1/2}$.

Given that conductive heat flow across the boundary layer is $Q \sim \lambda\Delta T/\delta$ (where λ is thermal conductivity and ΔT is the temperature drop across the thickening boundary layer, compare with eq. 4.5), seafloor heat flow decreases with lithospheric age as $1/\sqrt{\text{age}}$. Assuming that the temperature at the basement surface is 0 °C, the surface heat flow Q (age) is:

$$Q(\text{age}) = \lambda \; T_m/\sqrt{\pi\chi \; \text{age}}. \tag{4.27a}$$

An empirical relationship for heat flow observations from oceans younger than ~80 Ma yields:

$$Q = 473/\sqrt{\text{age}}. \tag{4.27b}$$

Thus, variations in heat flow and bathymetry with ocean age jointly reflect changes in oceanic geotherms with age. The basal temperature estimated from the fit with observations is $T_m \sim 1335–1370\,°C$.

Mantle potential temperature

The basal temperature T_m, which is an important parameter which defines variations of heat flow and bathymetry with ocean age in the cooling half-space model (eqs. 4.22, 4.25–4.27), is the potential temperature of the mantle (i.e. the temperature of the mantle if it were decompressed adiabatically to the surface). Starting from the early 1970s, the potential temperature of the oceanic mantle has been studied experimentally from the geothermometry of oceanic tholeiites from different spreading centers. These provide the lower bound temperatures necessary to generate olivine tholeiites as a primary melt from the mantle source rock, and these estimates are not sensitive to the composition of the pyrolite model (Table 6.2). The range of all temperature values determined for rocks from the Pacific and Atlantic Oceans varies from ~1200 °C to ~1450 °C (e.g. Fig. 6.4) with median values around 1290–1350 °C. For a standard set of thermodynamic parameters (Section 4.1), it yields the adiabatic temperature gradient

$$(dT/dZ)_s = T\alpha g/C_p \sim 0.5 - 0.6\,°C/km, \tag{4.28}$$

where T is the absolute temperature, g is the gravitational acceleration, α is the coefficient of thermal expansion ($\alpha \sim 3 \times 10^{-5}$ 1/K), and C_p is the heat capacity at constant pressure ($C_p \sim 1000–1200$ J/kg/K). With increasing pressure, the value of α decreases; however this effect is compensated by an increase in temperature with depth, so that the adiabatic temperature gradient is approximately constant in the upper mantle. The above conclusions are based on an assumption that the mantle potential temperature beneath oceanic spreading centers is the same as elsewhere.

In agreement with these conclusions, McKenzie and Bickle (1988) argue that if "normal" oceanic crust of uniform thickness 7.1 ± 0.8 km (which is an average thickness of oceanic crust unaffected by plumes or fracture zones, White et al., 1992) is formed by decompressional melting, mantle potential temperature should be $T_m \sim 1315 \pm 13\,°C$.

Cooling half-space model and ocean floor flattening

The cooling half-space (boundary layer) model successfully explains the observed heat flow and bathymetry for ocean floor younger than ~70–80 Ma. However, it fails to predict the flattening of the ocean floor topography, observed (apparently, see below) for oceans older than ~80 Ma (Fig. 4.21). Moreover, the flattening of the seafloor starts at different ages in different oceans (Marty and Cazenave, 1989) and is often asymmetric with respect to ridge axes (e.g. Morgan and Smith, 1992). Several mechanisms have been proposed to explain ocean floor flattening:

- heat flow perturbations at the lithospheric base (e.g. due to asthenospheric flow, small-scale convective instabilities, or hotspots) that provide additional heat from the mantle, balance heat loss through the lithosphere, and prevent it from fast cooling (Crough, 1975; Oldenburg, 1975; Heest and Crough, 1981);
- pressure forces in the asthenosphere underneath the oceanic thermal boundary layer (TBL) (Phipps Morgan and Smith, 1992);
- a pre-existing chemical boundary layer (CBL, formed by melt extraction and, as a result, dehydrated and chemically buoyant) which might control the thickness of the TBL by affecting secondary convection at its base (i.e. the CBL might effectively work as a conductive lid not involved in mantle convection) (Lee *et al.*, 2005).

It should be noted that theoretical and observational studies over the last decade cast serious doubt over whether ocean floor flattening takes place in reality:

- in the 3D Earth, "ocean floor flattening" can be an artifact of theoretical predictions, since the $\sqrt{\text{age}}$ predictions are based on 2D thermal boundary-layer theory (Bercovici *et al.*, 2000);
- two recent case studies (Korenaga and Korenaga, 2008; Adam and Vidal, 2010) suggest that "ocean floor flattening" is an artifact of field observations (see below).

Plate model

Traditional approaches

To account for departure of heat flow and bathymetry from the $\sqrt{\text{age}}$ law in old ocean basins, the "plate model" in which the base of the TBL is considered as an isothermal boundary has been proposed (McKenzie, 1967; Richter, 1973). In this model, the plate (oceanic lithosphere) is generated at a constant temperature beneath a mid-ocean ridge spreading with constant velocity. The "base of the plate" has no physical meaning, but the plate is assumed to have constant thickness to approximate constant heat flow, as observed in old oceans. Constant temperature at the base of the plate (oceanic lithosphere) as suggested by small variations in thickness of the oceanic crust (see discussion below) is maintained by small-scale convection in the upper mantle caused by instabilities growing at the base of the cooling lithosphere that become effective only below old ocean and regulate the plate thickness. The plate consists of two parts: a rigid mechanical upper layer on top of a viscous thermal boundary layer (Parsons and McKenzie, 1978). Cooling results in thickening of both of the layers, the thermal boundary layer becomes unstable at an ocean age of ~60 Ma, and small-scale convection creates the observed thermal and bathymetry patterns (Fig. 4.23). The model has a singularity in heat flow at the mid-ocean ridge, which can be removed by changing the boundary conditions.

In the case of constant thermal conductivity, the temperature distribution within the plate has an analytical solution, and when the spreading rate is >10 mm/yr the geotherm only depends on the age of the plate (McKenzie, 1967) and can be approximated as:

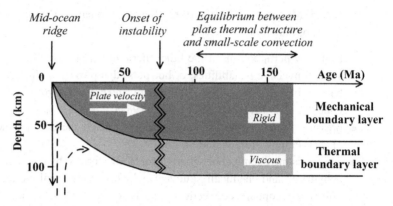

Simplified diagram showing the subdivision of oceanic plate into rigid mechanical boundary layer and viscous thermal boundary layer (modified after Parsons and McKenzie, 1978). Dashed lines – mantle flow.

$$T(\text{age}, z) = T_m(z/L + \Sigma \, 2/n\pi \sin(n\pi z/L) \exp(-\chi(n\pi/L)^2 \text{age})) \qquad (4.29)$$

where L is the plate thickness, z is the depth below the plate surface, and T_m is the basal temperature. Similarly, bathymetry variations with depth can be calculated for isostatically compensated plate and constant thermal expansion coefficient as:

$$D(\text{age}) = D_r + L/2 \; \alpha T_m \rho_m/(\rho_m - \rho_w)(1 - 8/\pi^2 \Sigma j^{-2} \exp(-\chi(j\pi/L)^2 \text{age})) \qquad (4.30)$$

where $j = 1,3,5$. The plate model does not explicitly provide the $\sqrt{\text{age}}$ dependence for heat flow and ocean floor topography:

$$Q(\text{age}) = \lambda T_m/L(1 + 2\Sigma \exp(-\chi(n\pi/L)^2 \; \text{age})), \qquad (4.31)$$

here the first term provides the asymptotic heat flow for old seafloor. The use of a Green's function approach allows us to express the asymptotic solution for young oceans away from the ridge crest in a form with explicit $\sqrt{\text{age}}$ dependence (Parsons and Sclater, 1977). The predictions of the cooling half-space model and the plate model are similar for young oceans, but differ for oceans older than 70–100 Ma (Fig. 4.21). In contrast to the cooling half-space model, the plate model predicts a near-constant ocean floor depth in the old oceans, as well as the age of the departure of bathymetry and heat flow from the $\sqrt{\text{age}}$ law.

A detailed analysis of the parameter space of the plate model shows that it is not possible to obtain all of the lithospheric parameters from data. Assuming that lithosphere density, thermal conductivity, and heat capacity are known, data on heat flow and bathymetry allow for calculation of plate thickness, temperature at its base, and thermal expansion coefficient α (Fig. 4.24; Table 4.7). A plate model with 125 ± 10 km-thick lithosphere, a basal temperature of $1350 \pm 275\,°C$, and $\alpha = (3.2 \pm 1.1) \times 10^{-5}\,°C^{-1}$ explains the observed age variations in ocean floor topography and heat flow in the North Pacific better than a cooling half-space model (Parsons and Sclater, 1977). For old oceans this model, however, still overestimates the topography and gives wrong predictions for heat flow. Stein and Stein (1992) further modified the plate model to yield a better fit to observations and suggested that a plate with 95 ± 10 km-thick lithosphere, a basal temperature of $1450 \pm 100\,°C$, and $\alpha = 3.1 \times 10^{-5}\,°C^{-1}$ provides

Table 4.7 Parameters specifying the plate model	
Constant	Estimate
$c_1 = 2\rho\alpha T \sqrt{\chi/\pi} / (\rho-\rho_w)$	$350 \pm 65 \ (\text{Ma})^{1/2}$
$c_2 = L^2/9\chi$	64–80 Ma
$c_3 = \rho L \alpha T/2(\rho-\rho_w)$	$3900 \pm 350 \ \text{m}$
$c_4 = \rho \, C_p \, T \sqrt{\chi/\pi}$	–
$c_5 = 0.22 \, L^2/\chi$	–
$c_6 = \lambda T/L$	$33.5 \pm 4.2 \ \text{mW/m}^2$

(after Parsons and Sclater, 1977). Parameters: T – temperature at the lithosphere (plate) base; L – plate thickness; α – coefficient of thermal expansion of the lithosphere; ρ and ρ_w – density of lithosphere (at $T=0\,°C$) and water; $\chi = \lambda/\rho c_p$ – thermal diffusivity of the lithosphere, λ – thermal conductivity, c_p – specific heat at constant pressure. Commonly assumed values are: $\rho = 3300 \ \text{kg/m}^3$; $\rho_w = 1000 \ \text{kg/m}^3$, $c_p = 1.17 \ \text{kJ/kg/K}$, $\lambda = 3.14 \ \text{W/m/K}$.

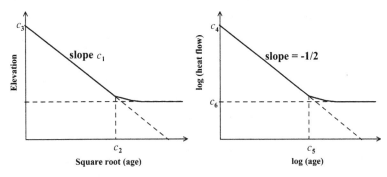

Fig. 4.24   Parameter space for the plate model (after Parsons and Sclater, 1977). See Table 4.7 for definitions of the constants c_1–c_6.

better fit to the observed topography and heat flow for ocean ages >70–100 Ma. Their model predicts that for oceans younger than 20 My, heat flow Q and bathymetry D can be expressed as:

$$Q_{oc} = 510/\sqrt{\text{age}}, \quad D = 2.6 + 0.365 \sqrt{\text{age}} \tag{4.32}$$

where the age of the ocean floor is in Ma, heat flow Q_{oc} is in mW/m^2, and bathymetry is in km.

Oceanic geotherms

The plate model is commonly used to calculate the average thermal structure of the oceanic lithosphere as a function of ocean age (Fig. 4.25). It also allows for speculation on melting conditions in the oceanic mantle (Fig. 4.26). In particular, the plate model of Stein and Stein

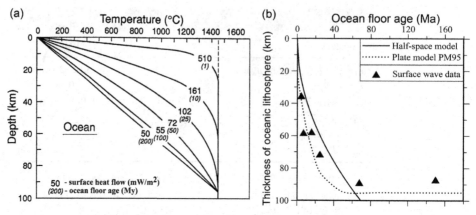

Fig. 4.25 Thermal structure of normal oceans. (a) Family of conductive geotherms for oceanic lithosphere calculated for plate model PM95 with plate thickness 95 km and temperature at the plate base 1450 °C (from Pollack *et al.*, 1993). When the temperature dependence of thermal conductivity is taken into account, temperatures in the oceanic mantle become lower than shown in (a). (b) Thickness of oceanic lithosphere predicted by a cooling half-space model and plate model PM95. Symbols – surface wave estimates of lithosphere thickness (see Fig. 3.121). For young oceans (<80 Ma), seismic data agree with both models. For old oceans, old seismic data do not support the half-space model, while recent data indicate that lithosphere beneath old parts of the Pacific Ocean can extend down to ~120–150 km (Fig. 3.99d).

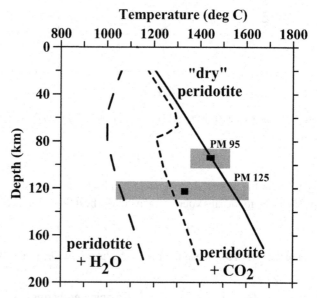

Fig. 4.26 Best estimates (black boxes) and range of uncertainties (gray boxes) for the temperature and thickness of oceanic plate for two plate models (based on plate models of Parsons and Sclater, 1977; Stein and Stein, 1992) compared with peridotite solidi: solid line – melting of a natural dry peridotite (Takahashi, 1986; Hirschmann, 2000); dashed lines – melting of peridotite saturated with H_2O and CO_2 (compilation by Wyllie, 1995).

(1992) with a 95 km-thick oceanic plate suggests dry peridotite melting in oceanic mantle. This result may be inconsistent with data on electrical conductivity which favor the wet composition of suboceanic mantle (Hirth *et al.*, 2001).

McKenzie *et al.* (2005) analyzed the effect of temperature-varying thermal conductivity $\lambda(T)$ and thermal expansion $\alpha(T)$ on oceanic mantle geotherms in the plate model. In the case of constant λ, any changes in model parameters produce changes in both the depth–age curve and in the heat flow–age curve. In the case of temperature-dependent conductivity $\lambda(T)$, different heat flow–age curves can correspond to the same depth–age curve, since the analytical solution uses a Fourier expansion of temperature variations in which only the odd terms contribute to the depth variations, but both odd and even terms contribute to the heat flow variations. When temperature dependence of thermal conductivity is taken into account, the predicted temperatures in the oceanic mantle become lower than Fig. 4.25 predicts (e.g. McKenzie *et al.*, 2005). A decrease in λ with a temperature increase reduces the temperature in the center of the plate compared to the analytical solution. Since temperature decrease causes decrease in α, the resultant effect is a greater amount of ocean floor subsidence. The latter, however, is counteracted by a significant reduction in plate thickness.

Alternative approaches

Alternatively, the thermal regime of normal oceanic lithosphere can be calculated if the thermal expansion coefficient is known (for example, from laboratory measurements on peridotite samples). An additional constraint on melting temperature (and thus on the potential temperature of the mantle) can be applied by requiring that the oceanic crust be produced by decompression melting of passively ascending mantle material at a mid-ocean ridge. In such a formulation, plate thickness remains the only free parameter in the system (McKenzie *et al.*, 2005).

In a general case, the thickness of "normal" oceanic crust z (in km) generated by decompressional melting can be parameterized as:

$$z = [T_\mathrm{m} - 1242]/10.32, \tag{4.33}$$

where T_m is the mantle potential temperature (in °C), and for a potential temperature increase of 12.5 °C crustal thickness increases by 1 km (McKenzie *et al.*, 2005). The plate model of Parsons and Sclater (1977) is more consistent with estimates of mantle potential temperature, since the potential temperature of 1450 °C (plate model of Stein and Stein, 1992) would generate by decompressional melting a 16 km-thick crust. This further implies that temperature variations in the oceanic mantle are very small since the standard deviation of 0.8 km in the average thickness of oceanic crust (7.1 ± 0.8 km) corresponds to a temperature variation of only ~10 °C. The latter explains the success of the plate model, which is based on an assumption of constant temperature at the base of the plate.

Testing half-space and plate models by observation

Carson and Johnson (1994) compared the predictions of the half-space and plate models with mean depth and depth to basement reported from the Deep Sea Drilling Project (DSDP) and Ocean Drilling Program (ODP). They concluded that:

- the cooling half-space $\sqrt{\text{age}}$ models can meaningfully fit the drilling and mean depth data for oceans younger than 73–81 Ma; for young (<~80 Ma) seafloor, cooling half-space models require a basal temperature of 1300–1370 °C; the minimum plate thickness compatible with best fitting cooling half-space models for young seafloor is ~120 km;
- plate models can also meaningfully fit the drilling and mean depth data for ocean ages <81 Ma; the best fitting plate models for young (<~80 Ma) seafloor require systematically higher temperatures of 1450–1470 °C and systematically thinner plates, 102–118 km, than derived from corresponding fits to the half-space model for young seafloor;
- no cooling plate model can explain variations in bathymetry with age over the entire range of ocean ages, 0–165 Ma; the misfit dramatically fails for ocean ages >81 Ma; the plate model of Stein and Stein (1992) minimizes the overall misfit at the expense of producing large misfits (above the limit of acceptable systematic misfit) over the entire range of ocean ages;
- failure of the plate model to explain the drilling data may indicate that the plate model is too simple to account for the full range of depth variations with age, and that the cooling history of old ocean floor cannot be approximated by extension of the cooling history of young ocean floor; depth versus age variations in old oceans can be explained by thermal perturbations superimposed on a "normal" half-space cooling trend.

Small-scale convection and geoid

Ocean floor flattening with age could not be explained by cooling half-space models and led to the development of plate models. However, plate models do not include any assumptions on physical mechanisms of flattening of old ocean floor. Proposed mechanisms for flattening included small-scale convection (McKenzie, 1967; Richter, 1973) and formation of oceanic plateaus and islands, since older oceans are more likely to have experienced anomalous evolution (e.g. Smith and Sandwell, 1997). For example, it was argued that all of the Pacific Ocean floor older than 80 Ma lies within 800 km of hotspots or hotspot tracks (Schroeder, 1984).

An alternative model of thermal evolution of old oceanic lithosphere has been proposed by Doin and Fleitout (1996). In contrast to the plate model, which assumes that the base of the oceanic lithosphere is at constant temperature, these authors assumed constant basal heating of oceanic lithosphere by small-scale mantle convection. The model agrees well with the plate model predictions for oceans younger than 80 Ma, provides a good fit for the bathymetry and heat flow for old-age oceans, but, in comparison with the plate model, predicts several hundreds of meters of subsidence in oceans older than 100 Ma and requires a thermal expansion coefficient much larger than assumed in plate models.

Another approach to modeling thermal evolution of oceanic lithosphere is based on the analysis of geoid data (e.g. Cazenave et al., 1983; Cazenave, 1994). Since cooling of the oceanic plate with age as it moves away from the spreading center leads to a change in its density, in theory one would expect geoid variations with ocean floor age. In practice, however, almost no correlation is observed between the two parameters, because the geoid signal is dominated by (i) long-wavelength components associated with density anomalies in the deep mantle and (ii) short-wavelength signals related to fracture zones and flexural anomalies in the shallow lithosphere. Later attempts (e.g. DeLaughter *et al.*, 1999) demonstrated that the application of spatial bandpass filtering allows for extraction of some age-dependent signal from the geoid. However, since isostatic geoid signals predicted from cooling half-space and plate models differ only in wavelength (for old ocean floor, the plate model predicts a zero slope of the geoid due to constant plate thickness, while the cooling half-space model does not), the filtered geoid data do not allow discrimination between the two models for ocean floor evolution (Hager, 1983). Seismic data also cannot reliably discriminate between the two models (Fig. 4.25b): while old interpretations of surface-wave data indicated a 90 km-thick lithosphere beneath old oceans, recent interpretations from the Pacific Ocean suggest that it may extend down to ~120–150 km, which would agree with the predictions of the cooling half-space model (eq. 4.25).

Ocean floor flattening: fact or artifact?

Two recent studies (Korenaga and Korenaga, 2008; Adam and Vidal, 2010), however, question the very seafloor flattening for old oceans and, as a result, question the necessity of using the plate model for the evolution of oceanic lithosphere. In fact, it has already been noted by Parsons and Sclater (1977) that there is no clear physical mechanism that determines the plate thickness, and the thickness of the slab is prescribed arbitrarily to fit the data on heat flow and bathymetry.

Korenaga and Korenaga (2008) argue that the analysis of bathymetry variations with age used in two popular plate models (Parsons and Sclater, 1977; Stein and Stein, 1992) was based on the entire data set, without excluding regions with "anomalous oceanic crust", such as hotspots and plateaus. As a result, for old oceans it is biased by data from anomalous regions because, due to isostasy, regions with anomalously thick crust produce topographic highs. Significant variations in the area of anomalous regions in different oceans may be the reason for significant differences in bathymetry versus ocean age variations between the oceans (Fig. 4.21).

Korenaga and Korenaga (2008) excluded data with a random signal, such as caused by crustal thickness variations, and limited their analysis of bathymetry–age data to the "normal seafloor", i.e. to regions unperturbed by hotspots and plateaus. They found that in this case the subsidence rate of ocean floor (the constant C_1 in Fig. 4.24; Table 4.7, eq. 4.26a) is 10% less, ~320 m/Ma$^{1/2}$. This, lower than conventional, value of the subsidence rate can be explained by the cooling half-space model, but with a reduced thermal expansivity of the mantle. The results of an earlier study (Nagihara *et al.*, 1996) also indicate that data on heat flow and isostatically corrected depth measurements follow the curve expected from the cooling half-space model (although with some offset in age).

Another recent study by Adam and Vidal (2010) suggests that ocean floor flattening is an artifact caused by confusing "*flow lines*" with "*age trajectories*". Age trajectories follow an age gradient and are usually used in analysis of the age dependence of bathymetry and heat flow. In contrast, flow lines represent a mantle convection pattern underneath the ocean floor and thus characterize thermal boundary cooling. Since the cooling half-space model assumes a constant plate velocity, these are flow lines that should be used for comparison with theoretical predictions of the cooling half-space model, but not age trajectories. Flow lines may differ strongly from age trajectories as an example from the Pacific Ocean illustrates. The difference is caused by the change in mantle convection pattern and in plate velocity that took place in the Pacific at ~47–50 Ma. Along the flow lines, no flattening is observed at old ocean floor ages in the Pacific Ocean, even far away from the ridge, while there is an apparent flattening along age trajectories. Along the flow lines, the bathymetry D shows a linear relation to the square root of the distance from the ridge $\sqrt{x}$ (compare with eq. 4.26):

$$D(x) = D_r + c_2\sqrt{x}, \tag{4.34}$$

where D_r is the ridge depth, and c_2 is the subsidence rate estimated to be in the range 0.5–3.5 m/m$^{1/2}$. In the Pacific Ocean, this relation is satisfied all along the plate, from the ridge to the subduction zone. If plate velocity is known, the distance from the ridge can be converted to the ocean floor age which, assuming a constant velocity of 9 cm/year for the Pacific plate, yields the subsidence rate between 200 and 900 m/Ma$^{1/2}$. Thus, Adam and Vidal (2010) conclude that there is no ocean floor flattening for old Pacific, and there is no need, in general, to invoke plate models to explain bathymetry in old oceans.

4.3.3 "Anomalous" oceans

Subduction zones, marginal basins, and accretionary prisms

A complicated thermal structure of subduction zones arises from their complex tectonic configuration. Horizontally, subduction zones can be roughly subdivided into forearc, arc, and back-arc (Fig. 4.27). Vertically they include the subducting slab and the overlying mantle wedge; the latter includes the convective asthenosphere and the conductive litho-spheric mantle (oceanic or continental) of the overriding plate. The mantle wedge (both its lithospheric and asthenospheric parts) can variably be infiltrated by fluids and melts. The thermal structure of subduction zones is highly variable (Peacock, 1990; Kelemen et al., 2003; Fukao et al., 2004; Pozgay et al., 2009) and depends on several parameters (Peacock, 2003) such as:

- the thermal structure of the subducting plate that depends on its age;
- the thermal structure (thickness) of the overriding plate;
- the angle of subduction;
- the convergence rate;
- the flow in the mantle wedge which depends on its rheology and coupling between the asthenosphere and the subducting plate;
- shear heating at the upper surface of the slab;
- processes of dehydration and metamorphism that affect mantle solidus;
- advective heating by raising melts;
- deformation and erosion.

Systematic studies of heat flow variations across subduction zones have been performed in the western Pacific. Between the trench axis and the volcanic arc, where oceanic lithosphere is returned to the mantle and cold lithospheric plate subducts downward, heat flow is very low, sometimes close to zero, and lithospheric temperatures in the subducting slab are

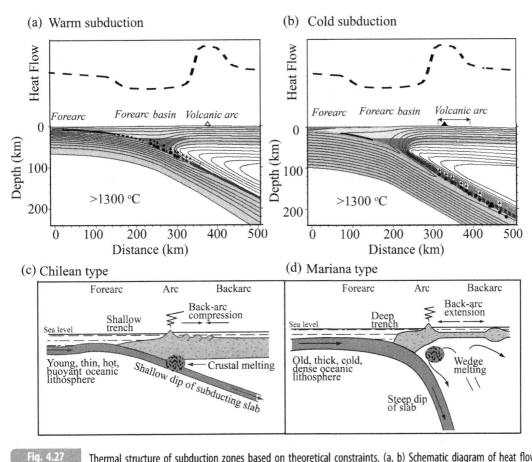

Fig. 4.27 Thermal structure of subduction zones based on theoretical constraints. (a, b) Schematic diagram of heat flow and temperature variations across warm and cold subduction zones. Oceanic plates are shown by gray shading. (c, d) Two end-member models of subduction zones based on the age of subducting slabs (modified after Uyeda and Kanamori, 1979). Thermal structures of the Chilean-type and Mariana-type subduction zones are significantly different.

depressed in comparison with adjacent oceanic mantle (Fig. 4.27). The thermal structure of the slab is asymmetric due to frictional heating at its upper surface. Melting of the mantle wedge above the subducting slab and associated volcanism result in high heat flow over the volcanic arcs behind the subduction zones due to upward advection of heat by magmas. These regions are also characterized by a large scatter in heat flow values.

Petrologic studies of mantle xenoliths from subduction zones worldwide (continent–ocean collision zones at the Kamchatka–Kurils arc, the Japan arc, and the Cascades (western USA); ocean–ocean collision zones at the Lesser Antilles (the Caribbean), the South Sandwich arc in the South Atlantic, the Izu–Bonin–Mariana arc, the Taiwan–Luzon volcanic arc, and the New Ireland island arc (in the Pacific Ocean) have revealed significant variations in their equilibration temperatures (Neumann and Simon, 2009). Most ultra-refractory harzburgites in oceanic forearcs have typically low equilibration temperatures, in the range 950–1050 °C; however the formation of magmas of such a composition requires high degrees of partial melting. Higher equilibration temperatures (>1100 °C) are determined for xenoliths from the continental arc of SW Japan, the Mariana back-arc rift, and the South Sandwich fracture zone; in addition to higher equilibration temperatures, these xenoliths also have fertile compositions.

Data on surface heat flow, seismic velocity, and xenolith thermobarometry (Fig. 4.28; for further details see Chapter 5) were used to examine the thermal structure of continental back arcs associated with subduction zones in the circum-Pacific and Southern Europe (Currie and Hyndman, 2006). These results suggest the presence of a thin (~60 km) lithosphere with uniformly high temperatures in the shallow mantle within the back arcs that extends over lateral distances of 250 km to >900 km and spreads over the back-arc widths. These authors conclude that

> a broad hot back arc may be a fundamental characteristic of a subduction zone that places important constraints on back-arc mantle dynamics. In particular, the thermal structure predicted for slab-driven corner flow is inconsistent with the observed uniformly high back-arc temperatures.

The preferred model of Currie and Hyndman (2006) includes a vigorous thermal convection in the back-arc upper mantle that brings the heat upward from the deeper mantle; it is facilitated by low viscosities due to fluids hydration from the subducting plate (Arcay et al., 2006). The predicted time-scale for thermal equilibration of the lithosphere in a continental back arc is ~300 Ma after subduction termination.

Similarly to oceanic subduction zones, zones of continental subduction are associated with volcanism and high heat flow over the volcanic arcs behind subduction zones due to upward heat advection by magmas. Some subduction zones (e.g. the northern part of the Andes and the Cascades in northwestern USA) have low heat flow; in these regions either no active arc volcanism is present or volcanism has migrated back from the trench.

Under a favorable stress regime, lithosphere extension associated with subduction zones can lead to the creation of new oceanic lithosphere in zones of back-arc spreading. Sclater et al. (1980) have analyzed data from marginal basins where both reliable heat flow data and age estimates were available. These included the Tyrrhenian Sea, the Balearic Basin (both in the Mediterranean), the Parece Vela and the West Philippine back-arc basins (both in the western Pacific), the Coral Sea, and the Bering Sea. This analysis demonstrated that for these basins the mean heat flow is similar to the major ocean basins of similar age and follows the theoretical $\sqrt{age}$ predictions. This implies that extensional processes similar to ocean spreading are responsible for the formation of back-arc basins. However, the depth of ocean floor in marginal basins is ~1 km deeper than expected for their age, perhaps because the small size of these basins may affect the efficiency of small-scale convection and basal heating (Louden, 1980; Watanabe et al., 1977; Park et al., 1990).

Heat flow measurements in accretionary prisms remain limited. The reported values are highly variable, both within a single prism and between different prisms, and probably reflect hydrothermal circulation in a highly porous water-saturated sedimentary layer (Davis et al., 1990; Foucher et al., 1990; Yamano et al., 1992). Some studies, however, suggest that heat flow in accretionary prisms can be lower than average (Watanabe et al., 1977).

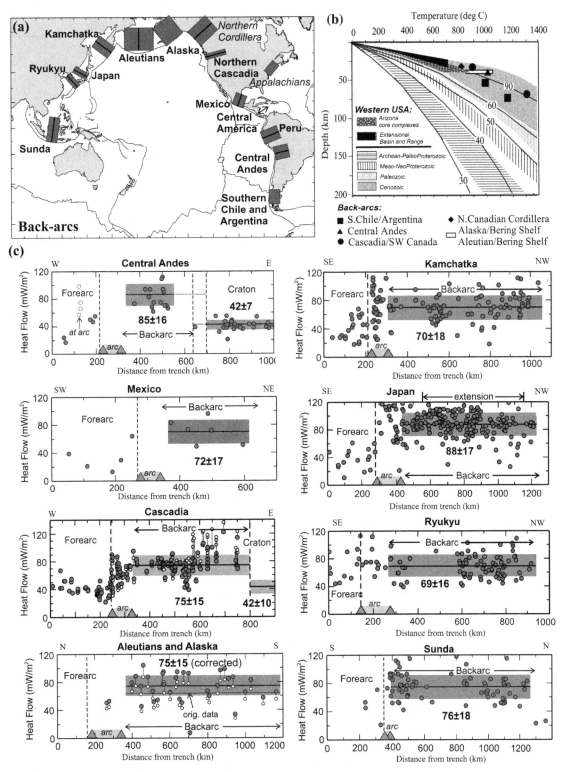

Fig. 4.28 Thermal regime of continental back-arcs. (a) Location map; (b) lithosphere temperatures in back-arcs (different symbols) based on mantle xenoliths P-T arrays plotted on top of typical continental geotherms (see Fig. 4.33 for details); the range of lithospheric temperatures in Cenozoic western USA is shown for comparison (data of Lachenbruch and Sass, 1977; Ketcham, 1996); (c) surface heat flow across different subduction zones (sources: (a) and (c) modified after Currie and Hyndman, 2006).

Ocean plateaus and hot spots

Several mechanisms have been proposed to explain the "anomalous" bathymetry in oceans (i.e. its departure from the $\sqrt{age}$ law):

(1) extra heating of the lithosphere by mantle plumes, shear heating, or vigorous secondary convection (Parsons and McKenzie, 1978; Schubert *et al.*, 1976; Davies, 1988);
(2) dynamic support from pressure-driven asthenospheric flow (Phipps Morgan and Smith, 1992);
(3) isostatic uplift of regions with anomalously thick crust which produces topographic highs.

Heat flow in oceanic regions with anomalous (non-square-root of age) bathymetry has been best studied in the Pacific: in the South Pacific Superswell (located between approximately 5°–30° S and 120°–150° W), the Darwin Rise (25° N–5° S, 160° E–160° W), and the Hawaiian hotspot swell. The presence of mantle upwellings with higher than normal mantle temperatures has been proposed for these regions to explain their shallow bathymetry. In particular, the thickness of the elastic lithosphere beneath the South Pacific Superswell has been estimated to be ~10 km thinner than beneath other hotspots and beneath "normal" regions of the Pacific Ocean (Calmant *et al.*, 1990).

However, the South Pacific Superswell and the Darwin Rise have normal heat flow that does not differ from other regions of the Pacific with similar lithosphere ages (Fig. 4.18) (Stein and Abbott, 1991). Similarly, heat flow measurements indicate that neither Hawaii, nor several other hotspots (Bermuda, Cape Verde, Crozet) have high heat flow (Courtney and Recq, 1986; Courtney and White, 1986; Detrick *et al.*, 1986; von Herzen *et al.*, 1989). Heat flow in these oceanic regions hardly differs from the values measured in ocean basins of similar ages and thus suggests that anomalous bathymetry is caused by dynamic support from the mantle, rather than by a thermal anomaly. This conclusion is supported by seismic tomography data that does not display low-velocity anomalies in the upper mantle beneath the Hawaiian Swell and the south-central Pacific (Fig. 3.99).

4.4 Thermal regime of continental lithosphere

4.4.1 Heat flow on the continents

Global observations

Data coverage of heat flow measurements on the continents still remains very uneven (Fig. 4.16). Along with densely sampled regions in Europe, North America, Asia, South Africa, and Australia, some large areas have very sparse data coverage (e.g. Arctic North America and north-central South America), while in other areas (e.g., north and central Africa, Antarctica and most of Greenland) heat flow measurements are still absent.

A large scatter of heat flow values is observed in tectonic provinces of all ages. In Precambrian regions, formed by accretion of Archean and Proterozoic terranes of different tectonic origin, it largely reflects complex, chemically heterogeneous crustal structure, particularly with respect to the distribution of heat-producing isotopes. In Phanerozoic regions, the diversity in surface heat flow reflects both crustal heterogeneity and variations in the thermal regime of the sublithospheric mantle.

The lowest heat flow values on the continents (20–30 mW/m^2) are reported for some of the Precambrian shields (e.g. the Archean Kola-Karelian province of the Baltic Shield, the Siberian craton, the northern parts of the Superior province in Canada, and the West African craton) (e.g. Sass and Behrendt, 1980; Duchkov *et al.*, 1987; Kukkonen and Joeleht, 1996; Levy *et al.*, 2010). The lower boundary on surface heat flow in these regions is set by mantle heat flow if almost no radiogenic heat is generated in the crust. Assuming that average crustal heat production is as low as in mafic granulites, 0.1–0.4 W/m^3 (Tables 4.3, 4.4) (e.g., Pinet and Jaupart, 1987; Rudnick and Fountain, 1995) and the crustal thickness is 40 km, it yields ~4–16 mW/m^2 for heat flow at the Moho in regions with surface heat flow of 20 mW/m^2.

Very low values of surface heat flow have been reported from some non-cratonic settings as well, e.g. in the 1500 km-long Magnitogorsk block in the Southern Urals where surface heat flow is ~25 mW/m^2 as compared to 40–50 mW/m^2 in the East European Platform. Possible explanations of such low heat flow anomalies include paleoclimatic variations (Fig. 4.3), unusually low crustal heat production, lateral groundwater heat transfer, or anomalously low mantle heat flow (e.g. associated with the Paleozoic subduction zones as proposed for the central part of the southern Urals) (Kukkonen *et al.*, 1997).

The highest heat flow on the continents (with average regional values of 120–130 mW/m^2) is observed in regions with advective heat transfer by ascending magmas, such as in Cenozoic zones of extension, lithosphere thinning, and volcanism. The upper limit on surface heat flow is apparently set by the start of crustal melting, since at this level of heat flow near-solidus temperatures are reached at the crustal base.

High heat flow associated with shear heating is expected at continental transform faults (e.g. Leloup *et al.*, 1999). However, heat flow measurements around the San Andreas fault in western North America do not display systematic variations in values as a function of distance from the fault (Lachenbruch and Sass, 1980) and recent studies suggest that heat flow near the San Andreas fault is dominated by conduction. The discrepancy between theoretical predictions and observations can be attributed to a complex interlink between fluid pressure, effective normal stress, melting, frictional stability, and other processes that control frictional heat generated during fault slip (see d'Alessio *et al.*, 2006 for details).

Age dependence of continental heat flow

In spite of limitations in data coverage and a large scatter in heat flow values, age dependence of continental heat flow had already been proposed in the 1960s (Lee and Uyeda, 1965). Further studies have supported this empirical relationship (Polyak and Smirnov, 1968; Vitorello and Pollack, 1980; Morgan, 1984) and shown that globally it holds, roughly, for both the crustal ages and the tectono-thermal ages of continental terranes (Fig. 4.29). Surface heat flow increases from 41 ± 11 mW/m^2 in the Archean terranes through 46 ± 15 mW/m^2 and 49 ± 16 mW/m^2 in the Mesoproterozoic and Neoproterozoic terranes, respectively, to >60 mW/m^2 in Phanerozoic regions (Pollack *et al.*, 1993b; Nyblade and Pollack,

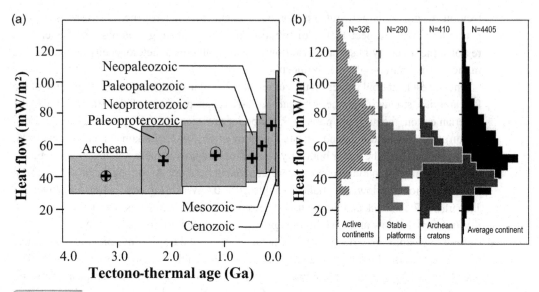

Fig. 4.29 (a) Observed continental heat flow versus tectono-thermal age (age of the last major tectonic or magmatic event at the measurement sites) averaged for seven age groups (gray boxes). Vertical dimension of boxes indicate ± one standard deviation of heat flow data from the mean (marked by crosses); horizontal dimension shows age range (Source: Morgan and Sass, 1984). A decade later update of heat flow database by all available borehole measurements at that time (Pollack et al., 1993) did not change results based on the earlier data set (circles). (b) Normalized histograms for continental heat flow measurements for three types of tectonic provinces based on the 3SMAC regionalization (Nataf and Richard, 1996). N – number of heat flow measurements (source: Röhm et al., 2000).

1993). This trend can roughly be approximated as $Q_o = 65–9*t$, where Q_o is the surface heat flow in mW/m^2 and t is the crustal age in Ga.

By analogy with the oceans, a simple model of a cooling thermal boundary layer has been proposed to explain the global age dependence of continental heat flow (Crough and Thompson, 1976). However, the model does not fit observations for the Precambrian terranes older than c. 1.5 Ga, for which it predicts nearly constant heat flow values.

At present, the age dependence of continental heat flow is commonly attributed to:

- the relaxation time of the lithosphere after a major tectono-thermal event,
- a systematic variation in crustal heat production with age,
- a systematic variation in lithospheric thickness and mantle heat flow with age,
- a combination of the above three factors.

Heat flow across the cratonic margins

As a result of the strong lateral heterogeneity of surface heat flow, the difference in heat flow values between tectonic provinces of dissimilar ages can significantly deviate from global averages. For example, in South Africa the contrast in surface heat flow between the Archean Kaapvaal craton and the surrounding Neoproterozoic mobile belts is significantly higher than suggested by global values, ~25 mW/m^2 versus ~8 mW/m^2 (Ballard and

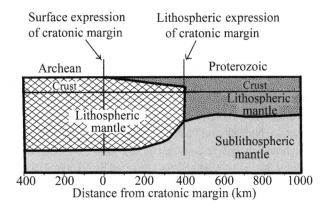

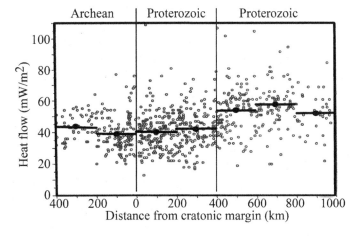

Fig. 4.30 Heat flow versus distance from the margins of Archean cratons for heat flow measurements from Precambrian terranes worldwid (lower part). Horizontal bars – averages over 200 km intervals (Sources: Nyblade and Pollack, 1993; Nyblade, 1999). The upper part displays a sketch of the corresponding lithospheric structure. Note that surface expression of cratonic margin as evidenced by geological and tectonic data may not coincide with the deep lithospheric boundary between Archean and younger terranes (see also Fig. 3.67).

Pollack, 1987). Similarly, a step-like change in surface heat flow by 20–30 mW/m² is observed across the Trans-European Suture Zone at the western margin of the East European craton and at the western edge of the North American craton (Fig. 4.18).

Nyblade and Pollack (1993) have analyzed worldwide data from the Precambrian terranes and argued that, although the transition from the Archean to Proterozoic terranes is marked globally by a pronounced increase in surface heat flow, the change in heat flow pattern does not necessarily correspond to the surface expression of the cratonic margin (Fig. 4.30). A regional pattern of decreased heat flow in Proterozoic terranes located within a 200–400 km-wide zone along the margins of the Archean cratons has been recognized at some Archean–Proterozoic boundaries: at the western margin of the Kalahari craton in South Africa (Ballard *et al.*, 1987), at the eastern margin of the Superior province in North America, and at the south-western margin of the Baltic Shield in Europe (Nyblade and Pollack, 1993).

Contrary to this, Jaupart and Mareschal (1999) do not observe any clear geographic trend at the eastern margin of the Canadian Shield, where in some regions (e.g. in the Abitibi province) heat flow increases from the craton margin to its center, and surface heat flow in the late Proterozoic Grenville province is the same as in the Archean terranes of the Superior province and in the early Proterozoic Trans-Hudson orogen (see Fig. 3.66 for a location map). Based on the Canadian heat flow observations, these authors argue that the regional heat flow pattern at the cratonic margins in South Africa solely reflects higher radioactivity of crustal rocks in the Proterozoic terranes. In contrast, Nyblade and Pollack (1993) propose that overthrusting of the Proterozoic crust onto the Archean lithosphere can explain low heat flow outside the margins of the Archean cratons (Fig. 4.30). This hypothesis is supported by seismic data of the LITHOPROBE transects across the Canadian Shield, which indicate that the boundaries between the terranes of different ages are not vertical, and that the age of the crust and the underlying mantle can be significantly different within a 100–200 km-wide zone along these boundaries. In particular, the middle Proterozoic crust of the Grenville province overthrusts the Archean mantle of the Superior province (Fig. 3.67), leading to a uniform surface and mantle heat flow across the lithospheric terranes of *apparently* different ages at the Superior–Grenville boundary. Additionally, atypically low heat flow observed in the Grenville province can be attributed to the time delay of thermal front propagation associated with the passage of the Grenville province over the Great Meteor and Monteregian hotspots at 180 Ma and 100 Ma, respectively (Morgan, 1983; Nyblade, 1999). For a lithospheric thickness of ~150–200 km, the time required for a thermal perturbation to reach the surface (eq. 4.18) and become seen in surface heat flow is comparable with the age of the hotspots. This observation is supported by seismic tomography models (which provide "snap-shots" of the mantle in contrast to the "inertial" heat flow image) that do not reveal the presence of a deep lithospheric keel beneath the Grenville province (e.g. van der Lee and Nolet, 1997a), in contrast with conclusions reached by Jaupart and Mareschal (1999).

4.4.2 Continental geotherms

Time-scale of thermal equilibration

Surface heat flow variations in continental regions can be interpreted in terms of the thermal structure of the continental lithosphere (Artemieva and Mooney, 2001). In stable continents, the thermal structure of the lithosphere can be approximated by the steady-state solution of the thermal conductivity equation. Note, however, that due to secular cooling of the Earth combined with radioactive decay of heat-producing elements, any steady-state thermal constraint is an approximation for the real Earth (Michaut and Jaupart, 2007) (see Section 4.1.4 for details).

The steady-state solution of the thermal conductivity equation is applicable to regions that have not experienced a major tectonic event for several hundred million years (Fig. 4.31). The exact time depends on the characteristic time of thermal equilibration which defines the time delay for a thermal front associated with a thermal perturbation at depth z to reach the surface and to become reflected in the surface heat flow (eq. 4.18). Figure 4.31 shows the estimated time-scale for the thermal diffusivity value widely used in the literature, $\chi = 10^{-6}$ m^2/s ~ 31.5 km^2/Ma. This implies that the steady-state approximation is valid for significant

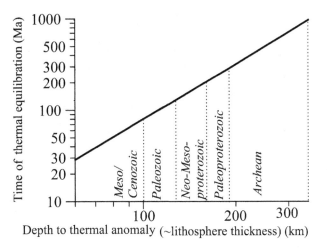

Fig. 4.31 Time-scale of thermal equilibration in the upper mantle (note logarithmic scales of the axes). The plot shows the minimum time required for a thermal perturbation in the upper mantle (at the base of the lithosphere) to be conductively transferred to the surface to become reflected in surface heat flow. Conductive heat flow measured in Cenozoic regions does not reflect the present thermal regime of the mantle beneath these regions.

parts of the continents which include old platforms (commonly with shelves) and many regions of Paleozoic tectono-magmatic activity where thermal equilibrium is expected to be re-established. In younger regions (<250 Ma), surface heat flow measured in boreholes may reflect the past thermal regime, but not the present thermal state of the mantle since, due to a slow rate of conductive heat transfer, thermal perturbation in the mantle has not yet reached the surface and is not yet reflected in surface heat flow. The lithosphere of these tectonically young regions is characterized by a transient thermal regime and the steady-state thermal approximation is invalid there.

The transient response of the lithosphere to a mantle thermal anomaly can be further illustrated by the passage of a hotspot underneath a continental lithosphere. The upper limit for a temperature anomaly associated with a plume is ~ 300 °C, compared to the ambient mantle temperature (e.g. McKenzie *et al.*, 1974; Gurnis, 1988; Schilling, 1991; also Fig. 4.39). An abrupt temperature increase, by 300 °C, at the lithospheric base causes a perturbation of its thermal regime and an increase in surface heat flow (Fig. 4.32). However, in a thick cratonic lithosphere this effect cannot be quickly detected. For example, if the cratonic lithosphere of the Grenville province passed over the Great Meteor and Monteregian hotspots at 180 Ma and 100 Ma, respectively (Morgan, 1983), the corresponding present-day surface heat flow increase could only be 5–8 mW/m^2, i.e. close to the uncertainty limit associated with heat flow measurements (Nyblade, 1999).

Conductive geotherms for stable continents

The temperature distribution in stable (i.e. in thermal equilibrium) continental lithosphere can be calculated from the thermal conductivity equation,

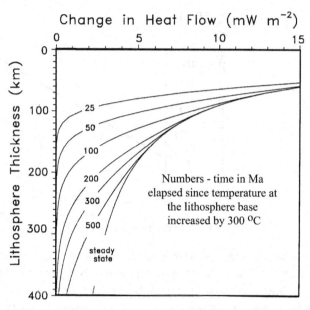

Fig. 4.32 Changes in surface heat flow associated with a step-like 300 °C temperature increase at the lithospheric base. Vertical axis – lithosphere thickness prior to thermal perturbation; numbers on the curves – elapsed time (in Ma) (after Nyblade, 1999).

$$\text{div } (\lambda(x, y, z) \text{ grad} T) + A(x, y, z) = \rho c(x, y, z)\partial T / \partial t \qquad (4.35a)$$

which in the steady-state case takes the form:

$$\text{div } (\lambda(x, y, z) \text{ grad} T) + A(x, y, z) = 0, \qquad (4.35b)$$

where T is temperature, $\lambda(x,y,z)$ is the thermal conductivity, and $A(x,y,z)$ is the radiogenic heat production. Boundary conditions at the top and the bottom include: (i) fixed temperature at the surface and (ii) either fixed *temperature* or fixed *heat flow* at the base. Additional boundary conditions (such as the symmetry of the temperature field) are set at the vertical boundaries.

The fact that the boundary conditions at the base of the lithospheric plate are unknown creates a fundamental problem for solution of the boundary problem (further discussed in the next subsection): the problem is ill-posed unless these conditions are assessed from independent data. The situation is similar to that of the oceanic thermal models, where the boundary condition at the base of the lithospheric plate on constant heat flow or constant temperature leads to two classes of models, the cooling half-space and the plate model, respectively.

An independent constraint on *temperature* at the lithospheric base can be derived, for example, from data on mantle melting temperature (solidus), while surface heat flow data $Q_0(x,y,0)$ provide a constraint on *heat flow* variations in the mantle. When the boundary condition at the base is known, the solution of Eq. (4.35b) is defined by lateral and vertical distributions of thermal parameters (radioactive heat production and thermal conductivity) within the lithosphere.

In many cases, the 1D approach to calculation of continental geotherms (temperature–depth profiles) is used since small-scale lateral variations of thermal conductivity and radiogenic heat production, in particular in the crust, are poorly known. In the 1D case, the temperature $T(z)$ at a depth z can be extrapolated from the surface temperature T_o and surface heat flow Q_o as:

$$T(z) = T_o + zQ_o/\lambda(z) - z^2 A(z)/2\lambda(z). \tag{4.36}$$

A family of conductive geotherms was first calculated by Pollack and Chapman (1977). Because of the simplicity of model assumptions, this family of conductive geotherms, commonly referred to as "reference geotherms", is widely used in literature for comparisons with other temperature estimates (Fig. 4.33). Table 4.8 summarizes the assumptions on which these geotherms are based. In short, the major assumptions are:

- a steady-state thermal regime regardless of the surface heat flow value;
- the concept of heat flow provinces (Section 4.1.4);
- generation of 40% of surface heat flow in the upper 8 km-thick crustal layer ("D-layer"), with the remaining 60% coming from below (and resulting from mid–lower crustal radioactivity, mantle radioactivity, and heat flow from the mantle);
- a globally uniform crustal thickness of 40 km;
- constant thermal conductivity of 2.5 W/m/K throughout the entire crust.

Data on surface heat flow variations together with data on the crustal structure and geological and tectonic information (interpreted in terms of lateral and vertical variations of thermal properties of crustal and mantle rocks) allow for calculation of *regional* continental geotherms. The surface heat flow database of Pollack *et al.* (1993) updated for newer borehole measurements formed the basis for a worldwide thermal model of the continental lithosphere (Artemieva and Mooney, 2001). It was constrained only by the most reliable heat flow values, based on quality criteria discussed by Pollack *et al.* (1993). In particular, heat flow data from shallow boreholes potentially affected by ground-water circulation and data obtained by non-conventional methods were excluded from the analysis. Additionally, the original heat flow data were analyzed, as far as possible, for potential regional perturbations and included paleoclimatic corrections to surface heat flow data due to climatic changes, such as Pleistocene glaciations (which in the northernmost areas can be as large as 15–20 mW/m^2, Fig. 4.3). The calculations were carried out for about 300 blocks with an area of $1° \times 1°$ to $5° \times 5°$ in size, where individual heat flow measurements in closely spaced boreholes were averaged for lithospheric blocks with similar crustal structure, ages, and tectonic setting.

In spite of significant differences between model assumptions (see Table 4.8 for details), conductive geotherms calculated globally from regional crustal and heat flow data in stable continental regions (Artemieva and Mooney, 2001) agree well with the family of reference continental geotherms of Pollack and Chapman (1977). Lithospheric temperatures fall between the reference geotherms for 30 and 40 mW/m^2 in the Archean–Paleoproterozoic terranes, and are close to the reference geotherms for 50 and 60 mW/m^2 in the Meso-Neoproterozoic and Paleozoic terranes (Fig. 4.33). The analysis also suggests that in the Precambrian terranes only ~30% of surface heat flow is generated in the upper crustal layer

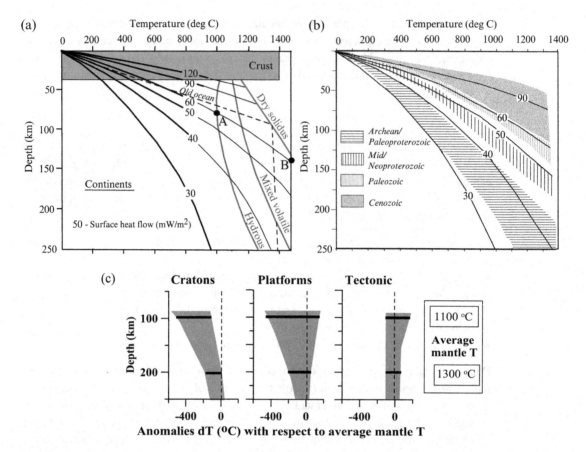

Fig. 4.33 Continental conductive geotherms. (a) A family of conductive geotherms versus surface heat flow (Source: Pollack and Chapman, 1977; Pollack et al., 1993). This set of geotherms is commonly referred to in the literature as "reference geotherms". Thin lines at higher temperatures indicate that non-conductive heat transfer can be important in the lower parts of the lithosphere. For comparison, dashed line shows a typical old ocean geotherm which corresponds to surface heat flow of ~50 mW/m². For the same surface heat flow, at any given depth continental temperatures are lower than oceanic due to radioactive heat generated in the continental crust. Gray lines – peridotite solidi for different volatile compositions. Note that, depending on the mantle composition, the thickness of thermal lithosphere defined by the intersection of a geotherm with mantle solidus, can be significantly different. For a 60 mW/m² reference geotherm, lithosphere thickness is ~80 km in the case of hydrous peridotite (A) and ~140 km in the case of dry peridotite (B). (b) Typical geotherms for continental regions of different ages constrained by heat flow data for stable continents and by xenolith data for active regions (see Table 4.8 for details, after Artemieva and Mooney, 2001). Thin lines with numbers – reference geotherms (Pollack and Chapman, 1977) as in (a) for different surface heat flow values. (c) Typical ranges of temperature differences between cratons, platforms, and tectonically active regions as compared to the average mantle (for average mantle temperatures of 1100 °C at a 100 km depth and 1300 °C at 200 km depth).

Table 4.8 Thermal parameters used in steady-state continental thermal models

	Pollack and Chapman, 1977	Rudnick *et al.*, 1998	Artemieva and Mooney, 2001
Crustal thickness	40 km	41 km	Variable, based on regional seismic data*
Heat production in sediments	Ignored	Ignored	0.7–1.3 $\mu W/m^3$ (based on regional lab data)
Heat production in the upper crust	40% of Q_o is generated in the upper D-layer ($D = 8$ km); this yields mean heat production in the D-layer: $A_D = 0.4\ Q_o/D$	14 km thick with 60% of A_{cr} (yields mean value of 0.7–1.2 $\mu W/m^3$)	0.4–6.7 $\mu W/m^3$ (based on regional lab data), 0.4–0.5 $\mu W/m^3$ at $z > 12$ km
Heat production in the middle-lower crust	0.25 $\mu W/m^3$	Middle: 13 km thick with 34% of A_{cr} (mean 0.4–0.75 $\mu W/m^3$) Lower: 14 km thick with 6% of A_{cr} (mean 0.07–0.12 $\mu W/m^3$)	Middle: 0.2–0.4 $\mu W/m^3$, Lower: 0.1 $\mu W/m^3$
Bulk crustal heat production, A_{cr}	Not assumed *a priori*, $0.01Q_o + 0.2$ $\mu W/m^3$ (yields 0.5–1.4 $\mu W/m^3$ for Q_o in the range 30–120 mW/m^2)	0.4–0.7 $\mu W/m^3$; preferred values: 0.4–0.5 $\mu W/m^3$ for $Q_o = 40$ mW/m^2 and 0.7 $\mu W/m^3$ for $Q_o = 50$ mW/m^2	Not assumed *a priori*, 0.3–1.0 $\mu W/m^3$
Mantle heat production	0.01 $\mu W/m^3$ at $z < 120$ km, 0.0084 $\mu W/m^3$ at $z > 120$ km	0.02–0.10 $\mu W/m^3$; preferred value 0.03 $\mu W/m^3$	0.01 $\mu W/m^3$ if $Vp < 8.3$ km/s, 0.004 $\mu W/m^3$ if $Vp > 8.3$ km/s
Thermal conductivity in the upper crust	2.5 W/m/K at $T < 500\ °C$ with $k(T)^{**}$ at $T > 500\ °C$	2.6–2.7 W/m/K	Sediments: 0.5–4.0 W/m/K, Upper: $k(T)^{**}$ in the range 2.5–3.0 W/m/K
Thermal conductivity in the lower crust	$k(T)^{**}$ at $T > 500\ °C$	2.6–2.7 W/m/K	Middle: 2.0–2.5 W/m/K, Lower: 2.0 W/m/K
Thermal conductivity in the mantle	–	$k(T)^{**}$	4.0 W/m/K
Treatment of high-heat flow regions	Assumed to be steady state	Not considered	Excluded from steady-state modeling; $T(z)$ based on xenolith data and transient models

* Subdivision of the crust into the upper, middle, and lower layers is based on Vp seismic velocities with the tops of the layers defined by 5.6 km/s, 6.4 km/s, and 6.8 km/s, respectively.
** Based on Schatz and Simmons (1972).
A_{cr} – bulk crustal heat production.

enriched by radiogenic elements, and the contribution of this layer increases in the younger crust to ~40% as assumed by Pollack and Chapman (1977).

Typical continental geotherms constrained by heat flow data for stable continents and by xenolith data for active regions indicate that the temperature difference between cratons, platforms, and tectonically active regions as compared to the average mantle is, correspondingly, $-310 \pm 200\,°C$, $-160 \pm 260\,°C$, and $+30 \pm 130\,°C$ at 100 km depth and c. $-90 \pm 70\,°C$, $-80 \pm 100\,°C$, and $-30 \pm 70\,°C$ at 200 km depth (for average mantle temperatures of $1100\,°C$ at a 100 km depth and $1300\,°C$ at a 200 km depth; the uncertainties indicate standard deviations of the calculated geotherms rather than the true uncertainties of the modeling) (Fig. 4.33c). These values are in a general agreement with mantle temperatures calculated from joint interpretation of gravity and tomography models (Deschamps et al., 2002).

Uncertainties in conductive geotherms

Because of the large uncertainty in the lateral and vertical distribution of thermal parameters in the lithosphere, a large set of conductive geotherms can be produced for any surface heat flow value. This makes it difficult to assess the accuracy of estimates of continental temperatures and lithospheric thermal thickness. Sensitivity analyses indicate that for a typical range of possible variations of heat production and thermal conductivity, continental temperatures are constrained with an accuracy of ~50–100 °C at a 50 km depth and ~100–150 °C at a 100–150 km depth (see Table 4.9 for details).

One of the possibilities for testing the accuracy of continental geotherms constrained by surface heat flow is to compare them with independent data, such as xenolith P–T arrays (see Chapter 5 for details and a discussion of the uncertainties associated with xenolith geotherms). For example, Rudnick et al. (1998) used mantle xenolith data from the Archean cratons of South Africa, Tanzania, and Siberia together with typical values of surface heat

Table 4.9 Sensitivity analysis for steady-state conductive continental geotherms

Change of model parameter	Temperature at $z = 50$ km	Temperature at $z = 100$ km	Lithospheric thickness	Mantle heat flow
Average crustal heat production 20% higher	9–13% (50–70 °C) lower	11–16% (100–130 °C) lower	15–30% (25–80 km) greater	8–10% (4–5 mW/m²) lower
Average crustal conductivity 10% higher	8% (30–60 °C) lower	5% (30–60 °C) lower	3–6% (5–10 km) greater	No effect
Mantle conductivity 17% lower	2–3% (10–15 °C) higher	8% (50–80 °C) higher	3–8% (10–15 km) lower	No effect
Surface heat flow 5% higher	7–8% (30–50 °C) higher	8–9% (50–90 °C) higher	10% (10–25 km) lower	2–3% (2–3 mW/m²) higher

(from Artemieva and Mooney, 2001; the model parameters are specified in Table 4.8)

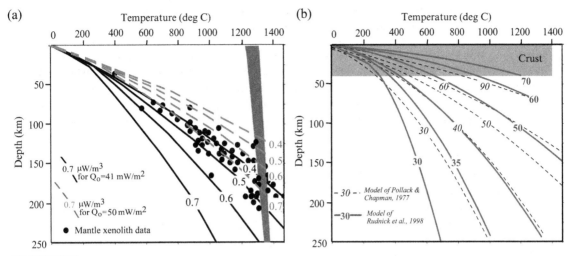

Fig. 4.34 Continental conductive geotherms calibrated by xenolith data from the Archean cratons (see Table 4.8 for model details, source: Rudnick *et al.*, 1998). (a) Geotherms corresponding to surface heat flow of 41 mW/m^2 (black lines) and 50 mW/m^2 (gray dashed lines) for a variety of bulk crustal compositions (with bulk crustal heat production of 0.4–0.7 μW/m^3, numbers on the curves) for mantle heat production of 0.03μW/m^3. Gray shading – mantle adiabat. (b) The same as in (a) but calculated for various values of surface heat flow (numbers on the curves) assuming bulk crustal heat production of 0.5 μW/m^3 (solid lines). Dashed lines – reference geotherms of Pollack and Chapman (1977). Note that two families of conductive geotherms agree only for a heat flow ~40 mW/m^2 (typical for Archean regions) that was used for calibration of crustal heat production in (a).

flow in the Archean regions to calibrate bulk crustal heat production and heat production in the lithospheric mantle. The determined values of thermal parameters were used to constrain the cratonic geotherms (Fig. 4.34). These geotherms match closely with conductive geotherms for regions with low surface heat flow (<50 mW/m^2) typical for the cratonic regions. Realistic geotherms (chosen to fit xenolith P–T arrays) require low, but non-zero heat production in the lithospheric mantle and low bulk crustal heat production which corresponds to the heat production of average granulites (Fig. 4.8). The heat production values preferred by Rudnick *et al.* (1998) are 0.4–0.5 μW/m^3 for the Archean crust and 0.03 μW/m^3 for the Archean lithospheric mantle:

- For higher values of bulk crustal heat production, geotherms do not intersect the mantle adiabat.
- Similarly, in the case that all surface heat flow results from crustal radioactivity sources, the temperature below the Moho is constant and never intersects the mantle adiabat.
- The shape of the actual distribution of heat production in the crust has a minor effect on the Moho temperature and no effect on mantle temperatures.
- In the case of no radiogenic sources in the crust, the calculated geotherms become similar to geotherms for old oceans.
- If, however, xenolith P–T arrays reflect the transient thermal regime in the cratonic lithosphere associated with magmatism, bulk heat production in the Archean crust can

be higher than the preferred values and the lithospheric thickness can be greater than predicted from the preferred heat production values.

An extension of the model of Rudnick *et al.* (1998) beyond the Archean crust, to continental regions with higher heat flow (>50 mW/m^2), displays a large discrepancy compared with the conductive geotherms of Pollack and Chapman (1977), and thus implies that bulk heat production in the post-Archean crust and the lithospheric mantle can be significantly higher than the values preferred by Rudnick and co-authors (Fig. 4.34). The discrepancy stems from the fact that, in the approach of Pollack and Chapman (1977), bulk crustal heat production A_{cr} is inexplicitly scaled by surface heat flow values Q_o as $A_{cr} = 0.01\ Q_o+0.2$ μW/m^3. For the range of considered surface heat flow values (30–120 mW/m^2) this yields A_{cr} ranging from 0.5 to 1.4 μW/m^3, while the preferred values of Rudnick and co-authors are fixed at 0.4–0.5 μW/m^3. From the other side, the formal solution of the steady-state thermal conductivity problem for regions with high heat flow as calculated by Pollack and Chapman (1977) may be misleading since these regions are likely to have a transient thermal regime of the lithosphere.

Mantle heat flow and mechanisms of surface heat flow variations in stable continents

The thermal conductivity boundary problem (Eq. 4.36) can be solved with conditions at the bottom boundary either for a constant heat flux (e.g. Jaupart and Mareschal, 1999), or for a constant temperature (e.g. Pollack and Chapman, 1977; Artemieva and Mooney, 2001). For the oceans, there are no obvious physical mechanisms to keep a constant heat flux at the plate bottom, while small variations in oceanic crustal thickness suggest that temperature variations at the base of the lithospheric plate are small (~15 °C) (McKenzie *et al.*, 2005). The situation is more complicated for the continental lithosphere. Assuming that the average potential temperature beneath the oceans and the continents is kept constant by large-scale mantle circulation, one should expect that a constant temperature at the lithospheric base (equal or close to the mantle potential temperature) rather than a constant heat flux from the mantle is the more likely situation beneath the continents. Because there are two options for the boundary conditions at the lithospheric base, lateral variations in surface heat flow of stable continents can be attributed to two end-member models with heat flow variations resulting primarily from:

- differences in crustal heat production (the case of a constant heat flux at the base) (e.g. Morgan, 1984; Jaupart and Mareschal, 1999), or
- lithospheric thickness (and thus mantle heat flow) variations (e.g. Ballard and Pollack, 1987; Nyblade and Pollack, 1993a).

Most of the practical approaches to calculation of lithospheric geotherms account for both of these effects (e.g. Artemieva and Mooney, 2001). Although the effect of thermal conductivity variations within the lithosphere on continental geotherms has gained some attention recently, its contribution to surface heat flow variations is insignificant (Levy *et al.*, 2010; see discussion in Section 4.1.2).

Precambrian regions with a steady-state thermal regime provide natural test sites for examining the relative contributions of crustal heat production and mantle heat flow in surface heat flow variations. In particular, worldwide analyses of surface heat flow variations across the Archean–Proterozoic boundaries indicate some heat flow increase from older to younger crust (e.g. Figs. 4.29, 4.30), while some geochemical evidence suggests that bulk crustal heat production in the Archean crust can be lower than in the Proterozoic crust (Table 4.4 and Section 4.1.3). A number of possibilities have been proposed to explain this possible age dependence of bulk crustal heat production:

(1) greater erosion of the Archean crust;
(2) selective survival of the Archean crust only with low crustal heat production;
(3) enrichment of the Proterozoic upper crust by radioactive isotopes during orogenic events;
(4) secular changes in crust-forming processes.

Although lower heat flow together with (possibly) lower bulk crustal heat production in the Archean terranes suggests that crustal heat production plays an important role in surface heat flow variations, other lines of evidence suggest a complex interplay between crustal heat production and mantle heat flow contributions to surface heat flow. Regional studies indicate that either one or the other mechanism can dominate on a regional scale. For example, thermal models in which surface heat flow variations are explained entirely by variations in crustal radioactivity cannot satisfy xenolith geotherms for South Africa, and variations in lithospheric thickness should be the principal factor controlling the surface heat flow pattern in the region (Nyblade and Pollack, 1993a,c). Ballard and Pollack (1987) have proposed that a thick lithosphere of the Archean Kalahari craton diverts mantle heat sideways to the surrounding Proterozoic mobile belts and further complicates the pattern of regional surface heat flow variations. This hypothesis is supported by recent mantle convection models that have focused on the insulating effect of continents. They demonstrate the similarity between trends in surface and mantle heat flow variations, interpreted as evidence that variations in crustal radioactivity are not the major factor controlling surface heat flow variations (Cooper et al., 2006).

In contrast, detailed analysis of the crustal structure and heat flow data in the Canadian Shield and in the South African craton (Jaupart and Mareschal, 1999) was used to argue that most surface heat flow variations in these provinces can be attributed to heterogeneous crustal heat production. This and similar studies suggest that mantle heat flow in the Canadian Shield apparently has a uniform value through Precambrian terranes of different ages (~12 mW/m^2), with no evidence for a diversion of mantle heat from the Archean terranes to younger adjacent regions (Lenardic et al., 2000) (Tables 4.10, 4.11). Indeed, the Canadian Shield is composed of a large number of small (sometimes less than 50–100 km across) distinct crustal terranes, which amalgamated at 2.72–2.69 Ga. Strong crustal heterogeneity leads to significant lateral large-amplitude surface heat flow variations with short wavelength (commonly less than 100 km), suggesting their crustal origin (Fig. 4.35).

However, with new, very low, heat flow data being accumulated in the northern parts of the Superior province (not shown in Table 4.10), the emerging picture is becoming more complicated. Very low values of surface heat flow (<30 mW/m^2) reported for large areas of the Siberian craton, some parts of the Baltic Shield, and the

Table 4.10 Heat flow in the main Precambrian–Paleozoic provinces of North America

Tectonic province	Tectono-thermal age (Ga)	Average heat flow (mW/m^2)
Superior province	2.7	42 ± 10
Trans-Hudson Orogen	1.9	42 ± 11
Grenville province	1.0	41 ± 11
Appalachians	0.4	57 ± 13

(source: Jaupart and Mareschal, 1999)

Table 4.11 Mantle heat flow in Precambrian and Paleozoic regions

Region	Moho heat flow (mW/m^2)*	Source**
Archean–Mesoproterozoic cratons	**10–30**	AM01
Canadian Shield (Superior province and	10–15	JM99
Trans-Hudson Orogen)	12–13	J98
	12–18	L10
	13–18 (Kapuskasing)	A87
	15–18	AM01
	18–25	RN99
Siberian craton (all)	10–22	AM01
Daldyn–Alakit terrane	10–15	AM01
- " -	10	MK05
Anabar Shield (Siberia)	~20	AM01
West African craton	10–15	AM01
Lewisian (Scotland)	11–16	WT84
Slave craton	12.4	MK05
Baltic Shield	12–15	AM01
Russian Platform	13–22	AM01
Kaapvaal craton	5–12 (Vredefort)	N91
	17	JM99
	17–25, best fit 18	RN99
	19–20	AM01
Western Australia	18–23	AM01
Central Australia	25–30	AM01
India	17–30	AM01
Sveco-Norwegian (Baltic Shield)	20–21	PJ87
South American craton	25–30	AM01
Northern China	25–30	AM01
Neoproterozoic–Paleozoic	**25–50**	AM01

* Heat flow at the base of the lithosphere is the same as at the Moho for heat production in the lithospheric mantle $A = 0$ μW/m^3 and ~1–3 mW/m^2 less than at the Moho for $A = 0.01$ μW/m^3.

** References: A87 = Ashwal *et al.*, 1987; AM01 = Artemieva and Mooney, 1999; 2001; J98 = Jaupart *et al.*, 1998; JM99 = Jaupart and Mareschal, 1999; L10 = Levy *et al.*, 2010; MK05 = McKenzie *et al.*, 2005; N81 = Nicolaysen *et al.*, 1981; PJ87 = Pinet and Jaupart, 1987; RN99 = Rudnick and Nyblade, 1999; WT84 = Weaver and Tarney, 1984.

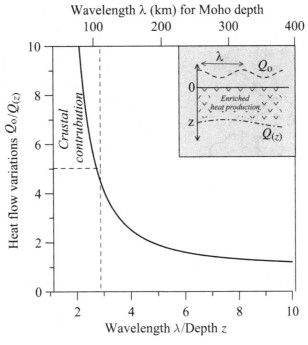

Fig. 4.35 The ratio between the magnitude of heat flow anomalies at the surface and at depth as a function of the wavelength of surface heat flow variations and depth to the anomaly (assumed to be produced by laterally sinusoidal distribution of heat production). To distinguish between surface heat flow anomalies of shallow and deep origin, as a rule of thumb, Jaupart and Mareschal (1999) suggest use of the wavelength of surface heat flow variations $\lambda > 2.8z$, which produces surface heat flow anomalies of magnitude 5 times higher than at depth z. In such an interpretation, surface heat flow anomalies with $\lambda < 110$ km are generated within the crust (for crustal thickness 40 km), while heat flow anomalies at the base of the cratonic lithosphere (at ~215 km depth) produce surface heat flow variations with $\lambda > 600$ km.

northern parts of the Canadian Shield require low mantle heat flow and, as a consequence, large lithospheric thickness in these regions (Fig. 4.36, Table 4.11) (Artemieva and Mooney, 2001; Levy *et al.*, 2010). In particular, a 6 mW/m^2 change in mantle heat flow in cold cratonic regions can be associated with variations in lithospheric thickness of ~100 km (Levy *et al.*, 2010) (Fig. 4.37). A simple estimate shows that heat flow at the Moho, Q_M, cannot be less than 10–12 mW/m^2 even in regions with a very low surface heat flow. For purely conductive heat transfer in the lithospheric mantle,

$$Q_M = \lambda [T_{TBL} - T_M]/[z_{TBL} - z_M] \tag{4.37}$$

where λ is the thermal conductivity of the lithospheric mantle, T and z are temperature and depth, respectively, and the indexes TBL and M refer to the base of the lithosphere and the Moho. Assuming a crustal thickness of 40 km, the lithospheric mantle extending

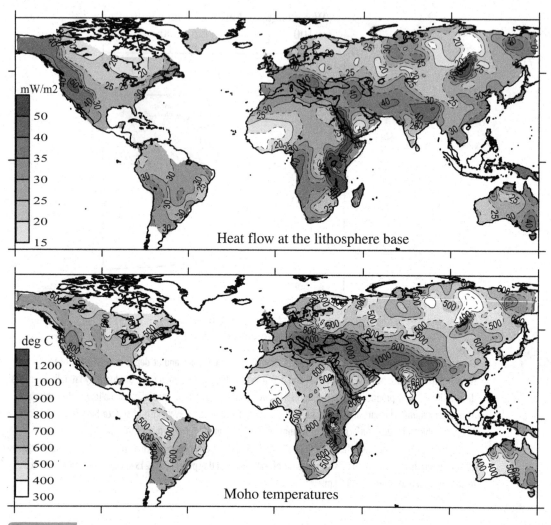

Fig. 4.36 Thermal regime of continental lithosphere constrained by heat flow data (Source: Artemieva and Mooney, 1999; 2001). The maps are constrained by a 15 deg × 15 deg interpolation; data coverage is shown in Fig. 4.16. (a) Mantle heat flow at the base of the lithosphere; (b) temperatures at the Moho (or at 40 km depth in regions where crustal thickness is poorly known).

down to the transition zone (410 km depth), mantle potential temperature of 1350 °C at the base of the lithosphere, and Moho temperature along a 30 mW/m^2 conductive reference geotherm, yields $Q_M \sim 11.5$ mW/m^2 for $\lambda = 4$ W/m/K and $Q_M \sim 10$ mW/m^2 for $\lambda = 3.5$ W/m/K.

Numerical models of mantle convection with floating continents (Lenardic and Moresi, 2001) resolve an apparent contradiction between two competing models of lithospheric thermal structure (controlled primarily by crustal heat production or by mantle heat flow variations) and demonstrate that they are the end-member solutions for layered and whole-

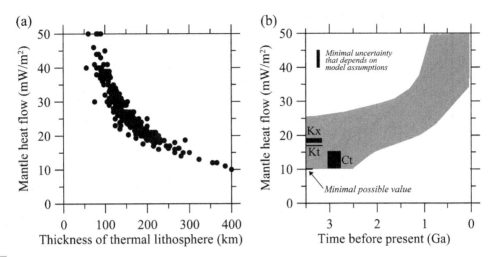

Fig. 4.37 Typical variations of mantle heat flow with lithospheric thickness (a) and with tectono-thermal age (b). Dots in (a) and gray shading in (b) correspond to the values calculated in global steady-state modeling of surface heat flow data (Source: Artemieva and Mooney, 1999; 2001). Other estimates of mantle heat flow: Kaapvaal craton (Kt) and Canadian Shield (Ct) thermal model (Jaupart and Mareschal (1999); Kaapvaal xenolith data (Rudnick and Nyblade, 1999). There is a significant uncertainty associated with the assumed values of thermal parameters. In particular, Jaupart and Mareschal (1999) assume four times higher heat production in the lower crust, which results in systematically lower values of mantle heat flow (on average, by 4–5 mW/m^2).

mantle convection models: the thermal resistance of the convecting mantle depends on the thickness of the upper thermal boundary layer and the system has a compensation mechanism for the former by adjusting the thickness of the latter. In the case of layered convection (i.e. thin convecting layer), the effect of lithospheric thickness on mantle heat flow is strong and large surface heat flow variations can be explained by small variations in relative lithospheric thickness (i.e. in the ratio of TBL to system depth). On the contrary, in case of whole mantle convection the dependence of mantle heat flow on lithospheric thickness is becoming weak.

Thermal evolution of the mantle and the "Archean paradox"

Heat generated by radioactive isotopes decreases exponentially with time, so that its present value is c. 4.5 times less than at 4.5 Ga and c. 2.5 times less than at 3.0 Ga (Fig. 4.38a). This means that the average mantle heat flow and mantle temperatures should have been significantly higher in the Archean than at present (e.g. Wasserburg *et al.*, 1964). In contrast, the lack of massive crustal melting in the Archean and Archean geotherms constrained by mineral assemblages in Archean high-grade terranes suggest that the Archean mantle temperatures were not significantly different from the present ones (Bickle, 1978; Burke and Kidd, 1978; England and Bickle, 1984).

The contradiction between certain lines of evidence that suggest that the Archean average mantle heat flux was significantly higher than at present and the evidence that Archean

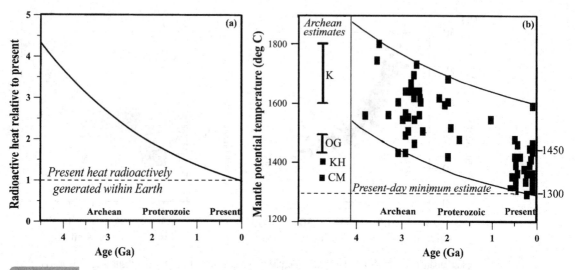

Fig. 4.38 Thermal history of the Earth. (a) Heat generated within Earth (normalized by its present-day value) by three major heat-producing isotopes: potassium, uranium, and thorium (based on estimates of van Schmus, 1995). (b) Potential temperatures of the upper mantle (hypothetical temperatures of mantle adiabatically brought up to the surface without melting) calculated for MORB-like ophiolite suites and greenstone belts (source: Abbott *et al.*, 1994). The upper and lower boundaries (exponential curves) approximate the shape of the radiogenic heat production curve (Wasserburg *et al.*, 1964). Potential temperature of the present-day mantle is commonly estimated to be between 1300 °C and 1450 °C. Left insert in (b): summary of mantle potential temperatures in the Archean, based on komatiite (K) (Nisbet and Fowler, 1983) and hydrous komatiites (KH) melting models (Grove and Parman, 2004), ophiolite suites and greenstone belts (OG) (Abbott *et al.*, 1994), and mantle convection models (MC) (Campbell and Griffiths, 1992).

continental geotherms were similar to modern ones produces the so-called "Archean paradox". This could easily be resolved if the ratio of oceanic to continental mantle heat flow was greater in the Archean due to greater heat loss through the oceans. The most obvious way to explain the paradox is to assume that heat loss due to the creation and subduction of oceanic lithosphere was greater in the Archean than at present (e.g. Burke and Kidd, 1978). This assumption can be fulfilled if plate velocities in the past were faster and ridge lengths were greater (e.g. Bickle, 1978). Assuming that (a) the Archean cratons were much smaller than present-day continents and (b) applying the present-day statistics of plate configuration (size distribution versus normalized area) to the past, Abbott and Menke (1990) concluded that the length of plate boundaries at 2.4 Ga was ~2.2 times greater than their present length (with >60 cratons). Note that both of these assumptions may be fundamentally wrong. Furthermore, mantle convection models that treat mantle convection as a thermal boundary layer phenomenon cannot explain the spatially disproportionate heat loss: spatial averaging over different, but with same lateral size, columns of convective cells will produce the same local heat loss out of a convective layer regardless of the location (i.e. oceanic versus continental settings) (Lenardic, 1998).

The presence of the continental crust affects mantle convection and helps to overcome the latter problem of spatially disproportionate heat loss by creating fundamentally different boundary conditions at the crust–mantle interface in continents and oceans (Lenardic, 1998). In oceans, the boundary condition at the crust–mantle interface is for a constant temperature, which is achieved by the involvement of oceanic crust in convective mantle overturns and by a large contrast in effective thermal conductivity of the oceanic crust and the mantle. If the crust–mantle boundary condition on Archean continents is for a near-constant heat flux (in particular, due to comparable thermal conductivities of the continental crust and the mantle), it produces spatially variable thermal conditions at the top of the convecting mantle (i.e. at the base of the TBL). As a result, Archean continental geotherms can become stabilized, despite higher convective vigor in the Archean, by forcing a greater proportion of the heat to be carried through the oceans.

On the other hand, following the "devil's advocate" arguments of Hamilton (2007), one should bear in mind that (also see Section 9.2.1):

- there is no geological proof for the existence of lithospherically distinct oceans and continents in the Archean; thus mantle convection models based on a different thermal regime for continents and oceans in the Archean may be fundamentally biased;
- the extremely mobile Archean continental crust "was incapable of behaving as the semi-rigid plates required, by definition, for plate tectonics"; thus thermal models that assume Archean plate tectonics may be misleading.

Recent mantle convection modeling based on back-tracking the thermal history of Earth from present-day conditions suggests that surface heat flux did not need to be significantly higher in the past than at present (Korenaga, 2006). Assuming that (a) plate tectonics have operated since the Paleoproterozoic and (b) secular evolution of mantle convection is affected by depleted cratonic lithosphere, the model predicts that in the Archean, when the mantle was hotter, plate tectonics was more sluggish, with slower plate velocities than commonly assumed. For the Urey ratio $\gamma = 0.15$–0.3 ($\gamma = H/Q$ and is the measure of the relative importance of internal heating H with respect to total convective heat flux Q, Christensen, 1985), the whole-mantle convection model predicts moderate secular cooling of the Earth with similar surface heat flow in the past and at present, but with the progressively increasing role of internal heating (due to radioactive heat generation) back in time (compare with Fig. 4.38a).

To add to the "paradox", petrologic and geophysical estimates of the Archean mantle temperatures differ significantly. In particular, mantle temperatures ~300–500 °C hotter than modern ambient asthenosphere are required for generation of Archean komatiites (Nisbet and Fowler, 1983). Some petrologic studies suggest, however, that these ultramafic lavas, that are almost entirely unique for the Archean and require eruption temperatures of at least as high as 1600 °C (and up to 1800 °C), were probably erupted by hotspots and are not representative of "normal" Archean mantle (e.g. Abbott et al., 1994; Fig. 4.39). Furthermore, recent petrologic studies that interpret Archean komatiites as products of hydrous shallow melting predict that the Archean mantle was only slightly hotter (~100 °C) than at present (Grove and Parman, 2004). Geochemical data from ophiolite suites and greenstone belts provide lower Archean temperatures than komatiite data and require that upper mantle potential temperature in the late Archean (2.8 Ga) should be only

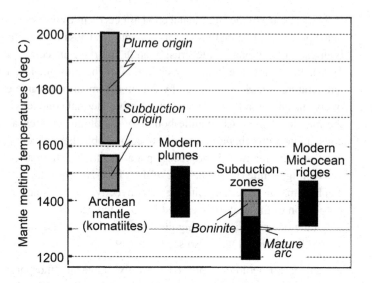

Mantle melting temperatures for different modern and Archean tectonic settings (data sources: Tatsumi *et al.*, 1983; Kinzler and Grove, 1992; Eggins, 1992; Baker *et al.*, 1994; Wagner and Grove, 1998; Falloon and Danyushevsky, 2000; Grove and Parman, 2004).

~130–200°C hotter than at present (Abbott *et al.*, 1994). More recent estimates based on mantle convection models indicate that the Archean mantle could, on average, be only less than 50 °C hotter than the present mantle (Campbell and Griffiths, 1992) (Fig. 4.38b).

Continental regions with transient thermal regime

Morgan (2001) argues that the transient thermal processes that operate in the continental lithosphere are, in general, similar to the processes that operate in the oceanic lithosphere. However, in contrast to the oceanic settings, the magnitude of these processes is less predictable due to thermal variations associated with the chemical heterogeneity of the continents.

Collisional orogens

Collisional orogens are characterized by a highly heterogeneous heat flow, with the highest values typically measured along the major fault zones. As for subduction zones, a steady-state approach is inapplicable to these tectonic structures. Thermo-kinematic models provide the best constraints on lithospheric temperatures in the orogens, such as the Alps (e.g., Royden *et al.* 1983; Davy and Gillet 1986; Bousquet *et al.*, 1997). However, they require detailed information on dynamic processes in the mantle, which are usually not completely understood. As a result, thermo-kinematic models are poorly constrained and detailed knowledge on lithospheric temperatures is unavailable.

A 2D model of the lithosphere of the Alps which takes into account the processes of crustal shortening and formation of crustal and lithospheric roots during subduction is an example of

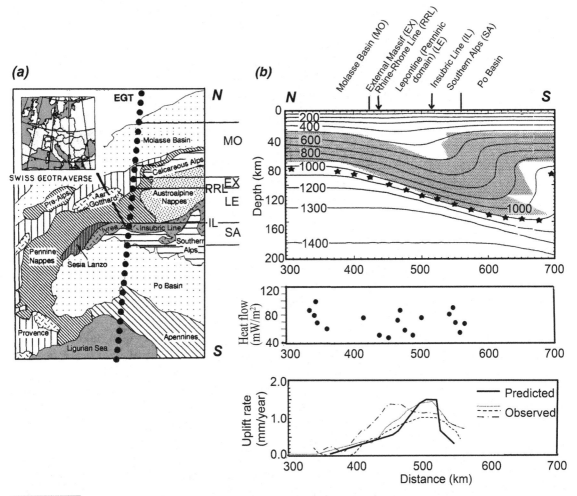

Fig. 4.40 Thermo-kinematic modeling for the Central and Western Alps. (a) Tectonic map showing the location of the European Geotraverse (EGT), (b) Top: cross-section along the EGT profile showing the predicted present-day lithospheric temperatures (asterisks – the lithosphere–asthenosphere boundary defined as the intersection of the geotherm with the mantle solidus; shaded area – the deformed lithospheric mantle derived from tectono-kinematic calculations); Middle: the present-day surface heat flow corrected for uplift and topography (the uncertainty is c. 20 mW/m²), Bottom: the present-day uplift rates derived from geodetic observations (dashed lines) and the exhumation rates at the surface predicted by the kinematic model (from Okaya et al., 1996).

an advanced thermo-mechanical model (Okaya et al., 1996). This model predicts that, although the crustal thickness across the Alpine region changes from ~30 km to the north of the orogen to ~55–60 km beneath the Alps and to ~30–34 km in the south beneath the Po basin, the Moho is an almost isothermal boundary with a temperature of ~500–600 °C, while the lithospheric thermal thickness gradually increases southwards from ~80 km to the north of the orogen to ~120–150 km beneath the southern Alps–northern Apennines (Fig. 4.40).

Extended regions

The Basin and Range Province in western USA is a classical example of an extended continental region (Gans, 1987). The region is characterized by anomalously high surface heat flow relative to the global continental average of 65 mW/m^2 (Pollack *et al.*, 1993): 92 ± 9 mW/m^2 in the presently active Northern Basin and Range and 82 ± 3 mW/m^2 in the presently quiescent Southern Basin and Range Province (Lachenbruch *et al.*, 1994). Although average heat flow in the northern and southern parts of the Basin and Range is similar on the province-wide scale (Fig. 4.18), the two regions differ both in topographic elevations (~1.0 km in the south and ~1.8 km in the north with an abrupt transition between them) and in Cenozoic extension histories. The average total Cenozoic extension may be similar in both regions; but while crustal extension is still on-going in the Northern Basin and Range and has roughly doubled its area in the Cenozoic (the ratio of area size prior to and after the extension $\beta = 2$), the Southern Basin and Range Province has been relatively inactive during the past 10–15 Ma (Hamilton, 1987).

The observed province-wide heat flow anomaly can be explained by a broad range of simple extension models that predict the time-scale of the process of *c.* 10–50 Ma (e.g. McKenzie, 1978). Within the frame of extensional models, the high heat flow in the Northern Basin and Range has been successfully explained by the Cenozoic extension-related upward influx of material in the lithosphere which advectively transfers the heat, increases the surface heat flow, and affects the thermal structure of the lithosphere. The first stages of tectonic extension and magmatism, when surface heat flow rapidly increases by non-conductive processes, are followed by tectonic quiescence, when heat transfer becomes dominated by conduction. Heat conduction, which leads to a decrease in surface heat flow, is a slow process; its earlier stages are controlled by the thermal structure of the lithosphere at the end of tectonic activity, while the equilibrium thickness of the lithosphere controls its later stages.

Figure 4.41a, based on the extensional model of McKenzie (1978), illustrates evolution of the extensional geotherms. An instantaneous extension (i.e. faster than required to redistribute heat by diffusion) of a thermally equilibrated lithosphere by a factor of β causes lithosphere thinning by factor β and an increase in thermal gradient and surface heat flow by the same factor β. To the first order, high surface heat flow observed in the Basin and Range Province supports this model even quantitatively (Lachenbruch *et al.*, 1994). For example, in the case where $\beta = 2$ (Fig. 4.41b), high heat flow is still close to its maximum value even after 15 Ma since termination of tectonism as observed in the Southern Basin and Range Province. However, a province-wide extension with $\beta = 2$ can also be achieved by an inhomogeneous extension with localized areas of high (e.g. $\beta = 10$) concentrated extension. Such areas will cool much faster and (neglecting lateral heat transfer) 15 Ma after the tectonism heat flow it will drop to <40% of its maximum value. Because of heat loss during the extension process, a uniform (as compared to instantaneous) lithosphere stretching at a constant rate requires slightly higher β to double the mantle heat flow (equations 15 and 16 in Lachenbruch and Sass, 1978).

Topographic elevation provides additional information for discriminating between various possible extensional models due to the buoyancy trade-off between extension and

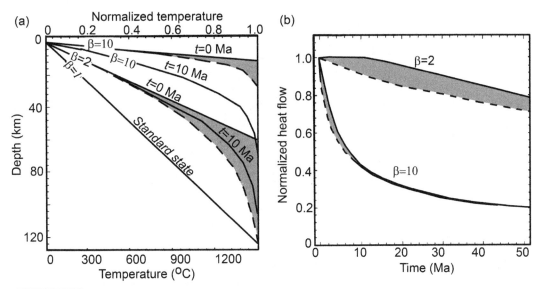

Fig. 4.41 Thermal model for extended regions (after Lachenbruch *et al.*, 1994). Radiogenic heat generation is neglected. (a) Lithospheric geotherms 0 Ma and 10 Ma after homogeneous stretching of a 125 km-thick lithosphere by a factor of β =2 or β=10 (solid lines). Dashed lines − steady-state geotherms for a uniform lithosphere extension that causes the same increase in heat flow as instantaneous stretching by factor β = 2 or β = 10. In the case where the lithospheric base is at constant temperature and extends uniformly at any rate so that heat flow increases twofold or tenfold, all geotherms fall within the gray areas. (b) Heat flow decay after lithosphere extension for the geotherms bounding gray regions in (a). Solid lines − heat flow for instantaneous stretching (β − stretching factor); dashed curves − for the steady-state case with a steady-state strain rate of 1 %/Ma (3×10^{-16} 1/s) and 25 %/Ma (10^{-14} 1/s) for β = 2 and β = 10, respectively.

magmatism (Lachenbruch and Morgan, 1990). This is, in particular, important in the Northern Basin and Range Province where high crustal extension ($\beta \sim 2$) is in apparent contradiction with high elevation and requires either the development of a compensating source of buoyancy during the extension or an extremely high elevation (~6 km above sea level) prior to the extension. Lithosphere delamination at the early stages of tectonism could produce topographic elevation despite stretching.

Metamorphic core complexes (MCCs), widely distributed in highly extended regions of western USA, provide additional information on the Cenozoic thermal evolution of the region (Fig. 4.42). The Arizona MCCs are interpreted to represent mid-crustal rocks that became exposed due to very fast (over a few million years) tectonic unroofing of 10–20 km of the upper crust. As discussed by Lachenbruch *et al.* (1994),

> such a massive removal of surface material should affect the surface heat flow in at least two ways: (1) it would expose warmer mid-crustal rocks, and for a time at least, cause a large local increase in heat flow, and (2) by removing the most radioactive (upper) part of the crust, it would cause a decrease in the steady state background heat flow, observable only after the initial warming decays.

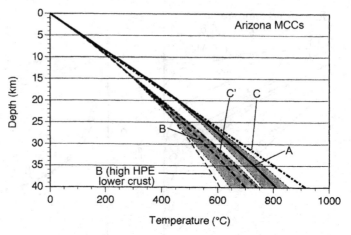

Fig. 4.42   Lithospheric geotherms A and B in Southern Basin and Range Province based on data from two metamorphic core complexes and assuming temperature- and pressure-dependent thermal conductivity and surface heat flow of 62 mW/m^2 (after Ketcham, 1996). In the case where mantle heat flow is 10 mW/m^2 lower, Moho temperatures will reduce by 235 °C. Shaded areas – estimated variability due to uncertainty in heat production (the corresponding distribution of heat production with depth for sets A and B is shown in Fig. 4.13(d–e), respectively). Geotherm C calculated for the lithosphere stretching model is shown for comparison (Lachenbruch and Sass, 1978). C′ – the same model as C but with constant thermal conductivity of 2.5 W/m/K.

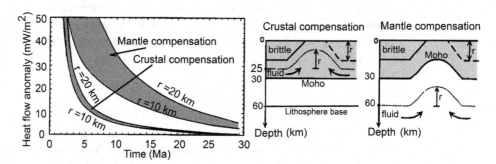

Fig. 4.43   Surface heat flow anomaly caused by an instantaneous unroofing with $r = 10$ km and $r = 20$ km in a 60 km-thick lithosphere at constant basal temperature (after Lachenbruch et al., 1994). In the case of crustal compensation, the compensating return flow occurs at 25 km depth. In the case of mantle compensation in a warm lithosphere with temperature gradient of 23 K/km, the return flow occurs at 60 km depth. Radiogenic heat generation is neglected.

The evidence for the latter effect is provided by spatial correlation between the elongated low heat flow anomaly and the belt of MCCs in the Southern Basin and Range Province. A 1D thermal model of instantaneous unroofing (Block and Royden, 1990) implies that removal of the upper crustal material is compensated by a lateral material influx that occurs either in a ductile lower crust or at the base of the lithosphere (Fig. 4.43). The mechanism of crustal compensation is consistent with relatively uniform crustal thickness in the region, since (unroofed) regions with MCCs have the same crustal thickness as their surroundings.

Global patterns

Using the global heat flow data of Pollack *et al.* (1993) supplemented by newer heat flow measurements, Artemieva and Mooney (2001) calculated regional lithospheric temperatures in stable continents and found that for the typical values of heat flow observed in crustal terranes of different ages they are close to the reference conductive geotherms of Pollack and Chapman (1977) (Fig. 4.33). The thermal state of the continental lithosphere is illustrated by a set of maps (Figs. 4.36 and 4.44). For stable continents, tectonically quiescent at least since the mid-Paleozoic, these maps are derived from numerous geotherms constrained as the downward continuation of surface heat flow (eq. 4.36). The values of thermal parameters (thermal conductivity and heat production) in the sedimentary layer and in the upper crust are constrained by regional measurements on rock outcrops and borehole data, supplemented by regional detailed seismic data on the crustal structure. Seismic data are also used to determine the thicknesses of the middle and lower crustal layers, for which thermal parameters are fixed at constant due to large uncertainty in the actual values (Section 4.1, Table 4.8). Consistent model assumptions for stable continents permit meaningful comparisons of the thermal structure of the lithosphere. Since steady-state modeling is inapplicable to tectonically active regions, such as Cenozoic orogens and extended regions, in these regions temperatures at 50 km depth are assumed to be in the range 900–1100 °C, and the lithosphere thermal thickness is assumed to be 60–80 km, in accord with numerous petrologic and regional transient geothermal models (e.g., Lachenbruch and Sass, 1977; Polyakov *et al.*, 1988; Mechie *et al.*, 1994; Decker, 1995; Okaya *et al.*, 1996; Le Pichon *et al.*, 1997; Currie and Hyndman, 2006) (Fig. 4.39b).

The major results can be summarized as follows.

The base of the crust is not isothermal. Temperatures at the Moho vary widely from 300–500 °C in the Archean–Paleoproterozoic cratons to 500–800 °C in Meso- and Neoproterozoic regions, where the crustal thickness averages 40–45 km. In the Archean–Paleoproterozoic regions the highest temperatures at the Moho are expected to be either in the regions with anomalously thick crust, such as in central Finland, or in the regions that have been affected by Phanerozoic tectonic activity (e.g. the Sino-Korean craton). In young tectonically active regions, temperatures at the base of the crust can be as high as 800–900 °C. Moho temperatures can vary significantly over short lateral distances, reflecting not only variations in the thermal regime of the crust, but also variations in the crustal thickness and composition. For example, across the Trans-European Suture Zone, which is the major tectonic boundary in Europe between the Precambrian East European Craton and the Phanerozoic Europe, temperatures at the crustal base change from 450–550 °C within the craton (where the Moho is at a 40–45 km depth) to 600–700 °C at the base of the 30–32 km-thick Variscan crust.

The depth where lithospheric temperatures reach ~600 °C can be interpreted as the proxy for the base of the elastic lithosphere in continental regions where this depth exceeds the crustal thickness (Fig. 4.44b). Brittle–ductile transition in olivine corresponds to the critical isotherm of ~600–700 °C, and flexural rigidity of old (>200 Ma) continental lithosphere is dominated by olivine rheology of the mantle (Chapter 8). Analysis of the depth–temperature correlation of the intraplate continental seismicity indicates that most of it terminates at

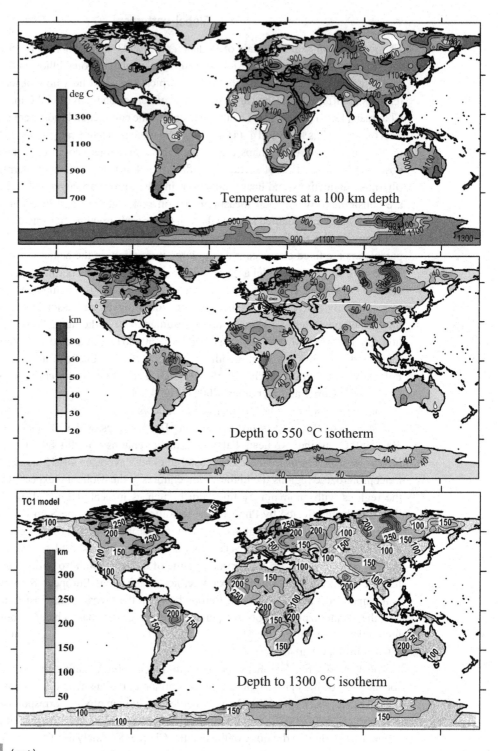

Fig. 4.44 (cont.)

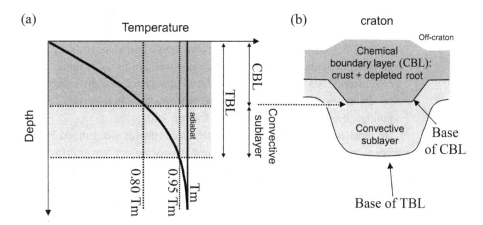

Fig. 4.45 Correlation between the thermal boundary layer (TBL), the chemical boundary layer (CBL), and the convective sublayer (CSL) (from Lee, 2006). The thickness of CSL is exaggerated. In this interpretation the base of the TBL is defined by the point at which the temperature becomes 0.95 of the ambient mantle temperature T_m.

lithospheric temperatures of 500–600 °C. The depth to 500–600 °C isotherm can also be considered as a proxy for the thickness of the magnetic layer (the maximal depth of crustal magnetization), since the Curie temperature for magnetite is close to 550 °C (Petersen and Bleil, 1982) (Section 5.4).

4.4.3 Thickness of thermal lithosphere

Thermal boundary layer

Heat flow from the deep interior is transferred upwards primarily by mantle convection. In the upper layers of the Earth the conduction mechanism of heat transfer starts to dominate; this upper part with mainly conductive heat transfer is termed the thermal lithosphere or TBL, the thermal boundary layer dominated by definition by conductive heat transfer. The transition from convection to conduction mechanisms is gradual and is responsible for the existence of the "convective boundary (sub)layer" in the upper mantle at the base of the TBL (Fig. 4.45a). Sometimes this convective boundary (sub)layer is considered as a part of the TBL (e.g. Cooper *et al.*, 2004; Lee *et al.*, 2005; McKenzie *et al.*, 2005).

Caption for figure 4.44 (cont.)

Thermal state of the continental lithosphere constrained by surface heat flow for stable regions, xenolith data for active regions, and statistical values for regions with no borehole data (Source: Artemieva, 2006). The step between the contours corresponds approximately to the uncertainty of the models. (a) Temperatures at 100 km depth shown with a 200 °C step in contours; (b) the depth to a 550 °C isotherm; the map is a proxy for the depth to the Curie isotherm, the depth to the brittle–ductile transition in olivine, and the thickness of the elastic lithosphere; (c) the depth to a 1300 °C isotherm shown with a 50 km step in contours; the map provides a proxy for the thickness of the thermal lithosphere. The interpolation used to construct these maps results not only in lateral smearing of short-wavelength anomalies, but also in a significant reduction in their amplitude.

Since large-scale seismic tomography samples the top of the convective mantle, litho-spheric thickness estimated from seismic tomography models commonly extends deeper than the TBL, down to the base of the convective boundary (sub)layer. The thickness of the convective boundary (sub)layer, which is 40–50 km in the cratonic regions (Jaupart *et al.*, 1998; Sleep, 2005), defines the amplitude of a systematic difference in lithospheric thick-ness estimates based on thermal and seismic tomography data, in particular in stable continental regions (Artemieva and Mooney, 2002).

In practice, the thickness of the TBL is commonly determined by the intersection of a lithospheric geotherm (actually, its linear downward continuation) with a mantle adiabat of $T_m \sim 1350$–$1400\,^\circ$C (which is mantle isentropic temperature at a depth of 100–150 km; the adiabatic gradient in the upper mantle is given by eq. 4.28). For surface heat flow $<30\,\text{mW/m}^2$ the conductive geotherm asymptotically approaches the mantle adiabat at depths greater than $\sim$300 km (Fig. 4.33) and the depth of their intersection (thickness of thermal lithosphere) cannot be reliably estimated. Since for typical upper mantle compositions the transition from high lithospheric viscosity to a lower asthenospheric viscosity (with effective values of $\sim10^{21}$ Pa s) starts at $T \sim 0.8T_m$, the thickness of the TBL (thermal lithosphere) is sometimes defined as the depth to a 1050–1100 $^\circ$C isotherm (e.g. Pollack and Chapman, 1977).

The transition from high viscosity lithosphere to a lower viscosity asthenosphere is controlled not only by temperature, but also by mantle composition, in particular by volatile content. Because of the significant difference in mantle solidi, the thickness of the TBL can differ significantly for dry and wet peridotite melting (Fig. 4.33a). Furthermore, mantle melting leads to the formation of a chemical boundary layer that is depleted in meltable components and dehydrated, and thus may have a viscosity two orders of magnitude higher than the upper mantle viscosity (Hirth and Kohlstedt, 1996). Lee *et al.* (2005) argue that the CBL, formed by mantle melting and dehydration, does not participate in large-scale or secondary mantle convection and forms a conductive lid both in oceans and continents that controls the thickness of the TBL and plays a significant role in its preservation (Fig. 4.45).

Temperature distribution in the lithosphere, required to estimate the lithospheric thermal thickness, can be calculated from several techniques:

(1) cooling models for normal oceanic lithosphere, in which thickness of the oceanic TBL is approximated by eq. (4.25), $L \sim 11\ t^{1/2}$;
(2) steady-state thermal conductivity models for stable continental regions constrained by surface heat flow measurements;
(3) transient thermal models for active continents and anomalous oceans;
(4) petrologic constraints on lithospheric geotherms based on mantle xenolith data, meta-morphic reactions;
(5) the depth to the Curie isotherm;
(6) conversion of seismic velocities, seismic attenuation, or electrical conductivity into temperatures constrained by laboratory measurements on physical properties of rocks at mantle P–T conditions;
(7) models of mantle convection that constrain mantle heat flow at the lithospheric base.

Some of these approaches are discussed in more detail in Chapter 5.

Global patterns for the continents

The age dependence of surface heat flow can, to first order, be recast as the relationship between lithospheric thermal thickness and age: statistically, regions with higher heat flow have thinner lithosphere. This transition is not, however, straightforward because of the non-uniform contribution of crustal heat production into the observed heat flow which results in significant scatter in age–thickness data (Fig. 4.46). Furthermore, since heat flow data coverage is very uneven and some of the regions are significantly oversampled in comparison to less studied regions, the best fit approximation for lithospheric thickness – the surface heat flow relationship may be misleading. For this reason, Fig. 4.47 shows only the upper and lower bounds as based on worldwide analysis of the thermal state of the continents (note that for regions with extremely low surface heat flow estimates of the TBL thickness are asymptotic). Although the exact depth extent of the cratonic lithosphere cannot be reliably assessed from thermal data (below ~300 km cold cratonic geotherms asymptotically approach the mantle adiabat so that the depth of their intersection cannot be determined), the depth to the 410 km mantle transition zone puts a natural limit on the base of the thermal lithosphere since the phase transition temperature at 410 km is expected to be close to the mantle potential temperature (Ito and Takahashi, 1989).

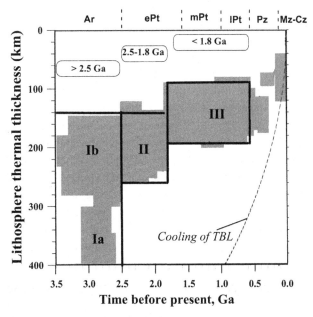

Fig. 4.46　Variations of lithospheric thermal thickness with tectono-thermal age. Gray shading shows lithospheric thickness estimates derived from surface heat flow (Source: Artemieva and Mooney, 2001). The Archean lithosphere has bimodal thickness distribution centered at ~350 and ~220 km. Key: Ar – Archean; ePt, mPt, lPt – early, middle, and late Proterozoic, respectively; Pz – Paleozoic; Mz-Cz – Meso-Cenozoic. Dotted line – theoretical curve based on cooling of thermal boundary layer (eq. 4.38).

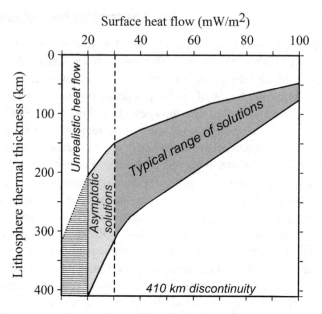

Fig. 4.47 Empirical relationship between surface heat flow and lithospheric thermal thickness. For each surface heat flow value, a large range of lithospheric thickness is possible, depending primarily on crustal heat production. For surface heat flow <30 mW/m² the conductive geotherm asymptotically approaches mantle adiabat and the depth of their intersection cannot be reliably estimated. The 410 km discontinuity limits the maximum possible thickness of thermal lithosphere.

The thickness of the continental lithosphere increases with tectono-thermal age (Fig. 4.46), but it does not follow the $\sqrt{age}$ pattern predicted by cooling of the thermal boundary layer (eq. 4.25). In contrast to the oceans where lithosphere thickness slowly increases with age in old oceans, old continents have deep-extending lithospheric roots (Fig. 4.46). A typical thickness of continental lithosphere is $c.$ 200 ± 50 km in the Paleoproterozoic terranes, 140 ± 50 km in Meso-Neoproterozoic terranes, and $100–120$ km in the Paleozoic regions (Artemieva and Mooney, 2001). The exceptions are the Archean cratons affected by Phanerozoic tectono-magmatic events (e.g. the Wyoming and the Sino-Korean cratons) where lithospheric thickness does not exceed $120–150$ km. In tectonically active Cenozoic regions (except for the collisional orogens where subducting lithospheric slabs can reach a depth of ~200 km) the lithosphere thickness is typically $60–80$ km.

Statistical analysis of thermal model constrained by surface heat flow

Lithosphere thickness–age correlation

A statistically significant correlation between the thermal state of the continental lithosphere and its age (Artemieva, 2006) is observed despite the large scatter of lithospheric thickness values (~100 km for any of the considered age intervals) (Tables 4.12, 4.13). This scatter is associated with uncertainties in the thermal model ($c.$ 25% for the lithospheric thickness)

Table 4.12 Statistics of thermal structure of the continental upper mantle

	n	T at 50 km (°C)	T at 100 km (°C)	T at 150 km (°C)	Depth to 1300 °C (km)
Archean average	79	505 ± 121	770 ± 174	1010 ± 201	219 ± 73
Archean I (<3.0 Ga)	35	430 ± 98	653 ± 142	867 ± 189	269 ± 80
Archean II (>3.0 Ga)	33	527 ± 67	809 ± 76	1079 ± 89	197 ± 23
Archean III – reworked	11	691 ± 77	1046 ± 83	Above mantle adiabat	135 ± 15
Paleoproterozoic	56	592 ± 98	871 ± 126	1135 ± 137	182 ± 34
Mesoproterozoic	20	649 ± 91	979 ± 128	1261 ± 111	153 ± 25
Neoproterozoic	42	781 ± 116	1154 ± 146	Above mantle adiabat	121 ± 24
Paleozoic	62	731 ± 135	1102 ± 181	Above mantle adiabat	130 ± 30
Meso-Cenozoic	–	900 to 1200	Above mantle adiabat	Above mantle adiabat	60 to 90

(Source: Artemieva, 2006)

Table 4.13 Geologic and tectonic ages and lithospheric thermal thickness for major Precambrian provinces

Precambrian cratons	Basement age (Ga)*	Cratonization age (Ga)	Major tectonic events (Ga)	Thermal thickness (km) (uncertainty c. 25%)
Australian craton				
Pilbara craton	**3.5–3.0**	3.0	2.0–1.7, 1.3–1.0	170–230 km
Yilgarn craton	**3.7–3.4** gneiss and 3.0–2.6 granites (3.7)	2.7–2.6		170–230 km
Hamersley and Naberru basins	Archean ?		**2.8–2.3**	No data
Gawler carton	**2.7–2.6**	1.5	1.9–1.8, 1.6–1.5	170–190 km
North Australian craton	**2.1–1.9 (2.5)**	1.85	1.3–1.1	130–200 km
Central Australian mobile belt	**1.9–1.4**	1.2–1.1	1.8–1.6, 1.5–1.3, 1.1–1.0, 0.9–0.5	140–160 km
South American craton				
Amazonian craton (Guyana Shield)	**3.4 and 2.9–2.6 (3.5)**	2.8–2.6	2.2–2.1, 1.85–1.55	No data
Mobile belts of Amazonian craton	**2.2–1.8**			No data
Central Brazil Shield	**3.0–2.8 (3.2)**	2.8–2.6	2.1–1.9, 1.6–1.1, 0.7–0.5	No data
Mobile belts of Central Brazil Shield	Archean–Paleoproterozoic		**1.4**	No data
Atlantic Shield (Saõ Francisco craton)	3.2 (3.4)	2.8–2.6	2.1–1.9, 0.6	190–200 km

Table 4.13 (cont.)				
Precambrian cratons	Basement age (Ga)*	Cratonization age (Ga)	Major tectonic events (Ga)	Thermal thickness (km) (uncertainty *c.* 25%)
Marginal belts of Atlantic Shield	**1.1**		0.7–0.5	110–160 km
Rio Apa craton	**1.6–1.0** (2.1)			No data
Indian craton				
Dharwar craton	**3.1–3.0**	2.6–2.5		180–200 km
Southern Granulite terrane	**3.3–3.0** (3.4)	2.6–2.5	2.1–2.0, 1.6–1.5, 0.9	180–230 km
Singhbhum craton	**3.4**	2.6–2.5	2.1	160–200 km
Aravalli craton	**3.4**	2.6–2.5	2.5–2.0, 1.8–1.5	160–200 km
Bhandara (Bastar) craton	**3.5–3.3**		2.2, 1.0	180–200 km
Cuddapah Basin	**2.0–1.5**		1.5–1.0	140–160 km
Narmada–Son Rift system			**0.6–0.5**	80–110 km
Siberian craton				
Aldan Shield & Stanovoy Ridge	**3.4–3.2**	3.1–2.9, 2.6–2.5	Partly at 2.0–1.8	140–190 km
Kolyma–Omolon and Okhotsk median massifs	**3.4**	?	1.6–0.5	No data
Anabar Shield	**2.9–2.7** (3.2)	2.6–2.5	1.9, partly at 0.9–0.5	190–350 km
Main Siberian craton	Mainly Archean (*c.* 75%), extensively reworked at **1.95–1.8 Ga** by epicratonic rifting	2.6–2.5	1.4–1.1, 0.8–0.5; Mz (Viluy basin)	150–350 km (120–140 km Viluy basin)
Pericratonic mobile belts	**3.2–2.5**		2.5, 1.9, 0.6	250–350 km
East-European craton				
Ukrainian Shield & Voronezh massif	**3.6–3.0**	3.2, 2.7–2.5	2.3–1.8, 1.4	170–230 km
Baltic Shield (Kola-Karelia & Lapland)	**2.9–2.7** (3.1)	2.7–2.6		200–300 km
SvecoFennian province	**2.0–1.8**			170–200 km
SvecoNorwegian province	1.75–1.5, extensively reworked at **1.1–0.9 Ga**		0.6–0.4	150–180 km
Transscandinavian Belt	1.8–1.6			110–140 km
East European platform	Archean (75%) and Paleoproterozoic (25%). Extensively reworked at **2.2–1.8 Ga** by widespread rifting	2.7–2.6	1.6–0.6	170–220 km
Volga–Urals anteclise	**~3.0**		1.6–0.6	200–220 km
Central Russia Rift System			**1.3–0.6**	170–200 km

Table 4.13 (cont.)				
Precambrian cratons	Basement age (Ga)*	Cratonization age (Ga)	Major tectonic events (Ga)	Thermal thickness (km) (uncertainty *c.* 25%)
North American craton				
Wyoming craton	**3.96–3.4**, extensively reworked in Meso-Cenozoic	?	2.8–2.6, 1.9–1.8, 0.06	130–150 km
Slave craton	**3.96–3.1** gneiss and **3.1–2.6** granites	2.7–2.6	1.0–0.8	Few data points (ca. 200 km)
Wopmay Orogen (Bear province)	**1.9–1.8**			140–180 km
NE Churchill province (Hearne/Rae cratons)	**2.9–2.7** (3.5)	2.7–2.6		No data
W Churchill province (Taltson and Queen Maud blocks)	2.9–2.7, extensively reworked at **1.9–1.8 Ga**	2.7–2.6		No data
Trans-Hudson orogen	**2.0–1.8**			160–200 km
Superior craton (north-central part)	**>3.35 and 3.0–2.7**	2.7–2.6		175–240 km
Superior craton (southern part and Ungava craton)	**3.0–2.7** (3.4)	2.7–2.6	**1.9–1.8**, 1.5–1.0	150–190 km
Nutak and Nain cratons (Labrador)	**3.9–3.4**		3.3–2.6	No data
Peripheral orogenic belts (Superior)			1.9–1.65, 1.3–1.0	No data
Penokean (Southern) province	**2.5–2.2** (3.5)		1.9–1.8	140–170 km
Yavapai (Central) province	**1.8–1.5**			140–170 km
Greenland & Lewisian	**3.82–3.7**	2.9–2.5		No data
Grenville province	**1.4–1.0**			140–200 km
Cathaysian craton				
Yangtze craton	Paleoproterozoic, extensively reworked in Meso-Cenozoic	1.85	1.05, 0.85–0.8	115–200 km
Orogenic belts at cratonic margins	**2.5–2.2, 1.86**			No data
Sino-Korean craton (Ordos, Ji-Lu nuclei)	**2.9–2.7** (3.57), extensively reworked in Meso-Cenozoic	2.6–2.5, 2.2–1.9	1.8–1.7, 1.5–1.4	115–200 km
Tarim craton	**Archean–Mesoproterozoic**	1.9, 1.0	1.5–1.4, 1.0, 0.85–0.6	No data
Median massifs (e.g. Junggar, Songliao)	**0.8–0.6**			140–180 km

Table 4.13 (cont.)				
Precambrian cratons	Basement age (Ga)*	Cratonization age (Ga)	Major tectonic events (Ga)	Thermal thickness (km) (uncertainty *c.* 25%)
African craton				
Kaapvaal & Zimbabwe cratons	**3.5–3.2** (3.64)	3.2–3.0, 2.7–2.6	2.7–2.6	180–200 km
Peripheral orogenic belts, South Africa (e.g. Kheis, Magondi)	**2.0–1.8**			180–200 km
Mid-Proterozoic mobile belts, South and Central Africa (e.g. Namaqua-Natal, Irumides, Kibarian)	**1.3–1.0** (2.0)			120–140 km
Tanzanian craton	**3.0–2.6**		2.1–1.9	180–250 km
Central Africa (Angolian & Kasai cratons)	**3.5–3.4** and 3.0–2.6	2.7–2.55	2.1–1.75, 1.3, Pz–Mz	No data
Central Africa (Congo craton)	**Archean**, Extensively reworked in Phanerozoic		1.0	No data
West African craton (Archean shields)	**3.0–2.9** (3.5)	2.7–2.55	2.0, 0.8–0.6	240–350 km
West African craton (Taoudeni Basin and Man shield)	Archean–Paleoproterozoic, extensively reworked at **2.2–1.9 Ga**		1.0	240–350 km
Benin–Nigeria Shield	(3.5–3.0)		2.75, 0.6	150–200 km
Trans-Saharan Belt and Tuareg Shield	? (3.5), Extensively reworked at ~0.6 Ga	3.0, 2.1–1.95	1.1, 0.6	No data
Pan-African (Central African and Mozambique) mobile belts	**1.1–0.6**		0.65–0.55	90–140 km
Arabian–Nubian Shield	**0.95–0.68**			120–180 km
Antarctica				
Antarctic craton	Archean ? (3.8–3.93)	Unknown		No data

(Source: Artemieva, 2006)

* Basement age refers to the oldest known major crust-forming event (see references in Goodwin, 1996), it is not the age of the oldest known rocks (shown in brackets). In bold – ages used to constrain the continental thermal model TC1 (Artemieva, 2006).

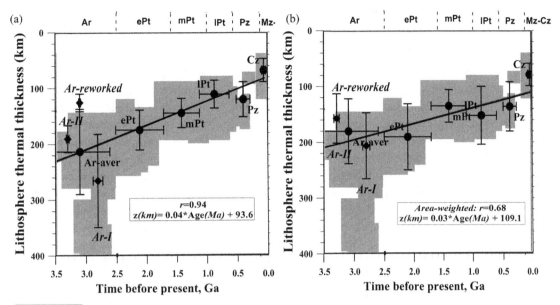

Fig. 4.48 Statistical analysis of variations of lithospheric thermal thickness with tectono-thermal age: (a) all continents; (b) area-weighted data for continents (Artemieva, 2006). Abbreviations are as in Fig. 4.46; Ar-I, Ar-II and Ar-average refer to young Archean (<3.0 Ga), old Archean (>3.0 Ga), and average values for all Archean terranes, respectively, excluding reworked Archean terranes.

and with a real scatter in lithospheric thickness values as a result of diverse tectonic evolution of terranes with similar ages.

If all lithospheric terranes are split into five groups by tectono-thermal age (the Archean (with all Archean terranes treated as one group), the early, middle, and late Proterozoic, and the Paleozoic, Table 4.12), the age–lithospheric thickness statistical correlation is surprisingly high ($r = 0.94$), and the average lithospheric thickness L can be approximated by a linear dependence as:

$$L[\text{km}] = 0.004^{*}t[\text{Ma}] + 93.6, \tag{4.38}$$

where t is the tectono-thermal age (Fig. 4.48a). This implies that, statistically, the minimal thickness of the continental lithosphere is ~90 km and it increases with age at a rate of ~4 km/100 My. This statistical relationship does not hold for some tectonic settings, such as modern zones of continent–continent or continent–ocean collision or Proterozoic terranes that are underthrust by Archean mantle. Furthermore, the statistics is significantly biased by the uneven distribution of borehole measurements over terranes of different ages: the correlation coefficient drops to, a still significant, value of $r = 0.68$ if the average values are area-weighted for terranes of different tectono-thermal age (Fig. 4.48b) and the lithospheric thickness roughly relates to tectono-thermal age as:

$$L[\text{km}] = 0.03^{*}t[\text{Ma}] + 109.1.$$

- Note that linear best-fits (eq. 4.38) do not have any physical meaning for the age-dependence of the thickness of the continental lithosphere, and are nothing more than mathematical approximations to the data.

Although the statistical analysis is based on tectono-thermal ages (the ages of the last major tectonic event) rather than the ages of the juvenile crust (i.e. the age of the major crust-forming event), such a distinction between geological and tectono-thermal age is speculative for many Precambrian terranes, where major crust-forming episodes were probably caused by large-scale thermal and tectonic events (e.g., superplumes or plate tectonic processes resembling those of the present). Moreover, some Archean–early Proterozoic terranes (e.g. the Wyoming, Sino-Korean, and Congo cratons) have undergone significant tectonic reworking in the Meso-Cenozoic. In the statistical analysis they are considered as a separate category of "reworked Archean cratons", since, having been formed under unique temperature and compositional conditions during the early evolution of the Earth, they are expected to have preserved, in part, the distinctive structure and composition of the lithosphere, different from that of Phanerozoic terranes (Artemieva, 2006).

To account for possible differences in the mechanisms of formation and later tectonic modification, the Archean terranes are subdivided into three groups: early Archean (older than 3.2–3.0 Ga), late Archean (3.0–2.5 Ga), and reworked Archean (with basement ages of 3.6–2.5 Ga). Surprisingly, unreworked Archean terranes display the opposite trend in variations of the lithospheric thickness with age: older Archean terranes (>3.0 Ga) have a lithospheric thickness of c. 200–220 km, while younger Archean terranes (<3.0 Ga) have lithospheric roots extending deeper than 250 km and perhaps as deep as 350 km (Fig. 4.49) (see next section for discussion).

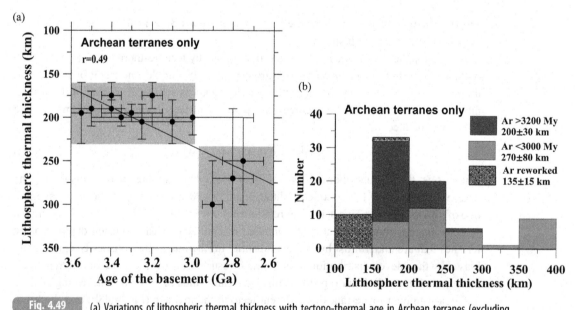

Fig. 4.49 (a) Variations of lithospheric thermal thickness with tectono-thermal age in Archean terranes (excluding reworked Archean cratons). (b) Statistics of lithospheric thermal thickness estimates for different Archean terranes. (Sources: Artemieva and Mooney, 2001; Artemieva, 2006.)

Global statistical thermal model

High statistical correlation between the ages and lithospheric geotherms, together with a $1° \times 1°$ database on lithospheric tectono-thermal ages (rather than the ages of the juvenile crust, Fig. 2.16), allows for constraining of the worldwide model TC1 of the thermal regime of the continents on a $1° \times 1°$ grid (Artemieva, 2006) (Fig. 4.44). The model is based on the following set of assumptions and simplifications.

(1) *Lithospheric geotherms*. Lithospheric temperatures in stable continental regions are constrained by high-quality surface heat flow data where available. In tectonically active continental regions (such as regions of Meso-Cenozoic extension, rifting, and volcanism) the thickness of the thermal lithosphere is assumed to be 60–100 km as indicated by xenolith geotherms. In modern zones of continent–continent or continent–ocean collision (e.g. the Andes, the Hellenic arc, the Alps) the lithospheric thickness is fixed at 220 km, in agreement with regional seismic tomography studies in continental collisional orogens which image the presence of subducted lithospheric slab in the upper mantle down to depths of 200–250 km. The Archean lithosphere of the Indian craton is assumed to underlain Tibet as suggested by recent seismic studies (e.g., Tilmann *et al.*, 2003 and Fig. 3.95c).

(2) *Archean cratons*. Based on lithospheric geotherms constrained by surface heat flow and xenolith P–T arrays for the cratonic mantle (see Chapter 5), Archean cratons are subdivided into early Archean, late Archean, and reworked Archean. Lithospheric thickness is assumed to be ~220 km for the early Archean terranes of Central and Southern Africa, South America, Labrador, and Greenland, and 300 km for the late Archean Baltic Shield, Siberia craton, and the northern and northeastern parts of the Canadian Shield. The lithospheric thermal thickness in reworked Archean cratons is assumed to be 140 km.

(3) *Lithospheric ages*. A critical assumption of the model is that the age of the continental lithospheric mantle is the same as the age of the overlying crust (see Section 2.2). This assumption is based on Re–Os dating of mantle-derived cratonic xenoliths which suggests that the crust and the lithospheric mantle of the cratons have been spatially and temporally linked since their formation (Pearson, 1999). High-resolution seismic transects have demonstrated that this may not always be true, in particular along the cratonic margins. For example, Archean lithosphere underlies Proterozoic crust at the margins of the Fennoscandian and the Slave cratons and at the transition between the Superior and the Grenville provinces (BABEL Working Group, 1993; Clowes *et al.*, 1998; Cook *et al.*, 1999). However, nowhere are lateral extensions of Archean blocks underneath the younger crust seismically traced further than 100–200 km from the terrane boundary observed at the surface, and the uncertainty associated with assigning lithospheric mantle the same ages as the overlying crust has little effect on the resolution of the TC1 model. A larger uncertainty is associated with poorly known basement ages in several regions with a thick sedimentary or ice cover (e.g., Greenland, Antarctica, north-central and western Africa, central Australia, Yangtze craton in China, Rio Apa and Rio de la Plata cratons in South America), where the model constraints should be treated with caution.

The resultant map of the lithospheric thickness constrained by thermal modeling in regions with heat flow data and statistical values in other continental areas is shown in Fig. 4.44 and on the book cover.

Bimodal thickness of Archean–Paleoproterozoic lithosphere

The bimodal distribution of lithosphere thermal thickness in Archean–Paleoproterozoic terranes (Fig. 4.46) is in accord with models of mantle convection that show that depleted cratonic lithosphere should have two equilibrium thicknesses, stable at ~350 km and unstable at ~220 km (Doin *et al.*, 1997). In the case of low activation energy of the upper mantle (expected for sublithospheric processes with a Newtonian rheology), $V^* < 2 \times 10^{-6}$ m^3/mol, a unique equilibrium lithospheric thickness is >450 km. However, the existence of the transition zone forbids lithosphere thickness in excess of 410 km. Lithosphere erosion from below or from the sides can occur when the heat flow supplied to the lithospheric base by sublithospheric small-scale convection exceeds the heat flow conducted through the lithosphere. In the case of thick (~350 km) lithosphere, small-scale convection at its base is sluggish and the basal part of the lithosphere is mainly destabilized by lateral erosion with a rate of ~0.4 km/Ma, so that in 2.5 Ga the width of the Archean lithospheric keel would decrease by ~1000 km (Fig. 4.50). In contrast, thinner (~220 km) lithosphere is in unstable equilibrium: it can either thicken or erode, but when the lithospheric thickness is close to the equilibrium value, its evolution should be slow. Numerical simulations demonstrate that the basal part of a ~220 km-thick depleted cratonal lithosphere is effectively eroded by strong sublithospheric small-scale convection with a rate of vertical erosion of ~0.16 km/Ma (Doin *et al.*, 1997). As a result, in ~1 Ga its thickness would be reduced to a stable value typical for thin, non-cratonic lithosphere.

Several observations related to the bimodal thickness of the Archean–Paleoproterozoic lithosphere should be mentioned:

- paleo-Precambrian cratons surrounded by Proterozoic mobile belts (as in South Africa, South America, western Australia, and India) have lithospheric thickness around 200–220 km, while the cratons without surrounding Proterozoic mobile belts (as in North America, Siberia, Europe, and West Africa) are characterized by thick lithospheric roots (250–350 km) (Fig. 4.44);

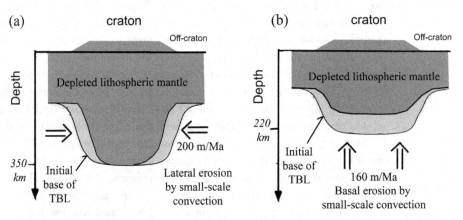

Fig. 4.50   Erosion of depleted cratonic lithosphere by small-scale sublithospheric convection (based on results of numerical modeling by Doin *et al.*, 1997). Thick (~350 km) lithosphere is eroded laterally, while thinner (~220 km) lithosphere is eroded vertically.

- the cratons with thick lithospheric roots were parts of Laurasia, while cratons with thinner lithosphere roots were parts of Gondwanaland;
- the cratons with thin (200–220 km) lithosphere are, in general, older (>3.2–3.0 Ga) than the cratons with thick lithosphere (Fig. 4.49a).

These correlations cannot easily be explained with the present knowledge on how the early lithosphere was formed, a question that still remains enigmatic. The widespread occurrence of metasediments with negative Eu anomalies (e.g. in Antarctica, Anabar Shield in Siberia, Greenland) suggests that relatively large fragments of the continental lithosphere already existed by 3.4–3.55 Ga (Kröner, 1991), while seismic reflection studies across the margins of the Archean part of the Canadian Shield (e.g. Calvert et al., 1995; Cook et al, 1999) and available paleomagnetic data provide cogent evidence that some form of plate tectonics already operated in the late Archean (Hale and Dunlop, 1984; Hale, 1987). The existing models of lateral and vertical growth of the continental nuclei based on diverse geophysical, tectonic, and geochemical evidence can be reduced to two types:

(1) the "Iceland type" model which includes plume-induced mantle melting, further differentiation and magmatic underplating (e.g. Ashwal et al., 1982), and
(2) the "accretion model" according to which the cratons grew by collision of oceanic terranes, shallow subduction, and underplating of slabs around the perimeters of pre-existing lithospheric fragments (Abbott and Mooney, 1995; Rudnick, 1995).

Windley (1995) distinguishes four different types of Archean crust, and one may expect that due to significant differences in their formation mechanisms, several different types of Archean lithospheric mantle may exist as well:

(1) cratons with intraoceanic tectonics followed by a stage of intracontinental tectonics (Kaapvaal);
(2) cratons formed at an active continental margin (Greenland, Labrador, NW Scotland);
(3) cratons formed at a passive continental margin (India and NE China);
(4) cratons formed by collision of island arcs (the Superior Province and the Slave craton).

While cratons of the first two types have a well-documented geological and tectonic history from late to early Archean, no early-mid-Archean history is known for the last two types. The spatial correlation between thin Archean roots and Proterozoic mobile belts suggests that the initial lithospheric thickness could have been the same in all of the Archean cratons, and the lower part of the roots could have been removed during adjacent Proterozoic orogenic activity. Alternatively, thick cratonic roots with ages of 2.9–2.6 Ga could be formed by Archean–Paleoproterozoic plate tectonic processes and were not an immediate product of mantle differentiation as the "Iceland type" model implies. Such thick lithospheric roots may remain stable with respect to mantle convection, preserving their original thickness since the Archean while gradually shrinking in lateral dimensions (Fig. 4.50).

Lithosphere preservation, thermal blanketing, and heat diversion

A collision of oceanic fragments with a continental nucleus could led to tectonic stacking, lithosphere thrust-thickening, and tectonic underplating, similar to the present-day Himalayas and Tibet. These tectonic processes could result in the formation of very thick crustal (>45–50 km) and lithospheric (in excess of 300 km) roots. This hypothesis is supported by data from the Baltic Shield, where a region of thick crust, atypical for Precambrian cratons and locally reaching 60 km, and thick lithosphere is located at the tectonic boundary between the Archean Kola-Karelian and the Paleoproterozoic SvecoFennian terranes in central Finland. Long-term preservation of crustal and lithospheric roots requires that they were thermally insulated from the mantle. If not, the lowermost part of the crust may have later transformed into eclogite facies and become seismically indistinguishable from the mantle or delaminate. The presence of a thick lithospheric root is supported by mantle xenoliths from the terrane boundary, which indicate a cold cratonic geotherm and show no evidence of shearing in xenoliths (commonly interpreted as being associated with mantle zones of reduced viscosity and asthenospheric flow) down to 240 km depth (Kukkonen and Peltonen, 1999). Extremely low surface heat flow in the area (30.2 ± 7.4 mW/m^2, $n = 16$) can be explained by either an abnormally low radiogenic crust, or by a very thick lithospheric root, or by a combination of both. In agreement with observations in Fennoscandia, a 300–350 km-thick lithospheric root produces a surface heat flow anomaly over an area of c. 1000 km across (Fig. 4.35).

The lateral extent of thick crustal and lithospheric roots can be relatively small and in the Baltic Shield it is largely restricted to the suture between the Archean–early Proterozoic terranes. Because of their small lateral size (a hundred to a thousand kilometers across, Fig. 4.50), such cratonic roots would not necessarily be resolved in seismic tomography models. However, they will still be efficient in diverting mantle heat sideways from the area with the thickest lithosphere. The latter process will lead not only to a reduction in surface heat flow above the root, but also to a heat flow increase in the adjacent regions, the mechanism initially proposed to explain the difference in surface heat flow between the Archean terranes and the surrounding Proterozoic mobile belts in South Africa (Ballard and Pollack, 1987). In central Finland, the belt of Proterozoic anorogenic magmatic activity (rapakivi granites) encircles a region with thick lithosphere and low surface heat flow from north, west, and south. This supports the hypothesis of an effective deflection of mantle heat sideways from the region with a thick lithospheric root (>250 km) which, during Proterozoic tectonism, enhanced mantle melting which produced anorogenic granites (Artemieva, 2006).

4.5 Heat flow balance of the Earth

Data on the heat balance of the Earth and the total heat loss provide critical constraints for models of planet evolution and boundary conditions for mantle convection models. Global heat loss has been examined in several studies (Williams and von Herzen, 1974;

Davies, 1980; Sclater *et al.*, 1980, 1981) and the estimated values range from 41.0 to 44.2 TW (10^{12} W). The latter value reported by Pollack *et al.* (1993) is commonly taken for the total heat loss.

Calculations of global heat loss are based on heat flow observations that are integrated over the Earth's surface. Since data are absent for significant parts of the globe, statistically averaged values calculated for well-sampled tectonic regions with certain ages and specific geological characteristics are assigned to areas with the same ages and similar tectonics. Pollack *et al.* (1993) distinguished 21 categories: 9 on continents and 12 in the oceans. Calculation of the mean heat flux of the continental lithosphere yields 65 mW/m^2. Although subdivision of oceanic lithosphere into categories is easier due to well-constrained ocean floor age, the calculation of the mean heat flux of the oceanic lithosphere is not as straightforward as for continents. The complications arise from two issues.

(1) Heat flow through oceanic lithosphere closely follows the square root dependence on ocean floor age (Parsons and Sclater, 1977). As discussed in detail in Section 4.3, thermal evolution of the oceanic plate (in particular <80 Ma) is well described by a cooling plate model (McKenzie, 1967), according to which oceanic plate is created at a mid-ocean ridge (age 0 Ma) and thickens as it cools and moves away from the ridge. This mathematical formulation does not allow integration of heat floor for all ocean floor ages because such an integration gives an infinite heat flux at the origin (the ridge). Several approaches have been proposed to overcome this difficulty. Sclater *et al.* (1980) modified the plate model following the assumption of Davis and Lister (1974): heat flow at the ridge is equal to the sum of the horizontal conductive heat loss and vertical heat loss by magmas. The error in such estimate of heat loss through the ocean floor is ~10–20% (Sclater *et al.*, 1980). An alternative approach was proposed by Hofmeister and Criss (2005) who attempted to ascertain the Earth's mean oceanic heat flux directly from compiled heat flow data. Their result, which yields the global heat loss value of 31 ± 1 TW, caused heated debates and is not generally accepted (Von Herzen *et al.*, 2005).

(2) The square root dependence of oceanic heat flow on ocean age may be obscured by a strong effect of hydrothermal circulation, which is particularly strong for young oceanic lithosphere (Fig. 4.22). A high scatter in heat flow values is observed near all spreading centers which leads to heat flow values significantly lower than predicted by the plate model. The reduction in measured heat flow is caused by seawater convection through oceanic crust. Most deep-sea sediments have low permeability. When the permeable crust is covered by a sufficient thickness of sediments, free water circulation in the crust still continues, but no heat is lost to the surface by advection, and heat flow measurements provide a reasonable estimate of deep heat flow (Lister, 1972).

Stein and Stein (1992) analyzed heat flow data from the oceans and concluded that, after the effect of hydrothermal circulation is excluded, heat flow in oceans older than 1 Ma can be approximated as $Q_{oc} = c / t^{1/2}$ where t is the age of ocean floor in Ma, heat flow Q_{oc} is in mW/m^2, and constant $c = 510$ for young oceans (see discussion in Section 4.3.2). These authors estimate mean heat flow in Quaternary oceans as 806.4 mW/m^2, in strong contrast to measurements that give a mean value of 139.5 mW/m^2 (Pollack *et al.*, 1993) (Fig. 4.19).

Using the statistical heat flow values for the continents and oceans older than 60 Ma and the preferred values of Stein and Stein (1992) for oceans younger than 60 Ma, Pollack *et al.* (1993) estimated that the global mean heat flow is 87 mW/m^2 and the global heat loss is 44.2 TW. Of the global heat loss at the Earth's surface, only ~30% of heat is lost through continents and shelves and ~70% through the deep oceans and marginal basins (and about 50% of this loss takes place in oceanic lithosphere, which is younger than 20 My, and 75% in oceanic lithosphere, which is younger than 80 My) (Sclater *et al.*, 1980).

A recent review of the heat loss of the Earth is provided by Malamud and Turcotte (1999) (Fig. 4.51). Of the heat loss from the mantle that constitutes *c.* 85% of the total heat loss, 58% is attributed to the subduction of the oceanic lithosphere and the remaining 42% are attributed to basal heating of the oceanic and continental lithosphere by mantle plumes (hotspots) and/or secondary convection. If there were no basal heating in the oceanic lithosphere, its thickness would have followed the cooling half-space model for all ocean-floor ages with a square-root relationship between the age and the ocean-floor topography (see Section 4.3 for a detailed discussion).

The mean basal heat flow through the oceanic lithosphere is estimated to be between 25 and 38 mW/m^2 (Sclater *et al.*, 1980). Other estimates of basal heat flow through the oceanic lithosphere are between 40 mW/m^2 (Doin and Fleitout, 1996) and 50 mW/m^2 (Stein and

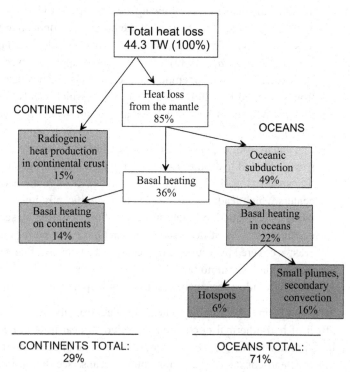

Fig. 4.51 Present heat balance of the Earth (source: Malamud and Turcotte, 1999). Total heat loss is based on the total global surface heat flow (Pollack *et al.*, 1993). Numbers – rounded percentage from the total.

Table 4.14 Heat flow balance of the Earth

Source of heat	Mean surface heat flow* (mW/m^2)	Heat loss $(TW)**$	Total global heat loss (%)
Continents (area $2.0 \times 10^8 \, km^2$)	65.0 ± 1.6	13.0 ± 0.3	29.3
Radiogenic heat production in continental crust	*34 ± 8*	*6.8 ± 1.6*	*15.3*
Basal heating of continental lithosphere	*31 ± 8*	*6.2 ± 1.6*	*14.0*
Oceans and marginal basins (area $3.1 \times 10^8 \, km^2$)	101.0 ± 2.2	31.3 ± 0.7	70.7
Subduction of oceanic lithosphere	*70 ± 8*	*21.7 ± 2.5*	*49.0*
Basal heating of oceanic lithosphere	*31 ± 8*	*9.6 ± 2.5*	*21.7*
Total global	86.9 ± 2.0	44.3 ± 1.0	100

(after Malamud and Turcotte, 1999)

* Sources: Sclater *et al.*, 1980; Pollack *et al.*, 1993

** $1 \, TW = 10^{12} \, W$

Stein, 1992). The mean area-averaged basal heat flow through the continental lithosphere is estimated as 25–29 mW/m^2 (Sclater *et al.*, 1980; Malamud and Turcotte, 1999).

Mantle plumes are expected to contribute ~5–7% to the total heat loss, and assuming a power-law correlation between the number of plumes (hotspots) and their size (buoyancy flux), the estimate of Malamud and Turcotte (1999) requires 5200 plumes. This extreme number of plumes received unexpected support from a recent global tomography study (Montelli *et al.*, 2004); however, probably not every local velocity anomaly (local upwelling?) should be considered as a plume. In contrast, calculation of plume buoyancy fluxes from swell magnitudes gives an estimate of plume heat flux of only ~1% of the global heat loss (Davies, 1988; Sleep, 1990). On the whole, ~65% of present global heat is lost by convection (which includes oceanic plate creation and continental orogenies), ~20% by conduction, and the remaining ~15% by radioactive decay of heat-producing elements in the crust. More details are provided in Table 4.14.

4.6 Thermal lithosphere: summary

Factors affecting heat flow data

Surface heat flow measurements provide the major constraint on the thermal state of the lithosphere. The measured parameters are (typically) depth variations of temperature gradient in a borehole and thermal conductivity of rocks. High quality scientific data requires that temperature equilibrium should be achieved in the borehole after drilling. This requirement is commonly not met when measurements are made in industrial boreholes.

The depth of the borehole is another critical factor for data quality. Shallow measurements are affected by temperature variations at the surface (e.g. climate change); the effect of the Pleistocene glaciation on heat flow measurements in shallow boreholes may amount to

10–20 mW/m^2. The amplitude of the paleoclimate effect on heat flow measurements decreases with borehole depth and becomes unimportant in drill holes deeper than 1.5–2 km. Groundwater circulation is another factor critical for heat flow quality; its effect is more severe in very shallow boreholes (e.g. in the coastal regions in Texas, USA), but it is difficult to assess its amplitude. Similarly, heat flow advection in shallow magmatic systems enhanced by crustal faults has a strong effect on heat flow measurements in tectonically active regions and leads to strong variability in measured values.

Even in the case when heat flow data are of high quality, several problems remain. First, data coverage by borehole measurements is very uneven. There are large areas both on continents and in oceans, where data are absent. Second, measurements in deep scientific drill holes indicate a very complex pattern of vertical variations in heat flow, inconsistent with theoretical predictions on heat flow decrease with depth. It is debatable which of the measured values is representative of surface heat flow. Importantly, heat flow measured in boreholes reflects not the present-day but the past thermal regime of the lithosphere. The reason being that conductive heat transfer is a slow process and a thermal anomaly propagates from the depth L [in km] to the surface over time $\sim L^2/31.5$ [in Ma].

Resolution problems

Surface heat flow is used to constrain lithospheric geotherms. Lateral resolution of thermal models depends critically on heat flow data coverage. The wavelength of surface heat flow variations puts an additional limit on lateral resolution of thermal models for the mantle, which is not better than 100–150 km for thermal anomalies produced below a 40 km depth (the continental Moho). Surface heat flow anomalies with lateral size <100 km originate in the upper 40 km (e.g. in the continental crust due to crustal radioactivity).

A severe problem in thermal modeling is amplitude resolution of thermal models (i.e. the range of possible temperatures at a fixed depth). On the continents, heat production plays a critical role in determining lithospheric temperatures. Several independent sets of data (such as laboratory measurements on near-surface rocks and overturned (exposed) crustal sections, geochemical constraints based on crustal/mantle chemical budgets, petrologic studies on crustal and mantle xenoliths, model constraints on heat flow–heat production relationships observed in many tectonic provinces), when used together, provide tight limits on bulk crustal heat production. However, direct data for the lower crust and lithospheric mantle is largely absent. As a result, heat production in the continental lithosphere is always a model assumption, and the true model uncertainty cannot be estimated. Temperature dependence (chiefly in the mantle and in the oceans) and compositional variability (in the continental crust) of thermal conductivity further affect thermal models, but the effect is much less important compared with heat production. Sensitivity tests for physically feasible ranges of heat production and thermal conductivity values indicate that amplitude resolution of thermal models is not better than 50–100 °C at any depth which is equivalent to c. 25% for lithospheric thickness. In continental regions with surface heat flow <30 mW/m^2, steady-state geotherms asymptotically approach the mantle adiabat at depths >250 km; as a result solutions are asymptotic and the true lithosphere thickness cannot be constrained; phase transition at 410 km depth limits the maximum thickness of cratonic TBL.

Major results and global trends

Surface heat flow is highly variable even at small distances. In general, higher heat flow is observed in young structures both on continents and in oceans. The mean heat flux is $65 \, \text{mW/m}^2$ on continents and $101 \, \text{mW/m}^2$ in oceans, with a global mean of $87 \, \text{mW/m}^2$.

In oceans, TBL thickens by conductive cooling. Oceanic regions where bathymetry follows $1/\sqrt{\text{age}}$ are termed "normal oceans" and they are typically younger than 80 Ma. In normal oceans, heat flow depends on ocean floor age as $\sim \sqrt{\text{age}}$, and the lithosphere thickens with age as $\sim 13\sqrt{\text{age}}$. In old oceans (80–160 Ma), bathymetry is more shallow than $1/\sqrt{\text{age}}$ dependence predicts (ocean floor flattening). The plate model, with a 95 km to 125 km-thick lithospheric plate, is commonly used to explain bathymetry in old oceans. A constant temperature at the base of the plate required by the model is assumed to result from small-scale mantle convection. Two recent studies (Korenaga and Korenaga, 2008; Adam and Vidal, 2010) reanalyzed bathymetry data in old oceans; they question the existence of seafloor flattening and, as a result, the necessity to use the plate model for the evolution of oceanic lithosphere.

In continental regions, lithosphere thickens with age but does not follow any simple relationship. Two major reasons for a complex thermal structure of the continental lithosphere are heterogeneity of distribution of crustal heat production (which is not correlated either with major rock chemistry or with tectonic setting) and a 4 billion years long tectonic evolution of the continental lithosphere. Several lines of evidence indicate that old cratonic lithosphere has been formed under unique conditions and, therefore, has a significantly different composition and tectonic structure (see Chapters 6 and 9) that cannot be explained by a simple conductive cooling of TBL. Many Archean cratons (usually older than 3.0 Ga) have a 200–250 km-thick lithosphere; some Archean terranes (typically younger than 3.0 Ga, such as in the East European craton, Siberian craton, West Africa, and in the northern and northeastern parts of the Canadian Shield) may have preserved 300–350 km-thick lithospheric roots. Thickness of continental TBL decreases to younger terranes, with typical values of $200 \pm 50 \, \text{km}$ in the Paleoproterozoic terranes, $140 \pm 50 \, \text{km}$ in Meso-Neoproterozoic terranes and 100–120 km in the Paleozoic regions. Some Archean cratons have lost the lower parts of the lithospheric mantle and the present lithospheric thickness in the Wyoming, Tanzanian, and Sino-Korean cratons does not exceed 120–150 km. In tectonically active Cenozoic regions (except for the collisional orogens), the lithosphere thickness is typically 60–80 km.

The temperature difference between cratons, platforms, and tectonically active regions as compared to the average mantle is, correspondingly, $-310 \pm 200 \, °\text{C}$, $-160 \pm 260 \, °\text{C}$, and $+30 \pm 130 \, °\text{C}$ at 100 km depth and $-90 \pm 70 \, °\text{C}$, $-80 \pm 100 \, °\text{C}$, and $-30 \pm 70 \, °\text{C}$ at 200 km depth. More details are provided in Table 4.15.

Table 4.15 Summary of thermal structure of the lithosphere for some tectonic provinces

Tectonic structure	Summary of thermal structure of the lithosphere
Normal oceans (<80 Ma)	Bathymetry $\sim 1/\sqrt{\text{age}}$; heat flow $\sim \sqrt{\text{age}}$; lithosphere thickness L is approximated by cooling plate or cooling half-space models with $L \sim 13\sqrt{\text{age}}$
Normal oceans (80–180 Ma)	Bathymetry more shallow than predicted by $1/\sqrt{\text{age}}$ and lithosphere thinner than predicted by $\sqrt{\text{age}}$. Some recent studies question these conclusions
Back-arc basins	Mean heat flow is similar to the major ocean basins of the same age and follows the theoretical $\sqrt{\text{age}}$ predictions
Mid-ocean ridges	A high scatter in heat flow values; mean heat flow values significantly lower than predicted due to seawater convection through the crust
Archean–Paleo-proterozoic cratons	Approximately steady-state thermal regime with very low heat flow. Moho temperatures 300–500 °C, lithospheric thermal thickness typically exceeds 200 km and in some terranes can be >300 km
Meso-Neoproterozoic cratons	Approximately steady-state thermal regime with low heat flow. Moho temperatures 500–700 °C, typical lithospheric thermal thickness 120–160 km
Paleozoic basins	Approximately steady-state thermal regime with moderate heat flow. Moho temperatures 600–700 °C, typical lithospheric thermal thickness 100–140 km
Collisional orogens	Transient thermal regime. Geotherms are depressed in the subducting slab
Extended crust	Transient thermal regime. Moho temperatures 700–900 °C, lithosphere thermal thickness typically 60–80 km

5

Thermal state of the lithosphere from non-thermal data

The previous chapter discussed the thermal regime of the lithosphere constrained by borehole heat flow measurements; the present chapter focuses on other methods to assess lithospheric temperatures. These include conversion of physical parameters measured in remote geophysical sampling, such as seismic velocities, seismic attenuation, and electrical conductivity, into the temperature of the crust and the upper mantle. The chapter also discusses magnetic methods for determining depth to the Curie point and lithospheric geotherms, while other approaches that provide additional information on the thermal state of the lithosphere (such as apatite fission track analysis, data from metamorphic core complexes, and data on temporal variations in magmatism) are excluded from the discussion.

Mantle xenoliths provide invaluable constraints both on lithosphere composition and its thermal structure, in particular in Precambrian terranes. Uneven worldwide coverage by high-resolution geophysical and mantle xenolith data makes comparison of different approaches for estimating mantle temperatures (or calibrating models) difficult in many regions (Table 5.1). The major reference books on petrologic studies of mantle xenoliths are Nixon (1987) and Pearson *et al.* (2003). Chapter 6 discusses petrologic studies of mantle xenoliths related to density of the lithospheric mantle, the chemical boundary layer, and the tectosphere, which closely links thermal and compositional anomalies in the cratonic lithospheric mantle.

5.1 Xenolith data

5.1.1 Xenoliths: advantages and disadvantages

Xenoliths and xenocrysts

Xenoliths (from the Greek word meaning "foreign rock") are rock fragments entrapped by magmas (such as basalts, kimberlites, lamproites, and lamprophyres) during eruption and transported to the surface. They range in size from small fragments to more than 1 m across, have a composition different from the host volcanic rock, and represent pieces of crustal or mantle rocks that were picked up by rising magmas either from the margins of a magma chamber or from the walls of magma conduit. In contrast, xenocrysts ("foreign crystals") are individual crystal inclusions within igneous rocks, such as diamonds within kimberlite diatremes.

Xenoliths and xenocrysts (together with orogenic peridotite massifs and ophiolites – mantle fragments exposed by tectonic activity) provide a direct, although non-uniform, sampling of the lithosphere at the time of eruption. This invaluable geochemical information on chemical and physical properties of the lower crust and the upper mantle forms a

Table 5.1 Geophysical and petrological data coverage for cratons

Cratons	Data on crustal structure[1]	Seismic tomography models	Heat flow boreholes[2]	Electrical conductivity data[3]	Mantle xenolith data[4]
North America and Greenland					
Superior province	Largely well known	Regional	Mostly at the southern margin, few in the NE part	Available, high-resolution	Available at the southern margin
Slave craton	Poorly known	Global	Only two boreholes	Available, high-resolution	Available
Northern margin of the Canadian Shield	No data	Global	Few boreholes	Unavailable	Available
Greenland	Poorly known	Regional	Only three boreholes	Unavailable	Available at the western margin
South America					
South American craton	Poorly known	Regional, poor ray coverage	Poor and patchy coverage, no data for large areas, quality variable	Unavailable	Mostly from the Guyana craton
Siberia					
Siberian Platform	Patchy coverage	Regional, poor ray coverage	Uneven coverage, particularly bad in the central part, good at the kimberlite fields and in the south	Available, largely old models	Available in the eastern part
Europe					
East European Platform	Good, except for central parts of the craton	Regional in south-west, global in northeast	Dense coverage with deep boreholes	Available, largely old models	Unavailable
Baltic Shield	Well known	Regional	Good coverage	Available	Available
Australia					
Australian craton	Patchy coverage	Regional	Mostly good coverage, but poor in Central Australia	?	Available in some locations, mostly at the margins

Africa				
South African craton	Well known	Dense coverage	Regional	Available, high-resolution	Available both in the craton and in mobile belts around it
West African craton	Largely unknown	Sparse coverage	Global	Unavailable	Unavailable?
Tanzanian craton	Good	Good coverage	Regional	Unavailable	Available, but can be affected by rift-related Cenozoic magmatism
India					
Indian Shield	Good	Dense coverage	Global, old regional	Unavailable	Available
China					
Sino-Korean and Yangtze cratons	Patchy coverage, does not exist in many regions	Largely good coverage, but of unknown quality	Regional	Unavailable ?	Available, but can be affected by Meso-Cenozoic tectonic activity

[1] Data on crustal structure is critical for accurate crustal corrections (see Fig. 3.38b for data coverage).
[2] See Fig. 4.16a for data coverage.
[3] See Chapter 7 for details.
[4] See Fig. 5.1 for data coverage.
[5] See Fig. 3.83a for ray coverage.

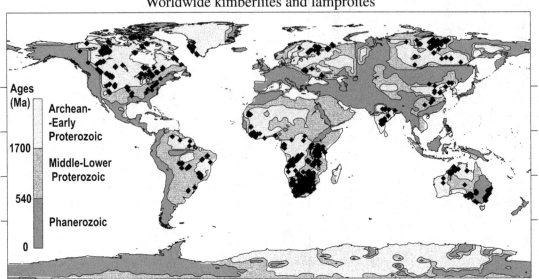

Worldwide kimberlites and lamproites

Fig. 5.1 Map showing worldwide distribution of kimberlites and lamproites (data source: www.consorem.ca) superimposed on a simplified map of crustal ages (source: Artemieva, 2006).

substantial part of present knowledge on the composition and thermal regime of the deep parts of the lithosphere (a review of lower crustal xenolith studies can be found in Rudnick and Fountain, 1995). In particular, mantle xenolith studies indicate that peridotite volumetrically dominates the composition of the upper mantle.

Mantle-derived xenoliths are abundant (at present ~3500 localities are known globally, see Canil, 2004 for references); however, their occurrences are largely restricted to certain tectonic settings, while large areas of the Earth remain un(der)sampled (Fig. 5.1). On the continents, peridotite xenoliths are available for many of the Archean–Proterozoic cratons of Africa, Canada (Fig. 5.2), Siberia (Fig. 5.3), Baltic Shield, Australia, India, China, and South America (Guyana craton). These xenolith data have significantly contributed to the knowledge of the composition and thickness of cratonic upper mantle and the mechanisms of formation, preservation, and destruction of lithospheric roots. Xenoliths from the Wyoming craton in North America and the Sino-Korean and Yangtze cratons in China provide data on the compositional and structural modification of the cratonic mantle during Proterozoic and Phanerozoic tectonism (Hearn, 2003). Spinel peridotites from xenoliths from the Phanerozoic mantle of the continents (e.g. the Vitim plateau in Siberia, Eastern Australia, the Pannonian Basin, French Massif Central, the Rhenish Massif) provide further constraints on the thermal regime of active continental regions and variations in mantle composition, related to the global evolution of the Earth. Some kimberlites contain eclogite xenoliths which are the products of high-pressure metamorphism of subducted oceanic basaltic crust.

Geochemical studies of mantle-derived xenoliths from different tectonic settings reveal a pronounced correlation between the age of the overlying continental crust and the mantle chemical composition, which is interpreted as a secular, irreversible trend from highly depleted (due to loss of iron during melt extraction) Archean continental lithospheric mantle

Fig. 5.2 Distribution of kimberlite clusters and fields in North America (after Heaman and Kjarsgaard, 2000) and the continental extensions of four Mesozoic and Cenozoic hotspots tracks (after Morgan, 1983). Gray line marks the profile discussed in Fig. 6.15.

to fertile Phanerozoic continental and suboceanic mantle (Griffin *et al.*, 1998a). This fundamental difference in major-element composition of the oceanic and cratonic lithospheric mantle had been recognized in the 1980s (Boyd, 1989; Fig. 3.105). The number of mantle-derived xenoliths from oceanic basins is very limited, although extensive xenolith suites are available from oceanic hotspots (Hawaii, the Canary Islands, Grande Comore, and Tahiti) and from the Ontong–Java oceanic plateau. Some xenoliths with metasomatized mantle wedge or mantle lithosphere compositions have been reported for several oceanic subduction zones (Indonesia, Japan, Lesser Antilles).

Limitations of xenolith data

Certain limitations related to geochemical studies of mantle-derived xenoliths should be mentioned:

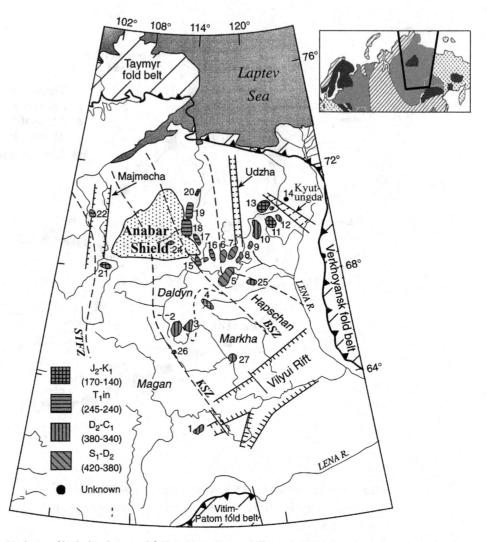

Fig. 5.3 Distribution of kimberlite clusters and fields in Siberia (Source: Griffin *et al.*, 1999a) plotted onto the tectonic map of Siberia with names of major terranes and rifts (Source: Rosen *et al.*, 1994). Abbreviations: SFTZ = Sayan-Taymyr fault zone; KSZ = Kotuykan shear zone; BSZ = Billyakh shear zone. Key to kimberlite fields: 1 – Malo-Botuobinsk (pipe Mir), 2 – Alakit, 3 – Daldyn (pipe Udachnaya), 4 – Upper Muna, 5 – Chomurdakh, 6–7 – Ukukit, 8 – Upper Motorchuna; 9 – Merchimden; 10 – Upper Molodo; 11 – Kuoika; 12 – Toluopka; 13 – Kuranakh; 14 – Lower Lena; 18 – Middle Kuonamka; 19 – Lower Kuonamka; 20 – Orto Irigakh; 21 – Kharamai; 22 – Dalbicha; 23 – Chadobez; 24 – Central Anabar; 25 – Severnei; 26 – Morkoka; 27 – Markha. The profiles along the kimberlite fields 1–13 are shown in Figs. 5.11 and 6.17.

(1) since xenolith occurrences are restricted to certain tectonic settings, it is unclear how representative "Nature's sampling" is, and if lithospheric mantle unsampled by xenoliths has the same composition as xenolith-sampled (see Section 6.3 for details);

(2) the small size of mantle xenoliths limits accuracy of estimates of their bulk composition and restricts studies of compositional heterogeneity of the mantle;

(3) chemical reactions between xenoliths and host magmas further complicate petrologic interpretations;

(4) xenoliths provide a snapshot of the thermal state of the mantle at the time of eruption; but this thermal regime (in particular, in cratonic settings) can be significantly perturbed from the steady state by active tectono-magmatic processes which caused magma eruption;

(5) compared to orogenic peridotite massifs, xenoliths lack a field relationship with surface tectonics. However, in contrast to peridotite (orogenic) massifs which tend to re-equilibrate during emplacement, xenoliths erupt fast and preserve the mineralogical and chemical signatures of their depth of origin;

(6) the maximum depth sampled by xenoliths does not exceed 250 km. This depth limitation, however, is apparently directly linked to the P–T conditions of kimberlite magma generation but not to the maximum depth extent of the lithospheric mantle. For example, some ultrahigh-pressure xenoliths from the Jagersfontein kimberlite in Kaapvaal, South Africa, and from Sierra Leone were derived from a depth of ~500 km (13–15 GPa) (Haggerty and Sautter, 1990).

5.1.2 Xenoliths and the boundary layers

Xenolith data are often used to constrain lithospheric thickness in various tectonic settings. Different lithosphere definitions (discussed further in detail) are employed for this purpose (Fig. 5.4).

Thermal boundary layer

Geothermobarometers (reviewed by Smith, 1999 and discussed in Section 5.2) provide a possibility for estimating equilibrium temperature and pressure (i.e. the depth of origin) of individual mineral grains in mantle-derived xenoliths at the time when they were entrained into the host volcanic rock. Such xenolith P–T arrays (commonly referred to as xenolith geotherms) provide an estimate of the lithospheric geotherm and of lithospheric *thermal thickness* (TBL) at the time of eruption of xenolith-carrying volcanic rocks (sometimes erroneously interpreted as reflecting the present-day, but not paleo-thermal regime of the mantle during magmatism). Similarly to thermal models, the base of the thermal boundary layer may be defined by the transition from a conductive to an adiabatic (isentropic) geotherm (Fig. 5.4).

In the petrologic community, xenolith P–T arrays are commonly compared with the family of conductive geotherms of Pollack and Chapman (1977) (Fig. 4.33), which provides a reference frame for comparison of the deep thermal regime of various tectonic settings. This family of conductive geotherms is based on a simple assumption that 60% of surface heat flow comes from the mantle. However, petrologic and thermal data indicate that average crustal heat production decreases with age and is the lowest in the Archean crust (e.g. Rudnick *et al.*, 1998; Tables 4.4 and 4.5), where only 30% of surface heat flow is generated in the crust (Artemieva and Mooney, 2001). Although the trend of

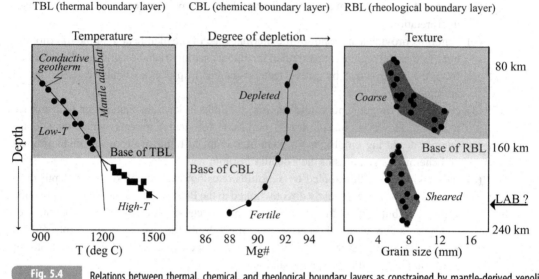

Relations between thermal, chemical, and rheological boundary layers as constrained by mantle-derived xenoliths. The bases of the TBL, CBL, and RBL may be at approximately the same depth which corresponds to the transition from low-temperature, depleted, and coarse peridotites to high-temperature, fertile, and sheared peridotites. The examples are based on xenolith data from southern Africa, the corresponding details are shown in Fig. 5.7b for TBL, Fig. 6.15a for CBL, and Fig. 3.9a for RBL. In this cratonic region, the base of the lithosphere (LAB) is expected to be at a depth of 200–240 km as indicated independently by seismic and thermal models (e.g. Figs. 3.98, 5.5) and thus it does not correspond either to TBL or to RBL bases as inferred from xenolith data. The transitional nature of the CBL base complicates comparison of CBL thickness with other data sets.

increasing mantle heat flow from Archean to Phanerozoic crust partly compensates for an increase in the crustal contribution to surface heat flow, any extrapolation of cratonic xenolith geotherms to surface heat flow along cold reference geotherms should be interpreted with caution.

Sometimes the lower part of the mantle section sampled by xenoliths exhibits a significant deviation (inflection) of xenolith P–T arrays from a conductive geotherm with a characteristic "kink" of the P–T array. This abrupt (almost step-like) change in mantle temperature gradient is often interpreted as the transition from a pure conductive to a non-conductive (adiabatic and convective) heat transfer regime. Thus, the depth of the transition from low-temperature (i.e. located along a conductive geotherm) to high-temperature peridotites is often interpreted as the base of the thermal boundary layer (thermal lithosphere) (Fig. 5.4). Other interpretations are possible as well (see Section 5.2.2 for discussion); they show that a kink on the xenolith geotherm may not necessarily mark the base of TBL.

Chemical boundary layer

Variations of major-element mantle composition with depth as indicated by geochemical studies of mantle-derived xenoliths from different tectonic settings provide information on the thickness of the *chemical boundary layer* (CBL). In most regions, mantle composition exhibits significant (commonly systematic) variation with depth, but no sharp compositional boundaries are observed (Fig. 5.4, see Section 6.1.4). In cratonic settings, however, the transition from low-T (with equilibrium $T < 1100–1200\,°C$) to high-T (with equilibrium $T > 1100–1200\,°C$) peridotites is often associated with a notable, although gradual, change in major-element mantle composition (from depleted in basaltic components to fertile mantle) (e.g. Fig. 6.15). Boyd and Gurney 1986) argue that high-T and low-T mantle peridotites have been formed under different physical–chemical conditions, and thus the boundary between them approximately corresponds to the petrologic lithospheric base. Alternatively, it may indicate the transition from the "intact" cratonic nucleus to the lowermost lithosphere metasomatized by melts and fluids from the convective mantle (Fig. 5.5).

As an example, the boundary at ~150 km depth between low-T and high-T xenoliths in the Williams kimberlite, Montana, USA, has been interpreted as the base of the ancient lithospheric keel, which apparently had been removed during the formation of the Rocky Mountains in the late Mesozoic (Carlson *et al.*, 1999). This conclusion is based on Os isotope studies which indicate that low-T samples have Archean–Paleoproterozoic Re depletion model ages (2.5–1.7 Ga), while high-T samples have Os isotope compositions similar to modern abyssal peridotites, and thus the high-T peridotites are recent additions to the lower parts of the lithosphere. Strong chemical stratification has been noted for the Slave craton (Kopylova and Russell, 2000). An important chemical boundary (with an increase in mantle fertility) identified between 160 and 200 km depth in the northern Slave craton can be interpreted as the boundary between the depleted Archean petrologic lithosphere and the lower lithosphere chemically and texturally modified by pre-kimberlitic magmatic activity (Fig. 5.6) (see Chapter 6). Alternatively, its geometry and sharpness (imaged by seismic receiver function analysis) suggests a structural origin, such as shallow paleosubduction (Bostock, 1997).

Ryan *et al.* (1996) propose that in cratonic regions the depth to the base of the petrologic lithosphere can be determined by the change from depleted (lithospheric) to undepleted (asthenospheric) trace-element signatures in garnet (see Section 5.2.3, Fig. 3.104). In some cratons the depth of this transition corresponds to mantle temperatures of 1200–1300\,°C at which the change from a conductive to an isentropic geotherm occurs, and thus the bases of the chemical (petrologic) and thermal lithospheres may coincide (Poudjom Djomani *et al.*, 1999). However, this is not necessarily the global pattern (see Chapter 6).

Rheological boundary layer

Two main populations of xenoliths, distinguished by their equilibrium temperatures and pressures, low-T and high-T peridotites, also differ by their texture (Fig. 5.4). The latter reflects differential stress conditions in the upper mantle at the time the xenoliths were sampled by ascending magmas. Low-T peridotites are usually coarse grained (grain size greater than 2 mm) and show a low level of lithosphere deformation over tens or hundreds

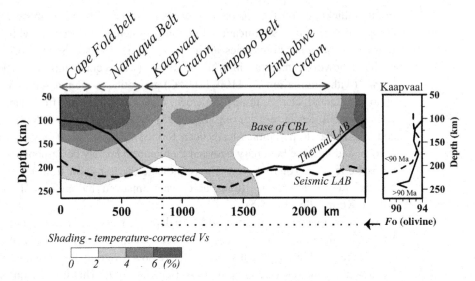

Fig. 5.5 Compositional profile through the lithosphere of southern Africa (modified from Artemieva, 2009). The cross-section shows compositional variations in the upper mantle (gray shading) constrained by absolute δVs velocity perturbations (surface-wave model of Shapiro and Ritzwoller, 2002) corrected for lateral temperature variations (thermal model of Artemieva and Mooney, 2001). Dark shading corresponds to fast (iron-depleted?) lithosphere, whereas white shading refers to fertile mantle. The transition from gray to white shading (2% temperature-corrected δVs anomaly) can be interpreted as the base of CBL. Black lines – lithospheric base as defined by a 2% δVs seismic (uncorrected for T-variations) anomaly with respect to $ak135$ continental reference model (dashed line) or by a 1300 °C isotherm (solid line). Right insert: depth distribution of Fo in Kaapvaal for mantle peridotites emplaced at > 90 Ma and < 90 Ma (data of Gaul $et\ al.$, 2000). The location of kimberlite is shown by dotted line.

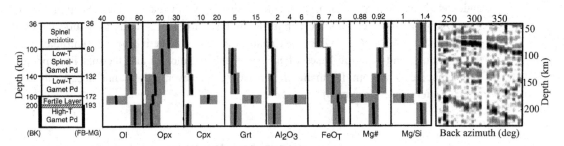

Fig. 5.6 Chemical stratification of the lithospheric mantle beneath the Slave craton (Source: Kopylova and Russell, 2000). Black lines in each panel show mean value determined for each type of peridotite (Pd), gray bars denote one standard deviation for each suite. Depth (left scales) based on Brey and Kohler (BK) and the Finnerty-Boyd and MacGregor (FB–MG) thermobarometry. Right panel: seismic P-to-S transmissivity (receiver function method) beneath the Slave craton as a function of depth and back azimuth (darker shades refer to negative values) (Source: Bostock, 1997).

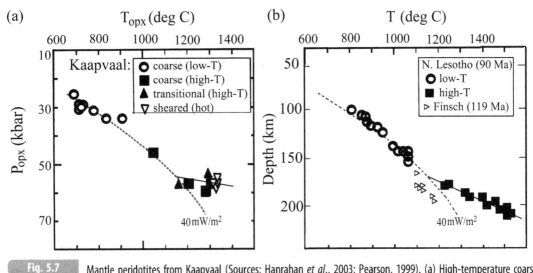

Fig. 5.7 Mantle peridotites from Kaapvaal (Sources: Hanrahan *et al.*, 2003; Pearson, 1999). (a) High-temperature coarse and sheared peridotites may fall very close in the P–T field; (b) the kink in the P–T array corresponds to the transition from low-temperature to high-temperature peridotites. However, the kink may not exist if other geothermobarometers are used (compare with Fig. 5.9d).

of millions of years (Fig. 3.104,d–e). In contrast, high-T peridotites are generally (but not always) finer grained (porphyroclastic) and highly deformed (sheared) (Nixon and Boyd, 1973) (Fig. 5.7). Porphyroclastic textures may be formed by shear heating or deformation during diapiric upwelling (Nixon and Boyd, 1973); however, their origin is not yet fully understood. Due to a fast rate of recrystallization and coarsening of olivine grains, these textures cannot exist for a long time at high mantle temperatures and thus are transient features acquired by xenoliths for a short time before their entrainment by ascending magmas (Pearson *et al.*, 2003).

Variations in texture of xenoliths brought to the surface from different depths in the mantle are often interpreted in terms of the *rheological ("mechanical") boundary layer* (RBL). High-temperature, sheared (and often metasomatized) peridotites are commonly interpreted as being associated with mantle zones of reduced viscosity and asthenospheric flow (compare with Fig. 3.117). Since mantle viscosity is strongly temperature dependent, the thickness of the RBL should be proportional to the thickness of the TBL, unless weak mantle rheology is due to the presence of fluids.

Os isotope results indicate that the transition from low-T granular xenoliths to high-T sheared xenoliths does not mark the base of the present lithosphere in the Kaapvaal and the Siberian cratons (Carlson *et al.*, 1999). In the Baltic Shield, high-T sheared peridotites were not found even in xenoliths of deepest origin (from ~240 km depth) which do not show the shearing associated with flow in the rheologically active boundary layer beneath the base of the conductive thermal boundary layer (Kukkonen and Peltinen, 1999). This implies that the thickness of the rheological boundary layer beneath the central Baltic Shield is greater than 240 km.

LAB and xenolith-based TBL, CBL, and RBL

Figure 5.4 based on xenolith data from southern Africa illustrates that the base of the lithosphere (the lithosphere–asthenosphere boundary, LAB) does not necessarily correspond to any of the TBL, CBL, or RBL bases inferred from xenolith data. Seismic and thermal models available for the Kaapvaal craton independently indicate 200–240 km as the depth to the LAB (Fig. 5.5), while xenolith data suggest that the TBL (defined by the kink on the geotherm) is ~180 km thick and the RBL constrained by grain size variations is ~140–160 km thick (Fig. 5.4). The position of the CBL base is somewhat arbitrary due to a gradual change in mantle composition with depth; however (particularly for young kimberlites, <90 Ma) it is shallower than a 200 km depth where Mg# decreases to ~90.

This conclusion is further illustrated by Fig. 5.5 which shows compositional variations in the upper mantle beneath southern Africa as constrained by joint analysis by surface-wave seismic tomography and thermal modeling. Compositional velocity anomalies (Artemieva, 2009) have been calculated by subtracting the elastic and anelastic effects of lateral temperature variations (temperature deflection from the $50\,mW/m^2$ continental geotherm) on Vs from observed seismic shear-velocities. The results illustrate that while there is a good agreement between the lithospheric thickness as constrained by seismic velocities (defined here by a 2% shear velocity perturbation uncorrected for T-variations) and mantle geotherms, the seismic and thermal LAB do not correspond to the base of the CBL in significant sections of the profile. The discrepancy between geophysically determined LAB and CBL is particularly strong beneath the Zimbabwe craton.

5.2 TBL and xenolith geotherms

5.2.1 Geothermobarometers

Major principles and assumptions

The idea of using mineralogical geothermobarometers to constrain ambient P–T conditions in the mantle sampled by xenoliths was first proposed by Boyd (1973) and further developed by O'Neill and Wood (1979), Finnerty and Boyd (1984), Brey and Köhler (1990) and many others (see Table 5.2 for summary). The approach is based on pressure and temperature dependence of the activity of exchange reactions (characterized by element diffusion, D) between coexisting (equilibrated) minerals. Since mantle peridotite is made of a small number of minerals (the abundant minerals include olivine, orthopyroxene, clinopyroxene, garnet, and spinel), exchange reactions are limited to a few mineral phases only.

Exchange reactions are experimentally calibrated for exchange equilibria. However, experimentally measured values of diffusion rates for cations in peridotites span over

Table 5.2 Some geothermobarometers used in xenolith studies

Method	Reference	Mineral system	Exchange reaction	P^* (GPa)	T^* (°C)	Accuracy**
Barometers						
BKN	Brey, Köhler and Nickel, 1990	Gt-Opx	Al	$1.0–6.0^{(2)}$	900–1400	0.5 GPa (15 km)
KB	Kohler and Brey, 1990	Ol-Cpx	Ca	$0.2–6.0^{(1)}$	900–1400	0.7 GPa (20 km)
NT00	Nimis and Taylor, 2000	Cpx-Gt	Cr	$0–6.0^{(2)}$	800–1500	0.3 GPa (10 km)
FB	Finnerty and Boyd, 1987	Ol-Cpx	Ca	$^{(1?)}$		?
P38	Grutter *et al.*, 2006	Cpx-Gt	Cr	$2.0–7.0^{(2)}$	600–1300	?
MC74	MacGregor, 1974	Opx-Cpx	Al-in-Opx	0.5–4.0	900–1600	?
Thermometers						
WS	Witt-Eickschen and Seck, 1991	Opx-Sp	Al/Ca	$1.5^{(1)}$	900–1400	20 °C
Ba	Ballhaus *et al.*, 1991	Ol-Sp	Fe-Mg$^{(4)}$	0.3–2.7	1040–1300	30 °C
BKN	Brey, Köhler and Nickel, 1990	Opx-Cpx	Ca-Mg$^{(3)}$	1.0–6.0	900–1400	60 °C
BKopx	Brey and Köhler, 1990	Opx	Ca-Mg$^{(3)}$	1.0–6.0	900–1400	62 °C
Ta	Taylor, 1998	Opx-Cpx	Ca-Mg$^{(3)}$	2.5–6.0	900–1400	60 °C
EG	Ellis and Green, 1979	Gt-Cpx	Ca-Mg $^{(3,5)}$ Fe-Mg $^{(4)}$	2.4–3.0	750–1300	?
Ha	Harley, 1984	Opx-Gt	Fe-Mg$^{(4)}$	0.5–3.0	800–1200	92 °C
LG	Lee and Ganguly, 1988	Opx-Gt	Fe-Mg$^{(4)}$	2.0–4.5	975–1400	96 °C
Kr	Krogh, 1988	Cpx-Gt	Fe-Mg$^{(4,5)}$	<3.0	600–1300	100 °C
OW	O'Neill and Wood, 1979	Ol-Gt	Fe-Mg$^{(4)}$	3.0–6.0	900–1400	180 °C
NT00	Nimis and Taylor, 2000	Cr-Di	En-in-Cpx	0–6.0	800–1500	30 °C

Abbreviations: Gt = garnet; Sp = spinel; Di = diopside; Opx = orthopyroxene; Cpx = clinopyroxene; Ol = olivine; En = enstatite

* pressure and temperature used in development and calibration of the method;

** 2σ accuracy in data reproduction;

$^{(1)}$ Sp-facies only;

$^{(2)}$ Gt-facies only;

$^{(3)}$ insensitive to Fe^{3+};

$^{(4)}$ depends on Ca-content;

$^{(5)}$ no systematic dependence on Mg# (this information may be incomplete since it was found only for a few methods).

Pressure–depth conversion: 1 GPa = 10 kbar = 10^9 kg/m sec^2. Pressure at the base of a 40 km-thick crust with density $\rho_{cr} = 2900$ kg/m^3 is $P = 1.16$ GPa; in the upper mantle pressure $P = g[\rho_{cr} z_{cr} + \rho_m(z-z_{cr})]$ and for upper mantle density $\rho_m = 3300$ kg/m^3 $P = 6.0$ GPa is achieved at a depth $z \sim 187$ km.

four and more orders of magnitude at single temperatures and are not well constrained for aluminium and chromium cations in mantle minerals (Canil, 2004). Mineral rims are more likely to be representative of equilibrium pressures and temperatures (Smith, 1999).

Reliable geothermobarometers should comply with several fundamental criteria.

(1) A critical assumption behind all geothermobarometer methods of estimating ambient temperature and equilibrium pressure of individual mineral grains in xenoliths (discussed by O'Reilly and Griffin, 1996 and Smith, 1999) is the assumption that equilibration has been attained in major elements in the minerals. In cases where this assumption is not valid, geothermobarometer techniques may give misleading constraints on xenolith P–T arrays and the corresponding geotherms. For example, evidence of mineralogical disequilibria has been found for xenoliths from the Siberian craton (Griffin *et al.*, 1996).

(2) Estimated P–T arrays should satisfy petrologic constraints imposed by plagioclase–spinel–garnet and graphite–diamond phase transitions. The transition of spinel peridotite to garnet peridotite in Cr-free, Al-bearing peridotites occurs at pressures of ~16–20 kbar (55–65 km depth). With increasing Cr/(Cr+Al) ratio, the phase transition depth shifts to pressures as high as 70 kbar (~220 km depth) (O'Neill, 1981). This dependence is employed, for example, in a widely used thermobarometer of Nimis and Taylor (2000). In cold cratonic lithosphere, the graphite–diamond phase transition occurs at ~130 km depth for $T = 800\,°C$ and deepens to ~190 km depth for $T = 1500\,°C$ (Kennedy and Kennedy, 1976).

A number of other geochemical criteria which depend on a particular exchange reaction may be critical for geothermobarometer reliability (see discussion in Nimis and Taylor, 2000).

Pressure constraints

The aluminium-bearing minerals of xenoliths (i.e. calcic plagioclase that is stable down to 25 km depth, spinel and garnet) are commonly used to constrain the depth of their origin. Studies based on garnet peridotite assume that all Al is within the pyroxene structure (specifically $MgAl_2SiO_6$) and rely on the Al content of orthopyroxene which is typically in the range between 0.7 and 2.0 wt% Al_2O_3 (Fig. 5.8e).

Most geobarometers are confined to garnet-facies samples; the number of geobarometers available for spinel-facies rocks is very limited (Table 5.2). Furthermore, the only quantitative geobarometer for spinel facies, that of Kohler and Brey (1990) is analytically challenging. For these reasons, geotherms in the shallow mantle within the spinel stability field are commonly poorly constrained by xenolith data. Since mineral exchange reactions used in geobarometry depend strongly on temperature as well, usually at the first stage the equilibrium T is estimated and the T value is next used to estimate the equilibrium pressure.

There is a fundamental difference between thermal arrays defined by Opx-Grt and Cpx-Grt barometry (Grutter and Moore, 2003). The temperature dependence of the Opx-Grt P–T arrays (such as MC74 and BKN) is of the order of 40–60 bar/°C and they follow steady-state conductive geotherms (e.g. of Pollack and Chapman, 1977). In contrast, the Cpx-Grt barometer NT00 of Nimis and Taylor (2000) has a temperature gradient of ~25–50 bar/°C. As a result, P–T arrays based on NT00 barometry transect conductive geotherms at a shallow angle (Fig. 5.8ab). In particular, at $T > 850\,°C$ the Cpx-Grt barometry (NT00) data for the Somerset Island fall parallel to the graphite–diamond equilibrium (and produce the apparent kink in the

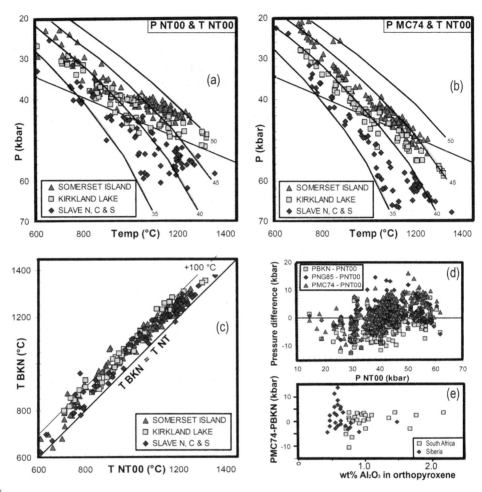

Fig. 5.8 Comparison of different geothermobarometers for garnet peridotite xenoliths (Sources: Grutter and Moore, 2003; Smith, 1999). Notations as in Table 5.2. (a–d) Thermobarometry for the same set of Canadian samples: (a–b) xenolith P–T arrays constrained by different geothermobarometers; black lines: reference conductive geotherms labeled in surface heat flow values and graphite–diamond phase transition. (c) Temperature difference between BKN and NT00 thermometry; (d) pressure difference between Opx-Grt and Cpx-Grt barometry; all pressures calculated at T(NT00); (e) pressure difference between Opx-Cpx and Opx-Grt barometry as a function of Al content of orthopyroxene for South African and Siberian xenoliths.

P–T array), while P–T arrays from the same xenoliths based on Opx-Grt barometry (MC74, PBKN and PNG85) intersect the diamond stability field at P ~ 50 kbar with no deflection of xenolith geotherm.

Temperature constraints

A detailed discussion of various thermobarometers can be found in Smith (1999). A popular thermometer OW is based on Fe–Mg exchange reactions between garnet and olivine

(O'Neill and Wood, 1979). However, temperature determinations based on the assumption that all iron in garnet is ferrous can differ by > 200 °C from the results in which the amount of ferrous iron is determined by Mossbauer spectroscopy (Canil and O'Neill, 1996). Similarly, for Fe–Mg exchange reactions between garnet and clinopyroxene the uncertainty in ferrous iron can result in > 300 °C difference in temperature determinations. Orthopyroxene-based thermometry may be less affected by bulk compositional variations and is potentially more accurate at temperatures below ~1000 °C. Two popular thermometers, BKN and NT00, produce temperature estimates with a small difference of ~50–100 °C, but BKN temperatures are systematically higher (Fig. 5.8c).

It is also important to remember that typically:

- xenoliths with temperatures between 900–1100 °C have equilibrated at ambient P–T conditions just before eruption;
- xenoliths with temperatures below 800–900 °C may not have equilibrated to ambient P–T conditions, since diffusion rate is temperature dependent;
- xenoliths with temperatures above 1100–1200 °C sample mantle affected by melt percolation, heat advection, and with transient thermal regime.

Xenolith geotherms: general remarks

A set of pressure–temperature pairs measured on xenolith samples from the same location forms a xenolith P–T array (xenolith geotherm) (e.g. Fig. 5.8a). The following remarks should be made:

- xenolith geotherms do not extrapolate to the surface temperature;
- the low-temperature subset of P–T data points probably provides the best estimate of the geotherm at the time of eruption;
- downward extension of the xenolith geotherm to its intersection with the mantle adiabat provides an estimate of the TBL even when no deep xenoliths are available;
- xenolith geothermobarometry may be inconsistent with estimates of the mantle adiabat based on studies of mid-ocean ridge basalts (MORB) (Bell *et al.*, 2003);
- in some regions where multiple magmatic episodes took place, xenoliths erupted at different times make it possible to reconstruct the evolution of mantle thermal regime in time;
- for tectonically active regions with a clearly transient thermal regime and significant heat advection by ascending magmas, a comparison of xenolith P–T arrays with the family of conductive geotherms does not have physical meaning and is a matter of practical convenience.

Uncertainties in xenolith P–T arrays

Temperature and pressure estimates obtained on the same rock by different geothermobarometers may differ by hundreds of degrees Celsius and tens of kilobars (Finnerty and Boyd, 1984; Smith, 1999; Grutter *et al.*, 2006) (Fig. 5.8). However, since most geothermobarometers are experimentally calibrated at temperatures between 900–1400 °C and pressures between 2–6 GPa (Table 5.2), most of them agree in this P–T range. At pressures and temperatures outside this range, the accuracy of P–T determinations for samples with mineral chemistries different from those used in calibration experiments cannot be assessed, since many thermobarometers are sensitive to bulk composition (Smith, 1999).

Uncertainties in temperature determinations based on Fe–Mg partitioning result from uncertainties in ferric/ferrous iron (Fe^{3+}/ Fe^{2+}) in pyroxene, spinel, and garnet. In iron-poor pyroxenes, like those from the cratonic settings with Mg# = 100 Mg/[Mg+Fe] of 91–94, the values of relative abundances of ferric and ferrous iron can be almost mean-ingless because of uncertainties in more abundant cations (such as Si) (Canil and O'Neill, 1996). For example, Fe^{3+}/ Fe^{total} in low-temperature peridotite garnets is between 0.02 and 0.07.

Some general observations based on experimental comparisons of different thermobar-ometers (see Table 5.2 for notations) are the following (Smith, 1999; Grutter and Moore, 2003; Preston and Sweeney, 2003):

- BKopx and BKN thermometers provide similar temperature estimates at T~900–1000 °C, but at T~1150 °C BKopx provides temperatures 100–150 °C lower than BKN;
- FB temperatures are 50–100 °C lower than BKN temperatures in the range 900–1200 °C; at higher temperatures T(FB) approaches T(BKN) and becomes hotter at $T > 1400$ °C;
- Kr and BKN temperatures are within 80 °C for the South African peridotites for all temperatures, but for the Siberian samples T(BKN) is 50–150 °C higher than T(Kr) at $T < 1000$ °C;
- The Ha thermometer, as compared to the BKN thermometer, provides higher temperature at $T < 800$ °C but lower temperature at $T > 1000$ °C;
- NT00 temperature values correlate well with temperatures calculated using Cpx-solvus thermometers (e.g. BKN) and give temperature 50–100 °C lower than BKN (Fig. 5.8c);
- the difference in equilibration temperatures determined with FB and NT00 thermometers reaches 200 °C but decreases with temperature increase;
- The LG thermometer, as compared to the BKN thermometer, provides temperature within 100 °C at $T > 1000$ °C but significantly higher temperature at $T < 1000$ °C;
- The Cr-barometer (Ryan et al., 1996) agrees with P38 at $P = 55 \pm 2$ kbar, but provides 6–12% higher values at lower pressures;
- Opx-Grt barometers yield pressures less than NT00 at pressure $P < 40$ kbar and higher than NT00 at high pressure, with absolute differences as large as ± 10 kbar;
- temperature gradients of Cpx-Grt barometers (NT00) and Opx-Grt barometers (MC74, PBKN and PNG85) are significantly different. As a result, at high temperatures Opx-Grt geotherms are parallel to cratonic conductive geotherms, while Cpx-Grt geotherms may orient sub-parallel to the graphite–diamond phase transition boundary and form a "kink" on the geotherm (Fig. 5.8ab).

5.2.2 Kink in xenolith geotherms

Xenolith P–T arrays from cratonic regions often display a characteristic kink at depths below 120–150 km, where mantle temperature gradient changes stepwise (Fig. 5.7). For example, xenolith geotherms for the Kaapvaal and Slave cratons and for the mobile belts in South Africa show kinks at depths of 220, 170, and 140 km, correspondingly, where they deflect from the conductive geotherm into the region of higher temperature. Boyd (1973) postulated that a sharp steepening of the shield geotherm from a normal temperature gradient of 2–4 °C/km at shallow depths to a high temperature gradient of > 10 °C/km below 140–200 km depth marks the base of the continental lithospheric plate. This conclusion has been questioned by different authors on several grounds (e.g. Green and Gueguen, 1974; Jordan, 1975a). Up to date, however, there is still no general agreement on the physical meaning of the kinks. Three new, principally different explanations have been proposed recently.

Kink as an artifact of thermobarometry

Recent experimental studies of cratonic mantle xenoliths with different geothermobarometers suggest that the presence or the absence of the kink often depends on the choice of thermobarometer. In particular, the kink in xenolith P–T arrays commonly appears in data based on Finnerty and Boyd (1987) thermobarometers but is absent in xenolith geotherms constrained by the BKN thermobarometer. For instance, for the African xenoliths from many different locations, P–T arrays based on P(MC)-T(FB) (see Table 5.2 for abbreviations) appear to be made of two linear branches which intersect at a "kink", while the same samples analyzed by the P(BKN)–T(BKN) thermobarometer combination show no kink. Similarly, for the Canadian data a kink is not observed for all thermobarometers (compare Figs. 5.8a,b and 5.9c).

Another concern was raised as early as 1975 (Wilshire and Jackson, 1975) when it was argued that high-temperature peridotites may represent unequilibrated mixtures of rocks with different compositions and are not likely to have been subject to their apparent equilibration temperatures for very long. Furthermore, Re–Os studies show no difference in Os model ages between high-T and low-T peridotites (Pearson et al., 1995; Carlson et al., 1997). These authors argue that high-temperature deformed peridotites are not derived from asthenospheric flow but represent pieces of the ancient lithosphere overprinted by later tectono-thermal events, and their fertile nature can be a recent phenomenon. This implies that *high-temperature peridotites should be included in the constraints on xenolith geotherms and the thickness of the TBL.*

Syn- or post-eruption thermal disequilibrium

The depths of the kinks, when they are present in P–T arrays, commonly correlate with the transition from undeformed (coarse) low-T to sheared high-T peridotites and with changes in peridotite composition: high-T peridotites are close in composition to the primitive mantle (Boyd, 1987). Thus, it is tempting to associate the boundary temperature between low-T and high-T mantle peridotite (~1100–1200 °C) either with the base of the thermal lithosphere (Finnerty and Boyd, 1987) or with the base of the rheological boundary layer (Boyd, 1973; Sleep, 2003).

An alternative explanation of the kink in xenolith geotherms implies that the lower part of the geotherm reflects a local thermal disequilibrium related to a thermal perturbation in the

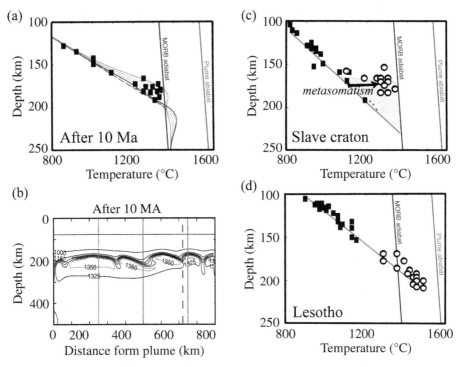

Fig. 5.9 Disequilibrium continental geotherms perturbed by plume–lithosphere interaction (after Sleep, 2003). (a) Mantle geotherms at 10 Ma after plume impact; different lines refer to different distances from the plume (shown in (b)). Symbols illustrate the scatter that might occur in a P–T array of xenoliths emplaced at slightly variable distances from the plume. (b) Potential temperature field at 10 Ma after plume impact. Different lines refer to mantle temperatures shown at steps of 250 °C. Vertical lines – positions of geotherms shown in (a); dashed vertical line – position of closely spaced geotherms in (a). (c–d) Symbols – xenolith P–T arrays (the BK method) from the Slave province (the Jericho pipe) and Lesotho. Solid symbols – coarse low-*T* xenoliths; open circles – sheared high-*T* xenoliths. Gray line – the inferred regional conductive geotherm determined by low-*T* xenoliths. In the Slave craton (c), high-*T* xenoliths lie hotter than this geotherm and require mantle heating proportional to the area of the gray triangle. In Lesotho (d), high-*T* xenoliths are at temperatures hotter than the MORB adiabat and are likely to sample the plume material.

mantle at the time of the magmatic event, while the upper part (constrained by low-*T* peridotites) refers to the unperturbed steady-state thermal regime (Green and Gueguen, 1974). Numerical models of mantle convection provide further insights into the possible physical origin of the kink on xenolith geotherms and illustrate features that may be observed in P–T studies of xenolith suites (Sleep, 2003). The results illustrate that the shape of transient geotherms depends on the distance from the plume and on the time elapsed since its interaction with the lithosphere. For example, xenolith samples erupted in a single event along-strike of a kimberlite dike may sample variable geotherms, causing scatter in the depth–temperature array (Fig. 5.9ab). As a result, random sampling of the lithosphere by kimberlites, as well as any systematic bias in the location of kimberlites may cause similar cratons to look different and obscure real differences between cratons (Sleep, 2003).

Xenolith data from two cratons, the Jerico pipe in the Slave craton and kimberlites from Northern Lesotho illustrate the complexity of interpretations of P–T arrays. Xenolith data from the Slave craton analyzed by BK thermobarometry displays a kink at ~170 km depth, that can be explained by significant (plume-related) local heating (Fig. 5.9c). In contrast, a linear P–T array that extends to high temperatures without any noticeable kink or curvature, as observed beneath Lesotho in South Africa, can be explained by ponding of plume material beneath a craton at ~90 Ma (Fig. 5.9d).

Steady-state inflected geotherms

Recent numerical simulations of mantle convection have demonstrated that there is no theoretical reason to assume that high-T peridotites reflect disequilibrium thermal conditions in the upper mantle (Lenardic and Moresi, 2000). Inflected geotherms can exist in thermal equilibrium in specific continental regions, such as transitions from a thick cratonic to a thin oceanic lithosphere (Fig. 5.10a).

In such regions, steep temperature gradients can be produced at a depth below the base of the thin lithosphere where vertical conduction is replaced by high horizontal heat flow in the mantle. The flowing upper TBL of the mantle convection cell moves from the oceanic region with high heat flow to the continental region with low heat flow. The uppermost layer of the convective mantle must adjust to the heat flow beneath the continent, thus forming the inner boundary layer and creating an inflected but still steady-state geotherm in the area of the ocean–continent transition. Similarly, inflected geotherms can exist at the cratonic margins, in areas with a steep change in the lithospheric thickness from cratonic to young continental lithosphere (Fig. 5.10b).

The third option for when an inflected steady-state geotherm is produced is the flow of a convective boundary layer below a plate with a step-like temperature discontinuity (Fig. 5.10c), although such conditions exist at the upper surface only temporally. Similarly to the first two cases, the inflected geotherm is produced by thermal inertia of the convective layer, which cannot instantly adjust its temperature over the full depth of the convective cell when convective flow encounters lateral heterogeneity. As a result, equilibrium thermal profiles of a non-classic shape are produced locally.

5.2.3 Garnet geotherms: CBL or TBL?

Griffin and co-authors (Ryan et al., 1996; Griffin et al., 1996, 1999a) use garnet T_{Ni} geotherms (temperature estimates from the Ni content distribution in xenocryst garnets) to analyze the structure of the continental lithospheric mantle. Their approach (illustrated in Fig. 5.11) is significantly different from classical geothermobarometry. At the first step, the temperature at the base of petrologic lithosphere is determined. This is defined as the

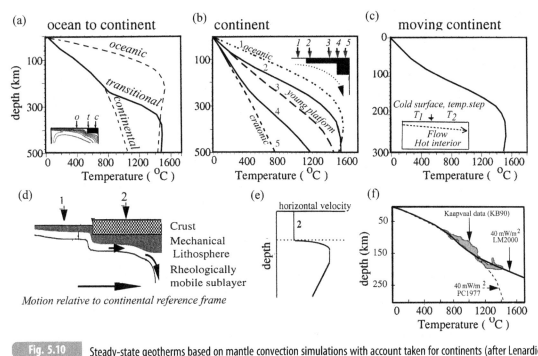

Fig. 5.10 Steady-state geotherms based on mantle convection simulations with account taken for continents (after Lenardic and Moresi, 2000). (a) Transitional equilibrium geotherms at the continent–ocean transition; (b) the same as (a) but with a step-like change in lithospheric thickness within the continent; (c) steady-state geotherm in the continental lithosphere with a temperature step at the upper cold free slip surface and with mantle flow within a hot interior. Inserts in (a–c) illustrate model configurations. (d) Cartoon illustrating subcontinental flow in the mantle and its effect on the lateral velocity profile near the edge of the continent (e). (f) Comparison of convection model predictions (LM2000) and conductive geotherm of Pollack and Chapman (1977) with Kaapvaal xenolith P–T arrays (Brey and Kohler, 1990 thermobarometry).

temperature above which garnets depleted in yttrium (a signature of the lithospheric mantle) disappear. Based on worldwide statistics, Y-depleted garnets are defined as those with less than 10 ppm Y. The base of the petrologic lithosphere defined in this manner may represent the upper limit of marked magma–wall interaction. In the next step, the thus determined T_{Ni} temperature at the base of the petrologic lithosphere is converted to depth along a conductive geotherm. This allows for the constraining of depth profiles of major (and trace-) element distribution in the lithospheric mantle and for calculating depth variations of density and seismic velocities. Griffin and co-authors have used this approach to map the olivine content and the thickness of the cratonic lithospheric mantle (e.g. Figs. 6.15 to 6.17; 6.24; 6.28).

In contrast to classical geothermobarometry, the definition of the lithospheric base by garnet geotherm combines features of both CBL and TBL definitions, since the depth at which Y-depleted garnets disappear is constrained from a conductive geotherm. Furthermore, this approach differs significantly from the way the lithospheric base is defined

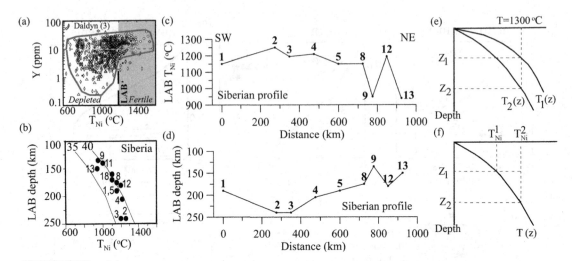

Fig. 5.11 Garnet geotherms and the Siberian lithospheric profile based on Ni-in-garnet thermometry. (a) An example of Y-T_{Ni} plot for the Daldyn kimberlite field in Siberia. The lithospheric base (LAB) is defined by the temperature above which garnets with less than 10 ppm Y disappear. The plot provides an estimate of temperature at the LAB. (b) Depth to LAB defined by Y-depleted garnets versus T_{Ni}. Temperature estimates for the LAB (a) are compared with the (assumed or independently calculated) conductive geotherm to obtain the depth of their origin. The lithospheric base defined in such a manner combines features of both CBL and TBL definitions. Solid lines – 35 and 40 mW/m^2 conductive geotherms; dots – kimberlite pipes. (c–d) SW–NE profiles across the Siberian craton showing Ni-temperatures at LAB (determined as in plot a) and the depth to LAB (determined as in plot b). Kimberlite pipes in (b–d) are numbered as in Fig. 5.3; lithological profile along the same line is shown in Fig. 6.17. (e–f) Sketches illustrating the difference between the thermal definition of the lithospheric base (e) and the definition of the lithospheric base based on Ni-in-garnet thermometry (f). (Data sources for (a–d): Griffin *et al.*, 1999a.)

in conventional thermal models (Fig. 5.11ef). According to thermal models, two regions have different lithospheric thicknesses because they have different geotherms that intersect a mantle adiabat (or a 1300 °C isotherm) at different depths. In contrast, as illustrated by garnet geotherms from the Siberian craton (Fig. 5.11b), two regions have different lithospheric thicknesses not because they have different geotherms, but because the temperature (and thus the depth) at which garnets become Y-depleted is different. All data for the Siberian craton based on Ni-in-garnet thermometry (Ryan *et al.*, 1996) fall close to the same conductive geotherm of ~40 mW/m^2, and thus according to a conventional definition of thermal lithosphere the lithospheric thickness along the Siberian kimberlite profile should be approximately the same and, in general, different from the lithosphere thickness defined by garnet-depletion. In particular, due to horizontal heat transfer short-wavelength fluctuations of the thickness of the thermal lithosphere are not possible, unless it were a non-steady-state feature immediately associated with kimberlite magmatism (Fig. 5.11cd). Note that conventional thermobarometry indicates that xenolith P–T arrays beneath all major kimberlite fields in Siberia fall close to 35–38 mW/m^2, with the coldest geotherm (~34–35 mW/m^2) beneath the Mir kimberlite pipe (e.g. Ashchepkov *et al.*, 2010) (Table 5.3).

Table 5.3 Summary of xenolith data for cratons

Location	Crustal age*	Age of emplacement	Lithospheric thickness (km) from xenoliths	Reference geotherm (mW/m²)		Reference
				**	***	
Africa						
Kaapvaal	ArI	<90 Ma (Group I), 119 Ma (pipe Finsch),	200–250	40	—	RN99
			180–200	40	—	G03
		1180 Ma (pipe Premier)	~180	39^	—	R96; ORG96
			>155–185	—	—	Gr06; Gr99
Lesotho	ArI		170–190	40	—	R96; BG86
Tanzania	ArII		~150	—	—	ORG96; LR99
West Africa	ArII	135–153 Ma, 800 Ma	>195–205	—	—	S04 (Guinea)
			>170	—	—	S04 (Sierra Leone)
			200	—	—	S04 (Liberia)
Namibia	ePt		RBL ~200	—	—	P94
			>125	—	—	Gr99
Mobile belts of S. Africa	lPt		~140	—	—	BG86
			>150	—	—	Gr99
Siberia						
Daldyn–Alakit terrane	ArII	345–360 Ma (pipe Udachnaya)	~240	~35	—	G99, B76
			>220	34	—	R96
Magan terrane	ArII	345–360 Ma (pipe Mir)	~220	~35	—	B76; G93; G96; P91
			200–250 ?	~35	—	A10
				40^	—	RN99; G99
Markha terrane (Muna)	ArII	345–360 Ma		38^	—	G99
Berekla-Olenek-Anabar	ArII	380–420 Ma, 240 Ma, 140–170 Ma		—	—	G99
S margin of the Siberian platform (Ingashi)	ArII-ePt	1268 Ma		—	—	
North America						
Slave craton	ArII	170 Ma	200–250	—	40	RN99
			200–220	35–38	—	OR01
			160–190 (North)	—	37–40	K99

Table 5.3 (cont.)

Location	Crustal age*	Age of emplacement	Lithospheric thickness (km) from xenoliths	Reference geotherm (mW/m²) **	***	Reference
Slave craton (Lac de Gras)	ArII	47–53 Ma	170–210 (North)	—	—	Gr99
			200 (Central)	—	—	P96
			180–200 (Central)	—	—	Gr99
			>230 (South)	—	—	K04, Gr99
Superior Prov. (Wawa)	ArII	140 Ma	200–250	37–40	—	Me03
				37–40	—	Kj96
					40	RN99
Superior Prov. (Kirkland Lake)	ArII	140 Ma		42	—	P98
Trans-Hudson Orogen	ePt			~40	—	Kj96
Saskatchewan	ePt	Cretaceous		42	—	Kj96
Wyoming craton	ArI	48–52 Ma	~140 ?		—	E88
				40^	—	R96
			>175		—	Gr99
SW Arkansas	mPt	106 Ma			—	
Midcontinent rift	Pt	90 Ma			—	
N Canadian Shield (Somerset Isl.)	Pt	99 Ma		44	—	SF99
Asia						
Sino-Korean craton	ArII	Pz	180	40^	—	ORG96
			140 ??		—	G98b, X00
Craton's core			>217		—	Gr99
Australia						
E Australia	Pz	Permian, Mz	~80–100	~90	—	ORG85
South America						
Guyana craton	ArII		160–180		—	G98a

	Crustal ages*		**| |**	Reference
Sao Francisco craton	ArI	89 Ma	—	34–40	R03
		85–80 Ma	—	39–50	RO3
Europe					
Baltic Shield (Karelia)	ArII-ePt	ePt, 589–626 Ma	220	36	PD99,
			200–240 (RBL > 240)	—	KP99
Baltic Shield (Archangelsk)	ArII-ePt	365–382 Ma	> 200	—	PD99, M03
Baltic Shield (Kola Province)	ArII-ePt	365–382 Ma	~150	—	PD99
Baltic Shield (SvecoFennian province)	ePt	Late Pz	~150	—	PD99
Baltic Shield (S Sweden)	ePt-mPt	Tertiary	<100	—	PD99
Baltic Shield (SW Norway)	Pz-lPt	Tertiary	<100	—	PD99
Ukrainian Shield	ArI		—	—	

* Crustal ages (see Table 4.13 for details): ArI – Archean I (> 3.0 Ga); ArII – Archean II (2.5–3.0 Ga); ePt – Paleoproterozoic (1.7–2.5 Ga); mPt – Mesoproterozoic (1.0–1.7 Ga); lPt – Neoproterozoic (540–1000 Ma); Pz – Paleozoic (250–540 Ma); Mz – Mesozoic (65–250 Ma); Cz – Cenozoic (<65 Ma).

** Reference conductive geotherms of Pollack and Chapman, 1977

*** Conductive geotherms calculated in the original publications (see references)

^ Ni-in-garnet thermometry

References: A10 = Aschepkov et al., 2010; B76 = Boyd et al., 1976; B84=Boyd, 1984; BG86 = Boyd and Gurney, 1986; E88 = Eggler et al., 1988; G96 = Griffin et al., 1996; G98 = Griffin et al., 1998a, 1998b; G99 = Griffin et al., 1999a; G03 = Griffin et al., 2003b; Gr99 = Grutter et al., 1999; Gr06 = Grutter et al., 2006; Kj96 = Kjarsgaard, 1996; K99 = Kopylova et al., 1999; K04=Kopylova and Caro, 2004; KP99 = Kukkonen and Peltinen, 1999; LR99 = Lee and Rudnick, 1999; M03 = Malkovets et al., 2003; Me03 = Menzies et al., 2003; ORG95 = O'Reilly and Griffin, 1985; ORG96 = O'Reilly and Griffin, 1996; OR01 = O'Reilly et al., 2001; PD99 = Poudjom Djomani et al., 1999; P91 = Pokhilenko et al., 1991; P94 = Pearson et al., 1994; P99 = Pearson et al., 1999; RN99 = Rudnick and Nyblade, 1999; R03 = Read et al., 2003; R96 = Ryan et al., 1996; S04 = Skinner et al., 2004; SF99 = Schmidberger and Francis, 1999; X00 = Xu et al., 2000.

5.2.4 Xenolith geotherms and TBL thickness

Figure 5.12 summarizes xenolith constraints on temperature distribution in the continental lithosphere in various tectonic settings. Following widely accepted practice in the petrologic community, xenolith geotherms are compared with conductive continental geotherms constrained by surface heat flow. The discussion to follow (Fig. 5.13) only includes xenolith geotherms constrained by classical geothermobarometry as described in Section 5.2.1, since garnet geotherms do not constrain pressure independently, but instead project temperature estimates from the Ni content distribution in xenocryst garnets onto a conductive geotherm to estimate pressure (depth).

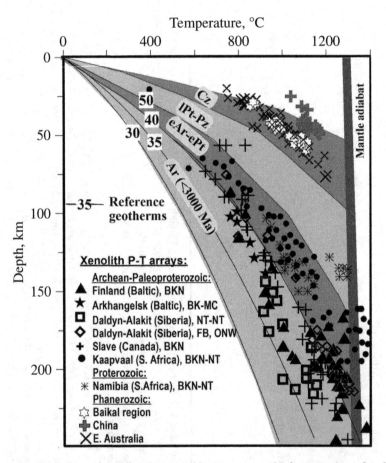

Fig. 5.12 A comparison of xenolith P–T arrays from different continental tectonic settings with reference geotherms from Pollack and Chapman (1977) and typical continental geotherms constrained by heat flow data (based on Artemieva, 2006). Abbreviations: Ar = Archean (< 3.0 Ga), eAr-ePt = Paleoarchean (> 3.0 Ga) and Paleoproterozoic (2.5–1.8 Ga); lPt = Neoproterozoic, Pz = Paleozoic, Cz = Cenozoic. For notations of thermobarometers used in xenolith studies see Table 5.2.

Xenolith constraints on mantle temperatures are of utmost importance in constraining the deep thermal regime of tectonically active regions where steady-state thermal models are invalid, while thermo-mechanical modeling requires detailed information on the crustal and upper mantle physico-chemical processes in these regions. Xenolith P–T arrays from different regions of Cenozoic extension, rifting, and volcanism from China, Australia, Siberia, Spitsbergen, and Europe are surprisingly similar and indicate high mantle temperatures, along 70–90 mW/m^2 conductive geotherms (Figs. 5.12, 5.13(i)). Assuming the MORB adiabat is representative of the convective upper mantle, xenolith thermobarometry yields a surprisingly uniform lithosphere thickness in tectonically active regions, ~60–80 km.

Cratonic geotherms constrained by xenolith geothermobarometry fall between 35 and 50 mW/m^2 conductive geotherms, with lower estimates of mantle temperatures for the Archean–Paleoproterozoic regions. Some cratonic regions (e.g. South Africa, Siberia) have been affected by several pulses of tectono-magmatic activity. Repeated eruptions that brought xenoliths to the surface during different tectonic episodes allow reconstruction of the evolution of paleolithospheric temperatures. For example, deep crustal xenoliths from South Africa suggest that the thermal regime of the lithosphere remained unchanged between 800–600 Ma and ~200 Ma (Schmitz and Bowring, 2003). It is likely, however, that a significant change in thermal regime occurred in the Mesozoic, as indicated by different xenolith P–T arrays below ~160 km depth from the Finsch pipe that erupted at 119 Ma and kimberlites from Northern Lesotho that erupted at ~90 Ma (Fig. 5.7b). Similarly, two generations of kimberlite eruptions in the SW São Francisco Craton, Brazil (at 89 Ma and at ~85 Ma) indicate significant change in the thermal regime of the lithospheric mantle (Fig. 5.13f). Kimberlites erupted at ~85 Ma clearly follow a 40–45 mW/m^2 conductive geotherm and show lithospheric temperatures ~ 250°C higher than xenoliths from the earlier eruption (~89 Ma) which follow a 35–40 mW/m^2 conductive geotherm. This thermal anomaly can be related to the Late Cretaceous opening of the South Atlantic and the break-up of Gondwana (Read *et al.*, 2003).

Xenolith geotherms from the off-craton Proterozoic mobile belts in South Africa and from the Archean cratons that have been significantly reworked in Phanerozoic (the Wyoming, Sino-Korean, and Tanzanian cratons) follow the cratonic conductive geotherm only down to ~120–150 km depth (Fig. 5.13b). At these depths, mantle temperatures reach 1000–1200 °C and increase over a small pressure (depth) interval. The deflection of xenolith P–T data from the conductive geotherm may be interpreted as the base of the upper mantle layer with a (pure) conductive heat transfer regime. Phanerozoic tectono-thermal events probably involved thermo-mechanical removal (delamination) or compositional transformation (refertilization by asthenosphere-derived fluids and basaltic melts) of the lower 80–140 km thick layer of these Archean keels (Eggler *et al.*, 1988; Griffin *et al.*, 1998b; Lee and Rudnick, 1999; Xu *et al.*, 2000).

5.2.5 Lithosphere thickness in Archean cratons

Two types of Archean xenolith geotherms?

Lithosphere thickness in the Archean cratons is a subject of great importance for global geochemical and geodynamical models of the Earth's early evolution, crustal and mantle

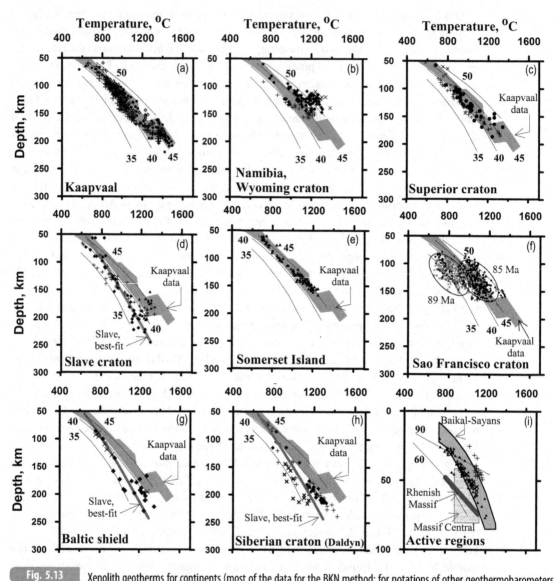

Fig. 5.13 Xenolith geotherms for continents (most of the data for the BKN method; for notations of other geothermobarometers see Table 5.2). Gray shading in (a–h): xenolith data from several locations in Kaapvaal (compilation of Rudnick and Nyblade, 1999); gray line in (d, g, h): best-fit for low-*T* peridotites from Slave craton (BKN data from Kopylova *et al.*, 1999; Kopylova and McCammon, 2003). Two geotherms are typical for the cratonic lithosphere: ~45 mW/m² (a–c, e, f) and ~37 mW/m² (d, g, h). Thin lines – conductive geotherms of Pollack and Chapman (1977), values are surface heat flow in mW/m². Note different depth scale of plot (i). For references and details see Artemieva (2006).

differentiation, longevity of the Archean lithosphere, and mantle–lithosphere convective interaction. It is also the subject of significant controversy and heated on-going debates.

Rudnick and Nyblade (1999) have compiled xenolith P–T arrays for four Archean cratons (Kaapvaal, Siberia, Slave, and Superior) and concluded that all of them fall onto the same mantle geotherm of ~40–42 mW/m². An extended analysis of xenolith P–T arrays from various Archean cratons does not support this conclusion (Artemieva, 2006). The analysis is based on a preferred choice of xenolith P–T arrays constrained (where available) by the BKN method, with the distinction of xenolith P–T data from different kimberlite clusters within a craton, and with account taken for different crustal and mantle densities (2.7 and 3.3 g/cm³, correspondingly) when converting pressures into depth values (unaccounted for by Rudnick and Nyblade, 1999). Figure 5.13 shows xenolith geotherms for different cratonic regions, which are compared with the family of reference conductive geotherms (Pollack and Chapman, 1977). P–T arrays from the Archean cratons fall into two distinctive groups:

(1) Xenolith geotherms from South Africa (Kaapvaal), South America, and the Superior Province follow a 40–45 mW/m² conductive geotherm (except for a high-temperature part which can be attributed to transient thermal effects, Sleep, 2003).

(2) Xenolith P–T arrays from the Slave craton, Fennoscandia (central Finland and Arkhangelsk region), and the Siberian craton (the Daldyn terrane) indicate a significantly colder conductive geotherm, 35–38 mW/m² (Table 5.3). There is, however, significant scatter of xenolith P–T data for the Siberian craton and the Baltic Shield, which can probably be attributed to mineralogical disequilibria, as observed in xenolith suites from Siberia (Griffin et al., 1996). In this case, xenolith P–T geotherms and associated lithospheric thermal structure should be interpreted with caution.

These results suggest that, because of the difference in xenolith geotherms between the two groups, the Archean cratons are likely to have two different lithosphere thicknesses. Small differences in references geotherms lead to large differences in the TBL thickness. Ignoring the existence of a rheologically active boundary layer with perturbed geotherm between the conductive TBL and the convecting mantle, a warmer 42 mW/m² geotherm intersects the mantle adiabat at ~220 km depth, while a colder 37 mW/m² geotherm – at ~300 km depth (the exact depth is difficult to constrain because at depths >300 km the geotherm asymptotically approaches the mantle adiabat). This conclusion is in general agreement with the results of thermal modeling (Fig. 4.46) and mantle convection models (Fig. 4.50) which indicate bimodal distribution of lithospheric thickness in depleted Archean cratons, 200–220 km and ~300 km (Doin et al., 1997; Artemieva and Mooney, 2001; Levy et al., 2010).

Do xenolith data ban a 300 km-thick lithosphere?

In accordance with the conclusion of Rudnick and Nyblade (1999), xenolith P–T arrays published for various cratonic settings (Table 5.3) are conventionally interpreted as indicating a globally uniform thickness of c. 200–220 km of cratonic lithosphere that has not undergone significant thermo-tectonic events since the Precambrian. This interpretation is constrained by the depth at which a 40–42 mW/m² xenolith geotherm intersects the MORB adiabat. As noted earlier, estimates of the mantle adiabat constrained by mid-ocean ridge basalts (MORB) may be inconsistent with xenolith thermobarometry (Bell et al., 2003).

A surprising similarity of thermal regime of the cratons as apparently indicated by xenolith P–T arrays, despite significant differences in the composition of the lithospheric mantle, tectonic evolution of the cratons, their size, and kimberlite emplacement ages can be easily understood if xenolith geotherms reflect a disequilibrium thermal regime of the cratonic lithosphere *during kimberlite magmatism*. In this case, xenolith geotherms carry information only on the P–T conditions of the eruption, but not on the ambient thermal structure of the lithosphere, making almost meaningless their comparison with *the snap-shots of the present state of the mantle* provided by seismic and electromagnetic studies. The very fact that kimberlite magmatism is so restricted both in time and in space (Table 9.5) suggests that it reflects not the equilibrium P–T conditions within the cratonic lithosphere, but rather some unusual localized thermal events in its tectonic history, such as plume–lithosphere interaction (see e.g. Fig. 5.9 and discussion in Section 6.3). The Re–Os studies of Carlson *et al.* (1997) show no difference in Os model ages between high-T and low-T peridotites. The authors conclude that the fertile nature of high-T peridotites is likely to be a recent phenomenon, and thus *the lithospheric base must lie below the depth range sampled by xenoliths in kimberlites*. The absence of shearing associated with flow in the rheologically active boundary layer in the deepest xenoliths from the Baltic Shield suggests that, at least locally, lithospheric thicknesses can be significantly greater than 240 km (Kukkonen and Peltinen, 1999; Poudjom Djomani *et al.*, 1999).

Another line of argument against a 300 km-thick cratonic lithosphere is based on the evidence (as seen in Fig. 5.13) that the deepest known mantle xenoliths usually come from a ~200–220 km depth with few samples from a ~240 km depth found in the Baltic Shield and in the Slave craton. *The absence of 'evidence for' is commonly interpreted as 'evidence against'*. However, the absence of known xenoliths derived from depths greater than 250 km within the lithospheric mantle does not prove that the cratonic lithosphere cannot be thicker than 220–250 km. The deeper layers of the upper mantle may not be sampled by xenoliths because kimberlite magmas from deeper mantle may not reach the surface or they are not generated deeper than 240–250 km. This view is supported by the petrologic studies of Brey *et al.* (1991) who suggest that xenoliths may not be representative of the entire cratonic lithosphere as a depth of c. 150 km is critical for generation of kimberlite magmas; at this depth they become highly fluid-charged which enhances their ability to capture and transport xenoliths to the surface.

Xenolith versus geophysical data

While numerous interpretations of xenolith P–T arrays from different cratons apparently favor a globally uniform thickness of 200–220 km of the Archean–Paleoproterozoic lithosphere, geophysical interpretations are less consistent and indicate significant variations in thickness of the cratonal lithospheric roots. A globally uniform thickness of c. 200–220 km of cratonic lithosphere is supported by electromagnetic studies from various cratonic regions (Fig. 7.38), but contradicts global and regional seismic tomography models, some of which show significant variations in cratonic lithosphere thickness from ~150–200 km to 300–400 km (e.g. Figs. 3.87, 3.90, 3.98, 3.101, 3.120, 3.122). However, the interpretations of seismic tomography models in terms of lithospheric thickness are non-unique and, among

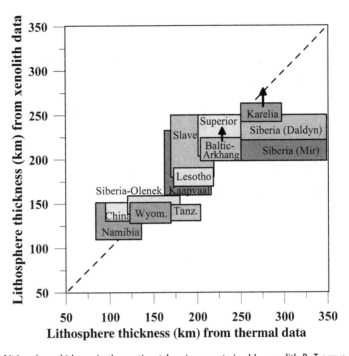

Fig. 5.14 A comparison of lithosphere thickness in the continental regions constrained by xenolith P–T arrays (Table 5.3) and surface heat flow (Artemieva and Mooney, 2001). Peridotites from the deepest known xenoliths from the Archean–Paleoproterozoic parts of the Baltic shield (Karelia and the Arkhangelsk region) do not show shearing, suggesting that the lithospheric mantle extends deeper than sampled by xenoliths (marked by arrows).

other factors, depend on the choice of reference models and the amount of velocity anomaly chosen to mark the lithospheric base (Sections 3.6.2–3.6.3). In agreement with high-resolution seismic tomography models, thermal modeling for the stable continental regions also indicates large differences in lithospheric thickness beneath different Archean cratons, with variations from ~200–220 km in South Africa, Western Australia, and the south-central parts of the Canadian Shield to ~300–350 km in West Africa, Baltic Shield, Siberia, and the northern parts of the Superior Province (Fig. 5.14) (Artemieva and Mooney, 2001; Levy *et al.*, 2010).

A direct comparison of lithospheric thickness estimates from xenolith, seismic, electro-magnetic, and thermal data should, however, be made with caution. The South African craton is probably the only region where all four sets of data exist and where all of the methods agree in estimating the lithosphere to be ~200–220 km thick (Fig. 5.5). Yet, in some cratons either high-resolution regional seismic models, or good quality electro-magnetic models, or reliable thermal constraints diverge from xenolith P–T constraints on the lithospheric thickness. The Archean–early Proterozoic part of the Baltic Shield is of special interest as seismic, thermal, electrical, and xenolith data are available there. In this region,

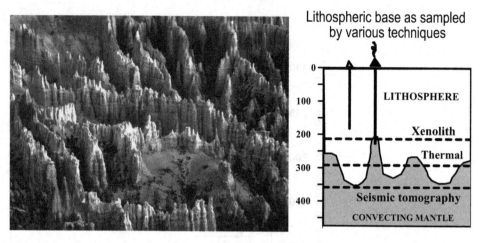

Fig. 5.15 The erosional pattern of the Bryce Canyon (Utah, USA) may be equivalent to the topography of the lithospheric base with short-wavelength significant variations in lithospheric thickness created by convective erosion. Different geophysical methods and petrologic xenolith-based data sample different depths in the upper mantle, leading to significant discrepancies in lithospheric thickness estimated by different methods.

(i) the electrical asthenosphere has not been found down to 220 km depth in high-resolution surveys (Jones, 1999),

(ii) seismic tomography and thermal models both indicate ~300 km-thick lithosphere (Figs. 3.120 and 4.44), regional high-resolution body-wave tomography suggests the lithosphere can be ~400 km thick (e.g. Sandoval et al., 2004), and

(iii) even the deepest xenoliths from 240 km depth have depleted composition and coarse texture (Poudjom Djomani et al., 1999; Kukkonen and Peltonen, 1999). In contrast to these diverse data sets, reinterpreted xenolith P–T arrays for the central Fennoscandian Shield in eastern Finland suggest only ~230–250 km thick TBL (Kukkonen et al., 2003), whereas earlier interpretations of the same data by the same group yielded values > 240 km for lithosphere thickness in the region (Kukkonen and Peltonen, 1999).

The apparent contradiction between lithosphere thickness estimates from geophysical and petrologic data reported for some cratons can be reconciled if the base of the lithosphere is highly heterogeneous at small scales (Artemieva and Mooney, 2002). The inverted topography created in the Bryce Canyon (Utah, USA) by surface erosion is an easy analog to think of (Fig. 5.15). In the case of an undulating lithospheric base with short-wavelength significant variations in lithospheric thickness, various techniques sample different depths in the upper mantle as the lithospheric base. The diffuse character of the seismic lithospheric base, especially pronounced in surface wave studies which would fail to resolve short-wavelength velocity variations, supports the model. Xenoliths sample the most shallow parts of the lower lithosphere where the pressure–temperature regime is favorable for producing kimberlite-type magmas. Thus, xenolith-based estimates of the lithosphere thickness are not representative of large-scale lithospheric thickness, but rather refer to a localized thinning of the lithosphere, much as a deeply eroded river gorge does not portray the average topography of a region. Seismic tomography and mantle convection models "average" the top of the convective mantle, i.e. the lowermost parts ("dips") of the lithospheric base (Fig. 5.15), whereas thermal data (due to laterally smoothing heat transfer) provides an averaged picture of lithospheric structure with values of lithospheric thickness intermediate between xenolith and seismic estimates.

5.3 Geotherms constrained by seismic data

5.3.1 Correlations between seismic velocity, attenuation, and temperature

Velocity–temperature correlations

The strong dependence of seismic velocity and elastic moduli on temperature is well known from laboratory studies (Fig. 3.3). Most measurements, however, were made at relatively low temperatures (<600–$700\,^{\circ}C$) where the temperature dependence of seismic velocities is approximately linear (eq. 3.7). Limited experimental data for $T > 1000\,^{\circ}C$ indicate that at near-solidus temperatures seismic velocities decrease rapidly even before any melting starts and both Vp and Vs show a pronounced drop with increasing percentage of melt (Murase and Kushiro, 1979; Murase and Fukuyama, 1980; Sato et al., 1989; Jackson, 1993).

The high sensitivity of seismic velocities to temperature variations provides important constraints on the thermal regime of the upper mantle. A large number of high-resolution global and regional seismic tomography models which have become available recently have been used to convert seismic velocities to temperatures. Such studies have been performed globally (Yan et al., 1989; Rohm et al., 2000), for the continents (Priestley and McKenzie, 2006), and on continent-scale for Europe (Furlong et al., 1995; Sobolev et al., 1996; Goes et al., 2000), North America (Goes and van der Lee, 2002; Godey et al., 2004), Australia (Goes et al., 2005); regional studies are available for South Africa (Kuskov et al., 2006), while beneath subduction zones mantle temperatures were estimated from seismic attenuation data (Sato et al., 1989; Roth et al., 2000; Nakajama and Hasegawa, 2003).

A pronounced geographical correlation observed in seismic tomographic models between velocity anomalies and surface heat flow further supports the idea that a significant proportion of seismic velocity variations may have a thermal origin. Shapiro and Ritzwoller (2004) propose to use thermodynamic constraints based on surface heat flow data and the thermal state of the upper mantle to improve seismic models by reducing their uncertainty (Fig. 5.16).

Knowledge of seismic velocity structure alone, however, does not allow for its unique interpretation in terms of mantle temperatures without using additional independent data. Starting from Jordan (1975a, 1979), Anderson and Bass (1984), Bass and Anderson (1984) it has been recognized that large-scale compositional variations in the upper mantle (such as in peridotite composition and Fe content), also play an important role in producing seismic velocity variations. This conclusion is supported by calculation of the density-to-velocity scaling factor which indicates significant compositional heterogeneity of the mantle

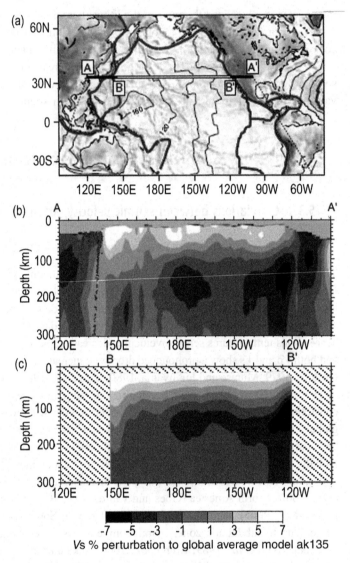

Fig. 5.16   Vertical cross-sections across the Pacific Ocean along the line AA′/BB′. (a) Location map; (b–c) seismic shear-wave velocity models: (b) with ad-hoc seismic parameterization; (c) with thermodynamic parameterization by the use of independent temperature profiles in the upper mantle (modified from Shapiro and Ritzwoller, 2004).

(Fig. 3.110; Godey *et al.*, 2004; van Gerven *et al.*, 2004; Artemieva, 2009; Simmons *et al.*, 2009). Theoretical constraints demonstrate that a +1% *V*s velocity anomaly can be produced either by a 100–150 °C temperature decrease or by a +4% δFe anomaly (Fig. 5.17). Other factors, such as the presence of partial melt or water and anisotropy also have a strong influence on seismic velocities and can significantly affect the results of velocity–temperature inversion (e.g. Sobolev *et al.*, 1996). In particular, the presence of water enhances anelastic relaxation and, by affecting *Q*, may reduce the seismic wave velocities by a few percent (Karato, 1995).

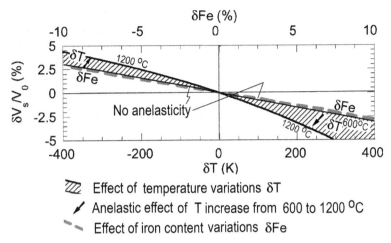

Fig. 5.17 The effect of temperature and iron-content variations in the mantle on shear velocities. The reference values are PREM for Vs and Fo89 for iron-content. Anelasticity significantly affects the temperature dependence of relative dVs anomalies (Fig. 3.6, eqs. 3.8–3.11). If its effect is ignored (thin black line), the temperature effect on Vs perturbations is underestimated for cold regions and overestimated for hot regions. The effect of Fe-depletion on dVs is of the same magnitude as the anelastic effect due to a 600 °C increase in temperature. The effects of temperature- and iron-content variations on density are shown in Fig. 6.14.

Attenuation–temperature correlations

Laboratory experiments demonstrate that at subsolidus temperatures, seismic attenuation $(1/Q)$ in olivine-rich mantle rocks increases exponentially with temperature T (eq. 3.8; Figs. 3.6, 3.8). The high sensitivity of $1/Q$ to temperatures was used to correlate seismic attenuation anomalies in the oceanic and continental mantle with temperature variations (Romanowicz, 1995, 1998; Roth *et al.*, 2000; Artemieva *et al.*, 2004; Yang *et al.*, 2007; Dalton *et al.*, 2008) and to calculate mantle geotherms from seismic attenuation (Figs. 3.124 to 3.129). Sato *et al.* (1988, 1989) were among the first to apply the correlation between temperature and seismic attenuation to constrain oceanic geotherms from seismic attenuation and laboratory data, and to argue that attenuation anomalies do not necessarily require the existence of partial melting. Other attempts to calculate temperatures from seismic attenuation data included studies on the crust of south-central Eurasia (Sarker and Abers, 1999) and the mantle wedge beneath northeastern Japan (Nakajima and Hasegawa, 2003).

The validity of conversion of seismic attenuation anomalies into mantle temperature performed by Sato and co-authors was questioned by Karato (1995). He argued that in Sato's experiments, anelastic relaxation was likely to be enhanced by water supplied by the dehydration of hydrous minerals (Karato and Spetzler, 1990). Since the enhancement of anelastic relaxation causes seismic velocity reduction, the low seismic wave velocities and high attenuation observed in back-arc basins and subduction zones are caused not only by high temperatures, but also presumably by high water fugacities. Thus, an analysis of seismic data (seismic velocities and attenuation) in terms of temperature profile should take into account the effect of water separately from that of temperature. In practice, this is not done in any of the published studies. A part of the problem arises from the fact that presently available laboratory studies of seismic attenuation do not examine the effects of water in detail (Sato *et al.*, 1989; Jackson *et al.*, 1992, 2000; Faul and Jackson, 2005). As a result, seismic attenuation (or velocity)–temperature conversions constrained by laboratory data need to be treated with caution. This is particularly important when a large amount of water may be present in the upper mantle, such as in (paleo)subduction settings (Fig. 7.19).

The opposite approach is to calculate theoretical models of upper mantle attenuation from the Arrhenius law (eq. 3.8) on the attenuation–temperature correlation, using independently constrained mantle temperatures (Artemieva *et al.*, 2004). A comparison between a theoretical attenuation model and seismic attenuation provides information on non-thermal effects on seismic Q, primarily caused by the presence of hydrogen and partial melt (uncertainties in seismic and theoretical models also contribute to the difference between the two models). The correlation between attenuation and temperature globally at any depth in the upper 200 km beneath the continents is less than 0.47 (Fig. 3.128), supporting the conclusion that effects other than temperature play an important role in producing attenuation anomalies in the upper mantle. Therefore, the attenuation–temperature correlation in the upper mantle is subject to significant uncertainty.

The attenuation values determined in laboratory experiments include only intrinsic (anelastic) attenuation, while global and regional seismological studies measure the sum of anelastic and scattering effects (eq. 3.6). A regional study of the Japan islands has demonstrated that scattering can account for 10–30% of the measured attenuation (Hoshiba, 1993) and thus can reduce the attenuation–temperature correlation. However, in the case of small *variations* of scattering in the upper mantle, the correlation between intrinsic (and total) attenuation and temperature will still hold.

The following discussion focuses only on velocity–temperature conversions.

5.3.2 Seismic constraints on temperatures: methodology and uncertainties

Velocity–temperature conversions are based on theoretical and experimental constraints on the elastic and non-elastic behavior of mantle rocks at high P–T conditions (e.g. Duffy and Anderson, 1989; Jackson *et al.*, 1990; Karato and Spetzler, 1990; Karato, 1993). However, seismic velocities can vary not only with temperature, but with mineral composition, phase changes, anisotropy, grain size, the presence of water, and with the onset of melting (see Chapter 3.1 and Figs. 3.2 to 3.8). Therefore, a significant proportion of velocity anomalies can be due to physical mechanisms other than thermal, as reflected in a low (< 0.60) correlation coefficient between shear seismic velocities and temperatures in the upper 200 km of the subcontinental mantle (Fig. 3.128). As a result, the conversion of seismic velocities into temperatures is subject to considerable uncertainty since non-thermal effects on seismic velocity variations are difficult to quantify. Furthermore, limitations associated with seismic models (see Section 3.2.3) propagate into additional uncertainties in velocity–temperature conversions.

Tomographic models

Major limitations and uncertainties "inherited" from seismic velocity models include:

- *Vertical and lateral resolution of seismic models.* In particular, seismic models based on fundamental-mode surface waves typically cannot resolve mantle structure deeper than 200–250 km and have a relatively low lateral resolution. In contrast, body-wave tomographic models smear the depth distribution of velocity perturbations (Figs. 3.79, 3.80).
- *Ray path coverage.* In teleseismic tomography it is globally very uneven (Fig. 3.77) and poor in some regions due to the fact that the majority of seismic events are restricted to the subduction zones and the areal distribution of seismic stations is highly non-uniform. On the continents, the worst ray path coverage in teleseismic tomography exists for Siberia, north-central-western Africa, large parts of South America, as well as Arctic and Antarctic regions (including northern Canada and Greenland). The problem is usually less important in regional tomographic models.
- *The regularization method and damping used in seismic inversion.* As a result, the amplitude of velocity perturbations can vary significantly between different seismic tomography models and their quantitative correlations are weak (Section 3.2.3).
- *Uncertainty specific to velocity–temperature conversions* (Fig. 5.18). Adjacent lithospheric blocks may have a strong contrast in seismic velocities (due to contrasts in composition, anisotropy, or grain size), whereas lateral heat transfer will prevent these blocks having a sharply localized contrast in temperatures (unless the thermal anomaly is young or associated with a very sharp contrast in thermal properties). As a result, strong horizontal temperature gradients produced by velocity–temperature conversions based on high-resolution regional seismic models are likely to be artifacts (such as a 800 °C temperature variation over a lateral distance of ~300–400 km at 100 km depth beneath south-eastern Europe in the interpretation of Goes *et al.*, 2000).

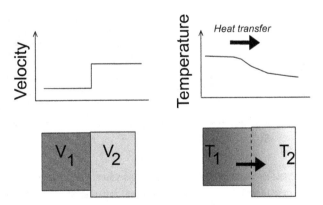

Fig. 5.18   Sketch illustrating an example of velocity-to-temperature conversion. Two adjacent blocks may have a sharp contrast in seismic properties, while lateral heat transfer smears the temperature contrast between them.

Relative and absolute velocities

A calculation of mantle temperatures from seismic velocities consists, in general, of the following steps. First, relative seismic velocity anomalies are converted to absolute velocities (unless the seismic model is in absolute velocities), since the velocity–temperature relation is strongly non-linear at near-solidus temperatures (Fig. 3.3). One of the largest problems arises, apparently, when the PREM model is used as the reference model in seismic inversion for the continents. The isotropic PREM has a 5% step in Vs velocity at 220 km depth, not necessarily required by continental data (see discussion in Chapter 3), and therefore a 220 km velocity discontinuity requires special consideration in velocity–temperature conversions for the continental regions, such as in the inversion by Röhm et al. (2000) (see below).

Effect of iron content

Weak, even at a regional scale, quantitative correlation of seismic velocity anomalies with surface heat flow suggests that factors other than temperature are important in producing velocity perturbations (Fig. 5.19). In the next step of velocity–temperature conversions, some assumptions on compositional anomalies in the upper mantle are made. In the simplest case, all observed seismic anomalies are assumed to result from variation in temperature and iron content. Using this first-order assumption, Röhm et al. (2000) estimated the contribution of iron-content variations to velocity anomalies beneath the continents. However, their results may be significantly biased since the values of iron depletion in the Archean cratons were derived from the data set of Jordan (1979) which includes both low-T and high-T Kaapvaal peridotites. Yan et al. (1989) interpret a global surface-wave tomographic model of Woodhouse and Dziewonski (1984) in terms of temperature anomalies (with respect to 1300 °C) or iron-depletion anomalies (with respect to Fo_{10}) (Fig. 5.20) (see Table 5.4 for summary of major model assumptions). Clearly, the approach does not allow for separation of thermal and compositional (iron-content) contributions to shear-wave velocity

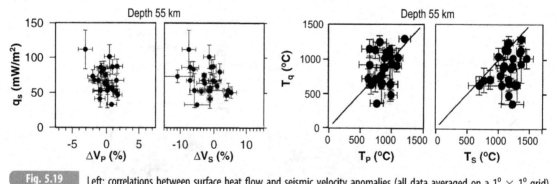

Fig. 5.19 Left: correlations between surface heat flow and seismic velocity anomalies (all data averaged on a 1° × 1° grid). Right: correlations between mantle temperatures extrapolated from surface heat flow using conductive geotherms (Pollack and Chapman, 1977) and temperatures derived from P- and S-seismic velocities. All data refer to the European upper mantle at 55 km depth (source: Goes et al., 2000).

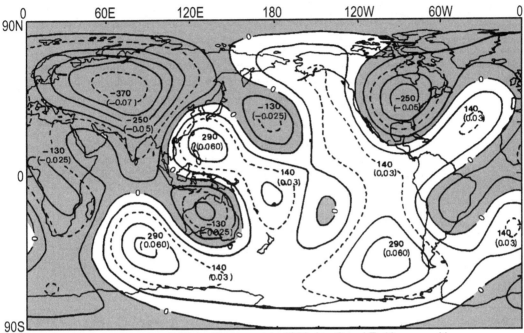

Anomalies are in: deg C (Fo) with respect to 1300 deg C and Fo10 composition

Fig. 5.20 Anomalies of temperature *or* iron content at 150 km depth constrained by a global surface wave tomographic model of Woodhouse and Dziewonski (1984). Positive velocity anomalies (with respect to PREM) in the upper mantle beneath the continents are attributed either to cold temperatures (given in deg C with respect to 1300 deg C) or to iron-depletion anomalies (given in brackets with respect to $Fo_{10} = (Mg_{90}, Fe_{10})SiO_4$), or to a combination of both, and the interpretation does not allow for their separation. No anelastic correction is introduced in the calculation. Anelasticity may reduce the amplitude of the inferred anomalies by up to ~50%. Redrawn from Yan *et al.* (1989).

perturbations which can have very close amplitudes (Fig. 5.17). The greatest negative temperature anomalies are associated with cratons and thus are well correlated with surface heat flow. In the case where all velocity anomalies have thermal origin, mantle temperatures at 150 km depth are expected to be ~900–950 °C beneath cratonic Eurasia, ~1050 °C beneath the Canadian Shield, and ~1150–1200 °C beneath cratonic Africa and Australia, with ~1440 °C along the mid-ocean ridges and up to +300 °C temperature anomalies in some parts of the Pacific Ocean. In contrast, if all velocity anomalies originate from iron anomalies, the composition of the cratonic mantle should be extremely depleted in Eurasia and North America. Note that this study did not account for anelastic effects on seismic velocity which, in regions of high attenuation, can reduce temperature anomalies derived from velocity perturbations by a factor of 2 (Sato *et al.*, 1989; Karato, 1993).

Effect of composition and anharmonicity

Different effects of pressure and temperature variations on elastic moduli and thermal expansion of single crystals produce anharmonic effects (due to the deviations of mineral

Table 5.4 Some velocity–temperature models		
Region and reference	Tomographic model	Assumptions in V–T conversion
Global (Yan *et al.*, 1989)	• Fundamental mode Rayleigh and Love waves with their first few overtones (periods > 135 s) and long-period body-wave data (periods > 45 s) (Woodhouse and Dziewonski, 1984); • Lateral resolution ~5000 km (up to degree l = 8 of spherical harmonics) • Velocity perturbations with respect to PREM	• Consider only 150 km depth slice • Composition: isotropic olivine, the only variable parameter: iron content • No anharmonicity • No anelasticity • No melting
Massif Central, France (Sobolev *et al.*, 1996)	• Teleseismic P-wave delay-time regional tomography model (in relative velocity anomalies) (Granet *et al.*, 1995)	• Compositional variations neglected • Vp, Vs, and density constrained by regional peridotite xenoliths • Account for anharmonicity • Anelastic correction[1] • Melting assumed
Europe (Goes *et al.*, 2000)	• Teleseismic body-wave Vp model (Bijwaard *et al.*, 1998); Relative velocities with respect to *ak135*; Lateral resolution ~60 km • Teleseismic body-wave Vs model (Marquering and Snieder, 1996)	• The approach of Furlong *et al.* (1995) • Independent inversion for Vp and Vs • Anharmonicity accounted for • Anelastic correction[1,2] for wave period of 20 s • Melting assumed
Global (Röhm *et al.*, 2000)	• Fundamental mode Love and Rayleigh surface-wave dispersion (period 40–150 s) plus a waveform data set (Trampert and Woodhouse, 1995); • Vertical resolution: 50–70 km in the upper 400 km • Lateral resolution ~2500 km (up to degree l = 16 of spherical harmonics) • Velocity perturbations with respect to PREM	• Archean cratons assigned a 42 mW/m^2 conductive geotherm, and $Vs(z)$ velocity profiles for other regions are scaled by perturbation from Archean $\delta Vs(z)$ profiles converted to $\delta T(z)$ from a 42 mW/m^2 geotherm • Two model compositions: primitive mantle (Mg# = 0.89) and cratonic mantle based on data of Jordan (1979) from both low-T and high-T peridotites (Mg# = 0.91) • No anharmonicity • Anelastic correction[1,2] for wave period of 20 s • No melting
North America (Godey *et al.*, 2004)	• Phase velocities of fundamental-mode Rayleigh waves (periods 40–150 s) • Average lateral resolution ~800 km, the best resolution (~400 km) in the western USA • Vertical resolution down to 260 km depth	• Iron anomalies are constrained by density-to-velocity scaling factor, calculated from a joint inversion of seismic model and geoid • Also considered variations in olivine and garnet • Anelastic correction[1,2] included

	Table 5.4 (cont.)	
Region and reference	Tomographic model	Assumptions in V–T conversion
North America (mostly USA) (Goes and van der Lee, 2002)	• Velocity perturbations with respect to PREM • Regional Rayleigh waves fundamental and higher mode model NA00 (van der Lee, 2002); Vertical resolution ~50 km; loses resolution below ~250 km depth • Global teleseismic body-wave Vp model (Bijwaard et al., 1998; Bijwaard and Spakman, 2000); the best vertical resolution ~50–100 km, minimum horizontal resolution ~100 km; relative velocities with respect to ak135; • Regional teleseismic body-wave Vp model (Dueker et al., 2001); the best vertical resolution ~50–100 km, minimum horizontal resolution ~50 km	• Partial melting ignored • The approach of Furlong et al. (1995) • Independent inversion for Vp and Vs • Anharmonicity accounted for • Anelactic correction[1,2] • Assumed three compositions: primitive mantle (Mg = 89), average garnet lherzolite (Mg = 90), depleted mantle (Mg = 93) • Partial melting ignored
Australia (Goes et al., 2005)	• Regional fundamental Rayleigh waves (40–200 s) and higher modes (20–125 s) (Simons et al., 2002); best resolution ~250 km laterally and ~50 km vertically; poor resolution for Western Australia (see Fig. 3.83); • Regional multimode Rayleigh wave phase velocities (40–150 s) (Yoshizawa and Kennett, 2004); best resolution ~400 km laterally and ~50 km vertically	• The approach of Furlong et al. (1995) • Anharmonicity accounted for • Anelactic correction[1] included • Assumed pyrolite composition • Partial melting ignored
Cratons and oceans (Priestley and McKenzie, 2006)	• Fundamental and first four higher modes Rayleigh wave SV-velocities (period of 50–160 s); • Velocity perturbations with respect to PREM (bad for continents); • Great circle approximation (invalid in case of strong lateral velocity gradients); • Poor resolution for some regions (e.g. Western Australia and Siberia, Fig. 3.83)	• Archean cratons assigned a ~40 mW/m2 conductive geotherm based on xenolith P–T arrays; • Oceanic geotherms calculated for the plate model; • Archean and oceanic geotherms used to calibrate velocity–depth profiles worldwide • Two compositions: primitive mantle (Mg# = 90) and melt-depleted cratonic mantle (Mg# = 92.5) • Anelactic correction[3] included • Partial melting ignored

Table 5.4 (cont.)		
Region and reference	Tomographic model	Assumptions in $V\text{--}T$ conversion
Kaapvaal (Kuskov *et al.*, 2006)	• Several regional Vp and Vs models	• Compositional variations constrained by regional peridotite xenoliths, also included are average garnet peridotite and primitive mantle compositions • Account for anharmonicity • Anelastic correction[1] included • Partial melting ignored

[1] Q model based on average experimental data and calibrated for wave period of 1 s to fit seismic observations with the old oceans geotherm (assumed to be a global average) and for mantle potential temperature of 1100 °C (Sobolev *et al.*, 1996).
[2] Q model based on experimental data for forsterite (Berckhemer *et al.*, 1982).
[3] Q model calibrated to fit seismic observations with the cratonic xenolith P–T arrays and with oceanic geotherms, taking into account the effect of grain size variations on Vs (experimental data of Faul and Jackson, 2005).

lattice vibrations from harmonic oscillations) (Duffy and Anderson, 1989; Chopelas and Boehler, 1989; Anderson *et al.*, 1992). Most velocity–temperature models that are constrained by petrologic data on crustal and upper mantle composition take these effects into account. For example, Sobolev *et al.* (1996) use regional mantle-derived peridotites from the Massif Central to calculate equilibrium mineralogical composition, density, and Vp and Vs seismic velocities which account for anharmonicity. In particular, with temperature increase (at constant pressure), the solubility of Al in pyroxenes increases leading to a decrease in volume fraction of spinel and garnet in ultramafic rocks. As a result, the overall density and seismic velocities of mantle rocks decrease (Fig. 6.12). To separate temperature and compositional effects of velocity variations, Furlong *et al.* (1995) propose the use of Vp and Vs velocities simultaneously in velocity–temperature conversions.

Mineral reactions, in particular at $P < 3$ GPa, may have a further effect on the temperature dependence of density and seismic velocities. In the deeper mantle, the effects of phase transitions should be included in velocity–temperature inversions. A self-consistent thermodynamic approach to calculate mantle temperatures from absolute P- and S-wave velocities and geochemical constraints has been used for the Kaapvaal craton (Kuskov *et al.*, 2006). The results (see below) show that calculated mantle temperatures depend strongly on bulk composition and on the proportion of stable phases at various depths, so that relatively small compositional differences lead to substantial variations in the inferred upper mantle temperature.

Effect of anelasticity

As a first-order approximation, mantle temperatures can be calculated from seismic velocities assuming linear dependence of velocity on temperature (eq. 3.17). If anelasticity is

Table 5.5 Temperature derivatives of elastic wave velocities for upper mantle olivine		
Q	$\partial \ln Vp/\partial T\,(\times 10^{-4}\ 1/K)$	$\partial \ln Vs/\partial T\,(\times 10^{-4}\ 1/K)$
50	-2.18	-2.32
100	-1.40	-1.54
200	-1.01	-1.15
300	-0.88	-1.02
400	-0.82	-0.96
∞	-0.62	-0.76

(From Karato, 1993).
For finite Q, temperature derivatives are the sum of anharmonic and anelastic effects. Anelastic effects for various Q values are calculated assuming $T = 1600\,K$ and $H^* = 500\,kJ/mol$. For $Q = \infty$, temperature derivatives are caused by anharmonic effects (values from Isaak, 1992).

ignored, the temperature effect on seismic velocities perturbations is underestimated for cold regions and overestimated for hot regions (Fig. 5.17). At seismic frequencies, a -5% anomaly of Vs at 100 km depth can be explained by a ~650 °C temperature anomaly if (eq. 3.17) is used, or by a ~325 °C temperature anomaly if the correction for anelastic effects is introduced (Karato, 1993). The anelasticity correction to the elastic part of seismic velocities can be introduced through seismic attenuation (eq. 3.11) which depends strongly on temperature (eq. 3.8) and the oscillation period (eqs. 3.9–3.10). Laboratory results on seismic attenuation in dry peridotite allows us to estimate Q_p from known geotherms in the vicinity of solidus temperatures. The relationship (3.17) allows introduction of anelastic corrections to Vs as well as Vp velocities. Taking the derivatives of eqs. (3.11, 3.8), one gets temperature derivatives of elastic wave velocities $\partial \ln Vp/\partial T$ and $\partial \ln Vs/\partial T$, which are the sum of anharmonic and anelastic effects. Table 5.5 lists their values for upper mantle olivine. Since for a given temperature, $Q_s < Q_p$, at any given depth $\partial \ln Vs/\partial T$ is significantly larger than $\partial \ln Vp/\partial T$.

Partial melts, fluids, grain size, and anisotropy

After assumptions on composition are made and the corrections due to anelasticity and anharmonicity can be introduced into velocity–temperature inversions, other non-thermal factors should be accounted for. These include the effects of melts, hydrogen, grain size, and LPO anisotropy on seismic velocities. An overview of these effects on elastic moduli is presented in Section 3.1 (see Figs. 3.8; 3.13; 3.14). Without repeating much of the text, the major aspects relating to velocity–temperature conversions are the following.

The amount of melt present in the upper mantle in different tectonic settings is not well constrained and the geometry of melt inclusions (which has a dramatic effect on seismic velocities) is hardly known. Thus, if the melt phase is present in the mantle, its effect is difficult to account for in velocity–temperature inversions. Typical assumptions used in inversions tend to underestimate the melt effect. For example, Sobolev *et al.* (1996) assumed the spherical geometry of melt inclusions, while Godey *et al.* (2004) assume that partial melting occurs locally and has negligible effect on the regional mean values of temperature and compositional variations.

Similarly, water (e.g. supplied by the dehydration of hydrous minerals) has a strong effect on seismic velocities and on attenuation, and an inversion of seismic data to mantle temperatures should take into account separately the effects of water and temperature. Most important are the indirect effects of water on seismic velocities due to enhanced anelastic relaxation and the possible change in preferred orientation of olivine with corresponding changes in seismic anisotropy (see Section 3.6.3). The effect of water is difficult to quantify due to the absence of systematic laboratory results on the effect of hydrogen on seismic attenuation.

Both seismic anisotropy and grain size variations may play an important role in seismic velocity variations, but at present their effects cannot be reliably quantified. For this reason, both effects are commonly ignored in velocity–temperature inversions.

5.3.3 Seismic constraints on temperatures: major results

Mapping temperature variations only

Velocity–temperature inversions have been applied globally and regionally to the calculation of lithospheric temperatures. These results provide useful constraints on lithospheric geotherms, especially for tectonically active regions, where any modeling of thermal structure of the mantle is highly uncertain (e.g. van Wijk *et al.*, 2001). A number of thermal models based on the conversion of seismic velocities to upper mantle temperatures are available for most of the continents. Such studies are limited to few research groups, and a summary of major assumptions and limitations associated with each of the approaches is provided in Table 5.4. As discussed earlier, the uncertainties of inversions include the uncertainties inherited from the seismic models used in the calculations and the assumptions of the velocity–temperature inversion itself.

Comparison of the results of velocity–temperature inversions with continental geotherms inferred from surface heat flow and xenolith data indicate a good agreement between the different approaches (Figs. 5.21, 5.22, 4.33). Röhm *et al.* (2000) calculated temperature variations in the upper mantle using the S-wave tomographic model S16RLBM (Woodhouse and Trampert, 1995). Since the S16RLBM model is constrained with respect to the PREM reference model, the velocity discontinuity at 220 km depth requires special handling. In the absence of a clear thermodynamic understanding of its origin (or its very existence under continents), its parameterization through a discontinuous variation of composition or rheology is not possible. To overcome this problem, the authors assume that all Archean cratons are characterized by a $42\,\text{mW/m}^2$ reference conductive geotherm. Under this assumption and assuming a cold mantle adiabat of $1200\,^\circ\text{C}$, it is possible to eliminate a $220\,\text{km}$ discontinuity for the velocity profile calculated from a $42\,\text{mW/m}^2$ conductive geotherm. Velocity–depth profiles for other tectonic regions are scaled by perturbation from the Archean velocity profiles and are converted to temperature perturbations from a $42\,\text{mW/m}^2$ geotherm. In the absence of independent data other than the

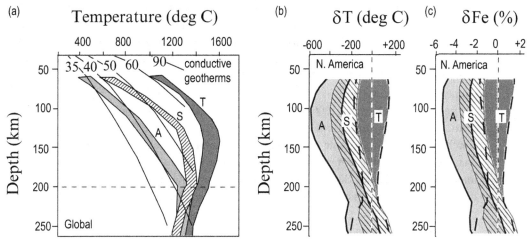

A = Archean craton; S = stable platform; T = tectonically active continent

Fig. 5.21 Vertical profiles for three different tectonic provinces of: (a) temperature variations assuming mg# = 0.89 and 0.91 (see details in Table 5.4), (b) temperature anomalies, and (c) iron anomalies with respect to XFe = 11% (Mg# = 0.89). All profiles are calculated from seismic velocities assuming that velocity anomalies are caused only by temperature and iron-content variations. (Sources: (a) Röhm *et al.*, 2000; (b–c) Godey *et al.*, 2004.)

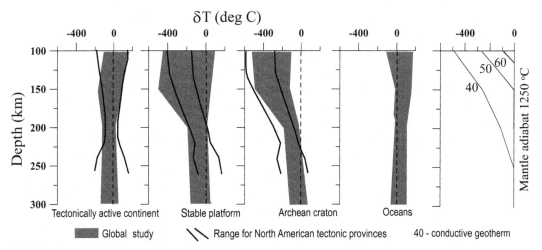

Fig. 5.22 Vertical profiles of temperature variations for different tectonic provinces based on a joint inversion of the S-wave tomographic model S16RLBM (Woodhouse and Trampert, 1995) and the geoid model EGM96 (Lemoine *et al.*, 1998) (gray shading). Reference temperature is 1250 °C (the results of Deschamps *et al.*, 2002). The results of Godey *et al.* (2004) for North America are shown by black lines (see Fig. 5.21). For a comparison, the right plot shows the difference between reference conductive geotherms and a 1250 °C mantle adiabat.

tomographic model, the study does not allow for separation of compositional and thermal effects, but allows for calculation of typical geotherms for three continental tectonic provinces (defined using the 3SMAC regionalization of Nataf and Richard, 1996). The effect of iron-content variations on inferred mantle temperatures is calculated for two compositions, with Mg# = 89 and 91, which provides the bounds on estimated mantle temperatures (Fig. 5.21a).

Priestly and McKenzie (2006) use oceanic geotherms derived from the plate model and cratonic geotherms constrained by xenolith P–T arrays from the Siberia, Slave, and Kola-Karelian cratons to calibrate shear-velocity perturbations (with respect to PREM) and convert them to mantle temperatures. Because of problems with the crustal correction, the tomographic model for Siberia may not be well constrained (the difference in the Moho depth in regional seismic data and in global crustal models is 5–10 km). Similarly to the study by Röhm et al. (2000), the approach relies heavily on the notion that xenolith geotherms are representative of the steady-state thermal regime of ancient continental nuclei. However, in contrast to the calculations of Röhm et al. (2000), no attempts are made to compensate for the 220 km velocity discontinuity when inverting relative velocity anomalies to mantle temperatures in the cratonic regions (see Figs. 3.85 and 3.86 and discussion in Section 3.6). The model also does not account for possible lateral variations in composition within the continental lithospheric mantle, which is assumed to be depleted cratonic and produced by a 20–25% melt removal with the residue composition characterized by Mg# ~92–92.5. For the oceans, temperature variations at 150 km depth beneath the entire Pacific region are within 20 °C of 1400 °C when anelasticity is accounted for. Analysis of seismic velocities for the East Pacific Rise indicates that the observed shear-velocity structure does not require the presence of melt and a 2.3% variation in shear velocity at 60 km depth at the eastern and western flanks of the Rise can be attributed to a 23 °C temperature difference at near-solidus temperatures when velocity–temperature dependence becomes strongly non-linear and anelastic effects are accounted for. By comparison, the elastic estimates of Dunn and Forsyth (2003) require a temperature difference of 350 °C to explain differences in velocity beneath the two flanks of the East Pacific Rise.

A velocity–temperature inversion for North America was performed by Goes and van der Lee (2002), following the approach used by Furlong et al. (1995) and Goes et al. (2000) for Europe. It includes independent inversion for V_p and V_s velocities with account taken of anelastic effects and it assumes three basic compositions of the upper mantle. As such, the approach does not allow for vertical mapping of compositional variations in the upper mantle. The results suggest a strong temperature anomaly in the upper 150 km of the mantle beneath the stable eastern USA with mantle temperature at 100 km depth of ~700–800 °C ($\delta T \sim 600$ °C) as constrained by the surface-wave velocity model. Beneath the active western USA, mantle temperatures at 100 km depth are close to the mantle adiabat. Using a similar approach and independently inverting two shear-wave velocity models for mantle temperatures, Goes et al. (2005) estimated geotherms for Australia. Again, in the absence of independent data or xenolith constraints on compositional heterogeneity of the mantle, seismic velocity perturbations are interpreted in terms of purely thermal anomalies. Importantly, seismic tomography models used in the velocity–temperature inversion have reduced resolution for the Archean cratons in Western Australia (Fig. 3.83). The results indicate that mantle temperatures beneath the cratonic part of Australia (both Archean and Proterozoic) follow a 40–45 mW/m^2 conductive geotherm with mantle temperatures along a 50–60 mW/m^2 conductive geotherm beneath the Phanerozoic regions.

Mapping thermal and compositional variations

Several studies aim to separate the effects of thermal and compositional variations in the cratonic upper mantle on velocity perturbations. These studies, in addition to seismic models, utilize independent data sets such as gravity data or regional petrologic data on the composition of the lithospheric mantle.

Deschamps *et al.* (2002) perform a joint inversion of the S-wave tomographic model S16RLBM (Woodhouse and Trampert, 1995) and the geoid model EGM96 (Lemoine *et al.*, 1998). Use of two independent data sets allows for separation of thermal and compositional anomalies in the upper 300 km; the latter are ascribed solely to iron-content variations (Figs. 3.19 and Fig. 5.22). The tectonic provinces are also defined using the 3SMAC regionalization. The results indicate that strong temperature anomalies (with respect to convective mantle) in stable continental regions are restricted to the upper 150 km with a fast decrease in amplitude between 150 km and 200 km depth. According to this study, below a 200 km depth, the temperature structure of the upper mantle beneath the continents and the oceans becomes indistinguishable. This specific depth is "inherited" from the velocity discontinuity at 220 km in the PREM model: a significant deviation of a 40 mW/m^2 reference conductive geotherm (expected for many cratonic regions) from the mantle adiabat extends deeper than 200 km depth (Fig. 5.22).

A similar analysis for the North American continent has been performed by Godey *et al.* (2004). They performed a joint inversion of a regional surface-wave tomographic model and geoid to separate thermal and compositional effects from shear velocities and density-to-velocity scaling factor. Compositional anomalies are interpreted in terms of iron-content variations, but the effects of other compositional parameters (garnet and olivine content) are examined as well. Since garnet contains iron, the global volume fraction of iron is kept constant, so that variations in garnet content effectively reflect variations in the aluminium content. The results indicate that a 2% increase in shear velocity can be caused either by a 120 °C temperature decrease, or a 7.5% iron depletion, or a 15% aluminium depletion. Variations in the fraction of olivine have a much weaker effect on shear velocities: a 50% depletion in olivine produces a 1% velocity perturbation (Fig. 6.14b). Depth profiles of temperature and iron-depletion variations in the upper mantle of North America are in overall agreement with the global results of Deschamps *et al.* (2002). The principal difference is that in the cratonic part of North America both temperature and iron-depletion anomalies persist down to at least 250 km depth (Fig. 5.21bc). Below this depth the model lacks resolution. Iron-content variations in the lithospheric mantle of the North American craton are in striking agreement with global xenolith data, which indicate a ~4% iron-depletion anomaly above 150–200 km depth reduced to ~2% at 200–240 km depth (Fig. 3.19). It is likely

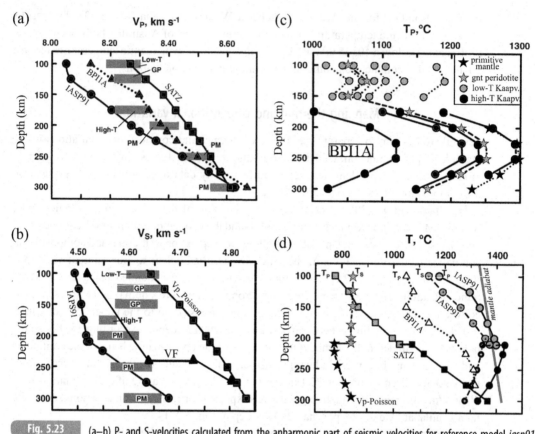

Fig. 5.23 (a–b) P- and S-velocities calculated from the anharmonic part of seismic velocities for reference model *iasp91* and regional seismic models for southern Africa: VF (Vinnik and Farra, 2002), BPI1A (Simon *et al.*, 2002), SATZ and Vp_Poisson (Zhao *et al.*, 1999). Gray boxes – Vp and Vs for low-*T* and high-*T* peridotites from the Kaapvaal craton (at 900 °C and 100 km depth, and at 1300 °C and 175 km depth, respectively) and Vp and Vs for compositions of average garnet peridotite (GP, depth 100–175 km) and primitive mantle (PM, depth > 175 km). (c) Effect of specific composition of garnet peridotite xenoliths from the Kaapvaal craton on the *T*p temperatures evaluated from the BPI1A model. For both low-*T* and high-*T* xenoliths, inferred temperatures vary by ~200 °C. Symbols refer to different kimberlite pipes. (d) *T*P and *T*S inferred from different P- and S-velocity models (BPI1A, IASP91, SATZ and Vp_Poisson) for three compositions: depleted garnet peridotite at depths of 100–210 km (GP, gray symbols), primitive mantle composition at depths of 210–300 km (PM, black symbols), and variable bulk composition with a gradual increase in fertility at depths between 200 and 275 km (BPI1A model, open triangles). Gray line – 1300 °C mantle adiabat. (Source: Kuskov *et al.*, 2006).

that the amplitude of iron depletion in the platform areas is overestimated compared with xenolith data.

Petrologic data from numerous kimberlite pipes in the Kaapvaal craton, South Africa, together with regional high-resolution seismic models were used to calculate mantle temperatures and to estimate the effect of the specific composition of garnet peridotite xenoliths on the temperatures evaluated from velocity models (Kuskov *et al.*, 2006).

Mantle composition was constrained by xenolith data and included petrologic data on garnet-bearing lherzolite xenoliths, average composition of garnet peridotites, and composition of primitive mantle. Phase equilibrium relations were computed by minimization of the total Gibbs free energy combined with a Mie–Grüneisen equation of state. Modeling included forward calculation of phase equilibria, seismic velocities and density, and inverse calculation of temperature with account for anharmonicity, anelasticity, and the effects of mineral reactions, which included calculations of phase proportions and chemical compositions of coexisting phases. The results indicate that the inferred temperatures can vary by $\sim$200 °C for both low-T and high-T peridotites (Fig. 5.23c). Furthermore, for models with a uniform bulk composition (irrespective of the composition model), mantle temperatures inferred from regional seismic models display a non-physical behavior with a negative temperature gradient at depths below $\sim$200–220 km. Neither inflexions of geotherms nor an anticorrelated behavior for T_P and T_S (mantle temperatures inferred from Vp and Vs, respectively) can be explained by a sharp change in composition from depleted garnet peridotite to fertile pyrolitic mantle. Temperature inflexions can be evaded if the mantle beneath the Kaapvaal craton is chemically stratified: the upper layer at 100–175 km depth consists of depleted garnet peridotite, while in the lower layer (at 200–275 km) mantle fertility increases continuously with a substantial gradual increase in FeO, Al_2O_3, and CaO content, and the lower layer does not differ from normal fertile mantle below 275 km depth.

5.4 Magnetic methods to determine lithospheric geotherms

5.4.1 Major principles

Magnetization of crustal and upper mantle rocks plays a critical role in paleomagnetic reconstructions, analysis of ocean floor ages, and ocean spreading rates. It is worth remembering that the ocean floor is dated by the magnetic method, which played a critical role towards widespread acceptance of plate tectonics (Heirtzler et al., 1968). Additionally, Curie temperature (a characteristic property of a ferromagnetic or piezoelectric material which is the reversible temperature above which the material becomes paramagnetic) may provide independent constraints on lithospheric temperatures. Only the latter aspect is discussed here.

The magnetic field of the Earth consists of contributions from the core and from the lithosphere (for review of terrestrial magnetism see Langel and Hinze, 1998; Purucker and Whaler, 2007). The magnetic field of the crust makes < 10% of the core magnetic field, but dominates at short (< 2600 km) wavelengths. At longer wavelengths, crustal signal may be distorted by overlap with the long-wavelength magnetic field from the core (Meyer et al., 1985).

Crustal magnetization consists of two components: remanent and induced magnetism. The Königsberger ratio Q is the ratio of remanent to induced magnetization in a rock; it ranges between 1 and 160 in oceanic basalts. Induced magnetism is caused by the present

Earth's core magnetic field and depends on the field strength (additionally affected by the interaction between the core field and the solar wind), the magnetic susceptibility of rocks, and the thickness of the magnetic layer ("magnetic crust"). A global model of induced and remanent lithosphere magnetization based on satellite data shows (i) long-wavelength magnetizations dominated by the continent–ocean contrast and (ii) short-wavelength enhanced magnetizations dominated by seafloor spreading. The remnant magnetization spectrum of the oceans has been modeled based on both non-satellite and satellite magnetic data (Cohen and Achache, 1994; Dyment and Arkani-Hamed, 1998). However, for the continents neither global nor regional remanent magnetization models exist because of incomplete knowledge on continental remanent magnetization. In practical modeling of continental magnetism, either the oceanic remanent model is used or it is assumed that induced magnetism dominates the continental magnetic field (Fox Maule et al., 2005).

Practical applications of magnetic methods are based on spectrum analysis of the magnetic field using forward and inverse methods. Forward methods are based on calculations of magnetic anomalies for bodies modeled as collections of simple geometric bodies (e.g. Bhattacharyya, 1966; Dyment and Arkani-Hamed, 1998). Inverse methods aim to estimate the depth to the magnetic bodies, their dimensions, and magnetization contrast. The depth to the magnetic source is often calculated in the spectral domain (e.g. Spector and Grant, 1970). The problem is extremely difficult since contributions from the bottom of the magnetic sources are, at all wavelengths, dominated by contributions from the top, and the top of the source is, in general, also unknown.

Until recently, studies of the lithospheric magnetic signal were based on regional aeromagnetic or global magnetic data such as provided by the Magsat satellite mission (launched in 1979). The launch of the Danish Ørsted high-precision geomagnetic field satellite in 1999 and the German CHAMP satellite in 2000 marked a breakthrough in studies of the magnetic field of the lithosphere (Olsen et al., 2000; Neubert et al., 2001; Mandea and Purucker, 2005). In satellite-based magnetic studies, with satellites typically orbiting at altitudes of a few hundred kilometers, the lithospheric anomaly field at (or above) the Earth's surface, depends on the magnetization of only a small lithospheric sector located directly beneath the observation point. Model resolution is, however, smoothed to at least a few hundred kilometers. In contrast, regional models based on aeromagnetic data provide high-resolution images of small-size lithospheric bodies (such as kimberlite pipes, impact structures (e.g. the Chicxulub), plutons, and ophiolites) which have a magnetic contrast with ambient rocks.

5.4.2 Calculation of lithospheric geotherms from magnetic data

The procedure for calculating lithospheric geotherms from satellite magnetic data typically includes the following steps and assumptions (e.g. Fox Maule et al., 2005):

- the magnetic field is represented in spherical harmonics (up to degree and order 90); the strong time-variable magnetic field in polar regions is controlled by combining data with spherical harmonic representations of the magnetic field with data from magnetically quiet times (as a result, satellite models of crustal magnetic fields are constrained);

- the core magnetic field is removed by high-pass (above degree 14) data filtering;
- the induced part of the magnetic signal is separated; in the oceans this is done by removing the remanent part (e.g. following Dyment and Arkani-Hamed, 1998); on the continents it is assumed that induced magnetization dominates (the assumption fails over the Bangui and Kursk magnetic anomalies);
- crustal magnetization is modeled by dipoles evenly distributed at the surface and the direction of the dipole is assumed to be given by the present core field;
- average crustal magnetization is proportional to the average magnetic susceptibility and the thickness of the magnetic layer;
- lateral crustal thickness variations dominate over magnetic susceptibility variations (which are assumed to be 0.035 and 0.0040 for the continental and oceanic crust, respectively);
- a crustal thickness model is used to account for the long-wavelength field component.

This approach allows modeling of the depth to the bottom of the magnetic layer, since magnetic methods for determination of the depth to the Curie isotherm build on the idea that average magnetization tends to zero below the Curie depth point. It is further assumed that:

- this depth corresponds to ~550–580 °C (the Curie temperature of magnetite);
- under certain assumptions (such as a steady-state thermal regime and thermal properties of the lithosphere) the depth to the Curie isotherm can be recalculated to surface heat flow values and lithospheric temperatures.

5.4.3 Curie temperature of mantle minerals

Curie temperature (Tc) is an intrinsic property of minerals which depends on their chemical composition and crystal structure; it determines the temperature (upon heating) at which (remanent or induced) magnetization of minerals disappears (ferromagnetic or ferrimagnetic minerals become paramagnetic). Lithospheric rocks consist of several mineral phases, and frequently their magnetization is characterized by a range of Curie temperatures (Fig. 5.24). The situation is similar to the melting of a multicomponent system, where solidus and liquidus temperatures provide the bounds on rock melting temperatures.

Magnetization of crustal and upper mantle rocks is controlled by ferromagnetic minerals of the iron–titanium oxide group and, to a much smaller extent, of the iron sulfide group, while iron hydroxides are also important in sedimentary rocks. Magnetite (Fe_3O_4) is the main ferromagnetic mineral in the lithosphere with Curie temperature $Tc \sim 575$–585 °C (see overview by Hurt et al., 1995) (Table 5.6; Fig. 5.25). The only iron oxide with ferromagnetic properties, maghemite (γ-Fe_2O_3) which is formed by the low-temperature oxidation of magnetite, is thermodynamically metastable and at $T > 250$ °C it converts to paramagnetic hematite (De Boer and Dekkers, 1996).

In general, the composition of the primary iron–titanium oxides varies with SiO_2 content of the rock. For this reason, the Curie temperature in titanomagnetites, which are the main magnetic minerals of igneous rocks, varies from less than 244 °C in basic rocks (gabbro,

Table 5.6 Curie temperature of some minerals		
Mineral	Composition	Tc (°C)
Iron	Fe	770
Magnetite	Fe_3O_4	575–585
Titanomagnetites	$Fe_{3-x}Ti_xO_4$	100–550
Magnesioferrite	$MgFe_2O_4$	440

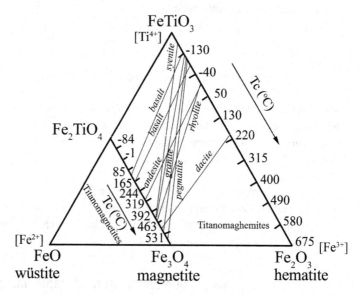

Fig. 5.24   Composition and Curie temperatures of primary Fe–Ti oxides from different igneous rocks, shown in the ternary system FeO (wüstite) – Fe_2O_3 (hematite) – TiO_2. Thin lines connect coexisting members of the (Fe_{2-y} Ti_yO_3) and (Fe_{3-x} Ti_xO_4) series (modified after Buddington and Lindsley, 1964).

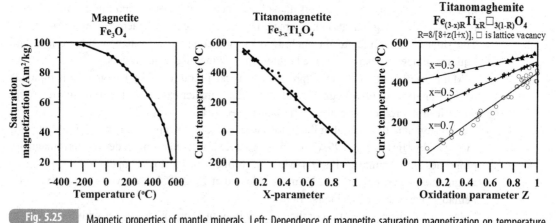

Fig. 5.25   Magnetic properties of mantle minerals. Left: Dependence of magnetite saturation magnetization on temperature. Middle and right: variations of Curie temperature in titanomagnetite and titanomaghemite as a function of composition (x-parameter) and oxidation (z-parameter) which is defined as the fraction of the initial Fe^{2+} ions in titanomagnetite converted to Fe^{3+} ions. (Data source: Hunt et al., 1995.)

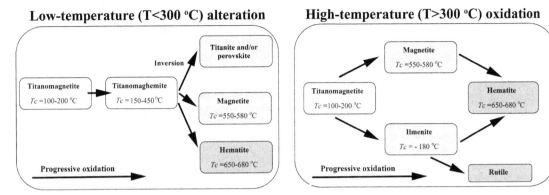

Low-temperature (T<300 °C) alteration

High-temperature (T>300 °C) oxidation

The effect of low-temperature alteration and high-temperature oxidation on the primary Fe–Ti oxides in basalt.

basalt) to 165–463 °C in intermediate rocks (dacite) and to 463–580 °C in acid rocks (e.g. granite). The Curie point of titanomagnetitesis is strongly dependent on the amount and type of titanium and iron oxides, and Tc varies from 100 to 550 °C (Nishitani and Kono, 1983) (Fig. 5.25). The ferric–ferrous iron transition is also important for preservation of magnetic properties within the lithosphere. It is characterized by oxidation parameter defined as the fraction of the initial Fe^{2+} ions in titanomagnetite converted to Fe^{3+} ions. The temperature range for the ferric–ferrous iron transition is 200–400 °C. This is sometimes used to explain the discrepancy between the depth to the Curie point determined by geothermal and magnetic methods in stable continental regions.

The effect of oxidation on the primary Fe–Ti oxides in basalt is illustrated by Fig. 5.26. In particular, low-temperature oxidation is typical for alteration in oceanic basalts (Fig. 5.27). Magnetic properties of rocks are strongly affected by hydrothermal circulation and meta-morphism. The former replaces magnetite by paramagnetic rocks (except for serpentinization of olivine-rich ultramafic rocks); the latter may have different effects depending on the P–T–t path of metamorphism and the composition of the protolith. Within the mantle, a significant magnetization can be associated with subducting slabs at temperatures below the Curie point (due to subsequent phase transitions metabasalt->eclogite->serpentine). Metallic alloys found in upper mantle xenoliths can produce significant magnetic anomalies down to depths of ~100 km (Purucker and Whaler, 2007). Because of the strong effect of chemical alteration on rock magnetization, primary and secondary Curie temperatures are distinguished. The primary Curie temperature characterizes the original composition of the primary magnetic minerals, while the secondary Curie temperature characterizes alteration of the original magnetic components of a rock which was subject to oxidation, reduction, or phase inversion. Comparison of the measured Curie temperature in a rock with its theoretical primary Curie temperature (e.g. Fig. 5.24) provides a measure of the degree of rock alteration.

5.4.4 Depth to the Curie point

Strictly speaking, it is unclear which Curie temperature, for magnetite or for titanomagnetite, should be used in interpretations of magnetic data since the concentration of Ti-oxides at

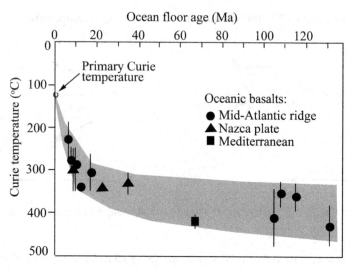

Fig. 5.27 The age dependence of Curie temperature in oceanic basalts based on data from DSDP drill holes in different locations. Error bars – range of measured Curie temperatures (modified after Petersen and Bleil, 1982). The observed pattern is unrelated to ocean plate cooling, but reflects the effect of low-temperature oxidation on Curie temperature of titanomagnetites which were separated from oceanic basalts.

different lithospheric depths is unknown. Commonly, the Tc of magnetite is used to determine the depth to Curie point, for the reasons that:

(i)   magnetite is the most abundant magnetic mineral in the crust,
(ii) magnetite and its compounds have the highest magnetization;
(iii) the Curie point of magnetite is the highest among the abundant magnetic minerals. This implies that the thus defined depth of the Curie point is the maximal possible depth for minerals to preserve their magnetization.

One should keep in mind that, due to compositional heterogeneity of the lithosphere, lateral and vertical variations in mineral concentrations should be expected. Chemical alterations, such as oxidation, further affect Curie temperature. As a result, the Curie depth determined for different regions may relate to different Curie temperatures and hamper their comparison.

Satellite magnetic data combined with the 3SMAC crustal model (Nataf and Ricard, 1996) were used to constrain the thermal regime of the lithosphere of Antarctica, where heat flow measurements are absent (Fox Maule et al., 2005). The first-order features of the model indicate: (i) the magnetic layer is thicker in East Antarctica than in West Antarctica; (ii) the highest lithospheric temperatures are calculated for regions of current volcanism where basal melting may occur; (iii) surface heat flow displays large lateral variations, 40–185 mW/m^2 (the model error is 20–25 mW/m^2).

A map of the depth to the 550 °C isotherm for the continents provides a geothermal estimate of the depth to the Curie point of magnetite (Fig. 4.44b). In Europe, magnetic data, in accordance with thermal data, show that craton-to-non-craton transition (the

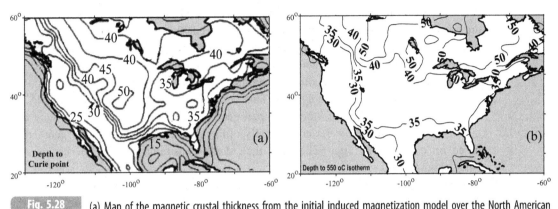

Fig. 5.28 (a) Map of the magnetic crustal thickness from the initial induced magnetization model over the North American region (modified from Purucker *et al.*, 2002); (b) depth to 550 °C isotherm constrained by geothermal data (Artemieva, 2006; www.lithosphere.info). The depth uncertainties of both maps are about 5 km.

Trans-European Suture Zone, TESZ, in central Europe) is a first-order feature in the lithospheric structure. In the Variscan Europe to the west of the TESZ the depth to the bottom of the magnetic layer is 29–33 km (Chiozzi *et al.*, 2005) and corresponds approximately to the seismic Moho. Similarly, in North America the thickness of the magnetic layer and the depth to 550 °C isotherm are >40 km in the Canadian Shield, become less than 30–35 km across the southern edge of the North American craton, and have values of 10–20 km in tectonically active regions of western USA (Fig. 5.28). The results for North America (Hinze and Zietz, 1985; Purucker *et al.*, 2002) illustrate that although thermal and magnetic estimates of the depth to 550–580 °C isotherm generally agree, there are significant short-wavelength differences that result from assumptions behind thermal and magnetic modeling.

5.5 Summary of non-thermal studies of lithosphere thermal state

Xenolith data

The advantage of xenolith data is that it provides the only available direct sampling of the deep sections of the lithosphere. The disadvantage is that the areal coverage by xenoliths is strongly non-uniform, and it is unclear how representative xenolith-based constraints are on the *in situ* unperturbed thermal state of the lithosphere.

There is a significant difference in xenolith geotherms constrained by different geothermobarometers (Fig. 5.8). Since most of them are calibrated at temperatures 900–1400 °C and pressures 2–6 GPa, the best consistency of xenolith P–T arrays is achieved at this P–T range. When different geothermobarometers are used on samples derived from the lithospheric mantle in the same location, pressure estimates may differ by 1.0–1.5 GPa and temperature estimates by >200 °C. The accuracy of P–T determinations on xenolith samples cannot be assessed since many thermobarometers are sensitive to bulk composition. There are no reliable geobarometers for pressures within the spinel stability field. Because of significant differences in temperature gradient, some geothermobarometers may produce a kink on the geotherm, while others (for the same data), do not. Geothermobarometry is based on the critical assumption of mineralogical equilibria in the system. This condition is not satisfied, for example, in xenoliths from some parts of Siberia. Garnet geotherms, widely used by Griffin and co-authors in their studies, have a conceptually different meaning than xenolith geotherms based on conventional approaches and their depth constraints should not necessarily agree (Fig. 5.11).

Xenoliths provide information on the thickness of TBL, CBL, and RBL (Fig. 5.4). The base of RBL can be constrained by variations in xenolith texture (the transition from coarse to sheared), CBL – by variations in major element composition (the transition from depleted to fertile), and TBL – by P–T arrays. Commonly the bases of all three boundary layers as constrained by xenolith data are at about the same depth. However, this depth may differ significantly from the lithospheric base determined by various geophysical methods. The kink commonly observed on xenolith geotherms should not necessarily correspond to the base of TBL; it can be an artifact of geo-thermobarometry, reflect thermal disequilibrium related to magmatism, or be a true steady-state feature produced by a sharp contrast in lithospheric thickness at the edge of a craton (Figs. 5.9, 5.10).

Xenolith P–T arrays from the Archean cratons apparently indicate that while some cratons have a 200–220 km thick TBL, others may extend down to at least ~300 km. Since the absence of evidence for cannot be interpreted as evidence against, the absence of known xenoliths derived from depths greater than 250 km within the lithospheric mantle does not prove that the cratonic lithosphere cannot be thicker than 220–250 km. Their absence may be related to the conditions of kimberlite magma generation and ascent.

Lithosphere temperatures derived from seismic data

Velocity–temperature conversions provide important constraints on lithospheric temperatures in regions with no (or low quality) borehole heat flow data, or regions where thermal modeling is subject to large uncertainty.

Velocity–temperature conversions are subject to a set of assumptions on the physical origin of seismic velocity perturbations. While temperature variations are assumed to be the major factor producing seismic velocity perturbations, non-thermal effects on seismic velocity variations such as anisotropy, the presence of melt and water, compositional and grain size variations may be equally important. However, most non-thermal effects are difficult to quantify and, except for the presence of melt, most of these factors are ignored in velocity–temperature conversions.

The effect of water on seismic velocity reduction can be dramatic, both direct and indirect through the enhancement of anelastic relaxation. This means that an analysis of seismic velocities and seismic attenuation in terms of mantle temperatures should take into account the effect of water, separately from the effect of temperature. However, currently, systematic laboratory results on the effect of hydrogen on seismic attenuation are unavailable, and there is no proper experimental basis for such corrections.

Compositional effects may be significant in producing seismic velocity anomalies, but velocity–temperature conversions cannot separate temperature and compositional effects unless an independent set of data (such as gravity) is used. Alternatively, assumptions on variations in the lithosphere composition are used to invert seismic data to temperatures.

Limitations associated with seismic models (such as model resolution in terms of vertical, lateral, and amplitude resolution) propagate into additional uncertainties in velocity–temperature conversions.

As a result, the uncertainty of temperature constraints from seismic data cannot be quantified and is not known. Additionally, horizontal heat transfer prevents the existence of sharp temperature contrasts in the lithosphere, and strong (several hundred degrees Celsius) short-wavelength lateral temperature contrasts are artifacts of inversion, unless they are caused by recent geodynamic processes.

Geotherms constrained by Curie temperature

Similarly to velocity–temperature conversions, magnetic data allow constraining of lithospheric geotherms in regions with no heat flow data (e.g. Antarctica). The depth where lithosphere magnetization tends to zero is interpreted as the depth where the Curie point of magnetite (550–580 °C) is reached. This depth–temperature pair constrains the lithospheric geotherm.

Lithospheric rocks have several Curie points associated with different minerals, and it is unclear which Curie point should be used in interpretations. Mineral alterations, in particularly oxidation, have a strong effect on Curie temperature. Therefore, the Curie depth determined for different regions may relate to different Curie temperatures. As a consequence, lithospheric geotherms may also be significantly different.

For these reasons, it is difficult to put uncertainty limits on lithosphere geotherms constrained by magnetic data. Additionally, separation of the induced part of the magnetic signal from the remnant part is subject to a significant uncertainty on the continents. Despite these limitations, a comparison with thermal models shows general agreement between the depth to Curie point and the depth to a 550 °C isotherm, but the details are different.

6   CBL and lithospheric density from petrologic and geophysical data

Mantle xenoliths provide invaluable information on lithosphere composition and thus on the structure and thickness of compositional boundary layer (CBL). This chapter focuses on the composition of the lithospheric mantle constrained by petrologic studies of mantle xenoliths (the major reference books are Nixon (1987) and Pearson *et al.* (2003)), lithosphere density from petrologic and geophysical data, compositional variations in the continental lithospheric mantle constrained by various geophysical methods, the thickness of the chemical boundary layer, and its correlation with thermal boundary layer.

6.1 Tectosphere and CBL

6.1.1 The tectosphere hypothesis

A significant difference in seismic velocity structure of the subcontinental and suboceanic mantle had been recognized as early as the 1960s from the analysis of dispersion curves for Love and Rayleigh waves which revealed differences in phase velocities out to periods of 300 s (e.g. Dorman *et al.*, 1960; Toksöz and Anderson, 1966; Kanamori, 1970; Dziewonski, 1971a). The early observations were used by MacDonald (1963) to argue against plate tectonics on the basis of the hypothesis that thermal and compositional differences between continents and oceans exist down to a ~500 km depth. In support of plate tectonics, Dziewonski (1971b) demonstrated that seismic dispersion data can fit if a significant velocity heterogeneity is restricted to the upper 200 km. The latter conclusion was challenged by Jordan (1975a) who compared (i) vertical travel-time differences for shear waves (ScS and multiple ScS phases) propagating through the upper 700 km of the oceanic and continental mantle and (ii) shear-wave travel times with free oscillation data (which provide estimates of the spherically averaged velocities, dominated by 2/3 by the oceanic structures). The ScS and multiple ScS phases were chosen because ScS travel-time curves are insensitive to radial variations in velocity and because travel time for the multiples is insensitive to the velocity structure in the vicinity of the source and receiver. The study showed that travel times for the continental mantle are:

- ~5 s less than for the oceanic islands of Hawaii, Samoa, Raratonga, and Bermuda (Fig. 6.1b) and 3–4 s less than for other oceanic regions;
- ~2–3 s less than the one-way vertical Vs travel times through the upper 700 km computed from free oscillation modes.

On the basis of these observations, Jordan (1975a) proposed that significant differences in both composition and temperatures should exist between oceans and continents down to at

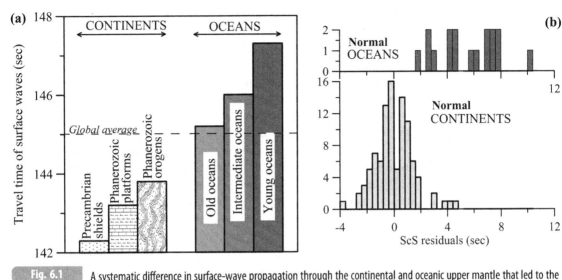

Fig. 6.1 A systematic difference in surface-wave propagation through the continental and oceanic upper mantle that led to the tectosphere hypothesis. (a) Travel time required for a shear wave to travel vertically from 700 km depth to the surface for three continental and three oceanic provinces; (b) histogram of ScS residuals for normal continents and oceans calculated by subtracting the average normal continental residual for each analyzed earthquake (Sources: Jordan, 1975b, 1979).

least a 400–500 km depth and perhaps as deep as 700 km. This proposition was further developed into the tectosphere hypothesis (Jordan, 1975b, 1978, 1979, 1981, 1988), according to which the tectosphere is:

(i)   the outer, at least 400 km-thick, layer,
(ii) it "translates coherently in the course of horizontal plate motions",
(iii) it is dominated by conductive heat transfer (with super-adiabatic thermal gradients),
(iv) and it is compositionally inhomogeneous so that "regions of lower potential temperature consist of material with intrinsically lower densities".

The depth extent of the tectosphere (the first assumption) is still a subject for debate. As written by Jordan (1975b, p. 1),

> It is apparent that the current thinking on the nature of the subcontinental mantle is dominated by the concept of thin lithospheric plates. This consensus maintains that continental plates are on the order of 200 km thick or less and that below this depth the subcontinental and suboceanic mantles are similar in composition and state. It has been argued that both the data relating to the temperature distribution and the seismic data are consistent with this model.

Thirty five years later, the concept of relatively thin (200–250 km thick) continental lithospheric plates is still alive, finding support in *some* seismic tomography and thermal models, and interpretations of mantle xenolith data. It coexists with *other concepts* based on seismic and thermal models which argue for 300–400 km-thick continental plates (see Chapters 3–5).

The controversy regarding the tectosphere thickness partly stems from the fact that the tectosphere hypothesis combines the concepts of MBL (mechanical boundary layer, the second assumption), TBL (thermal boundary layer, the third assumption), and CBL (chemical boundary layer, the fourth assumption). However, the bases of these three boundary layers should not necessarily coincide (as is discussed further in this chapter) and rarely do, and thus the tectosphere should rather be considered as a conceptual grouping of several interrelated boundary layers and not a single outer shell with a uniquely defined thickness.

The last of the tectosphere assumptions, which states the compositional heterogeneity of the lithospheric mantle, is based on the observation that stable continental regions do not produce any mass excess observed in the gravity field. This observation forms the basis for the "*isopycnic*" (i.e. equal-density) hypothesis (Jordan, 1978). His calculation of mineral composition, density, and seismic velocities of garnet lherzolites from whole rock compositions (without distinguishing low-T and high-T cratonic peridotites) demonstrated that density and seismic velocities beneath the oceans and the cratons are significantly different and are well correlated with the Fe/[Fe+Mg] ratio which reflects mantle fertility. Observation of the significant difference in the properties of oceanic and cratonic upper mantle provides support to the tectosphere hypothesis on the existence of the continental CBL where

> the large sub-lithospheric temperature gradients associated with deep continental root zones are dynamically stabilized by compositional gradients (Jordan, 1979).

Following the pioneering ideas of T. Jordan, numerous petrologic studies conducted since the early 1980s confirmed the existence of a fundamental difference between the upper mantle of continents and oceans, with a special volume devoted to the topic (Menzies and Cox, 1988). This was followed by many geophysical studies which examined the role of the continental CBL in maintaining stability and long-term survival of the cratonic lithosphere, its role in large-scale geodynamic processes, and its manifestation in different geophysical observables. This chapter focuses on the discussion of the last of the tectosphere assumptions (subcontinental CBL) and its petrologic, geophysical, and geodynamic significance.

6.1.2 Mantle melting and composition of the lithospheric mantle

Decompressional melting and MORBs

Compositional distinction between continental and oceanic upper mantle is produced as a result of mantle melting: the chemical boundary layer, CBL, is the residue left after melt extraction from the mantle. Experimental petrologic studies indicate that the major element composition of the residue depends on:

(i) the initial mantle composition,
(ii) the composition of the melt,
(iii) the amount of the melt extracted,
(iv) initial and final pressures of decompression melting.

Decompression-induced partial melting is the process thought to be responsible for generation of mid-ocean ridge basalts (MORB) and thus of the oceanic crust (Figs. 6.2, 9.10). According to the model (McKenzie and Bickle, 1988), solid material moves upwards

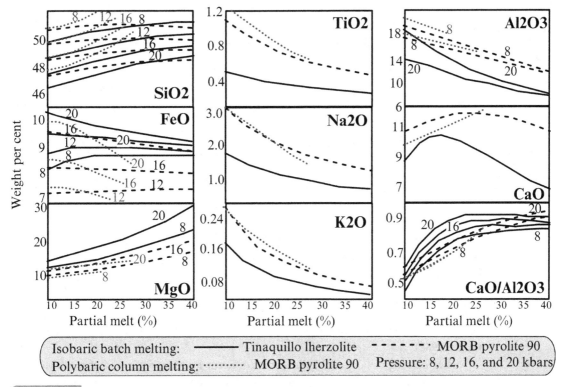

Isobaric batch melting: ——— Tinaquillo lherzolite - - - - - - MORB pyrolite 90
Polybaric column melting: ·········· MORB pyrolite 90   Pressure: 8, 12, 16, and 20 kbars

Fig. 6.2 Melt compositions for depleted lherzolite and MORB pyrolite for isobaric batch melting (when melt does not separate from the matrix) and polybaric column melting by which mid-ocean ridge basalts (MORB) are produced. The plots show the behavior of each oxide component in the melts as a function of extent of partial melting and pressure (labels on the curves); for some oxides the pressure effect is insignificant and is not shown. (Source: Niu and Batiza, 1991.) Melting experiments on fertile peridotite KR4003 at 3 to 7 GPa give slightly different variations of oxide abundances versus melt percent (Walter, 1998).

along the mantle adiabat and starts melting when it reaches solidus temperature; the latter is defined by mantle composition and composition of the volatile phase. A further rise of mantle material leads to its continuous melting, matrix compaction, and melt segregation, and the amount of melt increases with progressive material upwelling to a shallow depth (Fig. 6.3b). At low melt volumes, a rapid increase in melt fraction with an increase in the dimensionless temperature T' is caused by the cotectic melting of olivine and two pyroxenes.[1] The rate of melt fraction growth slows down when ~25% of melt is extracted; this corresponds to clinopyroxene exhaustion. A fast rate at $T' > 0.4$ corresponds to olivine melting as mantle temperature approaches liquidus. Adiabatic upwelling of the mantle with a potential temperature of ~1300 °C generates enough melt to produce oceanic crust with an average thickness of ~7 km. In regions with higher potential temperature (1450–1500 °C) basaltic melt produced by adiabatic

---

[1] Dimensionless temperature is defined here as $T' = [T - (T_l + T_s)/2] / [T_l - T_s]$, where T_s and T_l are the solidus and liquidus temperatures, respectively.

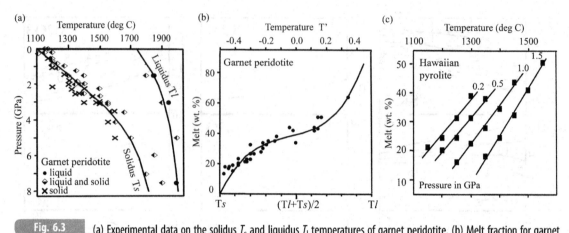

Fig. 6.3 (a) Experimental data on the solidus T_s and liquidus T_l temperatures of garnet peridotite. (b) Melt fraction for garnet peridotite as a function of dimensionless temperature $T' = [T-(T_l+T_s)/2] / [T_l-T_s]$. Symbols – experimental data. (c) Experimental data on the extent of partial melt in Hawaiian pyrolite as a function of pressure and temperature. (Sources: McKenzie and Bickle, 1988; Niu and Batiza, 1991.)

decompressional melting generates oceanic crust up to 25 km thick. A higher mantle potential temperature requires higher pressure at which melting begins (e.g. McKenzie & Bickle, 1988).

Studies of compositions of MORBs from different locations indicate that most MORBs can be explained by 10–20% decompressional melting with initial melting pressures of ~12–20 kbar. Significant differences exist, however, between fast- and slow-spreading ridges in (i) initial and final pressures of decompression melting and (ii) in the amount of the extracted melt. For example, at distances of 30–50 km away from the ridge axis of a fast ridge (the East Pacific Rise), mantle temperatures drop by 50–60 °C and the melt fraction decreases by 5–6% (Fig. 6.4). In contrast, beneath a slow ridge (the Mid-Atlantic Ridge) mantle temperature and major element chemistry across the ridge correlate with the bathymetry. While MORB compositions beneath the fast ridge can be explained by a narrow mantle upwelling, the local trend observed at the slow ridge is better explained by melting of multiple individual domains with different initial and final melting pressures (Niu and Batiza, 1991).

Residue composition and mantle depletion

The extraction of melt from fertile mantle peridotite (olivine, Ca- and Mg-pyroxene, and garnet or spinel) removes basaltic (pyroxene plus feldspar, with or without olivine) components (Fe, Si, Ca) and produces mantle residue depleted in magmaphil elements (Figs. 9.9, 9.10). The mantle residue has low Si and high Mg/[Mg+Fe] and low Mg/Si, Fe/Al, Cr/Al and Ca/Al ratios, typical for the continental lithospheric mantle (Fig. 6.5). While mantle density is sensitive chiefly to variations in iron and magnesium content, seismic velocities are controlled by the content of other elements as well (Fig. 3.24).

Under equilibrium melting conditions, the residue trends of partial melting of fertile peridotite have the following characteristics (Table 6.1, Fig. 6.6a,b) (Walter, 1998):

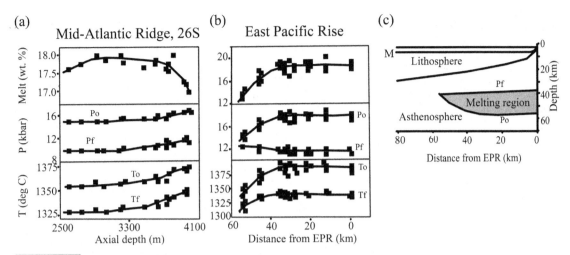

Fig. 6.4   The extent of partial melting beneath the Mid-Atlantic Ridge (26° S) and the Lamont Seamount near the East Pacific Rise (EPR); (c) the inferred region of melting based on (b). Po, Pf, To, Tf are initial and final melting pressures and temperatures, correspondingly. (Source: Niu and Batiza, 1991.)

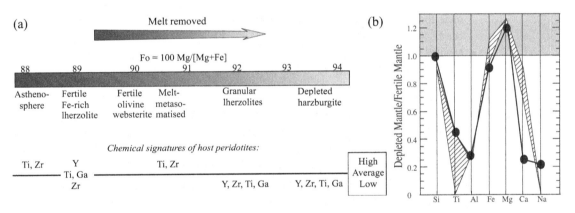

Fig. 6.5   The effect of melt removal on the upper mantle composition. (a) Variations in Fo olivine content and chemical signatures of mantle peridotites produced by melt depletion. (b) Major element abundances of depleted mantle normalized by fertile mantle. Dots: the composition of average continental garnet peridotite (McDonough, 1990). Mantle is assumed to be depleted by the extraction of the current volume of the continental crust (Taylor and McLennan, 1985). The amount of chemical fractionation in the mantle residue depends on the amount of mantle affected by the continent extraction; hatched areas are based on the assumption that 2% by mass of the mantle was affected. (Source for (b): Carlson *et al.*, 2005.)

- At low initial melting pressure (2–3 GPa), the residues can be strongly depleted in Al_2O_3, preserving FeO unchanged.
- Melting at higher initial pressure (> 3 Gpa) will produce residues (at constant MgO) with higher Al_2O_3, higher SiO_2, and lower FeO.

Table 6.1 Effect of mantle melting on oxide abundances

Oxide	Change in oxide abundance (wt%) caused by a melt fraction increase from $X = 10$ wt% to $X = 90$ wt%	Correlation between oxide abundance and melting pressure (at $P = 3$–7 GPa)
SiO_2	Increases from 45 to 48, may have a peak at $X \sim 50\%$	Negative
FeO	Decreases from 13 to 8	Positive
MgO	Increases from 16 to 34	Positive
CaO	Decreases from 11 to 5	Negative
Al_2O_3	Decreases from 13 to 4	Weakly negative with no dependence at $P > 5$ Gpa
Cr_2O_3	Increases from 0.25 to 0.55, may have a peak at $X \sim 60$–70%	No pressure dependence
TiO_2	Decreases from 1.7 to 0.2, weak dependence at $X > 30\%$	No pressure dependence
Na_2O	Decreases from 1.1 to 0.3	No pressure dependence
K_2O	Decreases from 1.0 to 0.1, weak dependence at $X > 20$–30%	No pressure dependence

(Source: Walter, 1998)

- At very high pressures of initial melting (7–10 GPa) and large (> 0.2) amount of extracted melt, the residues can be strongly depleted in Al_2O_3; if melt fraction is < 0.2 the residue can be slightly enriched in Al_2O_3.
- Melt fraction increases with the decrease of the pressure of final melting, producing residues with elevated MgO.

Petrologic studies of mantle peridotites from cratonic settings and abyssal peridotites (mantle fragments from modern ocean basins) indicate large differences in mantle composition which can be attributed to different melting pressures (Fig. 6.6c,d): the composition of cratonic peridotites is consistent with deep mantle melting at pressures > 3 GPa (in some cases as high as > 10 GPa), while abyssal peridotites can be generated by low-pressure (1–3 GPa) mantle melting. Extraction of a large amount of melt (20–40%) at very high pressures produces residue strongly depleted in basaltic components. The degree of melt extraction and mantle depletion is quantified by the iron content in mantle olivine $Mg\# = Mg/[Mg+Fe]$ or forsterite content $Fo = 100\,Mg/[Mg+Fe]$. Convective mantle has $Mg\# \sim 0.88$–0.89, fertile continental mantle has low Mg-number, ~ 0.90, while cratonic mantle strongly depleted by melt extraction has high Mg-numbers (> 0.92) with Fo in olivine ~ 92 and Fo in orthopyroxene ~ 93 (Fig. 6.5).

A remark on mineralogical nomenclature

Modal classifications of rocks were introduced in the mid 1970s and are based on the rock mineral content or mode. The modal proportions of mineral phases (in wt%) are calculated by mass balance between the composition of coexisting phases and the bulk composition of the rock. The advantage of using *modal compositions* comes from the fact that the number of mineral phases which the common mafic and ultramafic rocks contain is limited. This

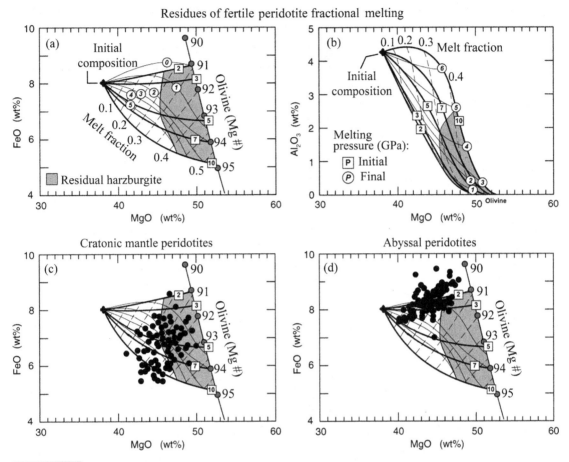

Fig. 6.6 Model residue compositions formed by fractional melting of fertile peridotite (Source: Herzberg, 2004). (a–b) Composition of residual harzburgite is shown by gray shading. Bold lines – initial melting pressures, light lines – final melting pressures, dashed lines – melt fractions. (c–d) Data for cratonic mantle peridotites (from Kaapvaal, Siberia, and Tanzania) and abyssal peridotites plotted on the top of (a).

limitation is known as Goldschmidt's mineralogical phase rule (1911) which says that the number of phases in a rock crystallized under equilibrium conditions does not exceed the number of independent chemical components. Igneous rocks normally contain no more than ten major and minor element components and thus cannot contain more than ten mineral phases. For example, the major element components of mantle peridotites are SiO_2, Al_2O_3, FeO, MgO, and CaO. Concentrations of these major element components define "pyrolite" (pyroxene + olivine) models, which represent the composition of a primitive mantle reservoir formed after the formation of the Earth's core (Table 6.2, also Table 2.2). All pyrolite models are based on the assumption that the primitive mantle is made up of mantle-derived melt and melt-depleted peridotite in such proportions that the final mixture would have mg# as in the least depleted known peridotite. The mixing proportion defines the

Table 6.2 Normative parameters for some pyrolite and peridotite compositions

	Pyrolite (Ringwood, 1966)	Pyrolite (Ringwood, 1975) minus 20% olivine basalt	Kimberlite xenolith PHN1611	Kimberlite xenolith PHN1569	Average continental garnet lherzolite
Garnet (% by mass)	12.3	1.7	10.4	3.1	6.0
Clinopyroxene (% by mass)	16.3	5.6	16.7	2.0	4.5
Orthopyroxene (% by mass)	13.5	17.7	9.5	39.7	22.5
Olivine (% by mass)	57.9	75.0	63.4	55.2	67.0
STP* density (g/cm^3)	3.397	3.341	3.415	3.309	3.353
STP* Vp (km/s)	8.265	8.277	8.217	8.300	8.290
STP* Vs (km/s)	4.802	4.826	4.763	4.868	4.833

* SPT = at surface (i.e. at room) pressure–temperature conditions
(Source: Jordan, 1981)

concentrations of other elements; for example a combination of primitive iron-rich basalt with dunite in proportions of 1:3 defines one of the early pyrolite models (Ringwood, 1975).

In contrast to modal composition, *normative composition* refers to the idealized mineralogy of a rock which is calculated from the known whole rock geochemistry of a rock sample. Such calculations are based on simplified mineral formulas and known phase relationships of minerals. Normative composition differs from modal composition, in particular for rocks that contain Fe and Mg and which may form numerous minerals with similar Fe and Mg ratios. For example, the composition of olivine $(Mg,Fe)SiO_4$ varies continuously from forsterite Fo (Mg_2SiO_4) to fayalite Fa (Fe_2SiO_4) and it is usually specified by reference to the Mg-rich end-member, so that a typical olivine from the cratonic mantle may be described as Fo_{92}. The common pyroxenes fall into two groups: orthopyroxenes and clinopyroxenes. Orthopyroxenes $(Mg,Fe)Si_2O_6$ also form a continues series with compositions ranging from Mg-rich enstatite En $(Mg_2Si_2O_6)$ to Fe-rich ferrosilite Fs $(Fe_2Si_2O_6)$ (Fig. 6.7).

6.1.3 Oceanic and cratonic trends

Since Bowen (1928), it has been recognized that, for the residues of peridotite partial melting, modal olivine and Mg# are correlated. However, it was not until the study of Boyd (1989) that the existence of a strong compositional difference between oceanic and continental lithosphere was recognized from petrologic analysis of mantle peridotites from different continental and oceanic locations. Continental peridotites analyzed by Boyd included abundant low-temperature coarse peridotites from the Archean Kaapvaal craton

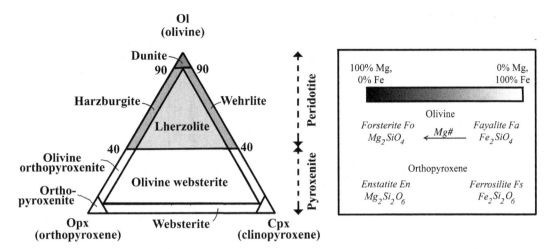

Fig. 6.7 Left: Modal classification of ultramafic rocks. Shaded areas refer to compositions of mantle peridotite which typically contain 40–60% of modal olivine. Right: nomenclature for some ferromagnesian rocks; Mg# = Mg/[Mg+Fe].

(South Africa), off-craton post-Archean peridotites from North America, Phanerozoic peridotites from Mongolia and Western Europe, as well as samples of oceanic lithosphere from abyssal peridotites, alpine peridotites, ophiolites, and oceanic basalts. These data allowed for a recognition of the "oceanic trend" typical of oceanic peridotites and modeled by shallow melting of a "pyrolite" or PUM (primitive, undifferentiated mantle) composition (Fig. 6.8). Further petrologic studies have demonstrated that most abyssal and ophiolite peridotites and many xenoliths from subduction zones lie along this trend. Furthermore, xenoliths from subcontinental mantle at the cratonic margins (post-Archean off-craton peridotites) and high-temperature cratonic peridotites also plot close to the "oceanic" trend. Thus, in fact, it is only low-temperature cratonic peridotites that do not follow the "oceanic" trend but plot at significantly higher mg-numbers at the same modal olivine content: peridotites from oceanic lithosphere are olivine rich with typical Mg#=90.5–91.5, while cratonic peridotites are enstatite-rich and have typical Mg#=91.5–93.5 (Mg#=100 mg#). This compositional difference defines the fundamental distinction between the cratonic low-temperature peridotites and peridotites from all other tectonic settings.

Boyd (1989) attributed the difference between oceanic and cratonic trends to different depths of mantle melting. Oceanic and Phanerozoic peridotites may have formed at low pressures (<2–3 GPa) as residues of pyrolite melting. Similarly, high-T peridotites are believed to be formed as the residues of melt extraction at 2–3 GPa. Although the origin of low-temperature cratonic peridotites remains enigmatic (it is difficult to find a process that can explain a coupling between high Mg# and high enstatite content), it is generally believed that they were formed as the residue of deep mantle melting at pressures greater than 5 GPa (Figs. 6.8 and 6.6c,d) (Walter, 1998).

Recent petrologic studies indicate that the difference between the continental and oceanic trends can be caused by differences in chemical differentiation during melt extraction or by processes unrelated to melting such as cumulate mixing and metamorphic or mechanical

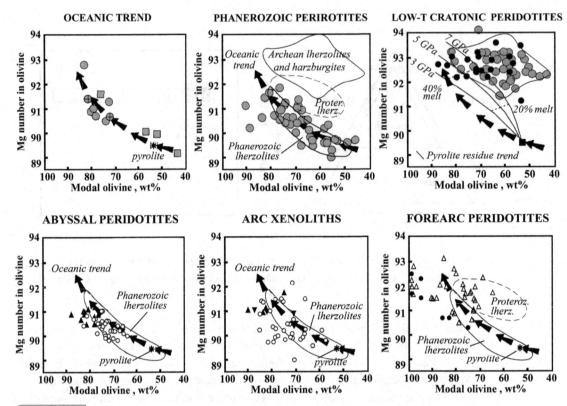

Fig. 6.8   Compositional trends for upper mantle peridotites from different tectonic settings (upper row: after Boyd, 1989; lower row: after Griffin *et al.*, 2003a). The plots also show the "oceanic trend" (Boyd, 1989) and typical ranges for the Archean, Proterozoic, and Phanerozoic peridotites from continental settings. The low olivine content of Archean peridotite xenoliths is related to high orthopyroxene content (compare with Fig. 6.13 for the Slave and East Greenland cratons). The plot for low-temperature cratonic peridotites also shows residue trends for melting of pyrolitic mantle (thin lines) labeled according to melting pressure and percent of melting (dashed lines) (Walter, 1998). Different symbols in some plots refer to samples from various locations.

sorting (for further details see Section 3.1.7 and Fig. 3.18). Residue oceanic peridotites are formed by one single melting event and their composition is characterized by Mg# alone. In contrast, the composition of continental peridotites is controlled by at least two processes (partial melting and the addition of Si-rich materials) and is characterized by two parameters, Mg# and Opx# (volume fraction of orthopyroxene) (Matsukage *et al.*, 2005). As a result of the second (unrelated to melting) process, low-*T* cratonic peridotite plot along the mixing trend between olivine and orthopyroxene and do not follow the residual trend for pyrolite melting (Fig. 6.9). For example, the average orthopyroxene component in Kaapvaal peridotites is too high to be explained as pyrolite melting residue. Other studies suggest that the average Siberian peridotites could be a residue of komatiite melt extraction from a pyrolitic source with 40% melting at ~6 GPa, while the average Kaapvaal peridotites could be a residues of 40% pyrolite melting at ~9 GPa (Walter, 1998).

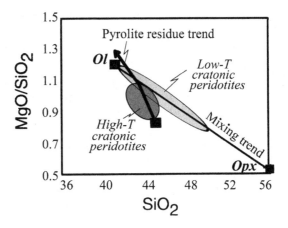

Fig. 6.9 MgO-SiO$_2$ (wt %) plot for cratonic peridotites (Source: Walter, 1998). High-*T* peridotites follow the residue trend for pyrolite melting; low-*T* peridotites clearly deviate from the pyrolite residue trend and plot along a mixing trend between olivine (Ol) and orthopyroxene (Opx).

6.1.4 CBL and compositional heterogeneity of continental mantle

Secular variations in composition of lithospheric mantle

The composition of the continental lithospheric mantle is dominated by depleted ultramafic rocks such as lherzolites (olivine + orthopyroxene + clinopyroxene + garnet or spinel), harzburgites (olivine + orthopyroxene), and dunites (just olivine), as evidenced by petrologic studies of mantle xenoliths and exposed orogenic massifs (Fig. 6.5). Numerous petrologic studies of mantle xenoliths from different continental tectonic settings have confirmed the existence of systematic variations in the composition of the lithospheric mantle with age (Boyd and McCallister, 1976; Boyd, 1989; Griffin *et al.*, 1998a; Boyd *et al.*, 1999; Lee and Rudnick, 1999; Gaul *et al.*, 2000). Depletion of the lithospheric mantle (primarily in Ca and Al, and to a lesser degree in Fe) increases with the age of the overlying crust: typically, the Archean lithospheric mantle has high Fo (93–94), low Ca and Al, the Proterozoic lithospheric mantle has intermediate Fo (91–92), Al, and Ca, while Phanerozoic lithospheric mantle is characterized by low Fo (~90) and high Al and Ca (Fig. 6.10a). A strong correlation between crustal age and composition of the lithospheric mantle suggests that they were formed during the same tectono-magmatic processes and have remained attached ever since.

Depleted Archean mantle is not only more depleted than younger lithospheric mantle, but it is depleted in a different way: in contrast to post-Archean continental and oceanic peridotites which show no correlation between Fe and Al, Archean peridotites with lower Al content also have lower Fe content. A similar pattern is observed for Cr–Al. Depleted Archean mantle contains rock types which are typically absent in post-Archean lithospheric mantle, low-Ca harzburgites and high-Cr garnets (Fig. 6.10c). Archean mantle has a distinct garnet signature: the mean Cr$_2$O$_3$ content in the garnet suites decreases, while the Y content increases from Archean to Phanerozoic mantle (Fig. 6.10b). These trends are interpreted as reflecting a decrease in average degree of mantle depletion with time from Archean to Phanerozoic,

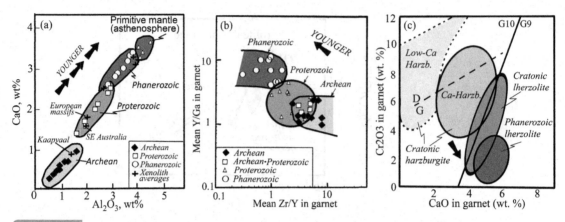

Fig. 6.10   Secular variations in mantle composition. Black arrows indicate the trend that corresponds to decreasing age. (a) Ca versus Al in oceanic and continental peridotites (Sources: Boyd, 1989; O'Reilly *et al.*, 2001). (b) Mean Zr/Y versus Y/Ga for lherzolitic garnets from different tectonic settings (Source: Griffin *et al.*, 1998a). (c) Cr$_2$O$_3$–CaO compositions for garnet peridotite xenoliths (Sources: Griffin *et al.*, 1998a; Grutter *et al.*, 2006). The field of low-Ca harzburgites which are characteristic for peridotite xenoliths from diamondiferous kimberlite pipes within the Archean cratons – after Sobolev (1974) and Gurney (1984). Dashed line marked G/D – proxy for the graphite–diamond transition; solid line G10/G9 (Gurney, 1984) separates garnets from diamondiferous and non-diamondiferous locations and follows closely the lherzolite–harzburgite boundary proposed by Sobolev (1974).

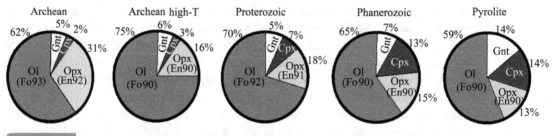

Fig. 6.11   Average model composition of the subcontinental lithospheric mantle (data of Griffin *et al.*, 1998a).

characterized by increasing (cpx+gnt) and cpx/gnt and decreasing Mg# (Fig. 6.11). To sum up, the Archean lithospheric mantle, as compared to the Proterozoic and Phanerozoic mantles, contains highly magnesian olivine and orthopyroxene and has less garnet and clinopyroxene. One of the most depleted known samples of the Archean mantle is from the cratonic keel beneath the Pilbara Craton which was sampled by kimberlites intruded ~1800 Ma ago (the Brockman Creek kimberlite) (Wyatt *et al.*, 2003).

Average model compositions of subcontinental lithospheric mantle calculated from a compilation of modal and compositional data on garnet peridotite xenoliths from different continental tectonic settings are summarized in Fig. 6.11. The evolution of the composition of the lithospheric mantle reflects global processes and is apparently closely linked to secular cooling of the Earth and possible changes in the style of global mantle convection. Numerical models of mantle convection are used to argue that layered mantle convection with periodic mantle overturns might have existed in the Archean time (Davies, 1995). Secular cooling of the mantle switched the style of mantle convection to the whole mantle (see Chapter 9) which had a dramatic effect on mantle melting and the composition of residue peridotites (Griffin *et al.*, 1998a).

Secular variations in mantle density and seismic velocities

Significant secular variations in modal composition are responsible for secular variations in seismic velocities and mantle density (Fig. 3.108). Mantle depletion results in a systematic decrease in the upper mantle density with age from essentially pyrolite density values in oceanic and Phanerozoic continental lithospheric mantle (Table 6.3) to a density deficit of ~2% in cratonic lithospheric mantle (Table 6.4). The density structure of the lithospheric mantle is discussed in detail in Section 6.2, while the effect of melt depletion on seismic velocities in the upper mantle is discussed in Section 3.1.7 (see also Figs. 3.20 to 3.24).

Jordan (1979) used experimental data on mineral composition, densities, and seismic velocities of two natural garnet lherzolites PHN1569 and PHN1611 to calculate normative (i.e. corresponding to standard pressure–temperature conditions, $P = 1$ bar, $T = 25\,°C$) densities and velocities of rock-forming minerals and end-member whole-rock compositions (Fig. 6.12). The results demonstrate that increasing garnet content produces an increase in both density and seismic velocities, whereas increasing iron content increases density but decreases seismic velocities. If changes in garnet and iron content are positively correlated, the effects of their variations on seismic velocities cancel each other. It should also be noted that shear velocities are much more sensitive to iron content variations than to variations in garnet content (Fig. 6.12) and thus are a good indicator of Mg# variations related with the degree of melt-extraction depletion of the mantle. Since Vp velocity in olivine is higher than in pyroxene, compressional velocities are also sensitive to the degree of silica saturation (Jordan, 1979) (compare with Fig. 3.21). Recent studies indicate that in the continental lithospheric mantle Vs also correlates with the orthopyroxene content (Fig. 3.24) (Matsukage *et al.*, 2005).

Table 6.3 Density of asthenosphere and suboceanic lithospheric mantle		
SPT* density, g/cm^3	Comments	References
3.37	Derived from seismic models	Lerner-Lam and Jordan, 1987
3.378	Calculated for pyrolite composition	Griffin *et al.*, 1998
3.38	Calculated for pyrolite or fertile peridotite compositions	Boyd, 1989
3.38–3.41	Calculated for garnet pyrolite	Irifune, 1987; Ringwood, 1975
3.39	Experimental data for high-T sheared Kaapvaal peridotites	Boyd and McCallister, 1976
3.39	Calculated for pyrolite composition	Poudjom Djomani *et al.*, 2001
3.397	Calculated for pyrolite composition	Jordan, 1979

* SPT = at surface (i.e. at room) pressure–temperature conditions

Table 6.4 Average density of subcontinental lithospheric mantle			
SPT* density, g/cm^3	Density deficit**	Comments	References
(3.30)	2.5%	Data from Iceland (basalt-depleted magmas)	O'Hara, 1975
Ar: 3.353	Ar: 1.3 % (reference density 3.397 g/cm^3)	Calculated for all Kaapvaal peridotites	Jordan, 1979
Ar: 3.33, mg# = 0.926 Pt: 3.34, mg# = 0.918 Ph: 3.36, mg# = 0.908	(Ar: 1.5%) (Pt: 1.1%) (Ph: 0.6%)	Based on isopycnic hypothesis and calculated from mg#	Jordan, 1988
Ar: 3.30	Ar: 2.7% (reference density 3.39 g/cm^3)	Experimental data for 2 low-T Kaapvaal garnet peridotites	Boyd and McCallister, 1976
Ar: 3.31–3.33	(Ar: 1.5–2.1%)	Calculated for two-phase harzburgite or four-phase garnet lherzolite	Boyd, 1989
(Ar: 3.324)	Ar: 1.7%	Calculated for low-T Kaapvaal peridotites	Pearson et al., 1995
Ar: 3.305	(Ar: 2.2%)	Calculated for garnet peridotites	Hawkesworth et al., 1990
Ar: 3.335	Ar: (1.4%)	Calculated for spinel peridotites	Hawkesworth et al, 1990
Ar: 3.310 Pt: 3.330 Ph: 3.354	(Ar: 2.1%) (Pt: 1.5%) (Ph: 0.8%)	Calculated from compositions of garnet peridotites	Griffin et al., 1998a
Ar: 3.31 Pt: 3.34 Ph: 3.37–3.39	Ar: 1.5% (reference density 3.36 g/cm^3)	Calculated from average mineral compositions	Griffin et al., 1999b
Ar: 3.31±0.016 Pt: 3.35±0.02 Ph: 3.36±0.02	(Ar: 2.1%) (Pt: 0.9%) (Ph: 0.6%)	Calculated from average mineral compositions	Poudjom Djomani et al., 2001

The values given in the original publications are in bold, derived values are in brackets. Density contrast, where non-reported in the original studies, is estimated for pyrolite SPT density of 3.38 g/cm^3. Abbreviations for ages: Ar = Archean; Pt = Proterozoic; Ph = Phanerozoic.

* SPT = at surface (i.e. at room) pressure–temperature conditions,

** calculated as the density contrast at $T = 1300$ °C between the cratonic root and the convecting mantle.

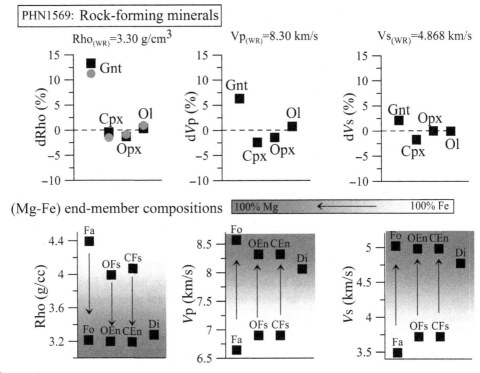

Fig. 6.12 The effect of variations in mineral composition on density and seismic velocities (based on data of Jordan, 1979). Upper panel: Densities and velocities of rock-forming minerals in natural garnet lherzolite PHN1569 normalized by whole rock (WR) values; black symbols – normative values (calculated for room P–T conditions), gray symbols – experimental values of Boyd and McCallister (1976); Gnt = garnet, Cpx = clinopyroxene, Opx = orthopyroxene, Ol = olivine. Lower panel: densities and velocities of end-member compositions at room P–T conditions; Fo = forsterite, Mg_2SiO_4; Fa = fayalite, Fe_2SiO_4; OEn = orthoenstatite, $MgSiO_3$; OFs = orthoferrosilite, $FeSiO_3$; CEn = clinoenstatite, $MgSiO_3$; CFs = clinoferrosilite, $FeSiO_3$; Di = diopside, $CaMgSi2O_6$.

There is a significant discrepancy between different petrologic and geophysical models on the quantitative effect of compositional variations on mantle seismic velocities (see also Section 6.3). For example, calculation of seismic velocities of garnet lherzolite from its whole rock composition gives values of $\sim$+1% for Vp and $\sim$+2.3 % for Vs velocity anomalies in the Archean mantle (Jordan, 1979). Using a similar approach, Schutt and Lesher (2006) estimate that melt depletion of fertile peridotite between 1 and 7 GPa has almost no effect on Vp and Vs so that

> "major element effects of melt depletion are thus insufficient to produce the high mantle velocities imaged beneath cratons or to cause significant velocity variations".

In contrast, Griffin *et al.* (1998a) argue that at the same temperature, the Phanerozoic mantle has Vp $\sim$5% slower and Vs $\sim$ 3% slower than the Archean mantle so that about 50% of amplitudes of seismic velocity anomalies observed in seismic tomography models can be ascribed to compositional variations.

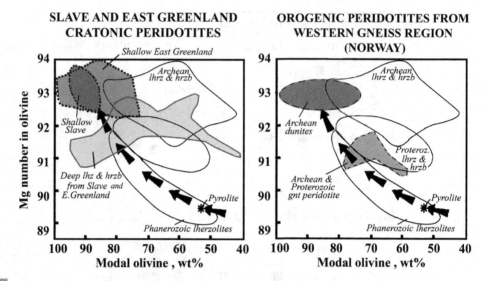

SLAVE AND EAST GREENLAND CRATONIC PERIDOTITES

OROGENIC PERIDOTITES FROM WESTERN GNEISS REGION (NORWAY)

Fig. 6.13 Left: Compositional trends for upper mantle peridotites from the Central Slave (Lac de Gras) and East Greenland cratons as compared to the "oceanic trend" (Boyd, 1989) and typical ranges for continental peridotites from the Archean, Proterozoic, and Phanerozoic settings. The shallow mantle beneath the Slave and East Greenland cratons is more depleted than a typical Archean lithospheric mantle (Source: Griffin *et al.*, 2003a). Right: Compositional trends for orogenic peridotites from the Western Gneiss Region (Norway). The Archean dunites are highly depleted. Less depleted and younger garnet peridotites are interpreted as dunites refertilized by post-depletion metasomatism (Source: Beyer *et al.*, 2006).

Note that the results reported by different petrologic groups for the effect of iron depletion on Vp velocity differ not only by amplitude but apparently also by sign (Fig. 3.21d). In particular, a limited xenolith data set from the Slave craton suggests that Vp has a strong inverse correlation with Mg#, implying that depleted Archean mantle should have Vp $\sim 0.5\%$ lower than Phanerozoic mantle (Kopylova *et al.*, 2004). In contrast, abundant peridotite data from various cratonic settings show either no Vp–Mg# correlation or a weak positive trend (Fig. 3.21bd; Lee, 2003). Although it is likely that regularities found for a very limited data set from the Slave craton are biased by scarce sampling, one cannot exclude the possibility that the lithospheric mantle of the Slave craton does not follow the patterns observed for other cratons, as indicated by a distinct compositional layering of the Slave lithospheric mantle (Fig. 6.13). Recent data from East Greenland suggest that highly depleted shallow (< 65 km) lithospheric mantle (Re–Os ages of 3.7–2.6 Ga) with high Mg# but low orthopyroxene/olivine content, similar to modern arc settings, may not be unique for the Slave shallow mantle (Bernstein *et al.*, 1998; 2006; Hanghøj *et al.*, 2001). Similar results were recently reported for the shallow mantle beneath the western terrane of the Kaapvaal craton (Gibson *et al.*, 2008).

Geophysical studies of the effect of melt depletion on seismic velocities are also controversial. A global study of the correlation between shear-wave seismic velocity anomalies in a global Rayleigh-wave tomographic model and mantle temperatures indicates that they are globally poorly correlated at depths of 100–200 km, and compositional effects may have a significant contribution (up to 50%) to the amplitude of Vs variations in the upper mantle

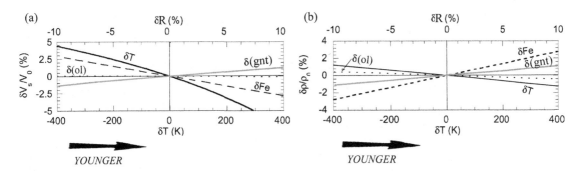

Fig. 6.14 Relative anomalies of S-wave velocity (a) and density (b) as a function of temperature anomalies δT (with respect to 1250 °C, solid lines) and compositional anomalies δR: volumic fraction of iron with respect to $X_{Fe} = 11\%$, i.e. mg# = 0.89 (dashed lines); volumic fraction of garnet with respect to $X_{gnt} = 15.3\%$ (gray lines); volumic fraction of olivine with respect to $X_{ol} = 61.7\%$ (dotted lines). Black arrows indicate a general trend of an increase in temperature and volumic fraction of iron with decreasing age. Note that anelasticity significantly affects the temperature dependence of relative Vs anomalies (Fig. 3.6). Density anomalies are based on a joint inversion of a seismic tomography model and geoid. (Based on results of Jordan, 1979; Yan et al., 1989; Deschamps et al., 2002; Artemieva et al., 2004; Godey et al., 2004). A 2% shear-velocity increase can be caused either by a decrease in temperature by 120 °C, or by a decrease in iron (depletion) by 7.5%, or by a decrease in aluminium by 15%. A 1% density increase can be explained either by a 3.5% iron enrichment, or by a 10% increase of aluminium, or by a 400 °C temperature decrease.

(Artemieva *et al.*, 2004). In contrast, a geodynamic modeling based on a joint inversion of shear-wave seismic tomography model S16RLBM (Woodhouse and Trampert, 1995) and the non-hydrostatic gravity anomalies derived from the geoid model EGM96 (Lemoine *et al.*, 1998) indicates that the compositional effect on seismic velocities is significantly smaller than the temperature effect: a Vs anomaly of 1% can be explained either by an iron depletion anomaly of -4% or by a temperature anomaly of about -50 °C (Fig. 6.14) (Deschamps *et al.*, 2002). Note that compositional effects other than iron depletion are not considered in this study. Clearly, a combination of several independent data sets on mantle velocity, density, and thermal structure are needed to estimate the magnitude of compositional variations of seismic velocities in the continental lithospheric mantle.

Secular variations in lithospheric temperatures (with the coldest geotherms typical of the Archean lithosphere and the hottest geotherms typical of young tectonically active structures) have opposite effects on *in situ* (i.e. at ambient mantle conditions and observed by remote geophysical methods) density and seismic velocities of lithospheric mantle (Figs. 3.108, 6.14). Density variations due to the secular trend in lithosphere depletion are compensated by secular variations in the thermal regime of the mantle (correlated with its tectono-thermal age): in young regions thermal expansion caused by high mantle temperatures compensates the high density of fertile lithospheric mantle and effectively masks density anomalies in gravity data (isopycnicity condition). In contrast, high STP (at standard P–T, i.e. at room, conditions) seismic velocities in depleted cratonic lithosphere are enhanced by low mantle temperatures due to the inverse correlation between seismic velocities and temperature (Fig. 3.2). This situation explains common interpretations of seismic anomalies in terms of hot–cold mantle temperatures, further provoked by the red–blue color code commonly used in seismic tomography images.

Depth variations of Mg# and CBL thickness

Regional studies of mantle-derived xenoliths indicate significant vertical and lateral compositional variations in the lithospheric mantle. A number of petrologic studies of cratonic peridotites from Siberia, Australia, Canada, and South Africa have been performed by Griffin and co-authors (e.g. Griffin *et al.*, 1996; 1999a; 2004; Gaul *et al.*, 2000). These studies were largely based on studies of major element compositions of peridotitic garnets coupled with temperature estimates from Ni content in garnet (T_{Ni}) (Ryan *et al.*, 1996; Chapter 5). The results were used to map the average distribution of forsterite content in mantle olivine as a function of depth (Fig. 6.15) by projecting temperature estimates for individual samples onto

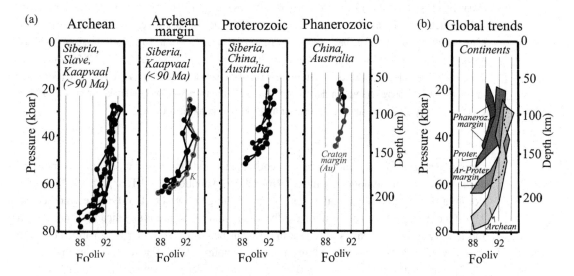

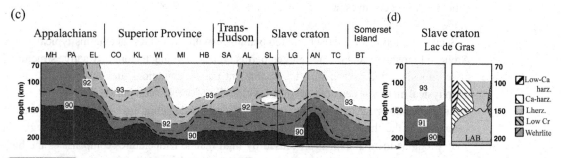

Fig. 6.15 Variations of forsterite content in mantle olivine with depth. (a) Stacked data for peridotites from different continental tectonic settings (compare with Fig. 3.19); (b) global trends based on stacked data; (c) a 2D profile through the lithosphere of North America showing the distribution of Fo content in olivine (see Fig. 5.2 for profile location), small letters – abbreviated xenolith locations; (d) other 1D interpretations of forsterite content versus depth for samples from the Central Slave Craton. LAB = lithosphere–asthenosphere boundary (here – the base of CBL). (Sources: (a) based on data of Griffin *et al.*, 1999a; 2003a; Gaul *et al.*, 2000; (c) Griffin *et al.*, 2004; (d) data of Kopylova and McCammon, 2003; Griffin *et al.*, 1999c.)

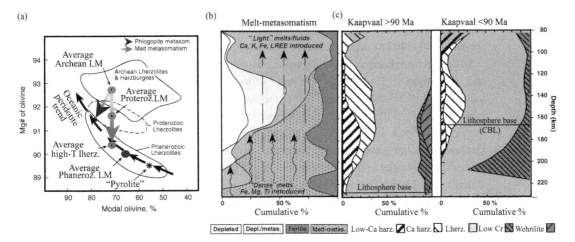

Fig. 6.16 Metasomatism of continental lithospheric mantle. (a) Mean compositions of Archean, Proterozoic, and Phanerozoic subcontinental lithospheric mantle (LM) and the mean composition of high-*T* sheared lherzolite xenoliths from kimberlites (compare with Fig. 6.8). Arrows show the effects of shallow phlogopite-related metasomatism (grayish-black) and melt-related metasomatism responsible for the composition of sheared xenoliths (gray arrow) which result in refertilization of the Archean lithospheric mantle. (b) Typical depth distribution of different types of metasomatism in Archean lithospheric mantle. (c) Kimberlites erupted at two different ages in Kaapvaal indicate a recent refertilization of the deep part of the cratonic lithosphere (below ~175 km). Rock types in lithologic sections are low-Ca harzburgite, Ca-harzburgite, lherzolite, low-Cr Phanerozoic lherzolite, and wehrlite. The tentative base of the lithosphere as proposed by Griffin and co-authors (strictly speaking, of CBL) is marked by thin lines. (Sources: Griffin *et al.*, 1996; 1999c; 2003a.)

the conductive geotherm to get the depth of their origin (see Fig. 5.11 for explanations). A similar technique was applied by the same group to map the depth distribution of mantle lithologies (e.g. Fig. 6.16c). The uncertainty of the constrained Fo-depth distribution depends on both the uncertainty of the Ni-in-garnet thermometer (see discussion in Section 5.2.1) and the uncertainty in the conductive geotherm; the latter is usually estimated either from xenolith/ xenocryst data or is taken from the family of reference conductive geotherms of Pollack and Chapman (1977). Thus, the constraints on CBL thickness obtained by this method are dependent on the constraints on the thickness of TBL through geotherms used to assign the depth of origin to xenocryst/xenolith samples. Accurate independent pressure (depth) estimates are possible only for garnet-bearing samples, which can be found only at certain depths sampled by xenocrysts/xenoliths. The shallow limit is controlled by the spinel–garnet peridotite phase transition which in cold cratonic mantle occurs at depths of 80–100 km; it is set by the presence of highly Al-depleted rocks in which garnet is absent.

The results (based on the Ni-in-garnet approach for the depth of xenolith origin) indicate that in the cratonic lithospheric mantle the mean values of Fo content of olivine, on average, decrease with depth and display an amazingly similar pattern for all Archean and all Proterozoic settings (Fig. 6.15, compare with Figs. 3.19, 3.106). In the Archean cratons, Fo ~ 93 at a 100 km depth and gradually decreases to Fo ~ 92 at ~120–170 km depth; below this depth some scatter is observed with Fo ~ 91–93. In some terranes of the Siberian and Kaapvaal cratons, a sharp decrease in forsterite content to Fo ~ 88–90 occurs in a narrow zone at ~210–240 km depth; however this pattern is not observed in the Slave data.

In the Archean Kaapvaal craton, young Group I kimberlites (erupted at or after ~90 Ma) and old Group II kimberlites (erupted before ~110 Ma) display significantly different values of Mg# below ~175–200 km depth, suggesting that recently (< 110 Ma) the low-ermost part of the lithospheric root has been refertilized (probably, melt-metasomatized), but not destroyed. Interestingly, the depth variation of Fe-depletion in young Kaapvaal kimberlites is the same as in the Siberian kimberlites that were erupted at the Archean/Proterozoic boundary (sites numbered 4–5 in Fig. 5.3). In Proterozoic terranes, Fo is ~ 92 at depths 70–130 km; below 130 km depth it decreases almost linearly with depth reaching Fo ~ 88 at a depth of ~160 km. Since Fo and Mg# quantify the degree of mantle depletion due to melt extraction (Fig. 6.5), the depth where the Fo content of olivine decreases to the value of primitive mantle (Fo ~ 88–89) may be interpreted as the base of the CBL (petrologic lithosphere) (compare with Table 2.2).

Thus, the thickness of the CBL, as defined by the forsterite content in mantle olivine, is apparently very uniform in both Archean and Proterozoic terranes and is ~210–240 km in the former and ~160 km in the latter. However, mapping depth variations of different lithologies in the cratonic lithospheric mantle by the approach similar to mapping olivine content (and also based on temperature estimates from Ni content distribution in xenocryst garnets, the so-called T_{Ni} geotherms) reveals a more complicated pattern than found by Fo- (or Y-, Zr-, TiO_2- contents in garnet) mapping (Griffin et al., 1999c). The lithological columns clearly indicate that there is no sharp lithological (petrologic) boundary that separates lithospheric mantle from the astheno-sphere (convecting mantle) (e.g. Fig. 6.16c), and that the base of CBL is transitional and can extend over some tens of kilometers.

Mantle metasomatism

The base of CBL defined as the depth where olivine content in the mantle decreases to the value of the primitive mantle may not necessarily mark the base of the cratonic lithosphere formed as the residue of mantle melting. Instead, it may mark a depth with a distinct increase in melt-related metasomatism. Such metasomatism, as for example associated with the eruption of the Karoo basalts, South Africa, and with the Siberian Traps, lowers the mean Fo content of olivine in the Archean lithospheric mantle and greatly reduces the proportion of subcalcic harzburgites, while increasing the proportion of fertile lherzolites. Thus, large-scale thermo-magmatic events may change dramatically the composition of the cratonic lithospheric mantle within the radius of several hundred kilometers from the eruption centers, not destroying most of the cratonic roots (Griffin et al., 2005).

Metasomatism, senso stricto, is a chemical change in a rock due to the addition and removal of materials to and from pre-existing lithologies (e.g. Bailey, 1982). Depletion (see Section 6.1.2) is the major metasomatic process within the upper mantle since it changes the bulk chemical composition of peridotitic rocks by melting and migration of incompatible elements into the basaltic liquids, and further material transfer from the upper mantle into the crust (Fig. 6.5). Two other processes have an important effect on further chemical alteration of the lithospheric mantle, melt-related metasomatism and "phlogopite" metasomatism (Fig. 6.16) (it is, however, debatable whether the former should rather be called a lithosphere–mantle interaction). Melt-related metasomatism has the opposite effect on mantle chemistry than the "depletion" trend (Fig. 6.10a), leading to a decrease in Mg# and an increase in Al and Cr content and thus producing a rock composition with higher density and lower seismic Vs velocity.

A recent petrologic study of garnet peridotite from orogenic massifs in western Norway indicates that most probably the Archean lithospheric mantle had a highly depleted composition (Beyer *et al.*, 2006) such as observed in xenolith suites from East Greenland (Bernstein *et al.*, 2006) (Fig. 6.13). Whole-rock Re–Os studies of dunite–harzburgite present in large volumes in the Norwegian orogenic massifs, indicate Archean ages (~3.0 Ga). These highly depleted rocks are interlayered with garnet lherzolites which represent zones of Proterozoic refertilization; their further alteration and tectonic deformation took place during the Caledonian (400–500 Ma) orogeny-related subduction of the Proterozoic crust surrounding the orogenic massifs.

Significant metasomatic modification of the lithospheric mantle can be produced by intensive regional magmatic activity, such as associated with large-scale trap magmatism in Siberia and South Africa. A comparison of lithological sections from the Kaapvaal craton shows that Group I kimberlites and Group II kimberlites (erupted before ~110 Ma) display significantly different lithological sections. Group I kimberlites erupted at or after a major metasomatic event (associated with the eruption of the Karoo Traps) which caused widespread enrichment of the Kaapvaal lithospheric mantle at ~90 Ma. These young kimberlites show a reduction in the proportion of depleted harzburgites and depleted lherzolites (compare with Fig. 6.10c) and chemical signatures of melt-related metasomatism in the lower lithospheric mantle and phlogopite-related metasomatism at shallow depths.

- It is important that xenolith suites from the south-western Kaapvaal Group I kimberlites (<90Ma), which provide the bulk of the samples and are the backbone of knowledge on the composition of the Archean lithospheric mantle, are strongly biased towards chemically altered composition (Griffin and O'Reilly, 2007).

Overall, similar patterns of metasomatism are observed in kimberlites from the eastern margin of the Siberian craton (Griffin *et al.*, 2005). Lithospheric mantle sampled by post- and pre-trap kimberlites is significantly different. Mantle metasomatism associated with the eruption of the Siberian Traps significantly modified the composition of the lithospheric mantle, especially in the depth range 80–130 km. At these depths, the mean Fo content of olivine is lowered, the proportion of subcalcic harzburgites is greatly reduced, whereas the proportion of fertile lherzolites is increased. Based on petrologic data from South Africa and Siberia, Griffin *et al.* (2005) argue that

- large-scale magmatic events can significantly modify the composition of the lithospheric mantle in regions extending over several hundred kilometers from the main magmatic centers.

Lateral variations in composition of cratonic lithosphere

Small-scale lateral variations in the composition of the lithospheric mantle are best established for the Canadian Shield, the Kaapvaal, the North China, and the Siberian cratons, where peridotite xenoliths are abundant (Fig. 5.1). Similar studies for the Baltic Shield are being started. The results of numerous petrologic studies worldwide indicate that each craton has a unique compositional structure of the lithospheric mantle which reflects a 2–3 Ga long and diverse history of its tectonic and geodynamic evolution. Furthermore, significant compositional variations exist not only within large cratons such as in Siberia (Griffin *et al.*, 1999a), but even within relatively small cratons, such as the Slave and the Karelian cratons (Fig. 6.17).

According to petrologic data from several kimberlite locations, the Slave craton in northern Canada has the blocky structure of lithospheric mantle with three compositionally distinct lithospheric domains separated by NW–SE oriented boundaries, whereas the crustal structure of the craton has mainly N–S divisions. An unusual stratification with a two-layered lithospheric mantle has been recognized beneath the central and northern parts of the Slave craton from studies of mineralogical and chemical composition of the peridotite xenoliths (Kopylova and Russell, 2000; Griffin *et al.*, 1999c). In the Northern Slave, the shallow level (down to 80–100 km depth) consists of depleted spinel peridotite, while the deeper garnet peridotite level (down to 200 km depth) shows a gradual reduction in Mg# with depth (Figs. 6.13, 6.15d). The deepest low-*T* peridotites have the lowest values of Mg# as compared to low-*T* peridotites from other Archean cratons. The depth with a sharp change in mantle composition approximately correlates with a sharp seismic boundary imaged by seismic receiver function analysis and interpreted as a shallow paleosubduction (Bostock, 1997) (Fig. 5.5). In the Central Slave, the shallow level (down to a ~145 km depth) consists of ~60% clinopyroxene-free harzburgite and 40% lherzolite, while the deeper layer sampled by xenoliths contains ~15–20% of harzburgite and 80–85% of lherzolite. The shallow layer beneath the Central Slave also has higher Mg# (Fo ~ 92–94) than the deeper layer (Fo ~ 91–92). The transition between the two compositional layers is very sharp, which apparently makes the structure of the Slave lithospheric mantle unique. The composition of the shallow layer is similar to some ophiolites from modern arcs, and it may have been formed in a similar tectonic setting during the accretion of the adjacent terranes to the ancient cratonic nucleus (the Anton terrane). The deeper layer is believed to originate from a plume head impact at ~2.6 Ga (Griffin *et al.*, 1999c). Recent data, however, indicate that the lithospheric mantle of East Greenland may have a similar compositional stratification (Beyer *et al.*, 2006; Fig. 6.13a).

In the Baltic Shield, petrologic studies of mantle-derived xenoliths and xenocrysts from two kimberlite provinces in Eastern Finland indicate striking differences in the composition of the lithospheric mantle (Fig. 6.17). The Archean nucleus of the Karelian craton has a relatively uniform lherzolite–harzburgite composition between 80 km and ~260 km depth. The lowermost part, ~30 km thick, may be melt-metasomatized. In contrast, xenoliths erupted at the SW edge of the Karelian craton (at the Archean–Proterozoic boundary) sample three compositionally distinct layers in the lithospheric mantle with boundaries at ~120 km and ~180 km depth. The deepest layer is interpreted either as Proterozoic underplating or as a part of the middle layer which was melt-fertilized (O'Brien *et al.*, 2003). A complicated lithological stratification of the lithospheric mantle at the southwestern edge of the craton may be caused by Neoproterozoic terrane accretion.

In Siberia, 65 kimberlites (Fig. 5.3) from several kimberlite fields across the northeastern part of the Siberian Craton were used to map the composition of the lithospheric mantle

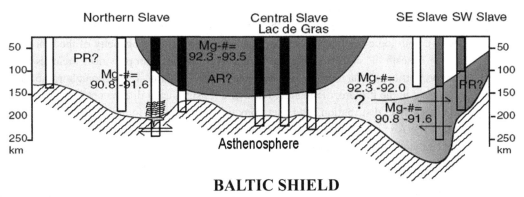

SLAVE CRATON

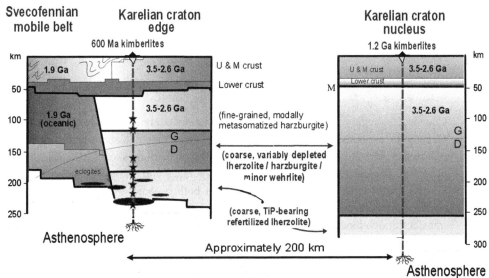

BALTIC SHIELD

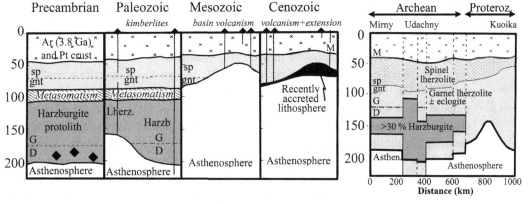

Fig. 6.17 Two-dimensional compositional transects through the lithospheric mantle of the Slave craton (Canada), the Karelian craton (Baltic Shield), and the Siberian craton. The North China transect reflects temporal evolution of the lithospheric mantle from Archean to Cenozoic. (Sources: Kopylova and McCammon, 2003; O'Brien *et al.*, 2003; Lehtonen *et al.* 2004; Menzies and Xu, 1998; Griffin *et al.*, 1999a.) The Siberian transect is the same as in Fig. 5.11. G/D = graphite–diamond transition; sp/gnt = spinel–garnet phase boundary; M = Moho; DM = depleted mantle; EM = enriched mantle.

along a 1000 km traverse through the Archean and Proterozoic provinces (Griffin *et al.*, 1999a). Petrologic sections show a strong compositional heterogeneity of the lithospheric mantle beneath the craton. Lithospheric mantle of the Archean provinces includes significant proportions of harzburgitic and depleted lherzolitic garnets, while harzburgitic garnets are absent in the Proterozoic mantle. Furthermore, significant differences in bulk composition, rock-type distribution, and metasomatic overprint are found for different terranes within the Archean provinces of the craton, sometimes over distances of ~100 km. This may indicate that differences in the structure and composition of the Archean and Proterozoic terranes were preserved during the Proterozoic assembly of the Siberian craton, and the width of the mixing zone between the Archean and Proterozoic mantle was probably less than 100 km. Lithospheric domains with different mantle composition are well correlated with crustal terrane boundaries defined by geological mapping and geophysical data, suggesting that the lithospheric sutures continue from the surface through the entire lithosphere down to, at least, 150–250 km depth.

6.1.5 The effect of CBL on TBL

Chemical boundary layers, both under the continents and the oceans, were called upon recently to explain the thermal structure of the continental and oceanic upper mantle (Cooper *et al.*, 2004; Lee *et al.*, 2005). Numerical models of small-scale convective processes demonstrate that pre-existing, dehydrated and melt-depleted, strong (in the viscous sense) CBL controls the thickness of both oceanic and continental TBL. If the chemical boundary layer does not participate in any secondary convection, vertical heat transfer through it occurs purely by conduction and the only part of TBL that participates in secondary convection is the one beneath the strong CBL (Figs. 1.2, 4.45b). In the oceans, the thickness of CBL is controlled by the potential temperature of the mantle at the time of mantle melting, while the thickness of TBL is controlled by conductive cooling and increases as the square root of seafloor age. This dependence may break for oceans older than ~80 Ma (ocean floor flattening, see Section 4.3.2). However, conductive cooling of the thermal boundary layer proposed to explain the thermal structure of the continents (Crough and Thompson, 1976) fails to explain a large thickness of continental lithosphere, where the thickness of TBL is significantly thinner than predicted by conductive cooling alone (Fig. 4.46). For the oceans, various mechanisms that involve secondary convection beneath the base of TBL have been proposed to explain flattening of old ocean floor. Lee *et al.* (2005) propose that a similar mechanism can control thickness of the continental, in particular cratonic, lithosphere. Similar to the oceans, the onset time of convective instability is controlled by the thickness of CBL: cooling of TBL leads to the situation where its thickness exceeds the thickness of CBL, and under this condition convective instability is no longer suppressed.

Numerical modeling of mantle convection with chemically distinct and non-deforming continents placed within the upper thermal boundary layer of a convecting mantle layer indicate a strong effect of the continents on the style of secondary convection: the thickness of a pre-existing strong CBL (which includes the crust and the depleted cratonic root) controls the thickness of the convective sublayer, and hence the total thickness

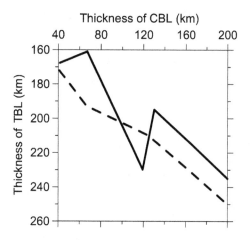

Fig. 6.18 The effect of chemical lithosphere thickness on thickness of thermal lithosphere for a radiogenically undepleted (solid line) and radiogenically depleted (dashed line) cratonic root. Thick chemically depleted cratonic lithosphere (CBL = crust + lithospheric root) controls the thickness of TBL by reducing small-scale convection beneath thick cratonic roots. (Source: Cooper *et al.*, 2004.)

of TBL. The ratio of TBL to CBL thickness decreases with increasing the thickness of the cratonic root (CBL) (Fig. 6.18), since thick cratonic roots reduce small-scale mantle convection beneath themselves (e.g. Doin *et al.*, 1997; Cooper *et al.*, 2004). Additionally, radiogenic enrichment of the continental crust plays a significant role in thermal coupling between stable continental lithosphere and the convecting mantle given that its depth distribution determines the local thermal structure of the continental lithosphere.

6.2 Xenolith constraints on mantle density

6.2.1 Method and uncertainties

Lithosphere is compositionally heterogeneous, both laterally and vertically, and both on large and small scales. Significantly different major element composition of xenoliths that sample lithospheric mantle poses the question of how representative is Nature's sampling (see Section 6.3) and of how large-scale physical properties of the lithosphere (such as density and seismic velocities) can be accessed from a local, small-scale, sampling of a highly heterogeneous at all scales crust and upper mantle. Laboratory measurements of the physical properties of natural peridotites may differ significantly due to compositional variations and may be misleading when directly extrapolated to large-scale (whole-rock) lithospheric properties. From the other side, major and minor element components of different lithospheric layers are well known and, combined with known phase relationships of minerals, provide information on modal mineralogical compositions of the crust and the lithospheric mantle.

Methods widely used in petrology to determine bulk physical properties of the lithosphere are based on analytical calculation of normative (i.e. approximate, see Section 6.1.2) mineral composition from whole-rock composition. This allows for calculation of normative density and normative seismic velocities, provided density and seismic velocities (or elastic moduli) of the constituting minerals are known (e.g. Jordan, 1979; Griffin *et al.*, 1998a; Poudjom Djomani *et al.*, 2001; Lee, 2003; James *et al.*, 2004) (Fig. 6.19). The errors in the values of normative parameters result from:

- uncertainties in thermobarometry of xenoliths;
- uncertainties in physical properties of the constituent minerals;
- assumptions behind calculations of whole-rock parameters from the properties of the constituent minerals;
- difference between normative and model compositions.

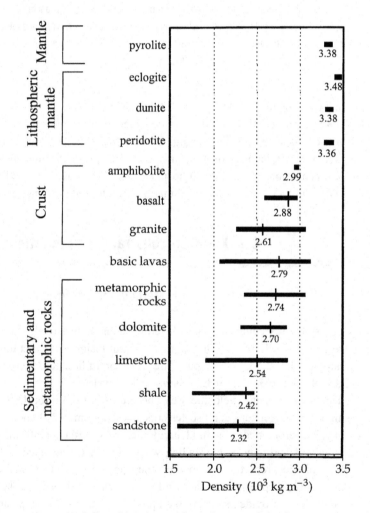

Fig. 6.19 Density of some rocks at room P–T conditions (mean value and a reported range).

Densities can be determined from mineral compositions by assuming the additivity of end-member molar volumes. Seismic velocities may be determined from end-member elastic parameters of constituent minerals under equilibrium P-T conditions at the mantle source by applying one of the averaging techniques widely used in calculations of elastic properties of composite materials. Many of these methods (such as the Voigt–Reuss–Hill approximation) are based on volume averaging of randomly oriented components and thus provide effective isotropic properties. The Hashin–Shtrikman method takes into account grain orientations and provides maximal and minimal bounds on elastic properties. Normative physical parameters are then commonly tested by comparing results with laboratory measurements on natural rocks from similar tectonic settings and with similar modal compositions. Thus, when using petrologic data on densities and seismic velocities, it is important to know whether they provide normative (i.e. model) parameters or experimental measurements on whole-rock samples.

6.2.2 Density of oceanic mantle

Density variations in the oceanic upper mantle are largely related to (i) variations in mantle temperature (which have an almost linear effect on density due to a weak temperature dependence of thermal expansion) and (ii) the amount of melt extracted. One of the first important studies on density structure of the upper mantle has been published by O'Hara (1975). He argued, in relation to Icelandic volcanism, that partial melting of the upper mantle will increase Mg# in the residuum and deplete the dense garnet phase. As a result, the depleted residue will be ~2.5% less dense than undepleted parental mantle. These results were further supported by experimental measurements on two natural garnet peridotites (Boyd and McCallister, 1976).

The amount of melt generated by decompressional melting in adiabatically upwelling asthenospheric mantle depends on the pressure of decompression melting and mantle potential temperature (Fig. 6.3). The effect of melting of MORB pyrolite and mantle peridotite on density of the mantle residue is illustrated in Fig. 6.20a. For melting of MORB pyrolite, the calculation includes an estimate of the amount of mantle material that must be depleted to produce a typical oceanic crust and the mean composition of the depleted residue (Hynes, 2001). The results indicate that the density of the depleted residual mantle (approximately corresponding to a 60 Ma old oceanic lithosphere) decreases with increasing melt fraction, largely due to a decrease in relative abundance of plagioclase feldspar (Fig. 6.20b), and stabilizes when Al is exhausted.

Compositional variations in the oceanic mantle should have a significant effect on mantle density as observed in gravity data, although compositional density variations are obscured by regional temperature variations (compare Figs. 6.21 and 4.27). Within the oceanic mantle, thermal effects dominate shear-wave and density heterogeneity as demonstrated by a joint inversion of shear-wave travel-time data and a set of geodynamic convection-related constraints (Simmons et al., 2009). However, according to the oceanic cooling plate model, temperature differences between oceanic lithosphere of different ages are restricted to the upper 130 km of the mantle, while global seismic tomography models indicate that the velocity anomalies beneath the oceans may persist down to at least a depth of 200–250 km with the most

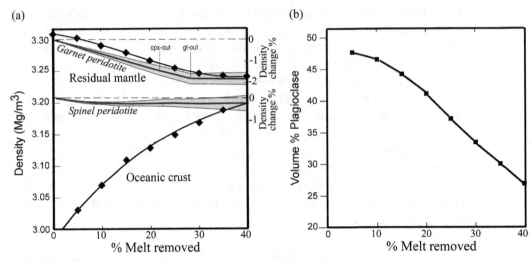

Fig. 6.20 The effect of melting on mantle density: solid lines with symbols – mean densities of oceanic crust and residual mantle for melting of MORB pyrolite-90 at a depth of 50 km and temperature of 860 °C (a) and the volume fraction of plagioclase in the oceanic crust as a function of the melt fraction by weight (b). Gray shading in (a) – density change in spinel and garnet peridotite at room temperature and pressure as a function of the percentage of melt removed and the error margin of the density calculation. Labels "cpx-out" and "gt-out" indicate the percentage of melting where clinopyroxene and garnet are completely consumed (Sources: Hynes, 2001; Carlson *et al.*, 2005.)

pronounced seismic velocity anomaly at 100–200 km depth (e.g. Fig. 3.116). The top of this seismic anomaly corresponds approximately to the depth of the G-discontinuity (~70 km depth), which may be a compositional boundary between the depleted layer from which the oceanic crust was extracted and the underlying fertile mantle (Gaherty *et al.*, 1996). This interpretation is supported by a joint inversion of seismic and gravity data which indicates a sharp change in the value of a relative velocity-to-density scaling factor at ~80 km depth, most likely associated with a transition from shallow dry harzburgite to wet peridotite (Fig. 3.110) (Kaban and Schwintzer, 2001). In general agreement with these results, Lee *et al.* (2005) propose that the pre-existing chemical boundary layer (formed during mantle melting and lithosphere extraction) controls the thickness of oceanic TBL. According to this model, the base of CBL (located at depths between 50 km and 110 km, depending on mantle potential temperature) is associated with a steep (up to two orders of magnitude) drop in viscosity from strong, high-viscous water-depleted CBL to convectively unstable, low-viscous, and wet mantle. Although no sharp change in mantle density occurs at any depth, the model predicts that the entire oceanic mantle older than ~20 Ma becomes negatively buoyant due to the undercompensation of thermal cooling by chemical buoyancy.

6.2.3 Density of continental mantle

Numerous petrologic studies of craton and off-craton mantle-derived peridotite xenoliths indicate that (at least, the Precambrian parts of) the continents are underlain by a distinct

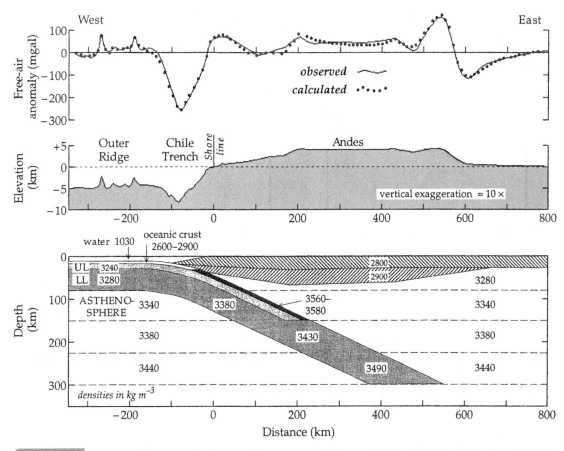

Fig. 6.21 Density structure across the Chilean subduction zone at 23°S constrained by seismic, thermal, and petrologic data (from Lowrie, 1997).

chemical boundary layer, depleted in magmaphil elements as compared to pyrolite mantle. Pyrolite melting and the transition from fertile to depleted peridotite is characterized by the loss of clinopyroxene and garnet (and then of orthopyroxene) (Fig. 6.5). Garnet density is ~15% higher (~3.75 g/cm³ at $P > 15$ GPa, Irifune *et al.*, 1986) than the density of olivine and orthopyroxene (Fig. 6.12). Thus, mantle melting reduces the density of depleted peridotite by ~2% as compared to fertile mantle at the same temperature due to garnet loss and an increase in Mg/[Mg+Fe] (Fig. 6.20). Because of a systematic increase of mg# with lithospheric age (Fig. 6.11), the density of the continental lithospheric mantle is also strongly correlated with age, as indicated by density calculations based on the mean mineral compositions of the lithosphere and experimental data on densities of end-member minerals (Table 6.4; Fig. 6.22a). At the same pressure–temperature conditions, the Phanerozoic mantle is more dense that the Archean mantle. In particular, at room conditions, lithosphere density is ~3.31–3.34 g/cm³ in the Archean–Paleoproterozoic regions, 3.36–3.37 g/cm³ in the Mid-Neoproterozoic regions, and 3.36–3.39 g/cm³ in the fertile Phanerozoic lithosphere

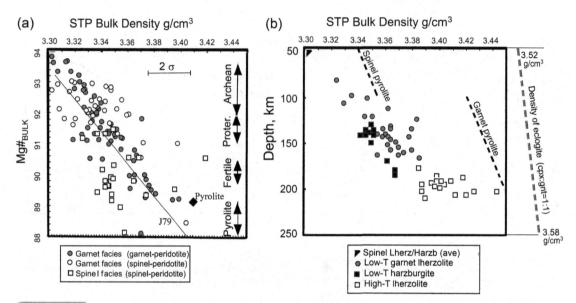

Fig. 6.22 Bulk normative densities of mantle peridotites (at room, STP, conditions). (a) Density versus Mg# = 100 Mg/(Mg + Fe) for garnet and spinel peridotites from the continental lithospheric mantle. The bar shows the largest estimated 2σ error. (Source: Lee, 2003.) Thin line – density of continental peridotites calculated by Jordan (1979). (b) Density versus depth estimated for typical compositions of the Kaapvaal peridotites along a typical cratonic geotherm. Dashed lines – theoretical estimates of densities of spinel and garnet pyrolites and eclogite (shown by gray dashed line outside of the plot with the density values at two depths). (Source: James *et al.*, 2004.)

of continents and oceans. The latter is believed to be formed by conductive cooling and thus has the composition of pyrolite mantle. The density of pyrolite mantle at room pressure–temperature conditions is estimated to be 3.37–3.41 g/cm³ (Table 6.3). The mean value of 3.39 g/cm³ is equivalent to an *in situ* density of 3.24 g/cm³ (assuming thermal expansion coefficient 3.5×10^{-5} °C⁻¹, the potential temperature of the upper mantle of 1300 °C, and neglecting the pressure effect).

Griffin *et al.* (2003) argue that if a depleted Archean mantle was formed in tectonic settings resembling modern mid-ocean ridges or arc-related collisional settings, it might follow the "oceanic trend" and thus may not have the same low density as the average or highly depleted Archean mantles with similar Mg#, because olivine is denser than orthopyroxene of similar Mg# (Fig. 6.23). If such an Archean mantle had ever been formed, it would have been unstable with respect to mantle convection due to its high density and would rarely survive until the present.

Because of a strong correlation between density and Mg# of the lithospheric mantle (Fig. 6.22), lithosphere depletion is often quantified by density deficit with respect to pyrolite mantle (asthenosphere). Typical estimated depletion values are 1.4–1.5%, 0.6–0.9%, and 0% for the Archean, Proterozoic, and Phanerozoic lithospheric mantles, respectively. The lowest value of depletion (1.3%) estimated for the Archean lithospheric mantle was reported by Jordan (1979) for the Kaapvaal craton since that study did not distinguish low-*T* and high-*T*

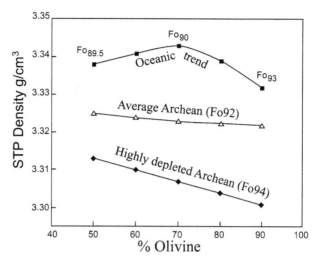

Fig. 6.23 Mantle density as a function of olivine content for the Archean lithospheric mantle of two different compositions (moderately and highly depleted) and for the Phanerozoic mantle which follows the "oceanic trend" (Fig. 6.8). For each curve, the ratio of orthopyroxene/clinopyroxene/garnet is constant as % olivine changes (from Griffin *et al.*, 2003a).

cratonic peridotites. Parameterization of the degree of depletion of the lithospheric mantle by its density deficit should be used with caution since this parameter provides an average characteristic, with density anomaly integrated over the entire column of the lithospheric mantle. The actual depth distribution of iron depletion and density deficit with respect to pyrolite mantle is significantly different in the upper and lower parts of the lithospheric mantle (Fig. 6.15), leading to significant variations of lithosphere density with depth. For example, high densities of the Archean Kaapvaal lithospheric mantle at ~ 200 km depth are clearly associated with melt-metasomatized, fertile composition (Fig. 6.22b). A similar pattern is observed for depth variations of normative seismic velocities in the Kaapvaal upper mantle calculated from mineral compositions derived from xenolith and garnet data (Fig. 6.24).

6.2.4 Isopycnicity

Thermal and compositional effects on mantle density

The combined effects of lateral variations in temperature and composition on density variations in the continental mantle may be expressed as:

$$\delta\rho/\rho = -\alpha\delta T + \partial(\ln\rho)/\partial R_{Fe}\,\delta R_{Fe} + \partial(\ln\rho)/\partial X_{gnt}\,\delta X_{gnt} \qquad (6.1)$$

where δT are temperature perturbations, $\alpha = 3.5 \times 10^{-5}\,1/K$ is the thermal expansion coefficient, and compositional variations may be expressed in terms of the iron molar ratio $R_{Fe} = X_{Fe}/(X_{Fe} + X_{Mg})$ and the molar fraction of garnet in the whole rock, X_{gnt}. A temperature anomaly of ~500 °C (which may exist at 100 km depth between stable and active regions, Fig. 4.33) will yield the high-density anomaly $\delta\rho/\rho = +1.75\%$ in cold cratonic regions. On the other hand, 10% fractionation of basalt from fertile subcontinental lherzolite corresponds

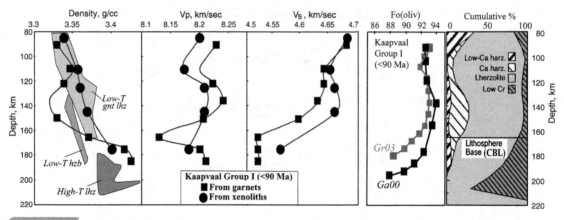

Fig. 6.24 Variations of normative density, seismic Vp and Vs velocities, and forsterite content in olivine with depth, and a lithologic cross-section based on Group I (< 90 Ma) kimberlites of the SW Kaapvaal craton and calculated from garnet data and a suite of xenoliths. Gray shading on the density plot: typical compositions of the Kaapvaal peridotites along a typical cratonic geotherm (compare with Fig. 6.22b). Lines labeled Ga00 and Gr03 on the Fo plot refer to data reported by Gaul *et al.* (2000) and Griffin *et al.* (2003a), correspondingly (other sources: O'Reilly and Griffin, 2006; James *et al.*, 2004; Griffin *et al.*, 1999c).

to $\delta R_{Fe} \sim 0.013$ and $\delta X_{gnt} \sim 0.056$ (Jordan, 1979), and for $\partial(\ln \rho)/\partial R_{Fe} = +0.32$ and $\partial(\ln \rho)/\partial X_{gnt} = +0.10$, compositional perturbations alone yield a density deficit $\delta \rho/\rho = -0.98\%$. This implies that the compositional heterogeneity due to basalt depletion in the continental lithospheric mantle should have a strong effect on ρ, producing $\delta \rho/\delta$ variations comparable in magnitude with thermally induced $\delta \rho/\delta$ variations (Forte and Perry, 2000). Such strong anomalies in mantle density if associated with differences in mantle depletion alone should be seen in global gravity data, in particular in geoid. However, they are not observed (MacDonald, 1963; Kaula, 1967; Shapiro *et al.*, 1999), and the question is, why?

Isopycnic hypothesis

A weak correlation between surface tectonics and long-wavelength geoid height anomalies (i.e. the absence of geoid anomalies over the cratonic regions with thick lithospheric roots and cold geotherms) as well as the near-constant values of the continental freeboard provide evidence for the existence of the continental chemical boundary layer with neutral buoyancy, so that the continental tectosphere and the oceanic mantle should have the same *in situ* density profile (Jordan, 1978, 1981, 1988). The gravitational effect of low temperature of cratonic upper mantle should be compensated by its low-density composition. This means that the relative compositional density decrease should be $\Delta \rho \sim -\alpha \Delta T$, where α is the thermal expansion coefficient. For the average temperature contrast $\Delta T \sim 300-500$ °C between the cratonic and the oceanic upper mantle and assuming $\alpha = 3.0 \times 10^{-5}$ 1/K, this yields an *average* $\Delta \rho$ of -0.9% to -1.5 % for a density deficit in the cratonic lithospheric mantle.

This idea was further developed into the isopycnic (i.e. equal-density) hypothesis according to which the *density deficit decreases with depth coherently with the decrease in the*

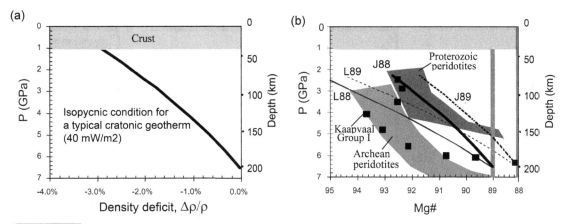

Fig. 6.25 (a) Compositional density deficit in the continental lithospheric mantle required to balance the thermal density difference between a cold cratonic lithospheric mantle that follows a 40 mW/m² conductive geotherm and the asthenosphere. (b) Compositional density in (a) converted into Mg# variations for two Mg# of the primitive mantle (solid lines Mg# = 88 and dashed lines Mg# = 89) and two values of thermal expansion a (bold lines marked J for $a = 3.0 \times 10^{-5}$ 1/K and thin lines marked L for $a = 3.9 \times 10^{-5}$ 1/K). Gray shading – typical mg# ranges for Archean and Proterozoic peridotites; symbols – average data for Kaapvaal Group I (< 90 Ma) kimberlites (compare with Fig. 6.15a). (Sources: Jordan, 1979; Lee, 2003.)

temperature contrast between cratons and oceans (Fig. 6.25a). Since the low-density structure of the cratonic lithospheric mantle is associated with its iron-depleted composition produced by the removal of basaltic magma, Jordan (1988) proposed a linear relationship between lithospheric density (in g/cm³ and normalized to standard pressure and temperature conditions) and Mg# (the degree of basaltic depletion) in the form:

$$\rho = 5.093 - 1.91\mathrm{Mg\#}, \tag{6.2}$$

which requires SPT densities of 3.33 g/cm³ for the Archean, 3.34 g/cm³ for the Meso-Neoproterozoic, and 3.36 g/cm³ for the Phanerozoic lithospheric mantle (assuming mg# of 0.926, 0.918, and 0.908, respectively, Table 6.4), or 1.5%, ~1.1%, and ~0.6% for their degrees of depletion, assuming that the primitive mantle has Mg# = 0.897 and $\rho = 3.38$ g/cm³. Figure 6.25b illustrates how density deficit in the cratonic lithospheric mantle corresponds to the iron-depletion number. Since the thermal expansion coefficient in the upper mantle (Fig. 4.1) and the upper mantle composition are not well known, the conversion is non-unique. Furthermore, due to the small-scale compositional heterogeneity of the cratonic lithospheric mantle, data from individual kimberlites would not necessarily follow generalizations such as given by eq. 6.2.

Tests of the isopycnic hypothesis

The validity of the isopycnic hypothesis is still debated. It is not clear whether a downward density increase due to compositional changes (i.e. decreasing depletion with depth) exactly

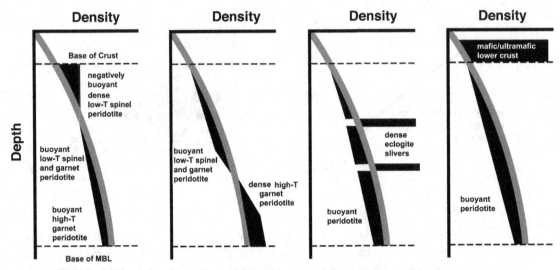

Possible density structures for a neutrally buoyant craton (Source: Kelly *et al.*, 2003). The cratonic mantle could be neutrally buoyant either if the density of the lithospheric mantle fits an isopycnic condition at every depth of the entire lithospheric column, or if the density versus depth distribution follows one of the four shown set-ups.

compensates a temperature increase with depth, in general, and at any depth in the litho-spheric mantle, in particular. Provided that the isopycnic condition is satisfied, five various scenarios for the neutral buoyancy of the cratonic upper mantle are possible (Fig. 6.26):

(1) density of the lithospheric mantle fits an isopycnic condition at every depth of the entire lithospheric column (not shown);
(2) negative buoyancy of dense shallow low-temperature peridotites is compensated by positively buoyant deeper low-temperature peridotites;
(3) positive buoyancy of shallow, highly depleted, low-temperature peridotites is compensated by negative buoyancy of deeper, more Fe-rich, high-temperature peridotites;
(4) positively buoyant depleted peridotites are compensated by layers of dense, basaltic material in eclogite facies;
(5) positive buoyancy of the entire cratonic mantle column is compensated by dense lower crust, for example if it contains abundant garnet granulites in the lowermost crust.

Studies of the cratonic peridotites from the Kaapvaal and Tanzanian lithospheric mantle support case 3 for maintaining neutral buoyancy of the cratonic upper mantle (Fig. 6.27). The upper part of the lithospheric mantle (down to ~ 140–150 km depth) is lighter than required by the isopycnic hypothesis, while the lower part of the lithospheric mantle (probably melt-metasomatized) is denser than predicted by the isopycnic hypothesis. When integrated over the entire lithospheric column, the temperature and compositional effects on mantle densities are balanced. Note that below 150 km depth, the Archean lithosphere is, however, more depleted than eq. 6.2 predicts for a 40 mW/m^2 geotherm (Fig. 6.25b).

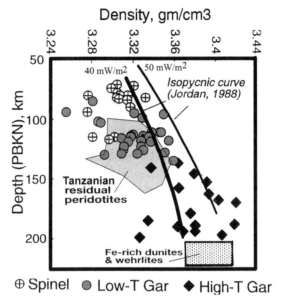

KAAPVAAL LHERZOLITES
TANZANIAN PERIDOTITES

Fig. 6.27 Densities of the Kaapvaal and Tanzanian peridotites plotted versus depth (pressure as determined by BKN geobarometer) (after Boyd *et al.*, 1999; Lee and Rudnick, 1999.) Solid lines – isopycnic curves for 40 and 50 mW/m^2 geotherms.

In the real lithospheric mantle mg# commonly decreases with depth, leading to an increase in STP density with depth (Fig. 6.25). Thus one should expect that

- while the isopycnic approximation holds when integrated over the lithospheric thickness, it still permits significant deviations in the upper and lower layers of lithospheric mantle, for example, to lower-than-predicted and higher-than-predicted densities, respectively, as is observed in the Kaapvaal and Tanzanian peridotite data. However, since heavy Fe-rich dunites may be undersampled by xenoliths, the distribution of lithospheric density with depth may further deviate from expression (6.2).

Using a large data set of mantle peridotites from different continental regions worldwide, Poudjom Djomani *et al.* (2001) calculated density–depth profiles for the lithospheric mantle of various ages (Fig. 6.28). They calculated normative (at room P–T conditions) densities from typical modal compositions of the Archean, Proterozoic, and Phanerozoic lithospheric mantle (with Fo93, Fo91, and Fo90, correspondingly) and approximated the asthenosphere by primitive mantle composition (with Fo90). In all cases the calculated density increases linearly with depth for all regions (except for hot Phanerozoic regions, such as Eastern Australia), and shows a good agreement with other data from the Archean cratons (e.g. Figs. 6.24; 6.27). A different approach has been undertaken by Lee *et al.* (2001) who calculated normative densities from Mg#, using the empirical relationship between density (in g/cm^3) and Mg# derived from the Tanzanian xenolith data:

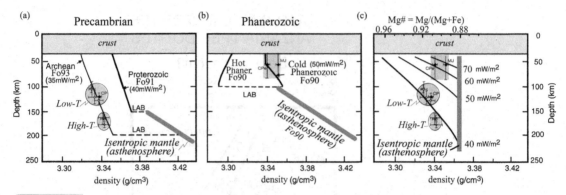

Fig. 6.28 Density profiles for the continental lithospheric mantle calculated from average mantle compositions. (a–b) Cumulative density (with account taken for column thickness) for lithosphere of different ages and asthenosphere (Source: Poudjom Djomani *et al.*, 2001). LAB (the lithospheric base) is defined by the depth where typical geotherms cross the mantle adiabat. (c) Isopycnic density profiles for geotherms corresponding to four surface heat flow values (Source: Lee *et al.*, 2001). Data points are for low-temperature and high-temperature xenolith suites from the Siberian, Tanzanian, and Kaapvaal cratons and for three locations in western USA (labeled MJ, CP, CPe). Pressures are based on the Brey and Kohler (1990) geobarometer for garnet peridotites and are constrained by the spinel–garnet transition for spinel peridotites (MJ and CPe). Asthenosphere is assumed to have an adiabatic gradient of 0.5 K/km and potential temperature of 1300 °C; thermal expansion is based on the polynomial expression (Fei, 1995) for the end-member minerals (a, b) or assumed to be constant, 2.7×10^{-5} 1/K (c).

$$\rho = 4.201 - 0.95mg\#. \tag{6.3}$$

Thus calculated, normative densities were recalculated to the upper mantle temperatures following four conductive geotherms. Note that while there is general agreement between the two approaches for the Archean lithospheric mantle (low-temperature and high-temperature cratonic peridotites agree with a 35–40 mW/m^2 cratonic geotherm), the predicted mantle densities are significantly different for younger (or warmer) lithospheric mantle. In particular, xenoliths from western USA (from the Mojave and the edge of the Colorado Plateau) fall onto a density profile predicted for a cold (50 mW/m^2) Phanerozoic region (Poudjom Djomani *et al.*, 2001), or onto density profiles for a 60–70 mW/m^2 conductive geotherm (Lee *et al.*, 2001). Note that density profiles for isentropic mantle are also significantly different in the two models (Fig. 6.28).

6.3 Are xenoliths representative of the cratonic upper mantle?

6.3.1 Sampling bias

The role of statistics and sampling in Earth Sciences has been discussed by Anderson (2007). He points out (p. 143) that the central limit theorem and the law of large numbers state that variably sized samples from a heterogeneous population will yield the same mean but will have variances that decrease as the sampling size or the volume of the sampled region increases. Small sample sizes are more likely to have extreme values than samples that blend components from a large volume

(here "the sample size" is the number of measurements that goes into the computation of each mean). An important aspect of the central limit theorem is that *regardless of the shape of the original statistical distribution in the natural system, sampling bias will always result in a normal distribution*. This means that petrologic data as well as geophysical methods sample over large areas of the mantle, average its material properties, and do not distinguish extreme values in the measured parameters. Thus, a homogeneous mantle (derived either from geophysical or petrologic sampling) does not imply that the source region is homogeneous, but is more likely to reflect sampling (averaging) over a large region.

Anderson (2007) further discusses the application of sampling statistics to volcanoes and the composition of the mantle that they sample. Since mantle xenoliths are brought to the surface by various types of magmas (basalts, kimberlites, lamproites, and lamprophyres) during their eruption, the discussion and conclusions of Anderson (p. 144–145) are directly applicable to xenolith sampling of the upper mantle. In particular,

- although magmas sample large volumes of the mantle, it is not necessary for the composition of the sampled region to correspond to the xenolith compositions (for example, pyrolite and picrite are hypothetical rock types that exist only as averages);
- sampling by magmas is biased not only by averaging (sampling method) but also by melting reactions (geochemical sampling processes), in which only definite components and compositions are possible under definite P–T conditions (in particular heavy dunites are likely to be undersampled by mantle-derived xenoliths);
- magmatic sampling is also biased by physical and tectonic processes that caused magmatism (geodynamical sampling processes) and defined the time of the eruption and stress and temperature conditions that caused magmatism.

The following discussion will show that geophysical studies based on three independent approaches indicate that xenolith data on densities and seismic velocities in the lithospheric mantle of the Archean–Proterozoic cratons may be non-representative of the "intact" (i.e. unsampled by mantle-derived peridotite xenoliths) cratonic mantle. Besides the sampling bias discussed above, the discrepancy between major-element petrologic models and geophysical observations can stem from at least two additional sources: (a) metasomatic modification of the cratonic lithosphere prior/during kimberlite magmatism; (b) underestimate of the effect of orthopyroxene on densities and velocities since many geophysical interpretations are based on parameterization of xenolith data solely in terms of iron depletion and do not take other compositional variations into account.

6.3.2 Why should xenoliths be representative?

Mantle xenoliths (rock fragments entrapped by magmas during their eruption and transported to the surface) provide unique direct sampling of the mantle and their role in our

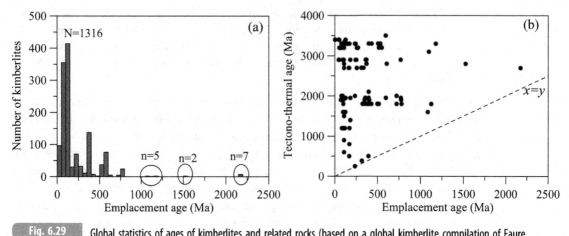

Fig. 6.29 Global statistics of ages of kimberlites and related rocks (based on a global kimberlite compilation of Faure, 2006 and a $1° \times 1°$ global compilation of tectono-thermal ages by Artemieva, 2006).

knowledge of upper mantle composition is crucial. However, from a geophysical perspective it is unclear why, in the first place, they should be representative of the cratonic lithospheric mantle: by definition, cratons are *stable* parts of the continents, while magmatism that brings xenoliths to the surface is not a process typical for stable regions. Xenoliths are also restricted both in space and in time (Figs. 5.1; 6.29) and, clearly, they do not reflect tectono-thermal conditions "normal" for stable continental interiors.

(1) Magmatic events that bring xenoliths to the surface require the presence of a thermal anomaly in the mantle (plume?). Such conditions do not reflect a steady-state thermal regime, typical for cratons.

(2) Kimberlites and related magmas are spatially localized; in many intraplate regions they are restricted to pre-existing weakness zones and thus sample lithospheric mantle with a potentially atypical structure.

(3) Magmas that bring mantle xenoliths to the surface can alter the *in situ* mineralogical composition of the lithospheric mantle during or prior to kimberlite emplacement. However, mantle xenoliths are believed to remain largely mineralogically unmodified during their transport to the surface; in particular, kimberlite magmas ascend with a rate of > 10 m/s and reach the surface from ~ 200 km depth in a time ranging from several hours to several days (e.g. Peslier *et al.*, 2006).

Orogenic peridotites, or peridotite massifs (sometimes also called alpine-type peridotites) provide complementary information on the composition of the lithospheric mantle. These peridotite bodies tectonically emplaced within the continental crust occur within high-pressure and ultrahigh-pressure metamorphic terranes of compressional orogens. In contrast to small-size mantle xenoliths, they represent large mantle volumes (with outcrop dimensions > 100 km^2) which permit the study of spatial and structural relationships of rock types in orogenic peridotites. Since orogenic peridotites are emplaced during collisional events, it is also questionable whether they should be considered as representative of the *in situ* structure and composition of the "intact" lithospheric mantle.

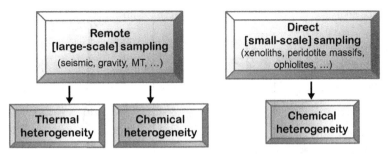

Fig. 6.30   Sketch showing the difference between remote geophysical and direct petrologic sampling.

There are significant differences in geophysical and petrologic sampling of the lithosphere and upper mantle and their results should be compared with caution (Fig. 6.30). Remote studies of the upper mantle are sensitive to *both thermal and chemical* heterogeneities and provide large-scale sampling of a large mantle volume. In contrast, petrologic data ("direct sampling") from mantle peridotites, orogenic peridotite massifs, and ophiolites are less affected by temperature variations and provide information on small-scale *chemical* heterogeneity of the upper mantle. Thus, a meaningful comparison of geophysical and petrologic data on mantle physical properties (such as density, seismic velocities, or electrical conductivity) requires, as the first step, a recalculation of *in situ* determined physical properties to laboratory P–T conditions. It is hardly possible to make a quantitative correction for different sampling scales of geophysical and petrologic data, so one should simply have this noteworthy difference in mind.

Several lines of evidence suggest that mantle-derived peridotites can be atypical of the unsampled (intact) lithospheric mantle. The following examples include a comparison of geophysical constraints and xenolith-based data on mantle densities, seismic velocities, and water content in the cratonic mantle.

6.3.3 Is the cratonic mantle dry or wet?

Experimental studies on deformation of olivine aggregates demonstrate that a wet (water-saturated) or a dry (water-free) composition of the upper mantle has a critical effect on mantle rheology (e.g. Karato *et al.*, 1986; Kohlstedt *et al.*, 1995; Mei and Kohlstedt, 2000a, b; Karato, 2008). Water depletion is often considered to be a key to the longevity of the cratonic lithosphere: a preferential partitioning of water into a melt phase during partial melting of the mantle should produce residue with a relatively low water content. A dry composition should increase the mechanical stiffness of the residue (the lithosphere) and thus should provide long-term stability for the continental tectosphere (Pollack, 1986). Numerical models of mantle convection with floating continents indicate that thick keels of Archean lithospheric mantle could have sustained convective instability and survived for 2–3 Ga only if the Archean mantle has (i) a distinct (depleted) chemical composition and (ii) has a water content of less than 100 ppm H/Si (e.g. Doin *et al.*, 1997; Lenardic *et al.*, 2003).

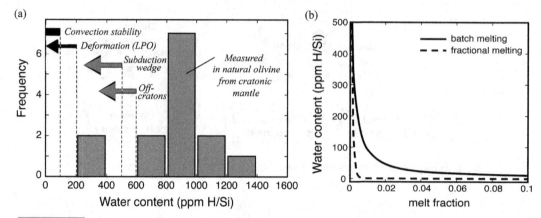

Fig. 6.31   (a) Water content measured in mantle olivines from different tectonic settings (gray) and geophysical estimates of water content in the cratonic upper mantle (black). (b) Water content left in the residue after partial melting of mantle. Melting process removes water and results in high water depletion in the lithospheric mantle. (Sources: Koga *et al.*, 2003; Peslier and Luhr, 2006; Li *et al.*, 2008; Katayama and Korenaga, 2009.)

While the first condition is undeniably confirmed by xenolith data, fulfillment of the second condition has not been indubitably proven so far, and the water contents of the oceanic upper mantle and continental lithospheric mantle are not yet well known. The presence of A-type fabric (Fig. 3.15a) of lattice-preferred orientation in mantle olivine from southern African xenoliths (Ben Ismail and Mainprice, 1998) has been interpreted as evidence that water-poor conditions (<200 ppm H/Si) prevailed during long-term plastic deformation of the cratonic lithosphere (Fig. 6.31) (Katayama and Korenaga, 2009). Experimental and theoretical studies of electrical resistivity of the upper mantle (the parameter that is highly sensitive to water presence) also indicate water-poor conditions in the continental lithospheric mantle (Fig. 7.38), in contrast to water-saturated oceanic upper mantle (e.g. Hirth *et al.*, 2000).

Contrary to geophysical constraints and theoretical expectations, mantle xenoliths from cratonic settings commonly contain hydrous minerals such as phlogopites, and thus indicate water-rich environments (Pearson *et al.*, 2003). Furthermore, laboratory measurements of water content in mantle-derived peridotites from South Africa and Lesotho indicate that olivines from the cratonic upper mantle contain more water than olivines from the mantle beneath continental mobile belts and even subduction wedges (Bell and Rossman, 1992; Kurosawa *et al.*, 1997; Peslier and Luhr, 2006). The cratonic xenoliths from the Kaapvaal craton contain ~800 ppm H/Si in mantle olivine, while the water content in mantle olivine from tectonically active regions is <600 ppm H/Si in continental off-craton settings and <500 ppm H/Si in mantle wedge in subduction zones (Fig. 6.31).

If the cratonic mantle has a higher water content than tectonically mobile regions, it should be mechanically weak and unstable with respect to convection instabilities. Thus, there is a clear contradiction between "direct sampling" by mantle xenoliths which indicates a wet mantle and "remote sensing" by various geophysical methods (such as depth variations of electrical conductivity in cratonic lithospheric mantle, see Chapter 7) which, together with theoretical models of mantle convection, favor (or even require) a dry mantle. This contradiction can be resolved if the present-day high water content in cratonic xenoliths has been acquired through localized metasomatism before/during volatile-rich kimberlite eruption (because of a high diffusivity of hydrogen, the water content of olivine can be easily modified), while the "intact" cratonic lithospheric mantle has a low water content, perhaps as low as at least one order of magnitude lower than the oceanic upper mantle (Katayama and Korenaga, 2009).

6.3.4 Mantle density from geophysical data

Gravity constraints on mantle density

Global modeling of gravity data corrected for regional variations in the crustal structure and lithospheric temperature suggests significant (non-thermal in origin) variations in density of the lithospheric mantle (Kaban et al., 2003). These results are in overall quantitative agreement with xenolith-based values of the upper mantle density, but indicate large density variations both within the cratonic regions and between different cratons. In that study, at the first step mantle residual gravity anomalies and mantle residual topography were calculated by subtracting the gravity effect of the crust from the observed gravity field. Comparison with a global seismic tomography model showed good positive correlation between the amplitudes of mantle residual gravity anomalies and seismic velocity anomalies in the upper mantle. By the sign of the residual gravity and residual topography the cratonic areas fall into two well-distinguished groups. The Precambrian Eurasia (the Baltic Shield, the East European Platform, the Ukrainian Shield, and the Siberian craton) has the largest positive residual gravity anomalies ($>70\,$mGal) and the most significant negative residual topography ($<-0.5\,$km), implying that it has a dense lithospheric mantle (either due to low mantle temperatures or due to heavy composition). In contrast, cratons of Western Australia, South America, India, and Southern Africa show negative mantle gravity anomalies and positive residual topography. This indicates that the average lithospheric density in these cratons is less than the upper mantle density beneath old ocean (taken in the study as the reference lithosphere). Other cratons have values of residual gravity and residual topography intermediate between the two groups.

In the next step, the effect of spatially differential thermal expansion due to lateral temperature variations has been removed from mantle residual gravity (density) anomalies in order to extract the effect of compositional density variations in the lithospheric mantle. Since gravity models provide estimates of density anomalies integrated over an unknown depth interval in the upper mantle (CBL), two models for the depth distribution of density deficit were considered for the cratonic lithospheric mantle. The first model was based on the assumption that the base of the CBL coincides with the base of the TBL (defined by the depth to a 1300 °C isotherm). The second model assumed that the base of the CBL is at a 200 km depth beneath all of the cratonic regions. The results show the following (Fig. 6.32).

(1) For the more realistic case of CBL = TBL, the *average* values of compositional density deficit are close to those predicted by xenoliths (Table 6.4), but display a significant scatter between 1.1% and 1.9% (a density deficit of $-49\,$kg/m^3 corresponds to ~1.3%).

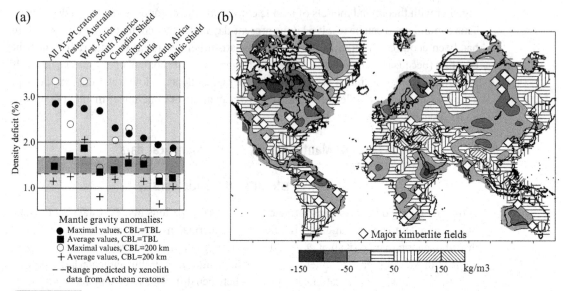

Fig. 6.32 Density anomalies in the upper mantle due to compositional variations (the temperature effect on mantle density is removed). The depth distribution of the anomalies within the CBL is unknown. (a) Average and maximal values of the density deficit in the lithospheric mantle of the Archean–Paleoproterozoic cratons due to compositional changes. The thickness of the CBL is assumed to be equal to the thickness of the TBL or to extend to 200 km depth. (b) Compositional density anomalies in the continental upper mantle. Note that all cratons have compositional density deficit; however, kimberlites do not sample cratonic mantle with the largest negative density anomalies. (Sources: Kaban et al., 2003; Kaban, personal communication.)

This scatter reflects, to some extent, the proportion of the Archean and the Proterozoic lithosphere within each of the cratons. However, a large scatter of density deficit within the Archean cratonic lithosphere is not correlated with crustal differentiation ages. It reflects strong compositional differences between various cratons which are not reflected in xenolith constraints on the bulk composition of cratonic mantle.

(2) For all of the cratons the *maximum* values of density deficit (which should characterize the Archean portion of each of the cratons) are significantly larger (1.7–2.8%) than implied by most petrologic studies of mantle xenoliths. However, the cratonic mantle with the highest amplitude of density deficit is not sampled by kimberlites (Fig. 6.32b). Either kimberlite magmatism is restricted to pre-existing weakness zones with mantle composition less depleted (more dense) than "intact" cratonic mantle, or kimberlite (and related) magmatism may refertilize the cratonic mantle during or prior to eruptions. The possibility of such a large-scale (over distances of several hundred kilometers) chemical modification of the cratonic mantle by melt-metasomatism is suggested by petrologic data from the Kaapvaal and the Siberian cratons (e.g. Griffin et al., 2005). In such a case, mantle metasomatism may cause a significant increase in iron content (see Section 6.1 and Fig. 6.16a) and lead to a density increase of the cratonic lithosphere (Fig. 6.33).

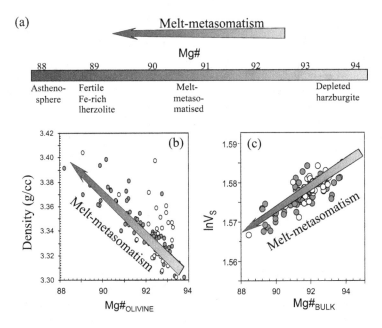

Fig. 6.33 Effect of melt-metasomatism on Mg# (a) and the corresponding changes in density (b) and natural logarithm of Vs seismic velocity (c). Symbols – STP density and seismic velocity versus Mg# for garnet-facies peridotites (data of Lee, 2003).

Mantle density from freeboard

The East European Craton is interesting for a freeboard analysis because it lacks surface relief despite huge amplitudes of topography at the top of the basement (20+ km), at the Moho (~30 km), and at the lithosphere–asthenosphere boundary (200+ km) with an amazing amplitude of *c.* 50 km in variations in the thickness of the crystalline crust (Figs. 3.37; 3.40). It is intriguing to examine how these huge variations in lithospheric structure are compensated to achieve near-zero variations in surface topography. High resolution regional seismic data on the crustal structure and high quality thermal data allow analysis of the relative contributions of the crust, the subcrustal lithosphere, and the dynamic support of the sublithospheric mantle in maintaining surface topography (Artemieva, 2007). The observed topography is the sum of (i) the crustal and lithospheric buoyancies and (ii) the dynamic topography caused by mantle flow due to density variations in the sublithospheric mantle. The buoyancies of the crust and of the lithospheric mantle depend on the thickness and *in situ* (i.e. accounting for lithospheric temperature variations) density of the crust and the lithospheric mantle, respectively.

Two model parameters, the density of the lithospheric mantle and the dynamic support of the sublithospheric mantle, cannot be assessed independently and require model assumptions. Numerous attempts to estimate the dynamic topography display a huge discrepancy, with values ranging from ~1.5 km to near-zero (e.g. Hager *et al.*, 1985; Gurnis, 1990; Forte *et al.*, 1993; Lithgow-Bertelloni and Silver, 1998). Because of the large uncertainty

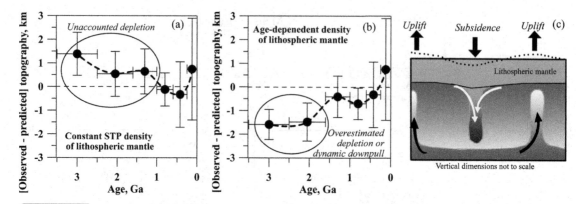

Fig. 6.34 (a–b) Unexplained topography of the East European craton (as the difference between the observed topography minus topography predicted by buoyancy modeling) for two end-member models of density of the lithospheric mantle (Source: Artemieva, 2007). Dots – average values (vertical error bars show standard deviations) calculated for six intervals (shown by horizontal bars) of tectono-thermal ages. (c) Cartoon of the effects of mantle dynamic support on surface topography. In the case of constant mantle density (a), the observed topography of the Precambrian parts is higher than model predictions. In the case of age-dependent SPT density of the lithospheric mantle (b), the predicted topography of the Archean–Paleoproterozoic parts of the craton is too low, either due to overestimated mantle depletion or due to the presence of a dynamic mantle downpull beneath the craton interior (c).

associated with the amplitude of dynamic topography, it is easier to make assumptions on the density structure of the lithospheric mantle, which depends on its composition and temperature. Given that the thermal effect on mantle density can be constrained independently from heat flow data, only compositional density variations remain unknown.

Two end-member models for density of the lithospheric mantle at room (STP) conditions include: (i) a model with constant SPT density equal to the SPT density of the primitive mantle (asthenosphere) so that all *in situ* lateral density variations in the lithospheric mantle are solely of thermal origin; (ii) a model with age-dependent SPT density of the lithospheric mantle with density values based on global petrologic studies of peridotites from mantle xenoliths (Table 6.4). These values correspond to 1.4–1.5% and 0.6–0.9% of the density deficit of the Archean–Paleoproterozoic and Meso-Neoproterozoic subcrustal lithosphere with respect to the underlying mantle. The Phanerozoic lithospheric mantle is assumed to have been formed by conductive cooling of the mantle and have the same composition as the latter, with 0% of the depletion.

Buoyancy modeling indicates that the observed topography cannot be explained for the end-member assumptions on density of the lithospheric mantle (Fig. 6.34). In the case of constant mantle density, the observed topography of the Precambrian parts of the craton is 0.5–1.5 km higher than model predictions. Since this model does not account for depletion of the Archean–Paleoproterozoic mantle, the heavy (with SPT density as of the fertile mantle), cold, and thick lithospheric root of the oldest parts of the craton would pull the surface topography downwards. In contrast, in the case of age-dependent SPT density of the lithospheric mantle, the observed topography of the Archean–Paleoproterozoic part of the craton is ~1.5 km lower than predicted. The end-member explanations for the lower-than-predicted topography include either overestimated mantle depletion or the presence of a dynamic mantle downpull beneath the craton interior, and the two effects cannot be reliably separated. The observed topography in the Archean–Paleoproterozoic parts of the craton can be matched if the density deficit in the lithospheric mantle is lower (~0.9%) than the assumed value based on global mantle xenolith data. Such an interpretation is supported by the presence of the Riphean rifts (sutures) in the central part of the East European Craton and the Devonian rifts in its southern parts (Fig. 3.69). Large-scale rift-related metasomatism could have reduced the iron depletion of the cratonic lithospheric mantle, increasing its density accordingly (Fig. 6.33). Such metasomatic modification of the cratonic mantle may explain, at least in part, the post-Devonian on-going subsidence of the vast area in the southern part of the East European Craton (Artemieva, 2003). On the other hand, the presence of a strong convective downwelling in the mantle beneath the craton interior can effectively divert heat from the lithospheric base, facilitating long-term survival of the Archean–Paleoproterozoic lithosphere. Indeed, a positive residual topography is observed at the cratonic margins, beneath the Norwegian mountains and the Urals, which still maintain significant elevation despite the Paleozoic age of both of the orogens.

6.3.5 Spatial correlations of seismic velocity anomalies and kimberlites

In the case where kimberlite-related melt-metasomatism causes an increase in iron content and a density increase in the cratonic lithosphere, it should also cause a decrease in Vs seismic velocity (Fig. 6.33). A comparison of Vs seismic velocity variations in the continental lithospheric mantle derived from body-wave and surface-wave tomography models and corrected for regional temperature variations indicates significantly different, non-thermal in origin, shear-velocity anomalies in the cratonic regions sampled and unsampled by xenoliths (Artemieva, 2009). Kimberlite provinces of the Baltic Shield, Siberia, and North America are clearly associated with a lower velocity mantle as compared to the "intact" cratonic keels. This is clearly illustrated by two parallel cross-sections through the East European craton (along 20E and 30E) which demonstrate a significant difference in the non-thermal (compositional) velocity structure of its lithospheric mantle (Fig. 6.35). The cratonic mantle that has been sampled by mantle xenoliths from kimberlite pipes in the Archean–Paleoproterozoic provinces of the Baltic Shield does not show the expected high Vs velocities, whereas the parallel profile reveals a high-velocity lithospheric mantle down to 200–250 km depth. It is appealing to ascribe the difference in velocity structure along the two profiles to the effect of kimberlite magmatism on the composition of the cratonic mantle. Such a pattern, also observed for other kimberlite regions, suggests that cratonic lithospheric mantle has been modified during or prior to the kimberlite magmatism and, as a consequence, *mantle composition constrained by xenoliths is likely to be non-representative of the unsampled "intact" cratonic mantle* (Artemieva, 2009).

Comparison of a body wave Vs tomography model (Grand, 2002) with kimberlite locations in South Africa indicates that *all of the best-studied xenolith suites*, which provide the bulk of the samples used in petrologic studies of the Archean lithospheric mantle, are *derived from the edges of the high-velocity cratonic nuclei* (Deen *et al.*, 2006). Similar pattern is observed in gravity modeling, which shows that xenoliths are derived from the

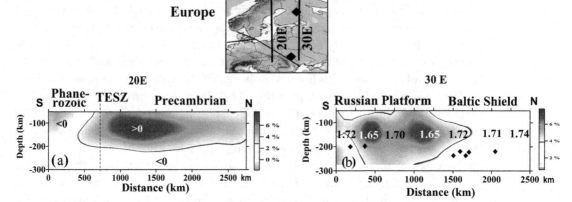

Fig. 6.35 Velocity cross-sections across Europe (along 20E and 30E) showing compositional heterogeneity in the upper mantle (modified after Artemieva, 2009). Seismic Vs velocities are derived from a surface-wave model (Shapiro and Ritzwoller, 2002) and a body-wave tomographic model (Grand, 2002) and are corrected for lateral and vertical temperature variations. Cross-section (b) also shows the Vp/Vs ratio in the upper mantle at 150 km depth (given by numbers) constrained by tomographic models of Bijwaard and Spakman (2000), Piromallo and Morelli (2003), Shapiro and Ritzwoller (2002). The locations of kimberlites are shown by diamonds. TESZ = Trans-European Suture Zone (shown by black line on the map). The cratonic mantle affected by kimberlite magmatism is less depleted.

edges of lithospheric terranes with the strongest density anomalies (Fig. 6.32). These xenoliths are strongly biased towards highly modified mantle material and represent the part of the lithospheric mantle strongly modified by metasomatic refertilization. Such a process produces rocks with higher density and lower shear-wave seismic velocity, while the more depleted material which constitutes *the bulk of the cratonic lithospheric mantle may be largely unsampled*. This implies that "intact" (or "pristime") Archean lithospheric mantle can be seriously underrepresented in petrologic studies so

> "that the widely accepted estimates of [the Archean subcontinental lithospheric mantle] composition are strongly biased" (Griffin and O'Reilly, 2007).

This conclusion is supported by recent petrologic studies from the Slave craton, East Greenland, and West Gneiss region in Norway which indicate that, at least some, cratonic peridotites can be unusually depleted (Fig. 6.13).

6.4 Summary

CBL from xenolith data

Melt extraction from the mantle produces a residue of peridotite partial melting (the lithospheric mantle) which is iron depleted and water depleted. Iron depletion is quantified by

Mg-number (Mg#). Melt extraction from the mantle produces an increase in mg#, which in turn leads to an increase in seismic velocities (particularly, in Vs) and a decrease in mantle density. Depending on melting pressures, temperatures, the amount of extracted melt, and the way the melt is extracted (batch or fractional melting), the composition of the residue can differ significantly. There is a strong compositional difference between the cratonic peridotites and peridotites from all other tectonic settings. The latter follow the "oceanic trend" of the correlation between modal olivine and Mg#. Compositional heterogeneity of the continental lithospheric mantle is also controlled by the addition of Si-rich materials to the residue and is characterized by the volume fraction of orthopyroxene (Opx#). Increasing Opx# causes a decrease in Vs (Fig. 3.24), but has no effect on density (Table 3.1).

The composition of the lithospheric mantle has a strong correlation with the age of the overlying crust; the average Mg# is highest in the lithospheric mantle of Archean cratons and lowest in Phanerozoic regions where it is close to the pyrolite values; Proterozoic lithosphere has intermediate Mg#. Significant lateral and vertical compositional variations in the lithospheric mantle exist on a scale of, at least, tens of kilometers.

Xenolith barometry allows for estimation of the depth of xenolith origin and vertical mapping of compositional variations in the upper mantle. The depth resolution depends on the choice of geobarometers (see Section 5.2.1). Pressure estimates may differ by 1.0–1.5 GPa (~30–45 km) for different geobarometers applied to the same data (the uncertainty is particularly high at $P > 6$ GPa, see also Table 10.1), with no reliable technique available for the spinel stability field (<60–80 km). With these uncertainties in mind, in the cratonic lithospheric mantle mg# is highest in the shallow mantle and gradually decreases with depth. In some cratons (Slave and East Greenland), the lithospheric mantle is compositionally stratified into two distinct layers. The base of CBL, if defined by the depth where Mg# becomes close to the pyrolite value, is transitional and may span over some tens of kilometers.

Mantle metasomatism (melt-related or "phlogopite"-metasomatism) can significantly modify the composition of the lithospheric mantle over a distance of several hundred kilometers from the centers of magmatism, as evidenced by xenolith data from the Kaapvaal and Siberian cratons. *Xenoliths from the Kaapvaal craton (Group I kimberlites), which are the backbone of the composition of the Archean lithospheric mantle, are strongly biased by chemical alteration caused by eruption of the Karoo traps.*

Density structure of the continental lithosphere

Uncertainties and resolution of gravity modeling

The density of the lithosphere is constrained by gravity data (available at high resolution) and models of the structure of the lithospheric layers. The latter are commonly constrained by seismic data, and velocity–density conversions are routinely used as a first-order approximation of the lithosphere density structure. Such conversions are non-unique and subject to great uncertainty (Fig. 3.25); compaction effects are important for calculating the gravity effect of the sedimentary cover. As a result, in some cases the uncertainty in gravity anomaly produced by a velocity–density conversion may exceed the range of observed gravity variations.

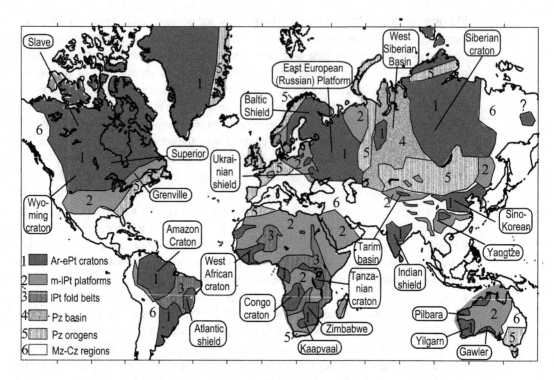

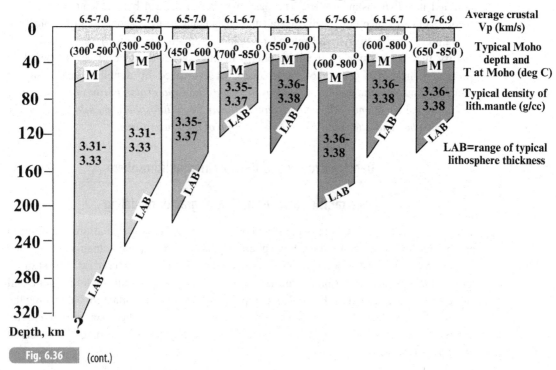

Fig. 6.36 (cont.)

Gravity modeling is non-unique. Density models for the lithospheric mantle require knowledge of the density structure of the crust, and their amplitude resolution depends on the quality of crustal model. Modeling should account for the spherical effects, and solutions for the flat Earth may lead to erroneous results. Importantly, gravity modeling only provides the average value of density variations in the lithospheric mantle, integrated over the entire column thickness. Indirect constraints on the depth of density anomalies can be made from the wavelength of observed gravity anomalies, which is about

$$\lambda \sim 2\pi z \qquad\qquad (6.4)$$

(where z is depth). This means that lateral resolution of a gravity model is no better than ~200 km when density variations have a subcrustal origin.

Lithosphere density from xenolith and gravity data

Xenolith data indicate that the density of the lithospheric mantle is correlated with its age and with Mg# (Table 6.4). This trend is explained by mantle depletion produced by melt removal. Gravity models, in general, agree with petrologic data, but indicate a broader range of average (non-thermal in origin) density variations in the cratonic lithospheric mantle. There is also significant scatter in the amplitude of mantle density anomalies between various cratons and within individual cratons. The amplitude of the strongest (non-thermal) negative density anomalies constrained by gravity is almost twice as large as the amplitude indicated by xenolith data. Kimberlites, apparently, do not sample the lithospheric mantle with the strongest negative density anomalies.

The density and thermal structure of stable continental lithosphere is summarized in Fig. 6.36 for different tectonic settings and geological ages. For some structures the geological (i.e. isotopic) age of the crust is much older than the tectono-thermal age (i.e. the age of the last major tectonic event). In particular, most basins with Paleozoic tectono-thermal ages were formed on Precambrian terranes and thus are parts of cratons. Archean cratons are subdivided into two groups with significantly different lithospheric thicknesses, ~200–220 km (South Africa, South America, India, Australia, probably also the Antarctic and Congo cratons) and 300–350 km (Siberia, the Baltic Shield, West Africa, the Canadian Shield). Since these cratons also show a significant difference in crustal thickness, positively correlated with lithospheric thickness, the density deficit in the lithospheric mantle of all Archean cratons required by free-board is the same (Artemieva and Mooney, 2001).

Caption for figure 6.36 (cont.)

Top: Tectonic provinces of the world based on geological ages and surface tectonics. The amount of data available for Antarctica is very limited and it is not shown on the map. Bottom: Typical cross-sections of stable continental lithosphere, showing ranges of crustal and lithospheric thickness, typical Moho temperatures, average crustal velocities, and bulk densities of the subcrustal lithosphere at room P–T conditions. Lithospheric thickness is based on various thermal models. Ar (I) – Archean cratons with a thick (300–350 km) lithosphere; Ar (II) – Archean cratons with a relatively thin (200–220 km) lithosphere; ePt and lPt – Paleo- and Neoproterozoic structures; Pz – Paleozoic structures; Pz (I) – the European Caledonides and Variscides and most of the Appalachians; Pz (II) – the Uralides and, probably, the northern Appalachians.

Because of significant differences in the lithospheric structure, Paleozoic orogens are also split into two groups. The first group, Pz(I), includes orogens with a thin (30–35 km) crust that have been significantly reworked by active tectonic processes since the Paleozoic. The second group, Pz(II), includes primarily the Uralides orogen which has remained intact within the continental interior since the Paleozoic and has preserved a thick crustal root (> 50 km).

Tectonically young structures, such as collisional orogens, continental rifts and regions of extended lithosphere are not shown due to a large diversity in their lithospheric structure. The oceanic upper mantle is also not shown; its STP density is close to the values estimated for pyrolite models (~3.38–3.39 g/cm^3), and at *in situ* conditions it is controlled largely by the thermal structure, which (for normal oceans) depends on the age of the ocean floor.

7 Electrical structure of the lithosphere

This chapter primarily discusses regional electromagnetic and magnetotelluric studies with a focus on electrical conductivity models of the lithospheric mantle for various tectonic settings. A short review of electromagnetic methods used in lithosphere studies is based largely on Vozoff (1986, 1991), Chave *et al.* (1991), Hermance (1995), Jones (1999), Jones and Craven (2004), Nover (2005), as well as on review papers published between 1992–2002 in several special issues of *Surveys in Geophysics* (volumes 13, 17, 18, 20, 23). Similarly to other chapters, it starts with an overview of laboratory measurements of electrical resistivity (conductivity) derived from numerous experimental studies. These data provide the basis for interpretations of electromagnetic (EM) and magnetotelluric (MT) results in terms of the lithospheric electrical conductivity structure. Regional conductivity models are presented for the oceanic and continental lithosphere.

7.1 Electromagnetic and magnetotelluric techniques

7.1.1 Electric conduction in rocks

Electrical properties of rocks, melts, and fluids play an important role in interpretations of remote geophysical surveys, since the electrical conductivity σ (or its inverse, electrical resistivity, $\rho = 1/\sigma$) is strongly sensitive to temperature, composition, and the presence of a fluid phase in the crust and the upper mantle. The sensitivity of electrical conductivity to variations in these parameters is significantly stronger than sensitivity of other geophysical characteristics, such as elastic moduli, density, and seismic velocities.

Four different processes of electrical charge transport are distinguished in crustal and mantle rocks (Nover, 2005).

- *Semiconduction* is the major process of electrical charge transport in mantle olivine. The process strongly depends on temperature and oxygen fugacity, which control the number and mobility of charge carriers (electrons). The effect of pressure is small in the absence of phase transitions. However, at high temperatures (>1200 °C) ionic conduction (with ions as charge carriers) becomes important in mantle olivine.
- *Electronic conduction* (with electrons or polarons as charge carriers)[1] is caused by the presence of an interconnected highly conducting phase (graphite or ores). For example, core samples from the KTB deep drillhole in Germany contain interconnected graphitic

---

[1] A polaron is a fermionic quasiparticle composed of a moving charge (electron or electron hole) that moves together with a region of lattice polarization and deformation which its movement caused (Landau, 1933).

veins and disconnected graphite crystals. The presence of even small volume fractions of carbon produces high conductivity anomalies, especially in the crust. The processes of electronic and electrolytic conduction are only moderately sensitive to changes in pressure and temperature.

- *Electrolytic conduction* occurs in fluid-saturated porous rocks. This charge transport process requires the presence of an interconnected pore network and its efficiency depends on rock porosity and permeability, the geometry of the pore space, chemical parameters of pore fluids, temperature, and surface interaction of the negatively charged mineral surface and water ions or dipoles. The process can produce electrical anisotropy associated with the rock fabric and the pore space geometry.

- *Conduction in interconnected partial melts* is important in volcanic areas, lower crust, and upper mantle; it significantly enhances bulk rock conductivity. The amount of melt and conductivity enhancement are controlled by temperature, pressure, composition, and the presence of fluids.

7.1.2 EM and MT methods

Data on electrical conductivity constrained by electromagnetic (EM) and magnetotelluric (MT) methods provide important information on the physical properties of the deep Earth. A large variety of EM and MT methods offers the possibility of studying tectonic structures at all scales, from local to regional, i.e. from a few meters to several hundred kilometers in size. Electromagnetic surveys are based on controlled-source studies (e.g. direct currents, power lines) and because of (usually) low source power these methods are used for studies of the continental upper and middle crust. However, ocean floor EM controlled-source experiments which use ocean-bottom cables can be used to constrain conductivity of the oceanic mantle.

Alternatively, the Earth's magnetic field is used as a source in MT surveys, providing the theoretical possibility of constraining conductivity profiles of the Earth at any depth. This possibility is based on the exponential decrease of the amplitude of electromagnetic waves with depth due to absorption. The skin depth δ (i.e. the depth in a homogeneous medium at which the amplitude decreases by exponent e from its surface value) depends on period T (in sec) as:

$$\delta = 0.503(T\rho)^{1/2} \tag{7.1}$$

(here δ is in km, $\rho = 1/\sigma$ is the electrical resistivity in ohm-m, and σ is electrical conductivity in S/m). This means that MT surveys based on low-frequency signals permit us to reach the deep layers of the Earth, and observations at increasing periods allow us to measure the Earth's conductivity at progressively greater depths. Information regarding the conductivity of the upper mantle (with a typical conductivity $\sigma \sim 0.01$ S/m) comes primarily from geomagnetic variations with periods of several hours (Fig. 7.1).

The MT method is based on the recordings of time-series of the two horizontal electric field components and the three magnetic field components. Variations in the magnetic field of the Earth induce electric currents, which generate an electric (telluric) field, the strength of

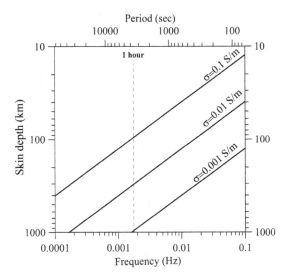

Skin depth (down from the surface) as a function of frequency (period) and the average rock conductivity.

which is proportional to the conductivity of a medium. Induced electric currents further produce variations of the magnetic field of internal origin, which depend on both the external field and the distribution of electrical conductivity within the Earth. Magneto-variational (MV) surveys (or geomagnetic depth soundings, GDS) are based on measurements of time variations of the components of the Earth's magnetic field alone. In polar latitudes, ionospheric EM sources cause additional difficulties in the interpretation of MT data (Osipova *et al.*, 1989).

7.1.3 Interpretations of magnetotelluric data

MT parameters

As the Earth's magnetic field is time-varying, simultaneous observations of three magnetic field components (H_x, H_y, and H_z) and two horizontal electric field components (E_x and E_y) at different frequencies are required in MT surveys to constrain conductivity (or its inverse, electrical resistivity $\rho = 1/\sigma$) variations with depth. The ratio of any two field components defines a complex impedance $Z_{ij}(\omega)$ as a function of the radial frequency of oscillations ω, e.g. $Z_{xy}(\omega) = E_x(\omega)/H_y(\omega)$, while the apparent resistivity is $\rho_{a,\,xy}(\omega) = 1/\mu\omega\,|\,E_x(\omega)/H_y(\omega)\,|^2$, where μ is the magnetic permeability. The phase difference between the electric and the magnetic field is: $\Phi_{xy} = 1/\tan[E_x(\omega)/H_y(\omega)]$; $\Phi_{xy} = 45°$ for a uniform medium, $\Phi_{xy} < 45°$ at the transition to a deeper more resistive layer, and $\Phi_{xy} > 45°$ at the transition to a deeper more conductive layer. Modern five-component instruments allow simultaneous magneto-variational and magnetotelluric surveys along a profile, which are inverted to 2D conductivity models for the crust and mantle. Most common conductivity models assume a layered Earth structure. During the last decade, codes to calculate magnetic induction in a laterally heterogeneous 3D Earth have been in development (e.g. Chou *et al.*, 2000).

The uniqueness theorem

The uniqueness theorem (Bailey, 1970; Weidelt, 1972) distinguishes EM and MT methods from other potential field studies. It states that only one model can fit **perfect data** for a 1D radially symmetric Earth with responses known for **all** frequencies. However, in the real Earth perfect data do not exist and data limitations (e.g. inaccuracy in response measurements, statistical errors, and limitations on the frequency range) lead to a loss of uniqueness of the inverse problem. The non-uniqueness means that there exists an infinite number of models which fit the **real data** equally well, and that the true value of conductivity cannot be resolved for any region within the Earth. To distinguish between different permitted conductivity models, EM and MT interpretations are often correlated with other geophysical and geological data (seismic and gravity interpretations, heat flow, and surface tectonics).

Inverse MT problems

Classical potential field theory interpretations are based on a comparison of the measured response of the Earth (e.g. the ratio of internal to external magnetic potentials) at different frequencies with theoretical response functions for assumed conductivity distributions. Such direct modeling of electromagnetic induction problems provides reasonably good starting models for a 1D distribution of electrical conductivity with depth; however their accuracy or uniqueness is unclear.

Another approach for determining the set of acceptable conductivity models directly from data is based on solution of the inverse electromagnetic induction problem. Different linear and non-linear algorithms have been proposed to constrain plausible models of conductivity, which both fit the data and are physically realistic. Some common approaches to estimate mantle conductivity from electromagnetic data are as follows.

(1) The Backus–Gilbert linearization (Backus and Gilbert, 1967, 1968) assumes that any model produced by the resolution kernels technique must be linearly close to the true conductivity distribution within the Earth. However, since this true conductivity distribution is *a priori* unknown, the technique may restrict the search to a space of conductivity models which are far from the best solution to the inverse MT problem.

(2) The extremal conductance inversions D+ and H+ are based on a search of the extreme solutions to the inverse problem using a non-linear inversion spectral expansion technique. The D+ and H+ models contain 1D inverse solutions for conductivity distribution within the layered Earth with the smallest possible misfit between the observed and the modeled surface inductive responses (Parker, 1980, 1983). Although models, in which minimized misfit is composed of delta-functions of conductivity, do not have direct physical interpretation, it has been noted that the locations of conductivity delta-functions are often correlated with seismic discontinuities (Schultz *et al.*, 1993). Extremal conductance inversions can be used to test lateral conductivity heterogeneities in the mantle (i.e. to test the validity of the assumption of one-dimensionality of the electric structure of the Earth). Lateral heterogeneity is not required by the data if for a set of joint inversions within the D+ space there exists a 1D solution which adequately fits all available response data from different sites.

(3) Two end-member approaches to estimate the distribution of resistivity (or conductivity) versus depth include models with a minimum number of layers or models with a minimum change of resistivity between multiple layers in an overparameterized problem. The latter approach, similar to the one used in seismic inversions, is called the Occam's inversion method (named after an English philosopher William of Ockham (1287–1347), best known for proposing the methodological principle (the so-called "Ockham's Razor") which states *don't multiply entities beyond necessity*"). The Occam's inversion method is used to find the "smoothest", or minimum norm, conductivity models to fit electromagnetic sounding data (Constable *et al.*, 1987; deGroot-Hedlin and Constable, 1990).

7.1.4 Resolution of MT methods

Spatial resolution of MT methods is lower than that of most seismic methods, since at the low frequencies used in MT studies the propagating electromagnetic waves are diffusive in nature (Jones, 1999). The horizontal resolution depends on the lateral sampling distance of measurements: conductivity anomalies with dimensions smaller than the sampling interval are poorly resolved. The vertical resolution of MT surveys is determined by the range of frequencies used and is typically *c.* 5–10% of the sampled depth. In the case of the proper spacing of measurements, the vertical and horizontal resolutions of MT studies are comparable.

There are, however, other limitations to the vertical resolution of the MT method. Commonly, MT models of mantle conductivity are based on a layered approximation. For the multi-layered Earth, problems arise in estimates of both resistivity values and layer thicknesses. The best resolved parameters are the *conductance* (i.e. the conductivity by thickness product) and the depth to the top of a conductive layer. This means that electromagnetic methods cannot reliably resolve the thickness and the conductivity (or its inverse, resistivity) independently, but only their product (e.g., Hjelt and Korja, 1993): a thin, highly conductive layer can produce the same MT response as a thick but more resistive layer. In the case when a conducting layer (e.g. lower crust) is sandwiched between two resistive layers (i.e. the upper-middle crust and the lithospheric mantle), it is possible to resolve both of the parameters of the conducting layer (conductivity and thickness) individually (Jones, 1999).

In general, because of the effect of screening, for deeper layers to be reliably resolved, their conductance should be higher than for the overlying layers. Thus, the presence of strong conductors in the crust creates additional complications in studies of mantle conductivity, as such conductors reduce the amplitudes of electromagnetic field at long periods (used for deep mantle sounding) and mask the conducting zones in the upper mantle (Haak and Hutton, 1986). As the lithospheric mantle is located between two well-conducting (and thus well-resolved) layers (the lower crust and the asthenosphere), its thickness is one of the best resolved model parameters with an error of 0.5% (Jones, 1999). However, the true resistivity of the lithospheric mantle is poorly resolved and is usually constrained with an error over an order of magnitude: the 1D MT method is insensitive to the resistivity of the resistive layers and only minimum bounds on their resistivity can be obtained as illustrated by the example in Fig. 7.2.

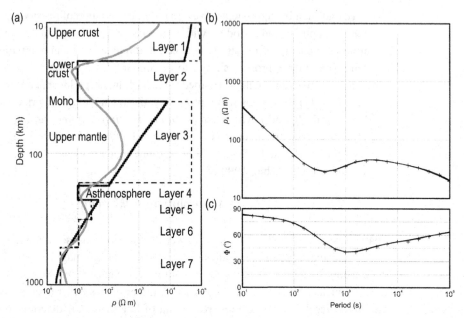

Fig. 7.2 (a) Theoretical reference resistivity model (bold black line) for the crust and upper mantle with high conductivity lower crust and asthenosphere, a 7-layer model (dashed line), and a smooth model (gray line). (b–c) Synthetic data for the apparent resistivity and the phase difference generated from the theoretical model (with 2% scatter). The 7-layer model and the smooth model both fit the synthetic data to within exactly one standard error on average, so that the true resistivity of the upper mantle cannot be resolved better than with an error of over an order of magnitude (from Jones, 1999).

7.2 Experimental data on electrical conductivity

7.2.1 Effect of lithology

EM and MT studies allow us to constrain conductivity–depth profiles for the Earth's crust and mantle. However, their interpretation in terms of composition, temperature, or physical state of the upper mantle would not be possible without laboratory studies of electrical conductivities (or resistivities) of different Earth materials (rocks, minerals, and fluids) at varying external conditions. The electrical conductivity of the upper mantle is essentially controlled by the conductivity of olivine, which comprises about 50–60% of the upper mantle. Experimental studies of olivine electrical conductivity demonstrate its strong dependence on a large number of physical and chemical parameters, which leaves a large degree of uncertainty in the interpretation of mechanisms responsible for geophysically observed conductivity anomalies. Major candidate mechanisms which can affect electrical conductivity of the upper mantle are reviewed in detail in this chapter and are summarized in Fig. 7.4. A review of different charge transport processes that control crustal conductivity can be found in Nover (2005).

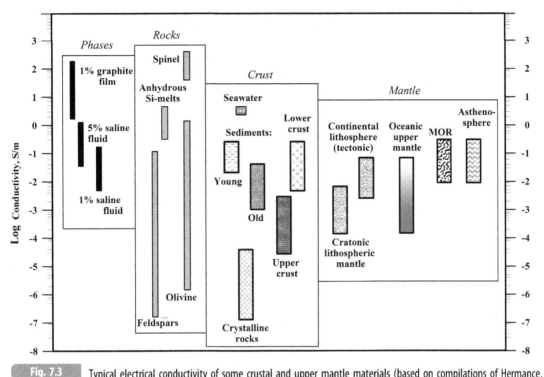

Fig. 7.3   Typical electrical conductivity of some crustal and upper mantle materials (based on compilations of Hermance, 1995; Tyburczy and Fisler, 1995; Jones, 1999; and data referred to in this chapter).

Most mantle rocks, including mantle olivine, are semiconductors (although graphite exhibits metallic conduction) and at low temperatures act as electric insulators given that their electric conductivity is very low. In ionic crystals, imperfections in the crystal lattice (point defects such as electron holes, electrons, lattice vacancies, interstitial ions) determine the electrical conductivity of minerals. In general, all of these defects are present in any mineral and contribute to its total conductivity, which is the sum of the conduction of all charge carriers:

$$\sigma = \sum c_i\, q_i\, v_i \tag{7.2}$$

with the summation over i, where c_i is the concentration of point defects of the i-type, q_i is its charge with respect to the neutral lattice, and v_i is its mobility (velocity of charge carrier per unit potential gradient).

The concentration of point defects depends on temperature, pressure, and the chemical reactions in the crystal lattice that produce defects of different types. Since even small compositional impurities in minerals can significantly change concentrations of defects and charge carriers, bulk conductivities of Earth's minerals can vary over orders of magnitude, and they are not well known for "pure" minerals due to the extreme sensitivity of defect concentrations to the content of minor and trace elements. Extensive experimental measurements indicate that conductivity of lithospheric materials varies by 7–8 orders of magnitude from highly conducting graphite and sulfides to highly resistant crystalline rocks (Fig. 7.3).

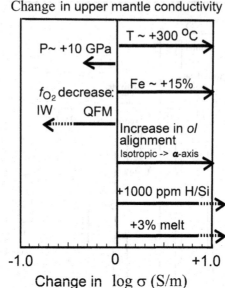

Change in upper mantle conductivity

Fig. 7.4   Summary of different effects on the electrical conductivity of the upper mantle.

7.2.2 Temperature dependence of conductivity in mantle rocks

Two conducting mechanisms in dry olivine

A temperature increase causes thermally induced semiconduction due to an increase in mobility of lattice defects and impurities. Both the mobility and the concentration of the charge carriers depend exponentially on temperature. The Nernst–Einstein relation (Mott and Gurney, 1948) links electrical conductivity σ to the diffusion coefficient D:

$$\sigma = ADcq^2/kT \tag{7.3}$$

where q is the electrical charge of charged elements, c is their concentration, k is the Maxwell–Boltzmann constant, T is temperature, and A is a numerical factor (close to unity). As diffusion is a thermally activated process, over a limited range of temperatures the conductivity of a rock can be described by an Arrhenius-type relation:

$$\sigma = \sigma_0 \, e^{-H/KT}, \tag{7.4}$$

where H is the activation enthalpy and σ_0 is a constant (the electrical conductivity at infinitely high temperature) which depends on the number and mobility of charge carriers.

Laboratory measurements of olivine conductivity at temperatures in the range 720–1500 °C show that at high temperatures a plot of logarithm of conductivity versus temperature deflects from a linear dependence as suggested by Eq. (7.4), indicating that conduction is more complex than a single conduction mechanism. Experimental data are better fit by:

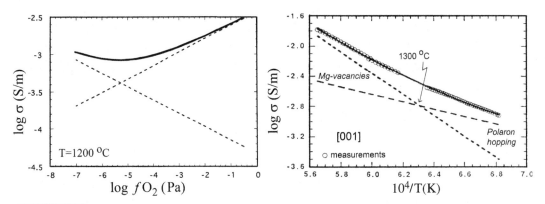

Fig. 7.5 Conductivity measured in San Carlos olivine (Fo89) as a function of oxygen fugacity and temperature (solid lines). Non-linear dependencies illustrate that total conductivity is the sum of two mechanisms (dashed lines) with different dependencies on oxygen fugacity and temperature. (Sources: Wanamaker and Duba, 1993; Shankland and Duba, 1990).

$$\sigma = \sigma_1 \exp(-H_1/kT) + \sigma_2 \exp(-H_2/kT), \tag{7.5}$$

where k is the Maxwell–Boltzmann constant and T is temperature (Constable and Duba, 1990). This parametric dependence indicates that there are two conducting mechanisms in dry olivine produced by two different thermally activated populations of defects with two different activation energies (Fig. 7.5). Low-temperature activation energy has values in the range 1.1 to 1.6 eV, while preferred values for higher-temperature activation energy are in the range 3.33 to 4.23 eV (Shankland and Duba, 1990; Constable et al., 1992). At a low-temperature regime (below 1200–1300 °C), the charge carriers are holes (polaron hopping associated with charge transfer between Fe^{2+} and Fe^{3+}), while at high temperatures conduction due to Mg vacancies dominates. Furthermore, the third conductivity mechanism with very low activation energy is likely to operate at low temperatures (640–720 °C) (Constable and Duba, 1990) (Table 7.1).

Experimental studies indicate that the temperature dependence of the electrical conductivity of other mantle minerals (namely, clinopyroxene and orthopyroxene) does not differ greatly from olivine conductivity. At any given temperature between 700–1500 °C, the ratio of their conductivities never exceeds 6 (Xu et al., 2000b).

"Standard olivine" models SO2 and SEO3

A strong consistency between different experimental studies on olivine allows us to constrain reference conductivity–temperature curves for the "standard olivine models" SO2 and SEO3 derived for dry subsolidus olivine (Fig. 7.6). The SO2 model is constrained by experimental data for single-crystal San Carlos olivine (Fo91.7) at constant gas mixture CO_2:$CO = 7$:1. The model has some kinetic problems because the samples are not entirely equilibrated with changes in temperature and oxygen fugacity. The SO2 model differs from the SEO3 mainly at the high-temperature end (>1200 °C) where the transition from polarons to magnesium vacancies conduction takes place (Constable et al., 1992; Constable, 2006).

Table 7.1 Activation enthalpy and σ_o constants for some mantle minerals (eq. 7.4)

Mineral	T (°C)	Log σ_o (S/m)	ΔH (eV)	Reference
Olivine (isotropic)	640–720	?	Very low	Constable and Duba, 1990
	720–1500	?	1.60 ± 0.01	
	>1500	?	7.16 ± 0.56	
Olivine (isotropic)	<1500	2.402	1.1–1.6	Constable et al., 1992
	>1500	9.17	3.33–4.23	
Olivine (isotropic)	1000–1400	2.69 ± 0.12	1.62 ± 0.04	Xu et al., 1998
Olivine [100]	1200–1500	?	1.11 ± 0.10^	Shankland and Duba, 1990
Olivine [010]	1200–1500	?	1.52 ± 0.02^	Shankland and Duba, 1990
Olivine [001]	1200–1500	?	0.97 ± 0.11^	Shankland and Duba, 1990
Olivine [100]	1200	2.27*	0.32**	Du Frane et al., 2005
Olivine [010]	1200	2.49*	0.56**	Du Frane et al., 2005
Olivine [001]	1200	2.40*	0.71**	Du Frane et al., 2005
Orthopyroxene	1000–1400	3.72 ± 0.10	1.80 ± 0.02	Xu and Shankland, 1999
Clinopyroxene	1000–1400	3.25 ± 0.11	1.87 ± 0.02	Xu and Shankland, 1999

^ for oxygen fugacity $f_{O2} = 10^{-4}$ Pa;
* for $f_{O2} = 0.1$ Pa;
** f_{O2} – independent values

The strong dependence of conductivity on temperature has been used to calculate theoretical profiles of electrical conductivity of the mantle from temperatures and surface heat flow (e.g. Dobson and Brodholt, 2000; Artemieva, 2006) and to decipher the thermal structure of the mantle from depth profiles of electrical conductivity (Sections 7.4–7.6). However, it appears that both reference conductivity–temperature olivine models may underestimate true electrical conductivity of the mantle (e.g. Gatzemeier and Moorkamp, 2005) (also Fig. 7.6).

7.2.3 High-temperature anisotropy of olivine conductivity

The mobility of the charge carriers, which is controlled by the geometry of the crystal lattice, is a tensor function which leads to anisotropy of conductivity (Fig. 7.7). The low-temperature activation energy shows essentially isotropic behavior (Shankland and Duba, 1990; Constable et al., 1992). However, at temperatures above 1200 °C, where the high-temperature conductivity mechanism (due to Mg vacancies) dominates, a significant intrinsic anisotropy of the orthorhombic olivine structure with directional dependence of conductivity along the principal axes is found in laboratory studies (Du Frane et al., 2005).

Anisotropic behavior of olivine conductivity at T > 1200 °C may confuse estimates of the activation energy and mantle temperatures from conductivity–depth profiles. At these temperatures, Mg and Fe can occupy two different crystallographic sites within the olivine structure thus determining the efficiency of conduction and resulting in different activation

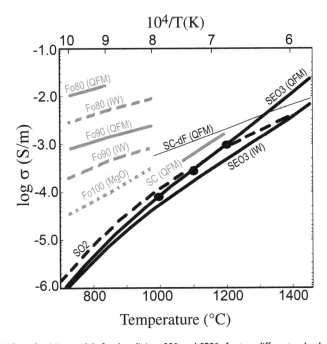

Fig. 7.6   Reference theoretical conductivity models for dry olivine, SO2 and SEO3, for two different redox buffers (black lines). Three experimental data points on which SEO3 is based are shown by dots (based on: Constable, 2006). Thin line marked SC-dF is based on the geometric mean of an anisotropic model of San Carlos olivine (Du Frane *et al.*, 2005). Experimental electrical conductivity for San Carlos olivine (Fo90) at 1 kbar (SC) and for synthetic olivines at 10 kbar with 80, 90, and 100 mol% Fo is shown for comparison (gray lines); equilibration redox buffers typical for the upper mantle are fayalite–magnetite–quartz (QFM), iron-wustite (IW) (data of Hintz *et al.*, 1981). Note a significant difference between olivine models SO2 and SEO3 and experimental data at $T < 1000\ °C$.

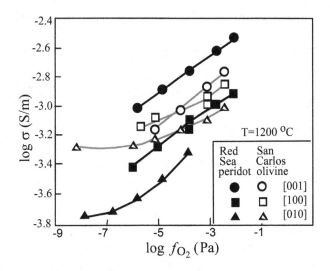

Fig. 7.7   Logarithm of electrical conductivity versus log oxygen fugacity measured along the principal directions for Red Sea peridot and San Carlos olivine at 1200 °C (data of Schock *et al.*, 1989; Hirsch and Shankland, 1993).

energy values for three crystallographic axes (Constable *et al.*, 1992). The difference between the activation energies along [010] and [001] crystallographic axes in the San Carlos olivine can be as large as *c.* 0.55 eV (Shankland and Duba, 1990) (Table 7.1).

Recently, experimental fully equilibrated data for a single-crystal dry San Carlos olivine at different temperatures and oxygen fugacity f_{O2} values was used to determine steady-state, f_{O2}-independent values of activation energies for each principal orientation of olivine. The results indicate that:

- the maximum azimuthal contrast in olivine conductivity at T < 1500 °C is ~$10^{0.4}$ S/m and the activation energies are in the range from 0.32 eV along [100] to 0.71 eV along [001] (Du Frane *et al.*, 2005).

The geometric mean of logarithmic conductivity along the three crystallographic axes is shown in Fig. 7.6. Karato (2006) argues that the anisotropy of conductivity can be weaker than expected from single crystal data if the grain-boundary effects are important. However, grain-boundary effects in iron-bearing olivine of grain-size between ~10 μm and ~1 mm are expected to be weak, although the existing evidence is insufficient for definite conclusions.

7.2.4 Pressure dependence of olivine conductivity

The pressure effect of electrical conductivity can be introduced through the activation volume V, and Eq. (7.4) transforms into:

$$\sigma = \sigma_0 \, e^{-(U+PV)/kT}, \tag{7.6}$$

where U is the activation internal energy and V is the activation volume. Thus, conductivity should decrease with pressure increase. In agreement with Eq. (7.4), an exponential dependence of conductivity on pressure was found in experimental measurements on synthetic olivines at room temperature and pressures up to 200 kbar (i.e. 20 GPa) (Lacam, 1983). However, recent laboratory measurements on olivine at temperatures 1000–1400 °C and pressures 4–10 GPa demonstrated that at upper mantle conditions the pressure effect on conductivity is weak and can be neglected (Xu *et al.*, 2000b) (Fig. 7.8).

7.2.5 Oxygen fugacity and redox buffers

In oxides, the chemical potential of oxygen determines the reactions of oxygen on the crystal lattice and thus controls the concentration of point defects and bulk conductivity. *Fugacity* (which has the same units as pressure) is a measure of the chemical potential of a mineral phase. It does not refer to a physical property of a phase or a rock, but is a tool used for predicting the final phase and reaction state of multi-component mixtures at various temperatures and pressures. In the equations of state, fugacity can be considered a "corrected pressure" for a real gas, defined in such a manner that the chemical potential equation for a real gas becomes similar to the equation for an ideal gas. In a multi-component system, the mineral phase with the lowest fugacity will have the lowest value of the Gibbs free energy and will be the most favorable.

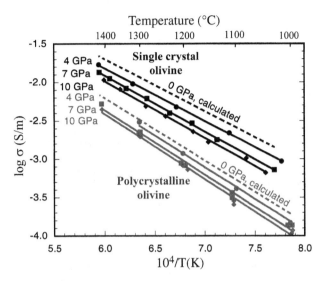

Fig. 7.8   Logarithm of electrical conductivity versus reciprocal temperature for single crystal and polycrystalline olivine as a function of pressure. The curves for zero pressure are calculated by fitting data for higher pressures to eq. (7.6) (modified after Xu et al., 2000a). In the upper mantle, the pressure effect can be ignored.

Oxygen fugacities (or the redox conditions, shorthand for reduction–oxidation reactions that are all chemical reactions in which the oxidation number of atoms is changed) control the relative proportions of the oxidation states of elements in mantle rocks. Three relatively abundant elements in the crust and upper mantle that occur in more than one oxidation state include Fe, Mg, and S. For example, Fe can exist either as ferrous iron (Fe^{2+}) or as ferric iron (Fe^{3+}). The Fe^{2+}/Fe^{3+} ratio of a rock determines its oxide mineral composition.

Buffer defines the situation where the concentration of the element forming the major phase remains fixed. Redox buffers are used widely in laboratory experiments to control oxygen fugacities. For example, the oxygen fugacity of a rock that contains pure minerals constituting a redox buffer is defined by one of the curves in Fig. 7.9. In mantle rocks, oxygen fugacity is commonly close to the FMQ (fayalite–magnetite–quartz) buffer. It means that the electron balance of the mineral assemblage is on average close to that given by the FMQ buffer even if not all of these minerals are present in a rock. In iron-bearing rocks, redox buffers control the proportions of Fe^{2+} and Fe^{3+} in the minerals. In mantle olivine, iron is present only as Fe^{2+}. Magnesium present in olivine stabilizes it to conditions more oxidizing than those required for fayalite stability. If redox conditions are more oxidizing than the magnetite–hematite (MH) buffer, most of the iron will be present as Fe^{3+} (e.g. in hematite), while at redox conditions more reducing than the iron-wustite (IW) buffer Fe^{3+} can still be present in some minerals (e.g. in pyroxene). The oxygen fugacity f_{O2} in the upper mantle lies between the IW and FMQ buffers and in the transition zone – near the IW buffer (Arculus, 1985; Hirsch, 1991). However, the variation between two oxygen buffers in the upper mantle does not have a strong effect on conductivity (0.2 to 0.8 log units).

By controlling the oxidation state of iron, oxygen fugacity controls the number of electrons that transport the electric charge. Olivine conductivity at oxygen fugacity close

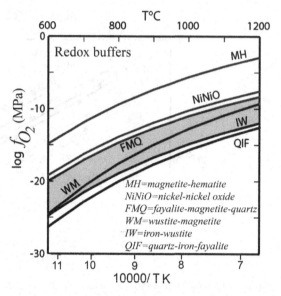

Fig. 7.9   Logarithm of oxygen fugacity versus temperature for common buffer assemblages at 1 bar pressure. The pressure effect on redox buffers is small. Gray shading – oxygen fugacity of the upper mantle (based on Frost, 1991). Oxygen fugacities versus pressure for several metal-oxide buffers are shown in Fig. 8.18.

to the FMQ buffer can be ~0.5 log units higher than near the molybdenum–molybdenum oxide (Mo–MoO$_2$) buffer (Xu *et al.*, 2000b). Experimental and theoretical studies have demonstrated that at low f_{O2} ($<10^{-12}$ MPa) olivine conductivity is generally independent of oxygen fugacity, but increases with fugacity at high f_{O2} (Fig. 7.9). This behavior is attributed to two different conduction mechanisms: at low f_{O2} – by electrons and at high f_{O2} – by polaron hopping of holes between Fe^{3+} and Fe^{2+} (Hirsch and Shankland, 1993).

7.2.6. Fe and Al variations in olivine

Electrical properties of olivine depend on the concentration of point defects. The parameters in eq. (7.4) that can possibly be iron-dependent are the pre-exponential factor σ_0, the activation energy, and the activation volume. Theoretical and experimental studies indicate a strong Fe-content dependence of σ_0 in olivine (Fe$_x$Mg$_{1-x}$)$_2$SiO$_4$ with a clear linear correlation between $\log \sigma_0$ and $\log x$ (x = Fe/(Fe+Mg)). The correlation between the activation energy and the iron content is not well known and can be modeled by a linear empirical trend, while no experiments reported a notable effect of iron on the activation volume (Vacher and Verhoeven, 2007) (see also eq. 8.31). For a multi-component system such as olivine, the general form of the conductivity can be written as (Tyburczy and Fisler, 1995):

$$\sigma = \sigma_0 \exp(-H/kT) \left(\mathrm{Fe}^x{}_{\mathrm{Mg}}\right)^m (f_{O2})^n (a_{\mathrm{SiO}_2})^p \tag{7.7}$$

where H is the activation enthalpy, k is the Maxwell–Boltzmann constant, T is temperature, σ_0 is a constant depending on the number and mobility of charge carriers, Fe$^x{}_{\mathrm{Mg}}$ is the

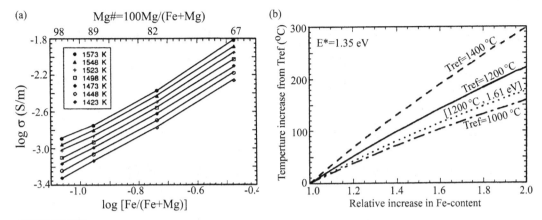

Fig. 7.10 (a) Logarithm of electrical conductivity as a function of iron content measured at different temperatures. (b) Relative temperature increase (from reference temperature) which produces a conductivity increase equivalent to that produced by an increase in iron content. Measurements are at oxygen fugacity for a gas mix $CO2:CO = 5:1$ (after Hirsch *et al.*, 1993). At high temperatures, variations in T and Fe cannot be separated.

fraction of Fe on Mg sites, a_{SiO_2} is the silica (or enstatite, $Mg_2Si_2O_6$) activity, f_{O2} is oxygen fugacity, and n, m, and p are constants that depend on the defect reaction stoichiometries.

Laboratory measurements of electrical conductivity in natural and synthetic olivines as a function of iron content have been made by Lacam (1983), Schock *et al.* (1989), Shankland and Duba (1990), Nell and Wood (1991), Hirsch and Shankland (1993), Hirsch *et al.* (1993), Farber *et al.* (2000), Vacher and Verhoeven (2007). They indicate a strong effect of iron content on olivine conductivity (Figs. 7.6; 7.10) which can be parameterized as (Hirsch *et al.*, 1993):

$$\sigma \sim [Fe/(Fe + Mg)]^{1.8}. \tag{7.8}$$

The experiments indicate that at temperatures below 700–750 °C and $f_{O2} > 10^{-17}$ MPa even minor amounts of Fe ($x = 1$ ppm) can significantly affect conductivity (Hirsch and Shankland, 1993). As a result of this extreme sensitivity, conductivities of forsterite Fo (Mg_2SiO_4) and fayalite Fa (Fe_2SiO_4) at the same temperatures can differ by at least five orders of magnitude (Lacam, 1983), while variations in the upper mantle composition between Fe-depleted cratonic lithosphere and Fe-enriched oceanic mantle can lead to a factor of two difference in mantle conductivity (Hirsch and Shankland, 1993). Since both *increase in mantle temperature and in iron content change olivine conductivity in the same direction*, it is worthwhile to compare the amplitude of these effects, since it has important geophysical implications in the interpretation of MT data such as possible distinction of compositional variations. The results of Hirsch *et al.* (1993) indicate that

- a 20% increase in fayalite content increases conductivity by 1.39; at near-solidus temperatures, the same conductivity increase can be achieved by a temperature increase of 40–70 °C only (Fig. 7.10b).

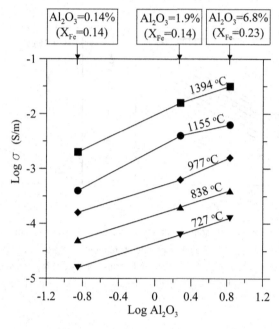

Fig. 7.11 The effect of aluminium on logarithmic electrical conductivity of orthopyroxene at different temperatures (based on experimental data of Duba *et al.*, 1976, 1979; Huebner *et al.*, 1979). Note that high-aluminous sample also has a high iron content.

Available experimental data do not permit discussion of the effect of metamorphic or metasomatic reactions of conductivity on the mantle. Boerner *et al.* (1999) proposed that hydrous modal metasomatism associated with paleosubductions can explain enhanced conductivity in the cratonic upper mantle. On the other hand, Jones *et al.* (2002) argue that metasomatism does not seem to result in a conductivity increase because hydrous mantle minerals such as phlogopite do not form an interconnected network. Infiltration of iron-rich basaltic magmas into cratonic lithosphere can result in metasomatic enrichment of depleted lithospheric root. However, this Fe-enrichment should not produce a significant increase in electrical conductivity.

Phase transition from olivine to spinel structure results in a discontinuous increase in electrical conductivity by ~2 orders of magnitude at the transition zone (Xu *et al.*, 1998). It is, however, still debated whether phase transition alone is sufficient to explain the observed jump in conductivity at ~410 km depth.

The effect of Al-content on olivine conductivity was discussed by Vacher and Verhoeven (2007). The only Al-bearing mineral for which experimental studies were reported for different Al_2O_3 content is orthopyroxene (Duba *et al.*, 1976, 1979; Huebner *et al.*, 1979). However, laboratory measurements reported by different authors were made at different iron-content, which complicates their comparison. The results apparently indicate that an increase in Al-content causes a significant increase in olivine conductivity, and the effect is stronger at low Al concentrations (Fig. 7.11). Similarly to Fe variations, a linear correlation between log σ_0 and

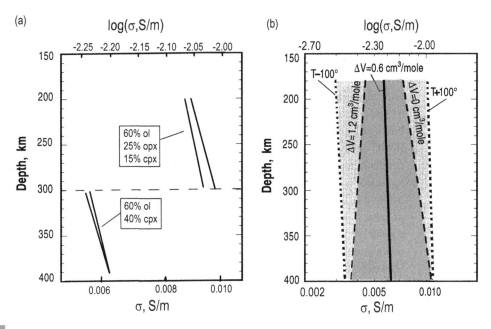

Fig. 7.12 Calculated electrical conductivity for the upper mantle. (a) Conductivity averages (Hashin–Shtrikman bounds) assuming different compositions above and below 300 km depth. (b) Conductivity calculated using eq. 7.6 and showing the effects of variations of ±100 °C in temperature (dotted lines) and ±0.6 cm^3/mole in activation volume (dashed lines). Solid line – conductivity calculated assuming olivine composition (with parameters similar to those of the standard olivine S02, Constable *et al.*, 1992) and for mantle temperature along a 1375 °C adiabat (based on Xu *et al.*, 2000a,b).

log(Al$_2$O$_3$% /Al$_2$O$_3$%ref) was found with the slope 0.53 ± 0.18. However, data scarcity precludes any definite conclusions on conductivity–Al$_2$O$_3$ correlation. Theoretical calculations of the upper mantle conductivity with a sharp compositional change at 300 km depth further illustrate that the effect of Al$_2$O$_3$ on conductivity variations (Fig. 7.12a) is comparable to the effects of variations in mantle temperature and the activation volume (Fig. 7.12b).

7.2.7 Effect of highly conducting phases

Two major conduction types within the lithosphere are electronic and ionic conduction; the former dominates in solid materials, while the latter dominates in fluids. At near-solidus temperatures, ionic conduction also becomes important in olivine. The bulk conductivity of a rock is sensitive to the presence of a number of conducting mineral phases, e.g. electronically conductive graphite, electrolytically conducting fluids, or, at high temperatures, ionically conductive partial melts in interconnected pores. Even small amounts of a conducting phase can significantly change bulk rock conductivity. When a conducting phase (with conductivity σ_f) is distributed within a non-conducting matrix, the bulk electrical conductivity σ_m depends on the connectivity and the amount of the conducting phase and is described by the empirical Archie's law (Hermance, 1979):

$$\sigma_m = C \, \sigma_f \eta^n. \tag{7.9}$$

Here η is the fractional porosity, n is between 1 and 2, and C is constant; the conductivity of a non-conducting phase has little effect on σ_m. In the case of more than one conducting phase, a modified Archie's law should be used to calculate bulk conductivity of a medium (e.g. Glover et al., 2000).

Trace amounts of free carbon (a few ppm) that form interconnected networks along grain boundaries can have a significant effect on bulk electrical properties of the Earth's mantle (Duba and Shankland, 1982). Graphite has been used successfully to explain lower crustal conductors (e.g. Mareschal et al., 1995), and its amount in the mantle, especially if CO_2 is included, is sufficient to affect bulk mantle conductivity down to a depth of 150–175 km where it enters the diamond stability field (Kennedy and Kennedy, 1976). Jones et al. (2003) propose that the presence of carbon in the lithospheric mantle can explain the mantle conductor beneath the Central Slave craton in North America since the bottom of the conductor coincides approximately with the depth of the graphite to diamond transition. However, to explain a mantle conductor by carbon presence one needs to explain why carbon does not affect conductivity at shallower depths.

The presence of an interconnected network of sulfides can also significantly increase mantle conductivity. Magnetotelluric studies across the Trans-Hudson orogen in the Canadian Shield reveal the longest so-far known crustal conductivity anomaly (Jones et al., 1993); the North American Central Plains conductivity anomaly could be associated with pyrite sulfides present in the crust (Jones et al., 1997). This mechanism may be valid for mantle conductors as well; the finding of interstitial sulfides in mantle xenoliths of the Kaapvaal craton (Alard et al., 2000) supports this hypothesis.

7.2.8 Partial melts

The influence of even a small amount of melt on conduction becomes significant when melt fractions form an interconnected network (Fig. 7.13; see also Section 3.1.6 and the next section on the effects of batch and fractional melting on water content in the upper mantle). Murase et al. (1977) were among the first to measure the electrical conductivity of mantle rocks as a function of the volume fraction of peridotite melt. They observed a steep increase in conductivity, when a substantial amount of melt was produced in the peridotite (at $T \sim 1170$ °C). These experimental results support the interpretation that the high-conductivity region in the mantle can be a region of partial melting in view of the fact that

- at $T \sim 1200$ °C, a volume fraction of melt of less than 3% can explain a conductivity increase by an order of magnitude.

Further experimental studies on olivine melting (Roberts and Tyburzcy, 1999; ten Grotenhuis et al., 2005) and on KCl aqueous solution within H_2O ice (which was considered as an analog of the Earth's mantle) (Watanabe and Kurita, 1993) confirmed earlier conclusions.

However, rock conductivity depends not only on the amount of melt, but on its geometry and connectivity (Fig. 7.14). For example, in the case when a highly conducting sulfide melt

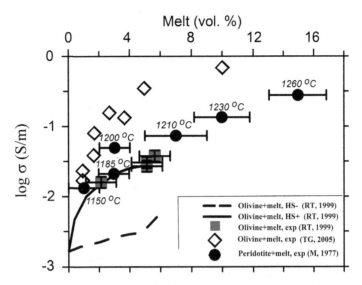

Fig. 7.13 Logarithm of electrical conductivity in peridotite and olivine as a function of melt volume. Circles – experimental data for peridotite melting, bars give possible errors in melt fraction (data of Murase *et al.*, 1977); diamonds – experimental data for melting of synthetic iron-free olivine rocks (ten Grotenhuis *et al.*, 2005); squares – experimental data for olivine melting; lines – theoretical calculations of Hashin–Shtrikman bounds. HS+ (solid line) models a well-interconnected network of melt and lies close to theoretical values for parallel conduction through interconnected melt network; HS- (dashed line) models isolated pockets of melt (data of Roberts and Tyburczy, 1999).

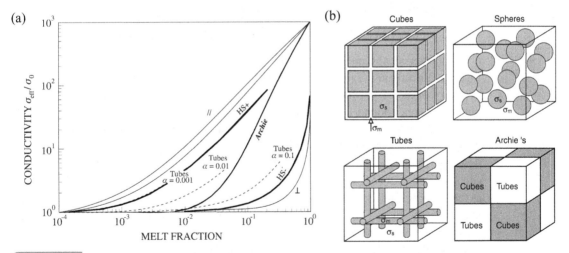

Fig. 7.14 (a) Theoretical models of the effect of melt concentration and geometry on rock conductivity (after Jegen and Edwards, 1998). σ_{eff} is the conductivity of a mixture of partial melt and solid rock, the conductivity of melt is taken to be 1000 times greater than of solid matrix σ_0. The symbols ⊥ and ∥ refer to models with thin layers of melt parallel and normal to the electric current flow, HS+ and HS- are the upper and lower Hashin–Shtrikman bounds, dashed lines refer to unconnected ellipsoidal melt inclusions with aspect ratios *a* (the ratio of the short axis to the long axis). Other curves are for Archie's law and for randomly oriented melt tubes. (b) Sketches of the melt distribution according to different geometric models (redrawn from ten Grotenhuis *et al.*, 2005).

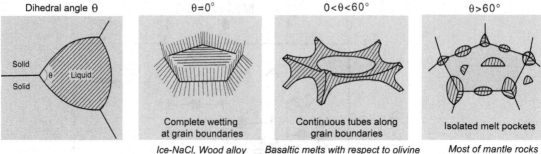

Dihedral angle θ   θ=0°   0<θ<60°   θ>60°

Solid

Solid   θ   Liquid

Complete wetting   Continuous tubes along   Isolated melt pockets
at grain boundaries   grain boundaries

Ice-NaCl, Wood alloy   *Basaltic melts with respect to olivine*   *Most of mantle rocks*

Fig. 7.15 The wetting angle between melt (shown by hatching) and solid in a melt–solid mixture. The geometry of melt distribution in solid matrix is controlled by the dihedral angle (after Karato, 2003).

is present in a peridotite matrix, a well connected melt can result in a bulk conductivity greater than 0.1 S/m while, with a completely disconnected melt, bulk conductivity is less than 0.03 S/m (Park and Ducea, 2003). Theoretical studies indicate that interconnectivity develops when the dihedral angle θ between the melt and the solid matrix is $\theta < 60°$ (von Bargen and Waff, 1986) (Fig. 7.15). Since salt–ice systems are characterized by complete wetting ($\theta = 0°$), it is dubious whether experimental studies on such systems are applicable to mantle conductivity. Sato *et al.* (1989) argue that for upper mantle rocks, the effect of partial melt on bulk rock properties is significantly weaker than determined for salt–ice or wood–alloy systems. Experimental studies (Daines and Richter, 1988) indicate that the dihedral angle in the olivine–basalt system with respect to mantle minerals is $\theta \sim 30$–$50°$ so that

- basaltic melt forms interconnected networks at melt fractions less than 0.005.

A comparison of experimental data on olivine Fo80–basalt partial melt at controlled oxygen fugacity (Roberts and Tyburzcy, 1999) with theoretical conductivity models for different geometry of melt inclusions indicates that experimental data are best fit by the Hashin–Shtrikman upper bound HS+ (Hashin and Shtrikman, 1962, 1963; Hashin, 1970) which approximates a model with a well-interconnected network of melt, or by one of the simplest theoretical models for parallel conduction (i.e. thin melt layers parallel to the electric current flow (e.g. Berryman, 1995)), or by Archie's law (eq. 7.9) with parameters $C = 0.73 \pm 0.02$ and $n = 0.98 \pm 0.01$ (Fig. 7.13). Theoretical models that refer to isolated melt pockets, such as the lower Hashin–Shtrikman bound HS- or the series model (thin melt layers perpendicular to electric current flow), apparently underestimate bulk conductivity by ~1 log unit in the experiments of Roberts and Tyburzcy (1999). The conductivity of the partially molten olivine samples analyzed by ten Grotenhuis *et al.* (2005) is best described by Archie's law (eq. 7.9) with parameters $C = 1.47$ and $n = 1.30$. Comparison with theoretical models has not been done for the peridotite melting experiments of Murase *et al.* (1977).

It is, however, unclear whether the geometry of melt distribution changes with increasing melt fraction. This question was addressed by ten Grotenhuis *et al.* (2005) who compared electrical conductivity measurements on synthetic, partially molten, iron-free olivine rocks with 0.01–0.1 melt fraction with theoretical predictions for different geometric models for

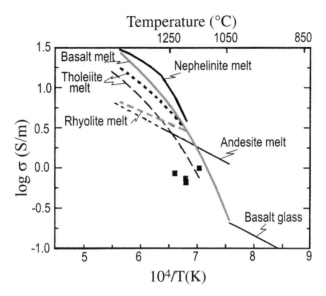

Fig. 7.16 Conductivity of melts of different composition versus temperature (redrawn from Roberts and Tyburczy, 1999). Squares – the Hashin–Shtrikman upper bound, assuming melt in equilibrium with Fo80.

melt distribution. Their results suggest that a gradual change in melt distribution can occur at melt fractions between 1% and 10% (Fig. 7.13): at melt fraction ~1% the melt is distributed in a network of triple junction tubes; an increase in melt fraction leads to the formation of grain-boundary melt layers, so that at 10% of melt the melt apparently occupies both a network of grain-boundary melt layers and triple junctions.

The bulk conductivity of a rock depends not only on the amount and geometry of melt distribution, but on melt composition since, depending on P–T conditions, oxygen fugacity, and the initial chemistry, melting of olivine mantle produces melts of different composition (Roberts and Tyburczy, 1999). These melts have significantly different conductivities (Fig. 7.16), which should affect the bulk conductivity of the upper mantle. Theoretical calculations for different melt compositions (for a melt fraction of 2–5 vol%) yield melt conductivity values in the range 0.55 S/m to 1.8 S/m and indicate that changes in temperature may balance the effect of changes in melt composition on mantle conductivity.

In summary, laboratory measurements on the melting of peridotite and olivine demonstrate that an increase in melt from 0 to 3% results in a conductivity increase by an order of magnitude (Murase et al., 1977; Roberts and Tyburzcy, 1999). Theoretical analysis indicates that the amount of melt beneath the mid-ocean ridges is c. 2% (McKenzie, 1985). Probably, the amount of melt away from MOR is even smaller, since interconnected melt is gravitationally unstable even at concentrations of <<1% (McKenzie, 1989) and migrates upwards. However, since the amount of melt in the upper mantle is poorly constrained and conductivity depends critically on the geometry of melt distribution in the solid matrix (e. g. Faul, 1997) which is largely unknown,

- synthetic conductivity-depth profiles which take into account the presence of melt remain rather speculative.

7.2.9 Role of water

Water in the mantle

Solubility of water

Large variations in water content are characteristic of the upper mantle (Ito *et al.*, 1983; Bell and Rossman, 1992; Stolper and Newman, 1994; Dixon *et al.*, 2002; Bell *et al.*, 2004; Hirschmann, 2006; Shito *et al.*, 2006). For fluid-saturated rock, the bulk rock conductivity at low temperatures is essentially controlled by the conductivity of a fluid. With increasing temperature, fluid conductivity increases, producing an increase in bulk rock conductivity unless pressure increase reduces the interconnectivity of fluid-bearing pores and cracks.

The amount of water (hydrogen) present in the mantle is characterized by two non-dimensional units: *the weight percent of water* (denoted as wt% H_2O) or *the H/Si ratio*. The latter is the ratio of the number of hydrogen atoms to the number of silicon atoms. For example, 1 wt% of water in forsterite Mg_2SiO_4 corresponds to 156 000 ppm H/Si (or $H/10^6$ Si, ppm denotes "part per million" or 10^{-6}). In a closed system (when the water content in the system is fixed and remains constant), the water content is directly related to water fugacity (for details see Section 8.1.2 and eq. 8.26).

Experimental studies demonstrate that significant amounts of water (hydrogen) can be stored in the upper mantle even in nominally anhydrous minerals (i.e. minerals like olivine and pyroxene that do not have hydrogen H in their chemical formula) (Bell and Rossman, 1992; Rossman, 2006). In general, the amount of hydrogen dissolved in nominally anhydrous minerals is small: olivines from natural peridotites have ~10–1000 ppm H/Si, and garnet peridotites usually contain more water than spinel peridotites. The amount of water that can be dissolved in nominally anhydrous minerals increases with increasing water fugacity, temperature, and pressure. In particular (Bai and Kohlstedt, 1992; Sisson and Grove, 1993; Kohlstedt *et al.*, 1995),

- the content of water dissolved in olivine can increase by a factor of ~100 if the pressure increases from 0.1 to 3.0 GPa (Fig. 7.17);
- water solubility in basaltic melt is significantly higher than in olivine, e.g. at a confining pressure of 300 MPa it is ~2500 times greater.

The hydrogen diffusion in a mineral has a characteristic time constant:

$$\tau_{\text{diff}} = d^2/\pi^2 D \qquad (7.10)$$

where d is the length-scale of diffusion process and D is the chemical diffusion coefficient. Thus *the amount of water measured in deformation experiments can be smaller than under geological conditions* and corresponds to partial, not complete, equilibrium conditions. Bell *et al.* (2003) note that

- experimentally determined water content values may underestimate actual water content by as much as a factor of three.

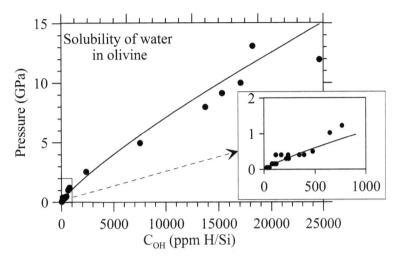

Fig. 7.17   Solubility of water in olivine as a function of pressure (based on data of Hirth and Kohlstedt, 1996). The best fit power line is the same on the plot and the insert. Water solubility in basaltic melt is up to a few orders of magnitude higher than in olivine.

Effect of melting on water content

Karato (1986) argue that partial melting leads to the removal of hydrogen (water) from such mantle minerals as olivine because water solubility in melt is 2–3 orders of magnitude greater than in mantle minerals. As a result, *even a small amount of melt can decrease water content in the rock matrix* (Fig. 7.18a). The processes of melt transport control how hydrogen (and other incompatible elements) is redistributed during melting.

In a multi-component system, such as the mantle, two types of partial melting processes are distinguished. In **batch melting** the melt transport is so slow that it stays in chemical equilibrium with the solid matrix and does not separate from the matrix. In contrast, **in fractional melting** melt can be easily (and quickly) removed from the solid matrix by gravity forces, and experimental studies indicate that this process dominates in the shallow mantle and beneath MOR. The **degree of melting** is the total amount of melt removed during fractional melting. The **melt fraction** is the amount of melt still present in the rock, and it equals the degree of melting only in the case of batch melting. For example, beneath mid-ocean ridges, the degree of melting is estimated to be ~10%, while the melt fraction can be significantly smaller. The ratio of hydrogen (water) concentration in the solid phase during and before melting, C/C_o, has different dependencies on the partition coefficient k (the ratio between the concentration of water in solid and melt) and the degree of melting Φ for batch and fractional melting (Karato, 2008):

$$C/C_o = k/[\Phi + (1 - \Phi)k] \quad \text{for batch melting,}$$
$$C/C_o = (1 - \Phi)^{(1-k)/k} \quad \text{for fractional melting.} \tag{7.11}$$

Estimates of the decrease in water content of olivine with decreasing depth beneath a mid-ocean ridge for both batch and fractional melting scenarios are provided by Hirth and

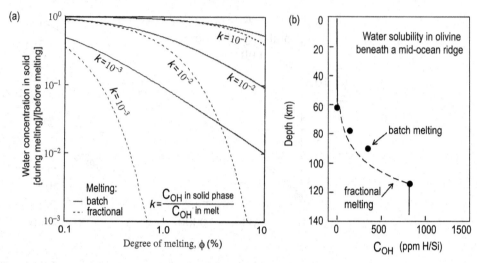

Fig. 7.18 Effect of partial melting on the amount of water in the upper mantle. (a) Water depletion by partial melting as a function of the degree of melting and the partition coefficient (the ratio between the concentration of water in solid and melt) (redrawn from Karato, 2008). (b) Water content in mantle olivine C_{OH} during partial melting beneath a mid-ocean ridge (redrawn from Hirth and Kohlstedt, 1996). Compare with Fig. 3.42.

Kohlstedt (1996). From mass balance calculations, they estimate that water solubility rapidly decreases with decreasing pressure (depth) from a value of ~810 ± 490 ppm H/Si (~50 ± 30 wt ppm of water) assumed for depths greater than 120 km (the MORB source) to 20–80 ppm H/Si at the top of the oceanic mantle (Fig. 7.18b), and at ~50 km depth mantle olivine with 810 ppm H/Si will be ~50% water-saturated.

In contrast to these results, Mierdel *et al.* (2007) argue that water solubility may have a pronounced minimum at depths below ~ 80 km beneath the oceans (in the asthenosphere) because at these depths, with increasing pressure, water solubility in Al-bearing orthopyroxene sharply decreases, while water solubility in olivine increases (Fig. 3.42). If, according to Rauch and Keppler (2002), the solubility of hydrogen in Al-orthopyroxene in the shallow upper mantle is significantly higher than in olivine, than it is the conductivity of orthopyroxene that controls the electrical conductivity of the shallow upper mantle (Karato, 2006; Dai and Karato, 2009).

Evidence for water in oceanic and continental mantle

Geochemical studies on basalts from mid-ocean ridges estimate the water content in the oceanic asthenosphere as ~0.01–0.02 wt%. Water is expected to play a particularly important role in subduction zones, where it is either stored in silicates as point defects (*c.* 0.1% in olivine and *c.* 1% in pyroxene (Bai and Kohlstedt, 1992; Bell and Rossman, 1992)) or as free water at a depth of *c.* 125 km. Garnets from some ultra-deep eclogites provide constraints on water content in subduction zone settings and indicate a concentration of hydrogen close to the water saturation limit. The amount of water in the mantle wedge above the subduction

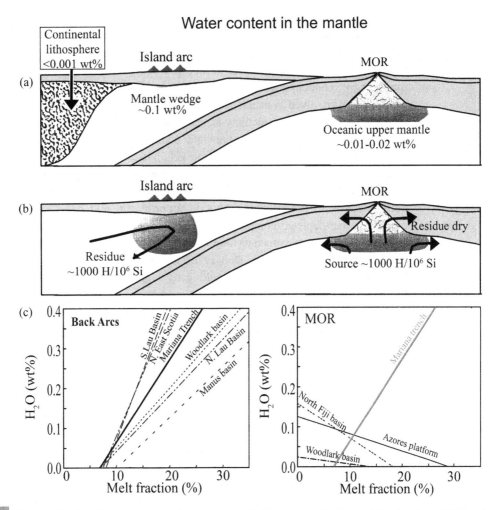

Fig. 7.19 A cartoon showing the water content in the upper mantle in different tectonic settings. (a) Based on Karato (2010) and showing dry continental lithosphere as compared to wet oceanic upper mantle (based on data from MOR and mantle wedge). (b) Based on Hirth and Kohlstedt (2003) and showing a fundamental difference in the role of water in the melting regions beneath arcs and ridges (darker shades refer to higher water content). Melting at MOR produces a dry residue, while melting in arcs produces a hydrous residue with water content similar to the MORB region. (c) Compilation of water content versus melt fraction for MOR and back-arc basins (from Kelley *et al.*, 2006). MOR = mid-ocean ridges; 0.1wt% of water in forsterite Mg_2SiO_4 corresponds to 15600 $H/10^6$ Si. Compare with Fig. 6.31.

zone is estimated to be as large as ~0.1 wt% (Figs. 6.31, 7.19a). Although dehydration reactions along the subducting slab can release more free water, if it leaves the stability field of a hydrous mineral, this water will either migrate upwards to the stability field of phlogopite or amphibole, or will partition into the melt and will be taken back to the lithosphere (Thompson, 1992; Constable and Heinson, 1993).

Figure 7.19b illustrates the fundamental differences in the role of water in the melting regions beneath arcs and ridges. During melting at mid-ocean ridges, water is extracted from the residue (Fig. 7.18). As a result, dry lithosphere with higher viscosity than in the oceanic asthenosphere is produced. In contrast, melting in arcs occurs with the influx of fluids from the subducting slab into the mantle wedge, and melting at back arcs produces a hydrous residue with water content similar to the MORB region (Fig. 7.19b,c). Downward water transport by subducting slabs can continuously supply the upper mantle with hydrogen which, being dissolved in mantle olivine, would facilitate the existence of a low viscosity, low velocity, and high conductivity asthenosphere.

In contrast to oceanic settings, the water content in the continental upper mantle is less well constrained (Karato, 2003a). Some geophysical evidence, in particular studies of mantle electrical conductivity, suggest that the continental lithosphere is "dry" as compared to "wet" oceanic mantle (e.g. Fig. 7.38); but due to the non-uniqueness of these interpretations they are only consistent with, but do not prove, the notion that the cratonic mantle is dry. The water-depleted composition of the cratonic lithospheric mantle is required by geophysical models to explain its long-term preservation and stability with respect to mantle convection (e.g. Pollack, 1986; Doin *et al.*, 1997).

While high degrees of mantle melting as suggested by petrological models for the formation of the continental lithospheric mantle would imply a high degree of water depletion, studies of kimberlite xenoliths from deep continental upper mantle indicate that it contains 0.001–0.01 wt% of water; the average concentration of water in cratonic peridotites is ~800 ppm H/Si (~0.005 wt% H_2O; Fig. 6.31a). Because the diffusion coefficient of hydrogen in olivine is very high, the water content in olivine can be easily modified. Thus, it is questionable whether the water content measured on xenolith samples reflects the original water content in the mantle.

The general issue of representativeness of xenolith sampling of the upper mantle is discussed in Section 6.3. In the case of water content, a plausible model for reconciling xenolith data with rheological requirements on high viscosity assumes that the bulk of the cratonic lithosphere has a low water content (<0.001 wt%) as a result of a high degree of mantle partial melting (Fig. 7.18a). Since volatile-rich magmatism, such as kimberlite magmatism, results in local metasomatic enrichment of the continental upper mantle associated with addition of incompatible components, including water, mantle-derived xenoliths brought to the surface by kimberlite-type magmas should have higher water content than the intact (*in situ*) cratonic mantle (Fig. 6.31a).

Conductivity of dry and wet olivine

Both theoretical and laboratory studies indicate that even a small amount of dissolved hydrogen significantly affects kinetic processes in silicates (Karato, 1990; Mackwell and Kohlstedt, 1990). Using the Nernst–Einstein equation (7.3) and extrapolating low-pressure laboratory measurements of olivine conductivity in the presence of hydrogen (20–80 ppm H/Si at $P = 300$ MPa) to the upper mantle conditions, Karato (1990) calculated electrical conductivity in wet olivine along the a-axis with the highest conductivity from the solubility and the diffusion coefficient of hydrogen (Fig. 7.20a). He argued that the presence of water in the upper mantle can explain the range of conductivity estimated from geomagnetic studies for the high-conductivity layer (interpreted as electric asthenosphere) without

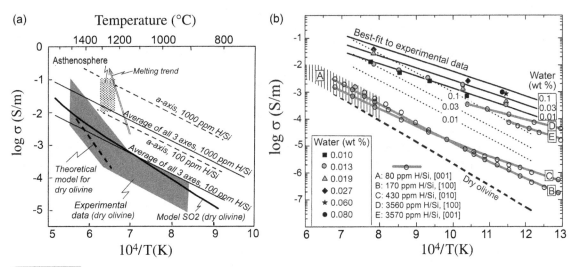

Fig. 7.20 Electrical conductivity in olivine as a function of inverse temperature, anisotropy, and water content. Plot (a): mostly theoretical calculations with a focus on the competing effects of hydrogen and anisotropy on conductivity (the discussion between Karato, 1990 and Constable, 1993). Conductivity of dry olivine conditions based on experimental data of Schock et al. (1989) (gray shading) and calculated from the diffusion coefficients of Mg^{2+} or Fe^{2+} (bold dashed line). Thin dashed lines – conductivity along the axis of highest conductivity (a-axis) calculated from experimental data on (anisotropic) hydrogen diffusivity (Karato, 1990). Thin solid lines – isotropic olivine conductivity calculated as an average of the series and parallel bounds; bold solid line – olivine conductivity of dry subsolidus olivine model SO2 (Constable, 1993). Melting trend based on data of Murase et al. (1977) (Fig. 7.13). Plot (b): comparison of experimental data with the predictions based on Karato's (1990) hypothesis (dotted lines; numbers near the curves – water content, wt %). Experimental data of Wang et al. (2006) for San Carlos olivine at $P = 4$ GPa and at different water content (wt %) with oxygen fugacity close to Ni–NiO buffer (filled symbols); solid black lines – multilinear regression from all data. Experimental data of Yoshino et al. (2006) for dry and wet olivine measured at 3 GPa along different axes is shown by gray lines labeled A to E. Dashed line – SO2 conductivity model (Constable et al., 1992).

requiring significant partial melting in the asthenosphere. These conclusions were challenged by Constable (1993) who argued that diffusivity along the a-axis in olivine is ~10 times greater than along the c-axis and ~100 times greater than along the b-axis. He calculated conductivity in wet mantle olivine for different mixing laws and for the isotropic mixture. His estimates imply that *the effect of hydrogen on conductivity is comparable to the effect of anisotropy*, and a high concentration of hydrogen can increase conductivity by 1–2 orders of magnitude as compared to the case of dry olivine.

Two recent experimental studies of conductivity of wet olivine further contribute to the earlier debate on the effect of water on the electrical conductivity of the upper mantle (Yoshino et al., 2006; Wang et al., 2006). Both of these experimental studies were made in Kawai-type multi-anvil presses at similar pressures (3–4 GPa) and thus their results are easily comparable (Fig. 7.20b). Both studies indicate that hydrogen enhances electrical conductivity. However, the slope and the magnitude of conductivity variations differ from predictions based on the Karato's (1990) hypothesis and constrained by the water diffusion coefficients of Kohlstedt and Mackwell (1998). These and earlier experimental data suggest (Huang *et al.*, 2005; Karato, 2006) that

- calculation of the electrical conductivity of a mineral from the solubility and diffusion coefficient of hydrogen may be somewhat misleading.

7.2.10 Other processes

Constable and Heinson (1993) discuss the grain-boundary effect among one of the major candidate mechanisms which can affect electrical conductivity of the upper mantle. Experimental studies of the electrical properties of olivine demonstrated that two distinct conduction mechanisms occur at different frequency ranges. At the low frequencies used in MT surveys, the electrical response is dominated (Roberts and Tyburczy, 1991) by:

- grain-boundary conductivity at $T < 1000\ °C$,
- grain-interior conductivity at $T > 1200–1250\ °C$.

However, no dependence of conductivity on the size of the grains has been found in experiments. When depth profiles of electrical conductivity constrained by MT studies are used to infer mantle temperatures, an unaccounted effect of grain boundaries on measured conductivity may lead to underestimated (by 50–130 °C) mantle temperatures (at $T < 1250\ °C$) (Roberts and Tyburczy, 1991).

An effect of strain on electrical conductivity of the mantle has been discussed by Constable and Heinson (1993). Since concentration of point defects in crystal lattice controls electrical conductivity, it has been suggested that deformation in the mantle associated with plate motions can increase the number of dislocations in olivine and thus produce an electric conductor. However, experimental studies indicate that high-temperature creep has little effect on electrical conductivity in olivine (Hirth, 1989).

7.2.11 Frequency dependence of electrical conductivity

Conductivity of the mantle is measured in a wide frequency range to ensure sufficient depth resolution. Therefore, frequency dependence of olivine is important in inversions of MT observations. Laboratory studies on dry and wet rocks have demonstrated frequency dependence of their conductivity: *conductivity increases with increasing frequency*. The effect is particularly significant at $T < 1100\ °C$ for frequencies $> 10^2$ Hz (Roberts and Tyburczy, 1991). Some frequency dispersion between 10^2 and 10^4 Hz is found in conductivity measurements on polycrystalline olivine at $T > 900\ °C$ (Wanamaker and Duba, 1993). Laboratory measurements on partially molten gabbro reveal a frequency-dependent electrical response at temperatures in the range 780 °C to 1200 °C and at frequencies between 10^{-2} and 10^4 Hz (Sato and Ida, 1984). These authors argue that conductivities in partially molten rocks measured at frequencies above 10 Hz are not directly applicable to MT interpretations and that the observed frequency dispersion can lead to misinterpretations of MT conductivity profiles in terms of mantle temperatures. Frequency dependence of conductivity can also lead to errors in estimates of activation energy values if these measurements are made at a single frequency (Roberts and Tyburczy, 1994).

7.3 Synthetic conductivity–depth profiles for the upper mantle

While electromagnetic studies of the Earth mantle typically report resistivity values, conductivity (which is the inverse of resistivity) is the parameter measured in laboratory studies. To facilitate a comparison of the results of EM and MT studies with laboratory data, the resistivity values reported in different publications have been converted here to conductivity values (most of them were digitized and plotted together to simplify comparison). Prior to discussion of the results of EM and MT surveys for the oceans and the continents, it is useful to examine some general patterns such as synthetic conductivity–depth curves for the upper mantle constrained by laboratory data.

Olivine is the most abundant and interconnected phase in the Earth's mantle down to the transition zone, where mantle composition changes to wadsleyite at a depth of $c.$ 410 km and further to ringwoodite at $c.$ 520 km depth. Laboratory measurements of olivine electrical conductivity at different upper mantle temperatures provide grounds for the so-called standard dry olivine models SO2 and SEO3 (Constable et al., 1992; Constable, 2006) (Fig. 7.6). The latter allow for calculation of synthetic conductivity–depth profiles in the upper mantle of spherically isotropic Earth, assuming variations of mantle temperature with depth are known (Artemieva, 2006). Temperature distribution in the upper mantle of different tectonic provinces can be derived from typical continental and oceanic geotherms (e.g. Pollack and Chapman, 1977; Parsons and Sclater, 1977) with isentropic temperature in the convective mantle. The resultant plot (Fig. 7.21) shows theoretical predictions for variations of electrical conductivity with depth for stable continents and old oceans. In spite of limitations related to model assumptions, the plot provides a background reference frame for comparison results of MT and EM surveys with laboratory data and theoretical predictions. Its details are described below.

Continental lithospheric mantle

For stable continents, synthetic conductivity–depth profiles for the lithospheric mantle are calculated assuming that the lithospheric temperatures follow the reference geotherms for surface heat flow of 37–50 mW/m^2 (Fig. 4.33) and the conductivity–temperature dependence follows the experimental data for isotropic synthetic olivine $Mg_{2(1-x)}Fe_{2x}SiO_4$ (Fo100 and Fo90) with oxygen fugacity buffered by IW (Hinze et al., 1981). A relatively high temperature gradient in the lithosphere (4–8 °C/km in stable continents and up to 20 °C/km in active continental regions) together with an exponential dependence of conductivity on temperature (eq. 7.4) predict fast growth in electrical conductivity with depth within the continental lithospheric mantle: by two and a half orders of magnitude over 100 km. Within the lithospheric mantle, at a fixed depth, the difference between the conductivity values calculated for 37 mW/m^2 and 50 mW/m^2 can be as large as two orders of magnitude. In

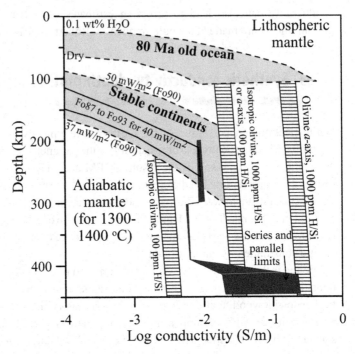

Fig. 7.21 Synthetic conductivity–depth distribution for the upper mantle. Conductivity in the lithospheric mantle is calculated for dry and wet (0.1 wt% H_2O) olivine for 80 Ma-old ocean and for dry Fo87–Fo93 olivine for stable continents as constrained by experimental data (Hinze *et al.*, 1981; Fig. 7.6) and typical continental and oceanic geotherms (Pollack and Chapman, 1977; Figs. 4.25 and 4.33). Hatching – conductivity in the sublithospheric mantle calculated for mantle adiabats of 1300 and 1400 °C from theoretical models of Karato (1990) and Constable (1993) for wet and dry anisotropic olivine (Fig. 7.20a). Dark shading – series and parallel limits on the upper mantle conductivity with depth-variable (dry) multi-component mantle composition (data of Xu *et al.*, 2000b and Fig. 7.12a). See Section 7.3 for explanations.

active regions a conductivity increase with depth should be even stronger. A conductivity profile for active continents (with lithospheric geotherms for heat flow of ~90 mW/m^2) for an olivine composition of Fo90 falls very close to the conductivity estimates for the oceans for Fo80 dry olivine composition.

Conductivity dependence on iron-content, though much weaker than temperature- or hydrogen-content dependence, can still be important in interpretations of regional electromagnetic models for the cratonic mantle. For the typical cratonic geotherm of 40 mW/m^2 a difference in conductivity for iron-enriched (Fo87) and iron-depleted (Fo93) compositions at the same depth becomes significant at depths >150 km.

Oceanic lithospheric mantle

For oceanic lithospheric mantle, conductivity profiles are calculated for lithospheric temperatures along the reference geotherm for a 80 Ma-old "normal" ocean (Fig. 4.25) and assuming that the lithospheric mantle has dry or wet (0.1 wt% H_2O) Fo80 isotropic olivine composition. The conductivity–temperature dependence is based on the experimental data of Wang *et al.* (2006) for wet olivine (Fig. 7.19) and on the SO2 model for dry olivine (Constable *et al.*, 1992). Because of a very high temperature gradient in the oceanic upper mantle, a significant increase in conductivity can be expected in the uppermost 50 km of the mantle. The presence of hydrogen has a strong (few orders of magnitude) effect on conductivity values, in particular in shallow mantle.

Adiabatic mantle and the transition zone

Since synthetic conductivity models SO2 and SEO3 are constrained by experimental or theoretical data for olivine at subsolidus temperatures, their extension to the convecting part of the mantle is, strictly speaking, inadequate and should be considered as a formal solution: the presence of even trace amounts of melt changes bulk rock conductivity by an order of magnitude (Fig. 7.13). For the sublithospheric upper mantle, conductivity profiles are based on the conductivity–temperature dependence calculated for isotropic and anisotropic olivine with a different hydrogen content (Karato, 1990; Constable, 1993), but with no account for possible melt, and assuming an adiabatic temperature gradient of 0.3 °C/km and mantle adiabats of 1300 and 1400 °C. Low adiabatic temperature gradient results in a very weak depth dependence of electrical conductivity in this part of the mantle. However, other factors, such as alignment of olivine crystals and presence of an interconnected network of a high conducting phase (dissolved hydrogen or partial melts) become critical in controlling conductivity variations in the convecting mantle. For this reason, the conductivity profile for the adiabatic mantle cannot be reliably constrained from the available experimental data.

Studies of mantle xenoliths indicate that a significant amount of water can be stored in the upper mantle minerals affecting grain-boundary diffusion and mantle conductivity. Solubility of water in olivine increases with pressure from 100 ppm H/Si at a depth of 5–10 km through 500 ppm H/Si at 30 km depth to 1300–1500 ppm H/Si at 100 km depth, and 10000–20000 ppm H/Si at 300 km depth (Fig. 7.17). Conductivity of isotropic olivine mantle highly saturated with hydrogen (1000 ppm H/Si) is by an order of magnitude higher than for the dry olivine model (100 ppm H/Si) (Fig. 7.21). Strong electrical anisotropy of olivine leads to a further, by an order of magnitude, increase in mantle conductivity (Karato, 1990; Constable, 1993).

Xu *et al.* (2000a, b) calculated conductivity–depth profiles for the upper mantle in the depth range 200–520 km with composition varying with depth, assuming 60% ol + 25% Opx + 15% Cpx at $z = 200$–300 km, and a two-phase mixture of 60% ol + 40% Cpx at $z = 300$–410 km and 60% wadsleyite + 40% Cpx at $z = 410$–520 km (Fig. 7.21). Among a set of models with different geometry of the mineral phases, the broadest bounds are given by the series/parallel model and thus they can be regarded as an uncertainty in conductivity due to an unknown mineral texture. Variations of ±100 °C in the mantle temperature profile lead to ±0.003 S/m uncertainty in conductivity (Fig. 7.12a), while variations in oxygen fugacity (which can vary between different tectonic settings) can produce ~0.5 log unit conductivity variations, and a 2% iron-content variation would produce a conductivity variation by a factor of 1.4. In the depth range 200–400 km, the conductivity values for a two-component olivine–clinopyroxene mixture fall between the estimates for isotropic and anisotropic dry olivine (100 ppm H/Si) (Hirth *et al.*, 2000). Below the 410 km transition zone, mantle conductivity in the model of Xu *et al.* (2000b) approaches the conductivity of wet aniso-tropic olivine, but can vary by an order of magnitude, probably due to a broad range of uncertainty caused by a large contrast in conductivity of the two mineral phases (wadsleyite and clinopyroxene) (Fig. 7.21). The amount of water present in the transition zone is still debated (Bercovici and Karato, 2003); the presence of hydrogen should have an additional strong effect on the electrical conductivity.

7.4 Regional electrical conductivity profiles in the oceanic upper mantle

7.4.1 Major marine experiments

The major methods for studying the conductivity structure of the mantle down to several hundred kilometers depth are magnetotellurics (MT) and geomagnetic depth sounding (GDS). Controlled-source electromagnetic sounding (CSEM) is usually limited to studies of the crust and the uppermost mantle due to the limitations on the penetration depth of the induced electric currents (Section 7.1.2). Some examples of CSEM marine studies are the RAMESSES marine experiment over the Reykjanes Ridge (part of the Mid-Atlantic Ridge) (Sinha *et al.*, 1997; MacGregor *et al.*, 1998), the PEGASUS (Pressure, Electromagnetic, Gravity, Active Source Underwater Survey) experiment in the northeast Pacific ocean (Constable and Cox, 1996), and electromagnetic studies of the axial zone of the northern East Pacific Rise (EPR) (Evans *et al.*, 1991) (Fig. 7.22). It has been recognized for more than 40 years that

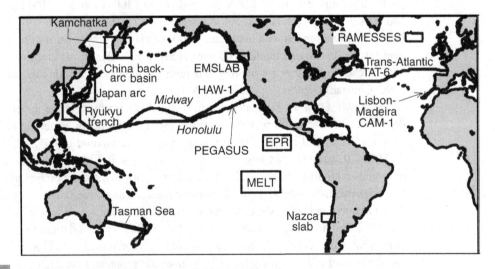

Fig. 7.22   Major marine electromagnetic experiments (bold lines and boxes).

geomagnetically induced currents in long conductors such as submarine telephone cables can be used to deduce the electrical properties of the Earth's crust and mantle (Runcorn, 1964). Although the first geophysical electromagnetic measurements at submarine telecommunication cables had already been taken in the 1970s and early 1980s (e.g. in the Atlantic Ocean from Rhode Island to France, Medford et al., 1981), the approach had significantly developed a decade later. The most important electromagnetic experiments include studies on submarine cables Honolulu–California (Heinson and Constable, 1992; Lizzaralde et al., 1995) as well as on other cables in the Pacific Ocean (Utada et al., 2003), and on cables Australia–New Zealand (Vanyan et al., 1995) and Lisbon–Madeira (Santos et al., 2003).

The conductivity structure of the oceanic mantle is also studied using natural electromagnetic fields induced either by external sources (ionospheric and magnetospheric currents) or by the oceanic dynamo (currents induced by an interaction of moving seawater with the Earth's magnetic field). The most significant marine experiments of this type include MELT (Mantle Electromagnetic and Tomographic Experiment) over the southern East Pacific Rise (Evans et al., 1999, 2005) and EMSLAB (with its on-shore extension EMRIDGE), which addressed the conductivity structure of the Pacific Ocean from the Juan de Fuca ridge to the subduction zone and further across the Cascades (e.g. Wannamaker et al., 1989). Heinson and Lilley (1993) combined ocean-floor and on-shore MT measurements to study mantle conductivity at the Tasman Abyssal Plain. Long-period MT data were used to examine lateral heterogeneity in the electrical conductivity structure of the mid-mantle in the Pacific Ocean and at the continents (Schultz and Larsen, 1990). Neal et al. (2000) complemented long-period MT with GDS data to compare the upper mantle structure beneath the Pacific Ocean and the North American continent. MT measurements at continental stations in Europe, supplemented by MT data from stations on the Mediterranean islands were used to estimate mantle conductivity of the Mediterranean region (Simpson, 2002a).

7.4.2 Major conductivity types in the oceanic mantle

It has long been recognized that considerable lateral heterogeneity of electrical conductivity may exist in the Earth's upper mantle. Indirect support for this hypothesis has been provided by global seismic tomography models, since both seismic velocities and electrical conductivity are sensitive to temperature variations, fluids, and partial melts and thus might be expected to exhibit a coherent pattern of anomalies in the mantle. A compilation of major results of marine EM and MT studies clearly indicates significant lateral variations in the conductivity of the oceanic upper mantle (Figs. 7.23 to 7.28). Although the total number of available conductivity models still remains limited and restricted to certain oceanic locations, it allows for grouping oceanic conductivity models into several types that correspond to the major marine tectonic settings: ocean basins, ocean margins, hotspots, and mid-ocean ridges (Fig. 7.23).

A comparison of conductivity profiles for different tectonic regimes indicates that *lateral heterogeneities of conductivity are restricted largely to the upper 200 km of the oceanic mantle*. Below this depth, all of the oceanic conductivity–depth curves (regardless of tectonic setting) are similar, suggesting an essentially homogeneous conductivity structure

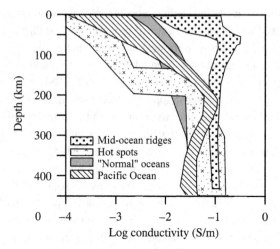

Fig. 7.23 Typical 1D electrical conductivity profiles for different oceanic settings (see Figs. 7.25 to 7.28 for details).

of the oceanic mantle below 200 km depth. However, this observation does not necessarily imply compositional homogeneity of the deep oceanic upper mantle since an increase in water content and olivine alignment have similar effects on the upper mantle electrical conductivity (Fig. 7.20a).

7.4.3 Age dependence of conductivity in oceanic mantle or the coast effect?

The strong temperature dependence of electrical conductivity (Fig. 7.6) together with high temperature gradients in the oceanic mantle and the strong age dependence of oceanic geotherms (Fig. 4.25) suggest that electrical conductivity in the oceanic upper mantle should be age dependent (Fig. 7.24). Indeed, an early 1D interpretation of three MT data sets for the Pacific Ocean included a high-conductivity layer (the "electrical asthenosphere") in the oceanic upper mantle, that decreased in magnitude and deepened with increasing ocean age (Oldenburg, 1981). The maximum conductivity values of $c.$ 0.1 S/m were achieved at depths of 60–80, 125–150, and 180–225 km for the ocean floor with ages of 1, 30, and 72 Ma, respectively. This correlation was generalized as evidence that in the Pacific Ocean the depth to a conductive layer and the amount of melt in the upper mantle are proportional to the age of the oceanic plate. Later interpretations of the same data have demonstrated, however, that (i) no 1D model can account for all three Pacific data sets, (ii) there exist a number of models in which conductivities are non-decreasing functions of depth, and (iii) the correlation between the depth to the high conductivity layer (HCL) and the ocean age, although still found, can be weak (Oldenburg *et al.*, 1984).

To check whether the HCL in the oceanic mantle is indeed required by the data, the same three MT data sets for the Pacific Ocean were reinterpreted a decade later (Heinson and Constable, 1992). This study demonstrated that the resistive coastlines around an ocean basin, which is electrically isolated from the deeper mantle, can cause significant distortion

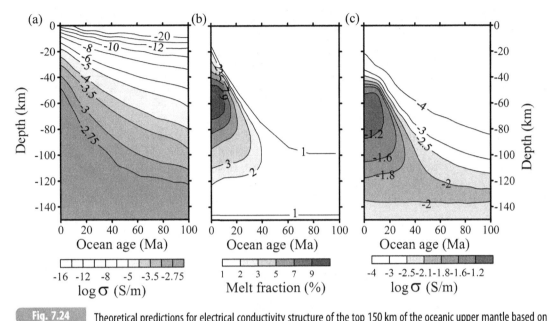

Fig. 7.24 Theoretical predictions for electrical conductivity structure of the top 150 km of the oceanic upper mantle based on the standard (dry, subsolidus) olivine conductivity model SO2 and typical oceanic geotherms (modified after Heinson and Constable, 1992). (a) Electrical conductivity of a pyrolite mantle with no volatiles and no melt fraction. (b) Expected melt fraction derived from the intersection of oceanic geotherms with the pyrolite mantle solidus. (c) Electrical structure of oceanic mantle with an interconnected melt fraction as in (b) modeled by Archie's conductivity law with an exponent $n = 1.5$ (eq. 7.9) and with the uppermost mantle conductivity set at 3×10^{-5} S/m.

of the electromagnetic response at the interior points of the ocean basin and, as a result, the presence of the HCL is highly dependent on data interpretation.

The coast effect has been debated by Tarits *et al.* (1993) who considered "leakage" models with horizontal and vertical electrical connections from an ocean basin to a continent. The study demonstrated that in the absence of electrical leakage paths, the seafloor electromagnetic field is essentially controlled by the geometry of the coastlines, but not by their physical properties. The presence of leakage paths can substantially reduce the effect of coastlines on the magnetotelluric response in the basin interior and can make the MT response of the basin interior close to the 1D case (Tarits *et al.*, 1993). Vertical electrical channels are observed worldwide at major subduction zones, faults, and sutures (Gough *et al.*, 1989; Wannamaker *et al.*, 1989), and consequent modeling demonstrated that the coast effect should not play a significant role in many marine MT studies (e.g. Lizzaralde *et al.*, 1995). However, further examination showed that even for a model with a highly conductive mid-mantle and extreme vertical leakage paths, ocean-floor MT responses can be far from the 1D case and thus the coast effect cannot be completely ignored (Constable and Heinson, 1993). A recent analytical calculation for a simple model, in which electric and magnetic fields are perturbed by a 2D sinusoidal topographic interface, indicates that accounting for the topographic effect of the coast is important due to the large conductivity contrast between seawater and the lithospheric rocks (Schwalenberg and Edwards, 2004). In

Table 7.2 Topography (coast) effect on magnetic and electric fields at land–seafloor transition

	Land surface	Ocean floor	Origin of distortion
TM, apparent resistivity	Distorted, frequency independent	Distorted, frequency independent	Galvanic
TM, phase	No effect	No effect	–
TE, apparent resistivity	No effect	Distorted, strongly frequency dependent	Inductive
TE, phase	No effect	Distorted, strongly frequency dependent	Inductive

particular, on land the ocean–continent transition distorts mainly the electric field, whereas in oceans it severely distorts both electric and magnetic fields (Table 7.2). Furthermore, a strongly frequency-dependent distortion of the TE mode (which suggests that the effect is inductive rather than galvanic) complicates MT data analysis and separation of the topographic effects (small-scale heterogeneity) from seafloor responses (large-scale mantle heterogeneity).

Upto the present, marine MT interpretations remain geographically limited, thus hampering definite conclusions whether electrical conductivity of the oceanic mantle exhibits any age dependence. Recent MT studies in the Pacific Ocean at the East Pacific Rise examined the conductivity structure of the young oceanic plate (with seafloor ages between 1.3 Ma and 4.5 Ma) down to 200 km depth (Evans et al., 2005). The results reveal an age-independent conductivity structure which is interpreted to reflect water content variations in the uppermost mantle, induced by melting beneath the spreading center, rather than thermal age-dependent variations predicted by conductive cooling models.

7.4.4 Pacific Ocean: fluids or anisotropy?

The conductivity structure of the upper mantle beneath normal oceans has been reported so far chiefly for the Pacific Ocean (Fig. 7.22). Utada et al. (2003) interpreted geomagnetic depth sounding and long-period MT data together with electric observations on abandoned submarine telecommunication cables all over the Pacific Ocean. Their 1D conductivity model (which has an unconstrained conductivity jump at the transition zone), however, refers not only to the Pacific Ocean but constrains a semi-global average conductivity profile for a quarter of the Earth, and as such may not be representative of a typical oceanic upper mantle (Fig. 7.25). In the depth range 100 km to 400 km, the model has very low conductivity values (10^{-3} to 10^{-2} S/m), which approximately follow a synthetic curve for dry olivine SO2 (Fig. 7.21). These values are significantly lower than typically reported in other regional oceanic models, except for two cases. In the depth range 100–200 km, the 1D conductivity model along the Lisbon–Madeira cable (Santos et al., 2003) shows the same conductivity values of 10^{-3} to 5×10^{-2} S/m as the quarter-global model of Utada et al. (2003). The similarity between the two profiles, the Lisbon–Madeira–Canaries and the quarter-global Pacific, in the upper 200 km suggests that at these depths the Pacific model

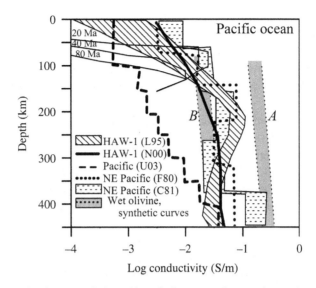

Fig. 7.25 Electrical conductivity profiles for the Pacific Ocean (data of Filloux, 1983; Chave *et al.*, 1981; Lizarralde *et al.*, 1995; Neal *et al*, 2000; Utada *et al.*, 2003). The model labeled U03 averages a quarter of the Earth and may be non-representative of oceanic mantle. Gray shading – synthetic conductivity curves as in Fig. 7.21 for wet anisotropic olivine and mantle adiabats of 1300 and 1400 °C (A: *a*-axis, 1000 ppm H/Si; B: *a*-axis, 100 ppm H/Si or isotropic olivine with 1000 ppm H/Si). Thin lines in shallow part – theoretical curves for a pyrolite mantle based on the SO2 model taking into account mantle melting for oceanic lithosphere of three ages (Heinson and Constable, 1992).

can essentially be dominated by conductivity at hotspots (e.g. the East Pacific Rise and maybe Hawaii, compare with Fig. 7.28). However, in contrast to the Pacific conductivity model, conductivity in the Lisbon–Madeira–Canaries region increases stepwise to 5×10^{-1} S/m at 200 km depth. In the depth range 200–400 km, low conductivities (5×10^{-2} to 10^{-2} S/m) as in the quarter-global Pacific model are observed only in a 1D average conductivity model for the Sakhalin Island in the north-western Pacific ocean (Moroz and Pospeev, 1995). However, the upper mantle beneath the Sakhalin Island, which is located at the suspected boundary between the Eurasian plate and the Pacific (or the North American) plate, can be highly heterogeneous so that, similarly to subduction zones, a 1D conductivity model may be misleading.

Other conductivity models calculated for the Pacific Ocean are rather consistent (Fig. 7.25). The PEGASUS, a controlled-source electromagnetic sounding experiment, has been carried out on 40 Ma-old lithosphere in the north-east Pacific Ocean (for locations see Fig. 7.22); its penetration depth was limited to the upper 100 km of the mantle (Constable and Cox, 1996). The preferred conductivity model has a highly resistive mantle in the depth range 10 to 40 km followed by a conductivity increase from $\sim 5 \times 10^{-5}$ S/m at 40 km to 10^{-3} S/m at 100 km depth. Such low conductivity values are compatible with a dry olivine geotherm (Fig. 7.24a).

Lizarralde *et al.* (1995) interpreted data from the submarine cable HAW-1, which extends over 4000 km across the Pacific Ocean from Point Arena (California) to Oahu (Hawaii)

(Fig. 7.22). These authors examined the coastal effect and the validity of 1D interpretations and concluded that the HAW-1 MT response is not consistent with the model of an electrically isolated ocean basin, implying that (i) there should be a substantial vertical leakage of electric current between the ocean basin and the deep mantle and that (ii) the uppermost mantle beneath the Pacific Ocean may not be so extremely resistive as proposed in early studies.

The results of Lizarralde et al. (1995) suggest the presence of a highly conductive zone (~0.1 S/m) between 150 km and 300 km depth. Note that a strong seismic shear-wave velocity anisotropy is also observed at 150–200 km depth in the central Pacific (Fig. 3.99c). Enhanced upper mantle electrical conductivity beneath the Pacific Ocean had previously been reported by Schultz and Larsen (1990). Earlier conductivity models for the Pacific Ocean (Filloux, 1983; Chave et al., 1981) also have a characteristic belly-shaped anomaly of increased conductivity with a maximal magnitude at ~200 km depth. The analysis of the HAW-1 MT data by Neal et al. (2000) shows some discrepancy with the earlier results and displays neither conductivity values smaller than 0.3–0.5 S/m nor the conductivity maximum at ~200 km depth (Fig. 7.25).

Enhanced conductivity in the Pacific upper mantle down to at least 300 km depth is similar to the conductivity models for the mid-ocean ridges (Fig. 7.23). For reasonable mantle geotherms, it is too high to be explained by solid-state conduction in dry olivine, and requires either an isotropic olivine mantle nearly saturated in hydrogen or water, or a combination of wet (undersaturated) olivine and mantle anisotropy. The high water content above a 200 km depth required by the conductivity model of Lizarralde et al. (1995) agrees with geochemical data on water content in the MORB source region (Bell and Rossmann, 1992). Alternatively, the presence of gravitationally stable melt in the mantle (e.g. 1–3% of melt of komatiite composition) can also explain the high conductivity layer centered at ~250 km depth (Lizarralde et al., 1995).

Lizarralde et al. (1995) favor interpretation of the electrical structure of the Pacific mantle as wet, depleted, and anisotropic in the upper 200 km and increasingly isotropic below 200 km depth with undersaturated concentrations of water (compare with Fig. 3.117). A significant seismic anisotropy in the upper mantle of the Pacific Ocean (Figs. 3.99, 3.100, and 3.116d) that can be explained by an alignment of the olivine a-axis in the direction of spreading has been revealed in seismic surface wave studies (Cara and Lévêque, 1988; Nishimura and Forsyth, 1989; Ekstrom and Dziewonski, 1998; Wolfe and Solomon, 1998; Nettles and Dziewonski, 2008). The most anisotropic is the upper mantle beneath the East Pacific Rise where maximum polarization anisotropy (up to 5–6%) is observed at depth intervals at 100–200 km depth, although some anisotropy is also observed in the shallow mantle below 50 km depth (see Section 3.6.2). In the case of an anisotropic mantle structure with lattice preferred orientation consistent with measured seismic anisotropy, data from the HAW-1 cable require orientation of the a-axis to be horizontal and parallel to the cable direction, i.e. also approximately parallel to the paleo-spreading direction of the Pacific plate. The study of Lizzaralde et al. (1995) could not, however, prove the presence of electrical anisotropy in the upper mantle of the Pacific Ocean. This question has been addressed in the MELT (Mantle Electromagnetic and Tomography) experiment (e.g. Evans et al., 1999, 2005).

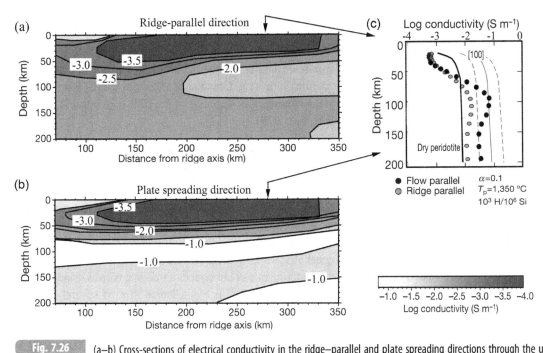

Fig. 7.26 (a–b) Cross-sections of electrical conductivity in the ridge–parallel and plate spreading directions through the upper mantle east of the southern East Pacific Rise. (c) 1D conductivity profiles through the cross-sections (a–b) at a distance of 250–300 km from the ridge axis (~4 Ma old) as compared to theoretical curves. Solid lines – isotropic and anisotropic (along olivine a-axis) conductivity models for a dry and wet (1000 ppm H/Si) peridotite mantle calculated for the geotherm for a 3 Ma-old ocean and for an adiabat with a potential temperature $Tp = 1350 \pm 150$ °C (the dashed lines show the effects of a variation in adiabat temperature of 150 °C) (after Evans et al., 2005). Compare with Figs. 3.99c and 5.16c.

The presence of a strong electrical anisotropy of the Pacific upper mantle has been proven in a number of recent studies based on the MELT experiment conducted at the mid-ocean ridge at 17° S at the East Pacific Rise (e.g. Evans et al., 1999, 2005). Two-dimensional MT interpretations revealed a strongly asymmetric electrical structure of the mantle beneath the region attributed to asymmetric spreading rates and the westward migration of the ridge axis. The MELT results indicate a significantly different electrical conductivity pattern in the ridge–parallel and ridge–perpendicular (plate spreading) directions below 60 km depth, which marks the transition from isotropic to transversely anisotropic electrical conductivity structure (Fig. 7.26). The high conductivity layer in the upper mantle at depths between 60 km and 120 km is significantly more conductive in the direction of plate spreading, which is expected to be the direction of dominant a-axis olivine alignment (Evans et al., 2005). It is also the direction of fast propagation of surface waves. The transition from the resistive to conductive mantle at ~60 km depth is attributed to the anisotropy of the hydrogen diffusion in olivine which is much higher in the [100] (a-axis) direction (Kohlstedt and Mackwell, 1998). Below this depth, partial melting causes water extraction from olivine into melt (Fig. 7.18) and, as a result, low viscous wet mantle material flows sideways from the Rise axis. These results are also broadly consistent with the asymmetry of seismic attenuation anomalies in the central Pacific (Fig. 3.126).

The geodynamic origin of the East Pacific Rise remains a subject for debate; it is likely that its origin is related to a large-scale anomaly in the mantle related either to a large-scale convective circulation or to a mantle plume (e.g. Grand *et al.*, 1997; McNutt, 1998; Gaboret *et al.*, 2003). A comparison of the conductivity structure of the EPR with oceanic hotspots follows in Section 7.4.6 (see also Fig. 7.29).

7.4.5 Mid-ocean ridges: mantle melting

Long-period MT interpretations have been reported for three mid-ocean ridges (MOR): the Reykjanes Ridge in the northern part of the Mid-Atlantic ridge (the RAMESSES experiment (Sinha *et al.*, 1997)), the Juan de Fuca Ridge in the northeast Pacific Ocean (Constable *et al.*, 1997; Jegen and Edwards, 1998), and the East Pacific Rise (the MELT experiment (Evans *et al.*, 1999, 2005)). Seismological observations in the East Pacific Rise demonstrate that the seismic velocity anomaly in the upper mantle is not concentrated in a narrow zone beneath the axis as observed in other mid-ocean ridges. It has a lateral size of several hundred kilometers, extends to depths greater than 100 km (The MELT Seismic Team, 1998), and is probably caused by a large-scale thermal anomaly in the mantle (plume?). Strictly speaking, two other locations are also not typical mid-ocean ridges but rather hotspot affected ridges, since the Reykjanes Ridge is located close to Iceland while the MT experiment at the Juan de Fuca Ridge was carried out on the Axial Seamount. However, upper mantle seismic velocity anomalies beneath these regions are localized beneath the axial zones, allowing us to consider their conductivity structure as being representative of mid-ocean ridges.

In the axial part of the Reykjanes Ridge, strong electrical anisotropy, probably related to the lattice-preferred orientation of olivine crystals by mantle flow, was determined in the upper 30–60 km. Despite a large difference in the conductivity profile in the uppermost 50 km, the Reykjanes Ridge and the Juan de Fuca Ridge have a similar conductivity versus depth structure beneath 60 km depth (Fig. 7.27). Both MT experiments consistently indicate the presence of a highly conductive layer (the electrical asthenosphere) below a 50–100 km depth. These high conductivity values (>0.1 S/m) cannot fit any model without melt; in both regions the melt fraction in the upper mantle was estimated to be between 1 and 10% (Constable *et al.*, 1997; Sinha *et al.*, 1997). However, since the electrical conductivity of the two-phase material depends on the geometry of melt distribution in the matrix and its interconnectivity (Schmeling, 1985b), a conductivity increase by an order of magnitude, as observed in the upper 100 km beneath MOR, can be caused by melt fractions varying by two orders of magnitude.

A ridge-parallel conductivity anomaly was found beneath the Juan de Fuca mid-ocean ridge at the 90 km-long Endeavour segment by vertical gradient electromagnetic sounding (Jegen and Edwards, 1998). A comparison with theoretical models for mantle conductivity with melt inclusions of different geometry and concentration suggests that melt fraction in the upper mantle upwelling zone may exceed 0.10 for Archie's mixing law with conductivity in the range of ~0.6 S/m. In the case of a well-interconnected melt network, such as sheet-like inclusions, melt fraction may be significantly lower, whereas

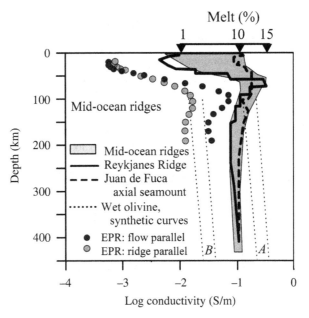

Electrical conductivity profiles for mid-ocean ridges (data of Constable *et al.*, 1997; Sinha *et al.*, 1997) and the East Pacific Rise (Fig. 7.26c). Melt fraction after Murase *et al.* (1977). Dotted lines A and B – synthetic conductivity curves as in Fig. 7.21 for wet anisotropic olivine and mantle adiabats of 1300 and 1400 °C (A: *a*-axis, 1000 ppm H/Si; B: *a*-axis, 100 ppm H/Si or isotropic olivine with 1000 ppm H/Si). Gray shading – generalized 1D model for MOR used for comparison in Figs. 7.23, 7.37.

in models with a crustal conductor the conductivity of the anomalous zone beneath the ridge is ~0.2 S/m.

Geochemical data from the MOR indicate that most lavas are derived from the depth above 60 km, although some melting occurs deeper, within the garnet stability field. Lava chemistry and crustal thickness in most of the MOR can be explained by anhydrous melting of peridotite (Klein and Langmuir, 1987). A comparison of conductivity profiles for the Reykjanes and the Juan de Fuca Ridges with experimental data on melting of dry peridotite suggests that conductivity values observed in the upper 100 km beneath MOR can be explained if the melt fraction is 10–15% (Fig. 7.27). The presence of water considerably lowers the melting temperature and can have a strong effect on melting in the MOR, leading to a greater melt production at a lower extent of melting (Thompson, 1992; Asimow and Langmuir, 2003). Experimental studies of solubility of water in olivine demonstrate that melting beneath a mid-ocean ridge can initiate at ~115 km depth, where olivine can contain about 10% of water (Fig. 7.18). Since mantle conductivity beneath the MOR approaches the conductivity of other oceanic regions at ~200 km depth (Fig. 7.23), this depth may indicate the depth of melt extent in the mantle beneath the MOR. However, an extraction of water from olivine into melt starts at a much shallower depth of 60–70 km (Hirth and Kohlstedt, 1996). This depth corresponds to the conductivity increase observed in MT studies beneath the MOR, in particular beneath the Reykjanes Ridge.

7.4.6 Hotspots: melt dehydrated mantle?

The number of conductivity models available for hotspots is very limited. Voltage measurements in the submarine telecommunication cable CAM-1 between Lisbon (Portugal) and Madeira island in combination with magnetic observation at the Canary Islands were used recently to investigate the conductivity structure of the entire mantle (Santos *et al.*, 2003). Although the inferred electrical structure does not refer solely on the Canary hotspot, the electrical conductivity profile is similar to the one from the axial part of the EPR and is significantly different from the conductivity–depth profile for the rest of the Pacific Ocean (Fig. 7.28). Similarly, the MT interpretations for the Society Island hotspot (near Tahiti) reveal atypical conductivity structure of the upper ~150 km of the mantle, with high resistivity values which are consistent with the dry olivine model (Nolasco *et al.*, 1998).

Thus, highly resistive mantle down to ~200 km depth is the characteristic feature of the oceanic hotspots and is consistent with the melt extraction and subsequent dehydration of the mantle. A comparison of mantle conductivity beneath oceanic hotspots with typical conductivity curves for other tectonic structures worldwide (Fig. 7.38) shows a striking

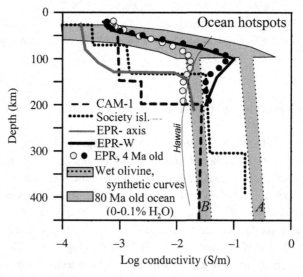

Fig. 7.28   Electrical conductivity profiles for ocean hotspots (1D profiles based on data of Nolasco *et al.*, 1998; Evans *et al.*, 1999, 2005; Santos *et al*, 2003). Abbreviations: EPR-axis and EPR-W – axial-eastern and western parts of the East Pacific Rise, respectively; EPR 4 Ma old – flow-parallel and ridge-parallel directions at 250–300 km distance from the ridge axis (Fig. 7.26c); CAM-1 – Lisbon–Madeira–Canary Islands. Thin line – conductivity profile for the Hawaii hotspot (Neal *et al.*, 2000). Gray shading – synthetic conductivity curves as in Fig. 7.21 for wet anisotropic olivine and mantle adiabats of 1300 and 1400 °C (A: *a*-axis, 1000 ppm H/Si; B: *a*-axis, 100 ppm H/Si or isotropic olivine with 1000 ppm H/Si) and in shallow mantle – for geotherms typical for a 80 Ma-old ocean for dry and wet (0.1 wt% H$_2$O) olivine.

similarity in the electrical structure of the cratonic mantle and hotspots, both requiring devolatilized depleted mantle composition down to at least 150–200 km depth.

A broad region of elevated seafloor topography in the East Pacific Rise (EPR) is associated with a large-scale seismic anomaly in the mantle. The latter has its origin in the deep mantle and can be produced by a large-scale buoyant mantle upwelling (a plume?) (Ekstrom and Dziewonski, 1998). Two-dimensional MT interpretations of the MELT electromagnetic experiment at the EPR (Evans *et al.*, 1999, 2005) show that the mantle beneath the ridge axis, as well as the mantle of the Nazca plate to the east of the ridge, is resistive down to 100–150 km depth and has a conductivity structure similar to the Society Island and the Canary Islands hotspots (Fig. 7.28). Conductivity in the axial zone remains consistent with the dry, depleted in melt and volatiles, olivine mantle down to at least a 200 km depth. Similarly to other hotspots, this can be attributed to mantle dehydration as a result of melt extraction. The presence of melt in small isolated pockets in the mantle beneath the EPR seems unlikely, since an interconnected melt network already forms at melt fractions of less than 0.5% (Daines and Richter, 1988; Faul, 1997).

In contrast to the axial zone beneath the EPR, high mantle conductivity values ($\sim 10^{-2}$ S/m) are observed to the west of the ridge axis already at depths of 40–60 km and increase to 0.5–0.1 S/m at 80–100 km depth. These conductivity values are significantly higher than expected even for water-saturated anisotropic olivine and may require the presence of 1–2% of interconnected melt in the mantle (Evans *et al.*, 1999). Thus, only the axial-eastern part of the EPR has a conductivity structure similar to oceanic hotspots consistent with dehydrated mantle. The asymmetric structure of shallow upper mantle beneath the EPR, clearly seen in seismic and electrical data, can be explained by mantle flow associated with a large-scale buoyant upwelling of mantle material which originates below the transition zone with a proposed origin beneath the South Pacific Superswell (Gaboret *et al.*, 2003). The ascending mantle flow may be horizontally channeled by the low-viscosity asthenosphere towards the EPR. The movement of the Pacific Plate above the ascending flow produces asymmetry in the mantle flow pattern due to a U-turn of shallow mantle flow west of the East Pacific Ridge (Fig. 7.29).

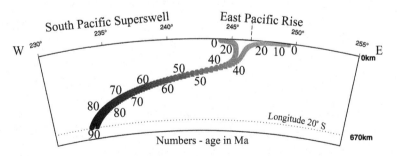

Fig. 7.29   Model of mantle flow associated with a large-scale buoyant upwelling of mantle material, which originates beneath the South Pacific Superswell and is horizontally channeled by the low-viscosity asthenosphere towards the East Pacific Rise, can explain the observed asymmetry in seismic and electrical structure beneath the East Pacific Rise (Source: Gaboret *et al.*, 2003).

Interpretations of long-period MT and GDS data for two locations in the Pacific Ocean with high heat flow (Hawaii and Midway Island) suggest a significantly different electrical structure of the upper mantle beneath the Hawaiian plume with a relatively high conductivity in the uppermost 150–200 km (Fig. 7.28). Since the Hawaiian plume is much younger and more active at present than other hotspots, there might be a correlation between the age of a hotspot and the degree of melt extraction from the mantle and its dehydration. Below a 200 km depth, the electrical structure of the upper mantle becomes similar beneath all oceanic structures and, as discussed earlier for the Pacific Ocean, can be explained either by an isotropic olivine mantle nearly saturated in hydrogen or water, or by a combination of wet (undersaturated) olivine and mantle anisotropy. Similar conductivities below 200 km depth in the mantle beneath mid-ocean ridges and hotspots may indicate, in agreement with modern models of magma genesis, the depth extent (200–250 km) of melt generation and extraction beneath these structures.

7.4.7 Subduction zones and continental margins

Strong lateral and vertical heterogeneity in electrical conductivity structure of the upper mantle has been reported for the Pacific margins. They include Kamchatka (Semenov and Rodkin, 1996), subduction zones beneath Japan (Fujinawa *et al.*, 1997), Izu-Bonin (Toh, 2003) and Ryukyu arcs (Shimakawa and Honkura, 1991) in the western Pacific, and the coasts of the northwestern USA (Wannamaker *et al.*, 1989; Heinson *et al.*, 1993), Argentina (Booker *et al.*, 2004) and Chile (Echternacht *et al.*, 1997) in the eastern Pacific. The data on the electrical heterogeneity beneath the Philippine Sea and the Mariana trench constrained by electric field variations from submarine cables crossing the Philippine Sea together with seismic Vp and Vs velocities and shear-wave attenuation were used by Fukao *et al.* (2004) to estimate mantle temperatures in the region. However, the results of seismic and EM tomography are inconsistent at the subduction zones crossed by the profile due to a significant difference in the sign of slab-related and mantle wedge-related anomalies. Some other studies of electrical conductivity structure of the upper mantle beneath other ocean margins include MT studies of SE Australia and the abyssal plain of the Tasman Sea (Kellett *et al.*, 1991; Heinson and Lilley, 1993) and the back-arc basin of China (Ichiki *et al.*, 2001).

Generalized models

A generalized 1D cross-section of the upper mantle electrical structure through some of the ocean margins is shown in Fig. 7.30; for the Nazca subduction the 1D profile refers to the region where the subduction angle changes to a steep one (Booker *et al.*, 2004). Because of a strong lateral heterogeneity of the mantle beneath active margins, 1D conductivity profiles for these regions represent a strong simplification and refer only to a particular location but not to the structure in general. The latter is better represented in 2D conductivity models (Figs. 7.32 and 7.33).

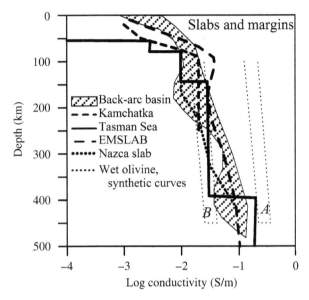

Fig. 7.30 Electrical conductivity profiles for ocean margins and subduction zones (generalized 1D profiles based on data of Heinson *et al.*, 1993; Heinson and Lilley, 1993; Semenov and Rodkin, 1996; Ichiki, 2001; Booker, 2004). Dashed lines A and B – synthetic conductivity curves as in Fig. 7.21 for wet anisotropic olivine and mantle adiabats of 1300 and 1400 °C (A: *a*-axis, 1000 ppm H/Si; B: *a*-axis, 100 ppm H/Si or isotropic olivine with 1000 ppm H/Si).

In spite of significant heterogeneity of conductivity in the uppermost mantle, all oceanic margins have a similar electrical structure below ~50 km depth with a high conductivity (*c.* 0.05–0.1 S/m) at around 100 km depth (Fig. 7.30). In 2D models, this conductivity peak is spatially correlated with the zone of arc magmatism. The conductivity profile for the Tasman Sea does not have a peak in conductivity but rather a smooth gradual conductivity increase between 50 and 150 km depth; however it also follows the general trend observed for oceanic subduction zones. High conductivity at 100–150 km depth requires both high water saturation and anisotropy of olivine in the mantle, or dehydrated mantle with ~5% of interconnected melt. Among different mechanisms responsible for an increase in conductivity in the Oregon subduction zone, Wannamaker *et al.* (1989) mention the possible presence of sulfides in the slab.

Geochemical studies indicate that, in subduction zones along active margins, water is released from oceanic slab by dehydration of the oceanic crust and sediments caused by pressure and temperature increase in the descending slab as it deepens into the mantle. This agrees with rock mechanics studies which suggest that earthquakes in subduction zones at depths of 50–300 km can be triggered by dehydration of hydrous minerals (e.g. serpentine) (Green and Houston, 1995; Kirby, 1995). Water released during dehydration of the slab decreases mantle solidus temperature and triggers partial melting in the mantle wedge, leading to arc-related magmatism. The degrees of melting in the mantle wedge can significantly exceed melting degrees achieved during normal decompressional melting

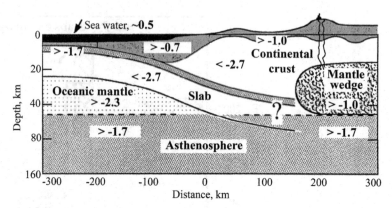

Fig. 7.31 Schematic 2D cross-section of conductivity variations across an active continental margin based on several regional studies in the Pacific Ocean (after Hermance, 1995). Numbers – log conductivity (S/m). Typical conductivity values are $2 \times 10^{-5} - 10^{-3}$ S/m for the upper crust, $3 \times 10^{-3} - 10^{-2}$ S/m for the lower crustal conductor, $\sim 10^{-3}$ S/m for the lithospheric mantle, 0.1–1 S/m for the HCL in the upper mantle ("electrical asthenosphere"), and 0.01–0.1 S/m for the deeper upper mantle. Not all features of this conceptual model are confirmed in regional surveys (compare with Figs. 7.32, 7.33, 7.43).

at mid-ocean ridges (Kelley *et al.*, 2006). Below a 150 km depth, conductivity in subduction zones gradually increases with depth. At these depths, the electrical structure of ocean margins resembles other oceanic regions (Fig. 7.23), and can be explained either by an isotropic water-saturated or by an anisotropic dehydrated olivine model, and probably at 150–200 km depth melting and dehydration in slabs ceases.

Regional studies

A generalized composite conductivity model for the upper mantle constrained from several electromagnetic studies across the Pacific Ocean and western North America illustrates lateral conductivity variations associated with ocean–continent transition (Fig. 7.31). In the real world, however, the conductivity structure of subduction zones is more diverse and displays either a dipping zone of high conductivity (e.g the Izu-Bonin arc and the subduction zone beneath Argentina), or a narrow dipping zone of high resistivity surrounded by a highly conductive mantle, however poorly constrained by the data (EMSLAB and the Lincoln line profiles across the Oregon subduction zone), or a combination of both (the Ryukyu trench) (Fig. 7.32). All of these conductivity anomalies are spatially correlated with the subduction zones; linear zones of high resistivity in the mantle are interpreted as traces of cold subducting slabs. The high resistivity layer in the uppermost 50–100 km of subduction zones is interpreted as cold and mechanically strong lithosphere, impermeable for ascending mantle fluids.

A 2D electrical conductivity model of the subduction beneath Argentina (Fig. 7.32a) clearly shows that a high conductivity zone in the mantle is associated with the subducting Nazca plate; contrary to the generalized model of Hermance (1995) cold descending

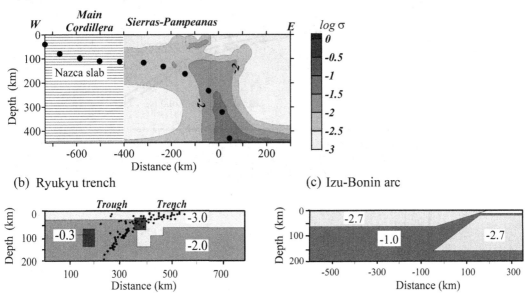

Fig. 7.32 Two-dimensional conductivity cross-sections across three subduction zones (based on the results of Booker *et al.* (2004), Shimakawa and Honkura (1991), Toh (1993)). Black circles: in (a) – position of the Nazca slab estimated from seismicity and in (b) – earthquake epicenters beneath the Ryukyu trench. Numbers – log conductivity (S/m).

slab is not seen (not resolved?) as a high resistivity anomaly in the model of Booker *et al.* (2004). Between 100 and 200 km depth, a strong low resistivity anomaly is located above the subducting slab and is probably produced by ascending diapirs (Schilling *et al.*, 1997). Similarly, in the Cascadia subduction zone, northwestern North America, a region of high electrical conductivity at 20–70 km depth is located above the slab but laterally in a much broader region than beneath Argentina (Fig. 7.33a); beneath Vancouver Island its top coincides with the seismic "E"-reflector in the mantle (Hyndman, 1988). In the Nazca subduction zone, the region of high conductivity in the almost vertically dipping part of the slab extends down to at least the transition zone. Beneath the Andes, to the east from the subducting Nazca slab, partial melting seems to occur down to a depth of at least 250 km and probably as deep as 400 km (Booker *et al.*, 2004).

Magnetotelluric and GDS studies at the active margin of northern Chile provide models of mantle conductivity across the magmatic arc and forearc regions (Echternacht *et al.*, 1997). A striking feature of the conductivity models is the presence of mantle conductor beneath the Western Cordillera with the top at ~20 km depth; its depth extent is not resolved by the data. Extremely high values of conductivity (0.5–2 S/m) require the presence of ~20 vol% of melt. The anisotropic pattern of conductivity observed beneath the Coastal Range has been modeled by a set of highly conducting vertical dykes with some percentage of melting. However, the presence of fluids in the system of coastal faults

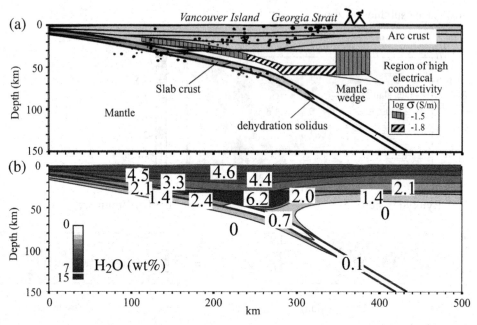

Fig. 7.33 Properties of the Cascadia subduction zone along a transect through southern Vancouver Island. (a) Seismicity and the region of high electrical conductivity (based on interpretations of Hyndman, 1988); (b) proposed maximum water content which shows that the slab mantle is nearly anhydrous, whereas the mantle wedge can locally reach 60–80% hydration (after Hacker *et al.*, 2003).

(the Atacama fault system) cannot be ruled out. Unlike other subduction zones (e.g. in southern Chile or of the Juan de Fuca plate at Vancouver Island, Fig. 7.33), no mantle conductor was identified beneath any part of the magmatic arc; this area spatially correlates with a region with no volcanism (the so-called Pica gap). Echternacht *et al.* (1997) speculate that the high conductivity of subducting slabs is partially associated with sediments on their tops. Since the Chilean profile crosses the margin in one of the driest places on Earth, sedimentary cover on the top of the slab can be missing given that the very arid environment of the Atacama desert is unfavorable for erosion; this could account for the invisibility of the slab in the conductivity profile.

7.5 Electrical structure of the continental upper mantle

7.5.1 Continent-scale conductivity patterns

Global induction studies using quiet time geomagnetic field variations demonstrate significant difference in upper mantle electrical conductivity between the continents (Fig. 7.34). Since the conductivity structure is controlled directly by the thermal state of the lithosphere, this difference reflects, primarily, the areal size of tectonic provinces of different tectono-thermal

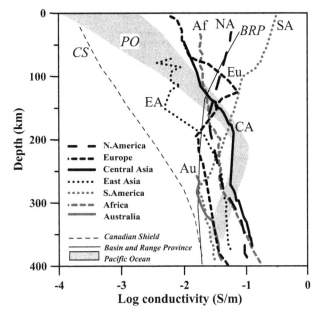

Subcontinental conductivity for seven continent-scale regions based on global induction studies (regression fitting) (data of Campbell and Schiffmacher, 1988). Regional conductivity profiles for three tectonic provinces (Precambrian craton, CS, region of Cenozoic extension and magmatism, BRP, and ocean, PO) are shown for comparison (data of Lizarralde et al., 1995; Neal et al., 2000).

ages within each of the continents; the presence of partial melt or volatiles in the mantle (controlled by regional tectonics) further contributes to regional conductivity variations.

Data for North America are from the stations located mostly in the USA, and although the corresponding conductivity-depth profile averages the craton and the tectonically active structures in the western part of the continent, it is clearly dominated by the latter (Fig. 7.34). Note that a significant difference in conductivity profiles exists not only for regions with different tectonic and thermal regimes, but even for tectonically similar structures (as in the Appalachians or in the Basin and Range Province) where conductivity at the same depth may vary by more than an order of magnitude (Fig. 7.35). These differences are discussed in detail later in this section. They clearly indicate that there are significant difficulties with MT models, related to data interpretation, model resolution, and to an extreme sensitivity of electrical conductivity to the presence (even local) of high-conducting phases.

The average European profile is characterized by a pronounced high conductivity layer at depths of c. 100–140 km (Fig. 7.34). For this continent data from the stations located in western Europe, Iceland, and northern Greenland were used (Campbell and Schiffmacher, 1988), and thus the European conductivity–depth profile reflects an average upper mantle structure for the North Atlantic Ocean and the Phanerozoic part of Europe.

By the geometry of station distribution, the South American conductivity profile is dominated chiefly by the mantle structure beneath the Andes. There, high conductivity values in the upper 150 km, probably associated with melts and entrapped fluids in the

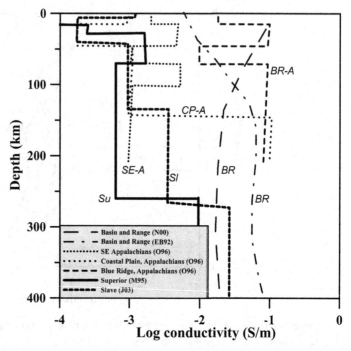

Fig. 7.35 One-dimensional conductivity profiles for various tectonic structures of North America which include the Precambrian Slave (Sl) and Superior (Su) cratons, the Paleozoic Appalachians orogen (SE-A, BR-A, CP-A), and the region of Cenozoic extension in western USA (BR) (based on data of Egbert and Booker, 1992; Mareschal *et al.*, 1995; Ogawa *et al.*, 1996; Neal *et al.*, 2000; Jones *et al.*, 2003).

Andean subduction zone, were found in a regional electromagnetic study of the Chilean coast (Echternacht *et al.*, 1997).

The stations in Central Asia were located approximately along the 70E–80E meridian and the low-conductivity anomaly at ~200 km depth is likely to be associated with the subducting Indian plate. In East Asia, the regional model reflects the conductivity structure of the subduction zones at the Pacific active margin, extending from the Aleutians and Kamchatka in the north to the Philippines in the south (compare with Fig. 7.30). In Australia, where the stations were distributed over the entire continent, the conductivity–depth profile reflects the average structure of the mantle beneath the entire continent. In Africa most of the stations were located south of the equator and the conductivity profile reflects a mixed signal from the cratonic mantle and the East African Rift.

7.5.2 Typical conductivity curves for different continental provinces

The results of electromagnetic and magnetotelluric studies of the last decades clearly indicate that significant differences in the conductivity structure of the crust and the upper mantle are correlated with tectonic settings and thermal regime. For example, the geoelectrical section across East Russia, which crosses active and stable tectonic regions, shows a strong heterogeneity of mantle conductivity down to at least 200 km depth (Fig. 7.36).

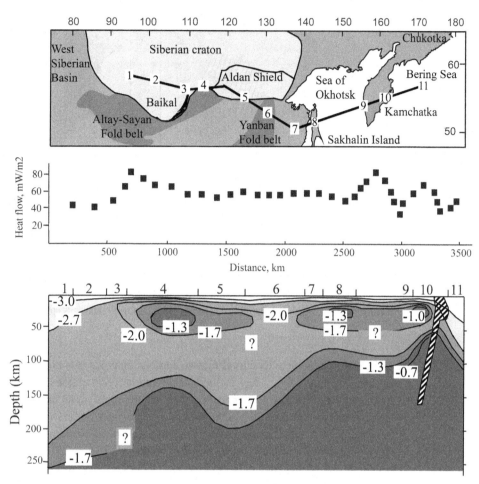

Fig. 7.36   Deep geoelectrical section across East Russia. Upper: profile location; middle: regional heat flow; lower: log conductivity (S/m); hatching – the focal zone beneath Kamchatka (modified after Moroz and Pospeev, 1995).

Zones of high conductivity are correlated with the Baikal rift zone and Vitim Plateau, the suspected plate boundary at the Sakhalin island, and the active margin at Kamchatka, whereas the Siberian craton has high resistivity of the upper mantle.

Clearly, the electrical structure of the continental crust and upper mantle is highly heterogeneous both laterally and vertically, and can be fully described only by 3D conductivity models. However, 1D profiles allow for generalizations of tectonic structures in similar tectonic settings and thus simplify their comparison. As discussed earlier, the choice of the inversion procedure (e.g. for a layered model which provides the best fit to the data, or with the Occam inversion which provides the smoothest possible solution to the data, see e.g. Fig. 7.38c) affects greatly the amplitude of recovered conductivity anomalies; most 2D interpretations include smoothing. In some cases, MT and EM data cannot be resolved reliably due to different types of near-surface distortions (such as in the high latitudes), and 1D models of the conductivity structure may be better constrained than 2D models.

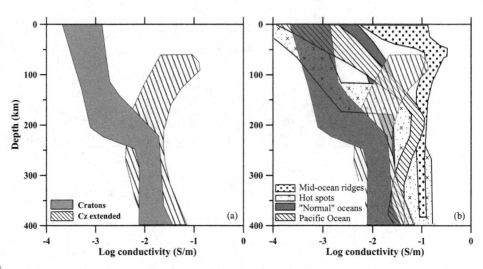

Fig. 7.37 (a) Generalized conductivity profiles for the cratons (see Fig. 7.38) and continental regions of Cenozoic extension (Fig. 7.41). (b) Comparison of typical conductivity of various continental and oceanic structures. Because of a large diversity in conductivity models for other continental regions, no generalization is possible for them.

Typical 1D conductivity curves for two contrasting types of continental upper mantle are presented in Fig. 7.37a. They include stable Precambrian cratons and tectonically active continental regions. The former are based on data reported for all cratons worldwide, while the latter incorporate data for a highly extended lithosphere of the Basin and Range Province in western USA and for the Phanerozoic lithosphere of western Europe, including the Pannonian Basin and the Central European Rift System (which extends from the Atlas mountains in Africa to the North Sea and includes, in particular, the Rhine graben). Below 80–100 km depth, the conductivity curves for the western Mediterranean and for Sakhalin island fall very close to the conductivity profile shown for the regions of Cenozoic extension. However, the conductivity structure of the upper 80–100 km beneath these regions differs by two orders of magnitude and cannot be generalized. Similarly, a significant diversity in the conductivity structure of other continental provinces (e.g. basins and orogens) precludes any 1D generalizations. In contrast to the significant difference in upper mantle conductivity above a 200 km depth, the conductivity structure of the continental upper mantle is relatively homogeneous at 200–400 km depths (Figs. 7.23 and 7.37a).

Comparison of conductivity curves for the continents and the oceans indicates that there is a fundamental difference between the cratons and normal oceans (Vanyan and Cox, 1983; Lizzaralde *et al.*, 1995; Hirth *et al.*, 2000), which extends down to a 200 km depth (Fig. 7.37b). Below this depth, the conductivity of the continental upper mantle becomes similar to mantle conductivity at the same depths beneath the oceans. This similarity suggests that, on a global scale, most differences in thermal and compositional (including melts and water) structure of the continental and oceanic upper mantle are concentrated above a 200 km depth. However, regional conductivity variations below a 200 km depth reach an order of magnitude suggesting that significant lateral variations in the mantle persist down to, at least, the transition zone.

7.5.3 Stable continents

Figure 7.38 shows a compilation of conductivity–depth profiles for different cratons of the world, based on data for the Slave craton, the Superior and Churchill provinces of the Canadian Shield, the Baltic Shield, the East European Platform, the Ukrainian Shield, the Siberian Shield, and the northern part of the Indian Shield. Some earlier results were excluded from the compilation since source field problems associated with high latitudes (Osipova et al., 1989) could cause misinterpretations of natural-source MT data, such as in the early studies of the northern parts of the Baltic Shield (Jones, 1984).

The electrical structure of the cratonic crust is highly heterogeneous and is commonly well correlated with major tectonic structures (e.g. Pajunpaa et al., 2002). Typically the shields have a highly resistive crust and are surrounded by narrow elongated highly conducting zones. In particular, reviews of crustal conductivity in Northern Europe and its correlation with tectonics and lithosphere evolution can be found in Korja et al. (2002). However, due to small-scale heterogeneity of crustal conductivity generalizations are hardly possible, and the following discussion focuses on the conductivity structure of the subcrustal lithosphere and upper mantle, which allows us to distinguish the characteristic patterns.

There is striking agreement between the general pattern of upper mantle conductivity for all cratonic regions; almost at any depth down to the transition zone the range of conductivity variations is less than 0.5 log units (Fig. 7.38). Even earlier results, which did not have sufficient resolution, show the same pattern. No other continental tectonic structure shows such consistency in conductivity–depth variations.

Experimental studies of conductivity–temperature dependence of olivine, which comprises about 50–60% of the upper mantle and may essentially control mantle conductivity, allows for calculation of theoretical conductivity–depth profiles in the mantle. Theoretical profiles calculated for dry subsolidus olivine model SO2 (Constable et al., 1992) and for typical cratonic geotherms of 37 and 50 mW/m^2 bound experimental data for all of the cratons in the depth range from 100–150 km to 220–280 km (Fig. 7.38) (Artemieva, 2006), thus confirming correctness of cratonic geotherms constrained by thermal and xenolith data. Agreement between theoretical predictions and model interpretations of electrical conductivity structure of the cratonic lithosphere also indicates dry composition of the cratonic mantle, although the presence of a trace amount of unconnected water cannot be ruled out (Hirth et al., 2000; Bahr and Duba, 2000).

Deflection of conductivity models for the individual cratons from the synthetic conductivity curve above 100–150 km depth can be attributed to the presence of highly conductive phases. Since for typical cratonic geotherms the graphite–diamond transition occurs at a depth of 120–170 km (Kennedy and Kennedy, 1976) and diamonds are abundant in the cratonic roots, it is reasonable to attribute an order of magnitude increase in conductivity of the uppermost mantle to the presence of graphite. Conducting graphite films were used to explain electrical anisotropy in the upper 100 km of the mantle beneath the Superior craton (Mareschal et al., 1995) (Fig. 7.38b). For a typical cratonic conductivity model, the base of

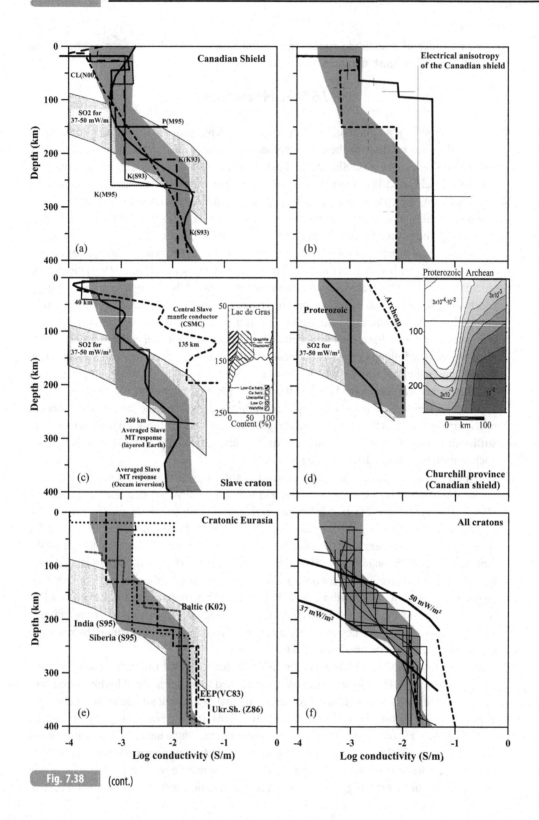

Fig. 7.38 (cont.)

the electrical lithosphere is expected to be at a depth of ~200–280 km. At this depth, the transition from conductive to adiabatic geotherm produces a step-like pattern in conductivity curves caused by a weak depth dependence of conductivity in the deeper, high-conductivity layer, commonly interpreted as the electric asthenosphere. In some cratonic regions (e.g. West Africa, central Finland, the East European Platform, the Canadian Shield, Figs. 7.39b, 7.47bc) no well-developed electrical asthenosphere has been identified down to a depth of ~300 km (Vanyan *et al.*, 1977; Ritz, 1984; Adam *et al.*, 1997; Korja, 1993, 2007). The conclusions of some of these models based on old interpretations may be questioned. Conductivity within the sublithospheric cratonic mantle closely follows theoretical constraints for conductivity of anisotropic wet olivine (Fig. 7.38f); however down to the transition zone the cratonic mantle still remains much drier than the oceanic mantle.

Superior Province: mantle anisotropy

Several long-period MT and GDS studies in the Archean Superior Province of the Canadian Shield (age 2.5–2.68 Ga) have demonstrated a relatively large scatter in estimates of the lithosphere thickness between different terranes (Fig. 7.38a). A uniform character of MT responses justifies 1D resistivity models for the Canadian mantle. For models of the layered Earth (commonly constrained by D+ Parker's analysis which provides the roughest, although the best possible fitting 1D layered model), a conductivity increase by an order of magnitude beneath the Kapuskasing structural zone is observed at depths of 205–260 km,

Caption for Fig. 7.38 (cont.)

Electrical conductivity profiles for different cratonic regions of the world (a–e); (f) and gray shading in (a–e): generalized cratonic conductivity based on data stacked for all cratons (after Artemieva, 2006). (a) Isotropic conductivity for the Superior Province of the Canadian Shield. Abbreviations: K–Kapuskasing, CL–Carter lake, P–Pontiac (data of Schultz *et al.*, 1993; Kurtz *et al.*, 1993; Mareschal *et al.*, 1995; Neal *et al.*, 2000; Ferguson *et al.*, 2005). (b) Anisotropic conductivity of the Superior Province (solid and dashed lines); thin lines – model uncertainties (data of Mareschal *et al.*, 1995). (c) Isotropic conductivity of the Slave craton. Solid lines – average MT response for the entire craton constrained for a layered model (thin line) and with the Occam inversion (bold line). Conductivity decrease by 0.5 log unit at 135–150 km depth closely correlates with the compositional boundary in the Slave mantle (insert, depth scale is the same, also see Figs. 6.15, 6.17). A sharp conductivity increase at ~260 km depth can be interpreted as the base of the electrical lithosphere. Conductivity structure of the Central Slave (dashed line) is significantly different, with a pronounced mantle conductor at 135 km depth (data of Wu *et al.*, 2002; Jones *et al.*, 2003). (d) An unusual conductivity structure of the Churchill Province, the Canadian Shield, where the Archean terrane is more conductive than the Proterozoic terrane. Two-dimensional profile is shown in insert (for the location see Fig. 7.39, profile C), numbers – conductivity in S/m, depth scale is the same (data of Boerner *et al.*, 1999). (e) Conductivity structure of the Baltic Shield (Korja, 1993; Korja *et al.*, 2002), the East European craton (Vanyan *et al.*, 1977; Vanyan and Cox, 1983), the Ukrainian Shield (Zhdanov *et al.*, 1986), Siberia (Safonov *et al.*, 1976; Vanyan and Cox, 1983; Singh *et al.*, 1995), and India (Singh *et al.*, 1995). The range of predicted conductivity variations is shown for comparison (see Fig. 7.21): light gray shading in (a, c–e) and thick solid lines in (f) – for continental geotherms of 37 and 50 mW/m^2 for the standard olivine model SO2 with 90% of forsterite; thick dashed lines in (f) – for mantle adiabats of 1300 to 1400 °C, along *a*-axis of anisotropic wet olivine (1000 ppm H/Si).

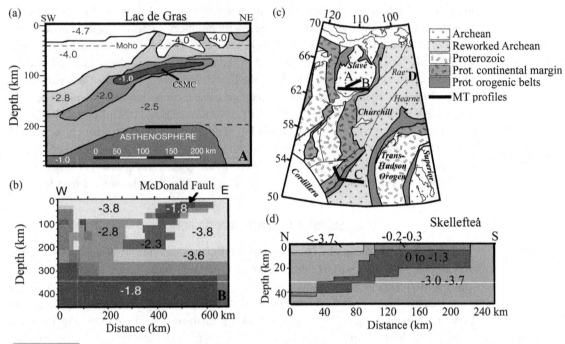

Fig. 7.39 (a, b) 2D conductivity structure of the upper mantle beneath the Slave craton (modified after Jones *et al.*, 2003). CSMC = Central Slave Mantle Conductor. (c) Location map for profiles (a–b) (modified after Boerner *et al.*, 1999). Conductivity structure of locations marked A–D is discussed in text. Profile C is shown in insert, Fig. 7.38d. (d) 2D conductivity structure of the Skellefteå area in Sweden (profile location in Fig. 7.47c). Numbers – log conductivity (S/m). Note different scales for the profiles. Compare with a conductivity model of a modern fault zone (Fig. 7.42).

and it is roughly coincident with the Lehmann discontinuity (Kurtz *et al.*, 1993; Mareschal *et al.*, 1995). In contrast, beneath the Pontiac Province, which is adjacent to the middle-Proterozoic Grenville Province, conductivity increases already at 100–150 km depth. Smooth conductivity models (the Occam inversion of long-period MT data (Schultz *et al.*, 1993) and joint interpretation of long-period MT and GDS data (Neal *et al.*, 2000)) smear differences in 1D conductivity structure between the Kapuskasing and the Pontiac provinces (Fig. 7.38a).

However, 2D studies indicate that these differences do exist and are expressed in the anisotropy of upper mantle electrical conductivity beneath the Superior Province, especially at 50–150 km depth (the base of the anisotropic layer was not well resolved) (Fig. 7.38b). Beneath the Kapuskasing Uplift, the most conductive direction is NW–SE (N20W), while a few hundred kilometers away, in the eastern Abitibi, the Pontiac Province and in the western Grenville, the most conductive direction is sublatitudal (N80E) (Mareschal *et al.*, 1995; Ji *et al.*, 1996). These directions roughly agree with the orientation of the fast axes of seismic anisotropy (Fig. 3.114). Electrical anisotropy in the upper mantle can be explained by the presence of a high-conducting phase, such as dissolved hydrogen or graphite films distributed within fractures or at grain boundaries and associated with metasomatism along major Archean shear zones (Kurtz *et al.*, 1993; Mareschal *et al.*, 1995). However, such a mechanism alone can be inefficient in explaining seismic anisotropy, and thus preferential alignment of olivine crystals in the upper mantle should additionally contribute to observed electrical anisotropy.

Slave craton: layered upper mantle

Long-period MT observations made within the frame of the LITHOPROBE project and interpreted as 1D, 2D, and 3D conductivity models, reveal an unusual electrical structure of the upper mantle of the Archean Slave craton (Jones *et al.*, 2003). Figure 7.38c shows three contrasting 1D conductivity profiles: an average conductivity for the entire Slave craton constrained by the Occam inversion which provides the smoothest possible solution to the data, an average conductivity calculated for the layered Earth which provides the roughest but the best possible fitting 1D layered model, and the Occam inversion of conductivity only for the central part of the Slave craton. Both average conductivity models show a two-layered structure of the upper mantle with conductivity jumps at ~135–150 km depth and at ~260 km depth. The latter is interpreted as the base of the electrical lithosphere. Jones *et al.* (2003) argue that the depth resolution of their model is very high and that the base of the electrical lithosphere beneath the Slave craton is at a depth of ~250–260 km in the southern part and shallows to ~210 km beneath the north-central parts of the craton. The contrasting structure of the south-eastern and north-central parts of the Slave craton has been also revealed in petrologic data (Section 6.1.4; Fig. 6.17).

A conductivity jump at mid-lithospheric depth is of a particular interest since it is not observed in any other cratonic region. The layered structure of the mantle lithosphere beneath the central part of the Slave craton (Lac de Gras area) is supported by petrologic studies of mantle-derived xenoliths (Griffin *et al.*, 1999c; Kopylova and Russell, 2000). The shallow layer has a higher iron-content (Fo_{92-94} as compared to the mean $Fo_{91.5}$ in the deeper layer) and lower concentrations of TiO_2, Y, and Zr in garnets (Fig. 6.15c,d). Its composition is interpreted as 60% harzburgite and 40% lherzolite (insert in Fig. 7.38c) and is similar to some highly depleted ophiolites from active convergent margins. At ~145 km depth mantle composition sharply changes to 15–20% harzburgite and 80–85% lherzolite and is interpreted as plume-related underplating of an existing lithosphere at ~2.6–2.7 Ga; the latter conclusion is supported by the finding of diamonds with an ultradeep (>760 km) origin (Griffin *et al.*, 1999c).

The central part of the Slave craton is also unique as it has a strong mantle conductor (the Central Slave Mantle Conductor, CSMC) which starts at depths of 80–120 km and is approximately 20–30 km thick (Figs. 7.38c; 7.39a). Its base correlates with a sharp transition from the ultra-depleted harzburgitic mantle layer to a less depleted lherzolitic mantle (Griffin *et al.*, 1999b) and with the phase transition graphite–diamond (Kennedy and Kennedy, 1976). The latter correlation suggests that the CSMC can be caused by the presence of carbon (Jones *et al.*, 2003). A problem with this explanation is an absence of carbon effect on the conductivity at shallower depths, since due to a very low thermal activation energy, conductivity of carbon is almost temperature independent (Constable and Heinson, 1993) and should affect bulk mantle conductivity even at low temperatures. Hydrogen diffusion can provide an alternative explanation for the conductivity anomaly, which can be attributed to trapped oceanic or arc-related paleoslab, probably during the Thelon–Talston orogeny.

Churchill Province: metasomatized lithosphere?

Recent MT studies in south-western Canada have revealed a contrasting conductivity structure in the Precambrian lithosphere of the Archean–Proterozoic Churchill Province with an unusual conductivity structure beneath the Archean part (Fig. 7.38d). While the conductivity structure beneath the Proterozoic part of the province is similar to typical cratonic conductivities, conductivity beneath the Archean terrane is one order of magnitude higher and is inconsistent with a dry olivine model (Boerner et al., 1999). These authors argue that variations in Fe content, an increased oxygen fugacity, or even fluids cannot explain the magnitude of the conductivity anomaly and its spatial correlation with the upper crustal provinces. Since the anomaly falls within the phlogopite stability field (depth ~80–190 km), hydrous modal metasomatism, probably related to the Proterozoic shallow subduction (Fig. 7.39c), could be responsible for the conductivity anomaly in the Archean lithospheric mantle. This conclusion is supported by xenolith evidence for a widespread mantle metasomatism under the Churchill Province. However, alternative mechanisms, such as olivine anisotropy or the presence of graphite or sulfides cannot be entirely ruled out.

7.5.4 Paleosubductions, suture zones, and craton margins

Precambrian plate tectonic processes are, at present, reliably distinguished seismically and mostly from observations of dipping mantle reflectors (Figs. 3.64 and 3.65). An analogy between the observed reflection geometries with reflections in modern subduction zones allows for the interpretation of them as images of paleosubduction zones. Similarly, by analogy with modern tectonic structures, electrical conductivity anomalies in the lithosphere are interpreted as evidence for ancient tectonism (Jones, 1993). Magnetotelluric studies over recent decades have been successful in revealing paleosutures in different continental settings, which have been interpreted as evidence for ancient plate tectonic processes associated either with paleosubductions or with collisional events.

North America

Systematic MT studies of the Precambrian large-scale sutures in the Canadian Shield were carried out within the LITHOPROBE project. In most cases the sutures are underlain by dipping conductors in the crust and the upper mantle. The mantle conductor (0.01–0.1 S/m) has been identified beneath the Great Bear magmatic arc. It extends down to at least 120 km depth and correlates spatially with a Proterozoic eastward subduction of oceanic lithosphere at ~1.8–1.9 Ga, interpreted from LITHOPROBE seismic data (Cook et al., 1999; Aulbach et al., 2004). The increased mantle conductivity beneath the Great Bear magmatic arc can be attributed either to carbon or to a high concentration of conductive minerals produced during the deformation and metamorphism (Wu et al., 2002). Fifty km eastwards, another pronounced mantle conductor (>0.01 S/m) extending down to at least 150 km depth has been identified beneath the Great Slave Lake shear zone at the southern margin of the Slave craton (Fig. 7.39c, location B). It appears to have nearly vertical geometry with an eastwards-dipping upper boundary and exhibits a sharp contact with more resistive mantle to the southeast. This sharp truncation of mantle anomaly has been interpreted as evidence for significant strike-slip movement of the mantle lithosphere. The origin of the enhanced conductivity beneath the Great Slave Lake shear zone is attributed to either hydrogen or carbon brought into the mantle during the Proterozoic subduction (Wu *et al.*, 2002). The McDonald Fault in the southern part of the Slave craton is another example of Proterozoic suture (Fig. 7.39b). It is expressed as a dipping at ~45°, 100 km-wide zone of an increased (by 1–2 orders of magnitude) conductivity from mid-crustal depths and down to ~200 km depth (Jones *et al.*, 2003). Conductivity anomalies associated with the McDonald and the Tintina strike–slip faults (Ledo *et al.*, 2002) are attributed to the Proterozoic Thelon–Talston orogeny and are explained by the presence of a highly conducting material (either graphite or sulfides) along fault zones.

The longest known linear conductivity anomaly, the North American Central Plains conductivity anomaly, has been identified by MT data (COPROD2 project) in the crust within the Paleoproterozoic Trans-Hudson orogen (Jones *et al.*, 1993). At depths ~8–25 km the conductivity values in the westwards-dipping anomaly are about two orders of magnitude higher than in the surrounding crust (1 S/m and 0.001–0.01 S/m, respectively) and can be associated with pyrite sulfides emplaced into the crust during Proterozoic westward-directed subduction (Jones *et al.*, 1997).

Regional MT studies in the Archean Churchill Province of the Canadian Shield identify a strong high conductivity anomaly in the crust at the boundary between the Rae and the Hearne domains (for the location see Fig. 7.39c, profile D). The crustal anomaly dips southwards and continues in the upper mantle as an interface between the more conducting Hearne mantle (log $\sigma \sim -3.8$–4.8) to the south and the more resistive Rae mantle (log $\sigma < -4.8$) to the north. Mantle conductivities beneath both of the Archean domains are low and are consistent with a dry pyrolite model. In alternative interpretations, the dipping crustal conductor is either a subdomain of the Rae footwall or a part of the Hearne hanging-wall. Jones *et al.* (2002) favor the latter model and propose that the leading edge of the Rae lithosphere has been underthrust under the Hearne crust by 150–200 km during the Neoarchean (*c.* 2.5–2.55 Ga) plate tectonic interaction. Further reactivation of the Neoarchean tectonic suture in the Proterozoic (~1.8 Ga) might have led to an overthrusting of the high pressure crust of the Hearne domain over low pressure crustal rocks of the Rae domain.

Europe

A strong crustal conductor (the Skellefteå conductor) has been identified near the northern part of the Bothnian Gulf of the Baltic Sea at the boundary between the Proterozoic and chiefly Archean domains (Fig. 7.39d). The Skellefteå conductor is similar to the crustal conductors between the Rae and Hearne domains in the Churchill Province of the Canadian Shield and beneath the McDonald Fault. The northward dip of the conductor beneath the Archean province suggests that it marks a zone of paleocollision, with the crust (and maybe subcrustal lithosphere) of the Proterozoic Svecofennian province underlying the crust of the Archean Kola–Karelian province over a distance of ~100 km. This conclusion is supported by high-resolution seismic reflection studies in the northern Bothnian Gulf along the BABEL lines B2, B3/4 (Fig. 3.64) which discovered a set of mantle reflectors at 50 to 80 km depths dipping northwards at 20° to 30° angle (BABEL Working Group, 1990, 1993). These reflectors are traced at horizontal distances over 100 km and are interpreted as relics of Proterozoic (1.8–1.9 Ga) tectonic processes related to the Svecofennian plate convergence, subduction, and accretion of Proterozoic terranes to the Archean nuclei of the Baltic Shield.

Several EM studies in Ireland and Scotland have aimed to reveal conductivity anomalies associated with the Iapetus suture – a Paleozoic collisional belt between Baltica and Avalonia which separates the Precambrian (>825 Ma) basement to the north from the 450–415 Ma Caledonian basement to the south. The suture was formed during the closure of the Iapetus Ocean and it is debatable whether the process included one or two subductions with northwestward or southeastward dippings. Two conductivity models from the northern UK suggest a southeastward-dipping conductivity anomaly, while the conductivity model across the Iapetus suture itself indicates a northwestward-dipping anomaly which correlates well with crustal reflectivity (Beamish and Smythe, 1986).

Australia

A continent-wide study of electrical conductivity structure in Australia obtained from simultaneously fitting results from multiple geomagnetic deep sounding surveys (Fig. 7.40) have revealed several large-scale linear conductivity anomalies (Wang, 1998;

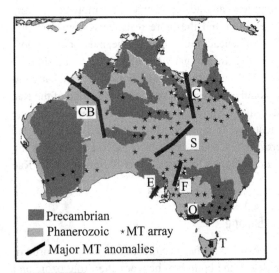
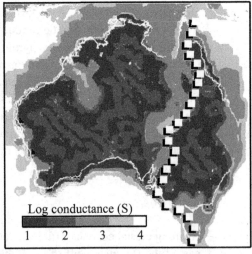

Fig. 7.40 Major electrical conductivity anomalies in Australia determined from regional magnetometer array studies (after Lilley et al., 2003). Solid lines in left map show orientation of major MT structures; C = Carpentaria, CB = Canning Basin, E = Eyre, F = Flinders, T = Tamar, S = SW Queensland. Right map: conductance anomalies (= product of conductivity by thickness). Dotted line marks the boundary between the Precambrian and the Paleozoic domains (the Tasman line).

Wang and Lilley, 1999). Similarly to other continents, these anomalies may define the location of the fundamental tectonic boundaries and can be caused by the emplacement of high-conducting phases (such as sulfides or graphite) along sutures during paleometamorphic events. A comparison with tectonic maps reveals that major conductivity anomalies in Australia are spatially correlated with Proterozoic orogenic belts and areas of Paleozoic tectono-thermal activity (Lilley *et al.*, 2003):

- the Carpentaria conductivity anomaly can be associated with the Carpentarian orogeny (1.8–1.4 Ga) which led to the formation of Mnt. Isa Orogen (see Fig. 3.82 for the location map). A detailed GDS study across the Carpentaria anomaly revealed a ~50 km-wide zone of enhanced conductivity (~1 S/m) within the upper 20–40 km of the lithosphere;
- the southwest Queensland anomaly spatially correlates with the Central Australian Mobile Belt formed during the Musgravian orogeny (1.4–1.0 Ga); this broad orogenic zone includes the Arunta–Musgrave blocks around the Amadeus Basin (with further deformation and metamorphism in the Arunta block at *c.* 1.1–1.0 Ga) and the Albany–Fraser belt (with earlier metamorphism at 1.7–1.6 Ga), which in its western part separates the Pilbara and the Yilgarn cratons;
- the Flinders anomaly follows the site of the Adelaidean orogeny (1.1–1.0 Ga) which in South Australia resulted in the formation of the Adelaide geosyncline and the Tasmania basin;
- the conductivity anomaly in the Canning Basin correlates spatially with the region of Paleozoic (particularly Devonian and Triassic–Jurassic) thermo-tectonic activity.

The conductivity anomalies in Australia encircle the Proterozoic cratonic core of the continent: the location of the Carpentaria, southwest Queensland, Flinders, and Eyre conductivity anomalies follows approximately the major tectonic boundary of the continent between the Proterozoic and Phanerozoic parts (the Tasman line), while the Phanerozoic Canning Basin at the western margin of the continent separates two major Archean–Proterozoic blocks. The large width (~500 km wide) of the zone with enhanced conductance along the Tasman line suggests that the Proterozoic plate collision has involved subduction of a significant amount of the oceanic lithosphere. This zone well correlates with a transition from fast cratonic to slow Phanerozoic mantle as seen in regional surface-wave tomography (Kennett, 2003).

Cratonic margins

Similarly to observations in Australia, a set of several linear conductors extending north–south over more than 1000 km along the Paleozoic Ural mountains has been found at the eastern margin of the East European craton (Avtoneyev *et al.*, 1992). The tectonic expression of the western margin of the East European craton is the Trans European Suture Zone (TESZ) (see Fig. 3.60 for location). The TESZ correlates with a 10–15 km decrease in the crustal thickness and with a 2–3% decrease in mantle seismic velocities across the cratonic margin (Zielhuis and Nolet, 1994; Cotte *et al.*, 2002; Janik *et al.*, 2002; Grad *et al.*, 2002, 2003). A low-velocity seismic anomaly has been attributed to the presence of fluids associated with ancient subduction at the cratonic margin (Nolet and Zielhuis, 1994).

However, no dipping structures are imaged in high-resolution regional seismic surveys. The recent EMTESZ–Pomerania project examined the electrical structure of the TESZ using long-period magnetotellurics (Brasse *et al.*, 2006). They found a typical low conductive mantle down to 300–400 km on the cratonic side of the TESZ and a highly conductive mantle with the top of the major upper mantle conductor at depths of 100–150 km beneath young structures of western Europe. The transition from the low conductive Precambrian mantle to highly conductive Phanerozoic mantle takes place over a distance of less than 200 km. Although the zone of enhanced conductivity in the crust, spatially correlated with the TESZ, is similar to the conductivity structure of other suture zones, electromagnetic data do not indicate the presence of a mantle conductor beneath the TESZ. The absence of a mantle conductor can, however, be apparent since thick sediments with conducting horizons within them restrict resolution of the deeper electrical structure beneath the TESZ. Nevertheless, it seems that, despite its tectonic name, the TESZ is not a suture *senso stricto*.

The conductivity patterns observed in Australia, the Urals, and across the TESZ are similar to observations in North America (Boerner *et al.*, 1996). There, electromagnetic studies of many of the Laurentian foredeeps show that conductivity anomalies correlate spatially with elongated graphitic belts contained within metamorphosed and deformed foredeep shale facies. Proterozoic foredeeps that encircle most of the cratons have similar stratigraphic characteristics and are interpreted as remnants of orogens formed during global-scale Proterozoic orogenic and accretionary processes. Thus, zones of enhanced conductivity can be associated with collisional processes at cratonic margins and their origin may be due to the presence of graphitic and sulphidic sedimentary rocks in the crust formed during deformation and metamorphism in an oceanic environment (Boerner *et al.*, 1996).

7.5.5 Extensional regions

Western USA

The number of mantle conductivity models for continental regions of lithospheric extension remains limited. The conductivity structure beneath the Southern Basin and Range Province in southwestern USA was studied with long-period MT data (Egbert and Booker, 1992) and with long-period MT data for the Tuscon (Arizona) station complemented with GDS data (Neal *et al.*, 2000). Although both models indicate a very conductive mantle beneath the region, there is a significant difference between the interpretations (Fig. 7.41). According to Neal *et al.* (2000), below ~200 km depth mantle conductivity follows a typical continental conductivity trend with the conductivity half an order of magnitude lower than beneath any oceanic structure (Fig. 7.37b). Enhanced conductivity in the uppermost 100–150 km can be explained by partial melting of the mantle; the depth range 150–200 km can require an anisotropic water-saturated or isotropic undersaturated composition of the mantle.

In contrast, the conductivity model of Egbert and Booker (1992) suggests relatively low conductivity in the upper 100 km, followed by a sharp (by an order of magnitude) conductivity increase at ~100 km depth which can be interpreted as the top of the electrical asthenosphere. Below ~100 km depth, very high conductivity values, similar to the Pacific Ocean and mid-ocean ridges, persist down to the transition zone (Fig. 7.41) and require an

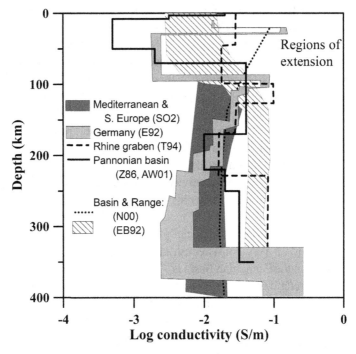

Fig. 7.41 Conductivity profiles for continental regions of Cenozoic extension and tectonic activity, including Western Europe, Central European Rift System, and the Mediterranean (models of Simpson, 2002a; Tezkan, 1994; ERCEUGT-Group, 1992), the Pannonian Basin (models of Zhdanov *et al.*, 1986; Adam *et al.*, 1997; Adam and Wesztergom, 2001), and the Basin and Range Province (models of Egbert and Booker, 1992; Neal *et al.*, 2000).

anisotropic mantle with some amount of water. These results are similar to the conductivity structure of the Blue Ridge extensional region in the southeast Appalachians (Fig. 7.35). A 3D MT study of the Basin and Range Province indicates a highly heterogeneous conductivity structure of the uppermost crust which can explain some discrepancy between the upper mantle conductivity models (Mackie *et al.*, 1996). However, in the uppermost mantle (at depth ~30 km) the model of Mackie *et al.* (1996) agrees better with the conductivity model of Neal *et al.* (2000) and reports mantle conductivity values (0.01–0.1 S/m) immediately below the crust. Furthermore, regional conductivity models for the Phanerozoic Western Europe show a surprising similarity with the conductivity–depth profile of Neal *et al.* (2000) (Fig. 7.41 and references to it).

Western Europe

Magnetotelluric interpretations for the Variscan Europe, the Central European Rift System, and the North German Basin consistently indicate that while the electrical structure of the crust is highly heterogeneous (e.g. Hoffmann *et al.*, 2005; Munoz *et al.*, 2005), the conductivity–depth profiles for different tectonic structures in western Europe show a concordant pattern below ~100 km depth (Fig. 7.41). At these depths, average conductivity models for western Europe and the Mediterranean, as well as regional conductivity models for the Pannonian Basin and the Rhine Graben show conductivity values similar to the model of Neal *et al.* (2000) for the southern Basin and Range Province. This conductivity–depth profile can be considered as typical for continental regions of active tectonics. Although a large scatter of conductivity values exists for the upper 50–80 km, all interpretations indicate the presence of a high conductivity layer at ~70–120 km depth, which can be interpreted as the electrical asthenosphere. Conductivity values below 200 km depth are similar to the cratonic values and require much drier mantle than beneath any oceanic structures.

Several extensional structures in western Europe were studied by MT methods. A sharp contrast in mantle conductivity structure beneath the Pannonian Basin and the adjacent regions was found in numerous MT studies (e.g. Adam *et al.*, 1997). These studies display a large diversity (one and a half order of magnitude) in conductivity values for the lithospheric mantle which can probably be attributed to its heterogeneous structure, but they are all very consistent in indicating that the thickness of the electrical lithosphere beneath the Pannonian Basin is ~70–90 km (Fig. 7.41). This result is supported by seismic interpretations of P-wave residuals (Babuška and Plomerová, 1992) and thermal modeling (Zeyen *et al.*, 2002). Strong negative residual gravity anomalies require the presence of low-dense, probably partially molten material in the upper mantle (Yegorova *et al.*, 1998). An almost two orders of magnitude conductivity increase at 70–90 km depth can also be explained by the presence of a significant amount of melt in the upper mantle, although mineralization and the presence of fluids in a subduction zone environment cannot be ruled out.

Data for the Rhine Graben require the presence of a highly conductive (0.01 S/m compared to the background conductivity of *c.* 0.001 S/m) layer at ~90–120 km depth (Tezkan, 1994). Although regional high-resolution P-wave and S-wave seismic tomography models have low mantle velocities at a depth of ~80 km within the Rhine Graben itself, absolute P-velocities do not reveal a low velocity anomaly in the upper mantle of the Rhine Graben down to a depth of ~280 km (Achauer and Masson, 2002). The lower crust and the upper mantle beneath the adjacent Rhenish Massif show strong electrical anisotropy. While the crustal conductor at 14–22 km depth has a N45°E orientation, the mantle conductor at 90–130 km has a west–east orientation (Leibecker *et al.*, 2002). The conductivity model for the upper mantle includes a background conductivity of 0.004 S/m (which agrees with the dry olivine model) superimposed with an anisotropic layer with a conductivity of 0.15–0.2 S/m in the west–east direction. A comparison with SKS shear-wave splitting data indicates that the preferred orientation of the olivine *a*-axis can explain both seismic and electrical anisotropy. However, LPO alone cannot explain the amplitude of conductivity anisotropy. Since the anisotropic layer appears to be below the base of the lithosphere, the presence of partial melt at these depths seems likely. This conclusion is supported by shear-wave tomography in the region which was interpreted as evidence for the presence of up to 1 vol% of melt in the upper mantle (Ritter *et al.*, 2001). Both olivine anisotropy and a small percentage of melt may be insufficient to explain high west–east conductivity, suggesting hydrogen can contribute to the enhanced conductivity as well.

Continental rifts

Several studies have addressed the electrical structure of other continental rift zones. A strong conductor, spatially correlated with a seismic low velocity zone, has been identified locally directly beneath the Baikal Rift zone at 15–45 km depth (Berdichevsky *et al.*, 1980), but information on the resolution of this model is unavailable. A high conductive layer (with $\sigma = 0.02$–0.1 S/m as compared to 10^{-3}–10^{-4} S/m above and below it) can be attributed to the presence of fluids and/or partial melts beneath the rift since the presence of graphite or sulfides would not cause a decrease in seismic velocities. Similarly, the conductivity profile across the Rio Grande Rift shows a 20 km-wide zone of increased conductivity (0.01–0.15 S/m) beneath the rift axis; the anomaly starts at 5 km and terminates at *c*. 30 km depth. However, the entire 100 km-wide zone beneath the rift has enhanced conductivity, 0.002 S/m in the upper 15–20 km and 0.03 S/m between 20 and 40 km depth (Hermance and Neumann, 1990).

In contrast to the "passive" continental rifts, such as the Rhine Graben and the Baikal Rift, the East African rift system has strong anomalies of increased conductivity in the shallow crust extending to at least lower crustal depths (De Beer *et al.*, 1982; Whaler and Hautot, 2006). These zones of high conductivity are correlated spatially with the magmatic segments within the rift. In the northern Main Ethiopian Rift, the region with the conductive material in the crust has high seismic velocity and high density in the upper to mid-crust and upper mantle as indicated by controlled-source seismic survey and gravity modeling (Mackenzie *et al.*, 2005; Cornwell *et al.*, 2006). Thus, seismic, gravity, and MT results concordantly indicate the presence of a mafic intrusion, and the MT interpretations suggest that it contains partial melt.

7.5.6 Orogens

Continental orogens exhibit a strongly heterogeneous 3-D conductivity structure and thus no typical 1-D conductivity-depth profile can be derived for them. This situation is similar to seismic studies, which show a significant diversity in lithospheric velocities beneath the continental orogens and their fundamentally 3-D seismic structure.

Alpine fold belt

The convergence of the European and the African plates that began at ~120 Ma resulted in plate collision and subduction at ~65 Ma; the consequent uplift of the Alpine orogenic belt started ~23 Ma. Numerous regional seismic tomography studies indicate a highly heterogeneous structure of the crust and the mantle of the Alpine fold belts such as the Betics, Pyrenees, Alps, Carpathians, Pontides, and Caucasus. Most geoelectric models for the region are very shallow (the first 5–15 km) with very few completed experiments to explore upper mantle conductivity. In particular, the deep electrical structure of the Alps has been addressed so far in two experiments only (Bahr *et al.*, 1993; Tarits *et al.*, 2004).

Magnetotelluric studies in Iberia started in early 1990s. They reveal a number of subvertical conductors in the crust of the Pyrenees, associated with the major fault zones and not distinguished in seismic reflection studies (Pous *et al.*, 1995a, 2004). Fluids are believed to be the major cause of crustal conductors in shear and fault zones. An important finding is a high-conductivity zone (0.3 S/m) at the contact between the Iberian and the European plates, parallel to the Pyrenean chain. By conductivity values, the European plate is estimated to be thicker than 115 km, while the Iberian plate is only about 80 km thick. The mantle conductor at the contact between the two plates extends over the entire orogen and correlates with the zone of low seismic velocities at the Axial zone of the orogen. The low velocity zone extends down to 80–100 km (Souriau and Granet, 1995); however its bottom cannot be resolved with existing MT data. High conductivity in the mantle can be explained by partial melting of the subducted lower crust of the Iberian plate, since mantle temperatures at this depth are expected to be less than 1300 °C (Pous *et al.*, 1995b). The presence of fluids along the subducting slab would facilitate melting and enhance conductivity. By analogy with the Alps (Austrheim, 1991), weak negative residual gravity anomalies in the Pyrenees are explained by eclogitization of the lower crust during its subduction (Vacher and Souriau, 2001).

A strong elongated arc-shaped conductor that encircles the entire Carpathian orogen had already been discovered in the 1960s. A large number of experiments have been carried out to try to understand its nature; but this has not been satisfactorily answered yet. Cross-sections across different parts of the Carpathians suggest a strongly heterogeneous mantle structure beneath the region. Electromagnetic studies indicate that the depth to the high conductivity layer in the mantle is ~70 km beneath the Eastern Carpathians. In contrast, the electrical lithosphere beneath the Western Carpathians can be as thick as 150 km. Thick lithosphere beneath this part of the Carpathians can be ascribed to the subducted Eurasian plate (?) as suggested by some seismic tomography models (Wortel and Spakman, 2000). The high-velocity body is imaged by seismic surveys and the main seismicity is localized in the depth range 60–180 km along a steeply dipping plane of the Vrancea zone (Fig. 3.93de). It is, however, surprising that conductivity profiles through the Vrancea zone do not show extension of strong conductivity anomalies below the Moho. Furthermore, by comparing MT and seismic data, Ernst *et al.* (2002) argue, that the conducting anomaly is within the sedimentary cover and is caused by fluids.

A number of recent MT experiments in Turkey targeted the electrical structure of active faults in the region with high seismicity rates (e.g. Gurer *et al.*, 2004). Studies on the Fethiye Burdur Fault zone in SW Anatolia and on the North Anatolian Fault zone reveal similar correlation between the conductivity structure of the crust, seismicity, and locations of the faults. In both cases, the faults are associated with dipping conductors in the crust, while the hypocenters are located in areas with high resistivity near the edges of the conductive zone (Fig. 7.42).

North American orogens

A long-period MT survey at 34–38° N latitude across the Paleozoic Appalachian orogen reveals two mantle conductors beneath the western and eastern margins of the orogen (Fig. 7.43). These conductors start at ~60–80 km and 140 km depth beneath the western and the eastern Appalachians, respectively (Ogawa *et al.*, 1996). Because of the shallow

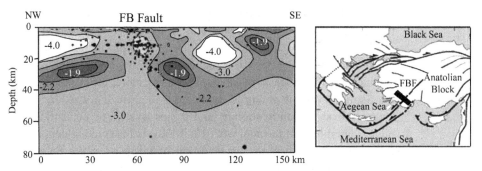

Fig. 7.42 Conductivity model for the Fethiye Birdur Fault Zone in SW Turkey (after Gurer *et al.*, 2004). Dots — seismicity. Numbers — log conductivity in S/m.

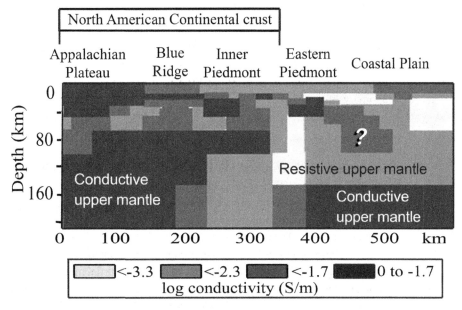

Fig. 7.43 Two-dimensional conductivity structure across the Appalachians (after Ogawa *et al.*, 1996). For 1D profiles see Fig. 7.35.

mantle conductor in the western part of the orogen, its 1D conductivity–depth profile is similar to the tectonically active structures of western Europe (Figs. 7.35 and Fig. 7.37). Two highly conductive blocks (log $\sigma < -1.7$) are separated by a 100–200 km-wide resistive mantle zone dipping northwestwards at ~45°. The resistive block, where conductivity is lower by almost an order of magnitude than in the adjacent mantle, is interpreted as a possible remnant of the late Paleozoic (Alleghanian) collision. The sub-horizontal lower crustal conductor that extends beneath most of the western Appalachians to the Blue Ridge is probably caused by the presence of conductive minerals (sulfides and graphite). Dipping crustal conductors beneath the Appalachian orogen are related to faults and major sutures.

A wideband MT survey across the Sierra Nevada showed that high resistivity values typical for batholiths are observed only in the upper 10–20 km of the lithosphere (Park et al., 1996). Below this depth, conductivity sharply increases by about two orders of magnitude from <-3 to >-1 log units. The highly conductive layer (0.1–1 S/m) extends down to at least 100 km depth. Furthermore, the conductivity structure beneath the Sierra Nevada is strongly asymmetric. Beneath the western side, the conductivity anomaly is significantly stronger and it dips westwards beneath the Great Valley. The highest amplitude of the anomaly (0.1–1 S/m) is observed at depths of 20–60 km. Since xenoliths from at least the upper 60 km of this part of the Sierra Nevada have high seismic velocities, the zone of enhanced conductivity cannot be attributed to partial melts. Park et al. (1996) propose that the conductivity anomaly beneath the Western Sierra can be explained by graphitic metasediments and/or dehydration of serpentine. Compared to the western part, the conductivity anomaly beneath the eastern part, where the highest topography is observed, has smaller magnitude (0.1–0.5 S/m); the zone with the highest conductivity is more shallow than beneath the western part (30–40 km versus 20–60 km) and it continues subhorizontally to the east beneath the Basin and Range Province. In contrast to the Western Sierra, xenoliths from the Eastern Sierra from depths of 40–70 km contain fractions of melt inclusions and the presence of 2–5% of melts in the upper mantle of the Eastern Sierra Nevada can explain the conductivity anomaly in the mantle.

Tibet

Broad-band and long-period MT data were collected across northern Tibet as part of the INDEPTH project (Wei et al., 2001; Li et al., 2003c; Unsworth et al., 2004). The conductivity profile shows a pronounced high conductivity anomaly (~0.1–0.2 S/m) below ~10–20 km depth extending down to the base of the crust (the Tibetan crustal conductor, TCC) (Fig. 7.44). Because of the presence of TCC, the conductivity of the upper mantle cannot be resolved by MT data; its maximum value is estimated to be ~0.01 S/m. This value strongly suggests elevated temperatures and perhaps partial melting of the mantle. The TCC, which underlies the entire north–south profile and terminates at the Kunlun Fault, may play an important role in the tectonic evolution of Tibet. Since the crust in northern Tibet is 60–70 km thick, TCC is probably located within the lower crust. This is consistent with seismic data which show low seismic velocities in the lower crust (Owens and Zandt, 1997). Low melt fractions (~2%) could explain both MT and seismic data (Unsworth et al., 2004). The rheologically weak mid-crustal layer should favor decoupling of the upper crust from the lower parts of the lithosphere. Clark and Royden (2000) argue that TCC could represent a channel of lower crustal flow transporting crustal material to the east and assisting the maintenance of mass balance in Tibet.

In the northern part of the profile (beneath the Qaidam Basin and the Kunlun Shan) where TCC terminates, a relatively resistive mantle ($\sigma < 0.01$ S/m) is required by MT data down to a depth of 150–200 km. This block is interpreted as the Asian lithospheric mantle. The apparent absence of low conductivity material south from the Jinsha River Suture suggests that the Asian lithosphere does not extend so far south.

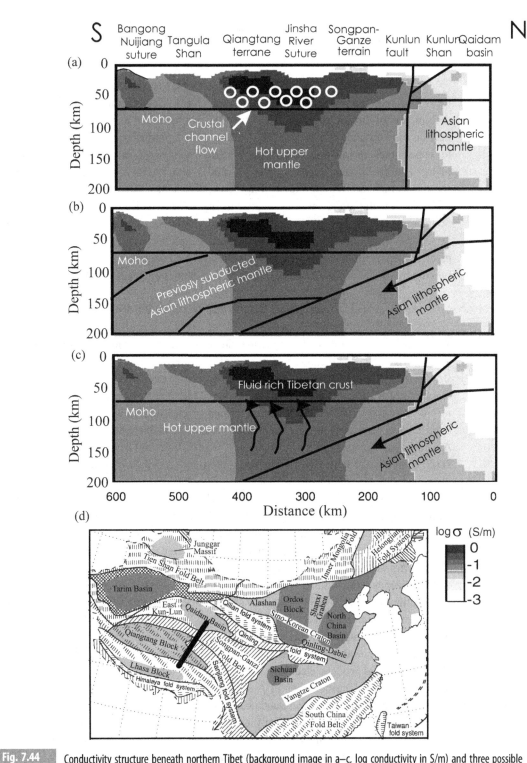

Fig. 7.44 Conductivity structure beneath northern Tibet (background image in a–c, log conductivity in S/m) and three possible tectonic interpretations of its crustal and upper mantle structures (after Unsworth et al., 2004). (a) Model with rheologically weak Tibetan lithosphere which leads to its delamination and consequent upward melt migration; (b) Model with cold, rheologically strong Tibetan lithosphere underlain by the subducting Asian lithosphere; (c) Model in which the subducting Asian slab is overlain by a wedge of hot upper mantle where melting occurs. Profile location is shown in (d).

Unsworth *et al.* (2004) attempted to discriminate between alternative geodynamic models for Tibet from the INDEPTH MT data (Fig. 7.44). Geodynamic models of continuum deformation of rheologically weak Tibetan lithosphere predict that a thickened lithosphere may become unstable and delaminate (Molnar, 1988), producing a conductivity anomaly in the zone of melt upward migration (model a). Alternatively, models with rigid cold lithosphere of the Tibetan Plateau (Tapponnier *et al.*, 2001) require that the Songpan–Ganze terrane should be underlain by the subducting Asian lithosphere (model b). This is apparently inconsistent with high lower crustal/upper mantle conductivities (0.03–0.1 S/m) beneath the Songpan–Ganze terrane. These high conductivity values require a few percent of interconnected melt fractions. However, if the subducting slab is overlain by a wedge of hot upper mantle where melting occurs (model c), the resistive lithosphere of the Asian plate would be masked by the conductive mantle above and would not be resolved by MT data.

7.5.7 Electrical anisotropy of continental mantle

The vector character of electromagnetic fields provides information on anisotropic conductivity. Two-dimensional models of electrical conductivity often show dipping conductive structures, which are sometimes described in terms of geoelectrical anisotropy. In most cases this anisotropy has a tectonic origin and, as compared to intrinsic anisotropy, is referred to as macro-anisotropy (Jones, 1999). Observations of electrical anisotropy in the continental mantle have been reported for the Canadian Shield, Baltic Shield, Central Europe, and Australia (Kurtz *et al.*, 1993; Mareschal *et al.*, 1995; Bahr and Duba, 2000; Leibecker *et al.*, 2002; Simpson, 2002b), but still remain limited in number. Several regional examples are discussed in detail in previous sections.

The anisotropy of conductivity is commonly attributed to the presence of an aligned highly conductive material (e.g. graphitized shear-zones as proposed by Duba and Shankland (1982) for the upper 150 km of the mantle) or to an anisotropic hydrogen diffusion. If water is present in the upper mantle, a large difference in hydrogen diffusivity along different axes of olivine crystals (Mackwell and Kohlstedt, 1990) would produce direction-dependent conductivity if olivine is partially aligned by asthenospheric flow (see also Sections 8.1.2–8.1.3). In the case of simple shear caused by the motion of an overriding lithospheric plate, the *a*-axes of olivine crystals should become aligned parallel to the direction of plate motion (Ribe, 1989). Since diffusivity along the *a*-axis in olivine is ~10 times greater than along the *c*-axis and ~100 times greater than long the *b*-axis, olivine alignment should produce electrical anisotropy of the lower lithosphere and asthenosphere. Surprisingly, geoelectric studies in Australia have shown a 27° discrepancy between the direction of highest electric conductance (electromagnetic strike) and the direction of the present-day absolute plate motion of the continent (Simpson, 2001).

A similar study performed in the Baltic Shield demonstrated that at 180 km depth the upper mantle beneath Fennoscandia is anisotropic (Bahr and Simpson, 2002). The authors interpreted this depth as corresponding to the sublithospheric mantle which, in contrast to results from other research groups, implies a less than 180 km-thick electrical lithosphere beneath the Baltic Shield (Fig. 7.47de). The results of electrical anisotropy analysis indicate some lateral variability of the electromagnetic strike direction beneath the Baltic Shield;

however they are in a reasonable agreement with azimuthal anisotropy constrained by shear-wave splitting (Vinnik *et al.*, 1992). The difference in electrical anisotropy patterns beneath the Australian and the Eurasian plates allows us to speculate on the roles of convection and plate motion mechanisms of asthenosphere deformation below fast-moving and the slow-moving lithospheric plates. The degree of olivine alignment is higher beneath the Baltic Shield than beneath Australia suggesting that convection, but not plate motion, plays the dominant role in mantle deformation (Bahr and Simpson, 2002).

The lattice-preferred orientation (LPO) of olivine crystals at the lithospheric base caused by asthenospheric flow produces both seismic and electrical anisotropy (Ribe, 1989). Therefore attempts have been made to correlate the two. Seismic data cannot resolve the depth where azimuthal seismic anisotropy originates, while the depth of the electrically anisotropic layer can be constrained by MT surveys. As a result, joint analysis of seismic and electrical anisotropy data can provide important geodynamical constraints on astheno-spheric flow. A study of mantle anisotropy beneath the north-central part of Australia supports the common mechanism for both of the observed anisotropies (Simpson, 2002b); it demonstrates a qualitative agreement between the direction of maximal electrical con-ductivity and the fast direction of SV-waves. The quantitative tests, however, show poor agreement between electrical conductivity values observed and calculated from predictions of seismic anisotropy for olivine LPO. The disagreement can be resolved if mechanisms other than hydrogen diffusivity contribute to electrical anisotropy or if Rayleigh-wave anisotropy in the Australian mantle is underestimated.

7.6 Electrical asthenosphere

7.6.1 High-conductivity mantle layer and electrical asthenosphere

The high-conductivity layer (HCL) in the upper mantle, where conductivity is 0.05–0.2 S/m and is an order of magnitude higher than in the lithospheric mantle, is usually termed "electrical asthenosphere". The conductivity jump across the lithosphere–asthenosphere boundary (LAB) is commonly explained by the presence of 1–3% melt fraction in the asthenosphere. The interconnected melt network, which forms at melt fractions <0.005 (Daines and Richter, 1988; Faul, 1997), significantly reduces the bulk electrical conductivity of the mantle. At near-solidus temperatures, interconnectivity of melt in partially molten mantle rocks increases significantly, which has a strong effect on electrical conductivity (Sato and Ida, 1984).

In some regions partial melt alone is insufficient to explain the observed anomalies of increased conductivity in the upper mantle. For example, beneath the Rhenish Massif the percentage of melt required by seismic data is significantly smaller than 5% of the melt required to fit observed electrical conductivities (Leibecker *et al.*, 2002).

Solid-state conductors (amorphous or graphitic carbon) have been suggested as other explanations of the HCL in the mantle challenging its interpretation in terms of "elec-trical asthenosphere". This mechanism, however, cannot explain the origin of the high-conductivity layer in the cratonic regions at depths below ~175 km since at these pressures graphite undergoes phase transition to diamond (Kennedy and Kennedy, 1976).

An alternative explanation of high conductivity in the upper mantle suggests the presence of water dissolved in anisotropic olivine mantle (Karato, 1990; Lizarralde et al., 1995). A large difference in conductivities of oceanic and continental mantles in the depth range 150 km to 250 km has been attributed to a higher water content in the oceanic mantle, where conductivity is too high to be explained by melting of a dry pyrolite mantle (Lizarralde et al., 1995; Hirth et al., 2000). However, this mechanism does not explain the limited depth extent of the HCL which is often less than ~100 km thick (Jones, 1999).

7.6.2 Thickness of electrical lithosphere

Oceans

The electrical lithosphere is defined as the layer above the high-conductivity layer (HCL) in the upper mantle. In the oceans, the conductivity of the upper mantle is clearly correlated with tectonic settings (Figs. 7.23, 7.37, Table 7.3). Beneath oceanic hotspots, a sharp conductivity increase at 200 km depth probably marks the base of dehydrated and melt-depleted upper mantle layer, from which hotspot magmatism is derived. Beneath mid-ocean ridges, the HCL with an estimated 10–15% of melt starts at a depth of ~50–70 km. Belly-shaped conductivity variations with depth observed in the Pacific and other oceans preclude definitive conclusions on the thickness of the electrical lithosphere there.

In "normal oceans", no age dependence of the conductivity structure of the oceanic upper mantle has been found. A conductivity–depth pattern based on theoretical predictions (Fig. 7.24) has not been observed to date anywhere in the oceans. These results are in a sharp contrast with seismic observations which indicate an age-dependent seismic velocity structure beneath the oceans down to 150–300 km depth (e.g. Figs. 3.100; 3.121; 3.125; 3.126). The inability of electrical conductivity data to reliably resolve the electrical lithosphere beneath the oceans can be associated with an extreme sensitivity of conductivity to factors other than the presence of partial melt in the mantle (primarily water).

Continents

In contrast to oceanic upper mantle, continents show a clear correlation between the depth to HCL (thickness of electrical lithosphere) and the lithospheric age (Fig. 7.45). In stable continental regions the top of the HCL is deeper than 200 km (note that in cratons HCL can be associated with solid-state conductors, and thus its top is not necessarily the LAB), while in young continents (except for collisional orogens with associated subducting slabs) electric lithosphere is less than 150 km thick.

These patterns are clearly seen in recently compiled data on the conductivity structure of the European upper mantle (Fig. 7.46). Active continental regions (e.g. the Rhine Graben) have a pronounced conductivity increase at a depth of ~100–120 km (Tezkan, 1994). In other regions, such as the Pannonian Basin, high conductivity anomalies are found at a much shallower depths, ~50–60 km (Adam and Wesztergom, 2001).

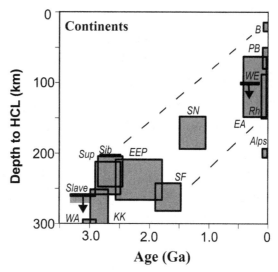

Fig. 7.45 Correlation between the depth to the high conductivity layer, HCL (commonly interpreted as thickness of electrical lithosphere) and the lithosphere age on the continents. Abbreviations: B = Baikal Rift; EA = Eastern Australia; EEP = East European Platform; PB = Pannonian Basin; Rh = Rhine Graben; KK, SF and SN = Kola-Karelian, SvecoFennian, and SvecoNorwegian provinces of the Baltic Shield; Sib = southern margin of the Siberian craton; Sup = Superior Province; WA = West Africa; WE = Western Europe. See references in the text.

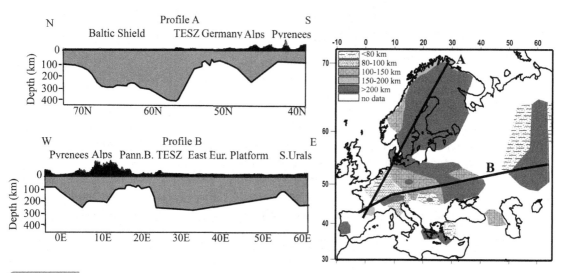

Fig. 7.46 Thickness of electrical lithosphere in Europe (based on compilations of Hjelt and Korja, 1993; Korja, 2007).

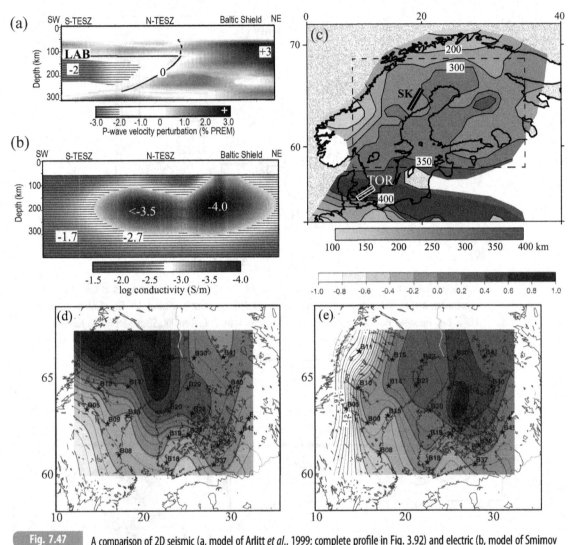

Fig. 7.47   A comparison of 2D seismic (a, model of Arlitt *et al.*, 1999; complete profile in Fig. 3.92) and electric (b, model of Smirnov and Pedersen, 2007) structure of the upper mantle along the TOR-array from North Germany via Denmark to southern Sweden. (c) Thickness of electric lithosphere (the depth to the top of the major high conductivity zone in the upper mantle) in the Baltic Shield based on the BEAR-array (model of Korja, 2007). SK = Skellefteå area in Sweden. Dashed line outlines the area shown in maps (d–e). (d–e) Electrical anisotropy at a depth of 180 km beneath the Baltic Shield shown by relative (with respect to the magnetic reference field measured at site B22, north of Skellefteå area) variations of the amplitudes of the *x*- and *y*- components of the geomagnetic field (from Bahr and Simpson, 2002).

   The most interesting result is the observation of the electrically resistive upper mantle in the Precambrian Baltic (Fennoscandian) Shield which has low conductivity values from the Moho to 300–350 km depth without any indication of electrical asthenosphere (Fig. 7.47c). These results are in contrast with the early interpretations of Jones (1984) which had source field problems associated with high latitudes (Osipova *et al.*, 1989), but agree better with recent seismic tomography results which indicate a seismically fast upper mantle beneath the Baltic Shield. Body-wave tomography (Sandoval *et al.*, 2004) shows high P-wave velocities down to possibly 400 km depth (although the model does not have sufficient depth resolution), while regional surface-wave tomography (Bruneton *et al.*, 2004) indicates fast velocities down to at least ~250 km depth. The discrepancy between P-wave and S-wave velocity models has been attributed to a possible vertical leakage in the P-wave tomography, to the edge effect associated with the small size of the main Archean block beneath the Baltic Shield, or to the presence of small-scale mantle downwelling. Shear-wave velocity models obtained by linearized inversion of observed dispersion curves are consistent with lithospheric thickness greater than 300 km (Pedersen *et al.*, 2009), while xenolith thermo-barometry from the same region indicates a lithospheric thickness of at least 220–250 km (Kukkonen and Peltonen, 1999). Long-period magnetotellurics along the TOR seismic tomography array across the transition from the Precambrian Baltic Shield to the younger Phanerozoic (Caledonian and Variscan) Europe to the south shows a sharp change in the conductivity structure down to ~300 km depth, which is in a broad agreement with P-wave velocity anomalies along the same profile (Fig. 7.47ab).

7.6.3 Correlations between electrical, thermal, and seismic LABs and the lithospheric age

In many continental regions, the estimated depth to the top of the asthenospheric electric conductor (HCL) is well correlated with the depth to the top of the seismic LVZ (e.g. Alekseyev *et al.*, 1977; Ryaboy and Derlyatko, 1984; Praus *et al.*, 1990; Adam and Wesztergom, 2001) (Fig. 7.48). The existence of such a correlation is suggested by laboratory measurements of seismic velocities and electrical conductivity in partially molten or fluid saturated rocks, since the presence of melts or fluids can explain both seismic LVZ and electrical HCL in the mantle (Schmeling, 1985b; Watanabe and Kurita, 1994). Furthermore, electrical conductivity can be correlated with seismic velocities of rocks (Carcione *et al.*, 2007).

The depth to the top of a partially molten layer in the mantle is essentially controlled by the upper mantle thermal regime and composition (the latter controls the solidus temper-ature). Assuming surface heat flow is representative of mantle temperatures (which is not necessarily true, especially in regions with a transient thermal regime, or an anomalous crustal radioactivity, or with an intense groundwater circulation affecting borehole heat flow measurements) and neglecting possible effects of compositional variations on mantle solidus, one would expect that the depth to the HCL (i.e. presumably the depth where partial melting of mantle material starts) should correlate with surface heat flow. Indeed, such an empirical statistical correlation between the depth to the top of the electrical asthenosphere and surface heat flow had been proposed for the continents long ago (Adam, 1976). A compilation of conductivity models published over the past three decades further suggests that surface heat flow can be used as a proxy for the depth of HCL (Fig. 7.49). However, because of the above mentioned assumptions on the direct correlation between surface heat flow and lithospheric geotherms (valid only as a rough, first-order proxy, see details in Section 4.3), and because factors other than melting strongly affect conductivity, this correlation should be considered with caution.

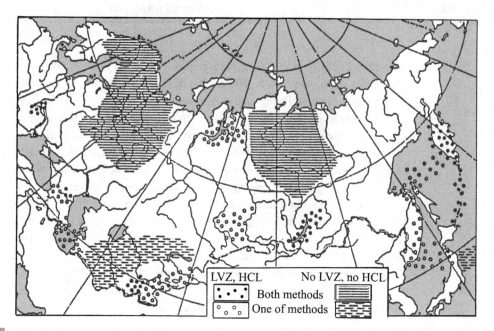

Fig. 7.48 Lateral extent of the asthenospheric layer in the upper mantle of Eurasia based on seismic and geoelectric studies. Dots – regions where either a seismic low-velocity zone (LVZ) or electric high-conductivity layer (HCL), or both have been identified in the upper mantle; hatching – regions where either both of the methods or one of the methods do not show the presence of the asthenosperic layer; white – regions with no constraints (after Alekseyev et al., 1977; Ryaboy and Derlyatko, 1984).

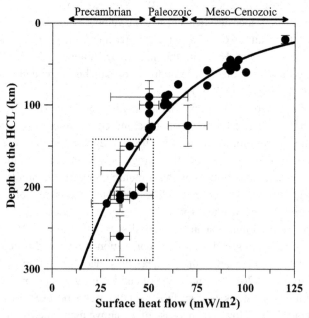

Fig. 7.49 Depth to high conductivity layer (commonly interpreted as the base of the electrical lithosphere) as a function of surface heat flow (after Artemieva, 2006). Globally on the continents the two parameters are clearly correlated. However, the correlation fails when only the Precambrian regions (where heat flow is typically <50 mW/m^2) are considered (outlined by dotted line).

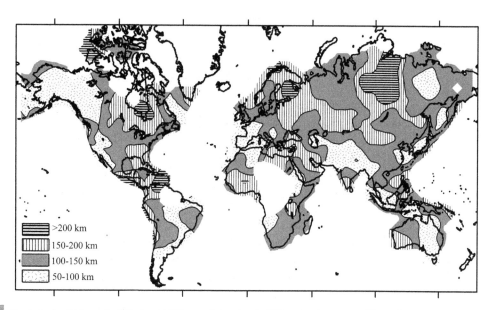

Fig. 7.50 Depth to the high conductivity layer (thickness of the electrical lithosphere) constrained by surface heat flow (averaged on a $10° \times 10°$ grid) and the relationship in Fig. 7.49. The constraints are inaccurate in subduction zones and in regions where heat flow is affected by groundwater circulation (e.g. Texas) or anomalous crustal radioactivity (e.g. Mnt. Isa orogen, N. Australia). Interpolation errors, especially at continental margins or in regions with scarce heat flow data, further contribute to the uncertainty of the constraints.

Figure 7.50 shows the map of the thickness of electrical lithosphere for the continents as predicted from surface heat flow values. The 50 km step between isolines corresponds approximately to the depth resolution of the top of HCL in electromagnetic models and the uncertainty in the conversion. Since surface heat flow is well correlated with the tectono-thermal age of the crust, the increase of the depth to the top of the asthenospheric conductor with age is similar to the general trend observed in geoelectric studies at the continents (Fig. 7.45): in Phanerozoic regions the top of the mantle conductor is typically more shallow (<100 km) than in Precambrian cratons (>150 km). Some regions have unreliable constraints due to uneven and, in some regions scarce, coverage of heat flow data. This is particularly the case in Africa, Arabia, and Brazil, as well as at the continental margins (see Fig. 4.16 for heat flow data coverage). In other regions (e.g. northern Australia, southern North America, and the Caribbean) artifact anomalies of electrical lithosphere thickness result from the effects of anomalous crustal radioactivity or groundwater circulation on surface heat flow.

As discussed earlier, electromagnetic studies do not prove that in the oceans the thickness of the electrical lithosphere increases with ocean floor age as might be expected from theoretical predictions, and a correlation between surface heat flow and the depth to the mantle conductor has not been found so far in the oceans. The lack of proven correlations precludes reconstruction of the depth to the HCL in the oceanic mantle from the age of the ocean floor, although it is tempting.

7.7 Electrical lithosphere: summary

Resolution of geoelectric models

The exponential decrease in the amplitude of electromagnetic waves with depth provides a theoretical possibility for constraining conductivity profiles of the Earth at any depth. Because of the frequency dependence of skin depth, MT surveys based on lower frequency signals allow us to reach deeper layers.

Vertical resolution depends on the range of frequencies used; typically it is ~5–10% of the sampled depth. In well-designed experiments, the vertical and horizontal resolutions of geoelectric studies are comparable.

The best resolved parameters are the conductance (the conductivity by thickness product) and the depth to the top of a conductive layer. Strong conductors in the upper layers reduce amplitudes of electromagnetic fields at long-periods and mask the underlying conducting layers.

Thickness of the lithospheric mantle is one of the best-resolved model parameters; however, the true resistivity of the lithospheric mantle is usually constrained with an error over one order of magnitude and only the minimum bounds on mantle resistivity can be obtained.

Factors affecting electric conductivity structure of the lithosphere

Electrical conductivity of the mantle is essentially controlled by composition (primarily of olivine), temperature, the presence of an interconnected network of a highly conducting phase (melt, fluids, graphite), and oxygen fugacity.

- Conductivity depends exponentially on temperature; at temperatures >1200 °C olivine conductivity is anisotropic. At upper mantle conditions, the pressure effect on conductivity can be neglected.
- The presence of a highly conducting phase (graphite, sulfides, fluids, melts) has a strong effect on conductivity. Five percent of melt (which forms the interconnected melt network) causes a 1.5–3 orders of magnitude increase in conductivity.
- The effect of olivine anisotropy on conductivity is comparable in magnitude to the effect of fluids.
- Because of variations in lithology, the conductivity of the lithospheric materials varies by 7–8 orders of magnitude. Variations in the upper mantle composition between Fe-depleted cratonic lithosphere and Fe-enriched oceanic mantle can lead to a factor of 2 difference in mantle conductivity. Variations in oxygen fugacity may produce ~0.5 log unit variations in mantle conductivity.
- The electrical conductivity of mantle rocks is frequency dependent.

Major results and global trends

Four different conductivity–depth profiles are characteristic of the oceans. The electrical structure of the Pacific mantle has a conductivity peak at 150–250 km depth which can be explained by depleted anisotropic composition in the upper 200 km of the mantle, with increasingly isotropic structure below 200 km, and with undersaturated concentration of water at all depths down to 300–400 km. The electrical structure of mid-ocean ridges requires asthenosphere at 50–100 km depth with a melt fraction in the upper mantle between 1 and 10%. Highly resistive mantle down to ~200 km depth is the characteristic feature of oceanic hotspots which is consistent with the dry olivine model due to melt extraction and the subsequent dehydration of the mantle. In normal oceans, conductivity is similar to the Pacific mantle down to 150 km. Below 200–250 km depth, all oceanic regions have similar conductivity.

The conductivity structure of the continental lithosphere is highly heterogeneous. In cratonic regions, lithosphere conductivity in the depth range between 100–150 km and 220–280 km is consistent with theoretical predictions for dry subsolidus olivine and geotherms of 37–50 mW/m^2. In young regions with extended crust conductivity at 100 km depth is ~2 orders of magnitude higher than in the cratonic mantle.

The electrical lithosphere is defined as the layer above the high-conductivity mantle layer (HCL) with conductivity, by an order of magnitude, higher than in the lithospheric mantle. This conductivity increase is commonly explained by the presence of 1–3% melt fraction in the asthenosphere (HCL). In many regions the depth to the top of the asthenospheric conductor is well correlated with the depth to the top of the seismic LVZ. In cratonic regions, solid-state conductors (amorphous or graphitic carbon) can produce HCL. Solid-state HCL is not necessarily the electrical asthenosphere, unless partial melt is also present.

In continents, the depth to HCL (thickness of the electrical lithosphere?), in general, increases with lithospheric age. In Phanerozoic regions the top of the mantle conductor is typically at 10–100 km depth and deepens to 150–250 km in Precambrian terranes. In some cratonic regions (Baltic Shield) no well-developed HCL has been found down to a depth of 300–350 km. In oceans, the correlation between the depth to HCL and ocean floor age is not confirmed by existing data.

Anisotropic hydrogen diffusivity along different axes of olivine crystals (if olivine is partially aligned by asthenospheric flow) produces direction-dependent electrical conductivity. Since the lattice-preferred orientation of olivine crystals caused by asthenospheric flow also produces seismic anisotropy, a joint analysis of seismic and electrical anisotropy data can provide important geodynamical constraints on asthenospheric flow at the lithospheric base.

Table 7.3 Summary of mantle electrical structure for major tectonic structures	
Tectonic structure	Summary of electrical structure of the mantle
Pacific Ocean	• Highly anisotropic down to 200–250 km depth. • Highly conductive at 150–300 km depth. *Alternative explanations of high conductivity at z < 300 km*: (a) an isotropic olivine mantle nearly saturated in hydrogen or water; (b) a combination of wet (undersaturated) olivine and mantle anisotropy; (c) 1–3% of gravitationally stable melt (of komatiite composition); (d) Preferred explanation: z = 0–200 km: depleted anisotropic with undersaturated concentration of water; z > 200 km: increasingly isotropic with undersaturated concentration of water; maybe drier than beneath other oceanic structures. • HCL at z = 150–300 km, no age dependence.

Table 7.3 (cont.)	
Tectonic structure	Summary of electrical structure of the mantle
Hotspots	• No HCL. • $z < 200$ km: Highly resistive, dry due to melt extraction and subsequent dehydration of the mantle. • $z > 200$ km: isotropic with undersaturated concentration of water.
Mid-ocean ridges	• Strong anisotropy in the upper 30–60 km associated with mantle flow in the spreading zone. • Highly conductive mantle down to $z \sim 300$–400 km. • HCL at $z = 50$–70 km; up to 10–15% of melt. • $60 < z < 150$ km: partially molten, water saturated. • $z > 200$ km: anisotropic with undersaturated concentration of water or saturated isotropic.
Active margins	• Strong lateral heterogeneity in the upper 200 km. • Highly conductive at $z < 150$ km and along the descending slab in most subduction zones; consistent with anisotropic water-saturated olivine model or/with partial melting. • $z > 150$–200 km; isotropic, undersaturated concentration of water.
Cratons	• HCL at $z = 200$–250 km; apparently not observed in the Baltic Shield. HCL may be solid-state (not necessarily asthenosphere). • $z < 200$–250 km: highly resistive; globally consistent with conductive geotherm of 37–45 mW/m^2. • Change from conductive to adiabatic geotherm (i.e. lithospheric base) typically occurs at 200–240 km depth. In some cratons no well-developed asthenosphere found down to 300 km depth. • $z > 200$–250 km: isotropic with undersaturated concentration of water, much drier than oceanic mantle down to the transition zone.
Extended continents	• HCL at $z > 70$–120 km. • $z > 200$ km: an isotropic with undersaturated concentration of water or saturated isotropic.
Orogens	• Strong lateral heterogeneity in the upper 200 km. • Thick electric lithosphere associated with subducting slabs.
Major continental suture zones	• Linear anomalies of enhanced crustal and mantle conductivity define the location of major tectonic boundaries; typically explained by graphite and sulfides emplaced during deformation and metamorphism in collisional and/or subduction environments.

HCL = high conductivity layer (commonly interpreted as electrical asthenosphere)

8 Flexure and rheology

This chapter offers an overview of lithospheric rheology as demonstrated by laboratory measurements of rock deformation, theoretical modeling of lithosphere elasticity, rigidity and flexure, seismic observations on the depth distribution of crustal and upper mantle seismicity, and models of upper mantle viscosity as constrained by field observations of postglacial rebound. Several excellent comprehensive monographs have been published recently on a number of these topics, including overviews of lithosphere rheology (Ranalli, 1995), deformation (Karato, 2008), flexure and isostasy (Watts, 2001). The reader is advised to consult these monographs for complete synopses of the methods, their geophysical applications and results, while this chapter provides only a general overview of the topics illustrated by recent research achievements in the field. The structure of this chapter follows the pattern of the other chapters: first, the basic parameters related to lithosphere rheology are introduced and discussed based on laboratory data; next, the theoretical background is briefly introduced with focus on model assumptions, limitations, and uncertainties; finally, the major research results are presented for different oceanic and continental tectonic settings and discussed within the global framework of lithosphere structure and evolution.

8.1 Rheology of rocks

8.1.1 Introduction to rock deformation

The goal of this brief introduction is to describe the principal physical concepts that are critical to understanding the rheology and deformation of the lithosphere. While the concept of viscosity goes back to Isaac Newton, the term rheology is very young and was first used in 1920 by E. C. Bingham, a professor at Lafayette College (Pennsylvania, USA). However, the concept of rheology is much older than the term and can be traced back to the pre-Socratic Greek school of philosophers, who used the expression "panta rei" to describe that "everything flows". Theoretical aspects of rheology are the relationships between deformations and stresses and the deformation/flow behavior of materials that cannot be described by theories of elasticity/classical fluid mechanics. In this regard, rheology unites the fields of plasticity and non-Newtonian fluid mechanics (fluid flow that cannot be described by a single constant value of viscosity) by studying deformation–stress relationships in materials

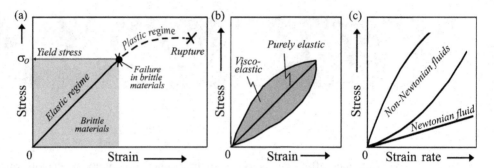

Fig. 8.1 Stress–strain relations during deformation of solid materials. (a) In a purely elastic material, strain increases linearly with stress and the slope of the curve is the elastic modulus E. When stress exceeds a yield point, the material experiences irreversible plastic deformation. (b) Purely elastic materials do not dissipate energy when a load is applied and subsequently removed, while a visco-elastic material does. As a result, the deformation of visco-elastic material is characterized by hysteresis in the stress–strain curve. (c) In a Newtonian fluid (purely viscous material), strain rate increases linearly with stress, while in non-Newtonian fluids the stress-to-strain rate response is non-linear.

that are unable to support a shear stress in static equilibrium. When a constant stress is applied to a material, its deformation can vary (Fig. 8.1):

- in a purely elastic material, strain increases linearly with stress;
- in a purely viscous material, strain rate increases linearly with increasing stress;
- in a visco-elastic material, neither the deformation strain, nor its derivative with time (strain rate) follow the stress.

Materials that fully recover when the external stress is removed have the suffix "–elastic", materials that do not recover fully when the external stress is removed have the suffix "–plastic", while materials that can creep have the prefix "visco-". In contrast to brittle materials which deform elastically, materials that undergo viscous deformation are termed ductile.

Elastic deformation

Elastic deformation, the end-member response to stress typical of solids, is characterized by a linear relationship between stress σ and strain ε (Hooke's law):

$$\sigma = E\varepsilon \qquad (8.1)$$

where E is the elastic modulus. Since, in general form, both stress and strain are tensors, several types of elastic moduli are recognized. *Bulk modulus K* is defined as the ratio of volumetric stress to volumetric strain and is used to quantify deformation of material in all directions under a uniform load acting in all directions (e.g. K gives the ratio of pressure to the resulting volume change). *Young's modulus E* is a one-dimensional form of the bulk modulus defined (in the case of extension) as the ratio of tensile (extensional) stress to tensile strain and it is used to quantify deformation along an axis when opposing (extensional) forces are applied along that axis. *Shear modulus μ* is defined as the ratio of shear stress to shear strain and it is used to quantify deformation of the shape at constant volume (i.e. material tendency to shear) under external opposing forces; viscosity is directly related to shear modulus. Two more parameters are commonly used to characterize elastic deformation of solids. *The Lamé parameter λ* allows simplification of the stiffness matrix in Hooke's law but has no physical interpretation. Together, parameters λ and μ fully parameterize the elastic moduli for homogeneous isotropic media, and are used in seismology to parameterize seismic velocities (see Section 3.1.1, eqs. 3.1–3.4, and Fig. 3.1). The dimensionless *Poisson's ratio ν* quantifies material deformation under extensional (or compressional) forces and (in the case of extensional forces) is the ratio of the contraction (or transverse strain) in the direction perpendicular to the applied forces to the extension (or axial strain) in the direction of the applied force; $\nu \leq 0.5$ in all materials with $\nu = 0.5$ in incompressible solids. The elastic properties of homogeneous and isotropic solids are fully described by any two elastic moduli; some relations between the elastic moduli are:

$$E = 3K(1 - 2\nu) = 2\mu(1 + \nu); \quad K = \lambda + 2/3\mu = \lambda(1 + \nu)/3\nu; \quad \nu = \lambda/2(\lambda + \mu). \quad (8.2)$$

In liquid, $\mu = 0$ and thus $E = 0$, $V_S = 0$ (since $V_S = \sqrt{\mu/\rho}$).

Plastic and elasto-plastic deformation

Plastic behavior is the other (in contrast to elastic behavior) end-member response to stress. It is characterized by irreversible deformation of material and occurs without fracturing. In the regime of plastic deformation the stress is independent of the strain rate. Steady-state plastic flow of rocks (p. 519) can be caused by dislocation motion, solid-state diffusion, and solution–diffusion–precipitation processes. A summary of the characteristics of brittle and ductile deformation as listed by Karato (2008) is shown in Fig. 8.2. Note that elastic and plastic deformation have essentially different sensitivities both to thermodynamic and material parameters.

A *yield stress* (or yield point) is the stress at which the transition from elastic to perfectly plastic deformation of a material occurs (Fig. 8.1a). Prior to the yield point, the material deformation is elastic and reversible, while once this elastic limit is passed, permanent and non-reversible deformation of material will occur.

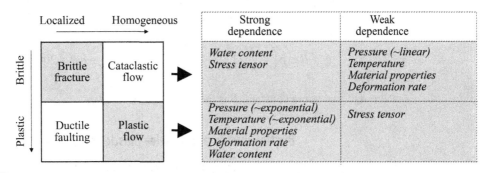

Fig. 8.2   Summary of characteristics of brittle and ductile deformation.

Table 8.1 Typical geological strain rates		
Tectonic setting	Strain rate (1/s)	Reference
Compressional orogens	3×10^{-15}	Cloetingh and Burov, 1996
Sedimentary basins	$10^{-17} - 10^{-15}$	Newman and White, 1997
Intracratonic basins	10^{-16}	Armitage and Allen, 2010
Plate interiors	$6 \times 10^{-20} - 10^{-17}$	Gordon, 1998

Plastic flow involves various microscopic mechanisms and cannot be described by one equation as elastic deformation. One of the most common forms of plastic flow is the power-law creep (described by the Arrhenius law which is characteristic of thermally activated processes). It is generally thought to be a good representation of the ductile deformation of the lower parts of the crust and lithosphere. It states that, at a constant strain rate, the ductile strength of materials decreases with increasing temperature as:

$$\dot{\varepsilon} = A_0 \sigma^n \exp\left(-\frac{H}{RT}\right) \tag{8.3}$$

where $\dot{\varepsilon}$ is the strain rate, σ is stress, an exponent $n > 1$ ($n = 1$ in the case of a Newtonian flow), A_0 is a constant dependent on the material, H is the activation enthalpy for the process, R is the universal gas constant, and T is the absolute temperature.

Experimental studies of deformation of dry polycrystalline olivine aggregates at stresses up to 5 GPa, temperatures up to 1600 °C, and strain rates of $10^{-6} - 10^{-4}$ 1/s (Goetze and Kohlstedt, 1973; Goetze, 1978) show that at low stresses, the strain rate–stress relation takes the form (**Goetze's criterion**):

$$\dot{\varepsilon} = A(\sigma_1 - \sigma_3)^n \exp\left(-\frac{H}{RT}\right), (\sigma_1 - \sigma_3) < 200 \text{ MPa} \tag{8.4}$$

where A is the stress constant, $n = 3$ is the stress exponent, $(\sigma_1 - \sigma_3)$ is the applied stress difference between the maximum (σ_1) and the minimum (σ_3) principal stresses, and it should be less than 200 MPa for eq. (8.4) to be valid. Strain rates used in the experiments of Goetze (1978) are significantly higher than geological strain rates (Table 8.1); but the analysis indicates that eq. (8.4) can be extrapolated to olivine deformation at low strain rates (Fig. 8.3). At $(\sigma_1 - \sigma_3) > 200$ MPa, **Dorn's law** provides a better fit to the observations:

$$\dot{\varepsilon} = B \exp\left(-\frac{H}{RT}\right)\left[1 - \left(\frac{\sigma_1 - \sigma_3}{A_d}\right)^2\right], (\sigma_1 - \sigma_3) > 200 \text{ MPa} \tag{8.5}$$

where A_d is the stress constant, B is the strain rate constant.

Brittle–ductile and brittle–plastic transitions

Brittle materials do not have a yield point; they fail while the deformation is elastic and thus do not undergo any plastic deformation (Fig. 8.1a, 8.2). A typical stress–strain curve for a

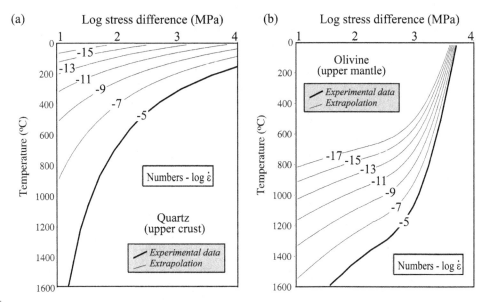

Fig. 8.3 Relation between the stress difference ($\sigma_1 - \sigma_3$) and temperature based on experimental studies of dry quartz and dry olivine at low strain rates (bold lines) and extrapolated to higher stresses using Goetze's and Dorn's laws (eqs. 8.4–8.5, thin lines) (based on Watts, 2001). Numbers on the curves – strain rates (in 1/s). Goetze's law is valid when $\log(\sigma_1 - \sigma_3) < 2.3$, and Dorn's law – when $\log(\sigma_1 - \sigma_3) > 2.3$.

brittle material will be linear as for pure elastic deformation. In contrast to brittle materials, ductile materials can be deformed plastically without fracture and they have a yield strength.

While at near-surface conditions most rocks exhibit brittle behavior, at greater depth plastic deformation dominates, and the same material can deform brittly at low temperatures, but ductilely at temperatures above the brittle–ductile transition. The transition between the two regimes of deformation takes place in the transitional zone (Duba *et al.*, 1990). Strength envelopes, like those in Fig. 8.4, show the maximum rock strength versus depth, and the strength of the lithosphere cannot exceed the strength of the constituting rocks. In each lithological layer (i.e. the layers of the crust and the lithospheric mantle), the strength increases with depth until the temperature is high enough for plastic flow to start. The transition between different deformation modes depends on frictional strength and plastic strength, and it is discussed further below.

Mohr–Coulomb failure criterion and Byerlee's friction law

The pressure of the brittle–ductile transition is predicted empirically by the intersection of the Mohr–Coulomb brittle failure criterion with Byerlee's frictional sliding law. The empirical ***Mohr–Coulomb failure criterion*** is a linear relation between shear strength τ and the normal stress σ_n at the point of brittle failure inside a material (when frictional sliding begins):

$$\tau = S_0 + \mu\sigma_n \tag{8.6}$$

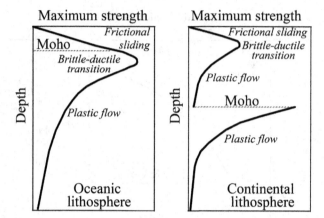

Fig. 8.4 Sketch showing maximum rock strength as a function of depth increases downwards for the oceanic and continental lithospheres (after Kohlstedt *et al.*, 1995). Compare with Figure 8.27.

where S_0 is the frictional cohesion of the material (or internal strength) and μ is the friction coefficient. Internal friction, as given by eq. (8.6), refers to the local slope of a failure criterion in shear/normal stress space: $\mu = \partial \tau / \partial \sigma_n$. Frictional strength provides a lower limit for rock strength, which is in particular important for understanding the deformation of large rock masses that contain joints and fractures. Lockner (1995) states that in the lithosphere, both rock fracturing and friction are macroscopic manifestations of the same processes, that include grain crushing, plastic yielding, crack growth and healing, and chemical reactions of dissolution, precipitation, and alteration.

Byerlee (1967, 1968) found experimentally that, in the upper crust, the criterion (8.6) that specifies the stress at which fracturing takes place can be simplified to (**Byerlee's friction law**):

$$\tau = 0.85\,\sigma_n, \quad \sigma_n < 200 \text{ MPa}$$
$$\tau = 0.5 + 0.6\,\sigma_n, \quad 200 \text{ MPa} < \sigma_n < 1700 \text{ MPa} \tag{8.7}$$

where the shear stress τ is given in MPa (Fig. 8.5a). Because at low stress $\mu = 0.85$, most faults in nature occur at a $30°$ angle to the direction of maximum principal stress. The empirical Byerlee's law becomes less accurate at temperatures above ~400 °C.

Pore pressure effectively reduces the normal stress throughout the rock mass:

$$\sigma_n{}^{\text{eff}} = \sigma_n - \alpha P_f \tag{8.8}$$

and equation (8.6) takes the form (**the Mohr–Coulomb law**):

$$\tau = S_0 + \mu(\sigma_n - \alpha P_f) \tag{8.9}$$

where P_f is the pore fluid pressure inside a rock and α is a constant related to pore geometry ($\alpha = 0$ for dry rock, $\alpha = 1$ for fracture strength of rock with well-interconnected porosity, $\alpha = 0.7$ corresponds to "wet" conditions, and $\alpha = 0.37$ – to hydrostatic conditions when the principal axes are equal). A material obeying the Mohr–Coulomb law experiences

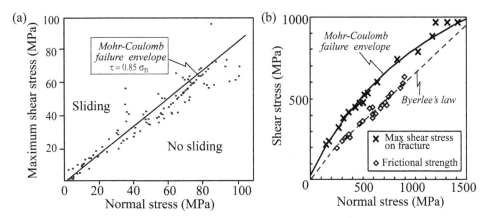

Fig. 8.5 (a) The Mohr–Coulomb rock failure criterion that shows maximum shear stress as a function of normal stress needed to cause sliding (eq. 8.6). Dots – experimental data for limestones, sandstones, granites, granodiorites, quartz monzonite, mylonite, gneiss, and gabbro (Byerlee, 1978); solid line – Byerlee's law. Because the slope of the curve is ~0.85, most faults in nature occur at a 30° angle to the direction of maximum principal stress. The relationship becomes invalid at temperatures higher than 400 °C. (b) Experimental shear stress – normal stress plot for Westerly granite showing the Mohr–Coulomb rock failure envelope and Byerlee's law (eq. 8.7) (after Lockner, 1995).

significant strengthening with pressure, in contrast to plastic deformation which has little or no pressure sensitivity. Besides, although Coulomb deformation can be described as plastic, the microscopic mechanisms responsible for plastic and Coulomb deformations are different (Lockner, 1995).

The shear stress τ and the normal stress σ_n can be expressed through the maximum (σ_1) and the minimum (σ_3) principal stresses. For a fault whose normal is 60° from the maximal principal stress, Byerlee's frictional sliding law (eqs. 8.7, 8.9) takes the form (Brace and Kohlstedt, 1980):

$$
\begin{aligned}
(\sigma_1 - aP_f) &\sim 4.9(\sigma_3 - aP_f), & (\sigma_3 - aP_f) &< 100 \text{ MPa} \\
(\sigma_1 - aP_f) &\sim 210 + 3.1(\sigma_3 - aP_f), & (\sigma_3 - aP_f) &> 100 \text{ MPa.}
\end{aligned}
\tag{8.10}
$$

The condition for rock failure is given by $(\sigma_1 - \sigma_3)$ (Figs. 8.3; 8.6). In the lithosphere, this corresponds to the difference between the vertical and the horizontal stresses, respectively. For example, when one of the principal stresses is oriented vertically, it is equal to the lithostatic pressure: $\sigma_V = P_1 = \rho g z$, where ρ is the average density of the overlying rock and z is the depth. In the case of a thrust fault, where the horizontal principal stress $\sigma_H > \sigma_V$, $\sigma_H = \sigma_1$ and $\sigma_V = \sigma_3$; in the case of a normal fault, $\sigma_V = \sigma_1$, and in the case of a strike–slip fault, $\sigma_V = \sigma_2 = (\sigma_1 + \sigma_3)/2$ (Fig. 8.6a). The corresponding relationships for each case can be found in Weijermars (1997), and application of Byerlee's law to natural faults is discussed by Hickman (1991).

In the case where the pore pressure is hydrostatic, $P_f = \rho_f g z$, where ρ_f is the density of the pore fluid. Fluid pressure is sometimes expressed through the parameter $\lambda = P_f/P_1$. The values of parameter $a = 0, 0.37, 0.76,$ and 0.94 (eq. 8.8) correspond to $\lambda = 0, 1, 2,$ and 2.5. Figure 8.6b

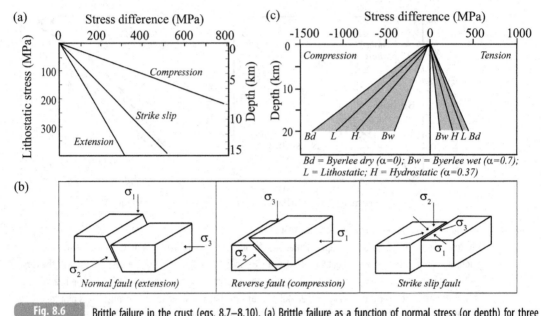

Fig. 8.6 Brittle failure in the crust (eqs. 8.7–8.10). (a) Brittle failure as a function of normal stress (or depth) for three geometries shown in (b) (redrawn from Stüwe, 2002). (c) Failure envelopes for the crust based on Byerlee's law (after Watts, 2001). Stress difference is the difference between the maximum (σ_1) and the minimum (σ_3) principal stresses (or, for the Earth, between the vertical and the horizontal stresses). a is a constant related to pore geometry; the cases of $a = 0$, 0.37, 0.76 correspond to $\lambda = 0$, 1, 2 where λ is the ratio of hydrostatic pore pressure to lithostatic pressure.

shows failure envelopes for the crust based on Byerlee's law (eqs. 8.9–8.10) for lithostatic and hydrostatic pressures and for dry rock composition.

Brittle–ductile–plastic transitions

The brittle–plastic transition occurs in two stages (Fig. 8.7). Following Rutter (1986), it is commonly accepted that the brittle–ductile transition (BDT) corresponds to a change in failure mode, while the brittle–plastic transition (BPT) corresponds to a change in the dominant deformation mechanism.

The brittle–ductile transition occurs as a change from localized deformation (brittle fracture) to distributed failure. Brittle fracture takes place when the strength is greater than the pressure, but less than the strength to cause frictional sliding (i.e. as established empirically under nominally dry conditions, the BDT occurs when the Mohr–Coulomb criterion for brittle failure intersects the frictional sliding curve, Byerlee's law, Fig. 8.5). In the case brittle failure is the only deformation mode, but strain is not localized macroscopically, the deformation is called cataclastic flow (Fig. 8.2). The transition from brittle to ductile regime is temperature dependent (and also strain-rate dependent, see Section 8.1.4). For quartz and feldspar-rich rocks, typical of the continental crust, the brittle–ductile transition occurs at temperatures approximately in the range 250–450 °C. In olivine under typical mantle conditions the brittle–ductile transition occurs at 600–750 °C.

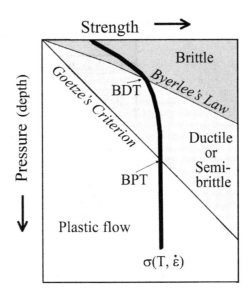

Fig. 8.7 Schematic relationships between rock strength and pressure/depth for a rock undergoing the brittle–ductile (BDT) and brittle–plastic (BPT) transition (after Kohlstedt *et al.*, 1995).

The brittle–plastic transition occurs as a change from brittle cracking to plastic flow (Fig. 8.7). Plastic flow occurs when the differential stress is about equal to or less than the effective confining pressure, i.e. it occurs at the intersection of the plastic yield strength curve with the line $\sigma_1 - \sigma_3 = \sigma_3'$ (Goetze's criterion, eq. 8.4). In the transitional zone, where both deformation mechanisms occur simultaneously, deformation is semi-brittle.

Visco-elastic and visco-plastic deformation

Visco-elastic materials are characterized by non-linear dependences of both the strain and the strain rate on the applied stress (creep). Their deformation behavior has elastic and viscous components that are modeled as linear combinations of these elements (e.g. the Maxwell model, the Kelvin–Voigt model, and the standard linear solid model). Elastic components are described by eq. (8.1); viscous components in case of a Newtonian fluid can be modeled by the stress–strain rate relationship (see eq. 8.3) such as:

$$\sigma = 2\eta\dot{\varepsilon}, \tag{8.11a}$$

where σ is the stress, η is the viscosity of the material, and $\dot{\varepsilon} = d\varepsilon/dt$ is the viscous part of the total deviatoric strain rate (note that eq. (8.11a) differs by a factor of two from the definition used by experimentalists: $\sigma = \eta\dot{\varepsilon}$). In the case of a non-Newtonian fluid, the general form of eq. (8.11a) is

$$\sigma^n = 2\eta\dot{\varepsilon}, \tag{8.11b}$$

where exponent n is a positive integer.

Viscosity, which describes viscous components, characterizes the material resistance to thermally activated plastic deformation. Thus, in contrast to purely elastic materials that do

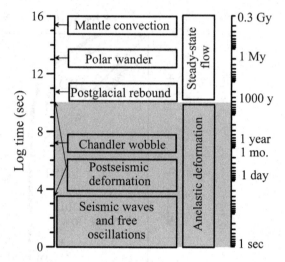

Fig. 8.8 Characteristic timescale of mantle deformation processes (based on Yuen *et al.*, 1982).

not dissipate energy when a load is applied and then removed, visco-elastic materials lose energy during the loading cycle. As a result, hysteresis is observed in the stress–strain curve, with the area of the loop equal to the energy lost during the loading cycle (Fig. 8.1b).

Visco-elastic deformation is time dependent: visco-elastic materials behave as elastic for short times and as viscous over long time-scales. For Earth materials, elastic behavior is an approximation, and their deformation is also time dependent (Fig. 8.8). For the Maxwell visco-elastic model (commonly used to approximate mantle deformation), the stress relaxes to 1/e of its original value over time t_{re} (visco-elastic relaxation time, or the Maxwell time):

$$t_{re} = 2\eta/E, \tag{8.12}$$

which yields $t_{re} \sim 1000$ years, assuming a dynamic viscosity of the asthenosphere $\eta \sim 10^{21}$ Pa s and $E = 70$ GPa. The visco-elastic relaxation time determines the deformation style of visco-elastic materials: they behave like elastic solids under short duration of stress (less than time t_{re}), but creep when the duration of the loading exceeds time t_{re}. In a general case of subsolidus dislocation creep, the stress-relaxation time depends strongly on temperature as (Turcotte and Schubert, 2002, p. 331):

$$t_{r2} = 3\eta_{eff}/E, \tag{8.13}$$

where t_{r2} is the time over which the stress relaxes to 1/2 of its original value σ_o and η_{eff} is the effective viscosity for dislocation creep at stress σ_o:

$$\eta_{eff} = \exp\left(E^*/RT\right)/\left(2\,C_1\sigma_0^2\right), \tag{8.14}$$

where E^* is activation energy (523 kJ/mol and 398 kJ/mol for dry and wet olivine, respectively) and constant $C_1 = 42 \times 10^4$ MPa^{-3} s^{-1} for dry olivine and 5.5×10^4 MPa^{-3} s^{-1} for wet olivine (Fig. 8.9). Note that the value of activation energy for wet olivine can be lower

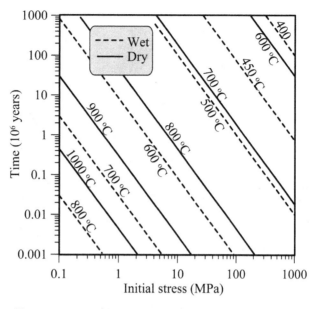

Fig. 8.9 Stress-relaxation time (the time over which the stress relaxes to 50% of its initial value) for dislocation creep of wet (dashed lines) and dry (solid lines) olivine as a function of the initial stress and temperature (numbers on the lines) (eqs. 8.13–8.14).

than assumed here. Experimentally determined values of these parameters are listed in Table 8.2a and discussed in Section 8.1.3.

8.1.2 Major mechanisms of non-linear deformation

Observations of rock microstructures indicate that in nature deformation can occur by several mechanisms (Groshong, 1988):

- granular processes including grain sliding and rotation;
- cataclastic processes including fracture and frictional sliding;
- crystal plastic processes including dislocation creep and pressure solution.

The present section discusses the mechanisms that can cause plastic flow of rocks. Three main mechanisms of non-linear deformation of Earth's materials (dislocation creep, diffusion creep, and Peierls creep) with significantly different dependences on thermodynamic and structural parameters are discussed below (Fig. 8.10). The present brief overview does not aim to provide a comprehensive review of rock deformation studies; such surveys can be found in a large number of publications (e.g. Carter, 1976; Frost and Ashby, 1982; Kirby, 1983; Groshong, 1988; Hirth and Kohlstedt, 2003; Karato, 2008).

Diffusion creep

Diffusion creep is caused by diffusion of atoms through the interior of crystal grains subject to stress; grain deformation caused by atom diffusion leads to rock strain. Diffusion creep is

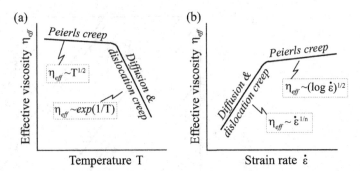

Schematic diagrams showing the influence of temperature (a) and strain rate (b) on the effective viscosity for Peierls and power-law creeps (after Karato, 2010). Peierls creep dominates at low temperatures or high strain rates $[\eta_{eff} \sim (T \log \dot{\varepsilon})^{1/2}]$, while power-law creep dominates at high temperatures and/or slow strain-rates $[\eta_{eff} \sim \dot{\varepsilon}^{1/n} \exp(H^*/nRT)]$. Experimental results obtained for one creep regime cannot be extrapolated to another regime due to different relations of effective viscosity on temperature and strain rate.

a Newtonian creep (i.e. it can be described by a single constant value of viscosity); it is dominant at low temperatures and low stresses, and its significance decreases with increasing grain size. Rocks that deform in this regime are composed of small grains and lack dislocation structures and lattice-preferred orientation (and thus are isotropic). Typical representatives are sheared olivine xenoliths, quartz mylonites, and calcite-dominated thrusts.

A generic form of dependence of the diffusion coefficient D on pressure and temperature follows the Arrhenius law characteristic of thermally activated processes:

$$D = D_0 \, \exp\left(\frac{-\left(E^{(\mathrm{diff})} + PV^{(\mathrm{diff})}\right)}{RT}\right) \tag{8.15}$$

where D_0 is the frequency factor (assumed to be temperature independent), $E^{(\mathrm{diff})}$ is activation energy, and $V^{(\mathrm{diff})}$ is activation volume for diffusion creep, and the term $PV^{(\mathrm{diff})}$ takes into account the pressure effect in reducing the number of vacancies and increasing the potential energy barrier between lattice sites (Turcotte and Schubert, 2002). The alternative expression (8.16) for the diffusion coefficient expresses it through the homologous temperature T_m/T where the pressure dependence of D is incorporated into the pressure dependence of the mantle solidus temperature T_m (Borch and Green, 1987):

$$D = D_0 \, \exp(-aT_m(P)/T). \tag{8.16}$$

The applicability of expression (8.16) was questioned by Karato (2010): since the solidus temperature of olivine is several hundreds of degrees Celcius higher than that of peridotite T_m, the rheological properties of peridotite and olivine would also be significantly different.

Since minerals are composed of several ions, the creep rate is determined by diffusion of the slowest ion along its fastest path. The diffusivity of a given ion also depends on the concentration of other defects; these include both intrinsic defects and extrinsic lattice point defects (i.e. impurities involving dissolved hydrogen, hydroxyl, and water). In particular, experimental studies demonstrate that diffusion flow depends strongly on point defect chemistry (e.g. Hobbs, 1981), because the concentration of point defects depends on the fugacities of other components. To the author's knowledge, the effects of point defect chemistry on diffusion creep have only been studied for Mg–Fe olivine (Nakamura and Schmalzried, 1983).

Two types of diffusion creep are distinguished: grain boundary diffusion and matrix diffusion. In **Coble creep** (Coble, 1963), grain boundary diffusion dominates over other dislocation mechanisms, and a generic expression for the creep rate is:

$$\dot{\varepsilon} = A_{\text{diff}} \frac{(\sigma_1 - \sigma_3)}{d^3} \exp\left(\frac{-\left(E^{(\text{GB})} + PV^{(\text{GB})}\right)}{RT}\right) \tag{8.17}$$

where the parameters A_{diff}, $E^{(\text{GB})}$, and $V^{(\text{GB})}$ are determined experimentally and are material dependent (the latter two are the activation energy and the activation volume for grain boundary diffusion creep). Equation (8.17) can be reduced to (Evans and Kohlstedt, 1995):

$$\dot{\varepsilon} = 44\delta D^{(\text{GB})} \frac{(\sigma_1 - \sigma_3)}{d^3} \frac{V}{RT} \tag{8.18}$$

where δ is the grain boundary width, $(\sigma_1 - \sigma_3)$ is differential stress, V is the molar volume, and $D^{(\text{GB})}$ is the boundary diffusion coefficient in the form (8.15). In the case where grain boundary diffusion creep is accompanied by grain boundary sliding, eq. (8.18) remains valid but will differ by the numerical prefactor.

In **Nabarro–Herring creep** (Nabarro, 1948, 1967; Herring, 1950), matrix diffusion dominates, and the strain rate in steady-state creep is directly related to stress and inversely related to grain size as:

$$\dot{\varepsilon} = 14dD^{(\text{L})} \frac{(\sigma_1 - \sigma_3)}{d^3} \frac{V}{RT} \tag{8.19}$$

where d is the average grain size, $(\sigma_1 - \sigma_3)$ is differential stress, and $D^{(\text{L})}$ is an effective matrix (lattice) diffusion coefficient. This creep mechanism is unlikely to occur at any conditions in the upper mantle, since it becomes effective only for grain size ≥ 1 m (Hirth and Kohlstedt, 2003).

Dislocation creep

The microscopic origin of dislocation creep is associated with dislocations, i.e. imperfections in the crystalline lattice structure. The general types of these imperfections are edge dislocations, screw dislocations, and mixed dislocations. Dislocations are specified by Burgers vector which is a measure of the relative movement of atoms (slip) that occurs when a dislocation line passes through the crystal lattice and by which the periodicity of the lattice is preserved.

The flow law used to describe dislocation creep has a power-law dependence of strain rate $\dot{\varepsilon}$ on the differential stress $(\sigma_1 - \sigma_3)$:

$$\dot{\varepsilon} = A_{\mathrm{dis}}(\sigma_1 - \sigma_3)^n \exp\left(\frac{-(E^{(\mathrm{dis})} + PV^{(\mathrm{dis})})}{RT}\right) \tag{8.20}$$

where n (~3.0–3.5) is a constant characteristic of the creep mechanism, the parameters A_{dis}, $E^{(\mathrm{dis})}$ and $V^{(\mathrm{dis})}$ are material dependent (the latter two are the activation energy and the activation volume for dislocation creep) and are determined experimentally. Commonly, the term $PV^{(\mathrm{dis})}$ is omitted from equation (8.20) when modeling the lithosphere deformation: for $V^{(\mathrm{dis})} = 17 \times 10^{-6}$ m^3/mol, the increase of stress is by a factor of ~2 over a depth range of 50 km (Kohlstedt et al., 1995).

During dislocation creep, the strain rate is proportional to dislocation density and velocity, both of which are stress dependent. As a consequence, stress and strain rates have a non-linear relationship so that the effective viscosity (strength at a given strain rate, eq. 8.11) is stress dependent (strain dependent) and relates to strain rate and temperatures as (Karato, 2010):

$$\eta_{\mathrm{eff}} \sim \dot{\varepsilon}^{1/n} \exp(H/nRT). \tag{8.21}$$

As a result, dislocation (or power-law) creep is dominant at relatively high stresses (> 10 MPa) and high temperatures and it is insensitive to the grain size (Fig. 8.10b; 8.11c). It produces lattice-preferred orientation (LPO) and is responsible for seismic anisotropy. In the case where the mantle with observed LPO is rheologically strong at present, anisotropy is frozen-in; otherwise, LPO is related to a recent deformation of a rheologically weak mantle region. Mantle-derived xenoliths from many continental regions show the presence of LPO, indicating that deformation of the lithospheric mantle takes place by dislocation creep. Dislocation creep of silicate rocks can be well approximated by eq. (8.20) when differential stress is < 1% of the shear modulus. However, this is not true for experimental deformation of olivine crystals when a single creep mechanism cannot explain deformation over the full range of lithospheric pressures and temperatures.

Some experimental studies suggest that at low stresses (< 1 MPa) and large grain sizes (> 1 cm) (i.e. at the conditions important for upper mantle rheology), strain rate may show a linear relationship with stress (**Harper–Dorn creep**, 1957). This implies that the dislocation density is independent of stress, but has a frozen-in value. In the case where this mechanism works in the real Earth as suggested by some studies (e.g. Wang et al., 1994), the extrapolation of power-law creep data to low stresses can be misleading. However, until the present there is no evidence that Harper–Dorn creep may be operating under geological conditions (Blum et al., 2002).

Peierls mechanism

At very high stresses (typically > 500 MPa) and low temperatures, the strain rate for plastic deformation increases approximately exponentially with differential stress and the power-law relationship in the form of eq. 8.20 becomes invalid (Fig. 8.10b). Dislocation creep in this high-stress regime, that may control deformation in heavily stressed, cold continental

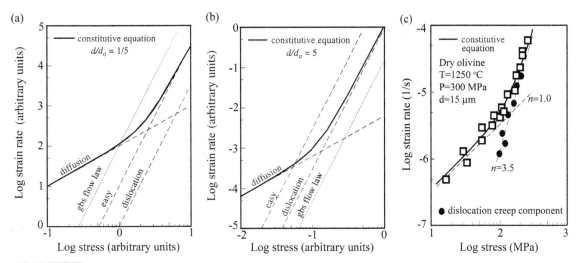

Fig. 8.11   Plots of strain rate versus stress illustrating transitions between different mechanisms of plastic deformation (from Hirth and Kohlstedt, 2003). (a–b) Schematic plots of strain rate versus stress for various grain size d/d_0. Solid lines – calculated using constitutive equation for steady-state deformation; dashed lines – flow laws for diffusion and dislocation creep, for olivine single crystals deformed on the easiest slip system (labeled "easy"), and for grain boundary sliding accommodated by a dislocation creep process (labeled "gbs"). (c) Experimental data on deformation of dry fine-grained olivine aggregates. Dislocation creep component has exponent $n = 3.5$; diffusion creep exponent $n = 1.0$.

lithosphere, is called Peierls creep (or power-law breakdown) (Guyot and Dorn, 1967). In the Peierls mechanism, the rate of deformation grows exponentially with stress which limits the stress magnitude, and the flow law in olivine is described by the following relation between the strain rate $\dot{\varepsilon}$ and the differential stress $\sigma = \sigma_1 - \sigma_3$:

$$\dot{\varepsilon} = \dot{\varepsilon}_0 \exp\left[-\frac{E^{(p)}}{RT}\left(1 - \frac{\sigma}{\sigma_p}\right)^2 \right] \qquad (8.22)$$

where σ_p is the Peierls stress (i.e. the lattice resistance to dislocation glide specified as the stress needed to move a dislocation without the help of a thermal activation mechanism), $E^{(p)}$ is the activation energy for dislocation glide. Effective viscosity in Peierls creep is related to the strain rate and temperatures as (Karato, 2010):

$$\eta_{\mathrm{eff}} \sim (T \log \dot{\varepsilon})^{1/2}. \qquad (8.23)$$

Constitutive equations for steady-state deformation

The following equations refer to the case of steady-state deformation, which means that steady-state distribution of defects (i.e. constant dislocation density, constant subgrain structures) is re-established after modifications in defect concentrations have occurred due to applied deviatoric stress. In general, the steady state can be defined only approximately because plastic deformation is sensitive to microstructures which always have a hierarchy in a real material. Figure 8.11 illustrates a general dependence of strain rate on stress in the dislocation and diffusion creep regimes.

A generic equation that describes deformation as a function of thermodynamic and structural parameters (Karato, 2008) is:

$$\dot{\varepsilon}^{(v)} = A \, f_{H_2O}^r \, (P, T) \, \sigma^n \, \left(\frac{b}{d}\right)^m \, \exp\left[-\frac{H}{RT}\right] \qquad (8.24)$$

In the case where Peierls mechanism is included, the equation has the form (Karato et al., 2001):

$$\dot{\varepsilon}^{(v)} = A \, \sigma^n \, \left(\frac{b}{d}\right)^m \, \exp\left[-\frac{H}{RT}(1 - \frac{\sigma}{\sigma_p})^q\right] \qquad (8.25)$$

where $\dot{\varepsilon}$ is strain rate (the meaning of superscript (v) is explained below), A is experimentally defined constant, $\sigma = \sigma_1 - \sigma_3$ is differential stress, σ_p is Peierls stress, q is a constant that depends on the mechanism of the dislocation glide, f_{H_2O} is water fugacity (see Section 7.2.5 for definition), r is a constant, d is grain size, n is stress exponent ($n = 1$ for diffusion creep and $n \sim 3$ for dislocation creep), m is the grain-size sensitivity exponent, b is the length of Burger's vector (length of the lattice distortion in dislocations, $b \sim 5$ nm), $H = E^* + PV^*$ is enthalpy (where E^* is activation energy, V^* – activation volume, P–pressure), R and T are the universal gas constant and temperature, respectively. Water fugacity can be replaced with water content $C_{OH} \sim f_{H_2O}$ when water content is fixed (closed system behavior). In a general case,

$$C_{OH} = f_{H_2O} \, A_{H_2O} \, \exp[-(E_{H_2O} + PV_{H_2O})/RT] \qquad (8.26)$$

where for F_{90} olivine $A_{H_2O} = 26$ ppm H/Si/MPa, $E_{H_2O} = 40$ kJ/mol, $V_{H_2O} = 10^{-5}$ m^3/mol (Hirth and Kohlstedt, 2003).

Experimental studies demonstrate that diffusion and dislocation creep are independent mechanisms of deformation and, at the first approximation, their strain rates are additive (Frost and Ashby, 1982). In contrast, both dislocation creep and Peierls creep involve the same dislocations which move under different pressure–temperature conditions. As a result, these deformation mechanisms are not independent and can be treated as alternative mechanisms: the one that produces the highest strain rate under given stress is considered to be active, while at the same time the other one is considered to be inactive. In this case:

$$\dot{\varepsilon}^{(v)} = \dot{\varepsilon}_{\text{diff}} + \max(\dot{\varepsilon}_{\text{disl}}, \dot{\varepsilon}_{\text{p}}), \qquad (8.27)$$

where $\dot{\varepsilon}_{\text{diff}}$, $\dot{\varepsilon}_{\text{disl}}$, $\dot{\varepsilon}_{\text{p}}$ are strain rates due to diffusion, dislocation, and Peierls mechanisms, respectively.

Another generic equation for creep, similar to eqs. (8.25, 8.27), has been proposed by Hirth and Kohlstedt (2003):

$$\dot{\varepsilon} = A_{\text{diff}} \left(\frac{\sigma}{\mu}\right)^{n(\text{diff})} \left(\frac{b}{d}\right)^{3} + A_{\text{GBS}} \left(\frac{\sigma}{\mu}\right)^{n(\text{GBS})} \left(\frac{b}{d}\right)^{2} \qquad (8.28)$$

where μ is the shear modulus which depends on temperature, pressure, and composition (e.g. on the mole fraction of iron in olivine). The subscript "diff" refers to grain boundary diffusion (Coble) creep in which diffusion is coupled with grain boundary sliding (Lifshitz, 1963). The subscript "GBS" refers to grain boundary sliding creep accommodated by a dislocation creep process. The parameters, A_{diff} and A_{GBS} are material dependent and are determined experimentally. Experimental studies indicate that in the diffusion creep regime, the stress exponent $n_{\text{diff}} = 1$, while in the grain boundary sliding creep regime the stress exponent n_{GBS} is determined from experiments.

Deformation maps for crustal and mantle rocks

Deformation maps are used when several deformation mechanisms operate; they provide a useful tool for analyzing which deformation mechanism dominates in a certain parameter space. Examples of deformation maps for some mantle and crustal rocks are shown in Figs. 8.12 and 8.13. These plots show how the aggregate strength depends on temperature, strain rate, grain size, and mineralogy, and which of the deformation mechanisms dominates under different physical conditions. Note that the deformation map for olivine (Fig. 8.12a) is shown not for the differential stress σ, but for its second deviatoric stress invariant $|\sigma| = \sqrt{\sigma_{ij}\sigma_{ij}/2}$. Its value depends on the experimental setup: in the case of uniaxial compression, $|\sigma| = \sigma/\sqrt{3}$ and $|\dot{\varepsilon}| = \dot{\varepsilon}\sqrt{3/2}$, while in case of simple shear $|\sigma| = \sigma$ and $|\dot{\varepsilon}| = \dot{\varepsilon}/2$. The effective viscosity used in calculations of the deformation map is $\eta = |\sigma|/2|\dot{\varepsilon}^{(v)}|$. Similarly, the deformation maps for plagioclase (Fig. 8.13a,b) are shown for shear stress $\sigma/\sqrt{3}$ and shear strain rate $\dot{\varepsilon}\sqrt{3}$.

Tables 8.2a,b summarize some experimental data on major parameters describing different deformation mechanisms that were used in the calculation of deformation maps. Note that for some of these parameters the reported values vary by several orders of magnitude due to differences in experimental setups. Additional information on the numerical values of rheological parameters for deformation equations (8.23–8.27) can be found in Karato *et al.* (1986); Green and Borch (1987); Borch and Green (1989); Gleason and Tullis (1995); Mei and Kohlstedt (2000a, b); Hirth and Kohlstedt (2003); Karato and Jung (2003); Rybacki and Dresen (2004); Li *et al.* (2004, 2006); Raterron *et al.* (2007); Katayama and Karato (2008); Kawazoe *et al.* (2009).

Viscous deformation: summary

Laboratory measurements of deformation indicate that the rheology (strength) of crustal and upper mantle materials depends on the following factors:

- temperature,
- pressure, pore fluid pressure,
- strain, strain rate, strain history,

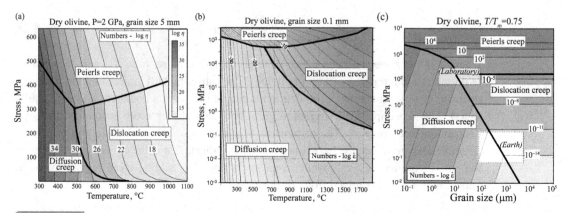

Fig. 8.12 Deformation mechanism maps of dry mantle olivine as functions of stress, temperature, strain rate, and grain size (a – from Elesin *et al.*, 2010; b – from Kameyama *et al.*, 1999; c – from *Karato*, 2008). (a) The effect of viscosity variations which are inversely proportional to strain rate $\dot{\varepsilon}$ (eq. 8.11), the vertical axis shows the second deviatoric stress invariant (see text for explanations). (b) The effect of strain rate variations. (c) The effect of grain size variations versus stress, Tm – solidus temperature. Light shadings in (c) show typical ranges of parameters in laboratory studies of rock deformation ("Laboratory") and in the lithospheric mantle ("Earth").

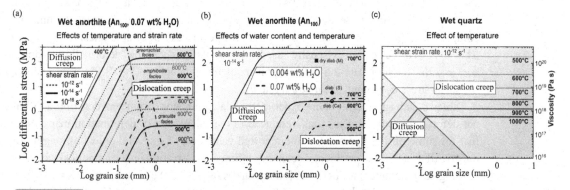

Fig. 8.13 Deformation mechanism map for plagioclase (wet anorthite) (from Rybacki and Dresen, 2004) and wet quartz (after Rutter and Brodie, 2004). Horizontal and vertical scales are the same in all plots. (a) The effects of temperature and strain rate on aggregate strength for shear strain rates of 10^{-12}, 10^{-14}, and 10^{-16} 1/s and temperatures of 500, 600, and 900 °C. The transition between grain boundary diffusion creep and dislocation (power-law) creep is shown by dashed and dotted lines. The approximate regions of greenschist, amphibolite, and granulite facies condition correspond to a shear strain rate of 10^{-14} 1/s. (b) The effect of (trace amounts of) water on shear stress for pure anorthite aggregates at temperatures of 700 and 900 °C and a shear strain rate of 10^{-14} 1/s. "Dry" rocks (broken lines) are considerably stronger than 'wet' rocks, in particular at lower temperature and in the diffusion creep regime. At similar temperature, the transition between regimes is shifted towards larger grain size for "wet" aggregates in comparison to "dry" rocks. Symbols – experimental data for natural Maryland diabase at 700 °C (labeled Ca and S) and dry Maryland diabase (labeled M). (c) The effect of temperature on strength of wet quartz for a strain rate of 10^{-12} 1/s.

- fugacity of water and other volatiles,
- grain size, mineralogy, chemical activities of the mineral components.

Because of the significant difference in thermodynamic conditions and material properties used in laboratory studies of rock deformation, on one side, and existing in the deep upper mantle, on the other side, the governing equations and deformation maps based on them should be considered as a first approximation only. A general summary based on the deformation maps and the constitutive equations presented in this section is the following (see Section 8.1.3 for details):

Diffusion creep:

- dominates at low temperatures and relatively low stresses;
- is characterized by high effective viscosity and low strain rate;
- has a relatively short transient time to re-establish steady-state distribution of defects;
- the strain rate linearly increases with stress, $\dot{\varepsilon} \sim \sigma$ (eq. 8.24);
- the strain rate increases significantly with grain size decrease, $\dot{\varepsilon} \sim d^{-m}$ ($m = 3$ for grain boundary diffusion, eq. 8.24) and is particularly important at grain sizes $d < 1$–5 mm (Fig. 8.12);
- viscosity decreases exponentially with increasing temperature and increases non-linearly with increasing grain size;
- strain rate slightly increases (viscosity decreases) with increasing Fe-content in olivine; a decrease of Mg# by 5 increases creep by a factor of ~5 (Zhou *et al.*, 2009; see Section 8.1.3);
- diffusion creep does not produce lattice-preferred orientation and thus anisotropy;
- diffusion creep can occur through grain boundary diffusion (Coble creep) and through matrix diffusion; the latter mechanism is unlikely to operate in the crustal and upper mantle.

Dislocation (or power-law) creep:

- dominates at relatively high stresses (> 10 MPa) and high temperatures; this deformation mechanism is likely to dominate in most of the lithosphere;
- is characterized by low effective viscosity and high strain rate;
- results in highly anisotropic deformation (at the level of grains) and in development of new grain-scale microstructures (i.e. formation of new grain boundaries, development of lattice-preferred orientation);
- has a long transient time to re-establish steady-state distribution of defects; grain-scale microstructures (lattice-preferred orientation and steady-state grain size) have longer time constants than dislocation microstructures;
- the strain rate increases non-linearly with stress $\dot{\varepsilon} \sim \sigma^n$ ($n = 3$–3.5, eq. 8.24);
- the strain rate is insensitive to the grain size ($m = 0$, eq. 8.24); the exception is dislocation creep of dry olivine near the transition from diffusion creep to dislocation creep (but there is no grain size dependence under hydrous conditions) (Hirth and Kohlstedt, 2003);
- viscosity decreases exponentially with increasing temperature (Fig. 8.10a), increases non-linearly with strain rate increase, but is insensitive to grain size;
- has strong dependence on the presence of trace amounts of water (Fig. 8.13b).

Table 8.2a Parameters for diffusion, dislocation (power-law), and Peierls creep (parameters refer to eqs. 8.24–8.27 and are used in calculations Fig. 8.12a)

	A ($Pa^{-n} s^{-1}$)	n	m	r	E^* (kJ/mol)	V^* (10^{-6} m^3/mol)	q	σ_p (GPa)	Reference
Diffusion creep									
Dry olivine	1.1×10^5	1	2.5	–	300	6	0	–	1
Wet olivine	6.6×10^4	1	2.5	0.7–1.0	240	0–20	0	–	1, 6
Dislocation creep									
Wet quartzite	4.0×10^{-18}	2.3	0	1.2	150	–	0	–	2
Anorthosite	2.1×10^{-23}	3.2	0	–	240	–	0	–	2
Diabase	8.0×10^{-25}	3.4	0	–	260	–	0	–	2
Dry olivine	1.3×10^{-12}	3.0	0	–	510 ± 30	14 ± 2	0	–	4, 5
Wet olivine*	2.0×10^{-14}	3.0	0	1.2	410 ± 40	11 ± 3	0	–	4, 5
Peierls creep									
Dry olivine	1.3×10^{12}	0	0	–	540	0	2	9.1	3
Wet olivine	6.3×10^{-5}	2.0	0	1.2	470	24	2	2.9	7

* closed system, ~1300 ppm H/Si

[1] Karato *et al.* (1986); [2] Ranalli and Murphy (1987); [3] Evans and Goetze (1979); [4] Karato and Jung (2003); [5] Mei and Kohlstedt (2000b); [6] Hirth and Kohlstedt (2003); [7] Katayama and Karato (2008).

Table 8.2b Flow law creep parameters for synthetic plagioclase aggregates (parameters refer to eqs. 8.24–8.25 and are used in calculations Fig. 8.13)

	H_2O (wt %)	$\log A$ ($MPa^{-n}\mu m^m s^{-1}$)	n	m	H (kJ/mol)	Reference
Diffusion creep						
An_{100}	0.004	12.1 ± 0.6	1.0 ± 0.1	3	467 ± 16	1
An_{100}	0.07	1.7 ± 0.2	1.0 ± 0.1	3	170 ± 6	1
An_{60}	0.3	1.1 ± 0.6	1	3	153 ± 15	2
Ab_{100}	0.2	3.9 ± 1.1	1	3	193 ± 25	3
Dislocation creep						
An_{100}	0.004	12.7 ± 0.6	3.0 ± 0.4	0	648 ± 20	1
An_{100}	0.07	2.6 ± 0.3	3.0 ± 0.2	0	356 ± 9	1
An_{60}	0.3	-1.5 ± 0.5	3	0	235 ± 13	2
Ab_{100}	0.2	3.4 ± 1.0	3	0	332 ± 23	3

An_{100} = anorthite; An_{60} = labradorite; Ab_{100} = albite; $q = 0$ in all cases.

[1] Rybacki and Dresen (2000); [2] Rybacki and Dresen (2004); [3] Offerhaus *et al.* (2001)

Peierls plasticity (or power-law breakdown) mechanism:

– dominates at relatively low temperatures and very high stresses (typically > 500 MPa); this deformation mechanism can be important in some cratonic settings and in cold slabs;

– is insensitive to the grain size;

– viscosity weakly decreases with increasing temperature (Fig. 8.10a), weakly increases with strain rate increase, and is insensitive to the grain size.

8.1.3 Selected experimental results on rock deformation

Several reviews on experimental studies of mantle rheology have been published recently (e.g. Karato and Wu, 1993; Kohlstedt *et al.*, 1995; Hirth and Kohlstedt, 2003; Karato, 2008, 2010). Dislocation and diffusion creep mechanisms are extensively studied in experiments for the most common crustal and upper mantle rocks, while Peierls creep received significantly less attention, recent experimental studies include Kameyama *et al.* (1999) and Karato *et al.* (2001). This section illustrates the effects of pressure, temperature, strain rate, grain size, water, partial melting, and mineralogy on the rheology of the Earth's rocks as measured in creep experiments (Fig. 8.14a,b). Before presenting the results, some issues relating to their uncertainties are worth a comment.

Uncertainties in experimental studies of rheological parameters

The reported values of some parameters that control deformation of the mantle vary by several orders of magnitude (Fig. 8.15a). For example, the values of activation volume range from 5×10^{-6} to 27×10^{-6} m³/mol; this corresponds to a more than 10 orders of magnitude difference in effective viscosity (at constant stress and temperature). A large scatter of rheological data exists for experiments at high pressures (> 1 GPa), which are technically more challenging than at low pressures. However, a large scatter in reported data exists even for low-pressure (< 0.5 GPa) experiments, to a large extent due to the incomplete characterization of water fugacity (water content). Furthermore, because of

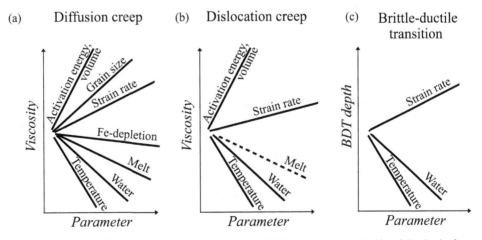

Fig. 8.14 Sketch illustrating the effects of some parameters on variations in lithosphere viscosity (a, b) and the depth of the brittle–ductile transition (c). Note that many effects are non-linear.

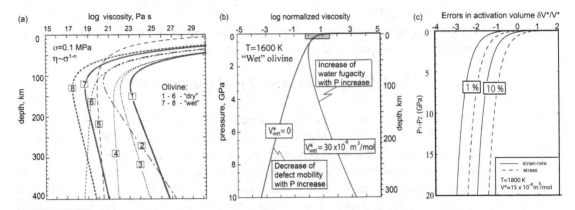

Uncertainties in experimental studies of deformation. (a) Viscosity versus depth relation for the upper mantle with a typical oceanic geotherm calculated from the power-law creep constitutive equation for olivine for the values reported in different experimental studies (data sources: 1 – Kawazoe *et al.*, 2009; 2 – Borch and Green, 1987; 3 – Karato and Rubie, 1997; 4 – Bussod *et al.*, 1993; 5 – Raterron *et al.*, 2007; 6 – Li *et al.*, 2006; 7 – for 0.007 wt% H_2O, Karato and Jung, 2003; 8 – for 0.07 wt% H_2O, Karato and Jung, 2003). (b) The range of viscosity uncertainties if only low-pressure laboratory data are used for extrapolation to the upper mantle under "wet" conditions. (c) Errors in the estimated activation volume δV^* based on measurements of creep strength (stress) or strain rate (both with errors of 10% or 1%) at pressures $P1$ and $P2$. Solid curves – for stress measurements (at constant strain rate), dashed lines – for strain rate measurements (at constant stress). Relative errors depend on the activation volume, V^* (after Karato, 2010).

extrapolation from laboratory to real Earth conditions, the accuracy of rheological constraints for the upper mantle is significantly lower than the precision of laboratory measurements. Some of the problems relating to experimental studies of plastic deformation and their interpolations are discussed below. The discussion follows Karato (2008, 2010) whose conclusions, due to the complicated nature of the topic and challenging requirements for experimental setup, may in some parts disagree with the conclusions of other researchers.

Experimental studies of plastic flow

According to Karato (2010), for experimental data on plastic deformation to be applied to geological conditions, deformation experiments must be performed: (i) on a single crystal or a polycrystalline aggregate (not on powder samples); (ii) at steady-state deformation conditions; (iii) for power-law creep rather than for Peierls mechanism; (iv) under control of water fugacity (water content); (v) with high accuracy of stress measurements; (vi) for a broad range of pressures, and *all* of these conditions must be satisfied during an experiment. Brief comments on these requirements are summarized below.

- When powder samples are used (as in the experiments by Li *et al.* (2004, 2006)), compaction and/or grain crushing complicate interpretations of creep experiments.

- Creep measurements must be made at steady-state deformation state which is the assumption behind the corresponding constitutive equations. However, power-law creep requires a long transient time to re-establish steady-state distribution of defects (stress-relaxation time for this plastic deformation is shown in Fig. 8.9). Furthermore, different creep mechanisms with significantly different transient times may operate simultaneously.

- Control of water content during the deformation experiment is important because even small traces of water have a strong effect on plastic flow (e.g. Fig. 8.13b). Since hydrous minerals are almost always present in natural rocks, their dehydration provides water into the sample (as was the case in experimental studies by Chopra and Paterson, 1981). Re-examination of some earlier experimental studies on nominally "dry" olivine (Green and Borch, 1987; Borch and Green, 1987, 1989) have demonstrated that it contained a significant amount of water. The opposite process, a decrease of water fugacity in undersaturated samples, can occur at high experimental temperatures due to partial melting. For this reason, water content must be measured both before and after deformation experiments. Unfortunately, in many experimental studies water fugacity is not clearly defined.

- The largest uncertainty is related to the pressures used in deformation experiments. Experimental studies of plastic flow at high pressures are hampered by the absence of reliable commercial apparatus. Stress values are not determined accurately in creep experiments at pressures ~4 to ~10 GPa. A significant amount of high-resolution rheological data was obtained at low pressures (<300–500 MPa); apparatus used for such measurements provides indisputable high quality in deformation experiments. However, these low pressures correspond to depths of less than 15 km and cannot be extrapolated to greater lithospheric depths. At low pressures, the pressure effect on the activation volume cannot be constrained reliably because the uncertainties in experiments are comparable with the pressure effect itself (due to exponential dependence of strain rate on pressure, the pressure effect on rheological parameters is weak at low pressures). Figure 8.15c, based on eq. (8.13), illustrates how errors associated solely with errors in strain rate or stress measurements (assumed to have a Gaussian distribution) propagate into errors in the estimated activation volume. With the increase in the span of pressure ($P1$ and $P2$ are pressures at which an experiment is conducted), the error in the activation volume decreases significantly, and high-pressure experiments provide higher quality (lower uncertainty) data on the activation volume than low-pressure experiments. When strain rate depends on grain size, the uncertainties increase.

- In the case of "wet" rheology, an additional complication arises from the fact that the rheological properties of wet materials depend not only on the pressure-dependent $\exp(-PV/RT)$, but also on the pressure-dependent water fugacity. The pressure effect is different for these two factors (Fig. 8.15b). At low pressures (<0.5 GPa), the fugacity effect dominates (because it depends almost linearly on pressure) and viscosity reduces with pressure increase. At high pressures (>1 GPa) and a large activation volume, the activation volume term becomes important; but its effect can only be determined with sufficient precision from data at $P > 1$–2 GPa. Thus, experimental results obtained from low-pressure studies cannot be extrapolated to lithospheric depths, because of (unresolved) pressure and water content effects. Note that the presence of volatiles other than water, e.g. carbon dioxide, may also have a strong effect on rheological properties (as, for example, its presence has on the mantle solidus); however, so far such experimental data do not exist except for some preliminary low-pressure reconnaissance (Keppler *et al.*, 2003; Shcheka *et al.*, 2006).

Extrapolation of laboratory data to real Earth conditions

Extrapolation of laboratory data to real Earth conditions carries with it significant uncertainties that are substantially greater than the accuracy of creep experiments. High-quality deformation experiments are conducted at $P < 0.5$ GPa (although some recent experiments have extended this range to pressures of up to ~17 GPa) and at temperatures up to ~1900 °C. Since laboratory temperatures are close to upper mantle temperatures, the major uncertainty comes from extrapolations of stress. Most laboratory measurements are made near the dislocation creep–diffusion creep transition, and the dominant mechanism may change with decreasing stress.

Furthermore, Fig. 8.12c shows that typical ranges of stress and grain size used in laboratory studies of rock deformation differ significantly from upper mantle conditions. Hirth and Kohlstedt (2003) note that extrapolation of stress by two orders of magnitude (e.g. from 100 MPa to 1 MPa) results in a ± 1 order of magnitude in viscosity uncertainty due to a ± 0.5 order of magnitude uncertainty in stress exponent. In dislocation creep, viscosity decreases non-linearly with increasing stress, and thus at laboratory time-scales experiments in this deformation regime should be carried out to high strain.

Another uncertainty is associated with grain size, which is particularly important in diffusion creep studies, where the strain rate non-linearly decreases (viscosity increases) with grain size increase (e.g. compare plots (a) and (b) in Fig. 8.11). In particular, a typical grain size used in plastic deformation experiments is ~ 0.01–0.1 mm (Fig. 8.12), while olivine grain size in mantle xenoliths is about 2–3 orders of magnitude greater (Fig. 3.9). As a result, in laboratory diffusion creep experiments rocks will deform 6–9 orders of magnitude faster than in the mantle. The correlation between grain size and deformation has been used in practice to estimate deviatoric stresses in continental mantle as a function of depth from known distributions of grain size variations with depth caused by dynamic recrystallization (Fig. 3.104d,e) (Mercier, 1980).

Effect of stress

Numerous experimental studies of plastic flow have examined the relationship between strain rate and stress. This relationship is expressed by the stress exponent n in the constitutive equations (8.24–8.25). An increase in differential stress leads to a transition from diffusion creep to dislocation creep (Fig. 8.11).

In the diffusion creep regime, plastic flow under dry conditions provides $n = 1.0 \pm 0.1$, i.e. strain rate depends linearly on the differential stress. Similar values are observed under water-saturated conditions. Slightly higher values reported in some studies probably indicate a contribution from dislocation creep.

Experimental studies indicate a significant scatter in the values of the stress exponent n in the dislocation creep regime, because much of the experimental data comes from experiments at the transition from diffusion to dislocation creep. For example, for the same experimental data (Fig. 8.11c), two different approaches give $n = 3.3 \pm 0.6$ and $n = 3.6 \pm 0.5$ (Hirth and Kohlstedt, 2003). For experimentally studied deformation of mantle rocks, the reported values of the stress exponent range, in general, between 3.3 and 3.7; similar values are determined for dry grain boundary sliding accommodated by the dislocation creep process. Experiments conducted at hydrous conditions yield n values from 3.0 ± 0.1 to 3.8 ± 0.2 (Mackwell *et al.*, 1985). The n values for polycrystalline samples are similar to those reported for single olivine crystals. For olivine crystals, the n values for the three major slip systems measured over a large range of pressure, temperature, and oxygen fugacity are similar and close to 3.5 (Fig. 8.11a,b). These data are very important for extrapolation of relatively high-stress experiments to low-stress conditions typical of the lithospheric mantle and the asthenosphere, since deformation experiments on single crystals can be performed at low stresses, low pressures, but at relatively high temperatures.

Temperature, pressure, and activation volume

A generic form of dependence of the strain rate on pressure and temperature follows the Arrhenius law (eqs. 8.16–8.19, 8.21, 8.23–8.24) and is illustrated by Fig. 8.16. A temperature increase produces an increase in the strain rate (or a viscosity decrease). The pressure

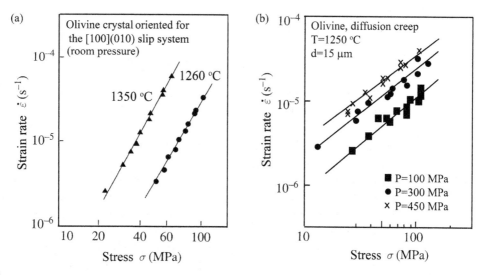

Fig. 8.16 Experimental data on plastic deformation of olivine. (a) Effect of temperature on stress–strain rate relationship in a single crystal of olivine (based on Bai *et al.*, 1991). (b) Effect of pressure on diffusion creep in olivine (data from Mei and Kohlstedt, 2000a).

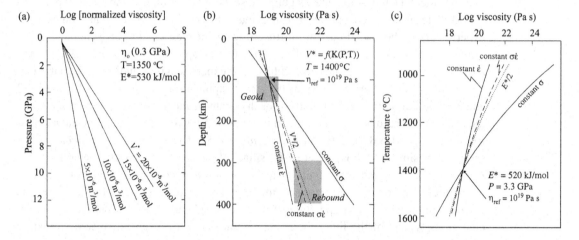

Fig. 8.17 Theoretical calculations showing the dependence of viscosity versus pressure (a), depth (b), and temperature (c) (after Hirth and Kohlstedt, 2003). (a) Viscosities calculated for constant values of activation volume and normalized by the value at $P = 0.3$ GPa. (b–c) Stress-dependent viscosity calculated relative to a reference viscosity of 10^{19} Pa s for constant $\dot{\varepsilon}$, constant σ (solid lines), and constant $\dot{\varepsilon}\sigma$ (dashed lines). The influence of depth (b) and temperature (c) of stress-independent viscosity (dotted lines labeled $V^*/2$ and $E^*/2$, respectively) is shown for comparison. Gray shadings in (b) – viscosity values in the upper mantle estimated from geoid and postglacial rebound.

effect on plastic flow is double: (i) through the term $\exp(-PV^*/RT)$ and the pressure dependence of the activation volume $V^*(P)$ and (ii) through the pressure dependence of fugacity $f_{H_2O}(P, T)$ (e.g. Fig. 8.15b).

Volume V^* depends on pressure as:

$$V^*(P) = -(\partial \ln \dot{\varepsilon}/\partial P)/RT + P\, \partial V/\partial P, \qquad (8.29)$$

thus estimates of V^* from the first term of eq. (8.29) in high-pressure experiments will underestimate V^*. Theoretical and experimental studies indicate that, despite a significant scatter in the reported values of V^*, the activation volume generally decreases with increasing pressure (depth) and increases with increasing melting temperature T_m. Although the pressure effect on viscosity based on theoretical predictions, where the elastic strain is expressed in terms of the pressure- and temperature-dependence of elastic moduli, provides a reasonable approximation to experimental data (Fig. 8.17a), it is only an approximation, since the calculations assume a constant value of activation volume for each of the lines.

For dislocation creep, the values of the activation volume are well constrained in experiments. For diffusion creep, the activation volume can be approximated as (Hirth and Kohlstedt, 2003):

$$V^* = E^*(\partial T_m/\partial P)/T_m. \qquad (8.30)$$

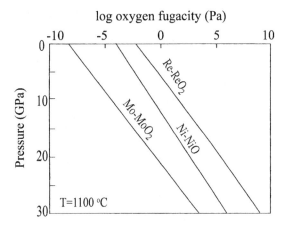

Fig. 8.18 Oxygen fugacities versus pressure for several metal oxide buffers (see Section 7.2.5 for explanation) (after Karato, 2008). Log oxygen fugacity versus temperature for common buffers is shown in Fig. 7.9.

However, the activation volume for diffusion creep in olivine is not yet well constrained, in particular under dry conditions.

Under wet conditions, the rheological properties of olivine are sensitive not only to $\exp(-PV^*/RT)$ but also to water fugacity $f_{H_2O}(P, T)$ (eq. 8.24) which increases almost linearly with pressure (Fig. 8.18). The interplay of these two factors, the pressure effect on the activation volume and fugacity, has been examined experimentally by Mei and Kohlstedt (2000a, b). At low pressures ($P < 450$ MPa), deformation increases with increasing pressure, probably due to increasing water solubility with pressure. At higher pressures, both effects, the increasing water solubility and the decreasing dislocation mobility, become important; as a result the dependence of strength on pressure becomes non-monotonic. The effect of water on plastic flow is discussed further below.

Upper mantle viscosity is strongly dependent on pressure and temperature due to high values of E^* and V^* for dislocation creep. As discussed above, the values of activation volume reported in the literature vary from 5×10^{-6} to 27×10^{-6} m^3/mol, leading to several orders of uncertainty in viscosity values for the upper mantle. The experimental data indicate that the pressure effect on viscosity can be as large (or even larger) than the temperature effect on mantle viscosity. For example, for the intermediate value of $V^* = 15 \times 10^{-6}$ m^3/mol, a four orders of magnitude change in viscosity can be achieved either by the pressure increase from 100 km to 400 km depth, or by a 400 °C temperature change.

Values of the activation energy E^* of mantle rocks (dunite, lherzolite, and olivine) reported in the literature usually vary from 310 ± 40 kJ/mol to 440 ± 80 kJ/mol for diffusion creep under both dry and wet conditions, with a representative value of 375 ± 50 kJ/mol (although values of 530–700 kJ/mol have been reported in some experiments). For dislocation creep, the representative value under dry and wet conditions is 520 kJ/mol, although values as low as 450 ± 100 kJ/mol have been reported for some lherzolite samples.

Mantle creep under constant stress or constant strain rate conditions results in different deformation that can be considered as end-member creep cases (Fig. 8.17b,c). In the case of

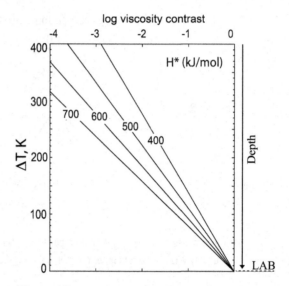

Fig. 8.19 The effect of temperature difference ΔT on the viscosity contrast between the continental lithospheric mantle and the surrounding mantle (after Karato, 2010). The viscosity contrast depends on activation enthalpy ($H^* = E^* + PV^*$). For a plausible range of activation energy and volume, the values of H^* range from ~400 kJ/mol to ~700 kJ/mol. The viscosity contrast is less than a factor of ~10 for a temperature difference of ~100 K which is expected in the lower portions of the continental lithospheric mantle. LAB = lithosphere–asthenosphere boundary.

constant stress, the P- and T-effects on mantle viscosity are the strongest and under constant strain rate – the weakest, while deformation at a constant rate of viscous dissipation ($\dot{\varepsilon}\sigma = $ const) results in intermediate change of mantle viscosity. Assuming $E^* = 520$ kJ/mol, under constant stress conditions a 100 °C temperature increase reduces mantle viscosity by one order of magnitude (Fig. 8.17c).

Figure 8.19 illustrates theoretical predictions of mantle viscosity variations, assuming that they are caused by temperature variations and variations in the activation enthalpy $H^* = E^* + PV^*$ only. Since the temperature difference between the lithospheric mantle and the convective mantle decreases when approaching the lithospheric base (e.g. Fig. 4.45), the viscosity contrast between the lower parts of the continental lithospheric roots and the convective mantle should be less than a factor of 10, in the case where all viscosity differences are produced by temperature variations alone. This result illustrates that low temperatures alone cannot explain the long-term stability of the continental lithosphere.

Effect of grain size

The effect of grain size on plastic flow is expressed by exponent m in the constitutive equations (8.24–8.25). In agreement with theoretical predictions, for grain boundary diffusion creep (the Coble creep) $m = 3.0 \pm 0.5$ (Fig. 8.11). For matrix diffusion $m = 2$; however, since the Nabarro–Herring creep mechanism is not important unless grain size is ~1 m, this

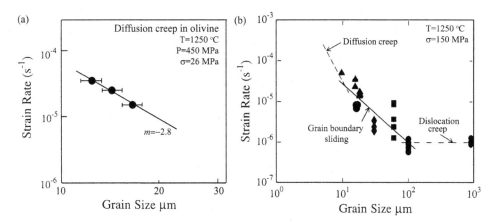

Fig. 8.20 Experimental results showing the effect of grain size on plastic flow in olivine (from Mei and Kohlstedt, 2000a,b; Hirth and Kohlstedt, 2003). (a) Diffusion creep in olivine under water-saturated conditions; the best fit gives a grain size exponent of $m \sim 2.8$. (b) Grain boundary sliding accommodated by a dislocation creep process; the best fit gives a grain size exponent of $m \sim 2$. The data are normalized to a constant stress using $n = 3.5$ after subtracting the component of diffusion creep from the total strain rate.

mechanism is unlikely to occur at any conditions in the upper mantle (Hirth and Kohlstedt, 2003), and the m-values for the Coble creep characterize diffusion creep in the mantle. The value of $m = 3.0$ is commonly used. Some uncertainty exists for diffusion creep at hydrous conditions because of a lack of experimental data.

A strong dependence of strain rate on grain size results in a non-linear viscosity increase with increasing grain size in the diffusion creep regime. As deformation maps (Fig. 8.12) indicate, the effect of grain size is particularly important at $d < 1$–5 mm. Laboratory measurements on rock samples indicate that most crustal rocks have a grain size of a few mm; smaller sizes (~ 10 μm) are observed in highly sheared regions and in fine-grained basalts that have undergone fast crystallization. Upper mantle rocks are expected to have a grain size of ~ 10 mm; however, this is influenced by the processes of dynamic recrystallization and grain growth. When dynamic recrystallization occurs, the grain size depends on stress, and the constitutive equations (8.24–8.25) should be modified.

Dislocation creep does not depend on grain size ($m = 0$). However, experimental studies of rock deformation near the transition from the diffusion to the dislocation creep regime indicate that grain boundaries significantly contribute to deformation of fine-grained rocks (Kohlstedt and Zimmerman, 1996). Grain boundary deformation is an important process in olivine and olivine + orthopyroxene because these minerals have a limited number of slip systems. If grain boundary sliding is accommodated by a dislocation creep process in dry olivine, deformation depends on grain size with $m \sim 1$–2 for grain sizes between 10 and 60 μm (Fig. 8.20b). This means that, due to grain boundary mechanisms, fine-grained samples can deform in the dislocation creep regime ~ 10 times faster than coarse-grained samples. This is true only in the case of dry rock; no grain size dependence is observed for dislocation creep in wet olivine (Hirth and Kohlstedt, 2003).

The transition between the deformation and dislocation creep regimes depends on grain size and activation volume. At intermediate values of activation volume for olivine, the transition between diffusion and dislocation creep in the mantle occurs at a differential stress of 1 MPa ($\eta = 10^{19}$ Pa s) if the grain size is 1 mm and at 0.1 MPa ($\eta = 3 \times 10^{20}$ Pa s) if the grain size is 6 mm. Because of significant scatter in the reported values of activation volume, the actual transition between the two deformation regimes in the mantle is not well defined. If the activation volume is close to zero, as suggested by some experimental data, diffusion creep will always dominate for a grain size 1–10 mm, values typical for the upper mantle (Fig. 3.9). However, this contradicts observations from ultradeep (> 300 km) xenoliths that show a strong LPO (Ben Ismail and Mainprice, 1998; Ben Ismail *et al.*, 2001).

Effect of compositional variations in the mantle

Compositional models of the upper mantle usually assume that the continental mantle consists of 60–80% mol fraction of olivine, 10–30% of orthopyroxene, 0–10% of both garnet and clinopyroxene (e.g. Walter, 1995). The oceanic mantle has less orthopyroxene, but more garnet and clinopyroxene, with 7–15% of each of these mineral phases (Figs. 5.6, 6.11). Since olivine dominates both in the continental and oceanic mantle, the rheological properties of the upper mantle, to a first approximation, are controlled by olivine rheology which is extensively studied in deformation experiments (Korenaga and Karato, 2008). Rheological properties of other mineral phases remain largely unknown. Because of the small volume fraction of garnet and clinopyroxene, their effect on mantle rheology can be ignored. Recent experimental studies of orthopyroxene deformation demonstrate that at low temperatures it is much weaker than olivine, but at high temperatures olivine and orthopyroxene rheologies are similar (e.g. Ohuchi and Karato, 2010). For crustal rocks, the best studied is the rheological behavior of quartz and plagioclase (Fig. 8.13), but the effect of compositional heterogeneity on plastic flow in these rocks has not been systematically examined so far.

Effect of Mg–Fe defects on diffusion creep

Early experimental studies of the effects of Fe and Mg on olivine rheology included deformation experiments of synthetic aggregates of olivine and orthopyroxene at 1500 K and 300 MPa confining pressure (e.g. Hitchings *et al.*, 1989). The results demonstrated that both olivine and orthopyroxene can be markedly weakened by the addition of magnetite but not by iron. This weakening is accompanied by an increased grain-boundary mobility and subgrain formation or recrystallization. A possible thermodynamic explanation is that the weakening effect of magnetite is due to an increase in point defect concentrations at the higher oxygen fugacity. However, when both iron and magnetite are added, only slight weakening is observed in the absence of other additions.

Zhao *et al.* (2009) studied diffusion creep in anhydrous polycrystalline olivine with grain sizes of < 50 μm in high-temperature, high-pressure triaxial compressive creep experiments. The samples of olivine $(Fe_xMg_{1-x})_2SiO_4$ had different Fe content (iron depletion) with

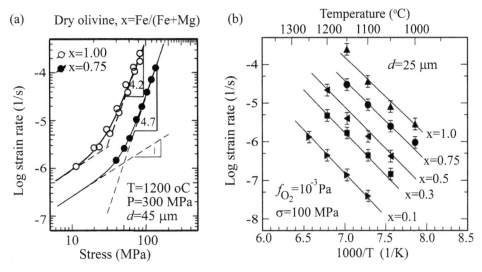

Fig. 8.21 Deformation experiments on dry olivine showing the effect of Fe-content (mole fraction of Fe defined as $x = Fe/(Fe + Mg)$) on grain-boundary sliding creep (after Zhao et al., 2009). (a) Logarithmic strain rate verses differential stress for $x = 1.0$ (open circles) and $x = 0.75$ (black circles). The dashed straight lines show power law relations for the two creep mechanisms; in the diffusional creep regime the stress exponent $n = 1$, in the dislocation-accommodated grain boundary sliding regime $n = 4.2$ for $x = 1.0$ and $n = 4.7$ for $x = 0.75$. (b) Experimental data (symbols) and empirical fit with $n = 3.9$ for all five compositions.

$x = Fe/(Fe+Mg) = 1.0$, 0.75, 0.5, and 0.3 ($Mg\# = 1 - x$). At differential stresses of 50 to 300 MPa, the confining pressure of 300 MPa, and temperatures between 1000 and 1200 °C, olivine deformation occurred primarily by the dislocation-accommodated grain boundary sliding (Coble) creep (Fig. 8.21). The effect of iron depletion on the strain rate is mainly through the effect of $Mg\#$ ($= 100\,mg\#$) on the activation enthalpy H^*, which based on experimental data can be expressed as:

$$H^* = H_o^*[1 - \alpha(1 - Mg\#/100)], \tag{8.31}$$

where H_o^* is the activation enthalpy for $Mg\# = 100$ and $\alpha = 0.093$. The results indicate that, under the same conditions (including fixed temperature), strain rate increases (viscosity decreases) systematically with increasing Fe concentration: the olivine samples with $x = 0.3$, 0.5, 0.75, and 1.0 creep ~15, ~75, ~365, and ~1480 times faster than olivine with $x = 0.1$, respectively. Since variations of iron concentration in the upper mantle are relatively small, the variation of $Mg\#$ by ~5 changes the creep by less than a factor of 5, making iron-depletion variations largely insignificant for upper mantle rheology.

Effect of water on mantle rheology

The presence of water in the upper mantle is discussed in detail in Section 7.2.9 (see also Figs. 7.17 to 7.19). In relation to deformation experiments it is important to note that,

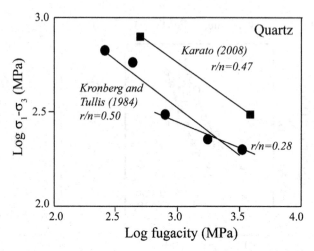

Fig. 8.22 Dependence of the creep strength of quartz on water content (from Karato, 2008). Symbols – experimental data for quartz; r/n – exponents for the constitutive equation (8.24); the effect of pressure on fugacity is not included.

because hydrogen diffusion is a slow process (eq. 7.10), experimentally determined water content may significantly underestimate actual water content and may correspond to incomplete equilibrium conditions (Bell *et al.*, 2002).

Because wet and dry creep have independent deformation mechanisms, the total strain rate may be approximated by (Karato and Jung, 2003):

$$\dot{\varepsilon}_{total} = \dot{\varepsilon}_{dry} + \dot{\varepsilon}_{wet} \tag{8.32}$$

where strain rate for wet and dry creep is expressed as (see eq. 8.24):

$$\dot{\varepsilon}_{dry} = A_{dry}\ \sigma^{n_{dry}} \left(\frac{b}{d}\right)^m \exp\left[-\frac{H_{dry}}{RT}\right] \tag{8.33}$$

$$\dot{\varepsilon}_{wet} = A_{wet}\ f^r_{H_2O}(P, T)\ \sigma^{n_{wet}} \left(\frac{b}{d}\right)^m \exp\left[-\frac{H_{wet}}{RT}\right]. \tag{8.34}$$

At water-rich conditions, the total strain rate is controlled mostly by wet rheology. The presence of even trace amounts of water has a strong effect on rheological parameters (Fig. 8.22). Although the amount of hydrogen (water) dissolved in nominally anhydrous minerals such as olivine is small (~10–1000 ppm H/Si), under anhydrous conditions hydrogen concentration exceeds the concentration of point defects in these minerals, having a strong effect on rock rheology (Mackwell *et al.*, 1985). Thus the effect of water on creep rate should depend on water concentration.

As discussed above, under hydrous conditions pressure has a double effect on rheology, through the pressure effects on the activation volume and on water fugacity. The latter effect is due to an increase in water solubility in nominally anhydrous minerals with increasing water fugacity. The effect of water fugacity on creep is specified by the water fugacity exponent *r* (eq. 8.34). An analysis of the global experimental data set for olivine (excluding the results of Borch and Green, 1987, 1989 with a poorly controlled water content) indicates

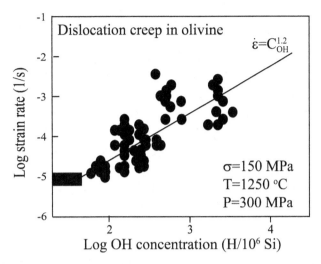

Dependence of the strain rate in olivine on water content (from Hirth and Kohlstedt, 2003). Symbols – experimental data from different authors normalized to a pressure of 300 MPa, constant stress and temperature for $n = 3.5$; black box – strain rate in dry olivine under the same P–T conditions.

that the best fit to data in the dislocation regime is achieved with $r = 1.2 \pm 0.4$ and $V^* = (22 \pm 11) \times 10^{-6}$ m^3/mol (Fig. 8.23). This means that strain rate increases approximately linearly with increasing water concentration. Experimental studies show that, in the diffusion creep regime, creep rates also increase with increasing water content (Mei and Kohlstedt, 2000a). The value of the water fugacity exponent $r = 1.0 \pm 0.3$, and the effect of water on the diffusion creep rate is through the effect of water fugacity on the concentration of point defects at the grain boundaries.

Viscosity profiles for a constant water content can be calculated by subtracting the effect of water from the activation parameters:

$$E^*_{\text{eff}} = E^* - E_{\text{H}_2\text{0}}; \tag{8.35}$$

$$V^*_{\text{eff}} = V^* - V_{\text{H}_2\text{0}}; \tag{8.36}$$

where $E_{\text{H}_2\text{O}}$ and $V_{\text{H}_2\text{O}}$ are given by eq. (8.26). Hirth and Kohlstedt (1996) estimated that in the presence of water, the viscosity of olivine aggregates decreases by a factor of up to 180 at a confining pressure of 300 MPa (Fig. 8.24a). As a result, the viscosity of the oceanic upper mantle in the MORB source region is 500 ± 300 times less than the viscosity of dry olivine. The dependence of water solubility in olivine on pressure and water fugacity has been used to estimate the melting depth range in the oceanic mantle (Fig. 7.18b) and to propose that wet to dry transition beneath the mid-ocean ridges may occur between 115 km and 60 km depth.

Role of partial melting

The effect of partial melting on deformation is important in plastic flow of the asthenosphere. The constitutive equation (8.24) can be considered as the power-law creep equation

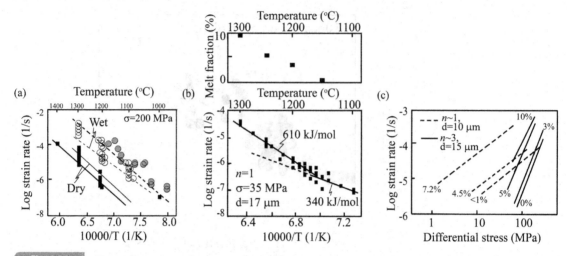

Fig. 8.24 Effect of water and melt on the creep strength of mantle rocks (from Hirth and Kohlstedt, 1996; Kohlstedt and Zimmerman, 1996). (a) Experimental data from different authors for deformation of olivine aggregates; circles and dashed lines – for wet olivine (open: at confining pressures $P = 0.3$ GPa, gray: $P = 1$ GPa, data of Borsch and Green, 1989); solid symbols and solid lines – for dry olivine ($P = 0.3$ GPa); thin lines – data of Karato and Wu (1993) adjusted using $n = 3.5$; (b) Deformation of spinel lherzolite (70% ol + 30% cpx) in the diffusion creep regime versus temperature; the upper plot shows the corresponding change in the melt fraction. The change in the activation energy at $T = 1160$ °C corresponds to the transition from subsolidus to hypersolidus temperature. (c) Diffusion (dashed lines) and dislocation (solid lines) creep in olivine at $T = 1250$ °C and confining pressure of 0.3 GPa as a function of melt fraction (numbers on the lines). A dramatic decrease in viscosity occurs when the melt fraction exceeds 5–7%.

for a melt-free rock. If melt is present, the equation must be modified to incorporate the dependence of the creep rate on melt fraction φ and the geometry of melt distribution; the latter is often characterized by the dihedral angle θ (Fig. 7.15). In an isotropic material, an interconnected network of melt forms along triple junctions when $0° < \theta < 60°$; when $\theta \rightarrow 0°$ the melt phase nearly or completely wets the grain boundaries. According to experimental studies, a 100 °C temperature increase reduces θ by ~10°. In diffusion deformation, the effect of melt depends critically on melt geometry. Diffusion through the melt phase is several orders of magnitude faster than through a grain boundary; as a result the strain rate increases dramatically in the diffusion creep regime when the melt totally wets the grain boundaries.

Under nonhydrostatic stress conditions, there is a significant difference in melt geometry during stress-controlled or strain-controlled deformation. In the case of stress-controlled deformation, melt films in grain boundaries and microcracks have preferential alignment parallel to the maximal principal stress. This pattern observed experimentally in deformation of granites at low effective confining pressures and some lherzolites was named as stress-induced wetting of grain boundaries (Jin *et al.*, 1998). In these experiments, fine-grained samples deformed in both the dislocation and diffusion creep regimes contained melt along all triple junctions.

In the case of strain-controlled deformation, melt is reoriented when a lattice-preferred orientation (LPO) or, at large strains, a shape-preferred orientation (SPO) develops. Because of crystalline anisotropy, LPO produces a preferred orientation of the melt channels; SPO results in elongated channels parallel to the long axis of the grain as triple junction tubes stretch out in this direction (Kohlstedt and Zimmerman, 1996). Because both LPO and SPO change very slowly during static annealing, the melt distribution will be effectively frozen-in to the microstructure.

Laboratory experiments on olivine + basalt and spinel lherzolite samples (most of them had grain sizes of less than 30 μm) suggest that a few percent of melt may have a strong effect on viscosity both in the dislocation and the diffusion creep regimes. For example, for olivine–basalt aggregates with typical values of $\theta = 35°$, the creep rate increases by a factor of two due to the presence of $\varphi = 0.03$ of melt. As illustrated by Fig. 8.24b, the major difference between the subsolidus and the hypersolidus regimes is in the value of the activation energy which increases by 1.8 when the melting starts. Experimental studies also indicate that melt has a much smaller effect on creep rate for coarse-grained than fine-grained samples: for example, in the case of 3% of melt, creep rate increases by a factor of 1.5 for $d = 17$ μm and by a factor of 4 for $d = 1$ mm (Kohlstedt and Zimmerman, 1996).

The effect of melting on olivine viscosity is shown in Fig. 8.25. When the amount of melt is small ($\varphi < 0.1$), the strain rate in both wet and dry conditions can be approximated by an exponential relationship

$$\dot{\varepsilon} \sim \exp(\alpha\phi), \tag{8.37}$$

or in case of constant stress by:

$$\eta \sim \exp(-\alpha\phi), \tag{8.38}$$

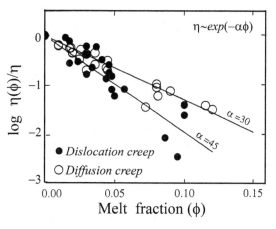

Fig. 8.25 Normalized viscosity versus melt fraction for deformation of olivine aggregates (from Hirth and Kohlstedt, 2003). For each data point, the viscosity is normalized by the viscosity of melt-free olivine aggregate deformed at the same conditions. The exponential relationship $\eta \sim \exp(-a\varphi)$ with $a = 30$–45 provides a good fit to data in both deformation regimes.

where $\alpha = 25$–30 for diffusion creep and $\alpha = 30$–45 for dislocation creep. In the case of dunite melting, the effect is weaker than for lherzolite melting, with $\alpha \sim 20$–25. The above proposed exponential relationships are empirical.

Experimental data on olivine + basalt samples indicate that a sharp decrease in viscosity occurs in both the dislocation and the diffusion creep regimes when the melt fraction exceeds 5–7% (Fig. 8.24c). In diffusion creep, the presence of 2% of melt reduces the strain rate by a factor of 1.1, while 7% of melt decrease strain rate by a factor of ~15. In dislocation creep, 3% of melt reduces the strain rate by ~1.5, while 10% of melt reduces it by a factor of 25. In the case of grain boundary sliding, a significant decrease in viscosity can occur at melt fractions as small as ~1% (Hirth and Kohlstedt, 2003). This effect is attributed to the presence of a large portion of melt-wetted grain boundaries that provide effective high-diffusivity paths and lead to a dramatic strain rate increase. The fraction of melt-wetted grain boundaries increases with melt fraction.

It is worth noting that experimental studies on lherzolite samples do not show a significant decrease in viscosity at melt fractions > 4–5% as observed in olivine + basalt samples (Kohlstedt and Zimmerman, 1996). In lherzolite samples, the strain rate increases smoothly with an increase in temperature and melt fraction. The difference in plastic flow in olivine and lherzolite can be explained by differences in the wetting geometry due to different mineralogies.

Significant controversy on the effect of melting on mantle rheology comes from the fact that extraction of melt from the mantle leads to water depletion. The interplay of the two processes (melting and dewatering) may have strong implications for the long-term preservation of deep continental lithospheric roots. The degree of water depletion depends on the mode of melt extraction, and it is significantly stronger for fractional melting than for batch melting (Fig. 7.18a). As discussed earlier, water depletion results in significant hardening of a rock, and thus partial melting may increase mantle viscosity if the retained melt fraction is small (0.1 wt%) (Karato, 1986). At a fixed temperature, the degree to which viscosity changes due to dewatering (the degree of hardening) depends on the initial water content, the degree of water removal, and the activation volume for dry olivine (Fig. 8.26). For example, a three orders of magnitude viscosity increase (without any change in temperature) can be achieved for a low degree of dewatering ($\sim 10^{-3}$) only if the activation volume is larger than $\sim 17 \times 10^{-6}$ m^3/mol. For small activation volumes ($\sim 5 \times 10^{-6}$ m^3/mol), the degree of hardening is always small. In such a case, long-term stability of the continental lithospheric roots requires a relatively large temperature difference between the continental roots and the surrounding mantle. If activation volumes are larger than $\sim 10 \times 10^{-6}$ m^3/mol, the water effect alone may result in a significant mantle strengthening, and there is a broad range of combinations of degree of water depletion and temperature difference between the surrounding mantle and the continental roots for the preservation of continental roots (Karato, 2010).

The interplay of the two processes (melting and dewatering) may also be important in controlling asthenosphere rheology both beneath continents and oceans. In particular, it is proposed that melting in the asthenosphere can be related to a zone with a minimum of water solubility in mantle materials (Fig. 3.42) so that excess water forms a hydrous silicate melt (Mierdel et al., 2007).

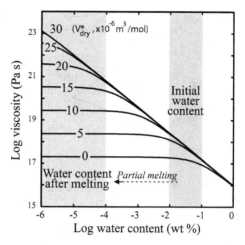

Fig. 8.26 Effect of water depletion due to partial melting on olivine viscosity (after Karato, 2010). The initial water content corresponds to the oceanic asthenosphere or mantle wedge above the subduction zone (shown by gray vertical band). Partial melting reduces the water content and increases the viscosity; the resultant water content is expected to correspond to cratonic lithospheric mantle (gray area on the left). Numbers on the curves – activation volume for dry olivine corresponding to the entire range of values reported in the experiments. The plot is calculated for $P = 7\,\mathrm{GPa}$ (~200 km depth), $T = 1430\,°C$, and $\sigma = 0.1\,\mathrm{MPa}$.

8.1.4 Lithosphere strength

Qualitative features of strength envelopes

Yield strength (failure) envelopes

Deformation laws described in Sections 8.1.2–8.1.3 allow for the constraining of strength profiles for the lithosphere (the so-called Brace–Goetze lithosphere profiles, or yield strength envelopes, YSE) (Fig. 8.27). The strength here has the units of stress and means the stress that leads to brittle fracture or viscous flow (defined accurately, lithosphere strength is the total force per unit width required to deform the lithosphere at a given strain rate). The profiles consist of linear parts that correspond to brittle deformation and curved parts that correspond to plastic flow. Integrating the area under the strength profile yields the vertically integrated lithosphere strength B which in the 2D case is expressed as:

$$B = \int_0^\infty \sigma(x, z, t, \dot{\varepsilon})\,\mathrm{d}z \qquad (8.39)$$

where the z-axis is oriented downwards. It is worth noting that yield strength envelopes are no more than approximations to lithosphere deformation since in the real Earth the strain–stress dependence is also time-dependent with characteristic time-scales.

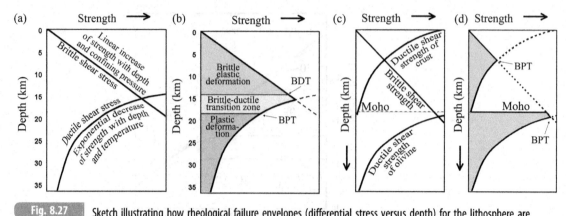

Fig. 8.27 Sketch illustrating how rheological failure envelopes (differential stress versus depth) for the lithosphere are constrained. (a–b) Strength profiles for continental crust; (c–d) yield strength envelopes for the crust and the lithospheric mantle. Rheology of shallow layers is controlled by brittle shear strength (see Fig. 8.6) which increases linearly with depth and confining pressure. Rheology of the deeper layers is controlled by ductile shear stress which is controlled by a number of thermodynamic and structural parameters and decreases with depth. Each curve in this regime corresponds to a fixed strain rate. At high strain rates, the uppermost mantle may deform brittle. In some cases, lithosphere decoupling may occur (d). Integrating the shaded area in (b) and (d) yields the vertically integrated lithosphere strength. BDT = brittle–ductile transition; BPT = brittle–plastic transition.

At a given depth, the strength of a rock is determined by the deformation mechanism that requires less stress; this leads to a "Christmas-tree" pattern of rheological profiles. In the brittle deformation regime, rock strength is specified by the Mohr–Coulomb law (eqs. 8.6–8.10) and increases linearly with depth as the normal stress increases (i.e. it also depends on the average density of the overlying lithospheric layers); in this regime rock strength depends also on the pore fluid pressure and pore geometry inside a rock, but is insensitive to mineralogy (see Section 8.1.1). In brittle deformation, the rock strength is different for extension and compression (e.g. Fig. 8.6), while such a difference does not exist for ductile deformation. Since, in the first approximation, the rock fails when its rheologically weakest mineral phase fails, and the weakest mineral in the crustal rocks is quartz, crustal rheology is commonly approximated by deformation of quartzite.

While confining pressure increases with depth suppressing the growth of dilatant micro-cracks and increasing the brittle strength, the ductile strength decreases with depth due to temperature increase. The transition from brittle deformation to plastic flow occurs within the brittle–ductile transition zone (Fig. 8.27); its existence is related to the pore pressure in lithospheric rocks (eqs. 8.9–8.10). There are no constitutive equations for semi-brittle rock behavior. Thus, the transition between brittle and ductile deformation regimes is sometimes defined by the depth where the effective pressure is 2.5 times less than the frictional strength (Kirby, 1980). Kohlstedt *et al.* (1995) propose that the onset of semi-brittle flow is when the yield or flow stress is less than 5 times the frictional stress. Another approach (Chester,

1988) relates the total strength in the semi-brittle regime to the creep strength σ_c and the fracture strength σ_f as:

$$\sigma_1 - \sigma_3 = \phi\sigma_c + (1 - \phi)\sigma_f \qquad (8.40)$$

where ϕ is a mixing parameter that depends on the rock material. Since the strength profiles have a maximum at the depth where the brittle–ductile transition (BDT) occurs, this zone is the strongest part of the crust where many earthquakes occur. In the ductile regime, the strength curves are calculated for a fixed strain rate (Fig. 8.17b,c) so that, at a given depth, a higher strain rate will produce a curve with higher strength. This implies that, in such an approach, the depth of the brittle–ductile transition is strain rate dependent. The BDT depth also depends on temperature gradient and composition (in particular, the presence of melts and fluids) (Fig. 8.15c). For slow deformation (low strain rate) and/or high heat flow (temperatures), the BDT is shallow, while for fast deformation and/or low heat flow it is deeper. For strain rates of 10^{-17}–10^{-14} s^{-1} typical of the continental lithosphere, the brittle–ductile transition occurs at ~300–400 °C in quartz and at 600–700 °C in olivine.

In the creep deformation regime, the strength is specified by the Goetze–Dorn laws (eqs. 8.4–8.5) which have an exponential dependence of strength on temperature. Since temperature in the lithospheric mantle increases with depth approximately linearly, the strength decreases (viscosity increases) with depth exponentially (Fig. 8.27). Because of large regional variations in lithospheric geotherms and a strong temperature dependence of strength, the rheological profiles for different tectonic settings differ dramatically. Depending on thermodynamic and structural parameters (activation volume, activation energy, pressure, grain size, the presence of hydrogen and melts), a large range of strength profiles is possible for the oceanic and continental crust[1] and upper mantle.

Since olivine dominates mantle composition, its rheology is commonly used to approximate mantle deformation. The calculations of strength envelopes typically assume that viscous deformation of the lithosphere is dominated by the dislocation creep mechanism, and diffusion creep is disregarded. Note that at high strain rates, a low-temperature uppermost mantle may deform brittly (Fig. 8.27). When the temperature of creep activation in the lower crust is lower than temperature at the Moho, lithosphere decoupling may occur.

Stress envelopes for lithosphere flexure

Strength profiles discussed above include only brittle and ductile deformation; however, they do not account for the elasticity of the lithosphere. In the alternative approach that takes into account viscous, elastic, and brittle deformation, lithosphere rheology is characterized by a stress state (Fig. 8.28), but not by a failure envelope, and it does not reflect the strain rate as the curves for viscous deformation do in the yield strength envelopes (Fig. 8.27).

---

[1] Not all published YSEs are correct. In the publication by Brace and Kohlstedt (1980), the parameter A required in creep calculations (eqs. 8.24–8.27 and Table 8.2) was misprinted for the dry quartz rheology, 5×10^6 instead of 5×10^{-6}. Several studies (Burov and Diament, 1992, 1995; Burov and Poliakov, 2001; Burov et al., 2001) that used the wrong A value estimated the crust (or upper crust) to be about three orders of magnitude softer than when the correct value is used (for discussion of these results see Ranalli, 2003; Burov, 2003).

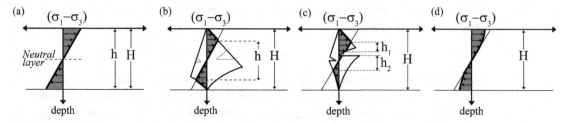

Rheology of the lithosphere of thickness H that takes into account viscous, elastic, and brittle deformation during lithosphere flexure (after Ranalli, 1994). Horizontal axes: compressive stresses are shown towards the right, tensional stresses – towards the left. Thick lines – strength profiles for brittle and ductile deformation. Straight thin lines (the "fibre" stress) that go from positive to negative stresses show lithospheric layers that are under compression (northeastern sectors) and under extension (southwestern sectors); they intersect the zero stress line in a stress-neutral layer that exists in the middle of the bent lithosphere. The shaded area shows the distribution of the differential stress with depth. The material fails when the strength envelope (compression positive) is less than the fibre stress. Failure reduces the thickness of the load-bearing lithospheric layer to h. (a) Elastic lithosphere; the whole plate thickness is load-bearing, $h = H$. (b) Flexure of rheologically layered lithosphere either with no ductile layer in the lower crust (continuous envelope) or when the softening of the lower crust is insufficient to cross the strength distribution (dotted envelope), $h < H$. (c) Flexure of rheologically layered lithosphere where the ductile zone in the lower crust crosses the strength envelope. The lithosphere is decoupled into two load-bearing layers of thicknesses h_1 and h_2, and the stress gradient is discontinuous across the decoupling zone. (d) In the case of non-linear work-hardening plasticity, the stress increases away from the neutral axis at a decreasing rate which is less than in the elastic case.

When a lithospheric plate is bent, a stress-neutral layer exists in the middle of the bending plate. The part of the plate above this layer is under compression (in the case of downward bending), while the part below this layer is under extension. In the case of elastic flexure, the whole lithospheric plate of thickness H supports the load. The mechanical thickness H of the lithosphere is defined as the depth to the level where the yield stress becomes smaller than a limiting value (see also next section). When the yield strength becomes less than the limiting value (the fibre stress), the material fails.

In the case of material failure, the thickness of the load-bearing lithospheric layer h (the "elastic core") becomes smaller than the mechanical thickness of the lithosphere H. The thickness of the load-bearing lithospheric layer depends on temperature, strain rate, plate curvature and composition. If the strength of the ductile lower crust is lower than the fibre stress, the lithosphere is decoupled into two load-bearing parts (Fig. 8.28c). If the lithosphere rheology is non-linear (see Section 8.1.3), the fibre stress also changes with depth non-linearly (Fig. 8.28d). When the elastic core vanishes, the lithospheric plate has reached the limit bending moment or the limit plate curvature.

In general, $h \leq H$; thickness of the elastic core h provides the minimum thickness of the elastic layer $Te(\min)$, while H provides the maximum thickness $Te(\max)$ which is the thickness of the entire lithospheric plate (Fig. 8.29). Thus

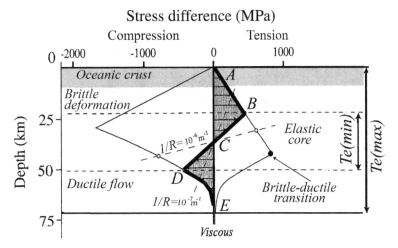

Fig. 8.29 Mechanical thickness of the lithosphere for an idealized case of oceanic lithosphere flexure. Thin solid line – the yield stress envelope calculated for 80 Ma old oceanic lithosphere with geotherms based on the cooling plate model. The brittle behaviour is modeled by Byerlee's law, the ductile behaviour – by an olivine power-law for dry olivine rheology. Thick solid line shows the stress difference for a load which generates a moment $M = 22 \times 10^{16}$ N/m and the plate curvature $1/R = 5 \times 10^{-6}$ 1/m. Dashed lines show results for the values of $1/R = 10^{-6}$ 1/m and $1/R = 10^{-7}$ 1/m which bracket observations at a deep-sea trench – outer rise systems. The bold line shows an example of a bending stress profile. The lithosphere exhibits brittle failure from surface A to depth B, linear elastic behaviour from B to D and ductile failure below depth D; depth C corresponds to the depth of no elastic strain. The load is supported partly by a strong elastic layer and partly by the brittle and ductile strength of the lithosphere; minimum elastic thickness corresponds to the thickness of the elastic core, $h = Te(\min)$ and the maximum – to the thickness of the entire elastic plate, $H = Te(\max)$ (after Watts and Burov, 2003).

- elastic models of lithosphere flexure can provide only the lower limit of the mechanical thickness of the lithosphere (and its flexural thickness)

because they overestimate the stress in the yielding parts of the lithosphere (Ranalli, 1994). Note, however, that some studies equate the mechanical thickness H with the flexural thickness of the lithosphere Te, $H = Te$ (see eq. 8.48 for the Te definition). This issue is discussed in detail in Section 8.3.

Lithosphere rheology

As discussed in Section 8.1, the lithosphere rheology is controlled by a large number of thermodynamic and structural parameters. As a result, rheological models and lithospheric strength profiles constrained for different tectonic settings are very different. For this reason, only a brief overview of a typical lithosphere rheology is presented here for some selected tectonic structures. Regional examples and detailed discussion follow in Section 8.3, after the concepts of flexural lithosphere and regional isostasy are introduced.

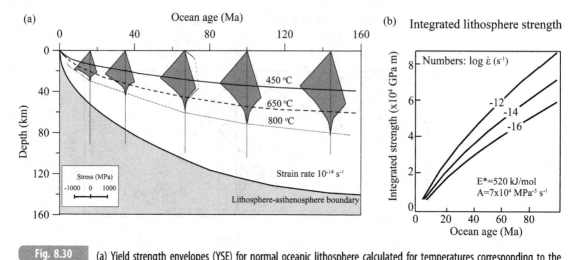

Fig. 8.30 (a) Yield strength envelopes (YSE) for normal oceanic lithosphere calculated for temperatures corresponding to the cooling plate model (YSE adopted from Watts (2001); assumptions on dry or wet rheology used in the calculations are unspecified). Thin line for 73 Ma-old lithosphere shows the estimate of Kohlstedt *et al.*, (1995) for dry 60 Ma-old oceanic lithosphere (Fig. 8.32a). Thin lines show the depth to 450, 650, and 800 °C isotherms; bold line marks the base of the thermal lithosphere. (b) Integrated strength of the oceanic lithosphere of different ages as a function of strain rate (from Stüwe, 2002). Some rheological parameters used in the calculation are indicated on the plot. Both plots (a–b) apparently refer to dry oceanic mantle.

Oceanic lithosphere

Since temperature has one of the strongest controls on lithosphere rheology, the rheology of the oceanic lithosphere, where the oceanic geotherms and lithospheric thickness are age-dependent in normal oceans, can be directly linked to ocean floor age (Fig. 8.30). The relatively homogeneous composition of the oceanic lithosphere makes such calculations straightforward. The major uncertainty arises from assumptions on the amount of water in the oceanic mantle. Based on the results from mid-ocean ridges that indicate a high degree of melting and as a result a high degree of water depletion, Kohlstedt *et al.* (1995) calculated strength envelopes for dry oceanic mantle. Following this approach, apparently a large number of strength studies for the oceanic lithosphere assumed that it has dry rheology. However, electromagnetic studies (Section 7.4) indicate that, away from mid-ocean ridges, the oceanic mantle can be hydrogen saturated. In this case, its strength will be significantly lower than predicted for dry rheology. Since, in the first approximation, the water content in the oceanic mantle away from mid-ocean ridges can be assumed to be the same, qualitatively the results based on dry rheology will remain valid. Note that large uncertainty over the value of the activation volume (see previous section) puts additional limitations on the validity of theoretical strength profiles.

Yield strength envelopes calculated for normal oceanic lithosphere show an expected dependence on ocean floor age and indicate that the integrated strength of the oceanic lithosphere increases with its age. In old oceanic lithosphere, the layer that can support loads is thicker than in young oceans, and it can support significantly larger loads than young oceanic lithosphere. This result agrees with the age dependence of the flexural thickness of the elastic lithosphere (Fig. 8.49b) discussed in Section 8.3.1.

Continental lithosphere

Owing to a large diversity in the structure of the continental lithosphere (thickness, structure, and mineralogy of the crust and the lithospheric mantle, the presence of melts and fluids) and in strong lateral heterogeneity of its thermal structure, simple strength models like for the oceanic lithosphere are not possible for the continents. Figure 8.31 illustrates some general cases and the effects of variations in stress regime, crustal thickness (assumed to have anorthosite rheology), and lithospheric temperatures on yield stress of the continental lithosphere. In the case of fluids in the upper crust, its brittle strength will be reduced as illustrated by Fig. 8.6c. The effect of crustal lithology on yield strength envelopes is illustrated by Fig. 8.32 and Fig. 8.33. *Relative strength*, defined as the ratio of lower crustal strength to upper mantle strength, is another possibility for characterizing the relative importance of lower

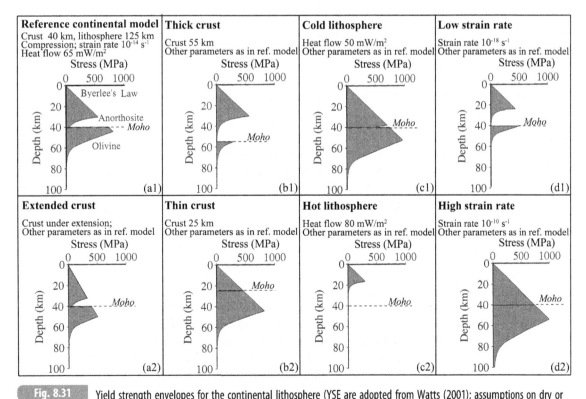

Fig. 8.31 Yield strength envelopes for the continental lithosphere (YSE are adopted from Watts (2001); assumptions on dry or wet rheology are unspecified). The plots show the effects of variations in the stress regime, crustal thickness (assumed to have anorthosite rheology), and lithospheric temperatures. The upper row corresponds to the cratonic lithosphere, while the lower row better reflects lithosphere strength in young tectonically active regions.

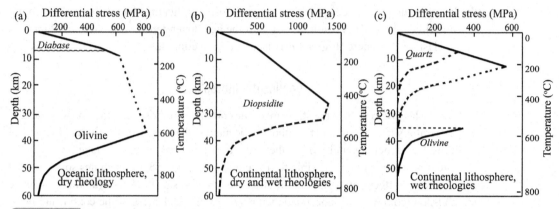

Fig. 8.32 Strength envelopes for the upper 60 km of the oceanic and continental lithosphere calculated for strain rate 10^{-15} s^{-1} (after Kohlstedt *et al.*, 1995). (a) Oceanic lithosphere with a geotherm for 60 Ma. Note that dry rheology is assumed; this can be the case only in regions with strong melting where strong dewatering occurs; such regions are unlikely to follow a geotherm for *normal* oceans. In the case of wet rheology, lithosphere strength will be significantly weaker. (b) Strength envelopes for diopsidite which can be used to represent the lower crust rheology. The curves are for a 60 mW/m^2 reference continental geotherm and are similar for wet and dry rheology. (c) Strength envelopes for continental lithosphere with temperatures along a 60 mW/m^2 reference continental geotherm. The crust is assumed to have wet quartzite rheology (two curves correspond to plastic flow with different parameters), the upper mantle is approximated by a wet olivine rheology. Dotted lines connect the brittle–ductile and the brittle–plastic transitions.

crustal and upper mantle contributions to the total lithospheric strength (Afonso and Ranalli, 2004). It may not necessarily correspond to the strength ratio of the lower crust and the upper mantle at the Moho depth; for example in the case of quartz- or plagioclase-controlled crustal rheologies, the local creep strength of crustal material can be less than the rock creep strength at the Moho, although the relative strength ratio is greater than 1.

Analysis of a wide range of parameters affecting strength of the continental lithosphere shows that, similar to the oceans:

- the continental lithosphere is stronger under compression than under extension;
- cold lithosphere is significantly stronger than hot lithosphere;
- the strength increases with a strain rate increase;
- fluids reduce crustal and mantle strength dramatically.

There is, however, an important qualitative difference between deformation of the continental and of the oceanic lithospheres: the continental lithosphere can experience mechanical decoupling of the upper lithospheric layers from the deeper layers due to the presence of an intermediate rheologically weak zone with ductile flow.

Yield strength envelopes for the continental lithosphere (Fig. 8.31) illustrate alternative first-order models of strength through continental lithosphere. In the *"jelly-sandwich"* model (a1, a2, d1), the strength envelope is characterized by a weak mid-to-lower crust squeezed between a strong upper crust and a strong mantle (composed dominantly of dry olivine). In particular, deformation of a thick crust or a crust with layered mineralogy (e.g.

Felsic granulite/dry peridotite

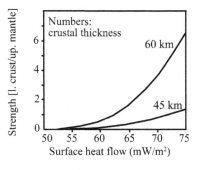
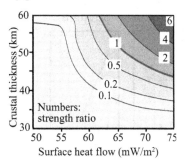

Mafic granulite/wet peridotite

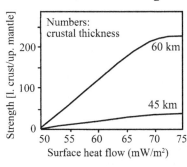
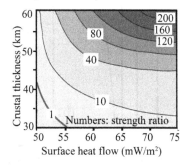

Fig. 8.33 Relative strength as the ratio of lower crustal strength to upper mantle strength (vertical axes on left plots and numbers on the curves on right plots) for selected compositions as a function of surface heat flow and crustal thickness (after Afonso and Ranalli, 2004).

quartz–diorite) can result in mechanical decoupling of the upper crust from the upper mantle by ductile flow in a weak zone. This mechanism is responsible, in particular, for the escape tectonics associated with crustal flow, or in some cases with flow in the uppermost lithospheric mantle, as observed in continental orogens (e.g. Chen and Molnar, 1983; Meissner *et al.*, 2002).

Cases with no decoupling are also possible (b2) when the strong part of the lithosphere includes mechanically coupled strong crust and upper lithospheric mantle. In the *"crème brûlée model"* (best represented by b1) a strong crust is underlain by a weak mantle (for example, due to high temperatures or water presence); the strength of the lithosphere is defined by dry and brittle crust. In the *"banana split" model* (c2) both the crust and the mantle are weak; this model is typical of major crustal fault zones that cut throughout the lithosphere, and weak lithosphere rheology is caused by a number of strain weakening and feedback processes. These models are further discussed in Sections 8.3.2–8.3.3.

The base of mechanical lithosphere

Similar to other lithosphere definitions, the base of the mechanical lithosphere (mechanical boundary layer, MBL) can be defined in a number of different ways (Fig. 8.34).

Defining the base of mechanical lithosphere

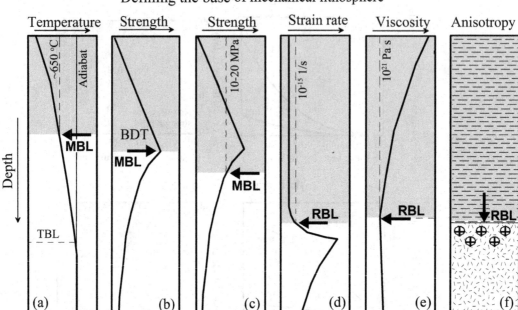

Fig. 8.34 Possible definitions of the base of mechanical lithosphere (a–c) and rheological lithosphere (d–f) (see Fig. 1.4). The base of MBL is defined (a) by an isotherm that approximately corresponds to the brittle–ductile transition (BDT, 600–750 °C for mantle olivine-dominated rocks); note that the base of thermal boundary layer (TBL) is much deeper; (b) by the depth of the BDT; (c) by the depth where the yielding stress or its vertical gradient becomes less than a particular absolute or relative value (e.g. 10–20 MPa or 1–5% of the lithostatic pressure for the yielding stress and 10–15 MPa/km for its vertical gradient). The base of RBL is defined (d) by the depth where the strain rate is less than a critical value (e.g. 10^{-15} s^{-1}); or by the depth where viscosity drops below a critical value (e.g. 10^{21} Pa s) (e), or where the change in the LPO anisotropy direction occurs (f and Fig. 3.117). In special cases this depth may correspond to a change from dislocation to diffusion creep (e.g. Karato, 1992) and ceasing of LPO anisotropy.

The situation is perhaps the closest to the definition of seismic lithosphere, since both mechanical (rheological) and seismic properties of the lithosphere are controlled by a broad range of physical and material parameters.

- The simplest and probably most often used approach is to associate the base of the mechanical lithosphere with a particular geotherm (Fig. 8.34a). This approach builds on the assumption that the base of the elastic lithosphere should correspond to the elastic–plastic transition (e.g. Bodine *et al.*, 1981). The physical grounds are formed by the exponential dependence of lithosphere plastic deformation (eqs. 8.25–8.26) on temperature. As a result, the depth to the brittle–ductile transition where the lithospheric strength is the greatest (Fig. 8.27) also depends on temperature (Fig. 8.34b). There is a large scatter in temperatures reported for this transition; the cited values are 250–450 °C for crustal quartz-bearing rocks and 600–750 °C for mantle olivine-dominated rocks (e.g., Meissner and Strehleau, 1982; Chen and Molnar, 1983). Note, however, that in the plastic deformation regime, the lithosphere strength also depends on the strain rate, which affects the depth of the brittle–ductile transition. The above cited temperatures refer to strain rates of 10^{-17}–10^{-14} s^{-1}, typical of the continental lithosphere. As discussed in Section 8.1.3, a number of other factors also have a strong effect on lithosphere rheology, the most critical ones being the presence of fluids and melts. This implies that assigning a certain isotherm to the base of mechanical lithosphere, although practically convenient, is a strong oversimplification that ignores many physical processes. As a result, one should not expect that the base of the mechanical lithosphere in different tectonic settings (characterized by different stress regime, and with different lithosphere structure and composition) is associated with the same isotherm. The possibility of lithosphere decoupling (Figs. 8.27; 8.31) further complicates the matter. However, in similar tectonic structures the depth to a particular isotherm provides, in the first approximation, the measure of the thickness of the mechanical lithosphere.

- Alternatively, the base of the mechanical lithosphere can be defined by the depth at which the yielding stress becomes less than a particular value (Fig. 8.34c). The choice of this value is somewhat arbitrary, and the situation is similar to seismic tomography where different seismic velocity values (absolute or relative) are chosen to indicate the base of seismic lithosphere. McNutt *et al.* (1988) associate the base of MBL with the depth where the critical yielding stress of 10–20 MPa. Burov and Diament (1995) define the base of MBL as the depth where the yielding stress becomes negligible (1–5%) compared to the lithostatic pressure at the mantle depth with temperature 700–800 °C. As Fig. 8.30 illustrates for oceans, choice of a particular value is somewhat arbitrary. However, regardless of the choice, the thickness of the mechanical boundary layer is always significantly thinner than the thickness of the thermal boundary layer, since the yield strength envelopes for normal oceanic lithosphere, as well as for the continental lithosphere, never extend deeper than the base of the thermal lithosphere.

- Since lithospheric strength is expected to decrease exponentially with depth, another possibility is to define the base of MBL by some minimum value of the vertical gradient of yielding stress (Fig. 8.34c). As with the definition based on critical yielding stress value, the choice of critical gradient value is arbitrary. Burov and Diament (1995) suggest defining the base of MBL by the value 10–15 MPa/km.

The base of rheological lithosphere

Two more definitions are directly related to depth variations in mantle rheology.

- During plastic deformation, the lithospheric strength depends not only on temperature, but also on the strain rate, and the base of the mechanical lithosphere can be identified with the depth where the strain rate is less than some critical value (Fig. 8.34d). Eaton *et al.* (2009) propose to use the value 10^{-15} s^{-1}. For small grain size, this value approximately corresponds to the transition from dislocation to diffusional creep (Fig. 8.12b). The latter can dominate in the deeper upper mantle (Karato, 1992). Dislocation creep produces lattice-preferred orientation responsible for anisotropy of the upper mantle. Thus, this definition attempts to link the base of the lithosphere to the depth where anisotropy (seismic or electric) vanishes or changes its direction (Fig. 8.34f). The upper mantle temperature at the depth specified by this definition should be close to the temperature expected at the base of TBL, and this definition specifies the base of RBL, not MBL.

- Similarly, one can define the base of the lithosphere by some critical viscosity value, since viscosity defines strength at a given strain rate (eq. 8.11). Because of a significant (up to several orders of magnitude) discrepancy in viscosity values reported for the lithospheric mantle and the asthenosphere, there is a large ambiguity in choosing the critical viscosity value as the definition of the lithosphere base (Fig. 8.34e). Commonly cited values are 10^{19}–10^{21} Pa s for dynamic viscosity of asthenosphere and up to 10^{25} Pa s for the lithospheric mantle (e.g. Lambeck *et al.*, 1990; Forte and Mitrovica, 1996). Again, this definition refers to the base of RBL, not MBL (see Fig. 1.4).

An additional complication arises from the fact that the deformation style (elastic or viscous) of a visco-elastic material depends on the duration of tectonic loading (eqs. 8.12–8.14). For an independently constrained stress relaxation time, this provides the opportunity to define the parameter space (temperature, rheology, stress) when the lithosphere behaves mechanically. It allows for estimation of physical conditions corresponding to the base of the mechanical boundary layer. The approach is discussed in Section 8.3.1 in relation to the mechanical thickness of the oceanic lithosphere. The relationships between different definitions of the base of MBL and RBL are non-unique and depend on a large number of physical and material parameters that determine the rheology of the lithosphere and the external tectonic forces applied to it.

8.2 Flexural isostasy: the approach

8.2.1 Lithosphere isostasy

Classical models of isostasy

The concept of isostasy (from the Greek words for "equal" and "standstill") has been introduced in Earth sciences to explain the existence of topographic variations by assuming that the lithosphere (the crust, in early studies) and the asthenosphere are in gravitational equilibrium (Dutton, 1882). According to this concept, the light lithospheric material floats on the top of the denser mantle, and the hydrostatic buoyancy principle of Archimedes is the simplest interpretation of isostasy. The Airy–Heiskanen and Pratt–Hayford compensation mechanisms are the two widely used end-member isostatic models for gravitational equilibrium between the crust and the mantle (Pratt, 1845; Airy, 1855; Hayford, 1917; Heiskanen, 1931) (Fig. 8.35a,b). They are built on the assumptions that isostatic compensation is:

- hydrostatic,
- local,
- achieved by variations in crustal thickness (Airy's model) or crustal densities (Pratt's model); these lateral variations in crustal structure produce gravity anomalies,

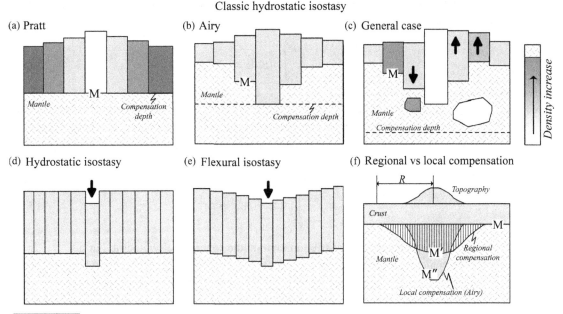

Fig. 8.35 Comparison of isostatic models. (a–c) Models of local hydrostatic equilibrium; darker shades correspond to higher crustal density; (a–b) the end-member models of Pratt and Airy with homogeneous mantle, (c) a general case, arrows show the directions of vertical adjustments caused by density inhomogeneities in the mantle, (d–f) difference between local hydrostatic isostasy and regional flexural isostasy, black arrows indicate topographic load; R is radius of regionality (in the case of crustal flexure $R \sim 250$ km, eqs. 8.43–8.46); M' and M'' – compensation depth (Moho) for regional and local equilibrium models (d–f modified after Stüwe, 2002; Watts, 2001). Dashed line in (b–c) – compensation depth.

- achieved at the base of the crust (Pratt's model) or at the base of the crustal root (Airy's model in its classical form), i.e. that:
  - all density inhomogeneities are within the crust;
  - the mantle is homogeneous;
  - crustal blocks can move vertically independently of each other (Fig. 8.35d).

Pratt's model is commonly used in calculations of isostatic subsidence of the ocean floor; since the crustal thickness of normal oceans is ~7 km worldwide, isostatic compensation is effectively achieved by density variations in the lithospheric mantle due to thermal cooling of lithospheric plates. On the continents, topography depends both on crustal thickness and density variations within the crust. Furthermore, due to a significant density heterogeneity within the mantle, the level of isostatic compensation can be well below the Moho (Fig. 8.35c). For example, gravity modeling indicates that in Eurasia variations in the crustal structure can account for no more than 85% of isostatic compensation (Artemjev *et al.*, 1994).

The concept of isostasy implies that at depth H, the level of isostatic compensation (which should not necessarily coincide with the base of the crust), the hydrostatic equilibrium is approximately achieved, so that the vertical sum of masses must be the same everywhere:

$$\int_t^H \rho(z)\mathrm{d}z = \text{const,} \tag{8.41}$$

where $\rho(z)$ is density (mass) distribution with depth and t is topography. This implies that the sum of the, so-called, anomalous masses produced by density anomalies within the crust and the mantle and located within a vertical column above the compensation level, equals zero:

$$\int_t^H [\rho(z) - \rho_0(z)]\mathrm{d}z = 0, \tag{8.42}$$

where $\rho_0(z)$ is the density distribution within the spherically symmetric Earth.

Tectonic and geodynamic processes modify the density structure of the lithosphere and the structure (thickness and density) of the crust, causing isostasy disturbances. Thus, isostatic models should be treated with caution, since many regions, in particular continental collisional zones, may not be in isostatic equilibrium. Gravity models that fail to achieve a complete local isostatic compensation indicate the presence of lithospheric stresses (Artemjev et al., 1972; Artyushkov, 1973).

If the detailed structure of the crystalline crust and the sedimentary cover are well constrained, the gravity anomalies caused by undulations of the basement and the Moho topographies and by low densities of sedimentary and crustal rocks can be calculated with high accuracy. The isostatic anomalies ("mantle anomalies") calculated by subtracting these crustal gravity anomalies from the observed gravity field, allow for evaluation of the gravity effect of density and structural inhomogeneities in the mantle below the level of isostatic compensation (Artemjev et al., 1984; Yegorova et al., 1998; Kaban et al., 2003).

Flexural isostasy

The concept of the elastic lithospheric plate is the alternative concept of isostasy which dates back to the works of Vening-Meinesz (1931, 1948) and which became particularly popular during the development of the theory of plate tectonics. In contrast to classical concepts of isostasy, flexural isostasy assumes (see Section 8.2.3 for more details):

- mechanical equilibrium of the lithosphere, i.e. compensation of topography is achieved not only hydrostatically by heterogeneities in mass distribution (due to density variations within the crust–mantle column above the compensation depth in the mantle), but also by elastic stresses in the lithosphere,
- mechanical interaction between neighboring crustal blocks with shear stresses between vertical columns (Fig. 8.35e),

- *regional* compensation (mechanical equilibrium of the lithosphere is approximated by an elastic plate overlying viscous mantle) (Fig. 8.35f).

Compared to Airy's model which predicts crustal thickening under a topographic load, the lithospheric flexure model predicts laterally broader thickening due to lithospheric bending (Fig. 8.35f). The idea of explaining topographic loads by *regional* isostasy through flexure of an elastic plate dates back to the 1920s (Putnam, 1922; Jeffreys, 1926; for a historical overview see Watts, 2001). An important result of these studies is the concept of *regionality* quantified by distance R (in the 3D case, the radius of the area) with departures from local compensation. Jeffreys (1926) approximated crustal bending under load by the flexure of a 2D beam of infinite length and estimated that

$$R = \frac{\pi}{\sqrt{2}} \beta \sim 2.22\beta; \tag{8.43}$$

$$\beta = [D/\Delta\rho g]^{1/4}, \tag{8.44}$$

$\Delta\rho$ is the density contrast between the crust and the mantle (the load and the beam), and D is the flexural rigidity:

$$D = ET^3/12. \tag{8.45}$$

Here E is Young's modulus and T is the beam thickness. Vening Meinesz (1931) modeled crustal compensation by 3D flexure of a circular plate with a concentrated load, and estimated:

$$R = 2.905\beta. \tag{8.46}$$

For $\Delta\rho = 500\,\mathrm{kg/m^3}$, $E = 10^{11}\,\mathrm{N/m^2}$, and crustal thickness $T = 40\,\mathrm{km}$, $\beta \sim 100\,\mathrm{km}$. This means that in the case of elastic flexure of the crust, departures from local compensation occur within a radius of $R > 250\,\mathrm{km}$ from the topographic load.

The degree of compensation Φ of the topographic load can be defined by the ratio of lithosphere deflection to its maximum (hydrostatic) deflection (in the latter case, the lithosphere has no rigidity and the topography is fully compensated). In the case of periodic loading of a thin elastic plate,

$$\Phi = (1 + \beta^4 k^4)^{-1}, \tag{8.47}$$

where $k = 2\pi/\lambda$ and λ is the wavelength of loading (Turcotte and Schubert, 2002, p. 123). This concept makes the basis for the coherence analysis of flexural isostasy (Section 8.2.4).

Both classical and elastic concepts of isostasy allow for determination of the density structure and rheology of the crust and the upper mantle. The classical (hydrostatic) concepts attempt to constrain only density inhomogeneities in the lithosphere. In contrast, the elastic plate approximation (which considers isostasy as a mechanical equilibrium of the lithosphere) attempts to constrain simultaneously the density structure of the lithosphere and its mechanical properties, as well as the external forces affecting it. Such a broad spectrum of independent parameters cannot be recovered from data without a set of model assumptions.[2] It is also worth noting that this approach, in contrast to classical isostasy, assumes that density inhomogeneities below the elastic lithosphere are absent (compare Fig. 8.35 c and f). Critical analysis of the practical applications of the concept of flexural isostasy can be found in Cochran (1980) and Artemjev and Kaban (1987, 1991).

8.2.2 Elastic thickness of the lithosphere

Definition

Elastic (or flexural) lithosphere is the layer that provides the isostatic response of the lithospheric plate to surface and/or subsurface loads and mechanically supports elastic (far-field subhorizontal) stresses induced by lithospheric bending (flexure). In models of lithosphere flexure, topography acts as an external load and variations in the crustal structure (due to variations in the topographies of the top of the basement and the Moho, density structure of the sediments and the crystalline crust) are considered as internal (subsurface) loads. *The elastic thickness of the lithosphere*, Te, is a measure of the mechanical strength of an elastic plate deflected by both topographic and internal loads.

In engineering, mechanical properties of an elastic bending plate (a beam) are characterized by flexural rigidity D_o defined as

$$D_o = MR, \tag{8.48a}$$

where M is the bending moment applied to the ends of the beam and R is the radius of beam curvature (in the case of lithosphere flexure, R is proportional to the wavelength of the topographic load). In geophysics, the parameters D_o and Te are related by the equation:

$$D_o = ETe^3/12(1 - v^2), \quad \text{uniform elastic plate,}$$
$$\text{or } Te^3 = 12(1 - v^2)M/Ek, \tag{8.48b}$$

where E is the Young's modulus, v is Poisson's ratio, and k is curvature. For Te values between 10 km and 100 km, flexural rigidity varies from $\sim 10^{22}$ Nm to $\sim 10^{25}$ Nm (assuming $v = 0.25$ and $E = 10^{11}$ N/m^2). The case where $D_o = 0$ corresponds to the classical Airy's local compensation model ($R = 0$, eq. 8.43). In a quasi-static approximation, eqs. (8.48) with some modifications are valid for any rheology.

The presence of weak zones in the crust, such as those associated with decoupling zones (Fig. 8.31) and faulting, significantly affects flexural properties of an elastic plate. McNutt *et al.* (1988) show that the presence of a decoupling layer at depth $Te/2$ (where Te is the thickness of the elastic plate) allows the layers above and below the decoupling zone to deform (bend) independently (the "jelly sandwich" model), so that the total bending

---

[2] Note that any gravity modeling is, in general, non-unique.

moment M supported by the plate and thus flexural rigidity of the plate become 4 times less than in the case of a uniform plate:

$$D = ETe^3/48(1 - v^2) = D_o/4, \quad \text{crust} - \text{mantle decoupling.} \tag{8.49a}$$

In general case of a multilayer elastic plate with n decoupled (detached) layers, the effective elastic thickness of the whole plate is:

$$Te^3(n) = \sum_{i=1}^{n} h_i^3, \tag{8.49b}$$

where h_i is the effective elastic thickness for each individual layer. For example, in case of a multilayer plate with equally strong layers and decoupled rheology

$$Te(n) = n^{1/3}h, \tag{8.49c}$$

which for a two-layer model yields $Te(2) = 1.26\,h$. However, in the case of a two-layer model with coupled rheology $Te = 2h$ (i.e. the base of the mechanically strong layer corresponds to the base of the lower layer).

Elastic thickness and mechanical thickness

The relationship between the elastic thickness of the lithosphere, Te, and the mechanical thickness of the lithosphere H (defined, for example, as the depth where the yield stress becomes smaller than a limiting value, see Fig. 8.32) is not straightforward, and depends on lithosphere rheology, temperature, strain rate, and plate curvature (see further discussion in Section 8.1.4 and Fig. 8.28). In general, the mechanical thickness of the lithosphere H is greater than the thickness of the load-bearing lithospheric layer (the "elastic core" in the yield strength envelopes). The two parameters are equal, $H = Te$ (elastic), only in the case of flexure of a single-layer elastic lithospheric plate, when the whole of the lithosphere thickness supports the load. In the case of a multilayered lithosphere rheology, the bending moment in eq. 8.48b is the sum of all the moments associated with the brittle, ductile, and elastic parts of the yield strength envelopes $M = M_{YSE}$ (Watts and Burov, 2003). In this case, D_o characterizes the mechanical properties of the lithospheric plate, M_{YSE} is determined by depth integration of the yield strength envelopes, and Te (YSE) provides not the actual thickness of the mechanical lithosphere H, but the thickness of the load-bearing lithospheric layer (the elastic core) which provides the integrated strength of the bending plate ($h = Te$ (YSE) in Fig. 8.28).

The relationship between Te (elastic) and Te (YSE) has been examined by McAdoo *et al.* (1985) (see also Section 8.3.1). Their analysis of yield strength envelopes constrained by Seasat altimeter data for nine major subduction zones shows that the value of Te (YSE) depends on the plate curvature $1/R$ and the bending moment M. Increasing the bending moment or increasing the plate curvature results in a decreasing *effective* elastic flexural rigidity and, as a consequence, in a decreasing *effective* elastic thickness, Te (YSE) < Te (elastic) (see Fig. 8.49b). For a 80 Ma-old oceanic lithosphere with a dry olivine rheology (activation energy $E^* = 525$ kJ/mol) subject to a uniform strain rate of 10^{-14} s^{-1} and for plate curvature $1/R = 10^{-8}$ m^{-1}, Te (YSE)/Te(elastic) ~ 1, while for $1/R = 10^{-6}$ m^{-1},

Te (YSE)/Te(elastic) ~ 0.5. When the lithospheric plate has reached the limit bending moment or the limit plate curvature, the elastic core vanishes and Te (YSE) $\rightarrow 0$ (Fig. 8.29).

8.2.3 Flexural isostasy: assumptions

The elastic plate model is a helpful approximation based on principles used in engineering; it provides a robust "reference" model for comparison of the mechanical and rheological behaviour of the crust and the upper mantle. In the real Earth, however, only part of the crust–upper mantle system satisfies approximation by an elastic plate. Flexure models of lithospheric bending are based on the critical assumptions (see also Section 8.2.1) that:

(1) isostatic equilibrium is maintained at regional scale,
(2) the lithosphere has elastic rheology and overlies inviscid fluid (asthenosphere).

The second assumption further implies that:

(3) static deformation of the lithosphere maintains its gravitational equilibrium over geological time-scales;
(4) lithospheric plate deflections are small and can be approximated by cylindrical bending of a thin infinite plate;
(5) horizontal stresses can be neglected (Turcotte and Schubert, 1982).

Finally, the approach assumes that there are no density inhomogeneities below the elastic lithospheric plate. The discussion below shows that many of these assumptions are rarely satisfied, while *all* of them should be true for the elastic plate model to be valid.

Regional isostasy

The first assumption, of regional isostasy, has been challenged by the analysis of lithosphere flexural rigidity for different tectonic regions of Eurasia (Artemjev and Kaban, 1986, 1991). Flexural compensation of a crustal load produces deviation from local isostasy over distances of $R > 250$ km (R is the radius of regionality). The results of high-resolution gravity modeling for Eurasia imply that the isostatic compensation of even large-scale continental structures with dimensions exceeding 100 km is primarily local. This conclusion can be illustrated by gravity models for the interplate Caucasus and the intraplate Tien Shan orogens, where detailed geophysical data allow calculation of the gravity effects of the sedimentary cover and variations in the crustal thickness with high resolution. In the Caucasus the isostatic equilibrium is achieved locally by crustal thickness and density variations beneath the mountain ridges and intramountain depressions; the isostatic equilibrium of the Tien Shan orogen is also local, but at least 50% of it is maintained locally by a reduced density of the upper mantle. These examples of local isostasy suggest that some continental tectonic structures (such as orogens) should be modeled by deviations from local isostasy rather than by models that assume regional isostatic compensation. This conclusion is, in particular, true for continental regions with a network of active deep faults, where elastic plate approximation can be invalid.

Elastic rheology and static deformation

The second assumption states that the lithosphere has an elastic rheology. It also implies that, from the rheological point of view, the base of the elastic lithosphere should correspond to the elastic–plastic transition (Bodine *et al.*, 1981) (Sections 8.1.1 and 8.1.4). Laboratory studies of rock deformations, however, indicate that on geological time-scales the lithosphere cannot be purely elastic: while the upper part of the crust is brittle, parts of the lower crust and the lithospheric mantle deform plastically (Fig. 8.30). A purely elastic model of flexure requires unrealistically high bending stresses and fails to explain the distribution of seismicity with depth and short-term rapid relaxation of the lithosphere after loading (Goetze and Evans, 1979; Kusznir and Karner, 1985). As eq. (8.49) indicates, the presence of decoupling zones in the lithosphere significantly reduces its flexural rigidity.

The fifth assumption on neglecting horizontal stresses is valid only for a perfectly elastic lithosphere that bends in a stable regime. In case of an inelastic lithospheric plate, strain softening and development of instabilities are possible, and thus the effect of horizontal stresses may become important (Cloetingh *et al.*, 1982).

Thin-plate approximation

The fourth assumption goes back to the studies of Timoshenko (1936) who has shown that plate flexure can be approximated by cylindrical bending if a purely elastic plate is loaded at the edges and its *horizontal dimensions are at least three times larger than its thickness*. Thus approximating elastic flexure of the continental lithosphere by a thin cylindrically bent elastic plate requires a plate size well in excess of 300 km. This requirement questions the validity of the approach when applied to relatively narrow convergent zones, such as the Alps and the Andes (e.g. Braitenberg *et al.*, 2002; Pérez-Gussinyé *et al.*, 2008). Furthermore, the validity of the thin-plate approximation to lithosphere flexure, in general, is a subject for debate (Comer, 1983; Wolf, 1985; Zhou, 1991). Following Timoshenko (1936), the approximation assumes that:

– lithosphere is perfectly elastic,
– lithosphere thickness is small compared to the curvature radius,
– vertical stress is small compared to horizontal stress components and can be neglected.

To test the validity of thin-plate approximation, the flexure of an elastic plate of arbitrary thickness has been examined analytically. According to Comer (1983), thin-plate and thick-plate solutions are close under the following conditions:

• small strain,
• linear boundary conditions,
• absence of gravitational forces.

If these conditions are satisfied (which is not the case in geophysics where gravitational forces are important and high strains are expected in many tectonic settings), the major difference between the thin-plate and thick-plate solutions is in the region immediately beneath the load where the thin-plate solution underestimates flexure by 5–10%. Wolf (1985) incorporated gravitational forces into a thick-plate model and found close agreement between thin-plate and thick-plate solutions in the case of long-wavelength loads. According to Watts (2001), thin-plate approximation may be satisfactory for many tectonic settings.

Effective elastic thickness *Te*

To avoid problems related to the assumption of a purely elastic lithosphere rheology, the term "*effective elastic thickness*" has been introduced for the real Earth. In this concept, *Te* refers to the thickness of an equivalent elastic plate that produces the same deflection under the same applied loads. Thus the lithosphere is treated as a "black-box" closed system for which, irrespective of the real strain and stress distributions during lithosphere deformation, only its integrated response to external forces is known (e.g. the vertically integrated lithosphere strength B given by eq. 8.39). This approach allows comparison of *Te* for different tectonic settings as the response function that relates the observed deflection of the plate under topographic loads (expressed, for example, in the gravity field) to the calculated deflection (for assumed mechanical properties of the lithosphere). It allows for modeling of lithospheric flexure by the bending of an inelastic plate with a complex rheology, such as multilayered non-linear brittle–elasto–ductile rheology (e.g. Burov and Diament, 1992).

Although the concept of effective elastic thickness does not provide insights into the "black-box", numerical studies indicate that *Te* depends on (Kusznir and Karner, 1985; Burov and Diament, 1995; Pérez-Gussinyé *et al.*, 2004):

- rheology of the crust and the upper mantle and the degree of their coupling, which are controlled by (e.g. Fig. 8.30):
  - thermal state of the lithosphere (i.e. its tectono-thermal age),
  - thickness and structure of the crust,
  - composition of the crust and the lithospheric mantle,
  - presence of fluids and melts;
- local curvature of the plate which is controlled by:
  - bending stresses (e.g. the dip angle of the subducting plate),
  - plate rheology,
  - distribution of internal and external topographic loads,
  - isostatic postglacial rebound;
- the presence of faults.

The concept of effective elastic thickness helps to avoid problems related to the assumption of a pure elastic lithosphere rheology, but it does not resolve the problem with the assumption on the existence of regional isostasy. Furthermore, classical gravity models, in contrast to the "black-box" lithosphere treatment by flexural models, are constrained by density heterogeneities in the crust and upper mantle and thus provide valuable information on the lithosphere structure (see Chapter 6).

8.2.4 Methods for estimation of effective elastic thickness Te

Flexural rigidity of the lithosphere can be estimated from joint interpretation of gravity and topography (bathymetry) data; the approach has been particularly successful when applied to island chains, seamounts, and continental regions of paleoglaciation. Flexural rigidity (and elastic thickness) is computed either by inverse or forward modeling. Standard techniques employ spectral methods that are based on the concept that the lithosphere responds as a mechanical filter to topographic loads: it suppresses high-amplitude short-wavelength deformations associated with local isostatic compensation (classical isostasy models) but passes long-wavelength deformation associated with regional isostatic compensation (flexural model). Since topographic loads (the input signal) produce gravity anomalies (the output signal), practical techniques examine the relationship between gravity anomalies and topography as a function of wavelength (i.e. of the horizontal dimension of the load). Separation of individual spectral components associated with different wavelengths allows computing of gravity anomalies produced by loads of any size. Thus, estimates of the elastic lithosphere are based either on spectral methods or on forward modeling when gravity and topography frequency spectra are compared with the predictions of local and regional isostatic models.

Two spectral methods, admittance and coherence analyses, are widely used in the estimates of D and Te (Timoshenko and Woinowsky-Krieger, 1959). The principal differences between these techniques are summarized below, since their application can lead to significantly different interpretations of geophysical data on lithospheric structure. As in other chapters, technical details are largely omitted in favor of a broader discussion of the results and their geophysical implications.

Admittance analysis

Gravitational admittance, or the transfer function, $Z(k)$, is a function of the wavenumber k ($k = 2\pi/\lambda = \sqrt{k_x{}^2 + k_y{}^2}$, where λ is the wavelength of the load in the horizontal dimension) and is defined as the ratio of the Fourier transform of the observed gravity, $\Delta g(k)$, to the Fourier transform of the surface topography, $H(k)$:

$$Z(k) = \Delta g(k)/H(k). \qquad (8.50)$$

This equation provides the "first-order" approximation for the dependence between the gravity and the topography, since in general the dependence between these two parameters in the wavenumber domain is non-linear, and the higher terms of expansion in k should be included (Parker, 1972).

Practical forward applications of admittance analysis include the following steps:

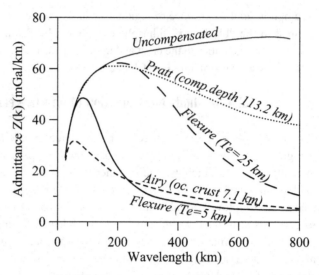

Fig. 8.36 Gravitational admittance calculated for different models of isostatic compensation of the ocean floor with 4.5 km bathymetry (modified after Watts, 2001): two flexure models with plate thickness $Te = 5$ km and 25 km, the Airy model for 7.1 km-thick oceanic crust, the Pratt model with crust of constant thickness and compensation depth at 117.7 km below sea level, and the model of uncompensated topography.

(1) calculation of $Z(k)$ for the assumed isostatic compensation model (e.g. Airy, Pratt, or flexural, with topographic or buried loads);
(2) calculation of the product $Z(k)H(k)$;
(3) inverse Fourier transformation of $Z(k)H(k)$ to obtain $\Delta g(k)$;
(4) comparison of calculated $\Delta g(k)$ with observed gravity anomalies and adjusting the assumed isostatic model.

The value of $Z(k)$ depends on the wavelength and the amplitude of the topographic load. Figure 8.36 illustrates gravitational admittance for several models of isostatic compensation. The Pratt model and the model with uncompensated topography predict the strongest gravity anomalies at long wavelengths, while the Airy model has the smallest anomalies at long wavelengths and small gravity anomalies at short wavelengths. The flexure models predict intermediate gravity anomalies, and the models with small Te tend to the Airy-type model, while flexural models with large Te approach the Pratt-type models. In all models of isostatic compensation, the admittance $Z(k)$ tends toward zero when the wavelength of topographic loads tends either to zero or to infinity.

Function $Z(k)$ modifies the *topography* to produce the observed gravity and thus it depends not only on flexural properties of the lithosphere (quantified by its flexural rigidity, eq. 8.48), but also on the density of the load and the density contrast (assumed to be uniform) between the load and the crust. To avoid dependence of the wavenumber function on density properties of the load, the *isostatic response function* $\varphi(k)$ is used. It modifies the *gravity effect of topography* to produce the observed gravity through linear convolution of $\varphi(k)$ and the load. Function $\varphi(k)$ is independent of the mean density contrast and depends only on the assumed isostatic compensation model: in the Airy-type model, on the thickness of the crust and, in the flexure model, on the flexural properties of the lithosphere. Similarly to practical application of admittance $Z(k)$, "experimental" $\varphi(k)$ is determined for the assumed isostatic compensation model and then compared with the "observed" parameters. The common approach is based on a Fourier analysis under the assumption that the observed gravity field is the sum of the gravity effects of the topographic load and the noise.

Gravity interpretations based on admittance or the isostatic response function aim to find a theoretical isostatic compensation model with *a priori* chosen parameters that will match the observed gravity anomalies. Obviously, the approach is a kind of inverse gravity problem, and its solution (as with any inverse gravity problem) is non-unique. This implies that a large number of density models with similar or nearly identical theoretical $\varphi(k)$ can be constrained for any tectonic structure. As a result, the solutions found within the *a priori* chosen narrow class of possible isostatic models, although matching observed gravity by theoretical predictions, may have nothing to do with reality. On the other hand, the admittance method has some advantages as compared to traditional gravity methods: it allows for fast testing of a large set of density models and has a higher sensitivity to model parameters.

Coherence analysis

Coherence γ^2 is the square of the correlation coefficient in the wavenumber domain between topography and gravity and indicates their statistical relationship. The coherence method introduced by Forsyth (1985):

– assumes that surface and subsurface loads are statistically uncorrelated;
– implies that both types of loads are compensated at the same density interface to which all loads are applied (i.e. that all internal loads are at a single depth);
– excludes the presence of internal loads with no topographic expression.

The validity of these assumptions is discussed in Section 8.2.5. Here surface loads include topography, while subsurface (internal) loads include any density deviations from a homogeneous lithospheric column (i.e. lateral variations in the thickness and density of sediments and the individual crustal layers, including Moho depth variations, as well as the presence of subcrustal density heterogeneities). Forsyth (1985) introduced *the loading ratio f* as the ratio between the weights of loads applied at the crustal base and at the surface, i.e. in the general case,

$$f = [\text{internal load}]/[\text{surface load}]. \tag{8.51}$$

Thus, f is a wavenumber-dependent parameter. The values of f typically range from 0.2 to 5. However, regional coherence analysis for the Canadian Shield indicates that it can be significantly higher in shield areas, ranging from $f \sim 10$ for wavelengths of surface loads > 500 km to $f > 100$ for wavelengths of surface loads less than 50 km (Wang and Mareschal, 1999).

Forsyth has found that while the admittance $Z(k)$ depends on the loading ratio, the coherence γ_B^2 calculated from the Bouguer gravity is not sensitive to it and thus better suits calculations of flexural properties of the lithosphere. This method has been used in almost all Te estimates since 1985. The practical application of Forsyth's method is to estimate Te by fitting the observed coherence with a coherence predicted for a model with initial surface and subsurface (internal) loading. The method was advanced by Lowry and Smith (1994) and termed "load deconvolution", because solving a pair of linear wave-number–domain equations allows for determination of both types of initial loads directly from observed gravity and topography. While for the oceans the admittance and the coherence between free-air gravity and topography are used to estimate flexural proper-ties of the lithosphere, for the continents the Bouguer gravity rather than free-air gravity is used since it accounts for the topographic loads above and below sea level and reflects density anomalies within the crust. The studies of McKenzie and Fairhead (1997) and McKenzie (2003) question the competence of this approach. This controversy, related to the use of either free-air or Bouguer gravity for the continents, is discussed in detail in Section 8.2.5.

The end-member relationships between topography and gravity explain the physical meaning of coherence (Fig. 8.37). They are based on the theoretical solution for bending of an infinite elastic plate that is subject to a periodic load. The solution for plate deflection $h(k)$ is given by:

$$h(k) \sim A\Phi(k) = A\left[1 + k^4 \frac{D}{\Delta\rho g}\right]^{-1} \tag{8.52}$$

where $\Delta\rho$ is the density contrast (assumed to be constant) between the mantle and the load, k is wavenumber ($k = 2\pi/\lambda$, where λ is the wavelength along the plate), D is flexural rigidity, and A depends on $\Delta\rho$ and the amplitude (height) of the topographic load. The parameter

$$\Phi(k) = 1/[1 + k^4\beta^4] \tag{8.53}$$

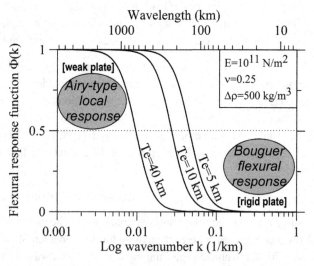

Fig. 8.37   Flexural response function (eq. 8.53) which characterizes the ratio of the output signal (topographic deflection) to the input signal (the load). The example is computed for bending of an infinite elastic plate under periodic load with wavelength $\lambda = 2\pi/k$. At $\lambda \to \infty$, the lithosphere is mechanically weak and its response to loads is local (Airy local compensation mechanism). At $\lambda \to 0$, the lithosphere appears rigid exhibiting regional compensation with flexural (Bouguer) response to loads.

characterizes the ratio of the output signal (topographic deflection) to the input signal (the load) and is called the *flexural response function* (compare with eq. 8.44).

As illustrated by Fig. 8.37, at long wavelengths ($\lambda \to \infty$) the lithosphere behaves as mechanically weak. Long-wavelength topographic loads produce large compensating roots (and thus large gravity anomalies) due to lithospheric flexure, while long-wavelength internal loads (caused by variations in crustal density and thickness) can produce surface topography (the Airy-type local compensation mechanism). This means that at long wavelengths, the observed Bouguer gravity anomalies and surface topography are strongly correlated and the coherence between them is high ($\Phi(k) \to 1$). In contrast, at short wavelengths ($\lambda \to 0$), the lithosphere behaves as mechanically strong regardless of its actual flexural rigidity; thus both surface and internal loads can be compensated by stresses within the plate. In this case, topographic loads will not necessarily produce significant local Bouguer gravity anomalies, while internal loads will produce gravity anomalies, but not topography. As a result, at short wavelengths the coherence between the Bouguer gravity anomalies and surface topography can be close to zero ($\Phi(k) \to 0$), i.e. they are statistically uncorrelated.

The wavelength of topographic loads, λ, which corresponds to the transition from coherent to non-coherent behavior of gravity anomalies and topography, reflects the wavelength for which surface and internal loads are supported by the finite flexural strength of the lithosphere and characterizes flexural rigidity of the elastic plate. Assuming $\Phi(k) = 0.5$ (dashed line in Fig. 8.37), eq. 8.52 yields:

$$\lambda = 2\pi/k = 2\pi(D/\Delta\rho g)^{1/4}, \; \Phi(k) = 0.5. \tag{8.54}$$

As before, $\Delta\rho$ is the density contrast at the interface where all load is compensated, and Te and D are given by eq. 8.48. Note that the choice of $\Phi(k) = 0.5$ is somewhat arbitrary, and in the case of $\Phi(k) = 0.4$ or 0.6 the corresponding λ will be 10% less or 10% greater than given by eq. 8.54, and the corresponding values of Te will be 13% less or 13% greater, respectively.

For a region with a non-uniform elastic thickness, the coherence method gives the weighted average value of λ for the entire region of interest. Practical application of coherence analysis is illustrated by Fig. 8.38 for two locations in western Canada. Note that in areas with thick elastic plates (e.g. continental shields), the difference between observed and predicted coherence becomes insensitive to variations in elastic thickness and the flexural parameters cannot be reliably defined (see Section 8.2.5 for details).

8.2.5 Limitations, problems, and uncertainties

Te uncertainties

Numerous studies of effective elastic thickness have demonstrated that Te is controlled by the rheology of the crust and the upper mantle, the distribution of loads, the presence of faults, the local curvature of the plate related to the bending stress and surface topography, and the isostatic postglacial rebound. Because of a large set of model assumptions and non-uniqueness of gravity modeling, the true Te uncertainty can hardly be estimated. According to Bechtel *et al.* (1990), the resolution of Te estimates is logarithmic,

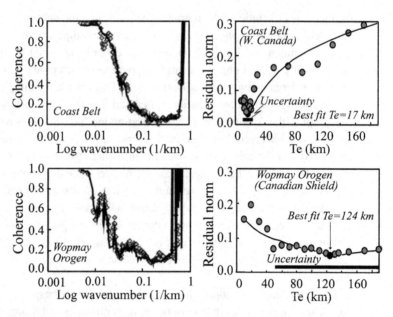

Fig. 8.38   Coherence plots and error functions for two sites in western Canada (after Flück *et al.*, 2003). The coherence plots (left) show the observed coherence between Bouguer gravity and topography (gray diamonds, with best fit predicted shown by black line). Function *Te* is defined by the transition from low to high coherence. Error plots (right) show the residual norm of the difference between observed and predicted coherence (gray circles, with best fit error function shown by black line) and the uncertainty in *Te* estimates (black bars).

and in spite of a huge range of parameters controlling effective elastic thickness of the lithosphere, Burov and Diament (1995) guesstimate the accuracy of most Te values to be ~25%. They suggest that a large part of the uncertainty results from variations in the assumed values of E and v, that typically lead to 10–20% scatter in Te estimates. For this reason, the uncertainty in estimates of D is less than for Te and should result mostly from neglecting external horizontal far-field stresses (Turcotte and Schubert, 1982) or displacements on faults (Bechtel *et al.*, 1990). In reality, the values of Te reported by different authors for the same tectonic structures differ by more than 25%, suggesting that this accuracy/ uncertainty estimate is far too optimistic (e.g. see Fig. 8.38, bottom right plot). Synthetic tests further demonstrate limitations of Te estimates (see discussion below).

The recovery of Te can be checked by calculating synthetic topography and gravity data for uncorrelated surface and subsurface mass loads placed on an elastic plate (e.g. Macario *et al.*, 1995; Pérez-Gussinyé *et al.*, 2004). Such recovery tests are performed with uniform (Fig 8.40a), periodic or non-periodic spatially variable synthetic Te values. In recovery tests with spatially varying Te, the synthetic input Te model often has a radial structure with a relatively strong region in the center and Te values decreasing outwards (Fig. 8.39). The results obtained with a wavelet method (which provides improved spatial resolution for recovering variations in Te as compared with traditional spectral methods) demonstrate limits of spectral methods in recovering spatial variations in Te:

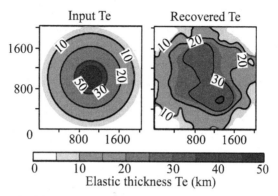

Recovered *Te* (right) from the input *Te* model (left) calculated by inverting local coherence with theoretical curves (after Audet and Mareschal, 2007).

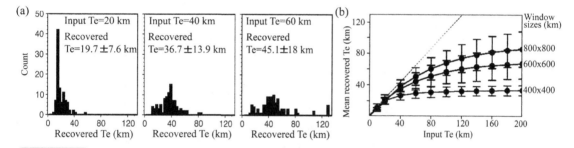

Uncertainties in *Te* estimates (from Perez-Gussinye *et al.*, 2008, 2009). (a) *Te* recovery as a function of the true (input) *Te* value based on tests with synthetic data for a window size of 600 × 600 km. (b) *Te* recovery as a function of the averaging window based on tests for synthetic topography and gravity data with no correlation at any wavelength. Circles – mean *Te* values based on 100 experiments, bars – standard deviations. Tests for synthetic data generated with initial loads correlated with $r = 0.66$ at wavelengths $\lambda \sim 100$ km plotted exactly on top of the *Te* lines indicating that, in the case of short-wavelength correlation, the *Te* estimates are not biased either for any window size or for the input *Te* value. Both (a) and (b) plots illustrate that the recovery of *Te* reduces with the increase in the true *Te* value. For large *Te* values (> 60–80 km), true *Te* values cannot be recovered even for large averaging windows (dashed line in (b) corresponds to the case of recovered *Te* equal to true *Te*).

- spectral methods can reliably recover high *Te* values only in the structures with wavelength longer than the transition wavelength from low to high coherence (Audet and Mareschal, 2007).

This conclusion is illustrated by an example when true *Te* varies from 50 km to 0 km over a distance of 1200 km, while recovered *Te* represents a weighted average of the true *Te* distribution. Thus, the accuracy of Te amplitudes recovered by gravity inversion depends strongly on the size of the study area as compared to the transition wavelength from coherent to non-coherent behavior of gravity anomalies and topography (Fig. 8.37). Tests with synthetic data also demonstrate limitations in lateral resolution: for example, the true (input) Te anomaly with a lateral dimension of 100 km is recovered as a ~300 km-wide Te anomaly (Pérez-Gussinyé *et al.*, 2009).

A large number of other factors further limit accuracy and resolution of flexural isostasy models and recovered Te values. They are discussed in detail in the rest of this section.

Uncertainties of Te estimation by spectral methods

Within the classical approach of spectral gravity analysis, practical applications of admittance and coherence analyses are based on standard spectral techniques which include the Fourier transform (FT) and the maximum entropy method (MEM). Fourier transform and MEM allow for estimation of the frequency characteristics of gravity admittance and coherence by using the wavelet transform (King, 1998; Stark and Stewart, 1997) and the multitaper technique (Thomson, 1982; Thomson and Chave, 1991; Percival and Walden, 1993). The isotropic Fourier transform allows us to estimate only one coherence (and admittance) for a data window. Since these parameters are wavenumber dependent, all spatial information is irretrievable. One of the first attempts to map Te variations in North America by calculating the coherence for 50 overlapping areas is based on this approach (Bechtel *et al.*, 1990); the resultant map of Te has low lateral resolution.

Application of the MEM or the windowed Fourier transform (WFT) (e.g. Lowry and Smith, 1994) allows us to overcome limitations in the isotropic Fourier transform method and to improve the spatial resolution of coherence analysis. The WFT approach is based on calculations of gravity admittance and coherence within moving areal windows; mapping Te values estimated for each moving window gives information on lateral variations of Te. Windowed Fourier transform has several limitations (Macario *et al.*, 1995):

- the window size should depend on lithosphere strength; strong plates with large Te require large windows because they have large transition wavelength from coherent to non-coherent behavior of gravity anomalies and topography (Fig. 8.37);
- WFT works best for data with a narrow wavelength band and it is not efficient for data with broadband signals;
- there is a trade-off between window size and lateral resolution: when windows are too small, they produce flawed spatial variations of Te, while very large windows smooth true Te structure by averaging spatially varying Te values.

The effect of window size on Te recovery has been examined in a number of tests calculated from synthetic topography and gravity data (Macario *et al.*, 1995; Wang and Mareschal, 1999; Ojeda and Whitman, 2002; Audet and Mareschal, 2007; Pérez-Gussinyé *et al.*, 2009). The results (illustrated by Figs. 8.40 and 8.41) can be summarized as follows:

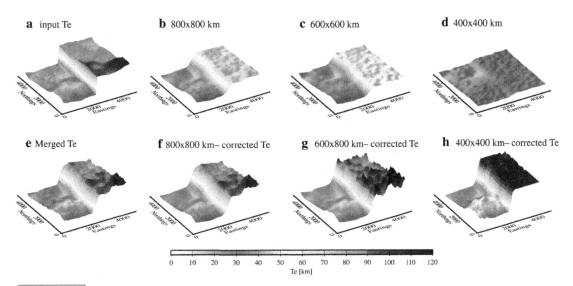

a input Te   **b** 800x800 km   **c** 600x600 km   **d** 400x400 km

e Merged Te   **f** 800x800 km– corrected Te   **g** 600x800 km– corrected Te   **h** 400x400 km– corrected Te

0   10   20   30   40   50   60   70   80   90   100   110   120
Te [km]

Fig. 8.41   Te recovery of the true (input) Te value (a) as a function of window size (b–d) based on tests for synthetic topography and gravity data with no correlation at any wavelength (from Perez-Gussinye et al., 2009). Lower panels (e–h) show Te recovery using sliding, overlapping windows with panel (e) showing the merged results for windows of three different sizes (f–h).

- All spectral methods have large errors when Te is high, because large multitaper window or wavelet sizes are required to resolve the flexural wavelength and the ratio [window size]/[flexural wavelength] decreases with increase in the true Te value.
- For large Te values ($>60–80$ km), true Te values cannot be recovered even for large averaging windows. This questions the principal ability of the WFT technique to recover large values of Te (Fig. 8.41).
- The number of outlier estimates (with recovered Te > 130 km) increases with the true Te increase. This results in a flat misfit curve between the observed (recovered) and predicted (input) coherence or admittance at high Te values, so that the best fit Te is effectively unconstrained (Fig. 8.38, bottom right).

These results have been obtained with multitapers, which are known to have an inherent bias: the spectral estimate at any given wavenumber k includes information from neighboring wavenumbers in the range $k \pm W$. Because of leakage from spectra at $(k-W)$ at wavenumbers k near the rollover, the estimated coherence will be larger than the true coherence, while the estimated admittance will be smaller than the true admittance. These effects can bias estimates of both coherence and admittance. Thus, Te estimates are biased toward low values when the observed multitaper admittance is compared to the theoretical admittance (McKenzie, 2003). The small size of averaging windows and large true Te values enhance this bias. For example, a test with synthetic data calculated for the true Te = 80 km yields a best-fit value of Te ~ 35 km unless the windows are larger than 2000×2000 km (Pérez-Gussinyé et al., 2009).

Estimation of *Te* using sliding, overlapping windows of different sizes can better recover both high and small *Te* values in a spatially variable *Te* structure. In this approach, as illustrated by Fig. 8.41, corrected *Te* are calculated using sliding, overlapping windows weighted with a Gaussian taper. The curves are centred on the *Te* value mid-way between the maximum *Te* recoverable with a given window size and the maximum *Te* recoverable with the next smallest window size. Merging the corrected results obtained with windows of three different sizes allows for improved recovery of the input *Te* structure.

The wavelet method provides improved spatial resolution for recovering variations in *Te* as compared with traditional spectral methods. In contrast to FT, where the coefficients depend on wavenumber only, wavelet coefficients depend both on wavenumber and spatial coordinates (Grossman and Morlet, 1984; Kirby, 2005). Thus 2D continuous wavelet transform allows for mapping of the directional variations of flexural rigidity (anisotropic *Te*) (e.g. Kirby and Swain, 2006, 2009; Audet and Mareschal, 2007). A further advantage of the wavelet approach is that the data need not be windowed and the full spectrum is recovered since the wavelet spans the whole study area. This is particularly advantageous for regions with large flexural wavelengths that cannot be recovered by the WFT (e.g., multitapers) method since the choice of window size in multitaper methods limits the bandwidth of the spectrum. The limitations of spectral methods in recovering spatial variations in *Te* are discussed above (Fig. 8.39).

Non-uniqueness of inverse gravity modeling

When discussing the flexural structure of the lithosphere, it is often forgotten that solution of the inverse gravity problem is intrinsically non-unique. This implies that flexural gravity models constrained

> for the a-priori chosen narrow class of possible solutions can confirm the ideas of their authors but have nothing to do with reality (Artemjev and Kaban, 1991).

In other words, it is always possible to constrain a large number of flexural and density models that will closely match the observed gravity field, but nonetheless will not match to the true density and flexural structure of the lithosphere. This situation is similar to seismic tomography where checkerboard tests are commonly used to verify tomography model solutions: although very popular, such tests can neither check the uniqueness of this solution, nor can they test the effect of errors in the theory as they are computed with the same theoretical model, assumptions, and simplifications as those used for the actual seismic tomography inversion. A detailed critical analysis of the assumptions on which the concept of flexural isostasy is based is provided in Section 8.2.3.

Correlation of internal and topographic loads

Forsyth's (1985) coherence method is based on the assumption that surface and subsurface loads are statistically uncorrelated. Estimates of *Te* are biased if this condition is not satisfied (as one may expect in magmatic provinces where mafic intrusions into the crust or crustal underplating are often correlated with topography) (Macario *et al.*, 1995). An extensive analysis of the effect of load correlation on Te estimates has been performed by Kirby and Swain (2009), who showed that wavelengths for which initial load correlation occurs can be identified from the real part of the complex Bouguer coherency. Pérez-Gussinyé *et al.* (2009) applied this approach to gravity data analysis for the East African Rift and concluded that the wavelength of correlation of surface and subsurface loads (associated with Cenozoic magmatism and volcanism) does not exceed 100 km and as such does not have a significant effect on Te estimates, since the waveband of load correlation is much smaller than the Te transitional wavelength.

Effects of crustal structure and sediment fill

Lateral variations in the thickness and density of sediments and individual crustal layers produce internal gravity loads and may have a significant effect on the results of gravity admittance and coherence. Furthermore, a thick sedimentary cover has a thermal insulation effect that may significantly weaken crustal rheology, leading to crust–mantle decoupling which occurs when bending stresses in the lower crust exceed the yield stresses of the lower crustal material.

Effect of crustal thickness variations

The importance of crustal thickness variations as a parameter controlling flexural strength of continental lithosphere has been examined by Lavier and Steckler (1997). They used a semi-analytical model for yield stress envelope parameterization with account taken of the effects of crust–mantle decoupling to analyze Te estimates for foreland basins and mountain belts. The results indicate that a change in the crustal thickness from 35 km to 40 km (with the geotherm fixed to correspond to a 200 km-thick continental lithosphere) leads to a hotter lower crust, so that the lithosphere with thicker crust decouples for a smaller load (or bending moment) than the lithosphere with the same geotherm but with a thinner crust. As a result, Te for a 40 km-thick crust model is 20 km less than for a 35 km-thick crust (Fig. 8.42).

A similar problem (the effect of crustal thickness variations on Te) but with a different approach has been addressed by Wang and Mareschal (1999). They performed sensitivity analysis for the coherence method which assumes that all internal loads (within the crust) are compensated at the same depth, while seismic data from the Canadian Shield indicate that lateral variations in Moho depth reach 10–15 km. Because of the low-relief topography of the shield, the ratio of internal to surface loads is regionally > 10 and locally (at wavelengths < 50 km) as high as 100. In contrast to Lavier and Steckler (1997), the analysis by Wang and Mareschal (1999) based on Bouguer gravity data indicates that, in cold cratonic lithosphere with no crust–mantle decoupling, a change in the compensating depth for internal loads (assumed to be at the Moho) from 40 km to 20 km results in a change of Te of less than 10% (which is equivalent to Te variations of 7–13 km for Te values reported for the Canadian Shield).

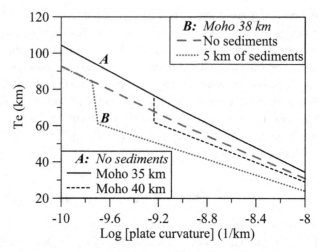

Fig. 8.42 Effects of crustal thickness variations (models A) and thickness of sediments (models B) on effective elastic thickness of the lithosphere Te (after Lavier and Steckler, 1997). For the same lithospheric geotherm and lithosphere rheology, thick crust or thick sedimentary cover cause crust–mantle decoupling and reduce Te (by ~20 km in models A and B calculated for a 200 km thick-lithosphere).

Effect of sediments for flexure models of orogens

Lavier and Steckler (1997) also examined the significance of sediment fill and found that it weakens the lithosphere and thus has a strong effect on estimated Te values. This weakening occurs, first, due to a significant density contrast between the sediments and the crystalline crust and, second, because the sediment cover thermally insulates the lower crust. At high bending stresses, the model with a 33 km-thick basement covered by 5 km of sediments (representative of the Zagros mountains) undergoes crust–mantle decoupling and has Te values 22 km lower than a similar lithosphere model with a 38 km-thick crust but without sediments (Fig. 8.42). The study of Lavier and Steckler (1997) has demonstrated that the thickness of sediment cover is likely to control the thickness of the effective elastic lithosphere Te in continental regions with a thin (35 km) crust, a thick (greater than 3–5 km) sedimentary layer, and cold conductive geotherm. Such rheological conditions are typical for some types of basins and continental margins. However, in young continental orogens with thick and hot crust (such as the Apennines, Eastern Alps, Western Alps), crust–mantle decoupling takes place regardless of the effect of the sediments.

Many studies of flexural rigidity of the continental lithosphere performed in the 1980s completely ignored the huge effects of sedimentary masses and crustal thickness variations in orogens (e.g. Kogan and McNutt, 1986; McNutt and Kogan, 1986; McNutt et al., 1988). As a result, lithosphere rigidity was significantly overestimated for many continental regions, such as most of the former USSR territories, including most orogens of Eurasia. These and similar unreliable results are excluded from the discussion in Section 8.3.2.

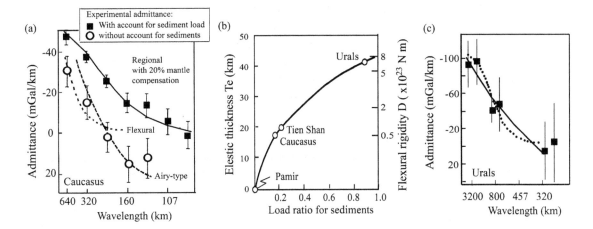

(a) Admittance for the Caucasus. Symbols and error bars – experimental admittance: open circles – based on topography and Bouguer anomalies; solid squares – based on corrected topography (after "compression" of sediments to density of 2670 kg/m^3) and on gravity anomalies corrected for anomalous density of sediments. Lines – theoretical admittances: short dashes – elastic flexure model ($Te = 25$ km), long dashes – Airy model ($Z_M = 45$ km) with account taken for the additional load of low-density sediments; solid line – model of local compensation, in which 70% is achieved by variations in crustal thickness, 20% – by density variations in the layer between the Moho and 100 km depth, and 10% is due to density contrast between the upper and the lower crust. Z_M – compensation depth (which in most cases is the Moho depth). (b) Dependence of flexural rigidity D on ratio f^* of the load produced by anomalous density of sedimentary masses to topographic load. (c) Admittance for the Urals. Symbols and error bars – experimental admittance; dotted line – theoretical admittance for flexure model with $D = 8 \times 10^{23}$ Nm, solid line – theoretical admittance estimated from eq. (8.55) for Airy-type model with account taken for anomalous density of sediments ($f^* = 0.9$) (after Artemjev and Kaban, 1991).

Problems related to the effect of thick sedimentary cover on flexural estimates are illustrated by examples from several orogenic belts in Eurasia, where thick layers of low-density sediments fill foredeeps and intermountain basins, as in the Caucasus and the Tien Shan orogens (Artemjev and Kaban, 1991). As in most orogens worldwide, the thickness of low-density sediments in the mountain regions of Eurasia is well correlated with topography. Topography adjusted for the anomalous density of sediments (i.e. when low-density sediments of the intramountain basins are "condensed" to a typical sediment density of 2670 kg/m^3) differs significantly from the real topography, and it is this corrected topography that has to be considered as the external load in flexural modeling. The results for the Caucasus (Fig. 8.43a) show a significant difference between gravity admittance calculated from real topography and Bouguer anomalies, on one side, and admittance calculated from corrected topography and gravity anomalies corrected for the anomalous density of sediments, on the other side. It is obvious that the use of uncorrected admittance, which ignores data on crustal density inhomogeneities, would lead to completely false conclusions about the model of compensation and would predict $Te \sim 25$ km.

Neglecting the effect of a thick layer of low-density sediments that are usually well correlated in thickness with the topography, and the corresponding gravity effect, has led to erroneous conclusions about the existence of a correlation between the elastic thickness of the lithosphere and the age of the continental orogens (McNutt, 1980; McNutt and Kogan, 1986). The anomalous-density load of sediments is equivalent to additional topography. Assuming that the total load of topography and anomalous masses is compensated by the Airy-type mechanism and the compensation depth Z_M is significantly larger in value than topography, the admittance for the Bouguer anomalies is:

$$Z_B^*(k) = 2\pi G \rho_0 [f^* - (1 + f^*) \exp(-2\pi k Z_M)] \tag{8.55}$$

where ρ_0 is the average density of basement rocks, G is the gravity constant, k is wavenumber, and f^* is the ratio of the load produced by anomalous density of sedimentary masses to the topographic load (Artemjev and Kaban, 1991). This equation allows for calculation of f^* which gives $Z_B^*(k)$ close to the admittance for the flexure model with a given value of flexural rigidity D (or Te). The dependence between D and f^* is shown by the solid line in Fig. 8.43b. The same figure shows the values calculated for different orogens in Eurasia (the results of McNutt and Kogan, 1986; Kogan et al., 1987). While the D value reported for the Pamir is correct (there are no low-density sediments in that region), the D (or Te) values reported for other orogens are erroneous. As Fig. 8.43b shows, the apparent D-values calculated by flexural models can be explained within regional Airy-type compensation models assuming that the ratio f^* of the load produced by anomalous density of sedimentary masses to the topographic load is c. 20–30% in the Caucasus and the Tien Shan and c. 90% in the Urals. An example for the Urals shows that the regional gravity model with $f^* = 0.9$ and with account taken for anomalous density of sediments, fits well with the observed admittance data (Fig. 8.43c).

Internal loads with no topographic expression

Usually Te for continental regions is calculated by Forsyth's (1985) method based on the Bouguer coherence (Section 8.2.4). The approach has been challenged by McKenzie and Fairhead (1997) and McKenzie (2003) who proposed instead the use of free-air gravity anomalies both for the continents and the oceans. The motivation of the authors was that the too large values of Te (up to 130 km) predicted for the continental shields by the Bouguer coherence seem to contradict expectations from mineral physics on the ability of the lithosphere to support elastic stresses over geological time-scales (for example, xenolith-based cratonic geotherms predict a temperature of ~1000 °C at a 130 km depth, Fig. 5.13). In contrast, the approach based on the free-air admittance gives systematically smaller estimates of Te (in some cases, by an order of magnitude less than the traditional approach based on Bouguer gravity) that never exceed 25 km even for the Precambrian shields.

Debate

In contrast to the traditional approach, McKenzie and Fairhead (1997) and McKenzie (2003) distinguish two types of internal loads (i.e. due to variations in crustal thickness and density) within the continental lithosphere (Fig. 8.44):

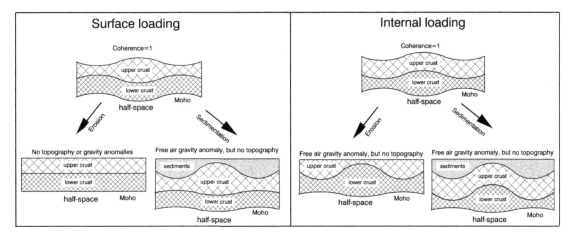

Fig. 8.44 Sketch illustrating how topographically expressed surface and internal loads can be modified by sedimentation and erosion into loads with no topographic expression, but with free-air gravity anomalies (except for the leftmost case) (after McKenzie, 2003).

- internal loads with topographic expression and
- internal loads with no topographic expression termed "noise" (in the signal processing sense) that are produced either by erosion or by sedimentation, and are especially important in continental shields with low topography.

The paper of McKenzie (2003) argued against load deconvolution methods, and particularly the coherence method, and caused a broad debate on the applicability of different approaches to flexural studies of the continental lithosphere. A detailed discussion of the controversy between Forsyth's (1985) and McKenzie's (2003) approaches can be found in Pérez-Gussinyé et al. (2004, 2009) and Kirby and Swain (2009). Some of the major points of the debate are summarized below, since they are critical to interpretations of continental flexural isostasy.

The major arguments of McKenzie and Fairhead (1997), further clarified in McKenzie (2003), are as follows.

- Forsyth's deconvolution method assumes that all loads produce topographic expressions (eq. 8.51).
- If substantial free-air anomalies (due to unexpressed internal loads) are observed in regions with almost flat topography, Forsyth's assumption (above) is not satisfied and the load deconvolution method cannot model the initial internal load. Close-to-zero coherence between free-air anomalies and topography at all wavenumbers is diagnostic of such a situation (Fig. 8.45).
- In such cases, the coherence between Bouguer anomalies and topography must also be low because of topographically unexpressed internal loads ("gravitational noise") in the gravity field.

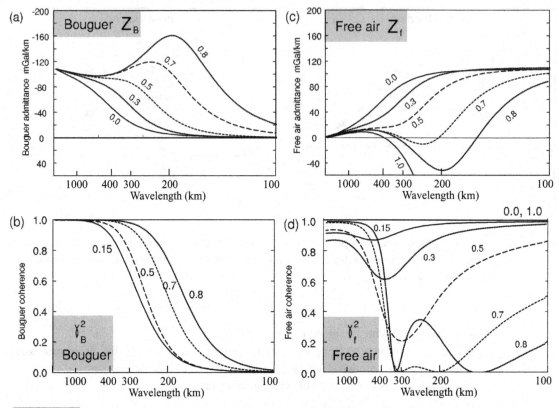

Fig. 8.45 Comparison of Bouguer and free-air admittance (top) and coherence (bottom) for a model with $T_e = 17$ km and with internal loads at the upper crust − lower crust interface (from McKenzie, 2003). Numbers − the fraction $F = f/(1 + f)$ of the internal load f at this interface to the total load (internal loads are due to density variations within the continental lithosphere; external loads are due to topography variations). Unaccounted internal loading reduces T_e by shifting coherence to shorter wavelengths (a, b). When most of the load is internal ($F \rightarrow 1$), the free-air admittance changes the sign (c) and thus it is diagnostic of the presence of topographically unexpressed internal loads. Free-air coherence (d) is low ($\rightarrow 0$) either when substantial free-air anomalies (produced by unexpressed internal loads) are observed in regions with almost flat topography ($F \rightarrow 1$) or when no internal loads are present ($F \rightarrow 0$).

The major conclusions of McKenzie's (2003) analysis are:

(1) In regions with eroded topography, the Bouguer coherence method will give the upper bounds for T_e estimates which may be many times larger than the true T_e values.

(2) The elastic thickness of continental lithosphere is less than 25 km worldwide.

(3) Surface loads rather than internal loads dominate stable continents, i.e. the loading ratio $F = f/(1+f)$ of the internal load f to the total load (note the difference with eq. 8.51) is close to zero in eroded, low-topography regions (however, unaccounted internal loading reduces T_e by shifting coherence to shorter wavelengths, Fig. 8.45b).

(4) Admittance is not biased by topographically unexpressed loads ("gravitational noise") as compared to coherence, and thus should be used in flexural isostasy studies.

(5) Flexural isostasy studies should be based on free-air anomalies, because the Bouguer anomalies cannot identify low free-air coherence at short wavelengths which are indicative of the presence of topographically unexpressed internal loads.

(6) The behavior of the free-air admittance is diagnostic of gravitational noise since it changes the sign when $F \rightarrow 1$ (most of the load is internal) (Fig. 8.45c).

These conclusions are based on the following assumptions (McKenzie, 2003):

- the internal loading interface is within the crust at 15 km depth, i.e. all density variations in the continental lithosphere are concentrated at mid-crustal depth (note that underestimating the depth of internal loading reduces the Te values recovered in the gravity inversion);
- the loading ratio f is assumed to have a uniform value (constant) over the entire region of study; in contrast to Forsyth's approach (eq. 8.51) it is not a wavenumber parameter because it is not estimated from gravity–topography data;
- the analysis is based on the multitaper spectral method with focus either on short-wavelength characteristics of free-air coherence and a free-air anomaly power spectrum, or on their average over a large bandwidth.

The application of McKenzie's (2003) approach to different continental regions worldwide gives low values of Te that are in remarkable contradiction to estimates based on Forsyth's (1985) method. This contradiction can, at least in part, arise from the non-uniqueness of the inverse gravity problem when theoretical solutions matching observed gravity are searched within the a-priori chosen class of possible isostatic models. Further, Kirby and Swain (2009) argue that while the load deconvolution method of Forsyth (1985) assumes that surface and internal loads are statistically uncorrelated (i.e. they must have a random phase difference when the complex admittance or coherence are analyzed), McKenzie's (2003) model requires that the initial loads must be correlated (i.e. have a phase difference of zero) to produce a flat surface topography after lithospheric flexure. These authors demonstrate that:

- "zero final topography" in McKenzie's (2003) model is a special case of a more general model,
- even 100% correlated initial loads do not always produce flat topography after flexure,
- initial correlation between surface and internal loads can bias Te estimates both upward or downward.

Synthetic tests

Several studies performed detailed numerical comparisons of Te estimates for the continental regions based on the load deconvolution method derived from the Bouguer coherence (Forsyth, 1985) and the free-air admittance uniform-f approach (with constant, a-priori fixed, ratio f of subsurface to surface loading, McKenzie, 2003).

Pérez-Gussinyé *et al.* (2009) compare the observed admittance/coherence functions calculated using the multitaper method with theoretical admittance/coherence curves calculated after McKenzie's (2003) approach, assuming that all subsurface loads are emplaced and compensated at the Moho. The a-priori fixed loading ratio f and Te of the theoretical admittance are varied to find a combination of these two parameters that would optimize the fit to the observed multitaper estimate of admittance. The results calculated for $f = 1$ using the coherence approach (Forsyth, 1985) and the traditional free-air admittance approach (McKenzie, 2003) demonstrate that:

– the bias and variance in the Bouguer coherence and the free-air admittance estimates increase for larger true Te and for smaller averaging windows;
– for a given size of averaging window, Te estimates based on the Bouguer coherence method have lower variance than the free-air admittance approach;
– because of a different variance of the two methods, the optimal window size for the admittance method is larger than for the coherence method, resulting in a lower spatial resolution due to smoothing (averaging of the spatially varying Te values).

Kirby and Swain (2009) treat both the admittance and coherency as complex quantities instead of using the real part of the admittance between gravity anomalies and topography in conventional spectral Te studies. Their results show that while the real parts provide estimates of Te, the imaginary parts indicate whether the inversion is biased by "gravitational noise" (topographically unexpressed loads).

They also examine the effect of gravity noise on the efficiency of Te recovery. Synthetic modeling with added gravity noise (generated by both surface and internal loads so that the final topography is zero) is performed for different structures of the North American continent. Two types of noise are modeled: type I (full-spectrum) noise generated by a load modeled as a random fractal surface with fractal dimension 2.5; type II noise generated by high-pass filtering of the load which produced the type I noise. The noise is added to the observed gravity, while the topography is left unaltered. Synthetic modeling is performed for plate models with a uniform Te and with Gaussian Te by two methods: the deconvolution method based on the Bouguer coherence (Forsyth, 1985) and the free-air admittance with constant loading ratio f (McKenzie 2003). The results of this modeling demonstrate the following (Kirby and Swain, 2009):

– In regions with very low-relief topography, even in the absence of added noise, the Bouguer coherence rollover is shifted towards longer wavelength than the theory predicts. When type I noise is added, the effect is even stronger, providing support to McKenzie's (2003) suggestion that the Bouguer coherence method can give unreliable Te estimates for flat regions with topographically unexpressed internal loads. However, the effect is weak when type II (high-pass filtered) noise is added to the gravity signal.
– Very-low-relief topography (such as in cratonic regions) can bias Te estimates towards high values, even in the absence of added noise; this overestimation affects both methods. When type I noise is added, the overestimation increases. In contrast to the conclusion of McKenzie (2003), the free-air admittance can become as biased as the Bouguer coherence. When type II noise is added, the bias is considerably less than in the case of full-spectrum noise; for the free-air admittance it is lower on average, but has larger extremes and apparently does not correlate with regions of low topography. This implies that the Bouguer coherence is almost unaffected by the presence of band-limited noise.

- The analysis of misfits (Fig. 8.46) indicates that for free-air admittance with uniform-f distribution the 99% Te confidence intervals almost always contain the true Te value for type II noise, but not for type I noise. Type I noise leads to overestimated Te in the Bouguer coherency method.
- Te calculated from the inversion of admittance generally has a better spatial resolution than Te calculated from coherence.
- Misfit analysis demonstrates that (depending on the gravity signal, surface topography, and the structure and geometry of internal loads) either the Bouguer coherence or the free-air admittance method provides more reliable and less biased results. In particular, flexural isostasy analysis for North America indicates that while the Bouguer coherence method provides better estimates for the north-central Canadian Shield, free-air admittance is more reliable for the Superior Province and the Midcontinent Rift.
- In contrast to McKenzie's (2003) conclusion number 3 (above), subsurface loading is important in most parts of the North American continent (see Section 8.3.2).
- In contrast to McKenzie's (2003) conclusion number 2 (above), large values of Te (> 100 km) are found for significant parts of the Canadian Shield. However, the Te values are less constrained than the F values (Fig. 8.46).

8.3 Lithosphere flexure and rheology

8.3.1 Flexure models of oceanic lithosphere

Lithosphere flexure under volcanic islands

The concept of lithosphere flexure was initially applied to island chains and seamounts, where it was very successful in explaining deviations in ocean bathymetry from the square-root of ocean floor age predictions (Fig. 4.21). The presence of volcanic islands causes flexural deformation of the oceanic lithosphere. If flexure of oceanic lithosphere is approximated by bending of an elastic plate, the maximum amplitude of plate deflection in direction z (downwards) (Turcotte and Schubert, 2002, p. 125–126) is:

$$w_o = V_o \alpha^3 / 8D \qquad (8.56)$$

where the flexural parameter α is:

$$\alpha = [4D/\Delta\rho g]^{1/4} = \sqrt{2}\beta, \qquad (8.57)$$

$\Delta\rho$ is the density contrast between the crust and the mantle (the load and the beam), and D is the flexural rigidity (eq. 8.48). The plate is assumed to be infinite in direction y, and to be thin compared to its width ($h \ll L$). The depression has a half-width of x_o, and the forebulge has a maximum height w_b at a distance x_b from the line load V_o:

$$x_o = 3/344\ \pi\alpha,\ x_b = \pi\alpha,\ w_b = \sim -0.0432\ w_o. \qquad (8.58)$$

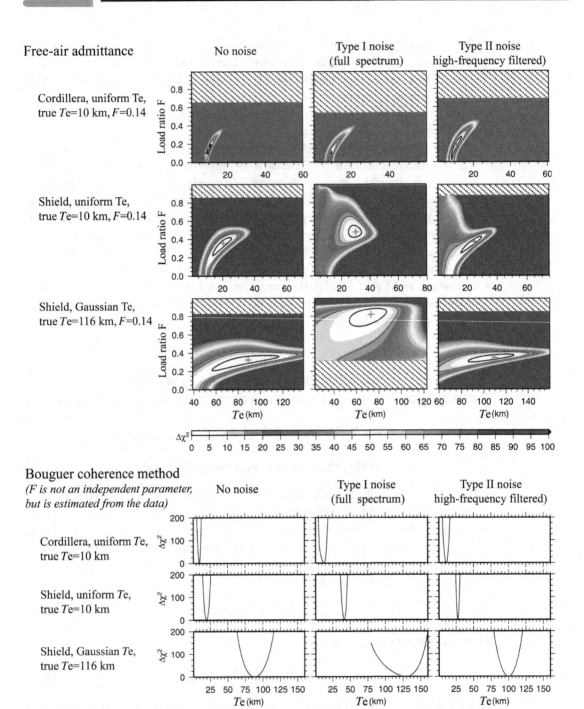

Fig. 8.46 Misfits χ^2 between the observed and predicted Bouguer coherency and free-air admittance with uniform-f distribution ($f = F/(1-F)$ where F is the ratio of internal to total load) at two locations in North America calculated for uniform Te and Gaussian Te plates (from Kirby and Swain, 2009). The boxes for free-air admittance misfits show misfit surfaces as a function of Te and F; crosses mark the best fitting Te and F values, thick and thin black contours show the 95% and 99.99% confidence limits. Curves for the Bouguer coherency show Te misfits. "Noise" refers to gravity signal produced by internal loads with no topographic expression.

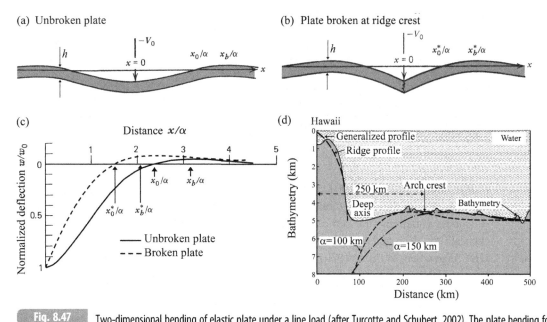

Fig. 8.47 Two-dimensional bending of elastic plate under a line load (after Turcotte and Schubert, 2002). The plate bending forms the depression with half-width x_0 and the forebulge which has a maximum height at distance x_b from the line load V_0. (a) Deflection of the elastic plate; (b) deflection of a broken elastic plate; (c) half of the theoretical deflection profile for cases (a, b) normalized by the maximum amplitude of deflection at $x = 0$ (eqs. 8.58–8.59); (d) observed bathymetry across the Hawaiian Ridge, deep and arch and theoretical profiles calculated for the broken plate (after Walcott, 1970). Parameter α depends on flexural rigidity (eq. 8.57).

In the case where the elastic plate is fractured along the line of the ridge (i.e. the case of a free plate edge) (Fig. 8.47), the maximum amplitude of plate deflection w_0^* is double the deflection amplitude of the unbroken plate and the maximum height of the forebulge w_b^* is larger, while the horizontal scales of the depression x_0^* and the forebuldge x_b^* are smaller than for the unbroken lithospheric plate:

$$w_0^* = 2\,w_0, \quad w_b^* = {\sim}\,{-}0.0670\,w_0 \sim 1.55\,w_b,$$
$$x_0^* = \pi\alpha/2 \sim 0.67\,x_0, \quad x_b^* = 34\pi\alpha = 0.75\,x_b. \tag{8.59}$$

This analysis has been applied to the Hawaiian Islands (Gunn, 1943), although the island load there is distributed over the area with a size of ~150 km (strictly speaking, not a line/point load). The plate thickness can be estimated by assuming that the volcanic load is applied in the center of the area, and that the distance from this center to the crest of the arch corresponds to x_b. The width of the arch in the Hawaiian Islands is commonly estimated to be about 500 km ($x_b \sim 250$ km) (e.g. Dietz and Menard, 1953); but accurate estimates are not possible because of the very gentle slopes on the outer edge of the arch. The predicted effective elastic thickness of the lithospheric plate is ~34 km in the case of an unbroken plate and ~49 km if the oceanic plate is fractured along the line of the ridge (Table 8.3). Figure 8.47d illustrates the results calculated for the case of a broken plate for two fixed values of flexural parameter α; they predict the amplitudes of the peripheral bulge to be

Table 8.3 Flexural parameters for bending of oceanic lithosphere

Location	x_b (km)	α (km)	D (N m)	Te (km)
Hawaiian Islands, unbroken plate	250	80	2.4×10^{23}	34
Hawaiian Islands, broken plate	250	106	7.26×10^{23}	49
Mariana trench	55	70	1.4×10^{23}	28

Based on eqs. 8.56–8.59, assuming the density contrast between water and mantle $\Delta\rho = 2300 \, \text{kg/m}^3$, $E = 70 \, \text{GPa}$, and $\nu = 0.25$.

600 m and 550 m for $\alpha = 100$ km and 150 km, respectively. These values are in reasonable agreement with the observations. By comparing model predictions with observed crustal thickness variations (the vertical Moho displacement by the load of the ridge is ~ 2 km beneath the deep and ~10.5 km beneath the ridge itself) and gravity variations across the Hawaiian chain, Walcott (1970) favored a model of a broken plate with $\alpha = 125$ km.

Ocean trenches

Lithospheric flexure at ocean trenches has been extensively studied starting from the early days of plate tectonics (e.g. Hanks, 1971; Watts and Talwani, 1974; Caldwell *et al.*, 1976; Parsons and Molnar, 1976; Turcotte *et al.*, 1978; Forsyth, 1982). Ocean trenches, a distinctive morphological feature of plate boundaries, form along convergent plate boundaries and mark the position at which the flexed oceanic lithospheric plate begins to subduct. Bending of the oceanic plate starts before subduction begins. Since the lithosphere is stronger in compression than in tension (Fig. 8.6), oceanic lithosphere has higher bending strength in a concave upward mode than in a concave downward mode (McNutt and Menard, 1982). Under the assumptions of flexure of a thin elastic plate, the half-width of the forebulge (which together with the height of the forebulge are the only parameters that can be measured directly) is $x_b - x_o = \pi\alpha/4$ (Fig. 8.48a). Since flexural parameter α is directly related to the flexural rigidity of the plate (eq. 8.57), and thus, to the elastic thickness of the oceanic lithosphere, the parameter $x_b - x_o$ allows for estimation of Te.

A comparison of model predictions with the bathymetric profile for the Mariana trench demonstrates that bathymetry can be explained by the bending of an elastic plate with $x_b = 55$ km, $w_b = 0.5$ km, and $\alpha = 70$ km, which corresponds to $Te \sim 28$ km (Table 8.3). Maximum bending stress of ~900 MPa is expected at a distance of 20 km seaward of the trench axis. It is unlikely that such high values of deviatoric stresses can be maintained at shallow depths; they are more plausible at larger depths with high lithostatic pressure.

A similar analysis for the Tonga trench, however, fails to explain a large curvature within the trench by elastic plate approximation. A model with $x_b = 60$ km, $w_b = 0.2$ km provides a fit to bathymetry data only from a distance of $x = -60$ km seaward of the trench axis (Fig. 8.48d). Fast bathymetry deepening in the direction of subduction can be predicted by plastic hinging that develops during deformation of a plate with elastic–perfectly plastic rheology. Plates with such rheology deform elastically at stresses lower than the yield stress and without limit (plastically) after the stress exceeds the yield stress. The transition between

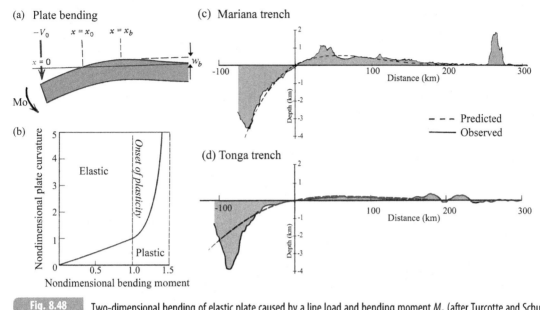

Fig. 8.48 Two-dimensional bending of elastic plate caused by a line load and bending moment M_0 (after Turcotte and Schubert, 2002). The plate bending forms the depression with half-width x_0 and the forebulge which has a maximum height at distance x_b from the line load V_0 bending moment M_0. (a) Sketch of elastic plate bending; (b) dependence of plate curvature on the bending moment for elastic–perfectly plastic rheology; the bending moment is normalized by its value at the onset of plasticity; (c–d) observed and predicted lithosphere bending at ocean trenches. Bathymetry at the Mariana trench can be explained by the bending of an elastic plate, while a large curvature in the Tonga trench can be predicted by plastic hinging, but cannot be explained by elastic plate approximation.

the two deformation styles is marked by a rapid increase in plate curvature referred to as a plastic hinge, which in the Tonga trench develops at $x = -60$ km.

The effective elastic thickness depends on plate curvature and the bending moment: increasing either bending moment or radius of the plate curvature results in a decreasing effective elastic flexural rigidity and effective elastic thickness (e.g. Fig. 8.29). Calculations of the maximum bending moment of the oceanic lithosphere indicate that it is loaded to its limit at most trenches (Fig. 8.49). The exception is the Izu-Bonin trench where the maximum bending moment is significantly smaller than the limit bending moment calculated for a flexure model of 80 Ma-old oceanic plate with dry olivine rheology subject to a uniform strain rate of 10^{-14} s^{-1}. Since the oceanic lithosphere subducting in the Izu-Bonin trench is significantly older (140 My) than used in modeling, it can be significantly stronger than predicted.

Normal oceans

Age dependence

The thermal regime plays a critical role in controlling lithosphere rigidity as both crustal and upper mantle rheologies are strongly temperature-dependent (see Section 8.1.4). Since the

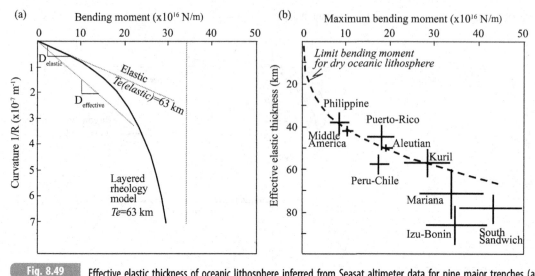

Fig. 8.49 Effective elastic thickness of oceanic lithosphere inferred from Seasat altimeter data for nine major trenches (after McAdoo *et al.*, 1985). (a) The relationship between *Te*(elastic) and *Te*(YSE) for ~120 Ma-old plate (eq. 8.48a). Increasing the bending moment results in a decreasing effective elastic flexural rigidity $D_{effective}$ and, as a result, in a decreasing effective elastic thickness. When the plate reaches the limit bending moment, $D_{effective}$ and *Te*(YSE) → 0. (b) The maximum bending moment of the oceanic lithosphere. Dashed line − the limit bending moment calculated for a flexure of 80 Ma-old oceanic plate with dry olivine rheology subject to a uniform strain rate of 10^{-14} s^{-1}. At most trenches, oceanic lithosphere is loaded close to its limit.

thermal state of normal oceanic lithosphere is controlled by its age (Fig. 4.25), flexural rigidity (and elastic thickness) of the oceanic lithosphere is expected to increase with ocean-floor age (Fig. 8.30) (Caldwell and Turcotte, 1979). For the same reason, flexural rigidity of the continental lithosphere is expected to depend on its tectono-thermal age (Kusznir and Karner, 1985), but because of the complicated thermal structure of the continental lithosphere, this trend is less apparent than in the oceans.

Numerous studies of elastic thickness of the oceanic lithosphere show that, indeed, it depends on ocean-floor age (and through it on the geotherm) (Caldwell and Turcotte, 1979). For normal oceans (i.e. parts of oceans where bathymetry follows the square-root-of-age relationship), typical values of *Te* are ~20 km for 20 Ma-old lithosphere and increase to 40–50 km in >100 Ma-old lithosphere (Figs. 8.30, 8.50b). For example, assuming that mantle temperature follows the plate model, McAdoo *et al.* (1985) estimated *Te* to be 40, 56, 64, and 68 km for 40, 80, 120, and 160 Ma-old oceanic plates, respectively, with a dry olivine rheology (activation energy $E^* = 525$ kJ/mol) subject to a uniform strain rate of 10^{-14} s^{-1}. These results are in general agreement with estimates of the lithosphere strength envelopes constrained by experimentally determined rheologies (e.g. Kohlstedt *et al.*, 1995), which show that for a 60 My-old oceanic lithosphere with dry rheology the transition from brittle deformation to plastic flow, accompanied by a rapid decrease of lithospheric strength, occurs at a depth of ~40 km (Fig. 8.32a).

This, usually unquestionably accepted, result has been challenged recently (Fig. 8.50b). Bry and White (2007) used a global database of bathymetric and free-air gravity profiles

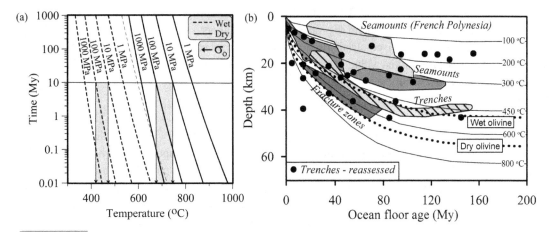

Fig. 8.50   (a) Relationship between temperature and time required for stress to relax to 50% of its initial value σ_0 (numbers on the curves) for dry (solid lines) and wet (dashed lines) olivine deformed by dislocation creep (eq. 8.60) assuming activation energy $E^* = 523$ kJ/mol and 375 kJ/mol for dry and wet rheologies, respectively. Thin dashed line — wet olivine with $E^* = 398$ kJ/mol at $\sigma_0 = 1$ MPa. Gray shadings show temperature ranges required for lithospheric stresses of ~10–100 MPa to relax in ~10 Ma (see eq. 8.60). (b) Compilation of effective elastic thickness Te for different oceanic structures as a function of ocean-floor age at the time of loading. Gray shadings — Te estimates grouped for four tectonic structures (based on compilation of Burov and Diament, 1995). Black dots — Te estimates for ocean trenches obtained by inverse modeling of free-air gravity 700 km-long profiles (based on Fig. 24c of Bry and White, 2007). Thin lines — geotherms based on the plate model (McKenzie et al., 2005), numbers on the curves — temperature in °C. Bold lines — theoretical predictions for the base of elastic lithosphere for dry and wet olivine rheologies (gray shadings in (a)). While compilation of Burov and Diament (1995) indicates that Te is correlated with ocean-floor age, the reassessed data for trenches (Bry and White, 2007) is inconsistent with a Te increase with plate age.

oriented perpendicular to oceanic trenches to reassess Te values. The influence of short-wavelength features not associated with lithospheric bending was removed by stacking adjacent profiles from 10-My bins; no prior assumptions were made about magnitude of load, size of bending moment, or whether the elastic plate is broken/continuous. The results do not show any consistent trend in increase of elastic thickness as a function of plate age. The authors conclude that, for ocean floor age between 20 and 150 Ma, Te ranges from 10 to 15 km, and this variation corresponds approximately to the depth to a 200 °C isotherm in the plate model. They further conclude that either Te is independent of plate age or that it cannot be determined with sufficient accuracy. However, the re-assessed pattern appears more complicated than the stated nearly uniform Te value of 10–15 km (Fig. 8.50b).

Temperature at the base of the elastic lithosphere

The following discussion disregards (but not rejects!) the results of Bry and White (2007) due to their significant scatter (Fig. 8.50b) and, for this reason, difficulty in assigning a characteristic isotherm to Te. The plot (8.50b) does not convince, however, that new Te values concentrate around the 200 °C isotherm as proposed by these authors.

In the case of visco-elastic rheology, the stress relaxation time t_{r2} over which the stress relaxes to its initial value σ_o (eqs. 8.13–8.14) depends strongly on temperature and the rheological parameters of the lithosphere:

$$t_{r2} = 3/2 \ \exp(E^*/RT)/[C_1 E\sigma_o^2] \tag{8.60}$$

where E^* is activation energy, R – universal gas constant, T – temperature, E – Young's modulus, and C_1 is the material constant. Because of the compositionally homogeneous (at least, in the first approximation) structure of the oceanic lithosphere, t_{r2} should be controlled primarily by temperature variations (i.e. the age of the oceanic plate) and the presence of fluids. Figure 8.50a illustrates this dependence for wet olivine ($E^* = 375$ kJ/mol) and dry olivine ($E^* = 523$ kJ/mol), assuming the same value of $E = 70$ GPa for both rheologies. The result for wet olivine with $E^* = 398$ kJ/mol (the value preferred by Turcotte and Schubert, 2002) is shown for comparison (see Table 8.2 for more details).

Assuming characteristic values of lithospheric stresses $\sim$10–100 MPa and stress relaxation time t_{r2} $\sim$10 Ma, significant stress relaxation for wet olivine is expected at $T > 425$–475 °C and for dry olivine at $T > 675$–750 °C (gray shadings in Fig. 8.50a). These isotherms can be used as proxies for the base of the elastic (mechanical) lithosphere for wet and dry olivine rheologies, respectively, *if the above assumptions on lithospheric stresses and stress relaxation time are correct*. The values for dry olivine are close to the estimates of the brittle–ductile transition temperatures in olivine that, at strain rates of 10^{-17}–10^{-14} s^{-1}, occurs at 600–750 °C (Section 8.1.4).

Indeed, the base of the elastic lithosphere in normal oceans is often identified with the depth to the 500–550 °C isotherm (Fig. 8.50b). Since this temperature range is mid-way between the temperatures of stress relaxation for dry and wet olivine, one may argue that the oceanic lithosphere has low hydrogen content. Clear deviations from normal oceans are seamounts with $Te \sim$ 10–30 km corresponding to 100–350 °C isotherms and, perhaps, large-offset fracture zones where the base of the elastic lithosphere corresponds to *c.* 600–650 °C isotherm. The former are examples of "anomalous oceans" clearly associated with pronounced thermal anomalies in the mantle. For these structures isotherms cannot be calculated within the frame of a cooling half-space or plate model, and plotting Te values on top of "normal" oceanic isotherms is misleading (Fig. 8.50b): the actual mantle temperatures can be several hundred degrees celcius higher than suggested by calculations based on ocean-floor age. As for the latter, one can question whether the model of elastic plate flexure is applicable to the deformation of a lithosphere which is highly fractured through its entire thickness. Alternatively, the results may be interpreted as evidence for dehydrated oceanic lithosphere in the large-scale fracture zones.

An analysis of relationships between Te and the age of the loads allows for speculation on the elastic relaxation time and thus on the long-term rheological behavior of the oceanic lithosphere (Watts *et al.*, 1980; McNutt and Menard, 1982). In particular, *assuming* that for normal oceans younger than $\sim$100 Ma the Te values cluster around a 500–550 °C isotherm (Fig. 8.50b) and assuming lithospheric stresses of $\sim$10–100 MPa, Fig. 8.50a yields an elastic relaxation time of $\sim$10–100 Ka for the oceanic mantle with wet rheology. The same line of argument apparently *forbids dry rheology of the oceanic mantle* since in this case it requires an unrealistically long elastic relaxation time, which exceeds the age of the Earth.

Alternative (classical isostasy) gravity models for the oceans

With a lot of attention drawn to flexural models of the oceanic lithosphere, gravity models based on traditional interpretations of regional isostasy are often neglected. However, these studies provide an alternative view on the regional structure of the oceanic lithosphere. A common problem with the traditional approach is that isostatic Airy-type models of local compensation require unrealistically large crustal thickness. This problem, can however be easily resolved if one assumes that crustal contribution to isostasy is 65–80% only and the rest of the compensation occurs in the mantle.

Figure 8.51 shows an example of flexural and regional isostatic models calculated for the deepwater Kuril Basin of the Okhotsk Sea, the northwestern Pacific Ocean. Although experimental admittance can fit a theoretical curve calculated for the model with a 30 km crust, such a model contradicts regional seismic data that indicate a crustal thickness of ~20 km (Sergeyev *et al.*, 1983). However, both seismic data on the crustal structure and gravity–bathymetry data can be explained by the Airy-type model if $\sim 1/3$ of the compensation is caused by a low density mantle at 50–70 km depth.

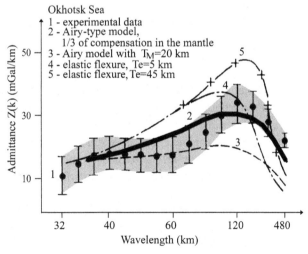

Fig. 8.51 Theoretical and experimental admittance for the Kuril Basin of the Okhotsk Sea (after Artemjev and Kaban, 1987). Gray shading and line 1: experimental data, lines 2–3: theoretical admittance for models of regional isostasy (2: for $T_M = 20$ km and assuming that $1/3$ of compensation is caused by low density mantle at 50–70 km depth; 3: for the classical Airy model with $T_M = 20$ km which agrees with the average crustal thickness in the region); lines 4–5: theoretical admittance for models of flexural isostasy (4: for the model with $Te = 5$ km; 5: for the model with $Te = 45$ km). T_M – average Moho depth, Te – elastic thickness.

Similarly, models of regional compensation can be constrained for other oceanic structures, including the Hawaiian ridge, assuming the presence of low-density mantle material beneath the region (Artemjev and Kaban, 1987). This does not imply that elastic flexure plays no role in maintaining isostasy in the Hawaiian region. However, a model of regional compensation, strongly supported by seismic data that indicates low seismic velocities in the upper mantle, will yield different parameters of the lithospheric structure than predicted by commonly accepted models of flexural isostasy of the region.

8.3.2 Flexure models of continental lithosphere

Lithosphere flexure under ice load

In the real Earth, deformation of the lithosphere during application or removal of a surface load (such as deposition of sediments in basins or melting of ice sheets) depends on both lithospheric rigidity and mantle viscosity. Thus, data on lithospheric flexure under known external loads allows for estimation of the range of possible values for both parameters. Studies of the Pleistocene–Holocene postglacial rebounds in Fennoscandia, North America, and Australia provide estimates of elastic thickness of the lithosphere and viscosities of the upper and lower mantle from data on sea-level changes, variations in paleoshorelines, and GPS studies of the rates of postglacial uplift (e.g. McConnell, 1968; Wolf, 1987; Lambeck, 1990; Forte and Mitrovica, 1996; Fjeldskaar, 1994; Lambeck *et al.*, 1998; Peltier, 1998; Kendall *et al.*, 2003; Sella *et al.*, 2007).

A number of rather diverse interpretations of the Pleistocene–Holocene postglacial rebound have been published for northwestern Europe (Fig. 8.52). Some of these models include the presence of a 50–75 km-thick low-viscosity, $\sim (1.3–2.0) \times 10^{19}$ Pa s, layer beneath the lithosphere (the asthenosphere) overlaying a 10^{21} Pa s upper mantle (e.g. Fjeldskaar, 1997). Other models either include a weak asthenospheric layer (Mitrovica and Peltier, 1993) or do not include it all (Lambeck *et al.*, 1998). The diversity of interpretations stems from the fact that, while a large set of Earth-model parameters describes the rebound with significant trade-offs between some of them (as for example, between the lithospheric thickness and asthenosphere viscosity), few models have examined the entire possible parameter space. Additionally, some parameters characterizing the ice-sheet may be particularly severe for model results, while most of the studies assume that the time evolution of the ice sheet geometry is perfectly known.

Lambeck *et al.* (1990, 1998) perform inversion of the observed sea-level variations in northwestern Europe since the time of the Last Glacial Maximum for glacial unloading of the lithosphere. They investigate the model solution space by including variations in ice-sheet thickness, lithosphere thickness, and mantle viscosity. The model accounts for the glacio-isostatic rebound, gravitational attraction of the ice, the effect of meltwater on global sea-floor loading, and localized water loading produced by the filling-in of the seafloor depression. The lithosphere is assumed to behave as an elastic plate on the time-scale of postglacial rebound (with viscosity $> 10^{25}$ Pa s), while the underlying mantle is modeled as a body with a linear visco-elastic Maxwell rheology. They find a better fit to data on sea-level variations if the ice thickness over Scandinavia at the time of maximum glaciation was only

(a)
Model of ice-sheet thickness (in m), 20 Ka

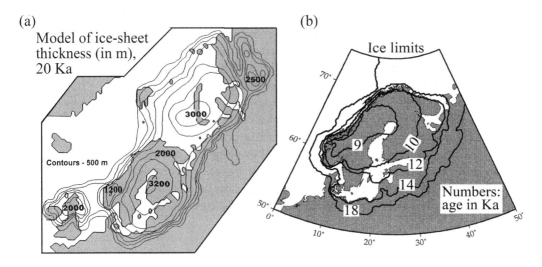

(b)
Ice limits

(c) Uplift rate and radial velocity

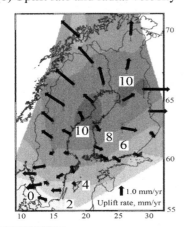

(d) Lithosphere thickness 65 km   L. mantle viscosity 20x10²¹ Pa s

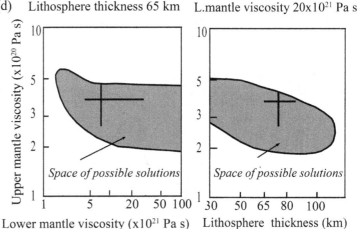

Fig. 8.52 Glacial rebound model for Fennoscandia (a – after Denton and Hughes, 1981; b, d – after Lambeck *et al.*, 1998; c – after Milne *et al.*, 2001, 2004). The maximum ice-sheet thickness at maximum glaciation (a) and ice limits at selected times during ice retreat (b). (c) GPS data on present-day vertical movements (gray shading) and horizontal velocity (arrows). (d) The parameter space for models best fitting the data from north-central Sweden that imply a maximum ice-sheet thickness of ~2000 m (but not as shown in (a)). The region for which the variance factor is less than 1 is shaded gray; error bars correspond to the best-fit model for the entire Fennoscandia with a lithospheric thickness of 75 ± 10 km, upper mantle viscosity of $(3.6 ± 1.0) × 10^{20}$ Pa s, and lower mantle viscosity of $8(-4/+22) × 10^{21}$ Pa s. The plots show the parameter space for upper and lower mantle viscosities for models with a lithospheric thickness of 65 km (left) and for upper mantle viscosity and lithospheric thickness for models with lower mantle viscosity of $20 × 10^{21}$ Pa s (right).

about 2000 m, instead of 3400 m (Fig. 8.52a) as assumed in earlier studies (e.g. Denton and Hughes, 1981). The data allows for a broad range of solutions within the possible parameter space (Fig. 8.52d). The best-fit solution to glaciation data for all of the region requires a $75 ± 10$ km-thick elastic lithosphere with upper mantle viscosity of $(3.6 ± 1.0) × 10^{20}$ Pa s and lower mantle viscosity of $8(-4/+22) \times 10^{21}$ Pa s. Similar results, with lithospheric rigidity corresponding to $Te \sim 80$ km, were obtained earlier (Wolf, 1987) based on sea-level variations in Fennoscandia, while other models based on similar data provide smaller values of the elastic thickness: $Te \sim 50$ km beneath the Bothnian Gulf of the Baltic Sea and $Te \sim 20$–25 km beneath the Norwegian Caledonides (Fjeldskaar, 1997).

Similarly, the estimates of elastic thickness of the lithosphere based on postglacial rebound in North America provide Te values in the same range, from ~ 20 km to ~ 90 km. The predicted present-day radial displacement rates, calculated for the VM2 mantle viscosity model (Peltier, 1998) and constrained by postglacial sea-level variations, show good correlation with recent GPS observations of the relative vertical movements in North America (Fig. 8.53). The glacial isostatic rebound has the strongest contribution to observed vertical movements. In overall agreement with theoretical models, GPS data indicate

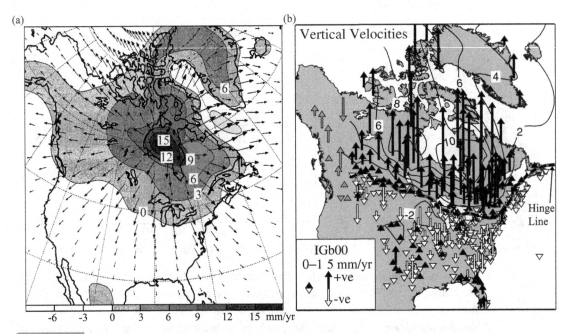

Fig. 8.53 (a) The present-day radial displacement rate (gray-shaded contours) and horizontal displacement rate (arrows whose length represents the magnitude of the tangential displacement rate and whose direction is indicated by the direction of the arrows) over Laurentia calculated for the VM2 viscosity model structure (from Peltier, 1998). (b) Relative vertical GPS site motions (from Sella *et al.*, 2007). Note large uplift rates around the Hudson Bay, and subsidence to the south. Black dashed line shows interpolated 0 mm/yr vertical "hinge line" separating uplift from subsidence. The vertical velocities show fast rebound (~ 10 mm/yr) near the Hudson Bay, the site of thickest ice at the last glacial maximum, which changes to slow subsidence (1–2 mm/yr) south of the Great Lakes. The "hinge line" separating uplift from subsidence is consistent with data from water-level gauges along the Great Lakes, showing uplift along the northern shores and subsidence along the southern shores.

horizontal motions outwards from the Hudson Bay with complex local variations, in particular in the far field, that are not entirely consistent with glacial rebound models. The discrepancy for horizontal motions can be caused by lateral variations in mantle viscosity and/or by inaccuracy in the ice-load model.

Crust–mantle decoupling in orogens

The range of Te values reported for different mountain belts span from 15–20 km in Pamir and the foreland basins of the Apennines and Southern Alps to 40–60 km in the Eastern Alps, Western Alps, Carpathians, Zagros, Tarim, Kunlun, and the Andes, reaching more than ~90 km in the Himalayas, Ganges, and perhaps in the Appalachians (Stewart and Watts, 1997). These huge variations reflect regional differences in lithospheric temperatures and crustal thickness, that are the major controls of the flexural strength of the lithosphere in continental collisional settings, and its coupled or decoupled behavior (Fig. 8.31). Since the thermal regime and crustal structure of the colliding lithospheric plates are significantly controlled by their age, some earlier studies proposed the existence of a correlation between the elastic thickness and the age of the continental orogens (e.g. McNutt, 1980; McNutt *et al.*, 1988). However, this correlation is an artifact of unaccounted low-density sediments, the thickness of which is well correlated with topography in most mountain regions thus biasing elastic thickness estimates. Additionally, because of the thermal insulation effect and additional load, the presence of thick sediments has a strong weakening effect on the lithospheric strength (see Section 8.2.5 and Fig. 8.42).

An analysis of Lavier and Steckler (1997) of the yield stress envelopes for foreland basins and mountain belts worldwide indicates that the presence of thick and hot crust in young continental orogens (such as Apennines, Eastern Alps, Western Alps) always leads to crust–mantle decoupling, regardless of the thickness of the sediments. As a result, the strength of the lithosphere in these orogens is low since the presence of a decoupling layer allows for independent deformation (bending) of the layers above and below the decoupling zone (e.g. eq. 8.49).

In orogens formed by collision of old and cold lithospheric plates (the Himalayas and Ganges), the weakening effects due to large crustal thickness, deep sediment fill, and/or large plate curvature are insufficient to cause lithosphere decoupling if the creep in the lower crust is controlled by plagioclase rheology. In such cases, the lithosphere deforms as a single plate and its flexural strength (elastic thickness) is controlled by the rheology of the lithospheric mantle. However, even in orogens formed by the collision of old and cold lithospheric plates the lithosphere can become permanently decoupled if the lower crustal creep is controlled by quartz rheology. When crustal rheology is stronger than quartz-diorite, crust–mantle decoupling is not possible unless sediments are present.

Similarly, in the collisional orogens with intermediate lithospheric temperatures (the Tarim, Kunlun, Zagros, Southern Alps, and the Andes) the crustal thickness and plate curvature alone are insufficient to explain the observed values of lithosphere rigidity. In these collisional settings, the presence of sedimentary cover plays a key role in weakening an otherwise strong lithosphere and decreasing the values of its effective elastic thickness by 10–30 km.

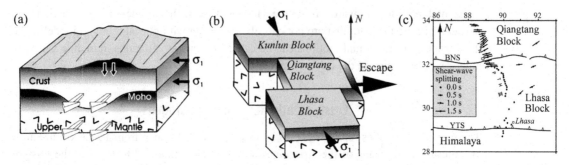

Fig. 8.54   (a) Sketch model of a young mountain belt with mountain-perpendicular stresses σ_1 (black arrows). Mountain-parallel creep and escape tectonics can occur in the ductile sections of lower crust and upper mantle, causing mountain-parallel anisotropy (white arrows). (b) Sketch model of the tectonic situation in central Tibet with eastwards extrusion of the Qiangtang Block. (c) SKS anisotropy in southern Tibet along the profile INDEPTH III. BNS = Bangong–Nujiang Suture, YTS = Yarlung Tsangpo Suture (after Meissner *et al.*, 2002).

The presence of a zone of ductile decoupling in the middle–lower crust or upper mantle of young mountain belts can result in crustal or uppermost mantle "tectonic escape" (or "lateral extrusion") tectonics under a mountain-perpendicular horizontal stress. This concept was first suggested for southern China by Molnar and Tapponier (1975) and has been extended later to eastern Asia (Burke and Sengör, 1986) and to the Alps (e.g. Ratschbacher *et al.*, 1989; Vauchez and Nicolas, 1991). Mantle tectonic escape is directly related to creep processes in the lithospheric mantle which produce a preferred orientation of olivine crystals and result in a "streamline-orientated" anisotropic viscosity with a 1–2 orders of magnitude difference between "shear" and "normal" viscosity (Christensen, 1987). Lattice-preferred orientation of olivine causes seismic anisotropy that has been measured beneath young orogens (Tibet, the Alpine chain, and young mountain ranges in the North and South Americas) for the seismic *Pn* phase that propagates in the uppermost mantle. The *Pn* anisotropy is mountain-parallel (i.e. it is aligned parallel to the structural axis of orogens), while tectonic stress and upper crustal structural features are usually mountain-perpendicular (Fig. 8.54c). Since the preferred orientation of the fast axes of olivine is in the creep direction, the anisotropy is interpreted as being caused by deformation during the most recent thermo-tectonic event, and indicates mountain-parallel ductile deformation in the uppermost mantle as a consequence of mountain-perpendicular compressive stresses (Meissner *et al.*, 2002).

Continent-scale models

North America

The results of continent-wide flexural studies are best illustrated by recent results for North America (Kirby and Swain, 2009) which include both the Bouguer coherence and free-air admittance methods (Fig. 8.55). As discussed in Section 8.2.5, sensitivity analysis indicates that the Bouguer coherence provides reliable results for the northern parts of the Canadian

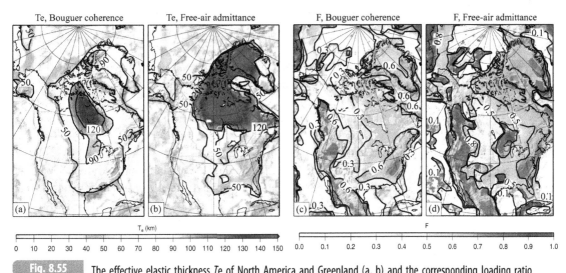

Fig. 8.55 The effective elastic thickness Te of North America and Greenland (a, b) and the corresponding loading ratio F (c–d), calculated from the Bouguer coherence and free-air admittance (modified after Kirby and Swain, 2009). Darker shading corresponds to larger values of Te and F (see Fig. 8.46 for model resolution).

Shield, while the results for the Superior Province and the Midcontinent Rift might be unreliable due to the presence of gravity noise. The major conclusions of the study are as follows.

The Canadian Shield and Greenland

The strength of the shield lithosphere is maintained by the upper mantle: both methods indicate a large elastic thickness of the lithosphere in the Archean–Paleoproterozoic terranes of the Canadian Shield with $Te > 120$ km; the region of thick mechanical lithosphere extends to north-central Greenland. The edges of the shield have a sharp and large decrease in Te values. The misfit in Te values for the Canadian Shield is 10–20 km for both methods. Importantly, the results indicate the presence of significant internal loads in most parts of the continent; the most significant values of the loading ratio F (the internal load to total load) are calculated for the stable cratonic part, with $F \sim 0.5$ for the Bouguer coherence and $F \sim 0.6$–0.8 for the free-air admittance. There are no confidence limits for F calculated from the Bouguer coherence since its value is calculated during the inversion; misfit for F calculated from the free-air admittance is < 0.05 for most of the shield (see Fig. 8.46 for model resolution).

Proterozoic and Paleozoic terranes

Within the southern parts of the Canadian shield and the Proterozoic parts of the North American craton including the Great Plains, Te values reduce to 50–90 km; similar values are calculated for the Appalachians lithosphere. The misfit in the Te values is 10–20 km for both methods. There is a large discrepancy in the results by the two methods for the Midcontinent Rift: while the Bouguer coherence indicates $Te \sim 100$ km, the free-air admittance method yields $Te \sim 37$ km. Importantly, the free-air admittance method has high uncertainty

(20–25 km) in the Great Plains and in the Midcontinent Rift. In the southern Appalachians, the loading ratio is $F \sim 0.7$–0.8 by both methods; the Great Plains have the lowest values of F, which nevertheless are not less than 0.2. The Midcontinent Rift has $F \sim 0.5$ ($f \sim 1$) for the Bouguer coherence and $F \sim 0.7$ ($f \sim 2.33$) for the free-air admittance. Misfit for F calculated from the free-air admittance is also high (0.25–0.30) for the Great Plains and the Midcontinent Rift. The results indicate that none of the methods can provide reliable estimates of Te and the loading ratio in the Superior Province, the southern Great Plains, and the southern Appalachians.

Tectonically young parts of the continent

Western parts of the continent including the Cordillera and the Basin and Range Province have $Te < 40$–50 km with uncertainties in Te values less than 10 km for both methods. The loading ratio is $F \sim 0.7$–0.8 by both methods for western USA and Cordillera with the misfit < 0.05 for the free-air admittance.

Other studies

Other studies (e.g. Bechtel *et al.*, 1990; Wang and Mareschal, 1999; Flück *et al.*, 2003) are in a general agreement with the results of Kirby and Swain (2009). Note that there is an overall improvement in the information on internal loads (related to lateral variations in the crustal structure) from the early to the more recent studies. As a consequence, there is significant divergence in model results for some of the tectonic regions. Since the crustal structure in the northern parts of the Canadian Shield is not well known, uncertainty in the model constraints, in general, increases northwards. The values of elastic thickness reported by different authors for the cratonic parts of North America range from more than 80 km in the Trans-Hudson Orogen and the Hearne Province to 50–100 km in the Slave craton and 50–60 km in the Hudson Basin. However, large values of elastic thickness cannot be recovered by flexural analysis (for resolution and uncertainties see Figs. 8.38 to 8.42). Some authors report a sharp decrease of coherence at long wavelengths in the southeastern parts of the Canadian Shield (the Abitibi and Grenville provinces) and in the northern Appalachians which cannot be explained by flexure of a thin plate (Audet and Mareschal, 2004b). Data on subsidence history of the intracratonic Williston Basin has been used to estimate its elastic thickness at the time of subsidence ($Te \sim 40$ km at 450 Ma) and at present (~ 80 km) (Ahern and Ditmars, 1985). Several authors report $Te \sim 20$–40 km in the Cordillera, except for the Proterozoic Wopmay Orogen in the north-eastern Cordillera where $Te \sim 90$ km. Earlier studies indicate a very weak lithosphere in the Basin and Range Province, with $Te < 10$ km. Overall, there is a general trend in lithosphere weakening from Precambrian to younger terranes; however, this trend is obscured by significant regional variations in lithosphere rigidity.

Flexural anisotropy

Although the mechanical behavior of lithospheric plates is commonly assumed to be isotropic, the possibility of anisotropic flexural response of the lithosphere was first suggested by Stephenson and Beaumont (1980). Flexural anisotropy expresses itself, for example, in directional variations of the coherence which may indicate a preferred direction of isostatic compensation. *The direction in which the coherence is maximal is interpreted as the direction in which the lithospheric plate is the weakest.*

Several studies have examined mechanical anisotropy of the elastic lithosphere in North America (Lowry and Smith, 1995; Audet and Mareschal, 2007; Audet et al., 2007). A clear correlation between seismic, electrical, and strength anisotropy suggests that they are caused by the same stress field that affected the entire lithosphere so that azimuthal variations of Te reflect deviatoric stress in the lithosphere. In an extensional environment, the mechanically weak direction correlates with the direction of minimum horizontal compressive stress, while in a compressive stress regime the minor Te axis (the weak direction) correlates with the direction of maximum horizontal compressive stress. Creep processes in the lithosphere that produce a preferred orientation of olivine crystals and the alignment of the olivine a-axis parallel to, or within 30° of, the direction of shear (Nicolas and Poirier, 1976; Nicolas and Christensen, 1987) result in *anisotropic viscosity* with "shear" viscosity along the a-axis of 1–2 orders of magnitude lower than "normal" viscosity in the perpendicular direction (Christensen, 1987). Since viscosity is proportional to strength (eq. 8.11), the lithosphere should be the weakest in the creep direction.

These conclusions are supported by the observed correlation between the directional variations of the coherence (strength anisotropy) and seismic and electrical anisotropy in the Canadian Shield: the average coherence increases (strength decreases) in the direction perpendicular to the direction of high electrical conductivity and fast seismic axis. Audet and Mareschal (2007) note that the mechanically weak axis is quasi-perpendicular to the main tectonic boundaries such as the Grenville Front. The correlation between the present-day horizontal displacement rate predicted from glacial rebound calculations and the vertical uplift rates as revealed by GPS data are, however, more striking (compare Figs. 8.53 and 8.56a). Note that the region with no strength anisotropy in the southeastern part of the Canadian Shield corresponds to the region of maximum vertical GPS motions. Strength anisotropy has been also reported for tectonically young provinces in western North America. While a weak Te anisotropy within the Basin and Range Province can be caused by fault orientation, a 50% strength reduction in the US Cordillera is likely to be associated with regional tectonic stress orientation which reduces Te in the direction of the stress axis. A notable exception is Yellowstone where strength anisotropy is not correlated with the stress regime.

Other continents

A significant number of flexural models have been published recently for all continents except Antarctica. The studies for some of the continents have limited resolution because reliable seismic data on the crustal structure are largely unavailable for most of Africa and South America and the quality of existing crustal compilations for Siberia is uncertain (e.g. Doucouré et al., 1996; Hartley et al., 1996; Petit and Ebinger, 2000; Ojeda and Whitman, 2002; Poudjom Djomani et al., 2003; Mantovani et al., 2005; Petit et al., 2008; Pérez-Gussinyé et al., 2009). Lack of uniform continental-scale gravity data coverage for significant parts of these continents further hampers flexural modeling. For some continents a number of lithosphere strength models have been published for selected tectonic structures. However, a significant difference in model assumptions limits the reliability of

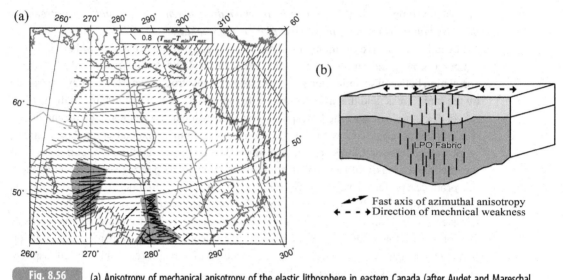

(a) Anisotropy of mechanical anisotropy of the elastic lithosphere in eastern Canada (after Audet and Mareschal, 2007). Direction of the bars shows the minimum strength direction, the length of the bars is proportional to $(Te_{max} - Te_{min})/Te_{max}$. Black bars on gray shading – the fast direction of seismic SKS anisotropy. (b) The relation between seismic and mechanical anisotropy in compressional regime (after Simons et al., 2003).

comparison of lithospheric strength between different studies. For this reason, the discussion below is limited to Australia and Europe for which continent-scale flexural models are available (Simons et al., 2000, 2003; Pérez-Gussinyé et al., 2004; Kirby and Swain, 2006).

Australia

In the lithosphere of Australia strength anisotropy is clearly correlated with tectonic provinces. A comparison of mechanical and azimuthal anisotropy at depths between 75 km and 175 km shows that, statistically, the weak axes and the fast seismic axes are orthogonal within the Precambrian parts of the continent (Fig. 8.57). The source of both types of anisotropy can be vertically coherent frozen-in deformation of the lithosphere such as created by large-scale collision (Silver, 1996) (Fig. 8.56b). Correlation further suggests the existence of a strong coupling of the crust and the upper mantle in the cratonic lithosphere, where the lithospheric strength is maintained mostly by the lithospheric mantle. The correlation vanishes below c. 200 km depth, where the pattern of seismic anisotropy becomes significantly different and is related to the present-day Australian plate motion (Fig. 3.115). The results of flexural modeling indicate that cratonic north-central Australia has a strong core with high $Te \sim 100$ km; the cratons of western Australia have small values of elastic thickness ($Te < 60$ km) as compared to other cratons worldwide. Analysis of the loading ratio (eq. 8.51) indicates that the Yilgarn craton has $f > 3$, i.e. that significant heterogeneity is present in its lithosphere.

Europe

Similar to North America, flexural analysis for the European continent based on a single approach indicates that the cratonic lithosphere is mechanically stronger than the lithosphere

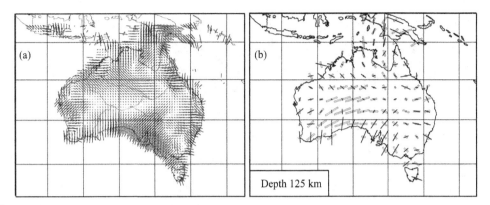

Mechanical anisotropy in Australia (from Kirby and Swain, 2006). (a) Direction of the bars shows the direction of minimum strength Te_{min} and their length is proportional to the anisotropy $(Te_{max} - Te_{min})/Te_{max}$. Gray lines – the crustal mega-elements. (b) Comparison of the axes of anisotropy in the direction of Te_{min} (averaged over a 300 km × 300 km area) (thin, black bars) with the fast axes of seismic anisotropy (the tomographic model of Kennett *et al.*, 2004) at a 125 km depth (thick, gray bars).

of younger Phanerozoic provinces (Pérez-Gussinyé and Watts, 2005). The results based on the Bouguer coherence method (assuming Young's modulus $E = 100$ MPa) indicate $Te > 60$ km in the Kola-Karelian province of the Baltic Shield and in all of the East European platform (60 km is the maximum value of the elastic thickness that could have been recovered, see Section 8.2.5 and Figs. 8.38 and 8.40). Similar values that agree with glacial rebound estimates (Fig. 8.52d) have been reported by Poudjom Djomani *et al.* (1999). The transition from the cratonic to Phanerozoic lithosphere is sharp, with a fast decrease in Te from > 60 km to < 40 km; this approximately coincides with the Trans-European Suture Zone at the western edge of the East European craton. Similar change in elastic thickness is observed at the southeastern edge of the SvecoFennian province of the Baltic Shield (see Fig. 3.60 for tectonic location). Within the Baltic Shield, Te values decrease northwestwards from > 60 km in the Archean Kola-Karelia and the Paleoproterozoic SvecoFennian of central Finland to 30–40 km in the ~ 1.0 Ga SvecoNorwegian province and to 10–25 km in the Norwegian Caledonides. Surprisingly, the tongue of strong lithosphere $(Te > 60$ km) extends through Denmark to the British Isles over the terranes of Avalonia accreted to the cratonic margin in the Paleozoic. The lithosphere of Phanerozoic Europe is weak, $Te \sim 10$–30 km. Throughout the entire continent, there is good overall agreement between the elastic thickness of the lithosphere and the depth to the 450–500 °C isotherm.

Global patterns

Overview

A compilation of recent estimates of the elastic thickness summarizes reliable constraints of Te for the continents (Fig. 8.58). This map reflects largely variations in the thickness of the mechanical boundary layer (MBL). Similar to the maps of seismic, thermal, and

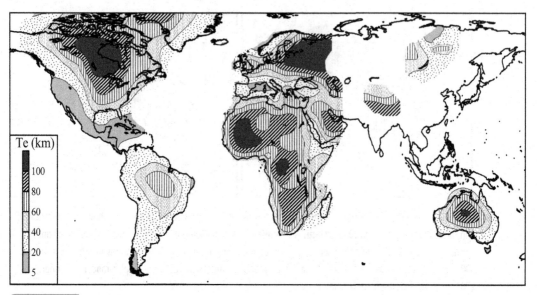

Fig. 8.58 Thickness of elastic lithosphere Te for the continents based on results of Tassara *et al.* (2007) for South America, Flück *et al.* (2003), and Audet and Mareschal (2004a) for Canada, Pérez-Gussinyé *et al.* (2009) for Africa, Pérez-Gussinyé and Watts (2005) for Europe, Poudjom-Djomani *et al.* (2003) and Petit *et al.* (2008) for Siberia and the Baikal region, Burov *et al.* (1993) and Royden (1993) for Central Asia, Swain and Kirby (2006) for Australia. The results for different continents should be compared with caution because of significant difference in the approaches used in regional studies.

electrical thicknesses of the lithosphere (Figs. 3.120, 4.44c, 7.50), a thick mechanically strong layer is typical of the Precambrian cores of the continents, while tectonically young structures have a smaller thickness of MBL. Likewise in seismic, thermal, and electrical approaches, the base of the MBL is diffuse because of a gradual change of lithosphere viscosity (and strength) with depth, and allows various practical definitions (Fig. 8.34). There is, however, an important difference between the Te thickness and the thicknesses of seismic, thermal, and electrical lithosphere. In contrast to other approaches, in general the effective elastic thickness Te does not correspond to a physical depth (such as, for example, the base of the MBL), but rather represents the integrated strength of all lithospheric layers (including the weak ones) that are supporting the loads. Therefore, Te is equal to the depth to the base of the MBL only in the case of mechanically coupled lithospheric rheology when the continental lithosphere deforms as a single plate, similarly to the oceanic lithosphere. The presence of decoupling zones in the lithosphere significantly reduces its effective elastic thickness (eq. 8.49).

Burov and Diament (1995) have analyzed the effect of crust–mantle decoupling on the strength of the continental lithosphere and concluded that due to the interplay of the effects of temperature and rheological properties of the crust and the lithospheric mantle, crust–mantle decoupling should always occur when the crustal thickness exceeds a critical value of 40 ± 5 km in the Precambrian lithosphere (this value decreases for younger structures). Their conclusion was supported by a comparison of Te estimates for different continental settings with lithospheric ages and geotherms. However, their result can be somewhat misleading:

(i) the continental geotherms calculated by these authors resemble oceanic geotherms and are the same for any continental structure older than 300 Ma (this explains the asymptotic behaviour of the critical for the decoupling process value of the crustal thickness);

(ii) an apparent asymptotic behaviour of effective elastic thickness for a strong (old and cold) lithosphere with $Te \sim 100$ km is due to the inability of flexural methods to resolve large Te values (see Section 8.2.5).

Large values of effective elastic thickness Te (up to 100–130 km which is the resolution limit of any flexural model, although some authors give even smaller resolution limiting values, ~60–80 km) reported worldwide for old stable regions of the continents rule out the possibility of crust–mantle decoupling in cratonic regions. Most of these results are reliable, and thus a large thickness of rheologically strong cratonic lithosphere is a robust feature. However, McKenzie and Fairhead (1997) argue that these values contradict expectations from mineral physics on the ability of the lithosphere to support elastic stresses over geological time-scales. Thus it is still an open question as to how lithosphere of such a thickness can remain mechanically stable over billions of years, dry rheology can be the clue.

Another not very well understood result is the very small effective elastic thickness reported for some tectonically active regions, since it is unclear how it can support tectonic loads. In particular, values less than 10 km have been estimated by several authors for volcanic and extensional provinces of western USA, including the Snake River Plain ($Te \sim 6$ km), the Northern Basin and Range Province ($Te \sim 4$–10 km), and the northern Rocky Mountains ($Te \sim 9$ km). Domains with the lowest Te correspond to the most extended crust.

Strength and loading time

It is commonly accepted that Te reflects fossil lithosphere equilibrium built up at the time of loading. This interpretation is challenged by a strong spatial correlation between the present-day pattern of flexural anisotropy in the Precambrian Canadian Shield and the pattern of present-day vertical velocities measured by GPS and caused by on-going postglacial rebound (Figs. 8.53b and Fig. 8.56a). Other studies, however, point out that strength anisotropy in the Precambrian cratons correlates with seismic anisotropy in the lithospheric mantle and with major terrane boundaries at the surface (Fig. 8.57) and thus advocate that Te values in the old continental lithosphere may correspond to its "fossil" rheological state (e.g. Simons et al., 2003; Audet and Mareschal, 2007).

In contrast, Pérez-Gussinyé et al. (2009) argue that the very fact that maps of Te for the continents are qualitatively similar to seismic tomography maps, that provide a snap-shot of modern lithospheric properties, indicates that Te in the cratonic regions reflects their modern rather than fossil thermal and compositional state (apparently this is not the case for the oceans where Te and seismic tomography are not well correlated (Kalnins and Watts, 2009)). Their further argument is that if loading occurs when the lithosphere is weak and no mass redistribution occurs afterwards, there is no need for stress to re-equilibrate, and the *Te* value can remain low even after consequent cooling and strengthening of the lithosphere. However, this hypothesis contradicts the results of Grotzinger and Royden (1990) who analyzed the sedimentary record in the Slave craton and concluded that although the present-day elastic thickness in the region is 100 ± 25 km, the craton had an elastic thickness of only 12 ± 4 km 1.9 Ga ago. Similarly, stratigraphic analysis of sediments in the intracratonic Williston Basin in North America indicates $Te \sim 40$ km at the time of subsidence ~ 450 Ma ago with a present value of $Te \sim 80$ km (Ahern and Ditmars, 1985). While it is easy to understand lithosphere strengthening in the intracratonic basin directly related to its cooling and thermal subsidence, it is not clear what could have caused an order of magnitude change in the elastic thickness in the Slave craton which stabilized at least 2.5 Ga ago.

Correlation between *Te* and tectonics

For the continents, the dependence of *Te* (or equivalently, *D*) on the tectonothermal age (i.e. the time since the last major tectono-thermal event) or, alternatively, on heat flow and lithospheric age, has been demonstrated in many studies. It can also be seen in the present compilation (Fig. 8.59), although the relationships are less pronounced than for the oceans

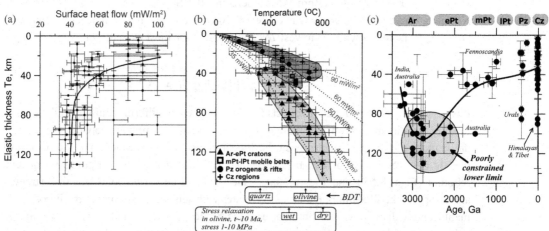

Fig. 8.59 Compilation of effective elastic thickness of continental lithosphere constrained by the Bouguer coherence method for tectonic provinces of different ages (symbols). A *Te* greater than 80–120 km cannot be reliably constrained by flexural models (e.g. Figs. 8.38 and 8.40). (a) *Te* versus surface heat flow. (b) *Te* versus lithospheric temperatures. Dotted lines – conductive geotherms for the continents (Pollack and Chapman, 1977; Artemieva and Mooney, 2001). Bottom boxes show typical temperatures of brittle–ductile transition (BDT) in quartz–rocks and olivine, and temperatures required for a stress of 1–10 MPa to relax over ~10 Ma (see Fig. 8.50a). (c) *Te* versus geological age, ages are based on the TC1 model (Artemieva, 2006). Ar = Archean, ePt = Paleoproterozoic, mPt = Mesoproterozoic, lPt = Neoproterozoic, Pz = Paleozoic, Cz = Meso-Cenozoic. Solid lines in (a, c) – approximations to *Te* values.

due to a complex and strongly heterogeneous structure of the continental lithosphere. For example, an analysis of elastic thickness for Canada shows correlation between Te and age only for the platform area, but not for the Canadian Shield, where Wang and Mareschal (1999) favor instead a correlation between Te and the composition of the crust. Within the platforms, Te also correlates with the thickness of sediments: while sediments weaken the lithosphere, low lithosphere rigidity amplifies basin subsidence. The analysis of Te in Fennoscandia relates the decrease of elastic thickness with tectonothermal age to secular variations in lithospheric mantle composition as derived from xenoliths data rather than to temperature variations (Poudjom Djomani *et al.*, 1999): the Archean–Paleoproterozoic lithosphere, which has the largest elastic thickness, has not only the coldest geotherms but also the most depleted composition. However, laboratory data suggest that the effect of iron depletion of cratonic lithosphere on elastic thickness is insignificant (Section 8.1.3).

Figure 8.59c based on a worldwide compilation of results constrained by the Bouguer coherence method for different continental regions indicates the existence of a broad global trend, showing a decrease of Te with time from the Precambrian to the Phanerozoic lithosphere. Several comments should be made in this connection.

First, a huge scatter of Te values (up to 50–100 km) is observed for the continental lithosphere of any age. For example, the values of Te reported for Cenozoic orogens vary from ~90 km in the Himalayas to ~20 km in the Apennines. Differences in the thermal regime are not the only cause of this scatter, although a strong correlation exists between Te and heat flow (Fig. 8.59a). Variations in the crustal structure and composition should play equally important roles in lithospheric flexure, in particular in young continental lithosphere. In young regions with high heat flow, Te better correlates with temperatures of 200–400 °C which correspond to the brittle–ductile transition in quartz-rich crustal rocks and to the base of mechanically strong crust; however higher temperatures should be expected for large Te values (Fig. 8.59ab). In older lithosphere (> 500 Ma), the mantle olivine component starts to dominate over quartz rheology. In these regions Te better correlates with temperatures of 400–800 °C. Temperatures of ~600–750 °C correspond to the brittle–ductile transition in olivine for typical strain rate values. Such temperatures are also required for relaxation of a 10–100 MPa stress during ~10 Ma: wet mantle rheology requires $T \sim 400$–500 °C, while dry mantle rheology requires $T \sim 700$–800 °C (Fig. 8.50a). Thus, high temperatures corresponding to large Te values in the cratons favor dry composition of the cratonic lithospheric mantle with mechanically strong lithosphere.

Second, the results of flexural studies suggest that the lithosphere of the young Archean cratons (2.5–2.9 Ga) is the strongest ($Te > 100$ km) and is stronger than the lithosphere of the old Archean cratons (> 3.0 Ga) where $Te < 100$ km (Fig. 8.59c). This conclusion is in agreement with the analysis of the thermal regime of the Precambrian lithosphere constrained by both surface heat flow and xenolith geothermobarometry, which indicates significantly colder geotherms in young Archean than in old Archean cratons (Artemieva, 2006) (Figs. 4.49, 5.13). However, the extent of the difference in lithospheric strength between young Archean and old Archean cratons cannot be constrained reliably because flexural isostasy analysis fails to resolve thick elastic lithosphere.

In contrast to the oceanic lithosphere, it is hardly possible to associate the depth corresponding to Te with any particular temperature (Fig. 8.59b). The major reasons are as follows:

- *Te* does not correspond to a physical depth but reflects the integrated strength of the lithosphere;
- possible crust–mantle decoupling has a strong effect in reducing *Te* values and complicates interpretations of effective elastic lithosphere since, commonly, there is no independent direct evidence whether the lithosphere is mechanically coupled or decoupled;
- similarly, the presence of lithospheric-scale faults may have a strong effect on lithosphere weakening and questions the validity of the thin elastic plate approximation in certain tectonic settings;
- inability of flexural methods to resolve large values of *Te* such as in the cratons can bias any correlations between *Te* and other physical parameters.

With these limitations in mind, the following principal conclusions can be made from the compiled results. The depth corresponding to *Te* is at 400–750 °C in Precambrian regions but spans the entire range of 100–800 °C in young lithosphere (including the orogens) (Fig. 8.60). Low *Te* values with associated low temperatures may correspond to decoupled lithosphere, while high *Te* values and higher temperatures are characteristic of regions with mechanically coupled lithosphere where flexural rigidity is dominated by olivine rheology of the lithospheric mantle.

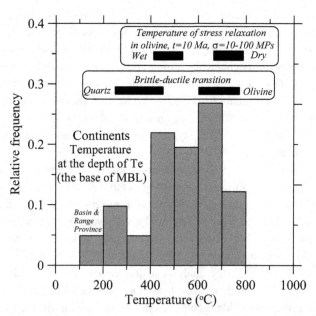

Fig. 8.60   Histogram of temperatures at the depth corresponding to the effective elastic thickness of continental lithosphere (constrained by the Bouguer coherence method only). Black boxes show temperature ranges for the brittle–ductile transition in quartz-rich rocks and mantle olivine and temperatures required for a stress of 10–100 MPa to relax over ~10 Ma in mantle with dry and wet olivine rheology. The results are biased by a non-uniform sampling of different tectonic provinces.

8.3.3 Seismogenic layer and *Te*

Seismicity is the most prominent manifestation of brittle deformation. Most of the continental earthquakes are confined to the uppermost brittle part of the crust, a seismogenic layer. According to studies of rock mechanics, in the continental regions seismicity is controlled by frictional sliding which, in turn, is controlled by lithosphere temperatures, strain rate, the amount of fluids, and crustal mineralogy (see Section 8.1.1). Frictional sliding is commonly confined to the upper ~15 km and becomes insignificant at depths greater than 40–50 km where ductile flow occurs. Although much deeper seismicity exists in the subduction zone environments, observations indicate that the mechanisms generating shallow and deep earthquakes are significantly different. For this reason, deep seismicity in subduction zones is not considered here.

It was noted some time ago that the thickness of seismogenic layer, *Ts*, which corresponds to the maximum depth of earthquakes, globally correlates with surface heat flow (Meissner and Strehlau, 1982; Chen and Molnar, 1983). Since lithosphere rheology is essentially controlled by temperature, and there is, in general, a broad correlation between *Te* and surface heat flow (Fig. 8.59a), one would expect that *Te* and *Ts* are also correlated. Furthermore, since both *Te* and *Ts* reflect the lithosphere strength, some authors propose that these two parameters are the same (Jackson and White, 1989; Maggi *et al.*, 2000). This would, however, imply that the strength of the lithosphere is entirely determined by brittle deformation only. Additionally, given that recent relocation of earthquake hypocenters indicates a remarkably uniform thickness of the seismogenic layer on the continents (~20 km, with rare deeper events in old tectonic structures), this implies that *Te*, with rare exceptions, is also ~20 km globally and that lithospheric loads are supported on geological times by the upper crust (assumed to be dry) rather than by the lithospheric mantle (assumed to be hydrous). This conclusion contradicts the results of flexural isostasy, and Watts and Burov (2003) argue that although both *Te* and *Ts* indicate lithosphere strength, they are not the same: while *Ts* reflects the stress level in the uppermost brittle layer, *Te* reflects an integrated strength of the lithosphere including both its brittle and ductile strength. Furthermore, while the time-scale of flexural processes (and *Te*) is ~0.1 Ma, the time-scale of seismic events (and *Ts*) is orders of magnitude smaller. A broad debate on the relationship between *Te* and *Ts* relates to the debate on a "jelly-sandwich" rheology (Ranalli, 1995; Watts and Burov, 2003; Afonso and Ranalli, 2004; Burov and Watts, 2006; Bürgmann and Dresen, 2008) versus a "crème brûlée" rheology (Jackson, 2002) (see Section 8.1.4 for a description of the two end-member rheological models).

A compilation of *Te* and *Ts* values from seismogenic regions worldwide confirms the existence of a global correlation between the two parameters. However, whereas the correlation is strong for the oceans, it is less unique for the continents (Figs. 8.61 and 8.62). In the oceanic lithosphere the depth distributions of *Te* and *Ts* values are statistically similar (although it does not necessarily imply that the two values correlate in each particular region); the vast majority of the values are < 50 km but extend significantly deeper than the base of the crust. *Ts* corresponds to the depth of the intersection of the moment–curvature curve with the brittle deformation field in yield strength envelopes (Fig. 8.61).

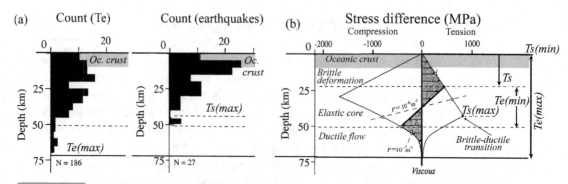

Correlations between thicknesses of elastic Te and seismogenic Ts layers in the oceans (redrawn from Watts and Burov, 2003). (a) Histograms of global compilation of Te and seismicity. (b) Te and Ts for an idealized case of oceanic lithosphere flexure. Thin solid line – the yield stress envelope calculated for 80 Ma-old oceanic lithosphere with geotherms based on the cooling plate model. The brittle behaviour is modeled by Byerlee's law, the ductile behavior – by an olivine power-law. Thick solid line shows the stress difference for a load that generates a moment $M = 22 \times 10^{16}$ N/m and curvature $r = 5 \times 10^{-6}$ 1/m. Both Ts and Te depend on the plate curvature. Two dashed lines show results for the values of r, 10^{-6} 1/m and 10^{-7} 1/m, that bracket observations at deep-sea trench – outer rise systems. See text for explanations.

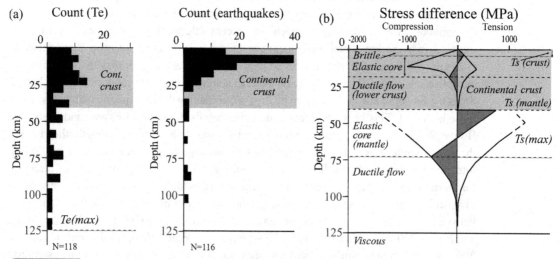

Correlations between thicknesses of elastic Te and seismogenic Ts layers in the continental lithosphere (after Watts and Burov, 2003). (a) Histograms of global compilation of Te and seismicity. (b) A general case of continental lithosphere flexure. Thin solid line – the yield stress envelope for a load that generates a curvature with $r = 7 \times 10^{-7}$ 1/m at a constant background strain rate of 10^{-15} 1/s. Geotherm corresponds to the lithosphere with a 250 km-thick thermal boundary layer. The crust has dry quartz rheology, lithospheric mantle is modeled by olivine with $E^* = 212$ kJ/mol (note that this value is c. 2 times less than indicated by laboratory data, Table 8.2). The brittle behaviour is modeled by Byerlee's law, the ductile behaviour – by an olivine power-law. Young's modulus $E = 80$ GPa, $v = 0.25$. In continental lithosphere the load is supported partly by the upper crust and partly by the lithospheric mantle (the "jelly-sandwich" model). Two main seismogenic layers (one in the uppermost crust and the other in the uppermost mantle) are separated by an aseismic region in the lower crust.

Minimum values of Ts correspond to the topographic surface while the maximum values of Ts correspond to the base of the brittle layer (the depth of brittle–ductile transition). However, neither minimum nor maximum Ts values correspond directly to minimum and maximum Te values. Since in the flexural isostasy approach the loads are supported partly by a strong elastic layer and partly by the brittle and ductile strength of the lithospheric mantle, minimum effective elastic thickness corresponds to the thickness of the elastic core, while the maximum Te values – to the thickness of the entire elastic plate. The reason why Te and Ts appear similar in the oceanic lithosphere is that the depth of brittle–ductile transition is approximately half the thickness of the strong elastic portion of the lithosphere (Watts and Burov, 2003).

On the continents, most seismicity is restricted to the upper crust, but in some regions (e.g. in the collisional orogens of Central Asia) earthquakes occur in the lithospheric mantle (Fig. 8.62). Effective elastic thickness in cratonic settings is significantly larger than the maximum depth of continental earthquakes. Thus, the relationship between Te and Ts is more complicated and less straightforward than in the oceanic lithosphere. In the continental lithosphere, crust–mantle decoupling may occur resulting in "jelly-sandwich" rheology with a rheologically weak lower crust being surrounded by a strong upper crust and a strong uppermost lithospheric mantle. Both of these strong layers have elastic cores that contribute to the strength of the lithospheric plate (eq. 8.49). (Note that there can be more than two strong layers.) However, according to Byerlee's law of brittle failure, brittle strength of a rock linearly increases with depth and, at zero fluid pressure, is ~0.65 of the lithostatic pressure at the corresponding depth (Section 8.1.1). This means that at 30–35 km depth (where most continental seismicity vanishes even in regions with cold and strong crust), theoretical limits on the crustal strength yield 0.6 GPa for tension and 2.0 GPa for extension and require unusually high dynamic forces to break the lithosphere. In accordance with these considerations, rheological models based on regional geotherms for the Charlevoix area in the intraplate stable interior of the Canadian Shield show that seismicity in this area is not controlled by the brittle–ductile transition in the crust, and the thickness of the brittle layer is larger than the maximum depth of earthquakes (~25 km) (e.g. Lamontagne and Ranalli, 1996). Although for realistic values of lithospheric stress Byerlee's law (eq. 8.7) limits the maximum depth of continental earthquakes to ~20 km, deeper earthquakes may occur in the brittle part of the uppermost lithospheric mantle in regions of high lithospheric curvature or high bending stress.

8.3.4 Lithosphere flexure: summary

- The concept of flexural isostasy is based on the assumptions that isostatic equilibrium exists at regional scale and that lithosphere deformation can be approximated by flexure of a thin elastic plate that overlies inviscid fluid (asthenosphere).
- Elastic thickness is commonly estimated by Bouguer coherence and free-air admittance. Both approaches are built on a large set of assumptions which have a strong effect on the uncertainties. In general, these uncertainties cannot be reliably estimated since the solutions (as in any gravity modeling) are non-unique. The accuracy of flexural modeling is

best when the detailed structure of the crust is well known. The approaches do not allow for estimation of large Te values (> 60–100 km).

- The effective elastic thickness of the lithosphere, Te, is a measure of the integrated strength of all lithospheric layers (brittle and ductile) that are supporting the loads and, in general, Te does not correspond to a physical depth. This characteristic of Te hampers, except in some special cases, its direct comparison with thicknesses of different boundary layers and lithosphere thickness in other definitions. The effective elastic thickness of the lithosphere, in general, is representative of the depth to the base of MBL. The latter can be defined by the depth to an isotherm of the brittle–ductile transition (i.e. the depth where the lithospheric strength is the greatest), or by the depth at which the yielding stress or its vertical gradient become less than a particular value.

- Te depends on lithospheric rheology (through elastic parameters – Young's modulus and Poisson's ratio) and its flexural rigidity. The thermal regime of the lithosphere has a critical control of elastic thickness, as both crustal and upper mantle rheologies are strongly temperature dependent. Since rheology is also controlled by composition, crustal thickness is a critical parameter in lithospheric flexure. A thick sedimentary layer works both as a thermal insulator and an additional load, and reduces lithosphere strength. The presence of water has a strong weakening effect on lithosphere strength.

- Deformation of a thick crust or a crust with a layered mineralogy can result in mechanical decoupling of the upper crust from the upper mantle by ductile flow in a weak lower crust. This process takes place only in the continental but not in the oceanic lithosphere. Mechanically decoupled lithosphere becomes weak. Lithosphere-scale faults also weaken the lithosphere, and one may question whether the flexural approach is valid in such cases.

- The elastic thickness of oceanic lithosphere increases with thermal age and is often identified with the depth to 500–600 °C isotherm. This conclusion has been questioned by a recent study, but the interpretations are not unique.

- In the continents, Te does not show a clear and unique correlation with a particular isotherm due to strong heterogeneity of the lithosphere and due to decoupling which is expected in many continental tectonic settings. For these reasons, the dependence of Te on the tectono-thermal age is less evident than for the oceans. Strong structural and compositional heterogeneity of the continental lithosphere, compared to the oceans, results in a large scatter in Te for continental lithosphere of any age. In general, young tectonic structures have small Te (except for some orogens) with temperatures of 200–400 °C at the corresponding depth (that is temperature of brittle–ductile transition in crustal quartz–feldspathic rocks), while cratonic lithosphere has $Te > 50$ km with temperatures of 400–800 °C at the corresponding depth (brittle–ductile transition in mantle olivine).

- In the continents, lithosphere strength is anisotropic. While the correlation of Te anisotropy with seismic anisotropy in Precambrian lithosphere favors its fossil origin, the correlation of Te anisotropy with directions of horizontal displacements and amplitudes of vertical displacements in regions of postglacial rebound suggests its recent origin.

- Elastic thickness and seismogenic thickness both reflect lithosphere strength, however in a different way and on different time-scales, and these parameters are not the same. Lithosphere strength is maintained not only by the brittle crust where most seismicity is concentrated, but also by brittle and ductile layers of the lithospheric mantle.

9

Evolution of the lithosphere

"When a fact appears opposed to a long train of deductions it invariably proves to be capable of bearing some other interpretation."

Sir Arthur Conan Doyle, *A Study in Scarlet* (1887)

In contrast to the previous chapters which discussed lithosphere *structure* based on different geophysical and geochemical data sets, this chapter focuses on lithospheric *processes*. It summarizes data on the lithosphere evolution, from its formation at the early stages of planetary differentiation to plate tectonics processes that lead to lithosphere modification and recycling. The primary focus is on the Precambrian evolution of the continental lithosphere: while the modern oceanic lithosphere is young, the Precambrian crust covers more than 85% of the Earth's geological history in time and occupies more than 70% of the continents in area. Structure and evolution of the Precambrian lithosphere have recently received much attention, and broad discussions and overviews can be found in a number of books (de Wit and Ashwall, 1997; Kemp and Hawkesworth, 2003; Patchett and Samson, 2003; Halliday, 2003; Benn *et al.*, 2006; Brown and Rushmer, 2006; Hamilton, 2007; Hatcher *et al.*, 2007; Condie and Pease, 2008; Santosh and Zhao, 2009).

9.1 Formation of the early lithosphere

9.1.1 Early lithosphere: geological and geochemical evidence

Preamble

Understanding the early evolution of the Earth is one of the fundamental problems in Earth sciences. The knowledge largely builds on recent progress in isotope geochronology and geochemistry and provides insights into the interplay of surface and deep processes that are both consequences and manifestations of the fundamental planetary heat loss. Creating a consistent global picture of the evolution and interaction of the lithosphere, deep mantle, atmosphere, and hydrosphere requires integration of data from a broad variety of Earth science disciplines, including geology, structural geology, tectonics, sedimentology, isotope geology and geochronology, geochemistry, petrology, and geophysics. Since much of the geological information about early Earth evolution is scarce and often restricted to small-size outcrops, regional and global synthesis and comparative analysis are the inherent components of structural and tectonic interpretations which, in turn, form the basis for geodynamic models and provide boundary conditions for them.

In spite of impressive progress in our perception of many processes in the early Earth evolution, there is a significant divergence in interpretations, concepts, and understanding of how the early Earth operated. For example, in the same volume Dewey (2007) and Hamilton (2007) provide two opposing views on the Precambrian evolution. It would have been equally naive of the author of this book to take a particular side in this discussion or to argue with world-renowned specialists. For this reason, this section avoids, as far as possible, highly debated arguments for or against (onset and style of) plate tectonics processes on the early Earth, but rather attempts to summarize the major broadly accepted facts, occasionally providing a word of caution or an alternative interpretation. For the sake of simplicity, Tables 9.1 and 9.2 summarize most of the information provided in this section. Following the convention used throughout the book, the age (i.e. time ago) is given in Ga or Ma, while duration of time intervals is abbreviated by Gy and My (gigayears and megayears). For the location of tectonic provinces discussed in the text see Fig. 9.1.

Chronology of early crust

A few words of caution

Many Archean tectonic settings are derived from geochemical data alone. Such interpretations can be inaccurate and inadequate for the following reasons (Hamilton, 2007):

- Application of tectonic discrimination diagrams based on geochemical data to *modern* oceanic tectonic settings (such as mid-ocean ridges, island arcs, and ocean islands) allows researchers to correctly classify samples with an accuracy of ~60% (Snow, 2006). Clearly, the accuracy of the same approach applied to ancient rocks is lower. For example, based on geochemical data, different studies interpreted the same volcanic assemblage in the Archean NW Pilbara craton as forearc, backarc, oceanic rift, and oceanic arc.
- Archean rocks share the same names as their modern analogs, but their composition is markedly different. The differences in composition between Archean and modern basalts are greater than between modern basalts from different tectonic settings, thus hampering their tectonic interpretations.
- The occurrences and associations of Archean tonalites and basalts that dominate early crustal and supracrustal assemblages, respectively, do not even have modern analogs. Tonalite–trondheimite–granodiorite (TTG) gneisses dominate the Archean middle crust. While modern tonalities form in mature or reworked island arcs which explains their elongated geometry, Archean tonalities form huge terranes such as the Superior Province.
- Although high-Mg Archean basalts do not have modern analogs, they are often interpreted as having originated in oceanic settings. However, there is no direct geological evidence (such as ophiolites or blueschists) for the Archean oceanic crust.

Oldest rocks and oldest crust

Most Archean rocks are dated by the uranium–lead isotope concentrations in zircon (see Chapter 2, Fig. 2.4). Zircon is a common trace mineral in most intermediate-to-felsic rocks. It preserves the age of grain crystallization, which is not reset by later erosion but could have

Table 9.1 The oldest dated or inferred geological events		
Feature	Age (Ga) [Reference]	Location
Formation of solar system	4.566 [B05]	The oldest presently dated meteorite
Final accretion of Earth	4.562 [M07]	The oldest dated basalts in the solar system (angrites from Sahara)
Formation of the Moon	~4.5 [LH95]	The oldest dated Lunar rocks
Core formation	*~4.5 [LH95]*	
Possible oldest crust	*4.5–4.4 [H05]*	*Western Australia*
Possible on-going mantle differentiation	*~4.4 [H05]*	
The oldest dated zircon on Earth	4.37 [H05]	Detrital zircons from the Jack Hills, western Australia
Intense meteorite impacting	4.2–3.9	
The oldest known rocks	4.03–4.00 [BW99]	The Acasta Gneisses, the western Slave Province, Canada
The oldest well-preserved crust	~3.9–3.8 [H78] with a single predecessor at 3.962 Ga	Metasediments in the Itsaq gneiss complex, southern West Greenland
The existence of hydrosphere	At least 3.6	Waterlaid sedimentary and volcanic sequences
The oldest well-documented supracrustal rocks	~3.60 [M06]	Volcano–sedimentary enclaves in 3.82 Ga gneisses, Akilia Island, southern West Greenland. The validity of this interpretation was questioned [W09]
The oldest well-documented volcanic extrusives	~3.5	Pilbara, western Australia and Barberton, South Africa
Start of global cratonization	3.6–3.3	Some cratons (e.g. western Australia, South Africa)
Possible start of some form of plate tectonics	*3.1–3.0 [D07]*	*Archean cratons*
The oldest subduction zone interpreted from seismic reflection data	*2.69 [LH00]*	*Superior province, the Canadian Shield*
Worldwide cratonization	~2.6–2.5	Cratons worldwide
Formation of K-granitoid rich crust	~2.6–2.5	Cratons worldwide
The first great rise of oxygen (the great oxidation event)	2.45–2.2 [F00], with a likely transient elevation in atmospheric and surface ocean oxygenation at 2.8–2.6 Ga [F09]	Cratons worldwide
The oldest obducted oceanic crust	1.90 (perhaps as old as ~2.5 Ga [K01])	Jormua ophiolite, Baltic Shield, Finland (questioned age: ophiolite from Dongwanzi, China)

Table 9.1 *(cont.)*		
Feature	Age (Ga) [Reference]	Location
The oldest known global glaciation	~0.95 [K92]	Cratons worldwide
Numerous ophiolites	0.9–0.8	
Possible start of modern plate tectonics	*~0.6 [D07]*	
The oldest preserved oceanic crust	~0.18	

Regular font: well-documented events; italics – speculated events.
Selected references: [B05] = Baker *et al.*, 2005; [BW99] = Bowring and Williams, 1999; [D07] = Dewey, 2007; [F00] = Farquhar *et al.*, 2000; [F09] = Frei *et al.*, 2009; [H05] = Harrison *et al.*, 2005; [H78] = Hamilton *et al.*, 1978; [H07] = Hamilton, 2007; [K92] = Kirschvink, 1992; [K01] = Kusky *et al.*, 2001; [LH95] = Lee and Halliday, 1995; [LH00] = Ludden and Hynes, 2000; [M06] = Manning *et al.*, 2006; [M07] = Markowski *et al.*, 2007; [W09] = Whitehouse *et al.*, 2009.

Table 9.2 Summary of major Precambrian tectonic events	
Age	Major tectonic and geological features
Archean (3.8–2.5 Ga)	**Paleontological marker**
	The beginning is marked by the first evidence of life ("chemical fossils").
	Typical structural styles
	Horizontal and subhorizontal;
	Dome-and-keel crustal tectonics.
	Major tectonic events
	Cratonization at *c.* 2.5–2.6 Ga (with a few exceptions at 3.3–3.6 Ga);
	Formation of epicratonic basins after termination of cratonization;
	LIPs (three at *c.* 2.7–2.8 Ga and one at 2.45 Ga);
	GDS (at *c.* 2.75 and 2.45 Ga).
	Unique phenomena
	Global cratonization;
	K-granitoid rich crust formed at *c.* 2.5–2.6 Ga;
	TTG made of plutonic rocks (although one Tertiary is known);
	Greenstone belts (mostly 3.5 Ga and 2.7 Ga; some exist in Paleoproterozoic).
	Documented for the first time
	LIPs, GDS;
	Formation of K-granitoid rich crust (~2.6–2.5 Ga).
	Unique composition of the crust
	60% granitoid gneiss;
	30% massive granitoid plutons (TTG);
	10% greenstone belts.
	Mean chemical composition of exposed Ar crust: Na-granodiorite.
	Prevailing crustal associations: medium- to high-grade gneiss-migmatite (metamorphic) terranes.

Table 9.2 (*cont.*)	
Age	Major tectonic and geological features
	Tectonic styles not observed in Archean terranes
	No known continental rifting (except for GDS);
	No evidence for lithospherically distinct oceans and continents (no known ophiolites or blueschists).
Paleoproterozoic (2.5–1.8 Ga)	**Paleontological marker**
	The beginning is marked by the earliest fossils.
	Typical structural styles
	Fold belts.
	Major tectonic events
	GDS (at 2.0–2.2 and 1.8–1.9 Ga);
	Mafic-ultramafic layered igneous complexes (at 2.7–1.9 Ga);
	Worldwide orogenesis at 1.8–1.9 Ga;
	Some greenstone belts.
	Unique phenomena
	Formation of the atmosphere;
	BIF (90% of all known);
	Worldwide orogenesis;
	No LIPs.
	Documented for the first time
	Redbeds (i.e. presence of free oxygen in the atmosphere);
	Ophiolites (the oldest at *c.* 1.9 Ga);
	A high proportion of stable-shelf sediments (i.e. uplift and erosion of nearby cratons);
	Wide-spread fold belts (i.e. compressional tectonics).
	Unique composition of the crust
	63% orthogneiss-migmatite;
	12% massive granitoid plutons;
	25% metasupracrustal rocks.
Mesoproterozoic (1.8–1.1 Ga)	**Typical structural styles**
	Transcontinental anorogenic belts;
	Pericratonic mobile belts.
	Major tectonic events
	Pericratonic mobile belts (as continental accretions and continent–continent collisions);
	Mid-continental (aulacogen-style) rifting that affected entire cratons;
	LIPs (at 1.27 Ga and 1.11 Ga);
	GDS (at 1.55–1.75 and 1.45 Ga, with a broad peak at 1.1–1.3 Ga);
	Transcontinental rifting-related anorogenic belts at 1.4–1.5 Ga;
	Broad platform subsidence.
	Unique phenomena
	Transcontinental anorogenic belts (70% at 1.4–1.5 Ga);
	Broad peak in GDS at 1.1–1.3 Ga;
	A strong correlation between the timing of GDS and LIPs episodes (i.e. assembly of a supercontinent).

Table 9.2 (*cont.*)	
Age	Major tectonic and geological features
	Documented for the first time
	Abundant, wide-spread redbeds (i.e well-developed atmosphere);
	Transcontinental thermal events ("mid-continental rifts");
	Large-scale anorogenic belts (i.e. the existence of a stable supercontinent).
	Unique composition of the crust in transcontinental anorogenic belts
	– Anorthosites,
	– Mangerites (pyroxene monzonite, charnockites),
	– Rapakivi granites.
Neoproterozoic (1.1–0.54 Ga)	**Paleontological marker**
	The end is marked by the global outburst of life.
	Typical structural styles
	Large interior basins;
	Mobile belts.
	Major tectonic events
	Pan-African thermo-tectonic events;
	Formation of the network of mobile belts;
	Large-scale subsidence (at average rate 5 m/Ma) and interior basins formation;
	Continuous GDS (at 1.1–0.54 Ga);
	LIPs (at 800 Ma and two at 723 Ma);
	Break-up of Rodinia (start at ~650 Ma);
	Widespread ophiolites (at 800–900 Ma).
	Documented for the first time
	Glaciations, four recognized as global events (tillites, 950–550 Ma);
	Blueshists.

GDS = giant dike swarms; LIPs = large igneous provinces; BIF = banded iron formations; TTG = tonalite–trondheimite–granodiorite gneisses

The table is based on summaries published in monographs of Windley (1984), Goodwin (1996), Condie (1997) updated for later publications.

been reset had zircon been in contact with hot dry melt. Some of the oldest dated zircons from ancient gneisses are (e.g. Hamilton *et al.*, 1978; Black *et al.*, 1986; Jacobsen and Dymek, 1988; Browring *et al.*, 1989; Harrison *et al.*, 2005; Bibikova *et al.*, 2008): 3.9–4.37 Ga in western Australia (Yilgarn craton) with population maxima at 4.35 and 4.1 Ga but with no gaps in age distribution; 4.03 Ga in northwestern Canada (Slave craton); 3.9–4.1 Ga in southern West Greenland and Labrador (Nutak and Nain cratons); 3.96 Ga in Wyoming craton; 3.93 Ga in Antarctica (Mount Sones, Enderby Land); ~3.7–3.8 Ga in Scotland (Lewisian craton); ~3.8 Ga in the northern part of the North China craton; ~3.3–3.7 Ga in the East European craton (Volga–Uralia domain); ~3.64 Ga in South Africa (Kalahari craton). All of these ages are older that the age of the oldest (~3.6 Ga, see below) supracrustal rocks (i.e. rocks deposited on pre-existing basement rocks), and the only known rocks older than 3.6 Ga are felsic gneisses. Thus some authors propose that continental terranes of 100–150 km in size already existed at >3.6 Ga (Kröner, 1981).

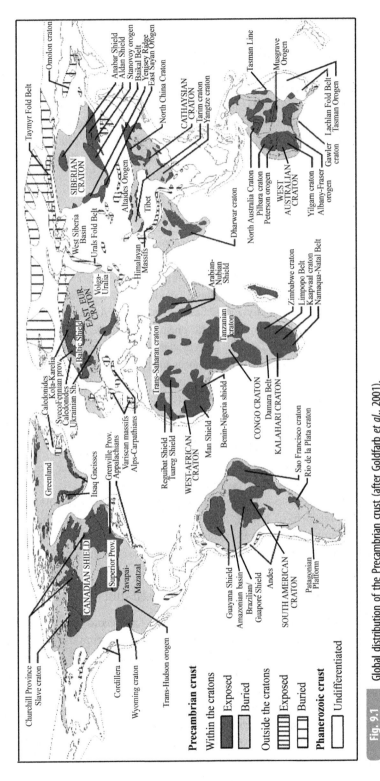

Fig. 9.1 Global distribution of the Precambrian crust (after Goldfarb *et al.*, 2001).

The age of the oldest crust, its volume, and composition are still debated. It is also uncertain whether the age of the oldest dated zircons reflects the age of the main rock mass, whether these zircons are remnants of a recycled protolith, or whether they were added by later magmatism. Commonly, the age of a rock is assigned as the age of its oldest abundant zircons. In particular, based on the geochemistry of detrital zircons from Australia Harrison *et al.* (2005) argue that a thick continental crust was formed globally within 70 million years of the Earth's formation (~4.4–4.5 Ga) and was rapidly recycled into the mantle. Taking the large uncertainties of isotope dating into account, Valley *et al.* (2006) have moderated this claim and proposed that while some crust with uncertain composition could have formed by 4.4 Ga, the oceanic crust formed by 4.2 Ga.

Greenstone belts

A vertical cross-section of the Archean crust typically includes thick supracrustal volcanic sequences that overlay the basement rocks of the middle and lower crust. Archean supracrustal assemblages, that make greenstone belts, commonly contain variably metamorphosed basalts (pillow basalt and high-magnesium basalts, including basaltic komatiites and komatiites) with minor interleaving sedimentary rocks. Seismic studies in Archean terranes indicate that all Archean supracrustal rocks overlay felsic gneisses (but never mantle rocks) with no clear relationship to the basement rocks. The correlation between greenstone belts and basement rocks becomes evident in Proterozoic provinces and progressively increases to younger terranes.

The Archean middle and lower crust is dominated by the relatively felsic tonalite–trondheimite–granodiorite (TTG) gneisses (e.g. Fig. 4.5, left section). The Archean TTG (which are high-pressure, second-stage melts of juvenile mantle-derived basalts) almost do not have the post-Archean analogs and the tectonic settings in which they were formed are uncertain. High-pressure TTG could have formed by slab melting at subduction settings, medium-pressure TTG – by melting at "orogenic" metamorphic conditions, and low-pressure TTG – at intraplate or oceanic plateau settings.

The origin of greenstone belts remains speculative (de Wit and Ashwal, 1997). Many authors interpret them as having formed by ancient oceanic spreading followed by sub-duction–accretion or by *in situ* arc-magmatism at island arc terranes; others interpret them as accretionary orogens formed by plate tectonic processes (e.g. Condie, 2007).

Rheology of Archean crust

Recognition of supracrustal rocks in the geological record is important because their presence indicates that the crust was sufficiently rigid to permit the rise of melts through it and to support the volcanics solidified at or near the surface. The oldest supracrustal rocks are ~3.6 Ga (perhaps even as old as 3.9–3.7 Ga) and are documented in southern West Greenland (Manning *et al.*, 2006). However, interpretation of these rocks as supracrustal has been questioned since the mafic–ultramafic gneiss could have been partly or wholly derived from igneous intrusions (Whitehouse *et al.*, 2009).

A number of authors argue that the Archean lower crust was hot and close to solidus temperature. Paleo- and Mesoarchean (3.8–3.6 Ga) gneisses from the Amitsoq province in southern West Greenland provide striking evidence for extreme ductile mixing of the lower crust (Fig. 5 in Hamilton, 2007). Note that this inference is not universally accepted since other researchers (Bickle, 1978; Burke and Kidd, 1978; England and Bickle, 1984) do not see evidence for massive crustal melting in the Archean high-grade terranes and conclude that the thermal regime of the Archean mantle was not significantly different from that of the present (Fig. 4.38b, insert). The proponents of hot Archean crust (e.g. Brown, 2008) propose that the rheology of the early lithosphere was either of a crème-brûlée type or that the ductile lower crust formed a decoupling zone between the upper crust and the upper mantle (Bailey, 2006). In the case of a crème-brûlée rheology, the lower portions of the lithosphere were unstable and could periodically delaminate. Decoupled rheology could result in continuous gravitational collapse or extrusion collapse, similar to expectations for the present-day Tibetan plateau. The principal difference in tectonic consequences between the two rheologies is the amplitude of mantle instability, as illustrated by numerical tests. Although these results are, most likely, quantitatively invalid for Archean tectonics due to the choice of numerical parameters (lithospheric geotherms, crustal and lithospheric thicknesses, the activation enthalpy for the crust and the lithospheric mantle), they nonetheless illustrate a qualitative difference between the two lithosphere rheologies (Fig. 9.2).

Lower crustal delamination may produce a flat Moho. A flat reflective Moho beneath, at least, some of the Archean terranes of the Canadian Shield and the Baltic Shield (Fig. 3.65 and Fig. 9.3) is similar to the reflectivity pattern observed in the Paleozoic and Cenozoic extensional crust of the Variscan Europe and the Basin and Range Province, western USA, where the lower crust and the lithospheric mantle have been delaminated (Fig. 3.59). A fairly uniform crustal thickness throughout a significant part of the Archean Kaapvaal craton (~38 km) could also have been caused by delamination (note that the composition of the Kaapvaal crust is highly heterogeneous as indicated by significant variations in Poisson's ratio, Fig. 3.37b). Lower crustal delamination could explain the general lack of a thick mafic basal layer observed in many of the Archean greenstone belts. It is worth noting that these patterns are not global. The recent high-resolution Finnish reflection experiment FIRE (Kukkonen and Lahtinen, 2006) has demonstrated that the crustal structure of the Archean terranes of the Baltic Shield is highly diverse (Fig. 9.3).

Cratonization and Archean dome-and-keel tectonics

The start of cratonization at ~3.6 Ga (which continued worldwide until ~2.5 Ga) approximately coincides with the age of the oldest supracrustal rocks. Deposition of thick volcanic and sedimentary sequences on top of the ancient gneisses could have resulted in a density inversion and thermal blanketing of the crust and the mantle. It caused sinking of supracrustal (volcanic and sedimentary) rocks into gneisses (or into the upper crust) and the diapiric rise of circular or elongated domiform batholiths between them (Hamilton, 2007). The result was the dome-and-keel tectonic style typical of Archean granite-and-greenstone terranes in which belts of supracrustal rocks occur as structural troughs between dome-shaped granite–gneiss–migmatite complexes (Fig. 9.4).

This mechanism can be illustrated by the Archean East Pilbara granite–greenstone terrane in western Australia. Granitic rocks are highly enriched in heat-producing elements

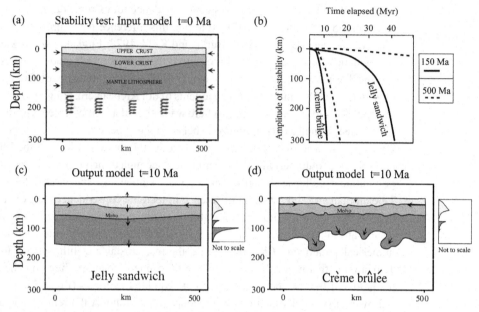

Numerical tests of mountain range stability for two types of lithosphere rheologies (after Burov and Watts, 2006). The failure envelopes associated with the jelly sandwich (crust–mantle decoupling) and crème-brûlée (weak lower crust and weak mantle) rheological models are shown in small inserts. Assumed rheologies are: upper crust – wet quartzite, lower crust – dry diabase and undried granulite, upper mantle – dry olivine for jelly sandwich and wet olivine for crème-brûlée. (a) Input model assumes a free upper surface and a hydrostatic boundary condition at the lower surface (shown by springs at the base). The stability test assumes a 3 km-high and 200 km-wide mountain range with a 36 km thick crust. At $t = 0$ the ridge is in isostatic equilibrium and has a zero topography. Horizontal compression at a rate of 0.5 cm/y is applied at the edges. (b) The displacement of the mantle root as a function of time tracked by a marker initially positioned at the base of the mechanical lithosphere (i.e., at the depth where the strength is 10 MPa) and at the 0 km axis. The thermal structure is modeled for 150 Ma (young and weak) and 500 Ma (old and strong) plates assuming a cooling half-space model with radiogenic heat production in the crust. (c–d) Crustal and mantle structure after 10 My have elapsed. Note that, although the tests illustrate the qualitative difference in rheological evolution of the lithosphere, the model parameters and quantitative results are likely to be invalid for Archean tectonics.

(Fig. 4.8), and a high contrast in heat production between granito-gneisses and supracrustals, in combination with the greenstone-over-granitoid density inversion, provides an efficient mechanism for the generation of crustal dome-and-keel structures (Sandiford et al., 2004). In East Pilbara, voluminous mafic-dominated volcanism and felsic plutonism started at ~3.515 Ga, the widespread tonalite–trondheimite–granodiorite (TTG) suites (now present in the middle crust) were emplaced at 3.47–3.43 Ga, and the formation of the early crust dominated by basaltic magmatism and the regional development of greenstone sequences was completed by ~3.335 Ga. As indicated by stratigraphic and structural relationships, voluminous granitoid plutonism started at ~3.325 Ga and the dome-and-keel crustal structure was formed between 3.325 Ga and 3.308 Ga. Rapid formation of granitic domes can be explained by a "conductive incubation" caused by the burial of radiogenic granitic crust beneath the accumulated greenstone sequences that regionally

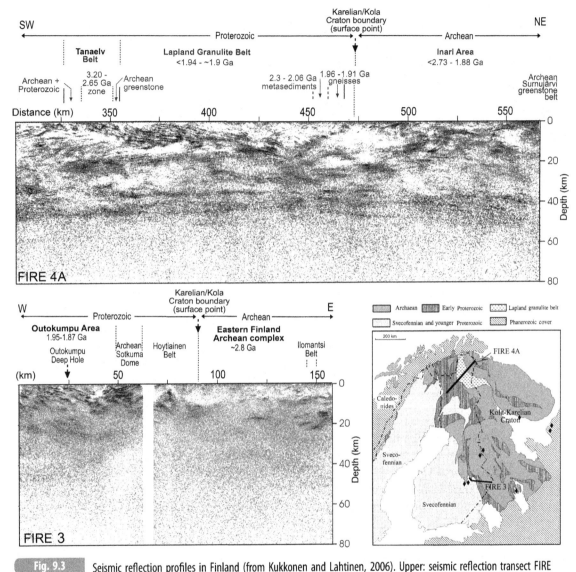

Fig. 9.3 Seismic reflection profiles in Finland (from Kukkonen and Lahtinen, 2006). Upper: seismic reflection transect FIRE 4A in northern Finland; lower: seismic reflection transect FIRE 3 in eastern Finland. Bottom right: location map of FIRE 3 and FIRE 4A seismic reflection transects; black diamonds show locations of diamond-bearing kimberlites and lamproites. The lower crust and Moho are distinctly reflective in northern Finland. In contrast, the lower crust of the Eastern Finland Archean complex is diffuse.

reach a thickness of 12–18 km (Fig. 9.4). The burial of radiogenic, granitoid-rich, TTG suites beneath the ~3.335 Ga basalts could have caused a ~200 °C temperature increase at 35 km depth. A corresponding 2–5 orders of magnitude effective viscosity decrease in the middle–lower crust may have triggered the formation of dome-and-keel crustal structure, further enhanced by the greenstone-over-granitoid density inversion.

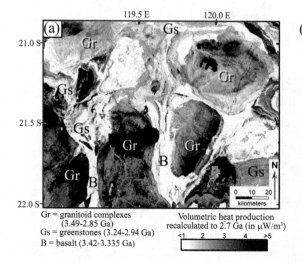

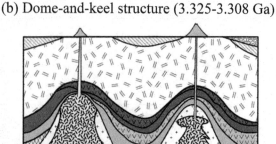

Gr = granitoid complexes
    (3.49-2.85 Ga)
Gs = greenstones (3.24-2.94 Ga)
B = basalt (3.42-3.335 Ga)

Volumetric heat production
recalculated to 2.7 Ga (in µW/m³)
<1   2   3   4   >5

Granitoid complexes
(3.317-3.307 Ga)   TTG (3.47-3.43 Ga)

Basalt (3.42-3.335 Ga)   Early crust (3.515-3.43 Ga)

Fig. 9.4 Crustal dome-and-keel structure of the East Pilbara granite-greenstone terrane (after Sandiford *et al.*, 2004). (a) Volumetric heat production recalculated to 2.7 Ga (the approximate time of cratonization) derived from the calibrated airborne radiometric data set. Dark colors (high radiogenic heat production) correspond to granitoid complexes. (b) Schematic crustal section showing the formation of dome-and-keel crustal geometry which initiated in East Pilbara at ~3.325 Ga. The deposition of basalt (prior to ~3.335 Ga) buried the earliest, highly radiogenic felsic crust. This led to heating of the middle–lower crust and voluminous granitoid formation. The process was amplified by the greenstone-over-granitoid density inversion, which led to sinking of greenstones into the thermally weakened middle crust.

It is worth mentioning that the tectonic origin of the dome-and-keel tectonics of the Archean greenstone belts is still debated, and the hypothesis that this tectonic style has been formed during cratonization is not unanimously accepted. Field studies of a granite–greenstone terrane in Brazil suggest that its dome-and-keel architecture is more than 0.5 Ga younger than the extrusion of the greenstone and that it originated in the Proterozoic during an episode of crustal extension that caused upward transport of hot basement rocks to the base of the supracrustals along a transcrustal normal fault system (Marshak *et al.*, 1992).

The time and style of cratonization varied between different cratons and even within the individual cratons. For example, the NE Pilbara craton (3.3–3.5 Ga) has a simple dome-and-keel style, the NW Pilbara (3.0–3.3 Ga) has both dome-and-keel and laterally disrupted styles, while the S Pilbara (2.5–3.0 Ga) has incomplete regionally continuous greenstone-belt type supracrustal sections with a limited number of batholiths. Similar successions of structural styles are typical of the Zimbabwe and Kaapvaal cratons, giving rise to tectonic models coupled with Pilbara Archean evolution (Zegers *et al.*, 1998). By the end of cratonization (which might have occurred ~0.4–0.6 Ga earlier in western Australia and southern Africa than in other Archean cratons), the lower crust could have lost its mobility, with rigid lithospheric plates gradually being formed. Granitic batholiths play a fundamental role in creating the compositional structure of the continental crust by delivering buoyant silica-rich and incompatible element-rich magmas to the shallow crustal levels.

Chronology of early continental lithospheric mantle

Although ancient crust is often restricted to small-sized outcrops, its dating is a relatively easy task compared to the dating of the lithospheric mantle. Several mechanisms deliver direct samples of the lithospheric mantle to the surface. These include:

● ophiolites;
● abyssal peridotites dragged from the ocean floor;
● massive peridotites brought to the surface in orogenic settings;
● spinel – and garnet–peridotite xenoliths brought to the surface by basaltic, lamproitic, and kimberlitic magmas;
● inclusions in diamonds.

While ophiolites and abyssal peridotites provide samples of the suboceanic lithosphere, massive peridotites and xenolith peridotites deliver fragments of the subcontinental lithospheric mantle. Isotope dating of mantle-derived xenoliths from cratonic settings and mineral inclusions in diamonds allow estimation of the age of the ancient continental lithospheric mantle. However, carbon and sulfur are incompatible elements in the mantle, and diamond inclusion ages are likely to reflect the age of mantle metasomatism rather than the mantle melting events that produced depleted lithospheric mantle.

In contrast to many other isotope systems that can be easily affected by infiltrating melts, the Re–Os system is of particular importance for determining the age of mantle differentiation: Os is compatible while Re is moderately incompatible during mantle melting, and the isotope system Re–Os is only weakly sensitive to later melt contamination (Carlson *et al.*, 2005). A recent compilation of Re–Os isotopic ages for the continental lithospheric mantle in different tectonic settings is shown in Fig. 9.5: T_{RD} and T_{MA} provide minimum and maximum bounds, respectively, on the time of melt depletion and the age of the lithospheric mantle. The

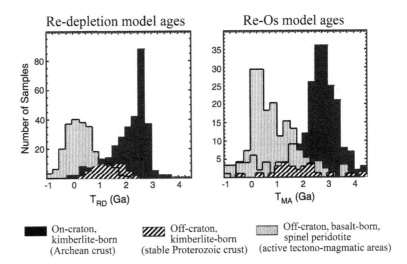

| Fig. 9.5 | Re-depletion and conventional Re–Os model ages for peridotite xenoliths from the continental lithospheric mantle (after Carlson *et al.*, 2005). T_{RD} and T_{MA} provide minimum and maximum bounds, respectively, on the time of melt depletion.

conventional Re–Os model ages T_{MA} indicate the time when a rock sample had the same isotopic composition as the fertile mantle (see Chapter 2 for details) and overestimate the true depletion time if a sample contains extra rhenium added during xenolith capture. Rhenium-depletion model ages T_{RD} are similar to T_{MA} but their calculation assumes that the Re/Os isotope ratio in an undisturbed sample is zero. The T_{RD} will underestimate the true depletion time for a low degree of melting when a significant amount of Re is left in the residue and thus the Re/Os isotope ratio is far from zero. With an increase in degree of melting, the Re/Os isotope ratio in the residue tends to zero, and the accuracy of T_{RD} increases.

The major results of Re–Os isotope studies of the continental lithospheric mantle demonstrate the following.

(1) There is a broad general correlation between the age of the crust and the age of the underlying lithospheric mantle: the older crust has an older lithospheric mantle. The median T_{RD} and T_{MA} ages are 2.4 Ga and 2.8 Ga, respectively, for the Archean provinces, 1.4 Ga and 2.1 Ga for stable Proterozoic crust, and 300 Ma and 700 Ma for young tectonically active regions (Fig. 9.5). Clearly, averaging over different tectonic regions worldwide smears regional patterns.

(2) At a regional scale, a pronounced correlation between the age of the crust and the age of the lithospheric mantle is well established for cratonic South Africa (Fig. 2.18). Similar correlations have been observed in the Siberian and Slave cratons and in East Greenland. The correlations are interpreted as first-order evidence that the crust and the lithospheric mantle have been formed by the same mantle melting events, and since then have remained mechanically coupled. The latter conclusion is supported by the fact that the Archean age of the lithospheric mantle has not been determined for any crustal province with geological ages younger than Archean. However, in general, it does not necessarily mean that such locations do not exist; simply they may not have been sampled by xenoliths. For example, if the Proterozoic Grenville Province were sampled close to the Grenville Front by mantle-derived xenoliths, they would be likely to show Archean ages similar to the neighboring Abitibi Province (Fig. 3.67). Archean–Paleoproterozoic lithospheric mantle has been identified in some provinces with young tectono-thermal ages (Section 2.2).

(3) A detailed analysis of vertical distribution of ages of the lithospheric mantle as sampled by the South African kimberlites and kimberlites from the Proterozoic Somerset Island (Arctic Canada) does not show any correlation between the depth and the age of the lithospheric mantle (Figs. 9.6 and 9.7). The basic conclusion from this observation is that the Archean and Proterozoic lithospheric mantle was not formed by vertical stacking or gradual accretion from below, unlike the oceanic lithosphere during its cooling (Fig. 9.8).

(4) Chemically stratified lithospheric mantle is recognized beneath the Slave craton in northern Canada (Chapter 6.1 and Fig. 6.17a). Although some authors speculate that the lower, more fertile, part of the lithospheric mantle is younger, no vertical profiles for age variations with depth, similar to those presented for South Africa and the Somerset Island, are available at present to support this hypothesis.

(5) The Archean lithosphere, or at least its lower portion, does not last forever. In several cratons (Wyoming, North China, Tanzania) the Archean lithospheric mantle exists only down to 140–150 km depth, and below this depth is underlain by the Phanerozoic

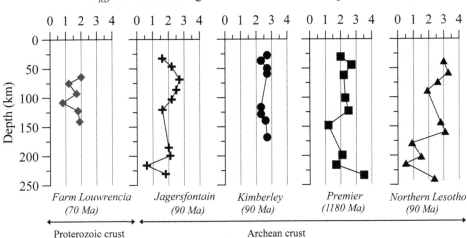

T_{RD} Re-Os modal ages of South African lithospheric mantle (Ga)

Fig. 9.6 Depth (constrained by mineral thermobarometry) versus Re-depletion ages for peridotite xenoliths from five kimberlite pipes in the South African craton (based on results of Pearson, 1999). The ages of kimberlites are 70–90 Ma except for the Premier pipe.

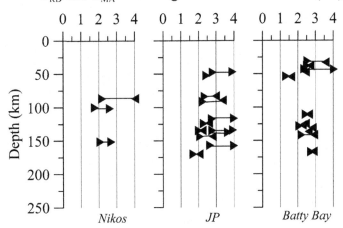

T_{RD} and T_{MA} modal ages for Somerset Island (Ga)

Fig. 9.7 Depth (constrained by mineral thermobarometry) versus Re-depletion and model ages for peridotite xenoliths from three kimberlite pipes in the Proterozoic Somerset Island (Arctic Canada) (based on results of Irvine et al., 2003). T_{RD} and T_{MA} provide minimum and maximum bounds, respectively. In some cases, the inferred ages exceed the age of the Earth (shown by arrows). There is no correlation between the depth and the age of the lithospheric mantle.

lithospheric mantle. Xenolith data from these cratons have been interpreted as evidence that the lower portion of the Archean lithospheric mantle has been lost during Phanerozoic major tectono-thermal events (Chapter 6.1 and Fig. 6.17d). In accordance with this inference, the Re–Os data from the northern Lesotho kimberlite pipe could indicate Neoproterozoic–Phanerozoic ages below 160–170 km depth, except for the deepest sample which can be from an older recycled protolith (Fig. 9.6).

Models of the early lithosphere formation

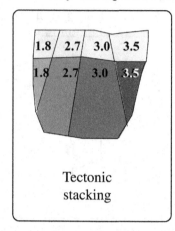

Instantaneous
formation

Tectonic
stacking

Growth from
below

Fig. 9.8   Sketch of three end-member models of lithospheric growth in the early terranes. Numbers – possible representative ages in Ga. The depth distribution of lithospheric ages constrained by isotope analysis of xenolith data does not support models of instantaneous or gradual lithospheric growth from below. The model of tectonic stacking of Archean terranes with similar ages of the crust and the lithospheric mantle is the one supported by the existing data.

To sum up, isotope dating of the Archean lithospheric mantle allows for the elimination of several possible model of early lithosphere formation (Fig. 9.8). Neither the models of instantaneous formation nor of gradual lithospheric growth from below are supported by geochemical data, at least for those cratons where such data are available. The model of tectonic stacking of Archean terranes with similar ages of the crust and the lithospheric mantle is the one supported by the existing isotope data on ages of the cratonic lithospheric mantle.

9.1.2 Models of early lithosphere formation

Preamble

Because of the short lifetime of the oceanic lithosphere, which is recycled to the mantle at a time-scale of ~200 Ma, the cratonic lithosphere of stable continents has received most attention in studies of the processes of lithosphere formation. These processes in relation to the formation of continental crust and continental lithospheric mantle are discussed in recent reviews (Bennett, 2004; Rudnick and Gao, 2003; Kemp and Hawkesworth, 2003; Lee, 2006). Owing to the complexity of geological and geochemical data on the structure and composition of Archean lithosphere and, often, to the non-uniqueness of data interpretations, no model of formation of early continental lithosphere can explain the entire observation set. Most of the evolutionary models look for present-day analogs of tectonic processes that occurred in the early history of the Earth and assume that lithospheric growth followed a few major universal scenarios (the principle of "uniformitarianism"). Although, undoubtedly, present-day processes provide the key clues to those that operated in the past, the extent to which the uniformitarianism approach is applicable to early processes of lithosphere growth and differentiation remains a matter of debate, since many Archean rocks are unique and do not have modern analogs.

Geological and geochemical data indicate significantly different compositions of the Archean and post-Archean crusts. Secular trends in the composition of the lower crust are difficult to distinguish due to difficulties with its sampling (Rudnick and Gao, 2003).

(1) Several types of rocks (such as TTG which dominates the earliest continental crust) are characteristic of the Archean crust and do not have post-Archean analogies.
(2) The crust at the Archean time was less mafic than the post-Archean crust. This is reflected in a systematic difference in the composition of Archean and post-Archean sedimentary assemblages (Taylor and McLennan, 1985, 1995; Condie, 1993).
(3) Felsic magmatic rocks exhibit a secular variation in composition (Kemp and Hawkesworth, 2003), related to (i) mantle cooling (for example, temperatures required for remelting basaltic rocks in the garnet stability field were rare in the post-Archean evolution); (ii) change in atmospheric conditions (crustal differentiation in the Archean may have been dominated by conditions when oxygen fugacity was less than in the post-Archean).

Different compositions of the Archean and the post-Archean crusts may be attributed to different tectonic processes that have operated at different stages of the Earth's evolution (Taylor and McLennan, 1995):

● different mechanisms of crustal growth on the Archean Earth;
● different nature of mantle sources for crust production;
● different processes of crustal differentiation from the mantle.

Geochemical constraints on early evolution

The continental crust is complementary to the depleted upper mantle for many trace elements (Taylor and McLennan, 1985; Hofmann, 1989). A comparison of trace-element abundances in the average continental crust and normal mid-ocean ridge basalts (N-MORB) demonstrates that enrichment of the continental crust in incompatible elements is mirrored by depletion in the same elements in N-MORB (Fig. 9.9). The complementary pattern of trace-element abundances makes the basis for the key concept in mantle geochemistry: the upper mantle is depleted (as compared to the primitive mantle) primarily due to extraction of the continental crust from it. Based on the geochemical arguments, McCulloch and Bennett (1998) propose that the mass fraction of the depleted mantle is 0.5 ± 0.15 of the whole mantle. Since it exceeds the entire volume of the upper mantle (that is 0.29 of the total mantle volume), the depleted mantle should fill not only the upper mantle, but also a significant portion of the lower mantle.

The depleted mantle is the source for mid-ocean ridge basalts (MORB): melting and extraction from it of MORB produces the chemically complementary oceanic crust (the melt) and residual depleted mantle (the residue) (O'Neill and Palme, 1998) (Fig. 9.10). Cooling of the thermal boundary layer, when the oceanic crust moves away from the mid-ocean ridge, produces oceanic lithosphere with the same composition as the surrounding mantle. In subduction zones a significant amount of the oceanic crust and the associated residual depleted mantle is recycled back into the depleted mantle. However, some of the most highly

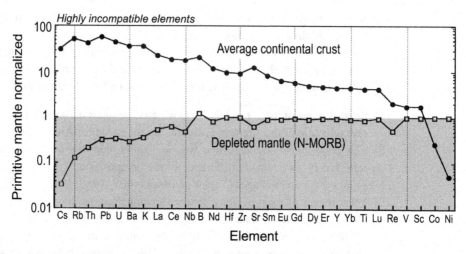

Fig. 9.9 Trace-element abundances in the average continental crust (Taylor and McLennan, 1985) and the depleted upper mantle). High enrichment of the continental crust by highly incompatible trace elements as compared to primitive mantle and depletion of normal mid-ocean ridge basalts (N-MORB) by the same elements makes the basis for assuming that the processes of continental crust formation and upper mantle depletion are complementary.

incompatible trace elements are brought back to the near-surface environment by island-arc magmatism which plays an important role in formation of the continental crust.

Secular trends in Nd, Hf, and Os isotope concentrations are commonly assumed to indicate progressive mantle differentiation by melt depletion (extraction) and corresponding growth of the continental crust (Figs. 2.6, 2.10 to 2.12). These elements did not participate in core formation, and the Bulk Silicate Earth (the mantle plus the crust) is thought to have chondritic relative abundances of rare earth elements. However, these elements fractionate during magmatic differentiation. Melt extraction from the primitive mantle causes light rare earth elements (LREE) such as Nd and Hf to partition into melt. As a result, the LREE-enriched continental felsic crust formed by mantle melting is relatively unradiogenic with $\varepsilon_{Nd} < 0$, while the upper mantle (melting residue) has radiogenic high-neodymium and high-hafnium ratios ($\varepsilon_{Nd} = +10+12$). The higher the mantle values of ε_{Nd} and ε_{Hf}, the more depleted the mantle composition and the greater the volume of produced continental crust. Both ε_{Nd} and ε_{Hf} increase with time from near-zero values typical of the primitive upper mantle to positive values that are already observed in the Archean mantle. Similarly, γ_{Os} systematically increases with time suggesting progressive mantle differentiation since, at least, Neoarchean. Some geochemical interpretations of Hf and Nd from the oldest zircons seem to require a very early (>4.3 Ga) differentiation of, at least, parts of the mantle (e.g., Bennett *et al.*, 1993; Amelin *et al.*, 1999; Albarède *et al.*, 2000).

For the known amount of early Archean crust, simple crustal extraction cannot explain the degree of depletion of the Archean lithospheric mantle observed in some interpretations of initial ε_{Nd} and ε_{Hf} compositions (Bennett, 2003). Similarly, while the composition of the Archean lithospheric mantle is consistent with a voluminous removal of komatiites and high-Mg basalts from the mantle, the volume of the Archean lithospheric mantle is too large to be complementary to the volume of these magmas if all of them are emplaced in supracrustal settings (Hamilton, 2007). Otherwise, if the present volume of (survived)

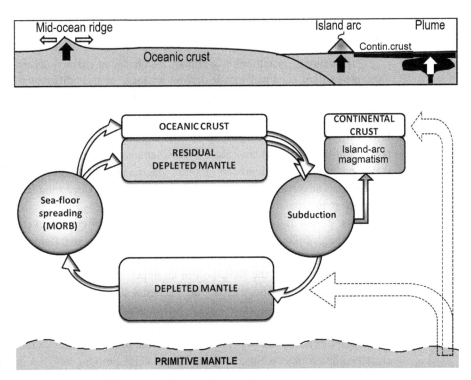

Fig. 9.10 Sketch illustrating major processes of crustal growth and recycling. Upper mantle is depleted as compared to the primitive mantle, primarily by extraction from it of the continental crust (dotted arrows); this key concept in mantle geochemistry is based on the complementary pattern of trace-element abundances (Fig. 9.9). Melting of the depleted mantle and extraction from it of mid-ocean ridge basalts (MORB) produces the oceanic crust (the melt) and the residual depleted mantle (the residue) which are chemically complementary. A significant amount of the oceanic crust and the associated residual depleted mantle is recycled back into the depleted mantle in the subduction zones. However, island-arc magmatism brings some of the most highly incompatible trace elements back to the near-surface environment, thus playing an important role in formation of the continental crust.

Archean felsic crust is representative of the degree of Archean mantle differentiation, the Archean lithospheric mantle is far too depleted. This controversy can be resolved if the Archean lithospheric mantle is complementary to the survived Archean crust plus the early (>4.0 Ga) ~100 km-thick continental crust that presumably was recycled into the mantle in the Hadean–Paleoarchean (Hamilton, 2007). Although this model can resolve the problem regarding lithosphere volume, it raises questions about what mechanisms could provide an exceptionally fast crustal growth rate at the early stages of the Earth's evolution.

Formation of early continental crust

While oceanic crust is formed by mantle melting at mid-ocean ridges, scenarios for continental crust generation are highly debated. In spite of the important role granitic (in the broad sense) rocks play in crustal evolution, their petrogenesis remains controversial, in part because of a large lithological and geochemical diversity of granites. Given that granitic magmas cannot be extracted directly from the upper mantle, three ways by which the granitic crust can be formed include:

"(i) fractional crystallization of primary basaltic liquid, (ii) mixing between partial melts of pre-existing crust and mantle-derived magmas, or their differentiates, and (iii) partial melting of young, mantle-derived mafic protoliths in the crust" (i.e. remelting within the crust) (Kemp and Hawkesworth, 2003).

Except for some specific conditions (such as can exist in migmatite terranes) even the basal portions of the thickened crust do not reach temperatures high enough (>800–900 °C) for partial melting and for significant granitic magma generation on time-scales of up to 100 My, and thus advection of mantle heat to the crustal base (for example, by underplating or interplating by mafic magmas) is required to produce voluminous granitic magmas (Clemens, 2006). If granitic magmas contain a component of juvenile mantle-derived material, granitic plutonism leads to the generation of new crust. Removal of large volumes of granitic magmas from the lower parts of the crust and their emplacement within the upper crustal levels results in the formation of an upper felsic crust enriched in heat-producing elements, and leaves mafic, partially dehydrated residue in the lower crust.

A comparison of variations in Nb/La, Sr/Nd, and Rb/Sr isotope ratios for average crustal compositions with the same rations for the primitive mantle, normal mid-ocean ridge basalt (N-MORB), ocean island basalt (OIB), and island arc tholeiite lavas, indicates that average crustal compositions do not correspond to any simple mixing between primitive mantle, or island arc tholeiites, or OIB (Kemp and Hawkesworth, 2003). In particular, because of a relatively high Mg# of the bulk continental crust, it cannot be produced by known tholeiite fractionation trends since such melting results in iron enrichment. Geochemical processes proposed for crustal growth include calc-alkaline differentiation similar to that observed in orogenic andesite suites or melting of high-magnesium andesites. The latter may be generated by partial melting of subducted oceanic crust with a subsequent melt contamination with the peridotites of the overlying mantle wedge (Kelemen, 1995).

The tectonic settings that most crust formation models assume include either plume-related magmatism, or compressive tectonic settings, including accretion of oceanic plateaus to continents and subduction related processes (Fig. 9.11). These scenarios are not mutually exclusive. Plume-related generation of a new crust provides a convenient explanation for the origin of the Archean TTG, since a consistent subduction signal is not observed in many greenstone belts. However, in other locations subduction related processes can explain compressive tectonic regimes at the depth of melting of the TTG and dislocation of TTG magmas from their voluminous ultramafic residues. The latter are recycled into the mantle by subduction of oceanic lithosphere. A compromise hypothesis (Kemp and Hawkesworth, 2003) assumes that, first, the basaltic protoliths to the TTG are formed in a plume-related environment and then, during subduction of this young mafic crust, they are remelted to form the silicic magmas. Alternatively, Zegers and van Keken (2001) have proposed that the earliest continental crust could have formed by a process similar to foundering of a continental arc root beneath the southern Sierra Nevada, western USA. Delamination of the lower part of an oceanic plateau protocrust that equilibrated at eclogite facies could have resulted in the production of the TTG suites and the formation of the earliest continental crust. Accretion of oceanic plateaus to continents also could

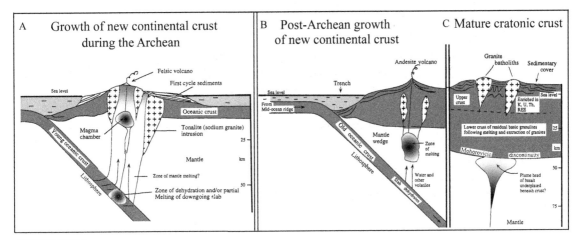

Fig. 9.11 Sketches showing the difference between Archean (A) and post-Archean (B, C) crustal development in subduction zone environments and when a plume impinges onto the base of the lithosphere. In Archean time, the subducting oceanic crust was younger and hotter, and melting occurred within the subducting slab. In post-Archean time, the subducting oceanic crust is old, and melting occurs within the mantle wedge (after Taylor and McLennon, 1995).

have played an important role in early Earth evolution, in particular because thick, buoyant Archean oceanic plateaus were difficult to subduct (in the case if subduction operated on early Earth). Crustal growth by this mechanism decreased with time and does not appear to be significant on modern Earth where most of the continental crust is generated at arc settings.

Formation of Archean lithospheric mantle

A multistage process of formation of the continental crust which involves production and differentiation of mafic magmas through a series of complex processes and further recycling of the continental crust to the convecting mantle at subduction zones makes the relationship between generation of the continental crust and the lithospheric mantle very complex. Since the composition of the average continental crust is, in general, complementary to the composition of the depleted upper mantle (Fig. 9.9), the processes responsible for crustal growth and formation of the lithospheric mantle should be essentially the same. Yet the origin of the Archean lithospheric mantle is not well understood because its unique composition requires unusual mantle melting conditions (see Section 6.1.2 for a detailed discussion). A ~30–50% melting at P~3–7 GPa can explain the composition of Archean harzburgites, the most depleted rocks in the Archean lithospheric mantle (Walter, 1998, 1999) (Fig. 6.6c). However, because the Al_2O_3–MgO systematics of cratonic peridotites is indistinguishable from Phanerozoic spinel peridotites (Bernstein et al., 1998), some authors challenge the requirement for high pressure melting and argue that highly depleted harzburgites could have been formed within a spinel stability field at P~1–4 GPa with their later transportation to 90–210 km depth (e.g. Canil, 2004; Herzberg, 1999).

The scenarios for the formation of the cratonic lithosphere are very diverse and, similar to the models of crustal growth, include models of lithosphere growth by tectonic and non-tectonic processes.

Models of lithosphere growth by tectonic processes:

- repeated cycles of differentiation and collisional thickening (Jordan, 1988);
- accretion of collided island arcs of depleted composition (Ashwal and Burke, 1989);
- imbrication of shallowly subducting (depleted) slabs (Helmstaedt and Schultze, 1989) and modification of this model, which assumes that subducting slabs are made of depleted material with trapped wedges of fertile mantle (Kusky, 1993);
- slab melting in subduction zones and uplift of Si-enriched magmas into the lithosphere (Barth *et al.*, 2001).

Models of lithosphere growth by non-tectonic processes:

- basal plume accretion (Campbell and Griffiths, 1992; Pearson *et al.*, 1995; Thompson *et al.*, 1996) and modification of this model which assumes episodic large-scale mantle avalanches (Condie, 1998).

These models can be reduced to three end-member scenarios of formation of lithospheric mantle (Table 9.3): (1) large-scale mantle melting caused by heads of mega-plumes or mantle overturns; the melt residues form the lithospheric mantle; (2) stacking/accretion of oceanic lithosphere; (3) arc accretion including accretion of large turbidite fans. The latter process may occur at convergent tectonic margins where subduction processes, that involve accretion of large submarine sediment fans (which contain large volumes of continental detritus), fragments of juvenile oceanic crust, and newly added continental crust produced by magmatic underplating, lead to the continental growth from marine successions. These three scenarios are not mutually exclusive and can be complementary, contributing to the lithosphere growth in the same locations. Crustal/lithosphere growth induced by mantle plume melting can be the oldest process since it does not require the existence of mechanically strong lithospheric plates. Associated magmatic underplating can further contribute to crustal/lithosphere growth. Table 9.3 summarizes evidence for and against these models as discussed by Lee (2006). However, these models cannot be reliably discriminated at present because of insufficient or controversial data on many of the early lithospheric properties.

9.2 Lithospheric processes in time

9.2.1 Plate tectonic processes and mantle dynamics

Mantle convection and mantle plumes

Introduction to mantle convection styles

Mantle convection transports heat by material advection, driven by gravity forces which cause ascent of material lighter than the surrounding mantle and descent of material that is denser. Density anomalies that drive thermo-chemical convection in the mantle can be produced by differences both in temperature (Rayleigh–Bénard instabilities) and in composition (Rayleigh–Taylor instabilities). Rayleigh–Bénard instabilities develop in a fluid that is

Table 9.3 Models of lithosphere formation

Observations	Large-scale melting caused by plume head or mantle overturns	Stacking of oceanic lithosphere (i.e. ~vertical stacking)	Arc (island-arc) accretion (i.e. ~ horizontal stacking)
Melting conditions			
Temperature conditions	30–50% melting requires potential $T > 1650$ °C (the hottest modern plumes have 1400–1600 °C)	High T are not required since melting is due to adiabatic decompression	
Pressure conditions	3–7 GPa	Requires that cratonic peridotites were partially formed within spinel stability field (1–4 GPa)	Cratonic peridotites could have been partially formed within spinel stability field (1–4 GPa)
Specific melting conditions	No	No	Within fluid-rich mantle wedge in a subduction zone; can explain anomalously Si-rich cratonic harzburgites
Absence of large volume of komatiites	Komatiites are produced at high-P melting (7–10 GPa), but they can be neutrally buoyant and sink	Low-T melting due to adiabatic decompression does not produce large volume of komatiites	
Geophysical signatures			
Absence of vertical age-stratification in cratonic lithospheric mantle	Yes	Can be explained only if formed in less than 0.5 Ga (uncertainty in Re–Os ages)	?
Positive buoyancy of cratonic lithospheric mantle	Positive	Negative, unless a hotter mantle and more vigorous convection in the Archean led to subduction of oceanic lithosphere younger than 10 Ma (at <10 Ma it is positively buoyant, Lee, 2003, Cooper et al., 2006)	Positive
Vertical stratification of mantle depletion (decreasing with depth)	Yes, Mg# = 91–92 at the top and 88 at the base	No, a periodic variation of fertility with depth is expected unless the oceanic lithosphere is chemically stratified and its lower part is removed during accretion	No, highly depleted mantle throughout the entire lithospheric mantle

	Table 9.3 (*cont.*)		
Observations	**Large-scale melting caused by plume head or mantle overturns**	**Stacking of oceanic litho-sphere** (i.e. ~vertical stacking)	**Arc (island-arc) accretion** (i.e. ~ horizontal stacking)
Secular trends (decrease of mantle depletion from Archean to Phanerozoic)	Yes, hotter Archean mantle → larger and hotter plumes, large mantle overturns	Yes, hotter Archean mantle → more vigorous convection, faster spreading rates, higher probability of stacking/accretion	
Geochemical signatures			
Presence of eclogites in kimberlites	?	Yes	Yes
Garnet trace-element signature (observed in cratonic peridotites)	Required	Not required	Not required ?
Two MgO groups of cratonic garnet peridotites	?	The low-MgO group has composition similar to modern basaltic oceanic crust (i.e. had oceanic protolith)	The high-MgO group has been interpreted to be formed at $P < 2$–3 GPa (Barth *et al.*, 2002) and represent primitive cumulates associated with continental arc magmatism
Proponents			
	Walter, 1998, 1999; Perason *et al.*, 1995b; Griffin *et al.*, 2003a	Sengör and Natal'in, 1996; Lee, 2006	Jordan, 1988; Abbott and Mooney, 1995; Herzberg, 1999; Lee, 2006
Model variations			
	Pearson *et al.*, 1995; Thompson *et al.*, 1996: Condie, 1998	Helmstaedt and Schultze, 1989; Kusky, 1993	Jordan, 1988; Ashwal and Burke, 1989; Rudnick, 1997

heated from below and cooled from above. These conditions are satisfied in the Earth's mantle which is heated from below by the core and is cooled from above at the surface; radioactive elements within the mantle provide additional internal heating of the mantle.

Internal heating as well as non-constant (e.g., temperature-dependent) viscosity of the mantle are critical for the generation of plate tectonics phenomena and mantle plumes (Davies, 1999; Schubert *et al.*, 2001). Numerical models demonstrate that the dominating pattern of three-dimensional convection with bottom and internal heating is the presence of upwelling cylinders (plumes) and long arcuate downwelling sheets (that resemble subducting slabs). The number of upwelling plumes, their size, and amplitude depend on the percentage of the total heat loss due to bottom heating; plumes transport the heat that enters the convective system from below, and in the end-member case of pure internal heating no plumes develop. With decreasing percentage of internal heating, there are fewer plumes but they transport more energy and merge with time into a smaller number of more vigorous plumes (Bercovici *et al.*, 1989). The structure of mantle convection is strongly time dependent and the upwelling and downwelling structures constantly evolve through time.

Effect of phase boundaries

The presence of two major phase transition boundaries (at 410 km and 660 km) has a strong effect on the mantle convection pattern. The phase change at 410 km depth from olivine to spinel is exothermic; it causes heat release and heating of the rock by the value $\Delta T = L/Cp \sim 90$ °C, where L is latent heat of phase reaction (~90 kJ/kg) and Cp is the specific heat at constant pressure (~1 kJ/kg/K). The phase change at 410 km depth enhances mantle convection and does not prevent subduction of downgoing lithospheric slabs (Fig. 9.12). The phase change at 660 km depth is endothermic and is associated with the transformation of spinel structures to perovskite and magnesiowüstite. It causes heat absorption that reduces rock temperature by $\Delta T \sim -70$ °C (for $L = -70$ kJ/kg) and blocks subduction of lithospheric slabs.

The effects of the 410 km and 660 km phase transition boundaries on mantle flow are opposite and their relative effects on mantle convection can be evaluated by the ratio (Schubert *et al.*, 2001):

$$P_{410}/P_{660} = [\gamma \Delta \rho / \rho]_{410} / [\gamma \Delta \rho / \rho]_{660} \tag{9.1}$$

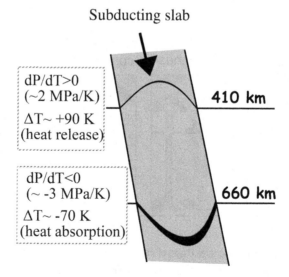

Subducting slab

dP/dT>0
(~2 MPa/K)

ΔT~ +90 K
(heat release)

410 km

dP/dT<0
(~ -3 MPa/K)

ΔT~ -70 K
(heat absorption)

660 km

Fig. 9.12   Sketch illustrating the effect of phase boundaries on slab subduction. Black lines illustrate the Clapeyron slopes.

$$\text{where } \gamma = (dP/dT)_c = L/T\Delta V = Q\rho_1\rho_2/T\Delta\rho \tag{9.2}$$

is the Clapeyron slope of the phase transition, L is the latent heat of the phase reaction from phase 2 to phase 1, T is the temperature, ΔV is the volume change of the phase transition, Q is the latent heat of the phase reaction per unit mass, ρ is density, and $\Delta\rho = \rho_2 - \rho_1$ is the density difference between the initial (lower) phase and the resultant (upper) phase. Equation (9.1) assumes that the following parameters are constant: density, coefficient of thermal expansion, and the thickness of the convective region. At 410 km the phase change, $\Delta\rho/\rho$ is ~5% and $\gamma = 1.5$–3 MPa/K; at 660 km discontinuity $\Delta\rho/\rho$ ~10% and $\gamma = -2$–4 MPa/K. Thus, P_{410}/P_{660} is in the range between ~0.2 and ~0.7. In the real Earth, the ratio is smaller because of a decrease of the coefficient of thermal expansion with depth in the transition zone. Thus, in mantle convection the effect of the 660 km discontinuity dominates over the effect of the 410 km discontinuity because the magnitudes of the density change and the Clapeyron slope of the phase transition are larger at the deeper discontinuity.

The presence of phase transition boundaries has led to the development of three classes of mantle convection models (Fig. 9.13a):

- whole mantle convection, in which the upwelling and downwelling structures cross the phase boundaries and there is a constant material circulation in the entire mantle volume with an approximately adiabatic temperature distribution with depth. Geochemical data, however, suggest the existence of isolated mantle reservoirs and thus do not support whole mantle convection (Allègre et al., 1983);
- layered mantle convection, in which the phase change at 660 km depth makes an impenetrable barrier for subducting slabs. In such a case, the upper mantle and the lower mantle should have two completely independent convective systems. The transition zone can also effectively work as a water filter (Bercovici and Karato, 2003). However, seismic tomography images suggest that at least some subducting slabs can penetrate into the lower mantle (van der Hilst et al., 1997) (Fig. 3.103);

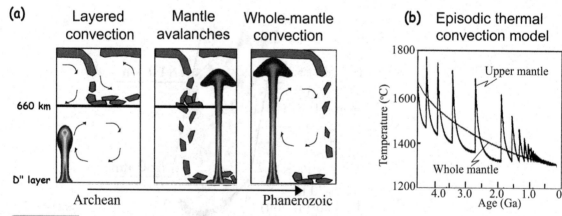

Fig. 9.13 (a) Sketch illustrating three classes of mantle convection models (modified after Condie, 1997). Dimensions not to scale. (b) Convective model of episodic thermal evolution in which convective overturns (marked by temperature peaks) are triggered only by the temperature difference between the upper and the lower convective systems (after Davies, 1995).

- hybrid models, in which the barrier at 660 km depth is strongly time dependent. The mantle avalanche model is an example of such a type of mantle convection (Tackley *et al.*, 1993).

In the case of layered mantle convection, four thermal boundary layers are, in general, formed in the system: one on the top and one at the bottom of each of the convective systems. The boundary layers at the interface between two convective systems (i.e. at around a 660 km depth) are of a particular interest. The boundary layer at the top of the transition zone is at a relatively high temperature and has a relatively low viscosity, whereas the boundary layer at the top of the lower mantle would have a relatively high viscosity and would be coupled more strongly to the interface (Davies, 1995, 1999). The efficiency of slab penetration through the transition zone depends on the age of subducting lithosphere (old and thick lithosphere is more likely to penetrate), the temperature difference between the upper and lower convecting systems, and the efficiency of heat transfer through the internal thermal boundary layers. The role of the latter in creating convective overturns has been examined by Davaille and Jaupart (1993) who studied convection in the case of a stiff, high-viscous boundary layer at the top of the lower mantle. Laboratory simulations complemented by numerical modeling indicate that the temperature difference between the upper and the lower convective systems and the corresponding change in viscosity trigger convective overturns in the mantle (Fig. 9.13b).

Mantle avalanches

Mantle avalanches are caused by the endothermic phase boundary at 660 km depth and do not appear in numerical models of mantle convection without this boundary. The mechanism of mantle avalanches is as follows. During the development of mantle convection, the 660 km discontinuity forms a barrier for whole-mantle circulation. Cold (and heavy) sheets of downgoing material cannot penetrate into the lower mantle but flow laterally and pile on top of the phase boundary. The exothermic phase boundary at 410 km depth promotes downflow of mantle material and its accumulation in the transition zone. When a significant amount of heavy downgoing material is accumulated, the pile sporadically (in time and space) breaks through the phase boundary and rapidly sinks into the lower mantle as massive cylindrical avalanches (Fig. 9.13ab). The latter spread out and fall into the lower mantle or at the core–mantle boundary as pancakes and effectively cool the core. Massive downflow of mantle material causes an upflow of hot mantle material to conserve mass balance and may trigger mega-plumes, which may have their origin near the core–mantle boundary. Passage of a plume through the endothermic phase boundary heats the plume material due to heat absorption during the phase transformation and gives birth to hotspots. As a result of downgoing mantle avalanches and upwelling mantle plumes, a two-layer mode of mantle convection sporadically becomes whole-mantle convection with a significant mass exchange through the phase transition barrier in both directions.

Mantle plumes

Both numerical modeling and laboratory experiments indicate that, under specific conditions (e.g. internal heating and non-isoviscous mantle rheology), plume-like structures of upwelling buoyant mantle material are typical for thermo-chemical convection and thus should play an important role in mantle dynamics and lithosphere evolution (Davies, 1999; Schubert *et al.*, 2001; Davaille and Limare, 2007). Mantle upwellings (plumes) are proposed to play a critical role both in lithosphere growth and reworking (Condie, 2001; Ritter and Christensen, 2007). For example, Phanerozoic large-scale tectono-magmatic processes, which may have their origin in mantle plumes, played an important role in erosion and metasomatism of the Archean lithospheric mantle in the Tanzanian craton in East Africa. On the other hand, some researchers question the very existence of mantle plumes in the real Earth. Arguments against the plume hypothesis can be found on the website: www.mantleplumes.org; these are motivated by the possibility of explaining most geophysical and geochemical observations by alternative, non-plume models. The present discussion does not aim to contribute to the debate on the existence of mantle plumes and is limited to selected geophysical and geochemical observations, such as a commonly proposed mantle (mega-) plume origin of continental rifting and supercontinent break-up, formation of large igneous provinces (LIPs) and giant dyke swarms (GDS), and komatiite extraction.

What does the mantle see?

The link between the plate motions at the Earth's surface and mantle convection is very complex. It raises a question similar to what was the first: the egg or the chicken? Holmes (1931) was among the first researchers to propose that continental motions can be driven by forces caused by thermal convection in the Earth's mantle. On the other hand, realistic numerical models of mantle convection demonstrate that the distribution of the continental landmass at the Earth's surface is fundamental in regulating temperature in the upper mantle and, through it, the style of mantle convection. Incorporation of continents into convection simulations modifies the style of mantle convection (e.g. Trompert and Hansen, 1998; Tackley, 2000). Davies (1999) suggests that the locations of convection upwellings and downwellings are influenced, or perhaps even controlled, by lithospheric structure rather than by properties of the deeper mantle. Recent results by Richards *et al.* (2000) show that changes in plate motions affect the dynamics not only of the upper mantle, but of the lower mantle and the D''-layer as well. Thus, while plate tectonic processes may be driven by mantle convection, mantle convection is affected by the geometry of the surface landmass (in other words, it "sees" the continents). Interaction of mantle convection with the continents is controlled by the following processes.

- Buoyancy of the continental lithosphere with respect to convective mantle prevents its large-scale subduction. Instead of subduction, plate tectonic processes cause aggregation of continents into supercontinents. Simple 2D numerical models with rigid and deformable continents suggest that the assembly time for supercontinents is most sensitive to the total surface area and is *c.* 400 Ma for the present continental area (Fig. 9.14).
- The aggregation of continents causes thermal insulation of the mantle, which leads to temperature increase and enlargement of the convection wavelength in the underlying mantle (Coltice *et al.*, 2007). The latter process also causes temperature increase below the continental lithosphere. Aggregation of sizable supercontinents (>3000 km) eventually triggers melting events and continental break-up (O'Neill *et al.*, 2007).

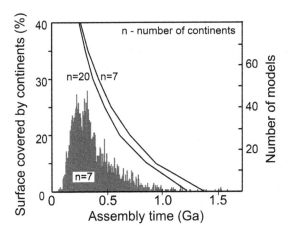

Fig. 9.14 Assembly time for supercontinents based on a simple 2D numerical model for colliding "continents" (based on results of Tao and Jarvis, 2002). Solid lines – median assembly times for models with 7 and 20 continental blocks demonstrate that the assembly time is most sensitive to the total surface area. Gray shading – frequency distribution of assembly times based on 2500 model runs with initially randomly distributed 7 continents with random initial velocities.

Large-scale rifting and break-up of supercontinents

Large-scale continental rifting can cause a supercontinent to split, giving birth to a new ocean, as it happens now at present in the Red Sea; the initial stages of continental break-up can be observed in the adjacent Afar Triangle in Ethiopia. Examples of continental rifting are also known for the Precambrian. The oldest known continental rifting event (~3.0 Ga) is documented by petrotectonic assemblages in the Kaapvaal craton; another well-studied Precambrian example is the Keweenawan rift in north-central USA.

The depth extent of the driving forces of active rifting is uncertain. Two depths are proposed as sources for continental rifting, the upper mantle (where the source is often associated with secondary convection (Griffiths and Campbell, 1991)) and the core–mantle boundary (Loper and Stacey, 1983). There is little direct evidence so far for a plume origin of continental rifting, and the role of plumes is mostly hypothesized based on large volumes of magmatic rocks associated with the rifting process (Anderson, 2007). Plate boundary forces can produce deviatoric stresses in the lithosphere causing lithospheric extension. In the case where the deviatoric tension is large enough and/or the lithosphere is relatively thin and hot, the lithosphere can fail (Kuznir and Park, 1984). Since lithosphere failure is accompanied by adiabatic ascent of hot mantle material into the weak zones of the lithosphere and associated magmatism (passive rifting), the consequences of this process are largely indistinguishable from active rifting, where either the upper mantle or the lower mantle is the driving force for rifting. Numerical models of mantle convection demonstrate that mantle reconfiguration events leading to aggregation of a large continental landmass can cause dynamic continental rifting due to mantle insulation by a supercontinent, and can be the ultimate cause of supercontinent break-up and related volcanism (O'Neill *et al.*, 2009).

Large igneous provinces

Large igneous provinces (LIPs), the term used to describe continental flood basalt provinces (LIPs located within large continents or at the edges of the cratons), volcanic passive margins, and oceanic plateaus, represent the largest known volcanic episodes on Earth (Mahoney and Coffin, 1997). The composition of continental flood basalts (CFB) and ocean island basalts (OIB) (hotspot magmas) is isotopically different from mid-ocean ridge basalts (MORB) and requires a geochemical reservoir (or its component) that has been isolated from the MORB source for more than 2 Ga (Anderson, 1994). The formation of LIPs is usually ascribed to plumes to explain large volumes of magma production observed in these regions and caused by massive mantle melting. However, some LIPs do not show evidence for plume melting and other mechanisms have been proposed to explain their origin. They include (but are not limited to) continental break-up, intense magmatism at jumping spreading centers, and secondary convection in the upper mantle. Global magmatic events (including LIPs) known throughout Earth history correlate with the ages of supercontinent formation; thus the thermal insulation effect of the supercontinents can play an important role in the formation of continental flood basalts ranging in age from the Archaean to the Mesozoic (Coltice *et al.*, 2007). Numerical models of mantle convection indicate that thermal insulation of the mantle by large landmasses (>3000 km) can cause elevated subcontinental mantle temperatures and associated magmatism; however for continents exceeding 8000 km in size lateral heat convection at these scales becomes inefficient, giving rise to the development of small-scale convecting systems under the continent (O'Neill *et al.*, 2009).

Small-scale edge-driven secondary convection which can develop at the cratonic margins has been proposed to explain some continental flood basalt provinces, such as the Siberian and Deccan trap provinces (King and Anderson, 1995; Korenaga and Jordan, 2002). Critical conditions for this type of secondary convection are (i) a significant (~100–150 km) step in the lithospheric thickness and (ii) warmer mantle beneath the cratons than under regions with thinner lithosphere. The former requirement is easily met at the cratonic edges, while the latter can be satisfied if the craton has sufficient size to cause efficient thermal blanketing of the mantle leading to an increase of its ambient temperature. A sharp change in the lithospheric thickness causes an interaction between the large-scale convective flow in the mantle and the small-scale edge-driven flow at the continental margin (Fig. 9.15). As the long-wavelength temperature anomaly weakens, the long-wavelength flow weakens with time and the small-scale edge-driven flow starts to dominate the flow pattern.

In many of the cratonic regions the models of secondary convection better fit the observations than the plume model since there is no evidence for significant lithosphere thinning during magma eruption. In particular, the Siberian traps are mainly formed by tholeiite basalts which suggest shallow subcrustal melting, while geophysical data indicate relatively low surface heat flow (38–40 mW/m^2) and lithospheric thickness >150 km, inconsistent with plume melting at ~250 Ma. If the lithosphere were thinned to 40–60 km during the magmatic event, its thermal cooling over 250 Ma would have produced a thickness of thermal lithosphere of no more than 115 km with the present-day surface heat flow of ~53 mW/m^2. Zorin and Vladimirov (1988) argue that since most eclogite xenoliths from the Siberian Platform correspond to tholeiitic basalts, and in some kimberlite pipes

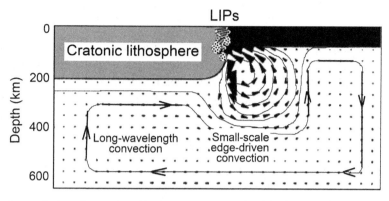

Fig. 9.15 Sketch showing mechanism of generation of large igneous provinces (LIPs) by edge-driven secondary convection. A critical condition for this type of secondary convection, warmer mantle beneath the cratons than under regions with thinner lithosphere (King and Anderson, 1995), can be achieved when the cratons have a sufficient size to cause thermal blanketing of the mantle.

the fraction of eclogites is as high as 50% of the total volume of mantle nodules, eclogites may form large lenses in the lithosphere of the Siberian Platform, so that due to gravity differentiation the lower part of the lithosphere can be formed mostly by eclogites and dunite–harzburgites. Small-scale secondary mantle convection can trigger partial melting of the lower lithosphere. High Fe content in Siberian eclogites should decrease their melting temperature by 150–200 °C and remove the so-called eclogite thermal barrier. As a result, eclogites should be more fusible than dunite–harzburgites and magnesian lherzolites. Given that the temperature difference between the solidus and liquidus of eclogites does not exceed 100 °C, a temperature increase at the base of the lithosphere can result in total melting of the eclogite lenses and can produce large volumes of tholeiitic magmas, such as observed in Siberia.

The Paraná Basin flood basalts in Brazil and the Etendeka flood basalts in Namibia are two parts of a huge LIP on the two sides of the Atlantic Ocean. In contrast to the Siberian and Deccan trap provinces, which were formed in less than 1 My, the Brazil–Namibia flood basalt province was erupted between 138 and 127 Ma (the major magmatic episode lasted for ~4–5 My). Although, the origin of this LIP is closely related to the opening of the South Atlantic (which started at ~150 Ma and continued until ~100 Ma when South America and Africa completely separated), the Tristan plume is assumed to contribute to generation of flood basalts (Wilson, 1992).

Phanerozoic examples of LIPs include the Ontong–Java plateau in the western Pacific, the Columbia River basalts in western USA, and the North Atlantic igneous province. Oceanic plateaus cover ~3% of the present-day sea floor, and Ontong–Java is the world's largest plateau. Most of the plateau basalts were erupted at ~122 Ma (with a smaller magmatic episode at ~90 Ma) producing an average crustal thickness of ~33 km which is in a sharp contrast to the normal thickness of oceanic crust (~7 km). Geochemical data suggest that the plateau could form by a mantle plume at a near-ridge setting (Tejada *et al.*, 1996); alternatively intense volcanism at an active ocean ridge or at a migrating triple junction can also explain most of the observations (Gladchenko *et al.*, 1997; Neal *et al.*, 1997).

Anorthosites

Edge-driven secondary convection at a sharp transition between thick cratonic and thin adjacent lithosphere can explain the tectonic origin of Proterozoic anorthosites (Artemieva, 2006). These enigmatic anorogenic magmatic rocks with a predominance of plagioclase feldspar (thus their name, which reflects that these rocks contain very little orthoclase) form *c.* 150 large intrusive massifs (Ashwal, 1993). Frequently (but not always) anorthosite massifs are enveloped by a large volume of anorogenic rapakivi-type granites; the latter could be formed by melting of host rocks in the course of magma ascent during anorthosite events (Sears *et al.*, 2005). The unique feature of massif-type anorthosites is their restriction in time and in space:

- all of them were emplaced in Proterozoic (Archean anorthosites have distinctly different composition and do not form large volume massifs) with the major peak between 1.73 Ga and 1.04 Ga when 58 out of 80 dated events produced 82.7% of the total known area;
- they seemingly form two global-scale "anorthosite belts", Laurentian and Gondwanian. The total number of massif-type anorthosites in the Gondwana continents is approximately half that in the Laurentia continents, but the statistics could be biased by a better exposure of basement rocks in the northern hemisphere.

Bimodal composition of Proterozoic anorthosite magmas has caused discussion on either crustal or mantle origin for their parental magmas: while the high positive ε_{Nd} determined in some anorthosite massifs rules out crustal rocks as the source material, melting of typical mantle peridotites fails to explain anorthosite-type compositions. Recent petrologic models favor a "hybrid" model with melting of both lower crustal and upper mantle material (Duchesne *et al.*, 1999). However, the tectonic processes responsible for generation of huge volumes of these anorogenic magmas (the size of the largest Proterozoic massif is 17800 km^2) remain enigmatic. An additional complication arises from the "oceanic" geochemical signature of massif-type anorthosites, in spite of them having been emplaced within the continental crust.

   None of the proposed tectonic models (a "global catastrophic anorthosite event", a moving diapir, or intracratonic rifting) or tectonic settings (e.g. the Tibetan-style or Andean-type convergent margins) is successful in explaining the entire set of their unique features (Herz, 1969; Emslie, 1978; Ashwal, 1993; Van Balen and Heeremans, 1998; Mukherjee and Das, 2002). The spatial and temporal restrictions of massif-type anorthosites suggest that their formation is closely related to a rapid growth of continents in the Paleoproterozoic; but no systematic study has been performed on a global scale. A rapid growth of continents at *c.* 1.95–1.80 Ga could have led to the formation of a supercontinent, which served as a thermal blanket and produced either a mantle superswell or a split of a supercontinent with the corresponding pulse of anorogenic magmatism (Hoffman, 1990; Zhao *et al.*, 2004). However, this hypothesis fails to explain spatial patterns of massif-type anorthosite occurrences, and in particular their small number within the Archean parts of the cratons and within the Gondwana-cratons.

   The statistical analysis of spatial and temporal distribution of Archean and Proterozoic anorthosites clearly shows (Artemieva, 2006) that:

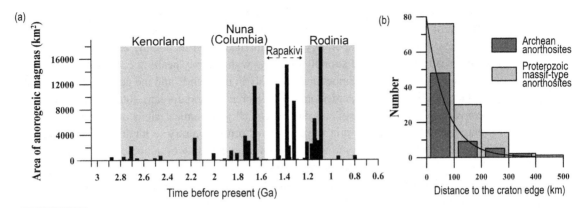

Fig. 9.16 Global occurrence of anorthosites in time (a) and space (b). (a) Area covered by Proterozoic massif-type anorthosites (data on both age and area are available for only 47 out of 133 known massifs), Archean layered anorthosites, and Proterozoic rapakivi-type granites. Shaded areas mark the time when supercontinents existed; their formation time correlates with a time of rapid lithosphere growth and with peaks in the anorthosite magmatism. Anorthosites emplaced at 1.5–1.3 Ga are commonly a part of large rapakivi-type complexes formed by rifting during the Mesoproterozoic pause in lithospheric growth. (b) Spatial patterns of global anorthosite occurrences. For Archean anorthosites – the distance to the nearest edge of the corresponding Archean craton; for Proterozoic anorthosites – the distance to the nearest edge of Precambrian province with geological age older than the anorthosite emplacement. Thin line approximates an exponential-type decay of the number of anorthosite complexes with the distance from a cratonic edge.

- the number of anorthosites decreases exponentially with distance from geologically defined cratonic boundaries (Fig. 9.16b): 85% of the Archean and massif-type Proterozoic anorthosites are found within 200 km of the edges of the Archean and Proterozoic crustal provinces. Thus their emplacement is caused not by processes within a supercontinent interior, but at the cratonic margins, and two global-scale "anorthosite belts" mark the edges of the Proterozoic supercontinent(s);

- anorthosite events correspond to Archean and Proterozoic peaks in crustal (lithosphere) growth (Fig. 9.16a). The prevalence of anorthosites in the cratons that later formed Laurasia can be related to higher lithosphere growth rates in the cratons of the Laurasia continents during the growth peaks than in the cratons of the Gondwana continents. An almost constant and low rate lithospheric growth in Australia and India (without a peak in Paleoproterozoic time related to formation of a sizable craton) can explain the small number of anorthosite events on the future Gondwana continents and their apparent absence in Australia;

- the termination of Proterozoic anorthosite events correlates with start of large-scale (in many cases, continent-scale) orogenies. Thus while a rapid growth of the continents at c. 1.95–1.80 Ga was the primary cause of the Precambrian anorthosite events, their termination is related to the initiation of continent-scale orogenies in the Meso- and Neoproterozoic.

Small-scale convection instability at the cratonic margins can explain the unique features of massif-type Proterozoic (and probably Archean) anorthosites (Artemieva, 2006), including the periodicity of Proterozoic anorthosite events which are ~44–46 Ma both at the western margin of the East European craton – the Baltic Shield and at the eastern margin of North America. According to laboratory and numerical experiments, the edge-driven upflow is intrinsically unsteady and strongly pulsating with velocity peaks at c. 10–100 Ma, depending on the amplitude of temperature anomaly in the mantle and the geometry of the model (Elder, 1976; King and Anderson, 1995; Korenaga and Jordan, 2002). The edge-driven convection model also explains the "oceanic" geochemical signature observed in most anorthosites emplaced within the continental crust: contrary to intuitive expectations, the edge-driven flow at the continental margin causes upflow of sublithospheric material that is directed from beneath the oceanic region towards the edge of the continent (Fig. 9.15).

The critical condition for the initiation of edge-driven convection, warmer mantle beneath the cratons than under regions with thinner lithosphere, can effectively be achieved when the cratons have sufficient lateral dimensions. This condition is met after periods of rapid lithosphere growth. The first peak in Proterozoic anorthosite events (at ~1.7–1.6 Ga) followed the formation of a supercontinent Nuna (Columbia), assembled at 1.9–1.8 Ga along global Paleoproterozoic collisional belts; the second peak (at 1.2–1.1 Ga, observed chiefly in North America) followed the early stages of formation of Rodinia at ~1.3–1.2 Ga. Non-synchronous peaks in lithospheric growth together with non-synchronous peaks in major anorthosite events imply that different cratons could have played different roles in the tectonics of the Precambrian supercontinents. Similarly, the emplacement of Archean anorthosites at c. 2.7 Ga apparently followed the period of rapid growth of Archean cratons, probably associated with the assemblage of the early protocontinents. The small volume of Archean anorthosites and their different petrologic characteristics could be consequences of the small size of the cratons in the Archean, which may have been insufficient to provide effective thermal blanketing of the mantle and maintain edge-driven convective upflow.

Anorthosite diapirism could have created lithosphere weakness zones during amalgamation of the supercontinents favorable for the consecutive development of belts of rapakivi-type magmatism. Anorthosites emplaced at 1.5–1.3 Ga are commonly a part of large rapakivi-type complexes which often possess features of layered mafic intrusions. Most of this type of magmatism took place during the Mesoproterozoic pause in lithospheric growth and it is apparently related to continental rifting (Fig. 9.16a). Initiation of continent-scale orogenies in the Meso- and Neoproterozoic transformed many cratonic margins into convergent plate margins which did not favor edge-driven small-scale convection. Secular cooling of the mantle led to the formation of a substantially thinner continental lithosphere in Phanerozoic time, and as a consequence, to a smaller step in lithospheric thickness at the cratonic margins and to emplacement of large igneous provinces rather than anorthosite massifs.

Giant dyke swarms

In contrast to large-scale rifting or large igneous provinces, some giant dyke swarms (GDS) univocally indicate an important role of plume–lithosphere interaction in the evolution of the continental lithosphere. Giant dyke swarms represent groups of genetically related (usually mafic) dykes that can be traced over hundreds of kilometers and are commonly associated with large igneous provinces. The radial structure of some GDSs (in the Paraná Basin in Brazil and the Mackenzie dyke swarm in North America) suggests that they radiate from a common "point" source such as caused by melting over a mantle plume head. The oldest known, the Ameralik swarm in southwestern Greenland, is ~3.25 Ga; but the best known giant dyke swarm due to its immense size is the Mackenzie dyke swarm (emplaced at 1.267 Ga). It extends for more than 2600 km from its source in Arctic Canada (the Coppermine continental flood basalt province) to the Great Lakes, has a maximum width of 1800 km, and covers an area of 2.7×10^6 km^2 (that is, greater than the area of the Ontong-Java plateau). The Mackenzie magmatic event is likely to be one of the largest LIPs in the geological record. Up to now, both the volume of the erupted magmas (generally tholeiitic) and the point source from which the dykes radiate cannot be explained by non-plume mechanisms.

Komatiites

Komatiites, ultramafic rocks with extremely high (more than 18%) MgO, are named after the Komati river in the Barberton greenstone belt, South Africa. Their extraction is an almost entirely Archean phenomenon and is commonly explained by mega-plume mantle melting (the only well-documented Phanerozoic example being the Gorgona island komatiites, extraction of which is probably related to the Galapagos hotspot). Geochemical data for komatiite formation are commonly interpreted as requiring very high melting temperatures (1600–1800 °C) (Figs. 4.38 and 4.39). To explain these data as well as to accommodate the difference in melting temperatures of komatiites and associated tholeiite basalts (~400–450 °C in the case of dry melting and ~200 °C in the case of wet melting, Grove et al., 1997), Campbell et al. (1989) have proposed that plume tails provide the local temperature increase required for komatiite generation, while plume heads with lower temperature due to the cooling effect of the surrounding mantle provide the source for associated basaltic magmas. Subsequent experimental studies significantly reduced melting temperatures required for generation of komatiitic magmas. While early studies indicated melting temperatures 300–500 °C higher than modern ambient mantle (Nisbet and Fowler, 1983), further studies reduced this value to ~50–100 °C, in particular in the case where magmas contain a significant amount of water (Grove and Parman, 2004). High-pressure experiments also demonstrate that komatiites can be produced not only by low-pressure, high-degree (>50%) melting, but by high-pressure, low-degree melting (Herzberg, 1995).

Modern plumes

The number of estimated modern mantle plumes varies significantly (Sleep, 1990; Montelli et al., 2004). Five criteria proposed to be characteristic of modern plumes include (Courtillot et al., 2003):

- the presence of a hotspot track,
- the presence of a large igneous province (LIP) at one of the ends,
- high buoyancy flux,

- high $^3He/^4He$ ratio,
- low seismic velocities in the mantle.

The number of "plume" criteria has been extended to twelve (Anderson, 2005) in order to include the parameters that characterize the seismic structure of the entire mantle and its thermal state. The controversy of the plume hypothesis is well illustrated by the example of Iceland: according to Courtillot *et al.* (2003), Iceland is a major hotspot, which satisfies at least four criteria, whereas according to Anderson (2005) Iceland is equal on "plume" and "plate tectonics" scores, and the major evidence for mantle plumes include a seismic low-velocity anomaly in the upper 400 km and extreme $^3He/^4He$ ratios that are among the highest on Earth. Recent surface-wave tomography indicates the presence of a significant, isolated (~1000 km in diameter), low S-wave velocity anomaly in the upper mantle of Iceland which may extend down to at least 600 km depth (Ritsema and Allen, 2003). A similar P-wave velocity anomaly persisting below the transition zone is observed in the body-wave tomography model (Bijwaard and Spakman, 1999). The observed S- and P-wave velocity anomalies (~5–10%) in the shallow mantle have been interpreted as evidence for a mantle plume. However, receiver function analysis of seismic data from Iceland shows a low-velocity zone only in the shallow mantle, centered at a depth of 100 km, and normal thickness of the transition zone; it challenges the conclusion of the very existence of the Iceland plume (Vinnik *et al*, 2005).

An elegant resolution of the debate on the existence of some modern plumes is provided by laboratory simulations of plume generation, evolution, and death in thermo-chemical convection (Fig. 9.17). These indicate that when a plume is dying, negative compositional buoyancy is no longer compensated by positive thermal buoyancy. As a result, a downward material flow develops along a still hotter-than-normal "plume channel", such as seen in Icelandic seismic tomography models.

Dying plumes start disappearing from the bottom up, sometimes even before reaching the upper boundary and they finally fade away by thermal diffusion. This sequence of events shows that time-dependence is a key-factor when interpreting present-day tomographic

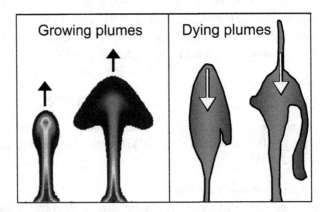

Fig. 9.17   Evolution of mantle plumes based on laboratory studies of thermo-chemical convection.

images of mantle upwellings. In particular, it could be erroneous to identify the depth of a present-day slow seismic anomaly with the depth of its origin, or to interpret the absence of a long tail as the absence of a plume. (Davaille and Vatteville, 2005).

These results have critical implications for the interpretation of mantle plumes based on seismic tomography images since it is often forgotten that mantle plumes evolve through time.

Subduction through time

Archean versus post-Archean tectonic processes

There is growing geological, geochemical, and geophysical evidence that some form of plate tectonics operated already in the early Archean (Calvert *et al.*, 1995; Condie, 1997; Percival *et al.*, 2004; Condie and Pease, 2008). However, massive mantle melting caused by heads of mega-plumes or mantle overturns may have been the dominant mechanism at the early stages of lithosphere formation. Arc-related magmatism is expected to become dominant with mantle cooling. Because of a secular decrease of mantle potential temperature (Campbell and Griffiths, 1992), the proportion of the crust generated by plume-related oceanic plateau basalts may have decreased from the Archean to the present day (Condie, 1994, 2000). In this case, a systematic difference between the bulk compositions of the Archean and post-Archean crust with respect to key chemical indices or elemental ratios should be expected. However, isotope analysis (in particular of Nb/La ratios which are sensitive to the proportion of plume component in volcanic rocks (Hofmann *et al.*, 1986)) shows that *the contribution of plume-related magmas to crust generation has not changed since the Archean* (Kemp and Hawkesworth, 2003).

The onset of plate tectonic-type processes, including subduction, provided new mechanisms other than basal lithospheric accretion by mantle plumes for the formation of submarine plateaus. Geological evidence from the well-preserved Kaapvaal craton indicates that continental fragments were amalgamated in the Archean by processes similar to modern plate tectonics (de Wit *et al.*, 1992), and accretion of oceanic terranes could have been an important global process of continental lithosphere growth in the Archean and Proterozoic (Percival and Williams, 1989). Hot mantle and depleted composition of Archean lithosphere resulted in buoyant subduction which is unique for Archean time. Buoyant subduction of oceanic slabs around the perimeters of pre-existing lithospheric fragments and collision of oceanic terranes (such as submarine plateaus and island arcs) with pre-existing continental margins contributed to lateral growth of the continental nuclei and lithospheric thickening (Abbott and Mooney, 1995; Rudnick, 1995).

Subduction on early Earth

Subduction (mantle downwelling) is an important component of mantle convection and manifestation of the operation of plate tectonics processes. A recent review of paleomagnetic, geochemical, and tectonic data pushes back the start of plate tectonics to at least 3.1 Ga (Cawood *et al.*, 2006). However, there is no generally recognized opinion on when plate tectonics began and whether its early style was like the modern style. Even a simple review of available data and interpretations deserves a book, and the reader is addressed to two recent summaries (Hatcher *et al.*, 2007; Condie and Pease, 2008). Leaving the debate aside, a brief (and inevitably incomplete) summary of highly controversial geological and geochemical evidence is as follows.

(1) There is no geological proof for the existence of lithospherically distinct oceans and continents in the Archean. Neither ophiolites (sections of the oceanic crust emplaced onto continental crust) nor blueschists (rocks formed by low-temperature metamorphism of basaltic rocks) are indisputably recognized in the Archean terranes or within the Paleoproterozoic orogens. The oldest well-documented ophiolite is ~1.9 Ga (in the Baltic Shield). The age has recently been pushed backwards to ~2.5 Ga based on dating of ophiolite in the North China craton (Kusky *et al.*, 2001), but this result remains controversial. Furthermore, a recent study based on the geochemistry of a sheeted-dyke complex within the Isua supracrustal belt (southwest Greenland) pushes the age of the oldest ophiolite as far back as ~3.8 Ga, by suggesting that it may represent oceanic crust accreted by sea-floor spreading (Furnes *et al.*, 2007).

(2) Important sites for growth of continental crust are convergent tectonic margins that involve accretion of large turbidite fans (Gray *et al.*, 2007). Neoproterozoic continental growth by accretion of submarine sediment fans is well established, for example, in the Damara orogen (~560–500 Ma), Namibia, from a distinct metamorphic zonation with belts of intermediate-temperature/intermediate-pressure metamorphism around the central zone that has undergone high-temperature/moderate-pressure metamorphism. Since locally derived coherent turbidites are known in many Neoarchean (3.0–2.5 Ga) greenstone terranes, they are sometimes used as evidence for Archean subduction (e.g. Percival *et al.*, 2004).

(3) The presence of both granulite-facies ultrahigh-temperature metamorphism, and medium-temperature eclogite high-pressure granulite metamorphism since 2.8–2.5 Ga may indicate the onset of a "Proterozoic plate tectonics regime" (Brown, 2008). Both types of extreme metamorphic conditions could have existed during amalgamation of the continental lithosphere into supercratons and, later, into supercontinents.

(4) The presence of eclogites in cratonic kimberlites that are not in equilibrium with harzburgite has been attributed to eclogite addition during subduction (McGreggor *et al.*, 1986). Later studies suggest kimberlitic eclogites can also be residues from the process of Archaean granitoid crust formation (Rollinson, 1997). Alternatively, crustal (lithosphere) delamination can produce a similar effect.

(5) Variations in oxygen isotope values and their correlations with major elements and radiogenic isotopes, such as observed in 2.57 Ga eclogite xenoliths from Udachnaya kimberlite pipe, Siberia, are interpreted as being originated in subducted oceanic crust (Jacob and Foley, 1999).

(6) Anomalies of isotope S^{33} found in diamonds from cratonic settings have been interpreted as having been formed in subduction settings in the oxygen-poor atmospheric environment already in the Archean (Pavlov and Kasting, 2002).

(7) Recent geochemical analysis of abundant Archean impact-melt spherules that provide indirect evidence on crustal composition at the time of impact has led to the conclusion that the Archean Earth was dominated by oceanic rather than by continental crust (Glikson, 2005). However, this interpretation as many other Archean geochemical interpretations, may be non-unique.

Post-Archean subduction

The post-Archean processes of lithosphere growth and recycling are in many regards different from the Archean processes. Mantle cooling changed the style of the subduction process: in post-Archean time subduction became steep because old and cold subducting lithospheric plates were no longer buoyant. This resulted in a change of mantle melting in the subduction zone environments: from slab melting in the Archean to mantle wedge melting in post-Archean time (Fig. 9.11); this, in turn, led to the change in composition of derived magmas. Strong compressional tectonics and development of continental subduction zones could form tectonic structures such as the modern subduction of the Indo-Australian plate beneath the Himalayas. Collisional growth of the continental lithosphere by accretion of oceanic terranes at convergent plate boundaries (such as in the modern Cordilleran and Appalachian orogens (Condie and Chomiak, 1996; ANCORP Working Group, 1999)) is similar to tectonic processes that operated in the Precambrian.

Subduction, together with lithosphere delamination, provides an efficient mechanism of lithosphere destruction and its recycling to the mantle (Section 9.3.4). While in Archean time the phase transition boundary at 660 km worked as an efficient barrier for whole-mantle convection and for sinking slabs into the lower mantle, secular cooling of the mantle made the transition zone more "transparent" for the slabs. Recent high-resolution P-wave and S-wave tomography images of the circum-Pacific subduction zones indicate that, although most of the slabs deflect at around 660 km depth, some positive seismic velocity anomalies that may relate to subducting slabs are traced deep into the lower mantle (Van der Hilst et al., 1997; Fukao et al., 2001) (Fig. 3.103). This result has important implications for mantle convection models (whole-mantle, layered, or hybrid) and provides indirect support for the hypothesis of episodic catastrophic overturns in the mantle caused by slab sinking (Condie, 1998). Slabs that reach deep mantle (either a compositionally heterogeneous layer at 1700–2300 km depth or the core–mantle boundary) can initiate large mantle plumes and massive mantle overturns (Davies, 1995; Van der Hilst and Karason, 1999).

9.2.2 Crustal growth and recycling

Crustal growth scenarios and crustal growth rate

Mechanisms of crustal growth are discussed in detail in Section 9.1.2. They include plume-related magmatism with associated magmatic underplating and subduction related processes (Fig. 9.11). Subduction processes at convergent margins that involve accretion of submarine turbidite fans, fragments of juvenile oceanic crust, and newly added continental crust produced by magmatic underplating lead to continental growth. A large number of quantitative

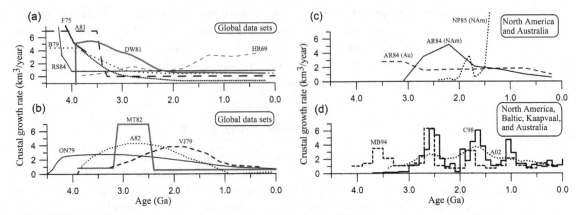

Crustal growth models presented as variations of crustal growth rate through time. References and comments: plot (a): global geophysical and geochemical models of A81 = Armstrong, 1981; B79 = Brown, 1979; DW81 = Dewey and Windley, 1981; HR69 = Hurley and Rand, 1969; F75 = Fyfe, 1978; RS84 = Reymer and Schubert, 1984; plot (b): A82 = Allègre, 1982 (global data for Nd and Sr isotopes); MT82 = McLennan and Taylor, 1982 (global analysis of sediments); ON79 = O'Nions *et al.*, 1979 (global analysis of Ar, Sr, Nd, Pb isotopes), VJ79 = Veizer and Jansen, 1979 (global analysis of Sr in sediments); plot (c): AR84 = Allegre and Rousseau, 1984 (Nd analysis of Australian and North American shales); NP85 = Nelson and DePaolo, 1985 (Nd analysis of 27 samples from North America, mostly from the Great Lakes and the Great Plains); plot (d): A00 = Abbott *et al.*, 2000 (geophysical models); C98 = Condie, 1998 (U/Pb isotope data, mostly for North America, Baltic Shield, and Kaapvaal); MB94 = McCulloch and Bennett, 1994 (Nd data for North America, Baltic Shield, and Australia).

models of crustal growth scenarios have been proposed starting from the late 1960s (Fig. 9.18). These models are based on the following general approaches:

(1) *Global-scale general isotope geochemistry (in particular, based on Sr, Nd, Pb, Hf, Ar, K isotopes) and trace element constraints.* Models of this type try to fit present-day isotope ratios and trace element compositions to the assumed starting values corresponding to the moment of the Earth's formation. The constraints are based on simple assumptions about melt extraction processes and assume compositional homogeneity of various reservoirs that approximate the Earth's structure (which is not necessarily the case in the real Earth, Fig. 9.19). For example, the two-reservoir model of O'Nions *et al.* (1979) in which the upper 50 km-thick layer (the "crust") is formed by mantle differentiation predicts fast crustal growth at 4.0–3.0 Ga with a gradual decrease in crustal growth rate afterwards. This prediction is typical for the models constrained by general isotope balance calculations (e.g. Allègre, 1982). The most extreme is the model of crust formation proposed by Armstrong (1981) and later advocated by Bowring and Housh (1995). This assumes that the continental crust of approximately present-day mass formed during the very early evolution of the Earth (by 3.5 Ga) and since then has remained unchanged. The latter conclusion, apparently, contradicts common-sense expectations since new crust is being formed today and has been formed both in the Precambrian and Phanerozoic as indicated by isotope ages of crustal rocks. Using free-

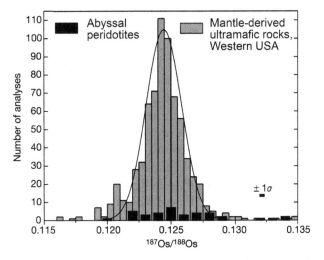

Fig. 9.19 Histograms of Os isotope compositions from mantle-derived ultramafic rock in western USA (from Meibom *et al.*, 2002). The measured data points form a Gaussian distribution (solid line). Data for abyssal peridotite samples from three different oceans are shown for a comparison. The Gaussian distribution shape is interpreted as the signature of a random mixing process between depleted and enriched domains in the upper mantle and demonstrates that the upper mantle is extremely heterogeneous. (see comment on p. 411)

board constraints, Armstrong (1981) proposed that, since the Archean, there has been a perfect balance between the amount of newly generated crust and the crust recycled to the mantle (the "steady-state" crustal model).

(2) *Regional isotope geochemistry (based on Sr, Nd, Pb, Hf, Ar, K isotope data)*. Early crustal growth models based on this approach are limited mostly to data from North America, with fewer constraints for Australia, Greenland, and the Baltic Shield. Since the sampled area is limited in size and often restricted to tectonic provinces of similar ages, the results show characteristic peaks in crustal growth that reflect the dominant crustal ages in the study areas (e.g. Nelson and DePaolo, 1985; Patchett and Arndt, 1986). Later studies based on the same approach have been constrained by large data sets with broader geographical coverage (Condie, 1998; Henry *et al.*, 2000). However, even these studies are heavily dominated by particular tectonic provinces, with other locations being seriously undersampled. For this reason, the conclusions of these studies are not necessarily robust and may change as new data become available. The most important conclusion is the growing evidence for episodic crustal growth with major peaks at 2.7–2.6, ~1.9, ~1.2 Ga; Nd isotope data from the Western Superior Province suggest additional peaks in zircon ages from juvenile crust at ~3.4 Ga and ~3.0–2.9 Ga and indicate that much of the post-Archean crust is contaminated with Archean subducted crustal material.

(3) *Areal distribution of radiometric ages on continents*. A few studies that followed this approach constrained crustal ages by Rb–Sr and K–Ar isotope systems (which can be affected by later tectonism) and the growth models were calculated from the total

surface area for each time slice (Hurley and Rand, 1969; Tugarinov and Bibikova, 1976). The results suggest that only ~25% of continents are older than 2.0 Ga and only 5–8% are older than 2.5–2.7 Ga.

(4) *Isotopic and chemical composition of sediments*. From studies of REE and Th abundances in the sedimentary clastic rocks, McLennan and Taylor (1982) concluded that the Archean crust makes up >70% of present crustal mass, and a large volume of cratonic sediments with ages of 2.4–2.0 Ga requires the existence of widespread continents at that time. An unusual type of crustal evolution with near-constant crustal growth rates from Archean to Phanerozoic was estimated from Nd isotope analysis of Australian shales and attributed to differential erosion (Allègre and Rousseau, 1984). A comparison of Nd model ages with stratigraphic ages shows that for post-Archean stratigraphic ages, the Nd model ages deviate from the equal age line (i.e. disagree with stratigraphic ages). This deviation has been attributed either to a small volume of crust that existed before 2.5 Ga (Condie, 1989) or to a vigorous crustal recycling in the Archean time (Veizer and Jansen, 1985).

(5) *Geophysical constraints (such as free-board consistency since the Archean)*. The number of geophysical models is very small as compared to geochemical models. One of the more important studies has been undertaken by Reymer and Schubert (1984) who estimated production rates for the continental crust based on the ages of juvenile granites in 17 magmatic arcs. The present-day crustal volume in the arcs constrained by seismic data and integrated over 4.5 Ga yields an average crustal growth rate of $1.72 \, km^3/year$. Assuming *(i)* a post-3.8 Ga crustal recycling rate of ~0.6 km^3/year (with all ancient, >3.8 Ga, crust being destroyed by vigorous convection), *(ii)* free-board consistency since the Archean, and *(iii)* taking into account secular cooling, they estimate that the net crustal growth since the end of the Archean is ~0.9–1.0 km^3/year.

Secular trends in crustal growth

The proposed end-member scenarios of the above described models of crustal evolution form three classes:

- progressive crustal growth models;
- steady-state (no-growth) models;
- models of episodic crustal growth.

Bennett (2003) argues that geochemical data alone cannot discriminate between these scenarios: although generation of the crust and mantle depletion are interrelated processes, the large number of variables that describe them does not allow for a unique solution. Detrital zircons older than 4.0 Ga with high $\delta^{18}O$ isotopic compositions are interpreted as evidence for the existence of both continental crust and surface water in the Hadean (Amelin et al., 1999; Wilde et al., 2001; Mojzsis et al., 2001). The main argument against models of a massive crustal production early in the Earth's history is the lack of any significant volumes of early Archean crust (Table 9.1), and geochemical arguments against the existence of significant volumes of the crust prior to 4.0 Ga have been summarized by Kramers (2007).

Vigorous recycling of the early crust into the mantle could explain the lack of ancient crust, if it existed; however if the early crust were felsic, it would be too buoyant to subduct completely. To resolve this problem, models with mafic–ultramafic and alkalic crust have been proposed (Chase and Patchett, 1988; Galer and Goldstein, 1991). These models would require much larger volumes of the crust to produce the required mantle depletion. The latter is possible because of high mantle temperatures in the early Earth which would have resulted in higher production rates of basalt crust than the present-day rates at oceanic spreading centers. In principle, models of the early basaltic crust can explain geochemical data; but (except for enigmatic >4.0 Ga detrial zircons) there is no direct supporting evidence.

In spite of the principal differences in the models of crustal growth, the following conclusions are generally accepted at present:

- increase of the crustal volume through time;
- existence of a significant amount of the crust by 2.5 Ga;
- importance of crustal recycling, in particular at the pre-2.5 Ga time;
- episodic crustal growth with a small number of peaks;
- global coincidence of major peaks at ~2.5–2.7 Ga and ~1.7–1.9 Ga.

The peak at ~1.7–1.9 Ga on the crustal growth curves (Fig. 9.18) coincides with the peak at ~1.9 Ga seen on all curves for the areal extent of the cratons (Figs. 2.17a and 9.21a) and corresponds to global Mesoproterozoic orogenic events (Table 9.2). Because of the large areal extent of areas affected by worldwide orogenies and statistically small variations in crustal thickness on the continents, the coincidence of the two peaks is not surprising.

The crustal "growth" models show, in fact, not only the growth of new crust but the balance between crustal production and crustal recycling. Thus, at face value, the curves of crustal growth except for those based on global-scale geochemical balance (Fig. 9.18), should be considered as the "survival" models of crustal evolution. Furthermore, depending on the tectonic setting during the convergence stage, the preservation potential of crustal rocks is significantly different (Fig. 9.22). At margins with retreating subduction zones, where extensional basins are formed, the preservation potential is greater than at non-accreting margins where subduction zones are advancing continent-ward, convey the continental crust into the mantle, and cause a high rate of crustal recycling (Table 9.4). This means that, since the formation of the first supercontinent, the preservation record of the continental crust is biased by tectonic styles (Hawkesworth et al., 2009). In accordance with this conclusion, high accretion-growth rates calculated for the Precambrian crust may be greatly overestimated since its volume can be significantly larger than commonly accepted (Kröner et al., 2006). For example, zircons with ages up to 2.75 Ga from the Arabian–Nubian Shield (600–870 Ma) may indicate the presence of an older basement; similar observations have been reported for many other Proterozoic terranes. Table 9.4 summarizes the reported rates of growth and recycling of the continental crust in different tectonic settings. For some processes, the reported values differ by an order of magnitude.

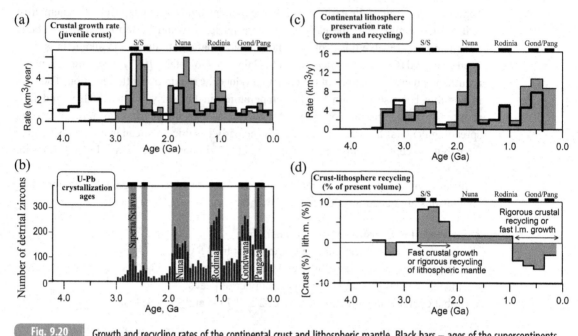

Fig. 9.20 Growth and recycling rates of the continental crust and lithospheric mantle. Black bars – ages of the supercontinents. (a) Growth rate of juvenile crust. Gray shading – based on U/Pb ages mostly from North America, Baltic Shield, Australia, and Kaapvaal (data of Condie, 1998); bold line – based on Nd data from North America, Baltic Shield, and Australia (data of McCulloch and Bennett, 1994). (b) U–Pb ages of 7000 detrital zircons (data of Campbell and Allen, 2008); they correspond to the age of crystallization of the parent igneous rock. Note a strong similarity to the age distribution of lithospheric volume in (c). Gray bars show the periods of supercontinent formation. (c) Preservation rate of the continental lithospheric mantle. Gray shading – calculations based on a $1° \times 1°$ global continental model TC1 of tectono-thermal ages and lithospheric thermal thickness (after Artemieva, 2006). Bold line – calculations based on body-wave global tomography model (Grand, 2002) assuming that the lithospheric base corresponds to a 2% velocity perturbation with respect to *iaspei* model. In both cases the lithospheric thickness is averaged over 0.2 Ga bins. Since the plots are based on present-day lithosphere thickness, they reflect both the rate of lithosphere differentiation from the mantle and the rate of its preservation. (d) Secular variations in crust-lithosphere recycling calculated as the difference between gray shadings in (a) and (c) (both normalized by their present-day volumes). Positive values indicate that normalized crustal volume exceeds normalized volume of lithospheric mantle either due to fast crustal growth or due to rigorous lithosphere recycling.

9.2.3 Global secular trends and episodicity of lithospheric processes

Growth and recycling of lithospheric mantle

Geophysical models of lithosphere thickness constrained by thermal and seismic tomography data can be used to calculate the volume of the continental lithosphere of different ages, assuming that the ages of the crust are representative of the ages of the underlying lithospheric mantle (Artemieva, 2006). The constrained curves reflect the combined effect of lithosphere growth and preservation, and show the volume of the lithospheric mantle of

Ages and lithosphere thickness: based on the TC1 model
Dashed lines: global averages

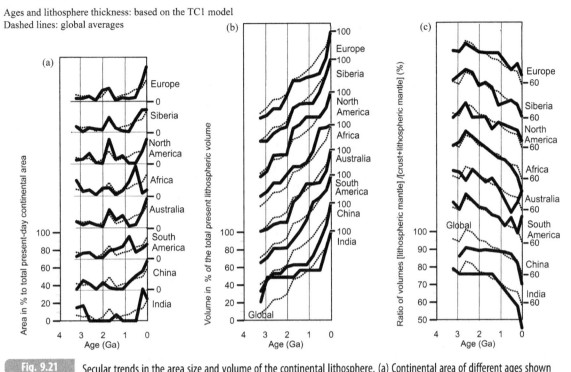

Fig. 9.21 Secular trends in the area size and volume of the continental lithosphere. (a) Continental area of different ages shown as percent of the present-day on-shore continental area; (b) volume of the continental lithosphere of different ages shown as per cent of the present-day total volume of the continental lithosphere; (c) the ratio of the volume of the continental lithospheric mantle of different ages to the total lithosphere volume (continental crust plus lithospheric mantle). Crustal volume for (c) is based on the CRUST 2.0 model; lithosphere volume is based on the TC1 thermal model. Bold lines in each plot show data for the individual continents, dotted lines – global averages. To simplify comparison, the curves for India are plotted with respect to the vertical axes shown on the left, while the curves for other continents are vertically shifted, preserving the scales. The results for continents with incomplete data (e.g. Africa) should be treated with extreme caution. The TC1 model is based not on geologic but on tectono-thermal ages (Artemieva, 2006), producing artifact trends of rapid lithosphere growth over the last 500 Ma.

different ages that has "survived" until present (Fig. 9.20). While young lithosphere has advantages in preservation due to its short exposure to recycling processes, the long-term stability of the cratonic lithosphere with respect to mantle convection due to its unique melt-depleted and hydrogen-depleted composition also favors its selective preservation.

Two periods in lithosphere evolution have distinctly different volumes of preserved crust and preserved lithospheric mantle (Fig. 9.20d). At 2.7–2.1 Ga, the proposed time of early super-continent formation, the normalized crustal volume exceeds the normalized volume of the lithospheric mantle. This suggests either rigorous recycling of the lithospheric mantle or a faster growth rate of the crust than that of the lithospheric mantle. The latter is possible, in particular, if different processes of lithosphere differentiation from the mantle operated on the Archean Earth. An excess volume of lithospheric mantle as compared to crustal volume in the Phanerozoic probably indicates an increasing role of plate tectonics processes in crustal recycling.

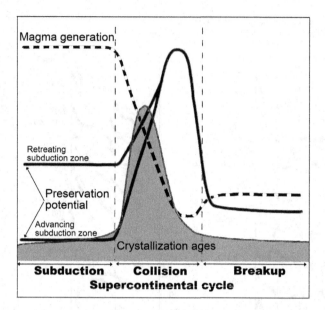

Fig. 9.22 The preservation potential of crustal rocks during the three stages of a supercontinent cycle (after Hawkesworth *et al.*, 2009). Dashed line – volumes of generated magma, solid lines – their likely preservation potential which depends on tectonic settings during the convergence stage (it is greater at the margins with retreating subduction zones where extensional basins are formed than at non-accreting margins where subduction zones are advancing continentward). Gray shading – preserved crystallization ages which reflect the balance between the produced magma volumes and their preservation potential.

Figure 9.21 illustrates secular variations in the areal size of modern continents and in volume of the continental lithosphere calculated for individual cratons by assuming the age of the lithospheric mantle is the same as the age of the crust. The continental lithosphere is defined as the area above the present-day sea level, excluding submerged areas with continental crust such as e.g. shelves and ocean plateaus. These assumptions as well as the uncertain age of the Antarctica craton introduce a significant bias into the global estimates. Lithosphere age (Fig. 2.16) and thermal thickness (Fig. 4.44c) are based on a $1° \times 1°$ global continental model TC1 (Artemieva, 2006). Estimates of the lithosphere volume are based on thickness of the thermal lithosphere defined by the intersection of continental geotherms with the 1300 °C mantle adiabat and, similarly to Fig. 9.20, show the volume of the lithosphere of different ages that has "survived" until present. The results suggest asynchronous lithosphere survival in different cratons (Fig. 9.21b).

The ratio of volume of the present lithospheric mantle to total preserved lithosphere volume (crust plus lithospheric mantle) shows a global general decrease from the end of the Archean to Phanerozoic time (Fig. 9.21c). This pattern is consistent globally and reflects the situation where a larger volume of the Archean lithospheric mantle with thick lithospheric keels is not matched by a similar trend in crustal thickness variations because thickness of Archean and post-Archean crust is similar. Some decrease in the ratio between the volumes of the crust and

Table 9.4 Growth rate and recycling rate of the continental crust

Process		Growth rate or recycling rate (km^3/y)	References
Growth	Crustal growth rate* (from global geochemical models)	1.25; 2.5 at a peak at 1.9 Ga	Patchett and Arndt, 1986
		1.0–1.5 at minima, 2.0–3.0 at peaks	Condie, 1998
	Crustal growth rate* (from global geophysical models)	1.6	Reymer and Schubert, 1984
		0.5–1.5 at minima, 2.0–4.0 at peaks	Artemieva, 2006
	Crustal growth by accretion of oceanic plateaus	3.7	Schubert and Sandwell, 1989
	Crustal growth by accretion of turbidite fans in subduction settings	From ~1–1.5 to ~13–18 km^3/My/km	Gray *et al.*, 2007 and references therein
	Modern production rate at MORB	20 km^3/y	Hoffman, 1989
	Crustal growth rate* in the absence of crustal recycling required to produce the present-day volume of continental crust	A constant rate of ~1.7 km^3/y over the entire Earth's history	Coltice *et al.*, 2000
Recycling	Sediment subduction	0.5–0.7	Rea and Ruff, 1996; Plank and Langmuir, 1998, and references therein
	Sediment subduction plus mechanical erosion of the crust at subduction zones	1.6	von Huene and Scholl, 1991
	Subduction of continental submarine plateau	2.5 (upper bound)	Schubert and Sandwell, 1989
	Modern subduction zones	**(1) Eroded forearc material:** Non-accreting margins: 1.3 (or ~40 km^3/Ma/km × 31,000 km) Accreting margins: 0.1 (or ~12 km^3/Ma/km × 11,000 km) **(2) Sediment subduction:** ~1.0 at all margins **Total: ~2.4–2.5**	Scholl and von Huene, 2007
	Crustal recycling rate (from models of Nd isotope secular evolutions)	0.8 ± 0.5	Albarède, 1998
		2.5	DePaolo, 1983

* Estimates are biased by preservation history

the lithospheric mantle for the Archean (pre-2.7 Ga) cratons may be artifact due to an underestimated lithosphere thickness or an overestimated crustal thickness. Alternatively, it may indicate that different processes of lithosphere differentiation and crustal accretion operated in the Paleo- and Mesoarchean or that much of the Archean lithosphere did not survive (Fig. 9.22).

Secular trends in lithosphere evolution

Secular evolution of the continental lithosphere is governed by secular cooling of the Earth (see discussion in Sections 4.1 and 4.4.2) which is caused by cooling of the planetary interior through the surface and by an exponential decrease in radioactivity within the Earth (Figs. 4.51 and 4.38a). For the oceanic lithosphere, which is very young compared to the age of the Earth, the effect of secular cooling is insignificant. In contrast, it is important for the continental lithosphere, the age of which spans from Archean to present. The most important evidence for high Archean mantle temperatures is provided by komatiites, magmas abundant in Archean. However, the quantitative estimates of how much hotter the Archean mantle was, remain uncertain: while some authors argue for close-to-solidus temperatures in the Archean lower crust, there is no clear geological evidence for a widespread crustal melting in the Archean.

In the continental lithosphere, secular cooling (through a decrease of mantle temperatures) controls (and manifests itself in):

- the thickness of the thermal boundary layer;
- the thickness and the composition of the chemical boundary layer; the latter is reflected in different compositions of the Archean and the post-Archean crusts and in secular trends in the composition of the lithospheric mantle (Fig. 6.11);
- some phenomena (such as global extraction of komatiites in the Archean and of anorthosites in the Archean–Proterozoic) that are unique for the Precambrian Earth (Table 9.2).

The processes through which secular cooling affects the lithosphere structure include both the processes of lithosphere formation and chemico-mechanical destruction.

- Different melting conditions in the mantle (together with the possible different nature of mantle sources and different processes of lithosphere differentiation from the mantle) allowed for generation of compositionally distinct, buoyant, thick cratonic lithosphere in the Precambrian.
- Long-term cooling of the thermal boundary layer on the continents could favor its further growth; but isotope age data from mantle-derived xenolith from Precambrian terranes do not indicate a significant growth of cratonic lithosphere from below (Fig. 9.6).
- Start of plate tectonics and the change of its style through geological evolution could have a significant effect on lithosphere growth and recycling rate at convergent boundaries (Table 9.4).
- Different style and vigor of mantle convection could affect both processes of lithosphere formation and its destruction (recycling).

Episodicity of lithospheric processes

An important feature of lithosphere evolution is the episodicity of a number of global-scale tectonic events. The processes responsible for these episodic events are closely linked to secular cooling of the Earth, formation of supercontinents, start and development of plate tectonics processes, and changes in the style and vigor of mantle convection due to mantle cooling.

Global episodic events include (McLennan and Taylor, 1985; Condie, 1998; Yale and Carpenter, 1998; Haggerty, 1999):

- global peaks in distribution of crustal ages at ~2.7–2.5, ~1.9, ~1.1 Ga (Fig. 9.20a,b) commonly associated with massive crustal production;
- episodic age distributions of granitoids and greenstone belts;
- episodic ages of large igneous provinces and giant dyke swarms;
- episodic age distributions of Archean and Proterozoic anorogenic magmatism (Fig. 9.16), including episodic ages of *major* kimberlite-type magmatic events (Table 9.5 and Fig. 6.26).

Except for the major kimberlite-type magmatic events, the ages of all other global episodic events are, in general, well correlated. Peaks in crustal growth (crustal ages) are often associated with supercontinent cycles. Similarly, other episodic global-scale geological/ magmatic events are commonly interpreted as manifestations of the existence (assemblage or break-up) of supercontinents (Courtillot *et al.*, 1999). The age of the oldest one is uncertain and the number of pre-Rodinia supercontinents is debated, while paleoreconstructions of reliably distinguished supercontinents are non-unique and problematic even for Gondwana (van der Voo and Meert, 1991). It is usually accepted that the first putative Archaean protocontinents Vaalbara (Zimvaalbara) formed probably at ~3.6 Ga (Barley *et al.*, 2005); this was followed by the Kenorland supercontinent which probably existed between ~2.8–2.7 Ga and ~2.1 Ga; the consecutive supercontinents include Nuna (Columbia) at ~2.0–1.5 Ga, Rodinia (formed at ~1.3–1.0 Ga with break-up at 0.8–0.6 Ga), Gondwana and Laurasia (assembled at 650–580 Ma, with the start of break-up at 170–120 Ma), and Pangea (formed at ~300–250 Ma).

Table 9.5 Major kimberlite-type (kimberlites and lamproites) magmatic events (updated after Haggerty, 1999)

Age	Location
~1.8 Ga	Western Australia (Pilbara)
~1.0 Ga	Africa, Western Australia, India, Brazil, Siberia, Greenland
~630–590 Ma	Baltic Shield (Finland)
~500–450 Ma	Northwest Russia (Arkhangelsk region), China, North America (Canada), southern Africa
~410–370 Ma	Siberia, North America (USA)
~200 Ma	North America (Canada), southern Africa, Tanzania
~120–80 Ma	Southern, central, and western Africa, India, Brazil, Siberia, North America (USA and Canada)
~50 Ma	North America (Canada), Africa (Tanzania)
~22 Ma	Northwest Australia (minor episodes)

Davies (1995) and Condie (1997, 2000) propose that global bursts of tectonic activity and massive volcanism could be caused by massive mantle avalanche events (Fig. 9.13b). They further hypothesize that layered mantle convection typical for pre-2.8 Ga (due to higher mantle temperatures and thus higher Rayleigh numbers) switched to present-style convection at ~1.3 Ga. The transitional period at 2.8–1.3 Ga, when the pre-Gondwana supercontinents were formed, was dominated by periodic mantle overturns caused by mantle avalanches. Alternative models with episodic massive volcanism and crustal growth (O'Neill *et al.*, 2007) suggest that high mantle temperatures of the early Earth were responsible for plate-driven episodic tectonics in the Precambrian. Lower lithospheric stresses caused by higher mantle temperatures could have led to rapid pulses of subduction interspersed with periods of relative quiescence. According to paleomagnetic data, documented pulses of rapid plate motion were coincident with peaks in the crustal age distribution.

9.3 The lithosphere is not forever

"*A diamond is forever*". This famous line, proposed by Frances Gerety in 1947 and used ever since by De Beers Companies in their advertising campaigns, was in 2000 named the best advertising slogan of the twentieth century. However, the lithosphere through which diamond-bearing kimberlite-type magmas erupt is not forever. Moreover, if the cratonal lithosphere were formed by cooling of the thermal boundary layer (similar to the oceanic lithosphere), its thickness should have been several hundreds of kilometers (Fig. 4.46). Hence, the cratonal lithosphere must be kept thin by convective heat transfer from the underlying mantle. Several mechanisms are particularly important in thermal, mechanical, and compositional destruction of the lithosphere and its recycling into the mantle (Fig. 9.23):

- thermal and convective erosion of the lithosphere by mantle upwellings, large-scale and localized small-scale mantle convection;
- thermo-mechanical delamination caused by density inversion (Rayleigh–Taylor-type gravitational instability);

MECHANISMS OF LITHOSPHERE DESTRUCTION

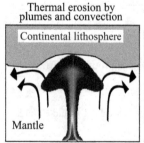
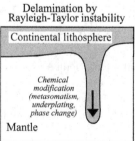
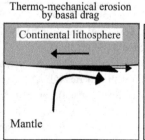
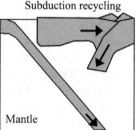

Fig. 9.23   Sketch illustrating major mechanisms of lithosphere destruction.

- thermo-mechanical erosion of the basal portions of the lithosphere by mantle drag due to relative motions of the lithospheric plates and convective mantle;
- thermo-mechanical recycling of lithospheric plates in subduction zones.

9.3.1 Thermal erosion by mantle convection and plumes

An interaction of an uprising anomalously hot mantle material with the lithosphere can result in thermal thinning of the lithosphere by conduction due to an increase of mantle heat flow at the lithospheric base (e.g., Crough and Thompson, 1976; Spohn and Schubert, 1982; Doin et al., 1996) or in mechanical thinning by delamination due to gravitational (Rayleigh–Taylor) instabilities in the lower lithosphere caused by density inversion (e.g., Bird, 1978, 1979; Houseman et al., 1981; Mareschal, 1983; Drazin, 2002). Each of these processes thins (destroys) the lithosphere, and prevalence of one over the other depends on the viscosity contrast between the lithosphere and the anomalous mantle (Sleep, 1994; Molnar et al., 1998). Low lithosphere viscosity favors the Rayleigh–Taylor-type instability, while thermal erosion plays a significant role in stable regions with cold, highly viscous lithosphere.

Erosion of Phanerozoic lithosphere

In the case where fertile Phanerozoic lithospheric mantle is heated by a mantle upwelling or mantle convection, a temperature increase in the lower part of the lithosphere (thermal boundary layer) will initiate its partial melting and a corresponding significant viscosity decrease (Figs. 8.10 and 8.25). The viscosity of a Newtonian fluid is expressed through the stress–strain rate relationship (Eq. 8.11): $\eta = \sigma/2\dot{\varepsilon}$, and the viscosity contrast between two materials with non-linear rheology depends on whether stress σ is constant or strain-rate $\dot{\varepsilon}$ is constant. During convective erosion, the lithosphere and the convective mantle are in contact without any constraints on strain rate, and the viscosity contrast must be considered at constant stress (Karato, 2010). In such a case, the temperature effect on viscosity change is the strongest (Fig. 8.17c). The presence of even trace amounts of water will further reduce the strength of the lower part of the lithosphere and make it mechanically unstable. Being close to the critical Rayleigh number in the lower part of the thermal boundary layer (Anderson, 1994), it cannot be isolated from the convective mantle, unless it is chemically buoyant and dry. As a result, the lithosphere will effectively become thinner (due to a temperature increase within the lithosphere, the intersection of the lithospheric geotherm with the mantle adiabat will occur at a shallower depth), and the viscous force exerted by the convection current in the nearby regions will remove the basal lithospheric material which will become part of the convective system.

Lithosphere thinning produces an isostatic uplift of lithospheric blocks leading to formation of high plateaus and domal swells; under certain thermo-mechanical conditions continental rifts may develop at the later stages of the process. The presence of fluids (water and carbon dioxide) in the lithospheric mantle has a dramatic effect on melting conditions by producing non-linear variations of solidus temperature with depth (Fig. 4.26). As a result, thermal erosion can cause non-monotonous (episodic) isostatic vertical movements when the melting front passes through highly fusible or refractory layers in the lithospheric mantle (Artemieva, 1989).

Longevity of cratonic lithosphere

In contrast to Phanerozoic lithosphere, the interaction of a mantle thermal anomaly with melt- and hydrogen-depleted cratonic lithosphere may not necessarily result in lithosphere thermal erosion and convective removal. Longevity and persistence of the cratonic lithosphere for ~3 Ga clearly demonstrates its unique stability with respect to mantle convection and, to some extent, to thermo-mechanical erosion by mantle plumes. Numerical simulations show that a 2–3 orders of magnitude viscosity contrast between the cratonic lithosphere and the convective mantle is required to sustain its preservation against convective erosion (Lenardic and Moresi, 1999; Shapiro *et al.*, 1999). Sleep (2003) argues that even a factor of 20 in viscosity contrast is sufficient for lithosphere stability in the case of weakly temperature-dependent viscosity and modern conductive heat flow through the chemical layer. In contrast, in the case of high heat flow and strongly temperature-dependent viscosity, few orders of magnitude viscosity ratios are required to maintain stability of the lithosphere (Doin *et al.*, 1997).

Experimental studies of olivine rheology indicate that the viscosity contrast is highly sensitive to the value of activation enthalpy (through activation volume) (Figs. 8.17a, 8.19); however, these parameters and, as a consequence, mantle viscosity values (Fig. 8.15) are determined non-uniquely (see discussion in Section 8.1). For plausible values of activation enthalpy, the viscosity contrast of ~1000 between the lower portions of the lithospheric mantle and the convective mantle can be achieved only when the temperature difference between them exceeds 200–400 °C (Fig. 8.19); this temperature contrast is too high to be realistic. The high viscosity contrast required for lithosphere stability with respect to mantle convection, however, can be achieved at a smaller temperature contrast between the convective mantle and the continental roots in the case of water-depleted (dry) rheology and relatively high values of activation volume V^*(Fig. 8.26). According to Karato (2010), for $V^* > 10^{-5}$ m^3/mol, there is a broad range of combinations of degree of water depletion and temperature contrast that can satisfy the necessary conditions for preservation of the continental roots and stability of the cratonic lithosphere, but a very dry mantle is required to maintain lithospheric strength at high temperature (Pollack, 1986). Dry composition can be achieved through early massive mantle melting and further isolation of the cratonic roots from mantle convection and subduction environments.

Numerical simulations of mantle convection with chemically distinct (water- and melt-depleted) continents residing within the thermal boundary layer of thermally convecting mantle demonstrate that only dry depleted continental roots are stable with respect to convective erosion and the dominant stabilizing factor is a relatively high brittle yield stress within the roots. The equilibrium thicknesses which such roots achieve during their interaction with convective flow are 200–220 km and 300–350 km (Doin *et al.*, 1997) (Fig. 4.50). These results accord with the simulations of Lenardic *et al.* (2003) which indicate that stability and longevity of continental roots during several convective overturns can be achieved if the chemically distinct, high-viscosity part of the lithosphere is at least twice the thickness of old oceanic lithosphere, and the effective mechanical thickness of the lithosphere is 100 km or more. Another possibility for preserving cratonic lithosphere arises when the cratonic upper and lower crust yields at higher stresses than the crust in adjacent continental terranes. In the latter case, the presence of continental lithosphere with a relatively low yield strength at the cratonic margins further enhances preservation of cratonic roots (Lenardic *et al.*, 2003). However, despite their chemical buoyancy and high viscosity, cratonic roots can be destroyed at contacts with subduction zones. Furthermore, chemical modification of depleted cratonic lithosphere associated with melt-metasomatism and fluid enrichment reduces its strength and makes it vulnerable to convective erosion.

In an alternative model (Polet and Anderson, 1995), inspired by numerical simulations of Gurnis (1988), the permanent, chemically distinct part of the cratonic lithosphere is only ~200 km thick. Since it is, nonetheless, significantly thicker than non-cratonic lithosphere, it affects the pattern of mantle convection so that cold mantle downwellings develop beneath the cratons. These cold downwellings, formed by constantly replaced mantle material, produce high-velocity cratonic roots seen in seismic tomography images. Although there is no xenolith sampling of the mantle structure below 240 km depth, the evidence based on the deepest xenoliths from the Baltic Shield, that are coarse-grained, non-sheared, and follow a very cold conductive geotherm, does not seem to support this model.

Modes of plume–lithosphere interaction

Similarly to convective erosion, plume impingement may also cause lithosphere erosion (destruction) (Davies, 1994). A recent analog modeling (Jurine *et al.*, 2005) demonstrates that plume–lithosphere interaction can have significantly different modes depending on the compositional and thermal contrasts between the lithospheric mantle and the plume (quantified by the ratio B between the intrinsic chemical density contrast and the thermal density contrast due to temperature differences). The viscosity contrast does not affect either the magnitude or the shape of the interface deformation, which are controlled by the B-ratio, but determines the time-scale of interface displacement. True plume penetration into the lithosphere occurs in the case of a weak compositional contrast between the lithosphere and the plume material ($B < 0.6$). In the case of a strong compositional contrast, such as between depleted cratonic roots and the upper mantle, true penetration of a plume into the cratonic lithosphere requires a temperature contrast of 300–400 °C (which corresponds to buoyancy fluxes higher than estimated for the present-day Hawaiian plume). Such a temperature contrast between an uprising plume and the cratonic lithosphere is sufficient to provide thermal buoyancy of a plume and to produce massive mantle melting and intensive magmatism, such as is observed in East Africa where the continental rifting develops in spite of the presence of depleted cratonic lithosphere. In contrast, a smaller temperature contrast between the uprising hot mantle material and the lower lithosphere, combined with depleted composition of cratonic lithospheric mantle, produces broad regional uplifts with no or scarce magmatic activity (Fig. 9.24).

The presence of abundant diamonds of superdeep origin (derived from a depth of at least 670 km) in the kimberlites from the Slave, the Gawler, and the Kaapvaal cratons (Davies *et al.*, 1999; Gaul *et al.*, 2000) suggests that lower mantle plumes have played

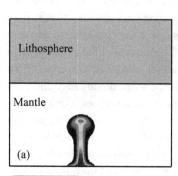 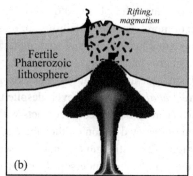 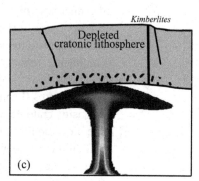

Fig. 9.24 Sketch illustrating the manner of plume–lithosphere interaction based on laboratory and numerical modeling (Jurine *et al.*, 2005). The manner of plume–lithosphere interaction is sensitive to the ratio between the intrinsic chemical density contrast and the thermal density contrast due to temperature differences between the lithosphere and the plume. In contact with depleted cratonic lithosphere, plume material spreads beneath the interface and its buoyancy force deforms the lithosphere–asthenosphere interface. In contact with Phanerozoic fertile lithosphere, plume penetrates into it. Vertical dimensions not to scale.

an important role in modification of the Archean lithosphere since the Precambrian. Higher heat flow from the core–mantle boundary (where most large plumes may originate) during the Archean may have led to a more frequent impingement of mantle plumes onto the base of the lithosphere, followed by massive magmatism and, in some cases, continental break-up (e.g. Courtillot *et al.*, 1999; Dalziel *et al*, 2000). Interaction of lower mantle plumes with the lithosphere of the Archean cratons could have resulted in the removal of a thick (up to ~200 km) lower portion of their roots; subsequent cooling of the thermal boundary layer could have led (as suggested by the layered lithospheric structure of some of the cratons (Gaul *et al.*, 2000)) to addition of a 70–80 km-thick layer of lithospheric material of the lower mantle paragenesis (as in the Slave craton). It is worth noting that, to date, superdeep diamonds are not known in the thickest (~300–350 km) parts of the Siberian cratonic lithosphere (the Daldyn–Alakit region). Their absence, interpreted as evidence of absence of lower mantle plume interaction with one of the present-day thickest cratonic lithospheres, suggests that all lithosphere formed in the Archean could have been at least 350 km thick.

Correlation between craton size and thickness of lithospheric roots

Phenomenological analysis based on available data on lateral dimensions for eight out of nine (except for Antarctica) cratons indicates (Artemieva and Mooney, 2002) that there is a strong correlation ($r = 0.87$) between the lithospheric thickness and the area of the Archean–Neoproterozoic cratons (Fig. 9.26).

- Large cratons have thicker lithospheric roots and are more stable with respect to mantle convection. This correlation only holds for the actual size of the Archean–Neoproterozoic block, but not for its relative size (i.e. the fraction of the lithospheric plate which the block with ancient crust occupies). The correlation is still strong ($r = 0.81$) when the area size of the entire Precambrian craton (Archean plus Proterozoic parts) is correlated with the lithosphere thickness in its Archean part. No correlation ($r = 0.18$) is found for the Meso-Neoproterozoic cratons where variations in lithosphere thickness are small.

- Extrapolation of the best fit line (Fig. 9.26a) to zero size of an Archean craton gives a lithospheric thickness of ~140–160 km, in agreement with geophysical observations of the thickness of the post-Archean lithosphere. This implies that tectono-magmatic reworking of the Archean lithosphere by lithosphere erosion tends to reduce its thickness to the same values.

- Extrapolation of the best fit line (Fig. 9.26a) to the total area of all preserved Archean terranes (which can be representative of the size of a hypothesized Archean super-continent) yields a lithospheric thickness of ~350–500 km. This value does not contradict geophysical constraints on the depth of the 410-km phase transition which is estimated to have occurred at 440–450 km depth in the Archean.

9.3.2 Rayleigh–Taylor instability and lithosphere delamination

The Rayleigh–Taylor instability (Rayleigh, 1883; Taylor, 1950) is a gravitational instability that develops at an interface between two viscous materials of different densities when the lighter material underlies the heavier material. A density inversion may have thermal or compositional origin and can be caused by pondering of hot mantle material beneath the lithospheric base, lithosphere thickening beneath orogenic belts, and phase transitions in the crust.

In collisional orogens formed by closure of an ocean basin (such as the Paleozoic Variscides in Europe and the Tethyan Fold belt – Himalayas in Asia Minor–Central Asia) continental or arc subduction can develop in the case of a strong lithosphere and high convergence rate. In contrast, weak lithosphere and low convergence rate favor develop-ment of the Rayleigh–Taylor instability. In the former case convergence is accommodated by internal deformation, while in the latter case – by pure shear thickening. In collisional settings, where the entire lithosphere is thickened as an orogen is formed, a cold and dense lithospheric root protrudes into the hotter asthenosphere and creates lateral temperature gradients. The latter produce thermal instability which drives convective flow that can remove (delaminate) the lower portions of the thickened lithospheric mantle. The presence of density inversion also causes foundering of negatively buoyant lithospheric material and lithosphere delamination. Detached lithospheric material is swept into descending drips or blobs. Pondering of hotter mantle material to maintain the mass balance causes adiabatic decompressional melting and triggers magmatism (Fig. 9.25).

Extensional collapse of overthickened crust and the uppermost lithosphere (Menard and Molnar, 1988) and gravitational collapse caused by formation of eclogites (Mengel and Kern, 1991) have been proposed as mechanisms responsible for the formation of thin crust in the European Variscides. Magmatic pulses documented in several continental regions and further supported by seismic data indicate that lithosphere delamination could play an important role in the Cenozoic tectonic evolution of the Colorado Plateau (Bird, 1978,

(a) Convergence-induced thickening   (b) Gravitational instability and thinning

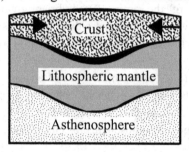

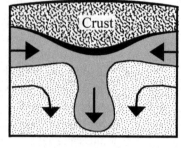

(c) Lithosphere extension   (d) Alboran Sea example

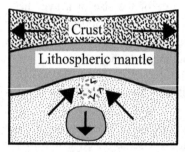

Fig. 9.25 Cartoon illustrating the principal stages in the conceptual model of Rayleigh–Taylor-like instability and lithosphere delamination (modified from Houseman and Molnar, 1997). Vertical dimensions exaggerated. (a) Lithospheric thickening by horizontal shortening, (b) thinning of the lithospheric mantle by Rayleigh–Taylor-like instability, (c) lithospheric extension, topographic collapse, and basin formation. Bottom right cartoon: north–south seismic cross-section from the Rif mountain belt (Morocco) to the Betic Cordillera (Iberia) across the Alboran Sea region of the western Mediterranean (modified from Molnar and Houseman, 2004; seismic model of Calvert et al., 2000). Regions of strong seismic velocity anomalies are shown by white (-2%) and black (+2%) shading. High-velocity anomaly at depths >100 km and shallow low-velocity anomaly are located directly beneath the Alboran Sea, which occupies the basin created by crustal thinning. Black dots – seismicity.

1979), the Sierra Nevada (Manley et al., 2000; Saleeby and Foster, 2004; Harig et al., 2008), the southern Puna plateau of the Andes (Kay and Kay, 1993), Tibet (Turner et al., 1993; England and Molnar, 1997), Himalaya (England and Houseman, 1989), Vrancea zone in the Carpathians (Fig. 3.93), and the Alboran Sea in the western Mediterranean (Houseman and Molnar, 1997; Turner et al., 1999). Seismic images from these regions often indicate the presence of, roughly cylindrical in shape, drip-structures between 100 and 300 km depth with high seismic velocities interpreted as delaminated lithosphere (Fig. 9.25).

The development of convective instability associated with mechanical thickening of cold, dense lithosphere beneath mountain belts has been examined numerically for Newtonian rheology (Houseman and Molnar, 1997; Molnar et al., 1998). The results suggest that for dry olivine rheology only the lower 50–60% of the lithosphere (with temperatures >910–950 °C) may be delaminated, while for wet olivine the delamination process may involve more than 80% of the lithospheric thickness (with $T > 750$ °C). Many mountain belts, built by crustal thickening, further collapse by normal faulting and horizontal crustal

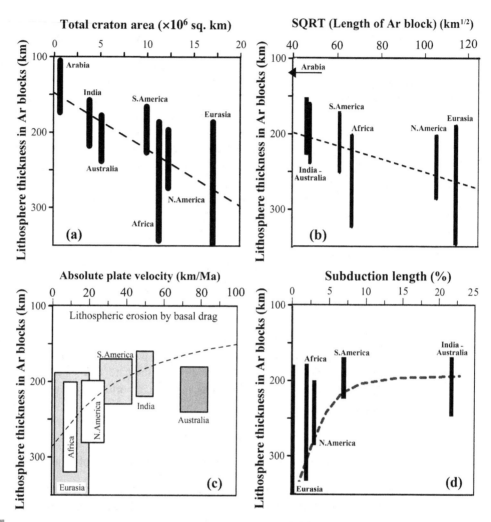

Fig. 9.26 Lithosphere erosion by basal drag caused by relative motion of lithospheric plates above convecting mantle (eq. 9.3) (after Artemieva and Mooney, 2002). The vertical dimensions of the bars and boxes indicate both the uncertainty and scatter in estimates of lithosphere thermal thickness; horizontal size of boxes in (c) shows the uncertainty in plate velocities. Dashed lines – best fits to the observations.

extension. The latter process, driven by removal of the lithospheric root that adds potential energy to the lithosphere, commonly does not start within a few My (and often >30 My) after the end of the crustal thickening.

9.3.3 Basal drag and mantle convection

The existence of thick lithospheric roots (extending down to at least 250–300 km) beneath the ancient parts of the continents poses questions on the style of the interaction between the lithospheric roots and mantle convection:

- do lithospheric roots affect mantle convection and plate motions? or
- are lithospheric roots eroded by mantle convection and plate motions?

The answer to the first question is provided by numerical simulations of mantle convection (Davies, 1999; Tackley, 2000) which indicate that the presence of thick lithospheric plates with dry and depleted composition affects the convection pattern, and thus can also have an effect on plate motions (Chapman and Pollack, 1974). On the other hand, as discussed above, mantle convection is an efficient mechanism for lithosphere erosion, unless lithospheric roots are melt- and hydrogen-depleted. In the case of depleted cratonal lithosphere, basal erosion by localized mantle convection aligned in rolls (basal drag) further restricts lithosphere thickness to ~225 km (Sleep, 2003) which, in the absence of thermo-mechanical basal erosion, could have been several hundred kilometers thick. Additionally, if the cratonal lithosphere had approached its present thickness in the Archean, its base should be at approximately steady-state with the heat supplied to the lithosphere from the mantle. The assumption of a 225 km-thick cratonal lithosphere, that formed the basis for the set of numerical simulations of Sleep (2003), is based on xenolith geotherms from some of the well-studied Archean terranes (e.g. the Kaapvaal craton and the Superior province). However, this assumption is not supported by the worldwide xenolith data set which indicates significantly different lithospheric geotherms in different Archean cratons with lithosphere thickness reaching 300 km and more in the Baltic Shield, the Siberian and Slave cratons (Fig. 5.13 and discussion in Section 5.2.4). In spite of the limitation with the observational premise, theoretical predictions indicate that basal drag on the thermal boundary layer, caused by plate motions, plays an important role in controlling the thickness of the lithospheric roots.

Correlation between plate velocity and thickness of lithospheric roots

The model of basal drag (caused by the relative movement of the lithospheric keel and the underlying mantle) suggests that the horizontal movement of lithospheric plates on atop of the convecting mantle produces simple shear at the base of the lithosphere, which leads to removal of its rheologically weak basal part (Sleep, 2003). In the case of stirring-dominated convection with vertical conduction and both horizontal and vertical heat advection, lithosphere thickness is proportional to the square root of the ratio of the craton length along the plate movement to the plate velocity:

$$\text{Lithosphere thickness} \propto \sqrt{\frac{\text{Craton length along the direction of plate motion}}{\text{Plate velocity}}}.$$

$$(9.3)$$

More precisely, the horizontal distance affected by the basal drag is more than the craton length along the direction of plate motion and is the distance to the nearest ridge axis in the direction of plate motion. However, the latter parameter is hard to estimate precisely. The proportionality constant in eq. 9.3 depends on the ratio between the temperature at the lithospheric base and the temperature change across the base of the thermal (chemical) boundary layer (Sleep, 2003).

Phenomenological analysis of lithospheric erosion by basal drag (based on a thermal model of lithosphere thickness) shows the following (Fig. 9.26) (Artemieva and Mooney, 2002):

- In agreement with eq. 9.3, lithosphere thickness in Archean cratons is strongly inversely correlated with the plate velocity ($r = -0.77$ including Australia and $r = -0.97$ without Australia): fast-moving Archean cratons have thinner lithosphere than slow moving cratons. Australia, which has the highest plate velocity, falls out of the general trend. Since the high velocity of the Australian plate may be a relatively recent phenomenon, it is likely that at present the lower portions of the Australian lithospheric mantle are subject to a strong shear which will erode it to a 150–170 km thickness if the Australian craton continues to move with such a high velocity (Fig. 9.26c).
- Lithospheric plates with a higher percentage of Proterozoic crust move faster (Stoddard and Abbott, 1996). This observation implies that the fast-moving Proterozoic keels are entirely within the low-viscosity asthenosphere which does not resist plate motions.
- Lithosphere thickness of stagnant plates (near-zero velocity) can be as large as 300–350 km. If Archean plates were moving slower than present-day plates (see discussion below), lithospheric thickness in the Archean may have been slightly larger than at present.
- In agreement with the basal drag model (eq. 9.3), lithosphere thickness in Archean cratons is well correlated with the length of the Archean block along the plate motion: shorter Archean cratons have thicker lithosphere. In the case where no Archean lithosphere is present within a craton, the Proterozoic cratonic roots are ~140–160 km thick, they are stable with respect to basal erosion by mantle drag and can preserve a lithospheric thickness of c. 150 km over hundreds of millions of years.

Constraints on Archean plate velocities

The above phenomenological correlations are based on present-day plate velocities and the present-day lithospheric thickness as only few constraints on paleo-velocities exist (Irving, 1977; Lithgow-Bertelloni and Richards, 1998), and all of them are for Paleozoic–Cenozoic time. Because of the high mantle temperature in the Archean, an early Archean lithosphere could have been underlain by an upper mantle with a viscosity 1–2 orders of magnitude lower than the present-day mantle (Davies, 1999) and plate velocities may have been significantly higher than the velocities of present-day plates. In such a case, lithospheric erosion by basal drag would probably have been different in the Archean than at present. Indirect constraints on paleo-velocities, however, suggest that they were even lower in the Archean than at present.

Based on the analysis of ancient subduction, Hargraves (1986) argues that if heat loss in the Archean was three times greater than at present, the ridge length in the Archean could have been 10–15 times its present length. In such a case, the Archean Earth (prior to the aggregation of the Archean supercontinent and/or after its dispersal) could have been covered by many slow-moving small plates. Similarly, Abbott and Menke (1990) estimate that at 2.4 Ga the length of plate boundaries was ~2.2 times greater than at present; to fit data on higher mantle heat production in the Precambrian this requires plate velocities ~16% slower than at present. An indirect constraint on paleo-velocities is provided by two-dimensional simulation of mantle convection with a constant-velocity upper boundary condition to account for the presence of lithospheric plates (England and Houseman, 1984). The modeling, used to explain the genesis of kimberlites, shows that lithospheric plates with low velocities (less than 20 km/Ma) favor the formation of mantle upwelling beneath them; the latter can lead to partial melting of the lowermost lithosphere and the generation of kimberlite magmas. Thus, the existence of kimberlites of different ages (Table 9.5) suggests that paleo-velocities in all of the continental plates were often slow, at least during the last billion years of Earth evolution. The absence of kimberlite-type magmatism prior to ~1 Ga can be interpreted as indirect evidence for high plate velocities (assuming rigid lithospheric plates already existed) in the Archean–Mesoproterozoic time.

The fate of the Archean lithosphere

Correlations between vertical and lateral dimensions of the Archean cratons and plate velocities indicate that basal drag and associated stirring at the base of the cratonic lithosphere have played an important role in the thinning of the cratonic lithosphere since the early Archean. The basal drag model (eq. 9.3) and the phenomenological correlations indicate that basal erosion of the lithosphere is weakest for Archean cratons with large area and slow plate velocities (Artemieva and Mooney, 2002). The large size of cratons is important for effective thermal insulation of the mantle and for an efficient deflection of heat from the deep mantle outside of the craton interior (Ballard and Pollack, 1987; Lenardic and Moresi, 2001). Although the basal drag model suggests that the long-term preservation of thick (300–350 km) Archean keels is possible only if the cratons have never experienced a period of high plate velocity, such a situation seems unlikely. However, due to higher mantle temperature in the Archean, the viscosity of the upper mantle could have been one or two orders of magnitude lower than today resulting in a much weaker basal erosion than today. These conditions would allow the preservation of thick lithospheric roots of Archean cratons even if Archean plate velocities were high.

Secondary convection at cratonic margins reduces the lateral size of Archean cratons with thick (~300 km) lithosphere (Fig. 9.27). The "critical size" of an Archean craton favorable to preserving lithospheric roots in excess of 200–250 km is $c.$ 5–$10 \times 10^6 \, \text{km}^2$ (Fig. 9.26a). This result, together with the conclusions of the convection simulation of Doin et $al.$ (1997), suggests that mantle convection erodes thick (300–350 km) lithospheric roots from the sides until their lateral dimensions decrease to the "critical size" (Fig. 4.50). Starting from that moment, the basal erosion of the lithosphere dominates until the new equilibrium thickness of 200–220 km (controlled by the thickness of the chemical boundary layer) is reached. Due to the viscosity–depth structure of the upper mantle, thinning of the Archean lithosphere will reduce its resistance to plate motion (Stoddard and Abbott, 1996). As a consequence, the craton will begin to move faster with respect to the underlying mantle, which, in turn, will enhance the basal drag and lithosphere erosion. The same can happen if a craton becomes part of a fast moving plate attached to a subduction zone (the case of present-day Australia).

Basal erosion of the lowermost part of the Archean lithosphere results in a lithospheric thickness that corresponds to the typical thickness of Proterozoic lithosphere (Fig. 9.26b). Thus, thinned Archean lithosphere becomes a candidate for tectonic reworking into

FATE OF ARCHEAN KEELS

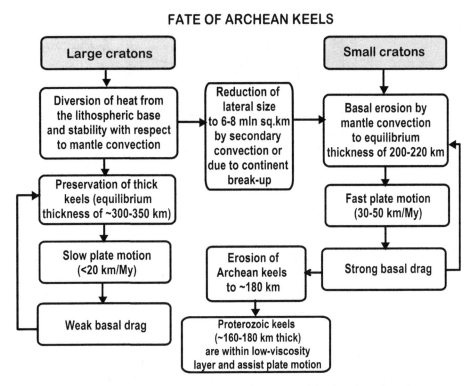

Fig. 9.27 Trends in evolution of Archean lithospheric keels. The fate of the cratonic lithosphere depends on the area size of the cratons. Large cratons are more stable with respect to mantle convection than small cratons, they move slower and preserve thicker lithospheric keels (300 km and more).

Proterozoic lithosphere, for example by strong deformation in collisional environments. This conclusion is supported by geological evidence which shows that Proterozoic lithosphere is often reworked Archean lithosphere. The absence of diamonds in the Proterozoic lithosphere can be explained by the removal of the lowermost part of the reworked Archean lithosphere by basal drag with its later growth by conductive cooling.

9.3.4 Lithosphere recycling in subduction zones

Figure 9.28 illustrates major plate tectonic forces that can act on the lithospheric plates and control lithosphere structure. The principal driving mechanisms of plate tectonics are subduction slab pull and mid-ocean ridge push, and the velocity of plates attached to subducting slabs is controlled chiefly by slab pull (Forsyth and Uyeda, 1975). Since continental lithosphere thickness is inversely proportional to square root of absolute plate velocity (Fig. 9.26c), one can expect that a similar correlation exists between lithospheric thickness and subduction pull (Artemieva and Mooney, 2002). An analysis of this correlation based on the comparison of lithospheric thermal thickness in the Archean cratons with fractional subduction length (i.e. the connectivity of a plate to a slab expressed as a percentage of the subduction length to the total plate boundary) suggests that, indeed, lithospheric thickness in

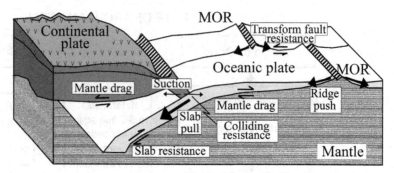

Sketch of major plate tectonic forces acting on the lithospheric plates (after Forsyth and Uyeda, 1975). MOR = mid-ocean ridge.

Archean cratons (in particular, maximal values) is inversely correlated with subduction length (subduction pull). Similar conclusions, although not well constrained because of non-uniform distribution of data, can be made from the analysis of correlations between lithospheric thickness in the Archean cratons and the effective ridge length (i.e. the ridge length exerting a net force, which means that the force is not canceled, for example, by a ridge on the opposite side of a plate). The results indicate that lithospheric thickness is controlled by two main driving forces of plate motion through their effect on plate velocity and, hence, on mantle drag. At the same time, thick Archean keels can slow down velocities of plates that are not attached to subducting slabs. Both of these conclusions hold only for the Archean cratons, but not for younger continents where the correlations are very weak.

Subduction is an effective mechanism of lithosphere recycling. Estimates show that, if the present convergence rate between the Ontong Java plateau and the Solomon Island arc is representative of the past, ~80% of the thickness of the oceanic plateau has been subducted and recycled into the mantle (Mann and Taira, 2004). With time, most of the oceanic crust is recycled at subduction zones. However, the situation is dramatically different for continental crust that is thicker, buoyant, and tectonically protected from subduction zones when located within the continental interior. Geochemical constraints based on $^{40}K–^{40}Ar$ isotope ratio indicate that during the Earth's entire history no more than 30% of the modern mass of the continents has subducted into the mantle, and 50–70% of the subducted material has been later reincorporated back either into the continental lithosphere as magmatic products or into the lower crust as metamorphic material (Coltice et al., 2000). In sharp contrast to isotope geochemistry estimates, geophysical data indicate that, if recycling rates in the past were the same as in modern subduction zones, a volume of continental crust equal to its present-day volume has been recycled since 2.5 Ga (Scholl and von Huene, 2007). This estimate provides a minimum bound since it does not include crustal material recycled at zones of continental collisions. It is worth noting that the rate of crustal recycling at non-accreting margins is about an order of magnitude higher than at accreting margins.

Table 9.4 summarizes available estimates for crustal growth and recycling in the subduction zones. Such estimates commonly do not consider recycling of the entire lithospheric plate, for which the recycling rate can be several times higher than for the crust. For example, about 2500 km of the Indian plate with a lithospheric thickness of ~200 km may have been recycled in the subduction zone beneath Tibet over the past ~50 My if the convergence rate (the part not absorbed by thrusting along the Himalayas front) was similar to the present-day rate of ~5 cm/year. The importance of lithosphere subduction in global mantle dynamics is illustrated by recent high-resolution P-wave and S-wave tomography images of the circum-Pacific subduction zones (Van der Hilst *et al.*, 1997; Fukao *et al.*, 2001) (Fig. 3.103).

10

Summary of lithospheric properties

"Evidence obtained under different experimental conditions cannot be comprehended within a single picture, but must be regarded as complementary in the sense that only the totality of the phenomena exhausts the possible information about the objects."

Niels Bohr

Originally, and following usual practice, my intention was to write a summary chapter. The topic of the book suggested several options for its structuring: (a) similar to the general structure of the book, by methods and techniques; (b) by major tectonic provinces; or (c) by definitions of the lithosphere and boundary layers. Now, having completed the rest of the book, is it is clear that none of these strategies will be appropriate. On one hand, it will result in significant repetition of what has been already written. On the other hand, lithospheric research remains a very active field with new data and new interpretations coming out on a regular basis. Many issues related to lithosphere structure and evolution are still subject to debate, and new results often provide unexpected evidence for or against highly speculative or well-established concepts. The goal of the book, in any case, is not to promote a certain line of interpretation, but instead to show the complexity of the subject and incompleteness of our present knowledge and understanding of many, sometimes fundamental, issues related to lithospheric processes. For these reasons, any short summary would be a dangerous oversimplification.

Therefore, this chapter should be considered as a reference tool to some parts of the book rather than a summary. It includes three tables and references to some key illustrations in the book.

Methods used in lithospheric research and lithosphere definitions

The methods used in lithospheric research are summarized in Table 10.1 which addresses the following questions:

(1) Which physical parameters do different techniques constrain?
(2) What is the resolution/uncertainty of these constraints (vertical, horizontal, and in amplitude)?
(3) Which characteristics of lithospheric structure affect the measured/modeled parameters?
(4) Which boundary layers may be characterized by the modeled parameters and how may they be interpreted in terms of lithospheric thickness?

The abbreviations MBL, TBL, RBL, and CBL stay for mechanical, thermal, rheological, and chemical boundary layers. Their definitions and relations to lithospheric structure and

Table 10.1 Summary of different methods used in lithospheric studies

	Method	Measured/ calculated property *	Resolution*/ uncertainty: Vertical	Resolution*/ uncertainty: Horizontal	Resolution*/ uncertainty: Amplitudes	Measured property is sensitive to:	MBL	TBL	CBL	RBL
Elastic lithosphere	Flexural studies	Flexural rigidity or elastic thickness T_e	• Due to a large set of assumptions and non-uniquenesses of solutions, cannot be quantified (Sections 8.2.3, 8.2.5). • ~25% in regions with known crustal structure. • Large errors for high T_e.	• In spectral methods, depends on averaging window; large windows are required to resolve large T_e; typically from 100 km to 1000 km	• Trade-off between the window size (leads to spatial averaging) and lateral resolution. • Cannot resolve $T_e > 60$–100 km	• Rheology (elastic-plastic-viscoelastic); temperature; strain rate; time scale of deformation. • For continents: composition of the crust and mantle; crust–mantle coupling; pore fluids, sedimentary and crustal thickness	(+)			+
	Glacial rebound	Mantle viscosity	n/a	n/a	• Trade-off between viscosity and lithosphere thickness	• Rheology		+		+?
Electrical lithosphere	MT and EM	Conductivity or resistivity from EM or MT surveys	• Sampled depth depends on period (eq. 7.1). • Vertical resolution ~5–10% of sampled depth. • High for lithospheric thickness	n/a	• The best resolved are the conductance (conductivity by thickness product) and the depth to the top of a conductor. • For lithospheric mantle: an order of magnitude for conductivity (resistivity) values	• Exponential dependence on temperature; interconnected network of highly conducting phase (graphite, melt, water) may change conductivity by several orders of magnitude; olivine composition; • oxygen fugacity; • anisotropy due to olivine LPO		+		+?
Seismic lithosphere	Seismic reflection	Crustal and mantle reflectivity	• ** Very high for TWT; low for depth: requires knowledge of velocity structure; decreases with depth if V_p is obtained from the same data	• ** Depends on the geometry of the experiment; high resolution for migrated data	n/a	• Geometry of seismic interfaces such as caused by: faults, lithologic and compositional layering, partial melts and/or fluids, anisotropy due to ductile flow			+?	+?
	Refraction/ wide angle reflection	Seismic velocities	• ** ~5% (~2 km for Moho). • Hidden layer problem	• ** Critically depends on the geometry of the experiment	• Crust 0.05–0.2 km/s; • mantle > 0.1 km/s; • poor resolution of LVZ; • may smooth sharp velocity interfaces	• Velocity structure variations such as caused by: temperature, composition, partial melts, fluids		+	+	
	Receiver functions	Depth–velocity product	• ** Depth resolution can be improved for known velocity structure • V_p/V_s using multiples improves resolution.	• ** Effectively, 1D model; sensitive to average structure below the station. • Dense spatial sampling may allow migration to improve resolution	• Depends on velocity contrast at interface. • Requires stacking of large volume of data	• Interfaces with strong seismic velocity contrast, e.g. due to composition, partial melts, fluids		+	+	
	Body-wave tomography ***	Body-wave P and S seismic velocities	• ** Teleseismic: not better than 50–100 km. • Regional: ~5 km for the crust	• ** Teleseismic: $1°$–$2°$ in regional studies; also high in global studies. Regional: up to 10–20 km	• Depends on damping and regularization. • Vertical averaging along the ray path. • Affected by crustal correction. • May depend on reference model	• Velocity structure variations, primarily due to temperature, partial melts, fluids		+	+?	

Table 10.1 (cont.)

	Method	Measured/ calculated property *	Resolution*/ uncertainty: Vertical	Resolution*/ uncertainty: Horizontal	Resolution*/ uncertainty: Amplitudes	Measured property is sensitive to:	MBL	TBL	CBL	RBL
Thermal lithosphere	Surface-wave tomography ***	Rayleigh and Love waves phase and group velocities	• ** Teleseismic: ~50 km in the upper mantle (controlled by resolution kernels). • The worst resolved depth is 300–400 km. • For fundamental modes, the maximum resolved depth is defined by f (e.g. ~200 km for 40 s). • Depends on ray coverage	• ** Teleseismic: in global studies is determined by spherical harmonic degree; commonly ~500–1000 km; Regional: up to 50 km. • Decreases with depth	• Depends on damping and regularization. • Horizontal averaging along the ray path. • Affected by crustal correction. • May depend on reference model	• Velocity structure variations such as caused by: temperature, composition, partial melts, fluids		+	+	
	Seismic anisotropy	Polarization and azimuthal anisotropy	• V_{SH}/V_{SV}: as in the corresponding tomographic approach; • SKS: no depth resolution. • Depends on ray coverage	• As in the corresponding seismic method	• As in the corresponding seismic method	• LPO fabric due to ductile flow; dynamic recrystallization; fluid-filled cracks; lithological layering (crust)				+
	Anelastic tomography (teleseismic)	Seismic attenuation	• ** Less than in surface-wave tomography	• ** Less than in surface-wave tomography	• Low resolution	• Extreme sensitivity to temperature, fluids, melts		+		
	Thermal modeling	Geotherms constrained by heat flow	• ~25% for lithospheric thickness. • Poorly constrained asymptotic solutions for lithosphere thickness >250 km	• Critically depends on borehole data coverage. • Depends on model setup >100–150 km for anomalies below Moho, <100 km for crustal anomalies	• Not better than 50–100 deg at any depth.	• Primarily heat production of the crust; • to a lesser extent thermal conductivity of the crust and the mantle	+	+		
	Thermal lithosphere from $V{\rightarrow}T$ conversion	Mantle temperatures from seismic velocities	• ** As in the corresponding seismic model	• ** As in the corresponding seismic model; artifact temperature anomalies can be produced in regions with a sharp velocity contrast	• Unknown because the effects of fluids, melts, anisotropy and composition cannot be quantified independently. • Also depends on the corresponding seismic model	• Velocity structure variations such as caused by: temperature, composition, partial melts, fluids, anisotropy. • Also anelasticity and anharmonicity	+	+?	+?	
Compositional lithosphere	Gravity modeling	Lithosphere density from gravity anomalies	• No depth resolution; can be roughly assessed from the wavelength of gravity anomalies	• $2\pi z$ (where z is depth). • No better than ~200 km when density variations have a subcrustal origin	• Difficult to assess because gravity modeling is non-unique. • Modeling is sensitive to velocity–density conversions which have a large (and unknown) uncertainty	• Density structure variations caused primarily by variations in temperature and composition			+	

					+
				+	

Xenolith data	Crustal and mantle xenoliths	Major element composition	• Pressure estimates may differ by 1.0–1.5 GPa for different geothermobarometers for the same data; • the uncertainty increases at $P > 6$ GPa. • There is no reliable technique for spinel stability field	• Effectively, 1D model	• Mineral equilibrium in the system; compositional variations; can be modified by melt-metasomatism	n/a
	Thermal lithosphere from xenolith P–T arrays	Xenolith geotherms calculated from various geothermobarometers	• Pressure estimates may differ by 1.0–1.5 GPa for different geothermobarometers for the same data; • the uncertainty increases at $P > 6$ GPa. • There is no reliable technique for spinel stability field	• Effectively, 1D model	• Temperature estimates may differ by > 200 °C for different geothermobarometers for the same data • Mineral equilibrium in the system	
	Lithosphere from xenolith texture	Fabric of peridotites (coarse versus sheared) and olivine grain size	• Pressure estimates may differ by 1.0–1.5 GPa for different geothermobarometers for the same data; • the uncertainty increases at $P > 6$ GPa. • There is no reliable technique for spinel stability field	• Effectively, 1D model	• Shearing due to asthenospheric flow	n/a

* Only major parameters are listed.

** Resolution is limited by Fresnel zone. Resolution of any seismic study cannot be better than the wavelength λ.

*** Except for anisotropy studies and anelastic tomography.

Notations: V_{SV} and V_{SH} – velocities of vertically and horizontally polarized shear waves.

Abbreviations: MBL, TBL, CBL, and RBL = mechanical, thermal, chemical, and rheological boundary layers, respectively; TWT = two-way traveltime; Te = effective elastic thickness; LPO = lattice-preferred orientation; n/a = non-applicable or non-available.

the lithosphere–asthenosphere boundary are discussed in detail in Chapter 1. They are further illustrated by the following figures:

Figure 1.4 compares different definitions of the lithosphere with MBL, TBL, RBL, and CBL.

Figure 1.5 illustrates relationships between different geophysical methods and different lithosphere definitions.

Figure 4.45 relates TBL to CBL as constrained by thermal data.

Figure 5.4 compares TBL, CBL, RBL as constrained by xenolith data.

Figure 8.34 compares MBL and RBL as constrained by lithospheric rheology.

Lithospheric structure in major tectonic provinces

It is important to remember that, in general, none of the geophysical or petrologic techniques can resolve the base of the lithosphere (regardless of its definition) with a resolution better than ~50 km. Resolution is limited not only by theoretical limitations, but also by a large number of model assumptions and the large variety of practical approaches used in data acquisition and interpretation. As a result, direct comparison of results coming from different techniques or from different groups using the same technique becomes meaningful only within the uncertainty range of *c.* 50 km (Table 10.1).

Lithospheric structure is summarized in Table 10.2; major tectonic provinces are outlined in Figs. 3.34 and 9.1. The parameters included in the table characterize the seismic structure of the crust and the lithospheric mantle and the thermal state of the lithosphere. They are further illustrated by the following figures:

Figure 3.32 shows typical crustal structure in some tectonic provinces;

Figure 6.36 provides typical seismic velocity and density cross-sections for continental lithosphere.

Figure 7.19 illustrates distribution of water in the upper mantle.

Physical characteristics of the continental lithosphere are summarized in Table 10.3.

Global variations in lithosphere structure are shown in the following maps

Figures 3.38 and 3.39 show the locations of major seismic profiles and global maps of crustal thickness variations and thickness of the sedimentary layer;

Figure 3.111 is a global map of azimuthal anisotropy; Fig. 3.118 shows global variations in seismic V_{SH}-V_{SV} at three depths, Fig. 3.117 is a sketch with interpretations of seismic anisotropy observations in continental and oceanic upper mantle; Fig. 3.120 shows three global maps of lithospheric thickness constrained by shear-wave seismic tomography models;

Figure 4.44 shows global maps of temperature at 100 km depth and lithospheric thermal thickness;

Figure 6.32 is a global map of variations in density of continental lithosphere constrained by gravity data;

Table 10.2 Summary of lithosphere structure in different tectonic settings

Typical values	Shields and platforms			Orogens			Extended crust		Shelves, passive margins	Oceans	
	Archean–Paleoproterozoic cratons	Proterozoic platforms	Paleozoic platforms and basins	Neoproterozoic orogens/mobile belts	Paleozoic orogens	Cenozoic collisional orogens	Paleozoic rifts and extended crust	Cenozoic plume-related regions and extended crust		Young "normal" oceans (20–80 Ma)	Old "normal" oceans (80–160 Ma)
Examples	(1) S.Afr., Western Australia (2) Baltic, Superior, Siberia; West Africa; Brazilian Sh.	EEP, Central N.Am., Central Australia	W. Siberia; Paris basin, North German Basin	S. and Central Africa	(1) Caledonides; Appalachians; (2) Urals	(1) Alps, Pyrenees, Carpathians; (2) Tibet	(1) Variscan Europe (2) Dnieper-Donets rift	W. USA; Massif Central, Pannonian Basin	Barents Sea		Western Pacific
Crustal thickness, km	(1) 32–40; (2) 40–50; locally 60	40–50, locally > 55	35–42, locally ~30	40–48	(1) 40–45 (2) 50–60	(1) 35–55 (2) up to ~80	(1) 30–32 (2) 38–42	25–32	25–35	5.6–7.1	5.6–7.1
Average V_p in the crust, km/s	6.5–7.0	6.5–7.0	6.1–6.7	6.1–6.7	(1) 6.5–6.7 (2) 6.7–6.9	(1) 6.3–6.7 (2) ?	(1) 6.1–6.5 (2) 6.5–6.8	6.1–6.5	6.3–6.8	6.5–6.6	6.5–6.6
Lithosphere thickness (seismic*), km	(1) 170–250; (2) 250–350	150–250	100–200	100–150	(1) 100–150 (2) 175–200	(1) 200–250 (2) 200–250	(1) 80–140 (2) ?	50–100	100–150	30–100	80–120 (150–200 ?)
Lithosphere thickness (TBL), km	(1) 170–250; (2) 250–350, commonly asymptotic solutions at z > 300	140–220	100–140	90–120	(1) 100–140 (2) ?	(1) 80–150, slab down to 150–220 (2) n/a	(1, 2) 100–150	40–100 (typically 60–70)	n/a	Plate model: 95–125 Half-space model: 60–115	Plate model: 95–125 Half-space model: 115–165
Surface heat flow, mW/m²	(1) 40–50; (2) 20–45	40–60	50–70	60–75	(1) 50–70 (2) 25–50	(1) 70–100 (2) ?	(1) 60–80 (2) 40–60	80–120	70–80	70–130	60–70
Moho temperature, °C	300–500	450–650	650–800	700–850	(1) 550–700 (2) 600–800	(1) 700–900 (2) ?	(1) 600–700 (2) 650–850	700–1000	~700	100–200	70–100
Grad T in the crust, deg/km	(1) 8–13 (2) 6–11	12–15	16–20	16–21	(1) ~15 (2) 10	(1) 12–20 (2) ?	(1) 18–22 (2) 16–20	28–40	~20	14–30	10–14
Grad T in the lith. mantle, deg/km	(1) 4.0–6.0 (2) 2.5–5.0	5–8	7–10	8–11	(1) 7–16 (2) 4–6	(1) 6–15 (2) ?	(1, 2) 5–10	10–22	~8–10	11–21	7–11

* from regional surface-wave tomography where available; n/a = non-available

Table 10.3 Summary of physical characteristics of the continental lithosphere

	Parameter	Estimate	References
Global	Total area of continental crust	2.104×10^8 km^2	MT99
	Between 80N and 80S (mean estimate)	1.60×10^8 km^2	A00
	On-shore continental crust	1.21×10^8 km^2	A06*
	Mass of the crust and mantle system	3.85×10^{24} kg	TM95
Crust	Volume of continental crust	7.575×10^9 km^3	TM95
		7.581×10^9 km^3	SS89
		7.76×10^9 km^3	RS84
		7.8×10^9 km^3	NR96, M98
	Mean crustal density	2835 kg/m^3	CM95
	Mass of continental crust (for density of 2835 kg/m^3)	2.08×10^{22} kg	TM95
		2.25×10^{22} kg	A06
	Mass of continental crust as percent of mass of the crust–mantle system	0.54%	TM95
	Mass of continental crust as percent of whole Earth's mass	0.35%	TM95
Lithosphere and lithospheric mantle	Average thickness of continental lithosphere	108 km	A06*
	Volume of continental lithosphere		
	On-shore continental lithosphere	32.1×10^9 km^3	A06*
	Between 80N and 80S (range)	$[24.8\text{–}34.1] \times 10^9$ km^3	A00
	Volume of lithospheric mantle		
	On-shore continental lithosphere	24.1×10^9 km^3	A06*
	Between 80N and 80S (range)	$[20.1\text{–}26.3] \times 10^9$ km^3	A00
	Ratio of crustal to total volumes for continental lithosphere	21.2%	A00
	Volume of continental crust as percent of volume of continental lithospheric mantle	26.8%	A00
	Mean density of lithospheric mantle (room P-T conditions)	3340 kg/m^3	G98
	Mass of continental lithospheric mantle	7.85×10^{22} kg	A06*
	Mass of continental lithosphere as percent of whole Earth's mass	1.68%	A06*
	Mass of continental lithospheric mantle as percent of Earth's mantle	2.04%	A06*

Data sources: A00 = Abbott et al. (2000); A06 = Artemieva (2006); CM95 = Christensen and Mooney (1995); G98 = Griffin et al. (1998a); NR96 = Nataf and Ricard (1996); M98 = Mooney et al. (1998); MT = McLennan and Taylor (1999); RS84 = Reymer and Schubert (1984); SS89 = Schubert and Sandwell (1989); TM95 = Taylor and McLennan (1995)

* The continental lithosphere is defined as the area above sea level, excluding submerged areas with continental crust such as e.g. shelves and ocean plateaus. The estimates of lithosphere thickness and its volume are based on a $1° \times 1°$ global thermal model (Artemieva, 2006).

Figure 7.50 is a map of predicted depth to high-conductivity layer (electrical lithosphere) constrained by heat flow data for the continents;

Figure 8.58 is a compilation of Te (effective elastic thickness) values reported for the continents.

Secular trends in lithospheric thickness variations are shown in the following figures

Figure 2.14 shows the time-scale of major global events related to the Earth's evolution; Figs. 2.15 and 2.22 are maps showing ages of the continental and oceanic crust;

Figure 3.108 is a sketch that illustrates secular variations in seismic velocities, density, and mg# of the lithospheric mantle; Fig. 6.15 summarizes variations of Mg# with depth for continental lithosphere of different ages;

Figure 3.88 shows 1D variations of seismic Vs velocity with depth in typical continental and oceanic structures and Fig. 3.89 compares 1D global averages of Vs versus depth for oceans and continents; Fig. 3.121 shows variations in thickness of oceanic lithosphere with age as constrained by seismic tomography;

Figure 4.19 are plots that show surface heat flow variations versus age for continents and oceans (data coverage is shown in Fig. 4.16); global continental trends are in Fig. 4.29; Fig. 4.21 illustrates variations of ocean heat flow and bathymetry with age; Figs. 4.25 and 4.33 show typical oceanic and continental geotherms; Fig. 5.13 shows typical xenolith geotherms; Fig. 4.46 shows variations in thickness of continental lithosphere with age as constrained by thermal models; Figs. 5.12 and 5.13 summarize xenolith geotherms from cratonic settings;

Figures 7.21, 7.23, 7.37 are plots that show typical profiles of electrical conductivity versus depth for different oceanic and continental settings; Figs. 7.45 and 7.49 illustrate global correlations between the depth to high-conductivity layer, lithosphere age, and surface heat flow on the continents;

Figures 8.50 and 8.59 show correlations between Te, geotherms, lithosphere age and surface heat flow for oceans and continents;

Figures 9.18 and 9.20 show various models of crustal growth rate over 3.5 Ga.

This summary, together with the index of the key figures, finalizes the overview of our present understanding of the lithosphere, its origin, structure, and evolution. In spite of impressive progress in lithospheric research over past decades and the continuous and growing interest in multidisciplinary studies of the crust and upper mantle of oceans and continents, much remains to be understood. Merging geophysical and geochemical field and laboratory observations and their joint interpretation, is becoming common practice, which advances our understanding of the lithosphere and its role in plate tectonics and global geodynamics. However, sometimes the more we know, the less we understand, with new information bringing new questions; some of the still unresolved controversies are discussed in this book. I hope this book will provoke interest and inspire further research in the exciting and still controversial field of multidisciplinary studies of the lithosphere.

References

Abbott, D. and Mooney, W. D. (1995). The structural and geochemical evolution of the continental crust: support for the oceanic plateau model of continental growth. *Rev. Geophys.*, **33** Suppl.: 231–242.

Abbott, D. H. and Menke, W. H. (1990). Length of the global plate boundary at 2.4 Ga. *Geology*, **18**, 62–66.

Abbott, D., Burgess, L., Longhi J., and Smith W. (1994). An empirical thermal history of the Earth's upper mantle. *J. Geophys. Res.*, **99**(B7), 13835–13850.

Abbott, D., Sparks, D., Herzberg, C., *et.al.* (2000). Quantifying Precambrian crustal extraction: the root is the answer. *Tectonophysics*, **322**, 163–190.

Achauer, U. and Masson, F. (2002). Seismic tomography of continental rifts revisited: from relative to absolute heterogeneities, *Tectonophysics*, **358**, 17–37.

Adam, A. (1976). Quantitative connections between regional heat flow and the depth of conductive layers in the earth's crust and upper mantle. *Acta Geod. Geophys. et Montanist. Acad. Sci. Hung.*, **11**, 503–509.

Adam, C. and Vidal, V. (2010). Mantle flow drives the subsidence of oceanic plates. *Science*, **328**, 83–85.

Adam, A. and Wesztergom, V. (2001). An attempt to map the depth of the electrical asthenosphere by deep magnetotelluric measurements in the Pannonian Basin (Hungary). *Acta Geol. Hung.*, **44**, 167–192.

Adam, A., Ernst, T., Jankowski, J., *et al.* (1997). Electromagnetic induction profile (PREPAN95) from the East European Platform (EEP) to the Pannonian Basin. *Acta Geod. Geoph. Hung.*, **32**(1–2), 203–223.

Afonso, J. C. and Ranalli, G. (2004). Crustal and mantle strengths in continental lithosphere: is the jelly sandwich model obsolete? *Tectonophysics*, **394**, 221–232.

Ahern, J. and Ditmars, R. (1985). Rejuvenation of continental lithosphere beneath an intracratonic basin. *Tectonophysics*, **120**, 21–35.

Airy, G. B. (1855). On the computation of the effect of the attraction of mountain-masses, as disturbing the apparent astronomical latitude of stations of geodetic surveys. *Phil. Trans. Roy. Soc.*, **145**, 101–104.

Aki, K. and Richards, P. G. (1980). *Quantitative Seismology: Theory and Methods*. San Francisco, W. H. Freeman.

Aki, K., Christofferson, A., and Husebye, E. S. (1977). Determination of the three-dimensional seismic structure of the lithosphere, *J. Geophys. Res.*, **82**, 277–296.

Alard, O., Griffin, W. L., Lorand, J. P., Jackson, S. E., and O'Reilly, S. Y. (2000). Non-chondritic distribution of the highly siderophile elements in mantle sulphides. *Nature*, **407**, 891–894.

Albarède, F. (1998). The growth of continental crust, *Tectonophysics*, **296**, 1–14.

Albarède, F. (2003). *Geochemistry: An Introduction*. Cambridge University Press, 248 pp.

Albarède, F., Blichert-Toft, J., Vervoort, J. D., Gleason, J. D., and Rosing, M. (2000). Hf–Nd isotope evidence for a transient dynamic regime in the early terrestrial mantle. *Nature*, **404**, 488–490.

Alekseyev, A. S., Vanyan, L. L., Berdichevsky, M. N., *et al.* (1977). Map of asthenospheric zones of the Soviet Union. *Doklady Akademii Nauk SSSR*, **234**, 790–793.

Allègre, C. J. (1982). Chemical geodynamics. *Tectonophysics*, **81**, 109–132.

Allègre, C. J. and Rousseau, D. (1984). The growth of the continents through geological time studies by Nd isotope analysis of shales. *Earth Planet Sci. Lett.*, **67**, 19–34.

Allègre, C. J., Lewin, E., and Dupre, B. (1988). A coherent crust-mantle model for the uranium-thorium-lead isotopic system. *Chem. Geol.*, **70**, 211–234.

Allègre, C. J., Staudacher, T., Sarda P., and Kurz, M. (1983). Constraints on evolution of Earth's mantle from rare gas systematics. *Nature*, **303**, 762–766.

Allis, R. G. (1979). A heat production model for stable continental crust. *Tectonophysics*, **57**, 151–165.

Allmendinger, R. W., *et al.* (1986). Phanerozoic tectonics of the Basin and Range – Colorado plateau transition from COCORP data and geologic data: a review. In: Barazangi, M. and Brown, L. D. (Eds.), 1986. *Reflection Seismology*. AGU Geodyn. Ser., v. **14**, 257–267.

Amelin, Y., Lee, D. C., Halliday, A. N., and Pidgeon, R. T. (1999). Nature of the Earth's earliest crust from hafnium isotopes in single detrital zircons. *Nature*, **399**, 252–255.

Ammon, C. J., Randall, G. E., and Zandt, G. (1990). On the nonuniqueness of receiver function inversions. *J. Geophys. Res.*, **95**, 15303–15318.

ANCORP Working Group (1999). Seismic reflection image revealing offset of Andean subduction-zone earthquake locations into oceanic mantle. *Nature*, **397**, 341–344.

Anderson, D. L. (1961). Elastic wave propagation in layered anisotropic media, *J. Geophys. Res.*, **66**, 2953–2963.

Anderson, D. L. (1994). Lithosphere and flood basalts. *Nature*, **367**, 226.

Anderson, D. L. (1995). Lithosphere, asthenosphere, and perisphere. *Rev. Geophys.*, **33**, 125–149.

Anderson, D. L. (2005). Scoring hotspots: the plume and plate paradigms, In Foulger, G. R., Natland, J. H., Presnall, D. C., and Anderson, D. L. (Eds), *Plates, Plumes, and Paradigms: GSA*, Special Paper 388, Boulder, CO, USA, 31–54.

Anderson, D. L. (2007). *New Theory of the Earth*. Cambridge University Press, 384 pp.

Anderson, D. L. and Archambeau, C. B. (1964). The anelasticity of the Earth. *J. Geophys. Res.*, **69**, 2071–2084.

Anderson, D. L. and Bass, J. D. (1984). Mineralogy and composition of the upper mantle. *Geophys. Res. Lett.*, **11**, 637–640.

Anderson, D. L. and Sammis, C. G. (1970). Partial melting in the upper mantle. *Phys. Earth Planet. Inter.*, **3**, 41–50.

Anderson, D. L., Ben-Menahem A., and Archambeau, C. B. (1965). Attenuation of seismic energy in upper mantle. *J. Geophys. Res.*, **70**, 1441–1448.

Anderson, R. N. (1980). Update of heat flow in the East and Southeast Asian seas In: D. Hayes (Ed.), *The Tectonic and Geologic Evolution of Southeast Asian Seas and Islands*. Geophysical Monograph Series, 23, pp. 319–326, Am. Geophys. Un., Washington, DC.

Anderson, R. N. and Hobart, M. A. (1976). The relation between heat flow, sediment thickness, and age in the eastern Pacific, *J. Geophys. Res.*, **81**, 2968–2989.

Anderson, O. L., Isaak, D. and Oda, H. (1992). High temperature elastic constants data on minerals relevant to geophysics. *Rev. Geophys.*, **30**, 57–90.

Anderson, O. L., Schreiber, E., and Lieberman, R. C. (1968). Some elastic constant data on minerals relevant to geophysics. *Rev. Geophys.*, **6**, 491–524.

Arcay, D., Doin, M.-P., Tric, E., Bousquet, R., and de Capitani, C. (2006). Overriding plate thinning in subduction zones: Localized convection induced by slab dehydration. *Geochem. Geophys. Geosyst.*, **7**, Q02007, doi:10.1029/2005GC001061.

Arculus, R. J. (1985). Oxidation status of the mantle: past and present. *Annu. Rev. Earth Planet. Sci.* **13**, 75–95.

Arlitt, R. (1999). Teleseismic body wave tomography across the trans-European suture zone between Sweden and Denmark. Ph.D. thesis, ETH, Zurich, 126 p.

Armitage, J. J. and Allen, P. A. (2010). Cratonic basins and the long-term subsidence history of continental interiors. *J. Geol. Soc.*, **167**: 61–70.

Armstrong, R. L. (1981). Radiogenic isotopes: the case for crustal recycling on a near-steady-state no-continental growth Earth. *Phil. Trans. R. Soc. London Ser.* A, **301**, 443–472.

Armstrong, R. L. (1991). The persistent myth of crustal growth. *Aust. J. Earth Sci.*, **38**, 613–630.

Artemieva, I. M. (1989). Influence of volatiles in the upper mantle on the dynamics of thermal thinning of the lithosphere. *J. Geodynam*, **11**, 77–97.

Artemieva, I. M. (1996). The dependence of transport properties of in situ rocks on pore fluid composition and temperature. *Surv. Geophys.*, **17**, 289–306.

Artemieva, I. M. (2003). Lithospheric structure, composition, and thermal regime of the East European craton: implications for the subsidence of the Russian Platform. *Earth Planet. Sci. Lett.*, **213**, 429–444.

Artemieva, I. M. (2006). Global $1° \times 1°$ thermal model TC1 for the continental lithosphere: Implications for lithosphere secular evolution. *Tectonophysics*, **416**, 245–277.

Artemieva, I. M. (2007). Dynamic topography of the East European Craton: shedding light upon the lithospheric structure, composition and mantle dynamics. *Global Planet. Change*, **58**, 411–434.

Artemieva, I. M. (2009). The continental lithosphere: reconciling thermal, seismic, and petrologic data. *Lithos*, **109**, 23–46.

Artemieva, I. M. and Chesnokov, E. M. (1991). Thermal characteristics of anisotropic media with inclusions. *Geophys. J. Int.*, **107**, 557–562.

Artemieva, I. M. and Mooney, W. D. (1999). Mantle heat flow in stable continental regions: A global study. *EOS Trans. AGU*, **80**(48), F967.

Artemieva, I. M. and Mooney, W. D. (2001). Thermal thickness and evolution of Precambrian lithosphere; a global study, *J. Geophys. Res.*, **106**, 16,387–16,414.

Artemieva, I. M. and Mooney, W. D. (2002). On the relation between cratonic lithosphere thickness, plate motions, and basal drag. *Tectonophysics*, **358**, 211–231.

Artemieva, I. M. and Thybo, H. (2008). Deep Norden: highlights of the lithospheric structure of Northern Europe, Iceland, and Greenland. *Episodes*, **31**, 98–106.

Artemieva, I. M. and Thybo, H. (2011). Crust and upper mantle of Europe, Greenland, and the North Atlantic region: an overview. *Submitted to J. Geoph. Res.*

Artemieva, I. M., Thybo, H., and Kaban, M. K. (2006). Deep Europe today: Geophysical synthesis of the upper mantle structure and lithospheric processes over 3.5 Ga. In: Gee, D. and Stephenson, R. (Eds.), *European Lithosphere Dynamics*. Geol. Soc. London Sp. Publ., v. **32**, 11–41.

Artemieva, I. M., Mooney, W. D., Perchuc, E., and Thybo, H. (2002). Processes of lithosphere evolution: new evidence on the structure of the continental crust and upper mantle. *Tectonophysics*. **358**, 1–15.

Artemieva, I. M., Billien, M., Lévêque, J.-J., and Mooney, W. D. (2004). Shear-wave velocity, seismic attenuation, and thermal structure of the continental upper mantle. *Geophys. J. Int.*, **157**: 607–628.

Artemjev, M. E. and Kaban, M. K. (1986). The free mantle surface – new possibilities to reveal subcrustal inhomogeneities from the structure of the Earth's crust. *J. Geodynam.*, **5**, 25–44.

Artemjev, M. E. and Kaban, M. K. (1987). Isostasy and cross-spectral method of its study. *Izvestiya. Phys. Solid Earth*, **23**(11), 935–945.

Artemjev, M. E. and Kaban, M. K. (1991). Isostatic processes and intracontinental orogenesis. *J. Geodynam.*, **13**, 77–86.

Artemjev, M. E., Bune, V. I., Dubrovsky, V. A., and Kambarov, N. S. (1972). Seismicity and isostasy. *Phys. Earth Planet. Inter.*, **6**, 256–262.

Artemjev, M. E., Babaeva, T. M., Mikhailov, V. O., and Voidetsky, I. E. (1984). Identification of mantle and lithospheric components of the gravity field by isostatic gravity anomalies. *Mar. Geophys. Res.*, **7**, 129–148.

Artemjev, M. E., Kaban, M. K., Kucherinenko, V. A., Demyanov, G. V., and Taranov, V. A. (1994). Subcrustal density inhomogeneities of Northern Eurasia as derived from the gravity data and isostatic models of the lithosphere. *Tectonophysics*, **240**, 249–280.

Artyushkov, E. V. (1993). *Physical Geotectonics*. Moscow, Nauka, 455 pp.

Artyushkov, E. V. (1973) Stresses in the lithosphere caused by crustal thickness inhomogeneities. *J. Geophys. Res.*, **78**, 7675–7708.

Ashchepkov, I. V., Pokhilenko, N. P., Vladykin, N. V., *et al.* (2010). Structure and evolution of the lithospheric mantle beneath Siberian craton, thermobarometric study. *Tectonophysics*, **485**, 17–41.

Aschepkov, I., Vladykin, N., Pokhilenko, N., *et al.* (2003). Clinopyroxene geotherms for the mantle columns beneath kimberlite pipes from Siberian craton. *Proc. 8th Int. Kimberlite Conf.*, Victoria, BC, Canada, June 2003: FLA-0355.

Ashwal, L. D. (1993). *Anorthosites*. Berlin, Springer-Verlag, 411 pp.

Ashwal, L. D. and Burke, K. (1989). African lithospheric structure, volcanism, and topography. *Earth Planet. Sci. Lett.*, **96**, 8–14.

Ashwal, L. D., Morgan, P., Kelley, S. A., and Percival, J. A. (1987). Heat production in an Archean crustal profile and implications for heat flow and mobilization of heat-producing elements. *Earth Plan. Sci. Lett.*, **85**, 439–450.

Ashwal, L. D., Phinney, W. C., Morrison, D. A., and Wooden, J. L. (1982). Underplating of Archean continents: evidence from the Bad Vermilion Lake anorthosite complex, Ontario. *Lunar Planet. Sci.*, **13**: 20–21.

Asimow, P. D. and Langmuir, C. H. (2003). The importance of water to oceanic mantle melting regimes. *Nature*, **421**, 815–820.

Audet, P. and Mareschal, J. C. (2004a). Anisotropy of the flexural response of the lithosphere in the Canadian Shield. *Geophys. Res. Lett.*, **31**, L20601, doi:10.1029/2004GL021080.

Audet, P. and Mareschal, J. C. (2004b). Variations in elastic thickness in the Canadian Shield. *Earth Planet. Sci. Lett.*, **226**, 17–31, doi:10.1016/j.epsl.2004.07.035.

Audet, P. and Mareschal, J. C. (2007). Wavelet analysis of the coherence between Bouguer gravity and topography: application to the elastic thickness anisotropy in the Canadian Shield. *Geophys. J. Int*, **168**, 287–298.

Audet, P., Jellinek, A. M., and Uno, H. (2007). Mechanical controls on the deformation of continents at convergent margins. *Earth Planet. Sci. Lett.*, **264**, 151–166.

Aulbach, S., Griffin, W. L., Pearson, N. J., *et al.* (2004). Mantle formation and evolution, Slave Craton: constraints from HSE abundances and Re–Os isotope systematics of sulfide inclusions in mantle xenocrysts. *Chem. Geol.*, **208**: 61–88.

Austin, N. J. and Evans, B. (2007). Paleowattmeters: a scaling relation for dynamically recrystallized grain size. *Geology*, **35**, 343–346.

Austrheim, H. (1991). Eclogite formation and dynamics of crustal roots under continental collision zones. *Terra Nova*, **3** (5), 492–499.

Avtoneyev, S. V., Ananyeva, Ye. M., Bashta, K. G., *et al.* (1992). Deep structure of the Urals from geophysical data. In: Beloussov, V. V., Pavlenkova, N. I., and Kvyatkovskaya, G. N. (Eds.) *Structure of the Crust and Upper Mantle of the (former) USSR. Intern. Geol. Rev.*, **34**, 263–278.

BABEL Working Group (1990). Evidence for early Proterozoic plate tectonics from seismic reflection profiles in the Baltic Shield. *Nature*, **348**, 34–38.

BABEL Working Group (1993). Deep seismic reflection/refraction interpretation of crustal structure along BABEL profiles A and B in the southern Baltic Sea. *Geophys. J. Int.*, **112**, 325–343.

Babuska, V. (1984). P wave velocity anisotropy in crystalline rocks. *Geophys. J.R. Astron. Soc.*, **76**, 113–119.

Babuška, V. and Plomerová, J. (1992). The lithosphere in central Europe – seismological and petrological aspects. *Tectonophysics*, **207**, 141–163.

Babuška, V., Montagner, J. P., Plomerova, J., *et al.* (1998). Age-dependent large-scale fabric of the mantle lithosphere as derived from surface-wave velocity anisotropy. *Pure and Appl. Geophys.*, **151**, 257–280.

Backus, G. and Gilbert, F. (1967). Numerical applications of a formalism for geophysical inverse problems. *Geophys. J. Roy. Astron. Soc.*, **13**, 247–276.

Backus, G. E. and Gilbert, J. F. (1968). The resolving power of gross Earth data, *Geophys. J. Roy. Astron. Soc.*, **16**, 169–205.

Bagley, B. and Revenaugh, J. (2008). Does the sharpness of the Gutenberg discontinuity require melt? *Eos Trans. AGU,* Fall Meet. Suppl., Abstract, U41F-01.

Bahr, K. and Duba, A. (2000). Is the asthenosphere electrically anisotropic? *Earth Planet. Sci. Lett.*, **178**, 87–95.

Bahr, K. and Simpson, F. (2002). Electrical anisotropy below slow and fast-moving plates: palaeoflow in the upper mantle? *Science*, **295**, 1270–1272.

Bahr, K., Olsen, N., and Shankland, T. J. (1993) On the combination of the magnetotelluric and the geomagnetic depthsounding method for resolving an electrical conductivity increase at 400 km depth. *Geophys. Res. Lett.*, **20**(24), 2937–2940.

Bai, Q. and Kohlstedt, D. L. (1992). Substantial hydrogen solubility in olivine and implications for water storage in the mantle. *Nature*, **357**, 672–674.

Bai, Q., Mackwell, S. J., and Kohlstedt, D. L. (1991). High temperature creep of olivine single crystals 1. Mechanical results for buffered samples. *J. Geophys. Res.*, **96**: 2441–2463.

Bailey, C. (2006). Crustal ductility and early continental Tectonophysics. In: Benn, K., Maraschel, J-C. and Condie, K. C. (Eds.), *Archean Geodynamics and Environments.* American Geophysical Union, Geophysical Monograph Series, **164**, 115–130.

Bailey, D. K. (1982). Mantle metasomatism. *Nature*, **296**, 525–530.

Bailey, R. C. (1970). Inversion of the geomagnetic induction problem. *Proc. Roy. Soc. London, Ser.* A, **315**, 185–194.

Baker, J. A., Bizzarro, M., Wittig, N., Connelly, J., and Haack, H. (2005). Early planetesimal melting from an age of 4.5662 Gyr for differentiated meteorites, *Nature* **436**, 1127–1131.

Baldridge, W. S., Keller, G. R., Haak, V., *et al.* (1995). The Rio Grande Rift. In: K. H. Olsen (Ed.), *Continental Rifts: Evolution, Structure, Tectonics*. Elsevier, Developm. Geotecton, v.25, 233–276.

Ballard, S. and Pollack, H. N. (1987). Diversion of heat by Archean cratons: a model for southern Africa. *Earth Planet. Sci. Lett.*, **85**, 253–264.

Ballard, S., Pollack, H. N., and Skinner, N. J. (1987). Terrestrial heat flow in Botswana and Namibia. *J. Geophys. Res.*, **92**, 6291–6300.

Ballhaus, C., Berry, R. F., and Green, D. H. (1991). High pressure experimental calibration of the olivine–orthopyroxene–spinel oxygen barometer: implications for the oxidation state of the upper mantle. *Contrib. Mineral. Petrol.* **107**, 27–40.

Balling, N. (1995). Heat flow and thermal structure of the lithosphere across the Baltic Shield and northern Tornquist Zone. *Tectonophys.*, **244**, 13–50.

Balling, N. (2000). Deep seismic reflection evidence for ancient subduction and collision zones within the continental lithosphere of northwestern Europe. *Tectonophysics*, **329**, 269–300.

Banner, J. L. (2004). Radiogenic isotopes: systematics and applications to earth surface processes and chemical stratigraphy. *Earth Sci. Rev.*, **65**, 141–194.

Barazangi, M. and Brown, L. D. (eds.) (1986). *Reflection Seismology*. AGU Geodyn. Ser., **v.13** (311 pp), **v.14**(339 pp).

Barley, M. E. Bekker, A., and Krapež, B. (2005). Late Archean to Early Paleoproterozoic global tectonics, environmental change and the rise of atmospheric oxygen. *Earth Planet. Sci. Lett.*, **238**, 156–171.

Barrel, J. (1914). The strength of the Earth's crust. *J. Geol.*, **v.22.**

Barruol, G. and Kern, H. (1996). Seismic anisotropy and shear-wave splitting in lower-crustal and upper mantle rocks from the Ivrea zone – experimental and calculated data. *Phys. Earth Planet. Inter.*, **95**, 175–194.

Barth, M. G., Rudnick, R. L. Horn, I. *et al.* (2001). Geochemistry of xenolithic eclogites from West Africa; Part I, A link between low MgO eclogites and Archean crust formation, *Geochim. Cosmochim. Acta*, **65**, 1499–1527.

Barth, M. G., Rudnick, R. L. Horn, I. *et al.* (2002). Geochemistry of xenolithic eclogites from West Africa, part 2: origins of the high MgO eclogites, *Geochim. Cosmochim. Acta*, **66**, 4325–4345.

Barton, P. J. (1986). The relationship between seismic velocity and density in the continental-crust – a useful constraint? *Geophys. J. Roy. Astron. Soc.*, **87**, 195–208.

Bass, J. D. and Anderson, D. L. (1984). Composition of the upper mantle: Geophysical tests of two petrological models. *Geophys. Res. Lett.* **11**, 237–240.

Bassin, C., Laske, G., and Masters, G. (2000). The current limits of resolution for surface wave tomography in North America, *EOS Trans AGU*, **81**: F897.

Beamish, D. and Smythe, D. K. (1986). Geophysical images of the deep crust – the Iapetus suture. *J. Geol. Soc.*, **143**, 489–497.

Beardsmore, G. R. and Cull, J. P. (2001). *Crustal Heat Flow, A Guide to Measurement and Modeling*. Cambridge University Press.

Bechtel, D., Forsyth, D. W. Sharpton, V. L. and Grieve, R. A. F. (1990). Variations in effective elastic thickness of the north American lithosphere, *Nature*, **343**, 636–638.

Beck, A. E. (1977). Climatically perturbed temperature gradients and their effect on regional and continental heat-flow means. *Tectonophysics*, **41**, 17–39.

Behn, M. D., Hirth, G and Elsenbeck, J. R. (2008). Implications of grain-size evolution on the seismic structure of the oceanic upper mantle. *Fall AGU Abstr.*, U41F-07.

Bell, D. R. and Rossman, G. R. (1992). Water in the Earth's mantle: the role of nominally anhydrous minerals, *Science*, **255**, 1391–1397.

Bell, D. R, Rossman, G. R., and Moore, R. O. (2004). Abundance and partitioning of OH in a high-pressure magmatic system: megacrysts from the Monastery Kimberlite, South Africa. *J. Petrol.*, **45**, 1539–1564.

Bell, D. R., Schmitz, M. D. and Janney, P. E. (2003). Mesozoic thermal evolution of the southern African mantle lithosphere, *Lithos*, **71**, 273–287.

Bell, D. R., Rossman, G. R., Maldener, J., Endisch, D., and Rauch, F. (2003), Hydroxide in olivine: a quantitative determination of the absolute amount and calibration of the IR spectrum: *J. Geophys. Res.*, v. **108**, doi10.1029/2001JB000679.

Ben Ismail, W. and Mainprice, D. (1998). An olivine fabric database: an overview of upper mantle fabrics and seismic anisotropy. *Tectonophysics*, **296**, 145–157.

Ben Ismail, W., Barroul, G., and Mainprice, D. (2001). The Kaapvaal craton seismic anisotropy: petrological analyses of upper mantle kimberlite nodules. *Geophys. Res. Lett.*, **28**, 2497–2500.

Benfield, A. E. (1939). Terrestrial heat flow in Britain. *Proc. Roy. Soc.*, **173**, 428–450.

Ben-Menahem A. (1965). Observed attenuation and Q values of seismic waves in the upper mantle. *J. Geophys. Res.*, **70**, 4641–4651.

Benn, K., Maraschel, J-C. and Condie, K. C. (Eds.) (2006). *Archean Geodynamics and Environments*. American Geophysical Union, Geophysical Monograph Series **164**, 320 pp.

Bennett, V. C. (2004). Compositional evolution of the mantle. In: *Treatise of Geochemistry*, v.2, Oxford., Elsevier-Pergamon, 493–519.

Bennett, V. C. and DePaolo, D. J. (1987). Proterozoic crustal history of the western United States as determined by neodymium isotopic mapping. *Geol. Soc. Am. Bull*, **99**, 674–685.

Bennett, V. C., Nutman, A. P., and McCulloch, M. T. (1993). Nd isotopic evidence for transient highly-depleted mantle reservoirs in the early history of the Earth. *Earth Planet. Sci. Lett.* **119**, 299–317.

Berckhemer, H., Kapfmann, W., Aulbach, E., and Schmeling, H. (1982). Shear modulus and Q of forsterite and dunite near partial melting from forced oscillation experiments. *Phys. Earth Planet. Interiors*, **29**, 30–41.

Bercovici, D. and Karato, S. (2003). Whole mantle convection and transition-zone water filter. *Nature,* **425**, 39–44.

Bercovici, D., Ricard, Y., Richards, M. A. (2000). The relation between mantle dynamics and plate tectonics: a primer. In: Richards, M. A. *et al.* (Eds.), *The History and Dynamics of Global Plate Motions*. AGU, Geophys. Monogr., **121**, 5–46.

Bercovici, D, Schubert, G, and Glatzmaier, G. (1989). 3-dimensional spherical-models of convection in the earth's mantle. *Science*, **244**, 950–955.

Berdichevsky, M. N., Vanyan, L. L., Kuznetsov, V. A., *et al.* (1980). Geoelectric model of the Baikal region. *Phys. Earth Planet. Inter.* **22**, 1–11.

Berman, R. G. and Aranovich, L. Y. (1996). Optimized standard state and solution properties of minerals: 1. Model calibration for olivine, orthopyroxene, cordierite, garnet, and ilmenite in the system $FeO-MgO-CaO-Al2Os-TiO_2-SiO_2$. *Contrib. Mineral. Petrol.*, **126**, 1–24.

Bernstein, S., Kelemen, P. B., and Brooks, C. K. (1998). Depleted spinel harzburgite xenoliths in Tertiary dikes from East Greenland: restites from degree melting. *Earth Planet. Sci. Lett.* **154**, 221–235.

Bernstein, S., Hanghoj, K., Kelemen, P.B., *et al.* (2006). Ultra-depleted, shallow cratonic mantle beneath West Greenland: dunitic xenoliths from Ubekendt Ejland. *Contrib. Mineral. Petrol.*, **152**, 335–347.

Berryman, J. G. (1995). Mixture theories for rock properties. In: T. J. Ahrens (Ed.), *Rock Physics and Phase Relations: A Handbook of Physical Constants*. Washington, American Geophysical Union.

Berthelsen, A. (1992). Mobile Europe. In: Blundell, D., Freeman, R., and Mueller, S. (Eds.), *A continent revealed. The European Geotraverse*: Cambridge University Press, 11–32.

Bessonova, E. N., Fishman, V. M., Shirman, M. G. *et al.* (1976). The tau method of inversion of travel times: 2. Earthquake data. *Geoph. J. Roy. Astr. Soc.*, **46**, 87–108.

Beyer, E. E., Griffin, W. L., and O'Reilly, S. Y. (2006). Transformation of archaean lithospheric mantle by refertilization: Evidence from exposed peridotites in the Western Gneiss Region, Norway. *J. Petrology*, **47**, 1611–1636.

Bhattacharyya, B. K. (1966). Continuous spectrum of the total magnetic field anomaly due to a rectangular prismatic body. *Geophysics* **31**, 97–121.

Bhattacharyya, J. G., Masters, G., and Shearer, P. (1996). Global lateral variations of shear attenuation in the upper mantle. *J. Geophys. Res.*, **101**, 22273–22290.

Bibikova, E. V., Bogdanova, S. V. Larionov, A. N. *et al.* (2008), New data on the Early Archean Age of granitoids in the Volga–Ural segment of the East European craton. *Dokl. Earth Sci.*, **419**, No. 2, 243–247.

Bickle, M. J. (1978). Heat loss from the Earth: a constraint on Archean tectonics from the relation between geothermal gradients and the rate of plate production. *Earth Planet. Sci. Lett.*, **40**, 301–315.

Bijwaard, H. and Spakman, W. (1999). Tomographic evidence for a narrow whole mantle plume below Iceland, *Earth Planet. Sci. Lett.*, **166**, 121–126.

Bijwaard, H. W., Spakman, W., and Engdahl, E. R. (1998). Closing the gap between regional and global travel time tomography. *J. Geophys. Res.*, **103**, 30055–30078.

Billien, M., Lévêque, J.-J., and Trampert, J. (2000). Global maps of Rayleigh wave attenuation for periods between 40 and 150 seconds. *Geophys. Res. Lett.*, **27**, 3619–3622.

Bijwaard, H. and Spakman, W. (2000). Non-linear global P-wave tomography by iterated linearized inversion, *Geophys. J. Int.*, **141**, 71–82.

Birch, F. (1943). Elasticity of igneous rocks at temperatures and pressures. *Geol. Soc. Am. Bull.*, **54**, 263–286.

Birch, F. (1958). Interpretation of the seismic structure of the crust in light of experimental studies of wave velocities in rocks. In: H. Benioff (Ed.), *Contributions in Geophysics in Honor of Beno Gutenberg*. New York, Pergamon, 158–170.

Birch, F. (1961). The velocity of compressional waves in rocks to 10 kilobars, part 2. *J. Geophys. Res.*, **66**, 2199–2224.

Birch, F., Roy, E. R., and Decker, E. R. (1968). Heat flow and thermal history and New England and New York. In: An-Zen E. (Ed.), *Studies of Appalachian Geology*. New York, Wiley, Interscience, 437–451.

Bird, P. (1978). Initiation of intracontinental subduction in the Himalayas. *J. Geophys. Res.*, **83**, 4975–4987.

Bird, P. (1979). Continental delamination and the Colorado Plateau. *J. Geophys. Res.*, **84**, 7561–7571.

Bjarnason, I. T., Wolfe, C. J. Solomon, S. C. and Gudmundsson, G. (1996). Initial results from the ICEMELT experiment: Body-wave delay times and shear-wave splitting across Iceland. *Geophys. Res. Lett.*, **23**, 459–463.

Black, L. P., Williams, I. S. and Compston, W. (1986). Four zircon ages from one rock: the history of a 3930 Ma-old granulite from Mount Sones, Enderby Land, Antarctica. *Contrib. Mineral. Petrol.*, **94**, 427–437.

Blackman, D. K. and Kendall, J. M. (1997). Sensitivity of teleseismic body waves to mineral texture and melt in the mantle beneath a mid-ocean ridge. *Phil. Trans. Roy. Soc. London, Ser.* A, **355**, 217–231.

Blichert-Toft, J. and Albarède, F. (1997). The Lu-Hf isotope geochemistry of chondrites and the evolution of the crust-mantle system. *Earth Planet. Sci. Lett.* **148**, 243–258.

Block, L. and Royden, L. H. (1990). Core complex geometries and regional scale flow in the lower crust, *Tectonics*, **9**, 557–567.

Blum, W., Eisenlohr, P., and Breutinger, F. (2002). Understanding creep – a review. *Metallurg. Mater. Trans.*, **33A**, 291–303.

Blundell, D., Freeman, R., and Mueller, S. (Eds.) (1992), *A Continent Revealed. The European Geotraverse.* Cambridge University Press, 275 pp.

Bodine, J. H., Steckler, M. S., and Watts, A. B. (1981). Observations of flexure and the rheology of the oceanic lithosphere. *J. Geophys. Res.*, **86**, 3695–3707.

Boerner, D. E., Kurtz, R. D., and Craven, J. A. (1996). Electrical conductivity and paleo-Proterozoic foredeeps. *J. Geophys. Res.*, **101**, 13775–13791.

Boerner, D. E., Kurtz, R. D., Craven, J. A. *et al.* (1999). Electrical conductivity in the Precambrian Lithosphere of Western Canada. *Science*, **283**, 668–670.

Bokelmann, G. H. R. (2002). Convection-driven motion of the North American craton: evidence from P-wave anisotropy. *Geophys. J. Int.*, **148**, 278–287.

Bolt, B. A. (1976). Hazards from earthquakes. *Naturwissenschaften*, **63** (8), 356–363.

Boltwood, B. B. (1907). On the ultimate disintegration products of the radioactive elements. Part II. The disintegration products of uranium: *Am. Jo. Sci.*, **23**, 77–88.

Booker, J. R., Favetto, A., and Pomposiello, M. C. (2004). Low electrical resistivity associated with plunging of the Nazca flat slab beneath Argentina. *Nature*, **429**, 399–403.

Borch, R. S. and Green, H. W., II. (1987). Dependence of creep in olivine on homologous temperature and its implication for flow in the mantle. *Nature*, **330**, 345–348.

Borch, R. S. and Green, H. W., II (1989). Deformation of peridotite at high pressure in a new molten cell: comparison of traditional and homologous temperature treatments. *Phys. Earth Planet. Inter.*, **55**: 269–276.

Boschi, L. and Ekstrom, G. (2002). New images of the upper mantle from measurements of surface wave phase velocity anomalies. *J. Geophys. Res.*, **107**, ESE-1.

Bostock, M. G. (1997). Anisotropic upper-mantle stratigraphy and architecture of the Slave Craton. *Nature*, **390**, 392–395.

Bostock, M. G. (1999). Seismic imaging of lithospheric discontinuities and continental evolution. *Lithos*, **48**, 1–16.

Bostock, M. G., Hyndman, R. D. Rondenay, S. and Peacock, S. M. (2002). An inverted continental Moho and serpentinization of the forearc mantle. *Nature*, **417**, 536–538.

Bott, M. H. P. and Gunnarsson, K. (1980). Crustal structure of the Iceland–Faeroe ridge: *Journal of Geophysics*, **47**, 221–227.

Bouhifd, M., Andrault, D., Fiquet, G., and Richet, P. (1996). Thermal expansion of forsterite up to the melting point, *Geophys. Res. Lett.*, **23**(10), 1143–1146.

Bousquet, R., Goffe, B., Henry, P., Le Pichon, X., and Chopin, C. (1997). Kinematic, thermal and petrological model of the Central Alps: lepontine metamorphism in the upper crust and eclogitisation of the lower crust. *Tectonophysics*, **273**, 105–128.

Bowen, N. L. (1928). *The Evolution of the Igneous Rocks*, Princeton, N.J., Princeton University Press, 332 pp.

Bowring, S. A. and Housh, T. (1995). The Earth's early evolution. *Science*, **269**, 1535–1540.

Bowring, S. A. and Williams, I. S. (1999). Priscoan (4.00–4.03 Ga) orthogneisses from northwestern Canada. *Contrib. Mineral. Petrol.*, **134**, 3–16.

Bowring, S. A., Crowley, J. L.; Ramezani, J., *et al.* (2007). High-precision U-Pb zircon geochronology: progress and potential. *Geochim. Cosmochim. Acta*, **71**, A117–A117.

Boyd, F. R. (1973). A pyroxene geotherm. *Geochim. Cosmochim. Acta*, **37**, 2533–2546.

Boyd, F. R. (1984). Siberian geotherm based on lherzolite xenoliths from Udachnaya kimberlite, USSR. *Geology*, **12**, 528–530.

Boyd, F. R. (1989). Compositional distinction between oceanic and cratonic lithosphere. *Earth Planet. Sci. Lett.*, **96**, 15–26.

Boyd, F. R. (1987). High- and low-temperature garnet peridotite xenoliths and their possible relation to the lithosphere–asthenosphere boundary beneath Southern Africa. In: Nixon, P. H. (Ed.), *Mantle Xenoliths*, Chichester, GBR, John Wiley and Sons, 403–412.

Boyd, F. R. and Gurney, J. J. (1986). Diamonds and the African lithosphere. *Science*, **232**, 472–477.

Boyd, F. R. and McCallister, R. H. (1976). Densities of fertile and sterile garnet peridotites. *Geophys. Res. Lett.*, **3**, 509–512.

Boyd, F. R., Fujii, T., and Danchin, R. V. (1976). A noninflected geotherm for the Udachnaya kimberlite pipe, USSR. *Carnegie Year Book*, 1975: 523–531.

Boyd, F. R., Pearson, D. G. and Mertzman, S. A. (1999), Spinet-facies peridotites from the Kaapvaal root, in: Gurney, J. J. Gurney, J. L. Pascoe, M. D. Richardson, S. H. (Eds.), *Proc. 7th Intern. Kimberlite Conf.*, v.1, 40–48.

Brace, W. F. and Kohlstedt, D. L. (1980). Limits on lithospheric stresses imposed by laboratory experiments. *J. Geophys. Res.*, **85**, 6248–6252.

Braile, L. W., Keller, G. R., Wendlandt, R. F., Morgan, P., and Khan, M. A. (1995). The East African Rift System. In: K. H. Olsen (Ed.), *Continental Rifts: Evolution, Structure, Tectonics*. Elsevier, Developm. Geotecton, v.25, 213–232.

Braitenberg, C., Ebbing, J. and Götze, H.-J. (2002). Inverse modelling of elastic thickness by convolution method – the eastern Alps as a case example, *Earth Planet. Sci. Lett.*, **202**, 387–404.

Brasse, H., Cerv, V., Ernst, T., *et al.* (2006). Probing electrical conductivity of the trans-European Suture Zone. *EOS*, **87**(28), 281–287.

Brey, G. and Köhler, T. (1990). Geothermobarometry in 4-phase lherzolites II. New thermobarometers, and practical assessment of existing thermobarometers. *J. Petrol.*, **31**, 1353–1378.

Brey, G. P., Kogarko, L. N., and Ryabchikov, I. D. (1991), Neues Jahrbuch fuer Mineralogie. *Monatshefte*, **4**, 159–168.

Brey, G., Köhler, T., and Nickel, K. G. (1990). Geothermobarometry in four-phase lherzolites I. Experimental results from 10 to 60 kb. *J. Petrol.*, **31**, 1313–1352.

Brown, G. (1979). The changing pattern of batholith emplacement during earthy history. In: Atherton, M. P. and Tarney, J. (eds.), *Origin of Granite Batholiths, Geochemical Evidence*. Shiva, Nantwich, UK, 106–115.

Brown, M. (2008). Characteristic thermal regimes of plate tectonics and their metamorphic imprint throughout Earth history: when did Earth first adopt a plate tectonics mode of behavior? In: Condie, K. C. and Pease, V. (Eds.), *When did Plate Tectonics Begin on Planet Earth?* Geological Society of America Memoirs, Sp. Paper **440**, 97–128.

Brown, M. and Rushmer, T. (Eds.) (2006). *Evolution and Differentiation of the Continental Crust.* Cambridge University Press, 553 pp.

Browring, S. A., King, J. E., Housh, T. B., Isachsen, C. E., and Podosek, F. A. (1989). Neodymium and lead isotope evidence for enriched early Archaean crust in North America. *Nature,* **340**, 222–225.

Brückl, E, Behm, M., Decker, K, *et al.* (2010). Crustal structure and active tectonics in the Eastern Alps. *Tectonics,* **29**, Article Number: TC2011.

Bruneton, M., Pedersen, H. A., Vacher, P., *et al.* and the SVEKALAPKO Seismic Tomography Working Group (2004). Layered lithospheric mantle in the central Baltic Shield from surface waves and xenolith analysis. *Earth Planet. Sci. Lett.,* **226**, 41–52.

Bry, M. and White, N. (2007). Reappraising elastic thickness variation at oceanic trenches, *J. Geophys. Res.,* **112**, B08414, doi:10.1029/2005JB004190.

Buddington, A. F. and Lindsley, D. H. (1964). Iron-titanium oxide minerals and synthetic equivalents. *J. Petrol.,* **5**, 310–357.

Budiansky, B. (1965). On the elastic moduli of some heterogeneous materials. *J. Mech. Phys. Solids,* **13**, 223–227.

Budiansky, B. and O'Connell, R. J. (1976). Elastic moduli of a cracked solid. *J. Solids Struct.,* **12**, 81–97.

Bullard, E. C. (1939). Heat flow in South Africa. *Proc. Roy. Soc.,* **173**, 474–502.

Bunge, H.-P. and Grand, S. (2000). Mesozoic plate-motion history below the northeast Pacific Ocean from seismic images of the subducted Farallon slab. *Nature,* **405**, 337–340.

Bürgmann, R. and Dresen, G. (2008). Rheology of the Lower Crust and Upper Mantle: Evidence from Rock Mechanics, Geodesy, and Field Observations. *Annu. Rev. Earth Planet. Sci.,* **36**, 531–567.

Burke, K. and Kidd, W. S. F. (1978). Were Archean continental geothermal gradients much steeper than those of today? *Nature,* **272**, 240–241.

Burke, K. and Sengör, C. (1986). Tectonic escape in the evolution of the continental crust, In: Barazangi, M. and Brown, L. (Eds.), *Reflection Seismology: the Continental Crust,* AGU, Washington. AGU Geodynamic Ser., **14**, 41–53.

Burlini, L. and Kern, H. (Eds.) (1994). Seismic properties of crustal and mantle rocks – laboratory measurements and theoretical calculations. *Surv. Geophys.,* **15**(5), 439–544.

Burmakov, Y. A., Chernyshev, N. M., Vinnik, L. P., and Egorkin, A. V. (1987). Comparative characteristics of the lithosphere of the Russian Platform, the West Siberian Platform and the Siberian Platform from seismic observations on long-range profiles. In: Kroner, A. (Ed.), *Proterozoic Lithospheric Evolution. AGU, Geodynam. Ser.,* v. **17**, 175–189.

Burov, E. (2003). The upper crust is softer than dry quartzite. *Tectonophysics,* **361**, 321–326.

Burov, E. B. and Diament, M. (1992). Flexure of the continental lithosphere with multi-layered rheology. *Geophys. J. Int.,* **109**, 449–468.

Burov, E. B. and Diament, M. (1995). The effective elastic thickness (Te) of continental lithosphere: what does it really mean? *J. Geophys. Res.,* **100**, 3905–3927.

Burov, E. and Poliakov, A. (2001). Erosion and rheology controls on synrift and postrift evolution: verifying old and new ideas using a fully coupled numerical model. *J. Geophys. Res.* **106**, 16461–16481.

Burov, E. B. and Watts, A. B. (2006). The long-term strength of continental lithosphere: "jelly sandwich" or "crème brûlée"? *GSA Today*, **12**, 4–10

Burov, E., Jolivet, L., Le Pourhiet, L., and Poliakov, A. (2001). A thermomechanical model of exhumation of high pressure (HP) and ultra-high pressure (UHP) metamorphic rocks in Alpine-type collision belts. *Tectonophysics*, **342**, 113–136.

Burov, E., Lobkovsky, L., Cloetingh, S., *et al.* (1993). Continental lithosphere folding in Central-Asia.2. Constraints from gravity and topography. *Tectonophysics*, **226**, 73–87.

Bussod, G. Y., Katsura, T., and Rubie, D. C. (1993). The large volume multi-anvil press as a high P-T deformation apparatus. *Pure Appl. Geophys.*, **141**: 579–599.

Byerlee, J. D. (1968). Brittle-ductile transition in rocks. *J. Geophys. Res.*, **73**, 4741–4750.

Byerlee, J. D. (1967). Frictional characteristics of granite under high confining pressure, *J. Geophys. Res.*, **72**, 3639–3648.

Byerlee, J. D. (1978). Friction of rocks. *Pure Appl. Geophys.*, **116**: 6I5–626.

Caldwell, J. G. and Turcotte, D. L. (1979). Dependence of the thickness of the elastic lithosphere on age. *J. Geophys. Res.*, **84**, 7572–7576.

Caldwell, J. G., Haxby, W. F., Karig, D. E. and Turcotte, D. L. (1976). On the applicability of a universal elastic trench profile, *Earth Planet. Sci. Lett.*, **31**, 239–246.

Calmant, S., Franchetau, J., and Cazenave, A. (1990). Elastic layer thickening with age of the oceanic lithosphere: a tool for prediction of the age of volcanoes or oceanic crust. *Geophys. J. Int.*, **100**, 59–67.

Calvert, A. J., Sawyer, E. W., Davis, W. J., and Ludden, J. N. (1995). Archean subduction inferred from seismic images of a mantle suture in the Superior province. *Nature*, **375**, 670–674.

Calvert, A., Sandvol, E., Seber, D., *et al.* (2000). Geodynamic evolution of the lithosphere and upper mantle beneath the Alboran region of the western Mediterranean: constraints from travel time tomography, *J. Geophys. Res.*, **105**, 10 871–10 898.

Campbell, I. H. and Allen, C. M. (2008). Formation of supercontinents linked to increase in atmospheric oxygen. *Nature Geosci.*, **1**, 554.

Campbell, I. H. and Griffiths, R. W. (1992). The changing nature of mantle hotspots through time: Implications for the chemical evolution of the mantle. *J. Geol.*, **92**, 497–523.

Campbell, W. H. and Schiffmacher, E. R. (1988). Upper mantle electrical-conductivity for 7 subcontinental regions of the Earth. *J. Geomagn. Geoelec.* **40**, 1387–1406.

Campbell, I. H., Griffiths, R. W., and Hill, R. I. (1989). Melting in an Archean mantle plume: Head it's basalts, tail it's komatiites. *Nature*, **339**, 697–699.

Canas, J. A. and Mitchell, B. J. (1978). Lateral variation of surface wave anelastic attenuation across the Pacific. *Nature*, **328**, 236–238.

Canil, D. (2004). Mildly incompatible elements in peridotites and the origins of mantle lithosphere. *Lithos*, **77**, 375–393.

Canil, D. and O'Neill, H. S. (1996). Distribution of ferric iron in some upper mantle assemblages. *J. Petrol.*, **37**, 609–635.

Cara, M. (1979). Lateral variations of S-velocity in the upper mantle from higher Rayleigh modes. *Geophys. J. Roy. Astron. Soc.*, **5**, 649–670.

Cara, M., and Leveque, J. J. (1988). Anisotropy of the asthenosphere – the higher mode data of the pacific revisited. *Geophys. Res. Lett.*, **15**, 205–208.

Carbonell, R., Perez-Estaun, A., Gallart, J., *et al.* (1996). A crustal root beneath the Urals: wide-angle seismic evidence. *Science*, **274**, 222–224.

Carcione, J. M., Ursin, B., and Nordskag, J. I. (2007). Cross-property relations between electrical conductivity and the seismic velocity of rocks. *Geophysics*, **72**, E193–E204.

Carlson, R. L. and Herrick, C. N. (1990). Densities and porosities in the oceanic crust and their variations with depth and age. *J. Geophys. Res.*, **95**, 9153–9170.

Carlson, R. L., Johnson, H. P. (1994). On modeling the thermal evolution of the oceanic upper-mantle – an assessment of the cooling plate model. *J. Geophys. Res.*, **99**, 3201–3214.

Carlson, R. W., Irving, A. J., and Hearn, B. C., Jr. (1999). Chemical and isotopic systematics of peridotite xenoliths from the Williams kimberlite, Montana: clues to processes of lithosphere formation, modification and destruction. In: Gurney, J. J., Gurney, J. L., and Richardson, S. H. (Eds.) *Proc. 7th Intern. Kimberlite Conf.* Red Roof Design, Cape Town, pp. 90–98.

Carlson, R. W., Pearson, G., and James, D. E. (2005). Physical, chemical, and chronological characteristics of continental mantle, *Rev. Geophys.*, **43**, RG1001, doi:10.1029/2004RG000156.

Carlson, R., Grove, T., de Wit, M., and Gurney, J. (1996). Anatomy of an Archean craton: a program for interdisciplinary studies of the Kaapvaal craton, southern Africa, *EOS Trans. AGU*, **77**, 273–277.

Carlson, R. W., Pearson, D. G., Shirey, S. B., and Boyd, F. R. (1997). Re-Os age constraints on the growth of lithospheric mantle beneath Southern Africa. *Eos, Trans, Am. Geophys. Union*; **78**, No. 46, Suppl., p. 747.

Carlson, R. W., Boyd, F. R., Shirey, S. B. *et al.* (2000). Continental growth, preservation and modification in southern Africa, *GSA Today*, **10**, 1–7.

Carter, N. L. (1976). Steady state flow of rocks. *Rev. Geophys. Space Phys.*, **14**: 301–360.

Carter, N. L. and Ave Lallemant, H. G. (1970). High temperature deformation of dunite and peridotite. *Geol. Soc. Am. Bull.*, **81**, 2181–202.

Castagna, J. P., Batzle, M. L., and Eastwood, R. L. (1985). Relationships between compressional-wave and shear-wave velocities in clastic silicate rocks, *Geophysics*, **50**, 571–581.

Catanzaro, E. J., Murphy, T. J., Garner, E. L., and Shields, W. R. (1969). Absolute isotopic abundance ratio and atomic weight of terrestrial rubidium. *J. Res. NBS*, **73A**, 511–516.

Cawood, P. A., Kröner, A., and Pisarevsky, S. (2006). Precambrian plate tectonics: Criteria and evidence. *GSA Today,* **16**(7), 4–11.

Cazenave, A. (1994). The geoid and oceanic lithosphere. In: Vanicek, P., Christou, N. T. (Eds.) *Geoid and Geophysical Interpretations*, Boca Raton, FL, CRC, 255–284.

Cazenave, A., Lago, B., and Dominh, K. (1983). Thermal parameters of the oceanic lithosphere inferred from geoid height data, *J. Geophys. Res.*, **88**, 1105–1118.

Cermak, V. and Rybach, L. (1982). Thermal conductivity and specific heat of minerals and rocks. In: Angenheister, G. (Ed.) *Landolt-Bornstein Numerical Data and Functional Relationships in Science and Technology*. Berlin, Springer, New Ser., Group V, 16, 213–256.

Chapman, D. S. and Pollack, H. N. (1974). "Cold spot" in West Africa: anchoring the African plate. *Nature*, **250**, 477–478.

Chapman, D. S. and Pollack, H. N. (1975). Global heat flow: A new look. *Earth Planet Sci. Lett.*, **28**, 23–32.

Chase, C. G. and Patchett, P. J. (1988). Stored mafic ultramafic crust and early Archean mantle depletion. *Earth Planet. Sci. Lett.*, **91**, 66–72.

Chauvel, C. and Hemond, C. (2002). Melting of a complete section of recycled oceanic crust: trace element and Pb isotopic evidence from Iceland. *Geochemi, Geophys., Geosyst.*, v.1, 10.1029/1999GC000002.

Chave, A. D. Constable, S. C., and Edwards, R. N. (1991). Electrical exploration methods for the seafloor, Vol. *2 of Electromagnetic Methods in Applied Geophysics*, Chap. 12, 931–966, Soc. of Explor. Geophys., Tulsa.

Chave, A. D., Von Herzen, R. P., Poehls, K. A. and Cox, C. S. (1981). Electromagnetic induction fields in the deep ocean north-east of Hawaii: implications for mantle conductivity and source fields, *Geophys. J. Roy. Astron. Soc.*, **66**, 379–406.

Chen, W. P. and Molnar, P. (1983). Focal depths of intracontinental earthquakes and their implications for the thermal and mechanical properties of the lithosphere. *J. Geophys. Res.*, **88**, 4183–4214.

Chester, F. M. (1988). The brittle–ductile transition in a deformation mechanism map for halite. *Tectonophysics*, **154**, 125–136.

Chiozzi, P., Matsushima, J., Okuboc, Y., Pasquale, V., and Verdoya, M. (2005). Curie-point depth from spectral analysis of magnetic data in central–southern Europe. *Phys. Earth Planet. Inter.*, **152**, 267–276.

Chopelas, A. and Boehler, R. (1989) Thermal expansion measurements at very high pressure, systematics and a case for a chemically homogeneous mantle, *Geophys. Res. Lett.*, **16**, 1347–1350.

Chopra, P. N. and Paterson, M. S. (1981). The experimental deformation of dunite. *Tectonophysics*, **78**, 453–573.

Chou, W., Matsumoto, R., and Tajima, F. (2000). Simultaneous modeling of time-domain magnetic induction using parallel computational schemes. *Comp. Phys. Comm.*, **138**, 175–191.

Christensen, N. I. (1965). Compressional wave velocities in metamorphic rocks at pressures to 10 kbar. *J. Geophys. Res.*, **70**, 6147–6164.

Christensen, N. I. (1979). Compressional wave velocities in rocks at high temperatures and pressures, critical thermal gradients, and crustal low-velocity zones. *J. Geophys. Res.*, **84**, 6849–6857.

Christensen, N. I. (1989). Pore pressure, seismic velocities, and crustal structure. In: Pakiser, L. C. and Mooney, W. D. (Eds.) *Geophysical Framework of the Continental United States.* Boulder, Colorado, Geol. Soc. Am. Memoir **172**, 783–800.

Christensen, N. I. (1996). Poisson's ratio and crustal seismology, *J. Geophys. Res.*, **101**, 3139–3156.

Christensen, N. I. and Crosson, R. S. (1968). Seismic anisotropy in the upper mantle: Transverse isotropy of olivine-rich ultramafic rocks, significance for upper mantle anisotropy, *Tectonophysics*, **6**, 93–107.

Christensen, N. I. and Fountain, D. M. (1975). Constitution of the lower continental crust based on experimental studies of seismic velocities in granulite. *Geol. Soc. Am. Bull.*, **86**, 227–236.

Christensen, N. I. and Lundquist, S. M. (1982). Pyroxene orientations within the upper mantle. *Geol. Soc. Am. Bull.* **93**: 279–88.

Christensen, N. I. and Mooney, W. D. (1995). Seismic velocity structure and composition of the continental crust: a global view. *J. Geophys. Res.*, **100**, 9761–9788.

Christensen, N. I., Medaris, L. G. Jr., Wang, H. F., and Jelinek, E. (2001). Depth variation of seismic anisotropy and petrology in central European lithosphere; a tectonothermal synthesis from spinel lherzolite. *J. Geophys. Res.*, **106**, 645–664.

Christensen, U. R. (1985). Thermal evolution models for the Earth. *J. Geophys. Res.*, **90**, 2995–3007.

Christensen, U. R. (1987). Some geodynamical consequences of anisotropic viscosity, *Geophys. J. Roy. Astron. Soc.*, **91**, 711–736.

Christoffel, E. B. (1877). Uber die Fortpflanzung von Stossen durch elastische feste Korper. *Ann. Mater.*, **8**, 193–243.

Claerbout, J. F. (1985). *Imaging the Earth's interior*. Oxford, Blackman Sci. Publ.

Clark, M., and L. Royden (2000), Topographic ooze: Building the eastern margin of Tibet by lower crustal flow. *Geology*, **28**, 703–706.

Clauser, C. (1988). Opacity – the concept of radiative thermal conductivity. In *Handbook of Terrestrial Heat Flow Density Determination*. Htinel, R., Rybach, L. and Stegena, L., (Eds), Dordrecht, Kluwer, 143–165.

Clauser, C. and Huenges, E. (1995). Thermal conductivity of rocks and minerals. In Ahrens, T. J. (Ed.). *A Handbook of Physical Constants: Rock Physics and Phase Relations* (Vol. 3), Am. Geophys. Union, Ref. Shelf Ser., **3**: 105–126.

Clemens, J. D. (2006). Melting of the continental crust: fluid regimes, melting reactions, and source-rock fertility. In: Brown, M. and Rushmer, T. (Eds.), 2006. *Evolution and Differentiation of the Continental Crust*. Cambridge University Press, 296–330.

Clitheroe, G., Gudmundsson, O., and Kennett, B. L. N. (2000). The crustal thickness of Australia. *J. Geophys. Res.*, **105**, 13697–13713.

Cloetingh, S. and Burov, E. B. (1996). Thermo-mechanical structure of European continental lithosphere: constraints from rheological profiles and EET estimates. *Geophys. J. Int.*, **124**, 695–723.

Cloetingh, S. A. P. L., Wortel, M. J. L., and Vlaar, N. J. (1982). Evolution of passive continental margins and initiation of subduction zones. *Nature*, **297**, 139–142.

Cloetingh, S., Ziegler, P. A., Beekman, F., *et al.* (2005). Lithospheric memory, state of stress and rheology: neotectonic controls on Europe's intraplate continental topography. *Quart. Sci. Rev.*, **24**, 241–304.

Clowes, R. M., Cook, F. A., and Ludden, J. N. (1998), Lithoprobe leads to new perspectives on continental evolution, *GSA Today*, **8**, 1–7.

Coble, R. L. (1963). A model for boundary-diffusion controlled creep in polycrystalline materials. *J. Appl. Phys.*, **34**, 1679–1682.

Cochran, J. R. (1980). Some remarks on isostasy and long-term behavior of the continental lithosphere. *Earth Plan. Sci. Lett.*, **46**, 266–274.

Cohen, Y. and Achache, J. (1994). Contribution of induced and remanent magnetization to long-wavelength oceanic magnetic anomalies, *J. Geophys. Res.*, **99**, 2943–2954.

Coltice, N., Albarede, F., and Gillet, P. (2000). K-40-Ar-40 constraints on recycling continental crust into the mantle. *Science*, **288**, 845–847.

Coltice, N., Phillips, B. R., Bertrand, H., Ricard, Y., and Rey, P. (2007). Global warming of the mantle at the origin of flood basalts over supercontinents. *Geology*, **35**, 391–394.

Comer, R. P. (1983). Thick plate flexure. *Geophys. J. Roy. Astron. Soc.*, **72**, 101–113.

Condie, K. (1989). Origin of the Earth's crust. *Palaeogeography, Palaeoclimatology, Palaeoecology* (Global and Planetary Change Section), **75**, 57–81.

Condie, K. C. (1993). Chemical composition and evolution of the upper continental crust: contrasting results from surface samples and shales. *Chem. Geol.*, **104**, 1–37.

Condie, K. C. (1994). Greenstones through time. In: *Archean Crustal Evolution* Condie, K. C. (Ed.). Elsevier, Netherlands, 85–120.

Condie, K. C. (1997). *Plate Tectonics and Crustal evolution*. Oxford, Butterworth-Heinemann, 282 p.

Condie, K. C. (1998). Episodic continental growth and supercontinents: a mantle avalanche connection? *Earth Planet. Sci. Lett.*, **163**, 97–108.

Condie, K. C. (2000). Episodic continental growth models: afterthoughts and extensions. *Tectonophysics*, **322**, 153–162.

Condie, K. C. (2001). *Mantle Plumes and Their Record in Earth History*. Cambridge University Press, 306 pp.

Condie, K. C. (2007). Accretionary orogens in space and time. In: Hatcher, R. D., Carlson, M. P., McBride, J. H., *et al.* (Eds.), *4-D Framework of Continental Crust*. Geol. Soc. Am. Memoirs, **200**, 145–158.

Condie, K. C. and Chomiak, B. (1996). Continental accretion: contrasting Mesozoic and Early Proterozoic tectonic regimes in North America. *Tectonophysics*, **265**, 101–126.

Condie, K. C. and Pease, V. (Eds.) (2008). *When did Plate Tectonics Begin on Planet Earth?* Geol. Soc. of Am. Memoirs, Sp. Paper 440, 313 pp.

Constable, S. (1993). Conduction by mantle hydrogen. *Nature*, **362**, 704.

Constable, S. (2006). SEO3: A new model of olivine electrical conductivity. *Geophys. J. Int.*, **166**, 435–437.

Constable, S. and Cox, C. S. (1996). Marine controlled-source electromagnetic sounding.2. The PEGASUS experiment. *J. Geophys. Res.*, **101**, 5519–5530.

Constable, S. C. and Duba, A. (1990). Electrical conductivity of olivine, a dunite, and the mantle. *J. Geophys. Res.*, **95**, 6967–6978.

Constable, S. and Heinson, G. (1993). In defense of a resistive oceanic upper-mantle: reply to a comment by Tarits, Chave and Schultz. *Geophys. J. Int.*, **114**, 717–723.

Constable, S. C., Parker, R. L., and Constable, C. G. (1987). Occam's inversion: a practical algorithm for generating smooth models from electromagnetic sounding data. *Geophysics*, **52**, 289–300.

Constable, S., Shankland, T. J., and Duba, A. (1992). The electrical conductivity of an isotropic olivine mantle, *J. Geophys. Res.*, **97**, 3397–3404.

Constable, S. C., Heinson, G. S., Anderson, G., and White, A. (1997). Seafloor Electromagnetic Measurements above Axial Seamount, Juan De Fuca Ridge. *J. Geomag. Geoelectr.*, **49**, 1327–1342.

Cook, F. A., van der Velden, A. J., Hall, K. W., and Roberts, B. J. (1999). Frozen subduction in Canada's Northwest Territories: Lithoprobe deep lithospheric reflection profiling of the western Canadian Shield. *Tectonics*, **18**: 1–26.

Cooper, C. M., Lenardic, A., and Moresi, L. (2004). The thermal structure of stable continental lithosphere within a dynamic mantle. *Earth Planet. Sci. Lett.*, **222**, 807–817.

Cooper, C. M., Lenardic, A., and Moresi, L. (2006). Effects of continental insulation and the partitioning of heat producing elements on the Earth's heat loss. *Geophys. Res. Lett.*, **33**, L13313.

Cordier, P. and Rubie, D. C. (2001). Plastic deformation of minerals under extreme pressure using a multi-anvil apparatus. *Mater. Sci. Eng.* A, 309–310, 38–43.

Cornwell, D. G., Mackenzie, G. D., Maguire, P. K. H., *et al.* (2006). Northern Main Ethiopian Rift crustal structure from new high-precision gravity data. In: Yirgu, G., Ebinger, C. J., and Maguire, P. K. H. (eds): *The Afar Volcanic Province within the East African Rift System.* Geological Society, London, Sp. Publ., **259**, 269–291.

Cotte, N., Pedersen, H. A., and TOR Working Group (2002). Sharp contrast in lithospheric structure across the Sorgenfrei-Tornquist Zone as inferred by Rayleigh wave analysis of TOR1 project data. *Tectonophysics*, **360**, 75–88.

Courtillot, V., Davaille, A., Besse, J., and Stock, J. (2003). Three distinct types of hotspots in the Earth's mantle. *Earth Planet. Sci. Lett.*, **205**, 295–308.

Courtillot, V., Jaupart, C., Manighetti, I., Tapponnier, P., and Besse, J. (1999). On casual links between flood basalts and continental breakup. *Earth Planet. Sci. Lett.*, **166**, 177–195.

Courtney, R. C. and M. Recq (1986). Anomalous heat flow neat the Crozet Plateau and mantle convection. *Earth Planet. Sci. Lett.*, **79**, 373–384.

Courtney, R. C. and White, R. S. (1986). Anomalous heat flow and geoid across the Cape Verde Rise: evidence for dynamic support from a thermal plume in the mantle, *Geophys. J. Roy. Astron. Soc.*, **87**, 815–867.

Couvy, H., Frost, D. J., Heidelbach, F., *et al.* (2004). Shear deformation experiments of forsterite at 11 GPa-1400&C in the multianvil apparatus. *Eur. J. Mineral.*, **16**, 877–89.

Crampin, S. (1981). A review of wave motion in anisotropic and cracked elastic-media. *Wave Motion*, **3** (4).

Crampin, S. (1987). Geological and industrial implications of extensive-dilatancy anisotropy. *Nature*, **328**, 491–496.

Crampin, S. (1977). A review of the effects of anisotropy layering on the propagation of seismic waves, Geophys. *J. Roy. Astron. Soc.*, **45**, 9–27.

Crough, S. T. (1975). Thermal model of oceanic lithosphere. *Nature*, **256**, 388–390.

Crough, S. T. and Thompson, G. A. (1976). Thermal model of continental lithosphere. *J. Geophys. Res.*, **81**, 4857–4862.

Currie, C. A. and Hyndman, R. D. (2006), The thermal structure of subduction zone back arcs, *J. Geophys. Res.*, 111, B08404, doi:10.1029/2005JB004024.

Dahlen, F. A. and Tromp, J. (1998). *Theoretical Global Seismology.* Princeton, NJ, Princeton Univ. Press.

Dahl-Jensen T., Larsen, T. B., Woelbern, I., *et al.* (2003). Depth to Moho in Greenland: receiver-function analysis suggests two Proterozoic blocks in Greenland. *Earth Planet. Sci. Lett.*, v. **205**, pp. 379–393.

Dai, L. D. and Karato, S. I. (2009). Electrical conductivity of orthopyroxene: Implications for the water content of the asthenosphere. *Proc. Japan Acad., Series B*, **85**, 466–475.

Daines, M. J. and Richter, F. M. (1988). An experimental method for directly determining the interconnectivity of melt in a partially molten system, *Geophys. Res. Lett.*, **15**, 1459–1462.

d'Alessio, M. A., Williams, C. F., and Bürgmann, R. (2006). Frictional strength heterogeneity and surface heat flow: implications for the strength of the creeping San Andreas fault, *J. Geophys. Res.*, **111**, B05410, doi:10.1029/2005JB003780.

Dalton, C. A., Ekstrom, G., Dziewonski, A. M. (2008). The global attenuation structure of the upper mantle. *J. Geophys. Res.*, **113**, B09303.

Dalziel, I. W. D., Lawver, L. A., and Murphy, J. B. (2000). Plumes, orogenesis, and super-continental fragmentation. *Earth Planet. Sci. Lett.*, **178**, 1–11.

Dana, J. D. (1896). *Manual of Geology*. NY, American Book Company, 1087 pp.

Darbyshire, F. A. (2005). Upper mantle structure of Arctic Canada from Rayleigh wave dispersion. *Tectonophysics*, **405**, 1–23.

Darbyshire, F. A., Larsen, T. B., Mosegaard, K., *et al.* (2004). A first detailed look at the Greenland lithosphere and upper mantle, using Rayleigh wave tomography. *Geophys. J. Int.*, v. **158**, pp. 267–286.

Davaille, A. and Jaupart, C. (1993). Transient high-Rayleigh number thermal convection with large viscosity variations. *J. Fluid Mech.*, **253**, 141–166.

Davaille, A. and Limare, A. (2007). Laboratory studies on mantle convection. In: *Treatise of Geophysics*, volume editor Bercovici, D., editor Schubert, G., Elsevier, 89–165.

Davaille, A. and Vatteville, J. (2005). On the transient nature of mantle plumes: *Geophys. Res. Lett.*, **32**, L14309, doi:10.1029/2005GL023029.

Davies, G. (1980). Review of oceanic and global heat flow estimates. *Rev. Geophys.*, **18**, 718–722.

Davies, G. F. (1988). Ocean bathymetry and mantle convection. *J. Geophys. Res.*, **93**, 10467–10480; 10481–10488.

Davies, G. F. (1994). Thermomechanical erosion of the lithosphere by mantle plumes. *J. Geophys. Res.*, **99**, 15,709–15,722.

Davies, G. F. (1995). Penetration of plates and plumes through the mantle transition zone. *Earth Planet. Sci. Lett.*, **133**, 507–516.

Davies, G. F. (1995). Punctuated tectonic evolution of the Earth. *Earth Planet. Sci. Lett.*, **136**, 363–379.

Davies, G. F. (1999). *Dynamic Earth: Plumes, Plates and Mantle Convection*. Cambridge University Press, 458 pp.

Davis, E. E. and Lister, C. R. B. (1974). Fundamentals of ridge crest topography, *Earth Planet. Sci. Lett.*, **21**, 405–413.

Davis, E. E., Hyndman, R. D., and Villinger, H. (1990). Rates of fluid expulsion across the northern Cascadia accretionary prism: Constraints from new heat flow and multichannel seismic reflection data, *J. Geophys. Res.*, **95**, 8869–8890.

Davies, R., Griffin, W. L., Perason, N. J., *et al.* (1999). Diamonds from the deep: Pipe DO-27, Slave Craton, Canada. In: *Proc. 7th Int. Kimberlite Conf.,* Red Roof Design, Cape Town, 148–153.

Davy, P. and Gillet, P. (1986). The stacking of thrust slices in collision zones and its thermal consequences. *Tectonics*, **5**, 913–929.

De Boer, C. B. and Dekkers, M. J. (1996). Grain-size dependence of the rock magnetic properties for a natural maghemite, *Geophys. Res. Lett.*, **23**, 2815–2818.

de Wit, M. J. and Ashwal, L. D. (Eds.), (1997). *Greenstone Belts*. Oxford, Clarendon Press, 813 pp., ISBN 0-19-854056-6.

de Wit, M. J., Roering, C., Hart, R. J., *et al.* (1992). Formation of an Archean continent. *Nature*, **357**, 553–562.

Debayle, E. and Kennett, B. L. N. (2000). Anisotropy in the Australasian upper mantle from Love and Rayleigh waveform inversion. *Earth Planet. Sci. Lett.*, **184**, 339–351.

Debayle, E., Kennett, B., and Priestley, K. (2005). Global azimuthal seismic anisotropy and the unique plate-motion deformation of Australia. *Nature*, **433**, 509–512.

DeBeer, J. H., Vanzijl, J., and Gough, D. (1982). The southern Cape conductive belt (south-Africa) – its composition, origin and tectonic significance. *Tectonophysics*: **83**, 205–225.

Decker, E. R. (1995). Thermal regimes of the Southern Rocky Mountains and Wyoming Basin in Colorado and Wyoming in the United States, *Tectonophysics*, **244**, 85–106.

Deen, T. J., Griffin, W. L., Begg, G., *et al.* (2006). Thermal and compositional structure of the subcontinental lithospheric mantle: Derivation from shear wave seismic tomography. *Geochem. Geophys. Geosyst.*, 7, Article Q07003

Deschamps, F., Snieder, R., and Trampert, J. (2001). The relative density-to-shear velocity scaling in the uppermost mantle. *Phys. Earth Planet. Int.*, **124**, 193–211.

deGroot-Hedlin C. and Constable, S. (1990). Occam's inversion to generate smooth, two-dimensional models from magnetotelluric data, *Geophysics*, **55**, 1613–1624.

DeLaughter, J., Stein, S., and Stein, C. A. (1999). Extraction of a lithospheric cooling signal from oceanwide geoid data. *Earth Planet. Sci. Lett.*, **174**, 173–181.

Denton, G. H., and Hughes, T. (1981). *The Last Great Ice Sheets,* New York, Wiley-Interscience.

DePaolo, D. J. (1983). The mean-life of continents – estimates of continent recycling rates from Nd and Hf isotopic data and implications for mantle structure. *Geophys. Res. Lett.*, **10**, 705–708.

DePaolo, D. J. and Wasserburg, G. J. (1976) Inferences about magma sources and mantle structure from variations of Nd-143/ Nd-144. *Geophys. Res. Lett.*, **3**, 743–746.

DePaolo, D. J. and Wasserburg, G. J. (1976). Nd isotopic variations and petrogenetic models. *Geophys. Res. Lett.*, **3**, 249–252.

DePaolo, D. J., Linn, A. M., and Schubert, G. (1991) The continental crustal age distribution; methods of determining mantle separation ages from Sm–Nd isotopic data and application to the southwestern United States. *J. Geophys. Res.*, **96**, 2071–2088.

Deschamps, F., Snieder, R., and Trampert, J. (2001). The relative density-to-shear velocity scaling in the uppermost mantle. *Phys. Earth Planet. Int.*, **124**, 193–211.

Deschamps, F., Trampert, J., and Snieder, R. (2002). Anomalies of temperature and iron in the uppermost mantle inferred from gravity data and tomographic models. *Phys. Earth Planet. Int.*, **129**, 245–264.

Deschamps, F., Lebedev, S., Meier, T., *et al.* (2008). Stratified seismic anisotropy reveals past and present deformation beneath the East-central United States. *Earth Planet. Sci. Lett.*, **274**, 489–498.

Detrick, R. S. and Watts, A. B. (1979). An analysis of isostasy in the world's oceans, 3. Aseismic ridges. *J. Geophys. Res.*, **84**, 3637–3653.

Detrick, R. S., Von Herzen, R. P., Parsons, B., Sandwell, D., and Dougherty, M. (1986). Heat flow observations on the Bermuda Rise and thermal models of midplate swells, *J. Geophys. Res.*, **91**, 3701–3723.

Deuss, A. and Woodhouse, J. H. (2004). The nature of the Lehmann discontinuity from its seismological Clapeyron slopes. *Earth Planet. Sci. Lett.*, **225**, 295–304.

Dewey, J. F. (2007). The secular evolution of plate tectonics and the continental crust: An outline. In: Hatcher, R. D., Carlson, M. P., McBride, J. H., *et al.* (Eds.). *4-D Framework of Continental Crust.* Geol. Soc. Am. Memoirs, **200**, 1–8.

Dewey, J. F. and Windley, B. F. (1981). Growth and differentiation of the continental crust. *Phil. Trans. Roy. Soc. London, Ser. A*, **301**, 189–206.

Dickin, A. P. (1995). *Radiogenic Isotope Geology.* Cambridge University Press, 490 pp.

Dietz, R. S. and Menard, H. W. (1953). Hawaiian Swell, deep and arch; subsidence of the Hawaiian Islands. *J. Geol.*, **61**, 99–113.

Divins, D. L. (2008). NGDC total sediment thickness of the world's oceans and marginal seas, *http://www.ngdc.noaa.gov/mgg/sedthick/sedthick.html*

Dixon, J. E., Leist, L., Langmuir, C. and Schilling, J.-G. (2002). Recycled dehydrated lithosphere observed in plume-influenced mid-ocean-ridge basalt. *Nature*, **420**, 385–389.

Dobson, D. P. and Brodholt, J. P. (2000). The electrical conductivity and thermal profile of the earth's mid-mantle. *Geophys. Res. Lett.*, **27**, 2325–2328.

Doin, M.-P. and Fleitout, L. (1996). Thermal evolution of the oceanic lithosphere: an alternative view. *Earth Planet. Sci. Lett.*, **142**, 121–136.

Doin, M.-P., Fleitout, L., and Christensen, U. R. (1997). Mantle convection and stability of depleted and undepleted continental lithosphere. *J. Geophys. Res.*, **102**, 2771–2787.

Dorman, J., Ewing, M., and Oliver, J. (1960). Study of the shear-velocity distribution in the upper manatle by Rayleigh waves. *Bull. Seismol. Soc. Am.*, **50**, 87–115.

Doucouré, C. M., de Wit, M. J., and Mushayandebyu, M. F. (1996). Effective elastic thickness of the continental lithosphere in South Africa. *J. Geophys. Res.*, **101**, 11,291–11,303.

Doyen, P. H. (1987). Crack geometry in igneous rocks: a maximum entropy inversion of elastic and transport properties. *J. Geophys. Res.*, **92**, 8169–8181.

Drazin, P. G. (2002). *Introduction to Hydrodynamic Stability.* Cambridge University Press, 238.

Drury, M. R. (2005). Dynamic recrystallization and strain softening of olivine aggregates in the laboratory and the lithosphere. *Geol. Soc., London, Special Publications*; **243**, 143–158.

Dueker, K. G. (1999). New seismic images of the Western US from crust to 660 km, *Seismol. Res. Lett.*, **70**, 2.

Du Frane, W. L., Roberts, J. J., Toffelmier, D. A., and Tyburczy, J. A. (2005). Anisotropy of electrical conductivity in dry olivine. *J. Geophys. Res.*, **32**, L24315, doi:10.1029/2005GL023879.

Duba, A. G. and Shankland, T. J. (1982). Free carbon and electrical conductivity in the Earth's Mantle, *Geophys. Res. Lett.*, **9**, 1271–1274.

Duba, A., Heard, H. C., and Schock, R. N. (1976). Electrical Conductivity of Orthopyroxene to 1400C and the Resulting Selenotherm. UCRL77655, *preprint. Proc. Lunar Sci. Conf. 7th,* 3173–3181.

Duba, A. G., Durham, W. B., Handin, J. W., and Wang, H. F. (Eds.) (1990). *The Brittle-Ductile Transition in Rocks: The Heard Volume.* Washington DC, *American Geophysical Union.*

Duba, A., Dennison, M., Irving, A. J., Thornber, C. R., and Huebner, J. S. (1979). Electrical conductivity of aluminous orthopyroxene. *Lunar Planet. Sci.* **X** (abstract), 318.

Duchène, S., J. Blichert-Toft, B. Luais, *et al.* (1997). The Lu–Hf dating of garnets and the ages of the Alpine high-pressure metamorphism. *Nature*, **387**, 586–590.

Duchesne, J.-C., Liégeois, J.-P., Vander-Auwera, J., and Longhi, J. (1999). The crustal tongue melting model and the origin of massive anorthosites. *Terra Nova* **11**, 100–105.

Duchkov, A. D., Lysak, S. V., and Balobaev, V. T. (1987). Thermal field of Siberia interiors. *Novosibirsk, Nauka*, 196 p. (In Russian).

Duffy, T. S. and Anderson, D. L. (1989). Seismic velocities in mantle minerals and the mineralogy of the upper mantle. *J. Geophys. Res.*, **94**, 1895–1912.

Dunn, R. A. and Forsyth, D. W. (2003). Imaging the transition between the region of mantle melt generation and the crustal magma chamber beneath the southern East Pacific Rise with short-period Love waves. *J. Geophys. Res.*, **108**, Article Number: 2352.

Durek, J. J. and Ekström, G. (1996). A radial model of anelasticity consistent with long-period surface-wave attenuation. *Bull. Seism. Soc. Am.*, **86**, 144–158.

Durrheim, R. J. and Mooney, W. D. (1991). Archean and Proterozoic crustal evolution: evidence from crustal seismology. *Geology*, **19**, 606–609.

Dutton, C. E. (1882). Physics of the Earth's crust. *Am. J. Sci.*, **23**, 283–290.

Dyment, J., and Arkani-Hamed, J. (1998). Contribution of lithospheric remnant magnetization to satellite magnetic anomalies over the world's oceans. *J. Geophys. Res.*, **103**, 15423–15441.

Dziewonski, A. M. (1971a). On regional differences in dispersion of mantle Rayleigh waves. *Geophys. J. Roy. Astron. Soc.*, **22**, 289–326.

Dziewonski, A. M. (1971b). Upper mantle models from "pure-path" dispersion data. *J. Geophys. Res.*, **76**, 2587–2601.

Dziewonski, A. M. (2003). Global seismic tomography: what we really can say and what we make up. *Abstract in "The Hotspot Handbook", Proc. Penrose Conf. "Plume IV: Beyond the Plume Hypothesis"*, Hveragerdi, Iceland, August 2003. *http://www.mantleplumes. org/Penrose/PenroseAbstracts.html*

Dziewonski, A. M. and Anderson, D. L. (1981). Preliminary reference earth model. *Phys. Earth Planet. Int.*, **25**, 297–356.

Eaton, D. W., Darbyshire, F., Evans, R., *et al.* (2009). The elusive lithosphere–asthenosphere boundary (LAB) beneath cratons, *Lithos*, **109**, 11–22.

Echternacht, F, Tauber, S, Eisel, M, *et al.* (1997). Electromagnetic study of the active continental margin in northern Chile. *Phys. Earth Planet. Inter.*, **102**, 69–87.

Egbert, G. D. and Booker, J. R. (1992). Very long period magnetotellurics at Tucson observatory: implications for mantle conductivity. *J. Geophys. Res.*, **97**, 15099–15112.

Eggins, S. M. (1992). Petrogenesis of Hawaiian tholeiites. *Contrib. Mineral. Petrol.*, **110**, 387–397, 398–410.

Eggler, D. H., Meen, J. K., Welt, F., *et al.* (1988). Tectonomagmatism of the Wyomimg Province. In: *Cenozoic Volcanism in the Southern Rocky Mountains Revisited; A Tribute to Rudy C. Epis*. Part 3. Colorado School of Mines Quarterly, v. **83**, No. 2, 25–40.

Egorkin, A. V., Zuganov, S. K., Pavlenkova, N. A., and Chernyshev, N. M. (1987). Results of lithosphere studies from long-range profiles in Siberia. *Tectonophysics*, **140**, 29–47.

Eken, T., Shomali, Z. H., Roberts, R., Hieronymus, C. F. and Bodvarsson, R. (2008). S and P velocity heterogeneities within the upper mantle below the Baltic Shield. *Tectonophysics*, **462**, 109–124.

Ekström, G. and Dziewonski, A. M. (1997). Three-dimensional velocity structure of the Earth's upper mantle. In: K. Fuchs (Ed.), *Upper Mantle Heterogeneities from Active and Passive Seismology*, 187–198. Dordrecht, Boston, London, Kluwer Acad. Publ.

Ekström, G. and Dziewonski, A. M. (1998). The unique anisotropy of the Pacific upper mantle, *Nature*, **394**, 168–172.

Ekström, G., Tromp, J., and Larson, E. W. F. (1997). Measurements and global models of surface wave propagation. *J. Geophys. Res.*, **102**, 8137–8157.

Elder, J. (1976). *The Bowels of the Earth*. UK, Oxford Univ. Press, 222 pp.

Elesin, Y., Gerya, T., Artemieva, I. M. and Thybo, H. (2010). SAMOVAR: a thermo-mechanical code for modeling of geodynamic processes in the lithosphere – application to basin evolution. In: F. Roure (Ed.), *ILP special Issue on Sedimentary Basins. Arabian J. Geosciences*, 3(4), 477–497.

Ellis, D. J. and Green, D. H. (1979). Experimental-study of the effect of Ca upon garnet-clinopyroxene Fe–Mg exchange equilibria. *Contrib. Mineral. Petrol.*, **71**, 13–22.

Emslie, R. F. (1978). Anorthosite massifs, rapakivi granites, and late Proterozoic rifting of North America. *Precambrian Res.*, **7**, 61–98.

England, P. and Bickle, M. (1984). Continental thermal and tectonic regimes during the Archean. *J. Geology*, **92**, 353–367.

England, P. and Houseman, G. (1984). On the geodynamic setting of kimberlite genesis. *Earth Planet. Sci. Lett.*, **67**, 109–122.

England, P. C. and Houseman, G. A. (1989). Extension during continental convergence with application to the Tibetan Plateau, *J. Geophys. Res.*, **94**, 17 561–17 579.

England, P. and Molnar, P. (1997). Active deformation of Asia: from kinematics to dynamics, *Science*, **278**, 647–650.

ERCEUGT-Group (1992). An electrical resistivity crustal section from the Alps to the Baltic Sea (central segment of the EGT), *Tectonophysics*, **207**, 123–139.

Ernst, T., Jankowski, J., Jo´z´wiak, W., Lefeld, J., Logvinov, I. (2002). Geoelectrical model along a profile across the Tornquist–Tesseyre zone in Southeastern Poland. *Acta Geophys Polonica,* **50**(4), 505–515.

Etheridge, M. A. (1983). Differential stress magnitudes during regional deformation and metamorphism: upper bound imposed be tensile fracturing. *Geology*, **11**, 231–234.

EUROBRIDGE Working Group and EUROBRIDGE'95 (2001). Deep seismic profiling within the East European Craton, *Tectonophysics*, **339**, 153–175.

Evans, B. and Goetze, C. (1979). Temperature-variation of hardness of olivine and its implication for polycrystalline yield stress. *J. Geophys. Res.*, **84**, 5505–5524.

Evans, B. and Kohlstedt, D. L. (1995). Rheology of rocks. *A Handbook of Physical Constants*, AGU Reference Shelf 3, AGU, 148–165.

Evans, R. L., Constable, S. C., Sinha, M. C., *et al.* (1991). Upper crustal resistivity structure of the East Pacific Rise near 13 °N, *Geophys. Res. Lett.*, **18**, 1917–1921.

Evans, R. L., Hirth, G., Baba, K., *et al.* (2005). Geophysical evidence from the MELT area for compositional controls on oceanic plates. *Nature*, **437**, 249–252.

Evans, R. L., Tarits, P., Chave, A. D., *et al.* (1999). Asymmetric electrical structure in the mantle beneath the East Pacific Rise at 178S. *Science*, **286**, 756–759.

Everett, J. D. (1883). *Report for British Association for 1882*, 72–90.

Exxon Production Research Company (1985). *Tectonic Map of the World*. Tulsa, USA, Am. Assoc. of Pet. Geol. Found.

Farber, D. L., Williams, Q., and Ryerson, F. J. (2000). Divalent cation diffusion in Mg_2SiO_4 spinel (ringwoodite), β phase (wadsleyite), and olivine: implications for electrical conductivity of the mantle. *J. Geophys. Res.*, **105**, 513–529.

Farquhar, J., Bao, H., and Thiemens, M. (2000). Atmospheric influence of Earth's earliest sulfur cycle. *Science*, **289**(5480), 756–758.

Farra, V. and Vinnik, L. (2000). Upper mantle stratification by P and S receiver functions, *Geophys. J. Int.*, **141**, 699–712.

Falloon, T. J. and Danyushevsky, L. V. (2000). Melting of refractory mantle at 1.5, 2 and 2.5 GPa under anhydrous and H_2O-undersaturated conditions: Implications for the petrogenesis of high-Ca boninites and the influence of subduction components on mantle melting. *J. Petroly*, **41**, 257–283.

Faul, U. H. (1997). Permeability of partially molten upper mantle rocks from experiments and percolation theory, *J. Geophys. Res.*, **102**, 10299–10311.

Faul, U. H. and Jackson, I. (2005). The seismological signature of temperature and grain size variations in the upper mantle. *Earth Planet. Sci. Lett.*, **234**, 119–134.

Faul, U. H. and Jackson, I. (2007). Diffusion creep of dry, melt-free olivine, *J. Geophys. Res.*, **112**, B04204, doi:10.1029/2006JB004586.

Faure, S. (2006). *World Kimberlites and Lamproites*. CONSOREM Database (Version 2006–1).

Faure, G. (1986). *Principles of Isotope Geology*. NY, Wiley, 589 pp.

Fei, Y. (1995). Thermal expansion. *A Handbook of Physical Constants*, AGU Reference Shelf 2, AGU, 29–44.

Feng, M., van der Lee, S., and Assumpcao, M. (2007). Upper mantle structure of South America from joint inversion of waveforms and fundamental mode group velocities of Rayleigh waves. *J. Geophys. Res.*, **112**, Article Number: B04312.

Ferguson, I. J., Craven, J. A., Kurtz, R. D., *et al.* (2005). Geoelectric response of Archean lithosphere in the western Superior Province, central Canada. *Phys. Earth Planet. Interiors* **150**, 123–143.

Filloux, J. H. (1983). Seafloor magnetotelluric soundings in the Mariana Island arc area. Part II. In D. E. Hayes (Ed.). *The Tectonic and Geologic Evolution of Southeast Asian Seas and Islands, AGU Geophys. Monogr.*, v. 27, Washington, DC, AGU, 255–265.

Finnerty, A. A. and Boyd, F. R. (1987) Thermobarometry for garnet peridotite xenoliths: a basis for upper mantle stratigraphy. In: *Mantle Xenoliths* Nixon P. H. (Ed.). Chichester, Wiley, 381–402.

Finnerty, A. A. and Boyd, F. R. (1984). Evaluation of thermobarometers for garnet peridotites. *Geochim Cosmochim Acta*, **48**, 15–27.

Fishwick, S., Kennett, B. L. N., and Reading, A. M. (2005). Contrasts in lithospheric structure within the Australian Craton – insights from surface wave tomography. *Earth Planet. Sci. Lett.*, **231**, 163–176.

Fjeldskaar, W. (1997). Flexural rigidity of Fennoscandia inferred from postglacial uplift. *Tectonics*, **16**, 596–608.

Fjeldskaar, W. (1994). Viscosity and thickness of the asthenosphere detected from the Fennoscandian uplift. *Earth Planet. Sci. Lett.*, **126**, 399–410.

Flanagan, M. P. and Shearer, P. M. (1998). Global mapping of topography of transition zone velocity discontinuities by stacking SS precursors. *J. Geophys. Res.*, **103**, 2673–2692.

Flück, P., Hyndman, R. D. and Lowe, C. (2003). Effective elastic thickness Te of the lithosphere in western Canada. *J. Geophys. Res.*, **108**(B9), 2430, doi:10.1029/2002JB002201.

Forsyth, D. W. (1985). Subsurface loading and estimates of the flexural rigidity of continental lithosphere. *J. Geophys. Res.*, **90**, 12,623–12,632.

Forsyth, D. W. (1982). Determinations of focal depths of earthquakes associated with the bending of oceanic plates at trenches, *Phys. Earth Planet. Inter.*, **28**, 141–160.

Forsyth, D. W. (1977). The evolution of the upper mantle beneath mid ocean ridges. *Tectonophysics*, **38**, 89–118.

Forsyth, D. W. (1975). The early structural evolution and anisotropy of the oceanic upper mantle. *Geophys. J. Roy. Astron. Soc.*, **43**, 103–162.

Forsyth, D. and Uyeda, S. (1975). On the relative importance of the driving forces of plate motion. *Geophys. J. Roy. Astron. Soc.*, **43**, 163–200.

Forsyth, D. W., Webb, S. C., Dorman, L. M. and Shen, Y. (1998). Phase velocities of Rayleigh waves in the MELT experiment on the East Pacific Rise. *Science*, **280**, 1235–1238.

Forte, A. M. (2007). Constraints on seismic models from other disciplines – implications for mantle dynamics and composition. *Treatise on Geophysics*, Chapter 1.23, 805–858.

Forte, A. M. and Perry, A. C. (2000). Seismic-geodynamic evidence for a chemically depleted continental tectosphere, *Science*, **290**, 1940–1944.

Forte, A. M. and Mitrovica, J. X. (1996). New inferences of mantle viscosity from joint inversion of long-wavelength mantle convection and post-glacial rebound data. *Geophys. Res. Lett.*, **23**, 1147–1150.

Forte, A. M., Dziewonski, A. M., and O'Connell, R. J. (1995). Thermal and chemical heterogeneity in the mantle: A seismic and geodynamic study of continental roots, *Phys. Earth Planet. Int.*, **92**, 45–55.

Forte, A. M., Woodward, R. L., and Dziewonski, A. M. (1994). Joint inversions of seismic and geodynamic data for models of three-dimensional mantle heterogeneity. *J. Geophys. Res.*, **99**, 21857–21887.

Forte, A., Peltier, W., Dziewonski, A., and Woodward, R. (1993). Dynamic surface topography: a new interpretation based upon mantle flow models derived from seismic tomography. *Geophys. Res. Lett.*, **20**, 225–228.

Foucher, J. P., Le Pichon, X., Lallemant, S., *et al.* (1990). Heat flow, tectonics, and fluid circulation at the toe of the Barbados Ridge accretionary prism, *J. Geophys. Res.*, **95**, 8859–8867.

Foulger, G. R., Natland, J. H., and Anderson, D. L. (2005), Genesis of the Iceland melt anomaly by plate tectonic processes. In Foulger, G. R., Natland, J. H., Presnall, D. C., and Anderson, D. L. (Eds.). *Plates, Plumes, and Paradigms:* Geological Society of America, Special Paper 388, Boulder, CO, USA, pp. 595–626.

Fountain, D. M. (1986). Is there a relationship between seismic velocity and heat production for crustal rocks? *Earth Sci. Plan. Lett.*, **79**, 145–150.

Fountain, D. M. and Salisbury, M. H. (1981). Exposed cross-sections through the continental crust: implications for crustal structure, petrology, and evolution. *Earth Planet. Sci. Lett.*, **56**, 263–277.

Fountain, D. M., Salisbury, M. H., and Furlong, K. P. (1987). Heat production and thermal conductivity of rocks from the Pikwitonei-Sachigo continental cross-section, central Manitoba: Implications for the thermal structure of Archean crust. *Can. J. Earth Sci.*, **24**, 1583–1594.

Fowler, C. M. R. (2004). *The Solid Earth. An Introduction to Global Geophysics.* Cambridge University Press, 2nd edn.

Fox, Maule C, Purucker, M, Olsen, N, and Mosegaard, K (2005). Heat flux anomalies in Antarctica revealed by satellite magnetic data. *Science*, **309**, 464–467.

Francis, T. J. G. (1969). Generation of seismic anisotropy in the upper mantle along the mid-ocean ridges, *Nature*, **221**, 162–165.

Frankel, A. (1982). The effects of attenuation and site response on the spectra of micro-earthquakes in the northeastern Caribbean. *Seism. Soc. Am. Bull.*, **72**, 1379–1402.

Frederiksen, A. W., Bostock, M. G., and Cassidy, J. F. (2001). S-wave velocity structure of the Canadian upper mantle. *Phys. Earth Planet. Int.*, **124**, 175–191.

Frederiksen, A. W., Miong, S.-K., Darbyshire, F. A., *et al.* (2007). Lithospheric variations across the Superior Province, Ontario, Canada: Evidence from tomography and shear wave splitting. *J. Geophys. Res.*, **112**, B07318, doi:10.1029/2006JB004861.

Frei, R. and Jensen, B. K. (2003). Re-Os, Sm-Nd isotope- and REE systematics on ultramafic rocks and pillow basalts from the Earth's oldest oceanic crustal fragments (Isua Supracrustal Belt and Ujaragssuit nunat area, W Greenland). *Chem. Geol.*, **196**, 163–191.

Frei, R., Gaucher, C., Poulton, S. W. and Canfield, D. E. (2009). Fluctuations in Precambrian atmospheric oxygenation recorded by chromium isotopes. *Nature*, **461**, 250–254.

Freybourger, M., Gaherty, J., and Jordan, T. (2001). Structure of the Kaapvaal craton from surface waves. *Geophys. Res. Lett.* **28**, 2489–2492.

Frost, B. R. (1991). Oxide minerals: petrologic and magnetic significance. *Rev. Mineral.*, **v. 25**.

Frost, H. J. and Ashby, M. F. (1982). *Deformation-Mechanism Maps; The Plasticity and Creep of Metals and Ceramics.* Oxford, Pergamon Press, 168 pp.

Froude, D. O., Ireland, T. R., Kinny, P. D., Williams, I. S., and Compston, W. (1983). Ion microprobe identification of 4,100–4,200 Myr-old terrestrial zircons. *Nature*, **304**, 616–618.

Fuchs, K. (Ed.), (1997). *Upper Mantle Heterogeneities from Active and Passive Seismology.* Kluwer Academic Publishers, 366 pp.

Fuchs, K. (1983). Recently formed elastic anisotropy and petrological models for the subcrustal lithosphere in southern Germany, *Phys. Earth Planet. Int.*, **31**, 93–118.

Fuchs, K. and Wenzel, F. (1997). Conservation of lithospheric DSS data. In: Fuchs, K. (Ed.), *Upper Mantle Heterogeneities from Active and Passive Seismology.* Kluwer Academic Publishers, 11–31.

Fujinawa, Y., Kawakami, N., Asch, T.H., *et al.* (1997). Studies of the georesistivity structure in the central part of the northeastern Japan arc. *J. Geomagn. Geoelect.*, **4**, 1601–1617.

Fukao, Y., Widiyantoro, S., and Obayashi, M. (2001). Stagnant slabs in the upper mantle and lower mantle transition region. *Rev. Geophys.*, **39**, 291–324.

Fukao, Y., Koyama, T., Obayashi, M., and Utada, H. (2004). Trans-Pacific temperature field in the mantle transition region derived from seismic and electromagnetic tomography. *Earth Planet. Sci. Lett.*, **217**, 425–434.

Furlong, F. P., Spakman, W., and Wortel, R. (1995). Thermal structure of the continental lithosphere: constraints from seismic tomography, *Tectonophysics*, **244**, 107–117.

Furnes, H., de Wit, M., Staudigel, H. *et al.* (2007). A vestige of Earth's oldest ophiolite. *Science*, **315**, 1704–1706.

Fyfe, W. S. (1978). The evolution of the earth's crust: modern plate tectonics to ancient hot spot tectonics? *Chem. Geol.*, **23**, 89–114.

Fyfe, W. S., Price, N. J., and Thompson, A. B. (1978). *Fluids in the Earth's Crust.* Amsterdam – Oxford – New York, Elsevier, 435P.

Gaboret, C., Forte, A. M., and Montagner, J.-P. (2003). The unique dynamics of the Pacific Hemisphere mantle and its signature on seismic anisotropy. *Earth Planet. Sci. Lett.*, **208**, 219–233.

Gaherty, J. B. and Jordan, T. H. (1995). Lehmann discontinuity as the base of an anisotropic layer beneath continents. *Science*, **268**, 1468–1471.

Gaherty, J. B., Jordan, T. H., and Gee, L. S. (1996). Seismic structure of the upper mantle in a central Pacific corridor. *J. Geophys. Res.*, **101**, 22291–22309.

Gaherty, J. B., Kato, M., and Jordan, T. H. (1999). Seismological structure of the upper mantle: a regional comparison of seismic layering. *Phys. Earth Planet. Int.*, **110**, 21–41.

Galer, S. J. G. and Goldstein, S. L. (1991). Early mantle differentiation and its thermal consequences. *Geochim. Cosmochim. Acta*, **55**, 227–239

Galitzin, B. (1917). *Zur Frage der Bestimmung der Ilerdtiefe eines Bebens.* New York, McGraw-Hill.

Gans, P. G. (1987). An open-system, two-layer crustal stretching model for the eastern Great Basin. *Tectonics*, **6**, 1–12.

Gao, S. Rudnick, R. L. Carlson, R. W., *et al.* (2002). Re-Os evidence for replacement of ancient mantle lithosphere beneath the North China craton. *Earth Planet. Sci. Lett.*, **198**, 307–322.

Gao, S., Rudnick, R. L., Yuan, H.-L., *et al.* (2004). Recycling lower continental crust in the North China craton. *Nature* **432**, 892–897.

Gao, S., Luo, T.-C., Zhang, B.-R., *et al*. (1998). Chemical composition of the continental crust as revealed by studies in East China. *Geochim. Cosmochim. Acta*, **62**, 1959–1975.

Gao, S., Davis, P. M., Liu, H., *et al*. (1997). SKS splitting beneath continental rift zones, *J. Geophys. Res.*, **102**, 22,781–22,797.

Gardner, G. H. F., Gardner, L. W., and Gregory, A. R. (1974), Formation velocity and density – the diagnostic basics for stratigraphic traps. *Geophysics*, **39**, 770–780.

Garfunkel, Z. (2007). Controls of stable continental lithospheric thickness: the role of basal drag. *Lithos*, **96**, 299–314.

Gatzemeier, A. and Moorkamp, M. (2005). 3D modelling of electrical anisotropy from electromagnetic array data: hypothesis testing for different upper mantle conduction mechanisms, *Phys. Earth planet. Int.*, **149**, 225–242.

Gaul, O. F., Griffin, W. L., O'Reilly, S. Y., and Pearson, N. J. (2000). Mapping olivine composition in the lithospheric mantle. *Earth Planet. Sci. Lett.*, **182**, 223–235.

Gibson, S. A., Malarkey, J., Day, J. A. (2008). Melt depletion and enrichment beneath the Western Kaapvaal Craton: evidence from Finsch peridotite xenoliths. *J. Petroly.*, **49**: 1817–1852.

Gladzcenko, T. P., Coffin, M. F., and Eldholm, O. (1997). Crustal structure of the Ontong Java plateau: modeling of new gravity and existing seismic data. *J. Geophys. Res.*, **102**, 22711–22029.

Gleason, G. C. and Tullis, J. (1995). A flow law for dislocation creep of quartz aggregates determined with the molten slat cell. *Tectonophysics*, **247**: 1–23.

Glikson, A. Y. (2005). Geochemical signatures of Archean to early Proterozoic Maria-scale oceanic impact basins. *Geology*, **33**, 125–128.

Glover, P. W. J., Hole, M. J., and Pous, J. (2000). A modified Archie's law for two conducting phases. *Earth Planet. Sci. Lett.*, **180**, 369–383.

Godey, S., Deschamps, F., Trampert, J., and Snieder, R. (2004). Thermal and compositional anomalies beneath the North American continent, *J. Geophys. Res.*, **109**, B01308, doi:10.1029/2002JB002263.

Goes, S. and van der Lee, S. (2002). Thermal structure of the North American uppermost mantle inferred from seismic tomography. *J. Geophys. Res.*, **107** (B3).

Goes, S., Govers, R., and Vacher, P. (2000). Shallow upper mantle temperatures under Europe from P and S wave tomography, *J. Geophys. Res.*, **105**, 11,153–11,169.

Goes, S., Simons, F., and Yoshizawa, K. (2005). Seismic constraints on temperature of the Australian uppermost mantle. *Earth Planet. Sci. Lett.,* **236**, 227–237.

Goetze, C. (1978). The mechanisms of creep in olivine. *Phil. Trans. Roy. Soc. London*, **288**, 99–119.

Goetze, C. and Evans, B. (1979). Stress and temperature in the bending lithosphere as constrained by experimental rock mechanics. *Geoph. J. Roy. Astron. Soc.*, **59**, 463–478.

Goetze, C. and Kohlstedt, D. L. (1973). Laboratory study of dislocation climb and diffusion in olivine, *J. Geophys. Res.*, **78**, 5961–5971.

Goldfarb, R. J., Groves, D. I. and Gardoll, S. (2001). Orogenic gold and geologic time: a global synthesis. *Ore Geology Reviews*, **18**, 1–75.

Goodwin, A. M. (1996). *Principles of Precambrian Geology*. London, San Diego, Toronto, Academic Press, 327 pp.

Gorbatsevich, F. F., Smirnov, Yu. P. (1998). Elastic anisotropy and paleostress along SG-3. In: *Kola Superdeep, 1998. Scientific Results and Research Experiments*. Orlov, V. P. and Laverov, N. P. (Eds.), Moscow, Technoneftegaz, 260 pp. (in Russian).

Gordon, R. G. (1998). The plate tectonic approximation: plate nonrigidity, diffuse plate boundaries, and global plate reconstructions. *Ann. Rev. Earth Planet. Sci.*, **26**, 615–642.

Gough, D. I., McKirdy, D. M., Woods, D. V., and Geiger, H. (1989). Conductive structures and tectonics beneath the EMSLAB land array, *J. Geophys. Res.*, **94**, 14099–14110.

Grad, M., Keller, G. R., Thybo, H., Guterch, A. and POLONAISE Working Group (2002). Lower lithospheric structure beneath the Trans-European Suture Zone from POLONAISE'97 seismic profiles. *Tectonophysics*, **360**, 153–168.

Grad, M., Bruckl, E., Majdanski, M., *et al.* (2009). Crustal structure of the Eastern Alps and their foreland: seismic model beneath the CEL10/Alp04 profile and tectonic implications. *Geophys.J. Int.*, **177**, 279–295.

Grad, M., Jensen, S. L., Keller, G. R., *et al.* (2003). Crustal structure of the Trans-European suture zone region along POLONAISE'97 seismic profile P4. *J. Geophys. Res.*, **108** (B11), Article Number: 2541.

Grand, S. P. (2002). Mantle shear-wave tomography and the fate of subducted slabs. *Phil. Trans. Roy. Soc.* London, Series A, **360**, 2475–2491.

Grand, S. P. (1994). Mantle shear structure beneath the Americas and surrounding oceans. *J. Geophys. Res.*, **99**, 11,591–11,621.

Grand, S. P. and Helmberger, D. V. (1985). Upper mantle shear structure beneath Asia from multi-bounce S waves, *Phys. Earth planet. Int.*, **41**, 154–169.

Grand, S. and Helmberger, D. V. (1984). The upper mantle shear structure of North America. *Geophys. J. Roy. Astron. Soc.*, **76**, 399–438.

Grand, S. P., van der Hilst, R. D., Widiyantoro, S. (1997). Global seismic tomography: A snapshot of convection in the Earth, *GSA Today* **7**, 1–7.

Granet, M., Stoll, G., Dorel, J., *et al.* (1995). Massif Central (France): new constraints on the geodynamical evolution from teleseismic tomography. *Geophys. J. Int.*, **121**, 33–48.

Gray, D. R., Foster, D. A., Maas, R., *et al.* (2007). Continental growth and recycling by accretion of deformed turbidite fans and remnant ocean basins: Examples from Neoproterozoic and Phanerozoic orogens. In: Hatcher, R. D., Carlson, M. P., McBride, J. H., *et al.* (Eds.), *4-D Framework of Continental Crust. Geol. Soc. Am. Memoirs*, **200**, 63–92.

Green, H. W. and Gueguen, Y. (1974). Origin of kimberlite pipes by diapiric upwelling in the upper mantle. *Nature*, **249**, 617–620.

Green, H. W. and Houston, H. (1995). The mechanics of deep earthquakes. *Ann. Rev. Earth Planet. Sci.*, **23**, 169–213.

Green, H. W., II. and Borch, R. S. (1987). The pressure dependence of creep. *Acta Metallurgica*, **35**: 1301–1305.

Green, R. W. E., Hales, A. L. (1968). The traveltimes of P-waves to 308 in the central United States and upper mantle structure. *Bull. Seismol. Soc. Am.*, **70**, 809–822.

Gregory, A. R. (1977). Aspects of rock physics from laboratory and log data that are important to seismic interpretation. *Am. Ass. Petrol. Geol. Mem.*, **26**, 15–27.

Gribb, T. T. and Cooper, R. F. (1998). Low-frequency shear-attenuation in polycrystalline olivine: Grain boundary diffusion and the physical significance of the Andrade model for viscoelastic rheology. *J. Geophys. Res.*, **103**, 27267–27279.

Griffin, W. L. and O'Reilly, S. Y. (2007). The earliest subcontinental lithospheric mantle. In: van Kranendonk, M. J., Smithies, R. H., and Bennett, V. C. (Eds.), *Developments in Precambrian Geology, Earth's Oldest Rocks*. Elsevier, v. 15, 1013–1035.

Griffin, W. L. and Ryan, C. G. (1995). Trace elements in indicator minerals: area selection and target evaluation in diamond exploration, *J. Geochem. Explor.*, **53**, 311–337.

Griffin, W. L., Kaminsky, F. V., Ryan, C. G., *et al.* (1996). Thermal state and composition of the lithospheric mantle beneath the Daldyn kimberlite field, Yakutia. *Tectonophysics*, **262**, 19–33.

Griffin, W. L., O'Reilly, S. Y., Ryan, C. G., Gaul, O., and Ionov, D. (1998a). Secular variation in the composition of continental lithospheric mantle. In: J. Braun *et al.* (Ed.), *Structure and Evolution of the Australian Continent*. AGU Geodynam. Ser., 26, 1–25.

Griffin, W. L., Zhang, A., O'Reilly, S. Y., and Ryan, C. G. (1998b), Phanerozoic evolution of the lithosphere beneath the Sino-Korean craton. In: Flower, M. F. J., Chung, S. L., Lo, C. H., and Lee, T. Y. (Eds.), *Mantle Dynamics and Plate Interactions in East Asia*, AGU Geodynam. Ser., v. 27, pp. 107–126.

Griffin, W. L., Ryan, C. G., Kaminsky, F. V., *et al.* (1999a). The Siberian lithosphere traverse: Mantle terranes and the assembly of the Siberian Craton. *Tectonophysics*, **310**, 1–35.

Griffin, W. L., O'Reilly, S. Y., and Ryan, C. G. (1999b). The composition and origin of subcontinental lithospheric mantle. In: Fei, Y., Bertka, C. M. and Mysen, B. O. (Eds.), *Mantle Petrology: Field Observations and High Pressure Experimentation: A Tribute to Francis R. (Joe) Boyd*. Geochem. Soc. Spec. Publ. No. 6, 13–45.

Griffin, W. L., Foyle, B. J., Ryan, C. G., *et al.* (1999c). Layered mantle lithosphere in the Lac de Gras Area, Slave Craton: Composition, structure and origin. *J. Petrol*, **40**, 705–727.

Griffin, W. L., O'Reilly, S. Y., Abe, N., *et al.* (2003a). The origin and evolution of Archean lithospheric mantle, *Precambrian Res.*, **127**, 19–41.

Griffin, W. L., O'Reilly, S. Y., Natapov, L. M., and Ryan, C. G. (2003b). The evolution of lithospheric mantle beneath the Kalahari craton and its margins. *Lithos*, **71**, 215–241.

Griffin, W. L, O'Reilly, S. Y., Doyle, B. J., *et al.* (2004). Lithospheric Mapping Beneath the North American Plate. *Lithos*, **77**, 873–922.

Griffin, W. L., Natapov, L. M., O'Reilly, S. Y., *et al.* (2005). The Kharamai kimberlite field, Siberia: Modification of the lithospheric mantle by the Siberian Trap event. *Lithos*, **81**, 167–187.

Griffiths, R. and Campbell, I. (1991). Interaction of mantle plume heads with Earth's surface and the onset of small-scale convection. *J. Geophys. Res.*, **96**, 18295–18310.

Griot, D. A., Montagner, J. P., and Tapponnier, P. (1998). Phase velocity structure from Rayleigh and Love waves in Tibet and its neighboring regions, *J. Geophys. Res.*, **103**, 21, 215–21, 232.

Groshong, R. H. Jr. (1988). Low-temperature deformation mechanisms and their interpretation. *GSA Bull.*, **100**, 1329–1360.

Grossman, A. and Morlet, J. (1984). Decomposition of Hardy functions into square integrable wavelets of constant shape. *SIAM J. Math. Anal.* **15** (4), 723–736.

Grotzinger, J. and Royden, L. (1990). Elastic strength of the Slave craton at 1.9 Gyr and implications for the thermal evolution of the continents. *Nature*, **347**, 64–66.

Grove, T. L. and Parman, S. W. (2004). Thermal evolution of the Earth as recorded by komatiites. *Earth Planet. Sci. Lett.*, **219**, 173–187.

Grove, T., de Wit, M. J, and Dann, J. (1997). Komatiites from the Komati type section, Barberton, South Africa. In: de Wit, M. J. and Ashwal, L. D. (Eds.), 1997. *Greenstone Belts*. Oxford, Clarendon Press, 438–453.

Grutter, H. S. and Moore, R. O. (2003). Pyroxene geotherms revisited – An empirical approach based on Canadian xenoliths. *Proc. 8th Intern. Kimberlite Conf.*, Victoria, BC, Canada, June 2003: FLA-0272.

Grutter, H., Apter, D. B., and Kong, J. (1999). Crust-mantle coupling: evidence from mantle-derived xenocrystic garnets. In: Gurney, J. J., Gurney, J. L., Pascoe, M. D., and Richardson, S. H. (Eds.), *Proc. 7th Int. Kimberlite Conf.*, 307–313.

Grutter, H., Latti, D., and Menzies, A. (2006). Cr-saturation arrays in concentrate garnet compositions from kimberlites and their use in mantle barometry. *J. Petrol*, **47**, 801–820.

Gu, Y. J. and Dziewonski, A. M. (2002). Global variability of transition zone thickness. *J. Geophys. Res.*, **107** (B7), 2135, doi:10.1029/2001JB000489.

Gu, Yu.J., Dziewonski, A. M., and Ekström, G. (2001). Preferential detection of the Lehmann discontinuity beneath continents. *Geophys. Res. Lett.*, **28**, 4655–4658.

Gung, Y., Panning, M. P., and Romanowicz, B. (2003), Global anisotropy and the thickness of continents *Nature*, **422**, 707–711.

Gunn, R. (1943). A quantitative study of isobaric equilibrium and gravity anomalies in the Hawaiian Islands. *J. Franklin Inst.*. **236**, 373–390.

Gupta, S., Rai, S. S., Prakasam, K. S., *et al.* (2003). The nature of the crust in southern India: Implications for Precambrian crustal evolution, *Geophys. Res. Lett.*, **30**(8), 1419, doi:10.1029/2002GL016770.

Gurer, A., Bayrak, M., and Gurer, O. F. (2004). Magnetotelluric images of the crust and mantle in the southwestern Taurides, Turkey. *Tectonophysics,* **391**, 109–120.

Gurney, J. J. (1984). A correlation between garnets and diamonds. In: Glover, J. E. and Harris, P. G. (Eds). *Kimberlite Occurrence and Origin: A Basis for Conceptual Models in Exploration*. Geology Department and University Extension, University of Western Australia Publication 8, 143–166.

Gurnis, M. (1988). Large-scale mantle convection and the aggregation and dispersal of supercontinents. *Nature*, **332**, 695–699.

Gurnis, M. (1990). Bounds on global dynamic topography from Phanerozoic flooding of continental platforms. *Nature*, **344**, 754–756.

Gutenberg, B. (1954). *Seismicity of the Earth and Associated Phenomena*. New York, Hafner, 310 pp.

Gutenberg, B. (1959). *Physics of the Earth's Interior*. New York, Springer.

Guyot, P., and J. E. Dorn (1967). A critical review of the Peierls mechanism, *Can. J. Phys.*, **J5**, p. 983.

Haak, V. and Hutton, V. R. S. (1986). Electrical resistivity in continental lower crust. In: Dawson, J. B., Carswell, D. A., Hall, J., and Wedepohl, K. H. (Eds.), *The Nature of the Lower Continental Crust*. Geol. Soc. London Sp. Publ., v. 24, 35–49.

Hacker, B. R., Abers, G. A., and Peacock, S. M. (2003). Subduction factory 1: Theoretical mineralogy, densities, seismic wave speeds, and H2O contents. *J. Geophys. Res.*, **108**, B1, 2029, doi: 10.1029/2001JB001127.

Hager, B. H. (1983). Global isostatic geoid anomalies for plate and boundary layer models of the lithosphere. *Earth Planet. Sci. Lett.*, **63**, 97–109.

Hager, B., Clayton, R., Richards, M., Comer, R., and Dziewonski, A., (1985). Lower mantle heterogeneity, dynamic topography and the geoid. *Nature*, **313**, 541–545.

Haggerty, S. (1999). A diamond trilogy: superplumes, supercontinents, and supernovae. *Science*, **285**, 851–860.

Haggerty, S. E. and Sautter, V. (1990). Ultradeep (greater than 300 kilometers), ultramafic upper mantle xenoliths. *Science*, **248**, 993–996.

Hale, C. J. (1987). The intensity of the geomagnetic field at 3.5 Ga: palaeointensity results from the Komati formation, Barberton mountain land, South Africa. *Earth Planet. Sci. Lett.*, **86**, 354–364.

Hale, C. J. and Dunlop, D. J. (1984). Evidence for an early Archean geomagnetic field; a paleomagnetic study of Komati formation, Barberton greenstone belt, South Africa. *Geophys. Res. Lett.*, **11**, 97–100.

Hales, A. L. (1991). Upper mantle models and the thickness of the continental lithosphere. *Geophys. J. Int.*, **105**, 355–363.

Hales, A. L. (1969). A seismic discontinuity in the lithosphere. *Earth Planet. Sci. Lett.*, **7**, 44–46.

Halliday, A. N. (2003). The origin and the earliest history of the Earth. *Treatise on Geochemistry*, Elsevier, Volume 1.

Hamilton, W. B. (2007). Earth's first two billion years – the era of internally mobile crust. In: Hatcher, R. D., Carlson, M. P., McBride, J. H., *et al.* (Eds.), *4-D Framework of Continental Crust.* Geological Society of America Memoirs, **200**, 233–296.

Hamilton, W. (1987). Crustal extension in the Basin and Range province, southwestern United States, In: *Continental Extensional Tectonics*, Coward, M. P., Dewey, J. F., and Hancock, P. L., (Eds.) *Geol. Soc. Spec. Publ. London*, **28**, 155–176.

Hamilton, P. J., O'Nions, R. K., Evensen, N. M., Bridgwater, D., and Allaart, H. (1978). Sm-Nd isotopic investigations of Isua supracrustals and implications for mantle evolution. *Nature*, **272**, 41–43.

Handler, M. R., Bennet, V. C., and Esat, T. Z. (1997). The persistence of off-cratonic lithospheric mantle: Os isotopic systematics of variably metasomatised southeast Australian xenoliths. *Earth Planet. Sci. Lett.*, **151**, 61–75.

Hanghøj, K., Keleman, P., Bernstein, S., Blusztajn, J., and Frei, R. (2001). Osmium isotopes in the Wiedemann Fjord mantle xenoliths: a unique record of cratonic mantle formation by melt depletion in the Archaean. *Geochem. Geophys. Geosys.* **2**.

Hanks, T. C. (1971). The Kuril Trench – Hokkaido Rise system: large shallow earthquakes and simple models of deformation, *Geophys. J.Roy. Astron. Soc.*, **23**, 173–189.

Hanrahan, M., Stachel, T., Brey, G. P., and Lahaye, Y. (2003). Garnet Peridotite Xenoliths from the Koffiefontein Mine, South Africa. In: *Proc. 8th Intern. Kimberlite Conf.*, Victoria, BC, Canada, FLA-0369.

Hansen, S. and Dueker, K. (2009). P- and S-wave receiver function images of crustal imbrication beneath the Cheyenne Belt in Southeast Wyoming. *Bull. Seismol. Soc. Am.*, **99**, 1953–1961.

Hargraves, R. B. (1986). Faster spreading or greater ridge length in the Archean? *Geology*, **14**, 750–752.

Harig, C., Molnar, P., and Houseman, G. A. (2008). Rayleigh-Taylor instability under a shear stress free top boundary condition and its relevance to removal of mantle lithosphere from beneath the Sierra Nevada, *Tectonics*, **27**, TC6019, doi: 10.1029/2007TC002241.

Harley, S. L. (1984). An experimental study of the partitioning of Fe and Mg between garnet and orthopyroxene. *Contrib. Mineral. Petrol.* **86**, 359–373.

Harrison, T. M., Blichert-Toft, J., Muller, W., *et al.* (2005). Heterogeneous Hadean Hafnium: Evidence of Continental Crust at 4.4 to 4.5 Ga. *Science*, **310**, 1947–1950.

Hart, S. R. and Dodd, R. T. (1962). Excess radiogenic argon in pyroxenites. *J. Geophys. Res.*, **67**, 2998.

Hart, S. R. and Zindler, A. (1986). In search of a bulk-Earth composition. *Chem. Geol.*, **57**, 247–267.

Hartley, R., Watts, A. B., and Fairhead, J. D. (1996). Isostasy of Africa. *Earth and Planet. Sci. Lett.*, **137**, 1–18.

Hashin, Z. (1970). Theory of composite materials. In: *Mechanics of Composite Materials*, 5th Symp. Naval Structural Mechanics, Wendt, F. W., Liebowitz, H., and Pertone, N. (Eds.), New York, Pergamon, 201–242.

Hashin, Z. and Shtrikman, S. (1962). A variational approach to the theory of the elastic behaviour of polycrystals. *J. Mech. Phy. Solids*, **10**, 343–352.

Hashin, Z. and Shtrikman, S. (1963). A variational approach to the theory of the elastic behaviour of multiphase materials. *J. Mech. Phy. Solids*, **11**, 12–140.

Hatcher, R. D., Carlson, M. P., McBride, J. H., *et al.* (Eds.), (2007). *4-D framework of continental crust*. Geol. Soc. Am. Memoirs, **200**, 643 pp., doi:10.1130/2007.1200.

Hawkesworth, C. J. and Kemp, A. I. S. (2006). Using hafnium and oxygen isotopes in zircons to unravel the record of crustal evolution. *Chem. Geol.*, **226**, 144–162.

Hawkesworth, C., Cawood, P., Kemp, A., Storey, C., and Dhuime, B. (2009). A matter of preservation. *Science*, **323**, 49–50.

Hawkesworth, C. J., Kempton, P. D., Rogers, N. W., Ellam, R. M., and van Calsteren, P. W. (1990). Continental mantle lithosphere, and shallow level enrichment process in the Earth's mantle. *Earth Plan. Sci. Lett.*, **96**, 256–268.

Hayford, J. F. (1917). Gravity and isostasy. *Science*, **45**(1163), 350–354.

Heaman, L. M. and Kjarsgaard, B. A. (2000). Timing of eastern North American kimberlite magmatism: continental extension of the Great Meteor hotspot track? *Earth Planet. Sci. Lett.*, **178**, 253–268.

Heaman, L. M., Kjarsgaard, B. A., and Creaser, R. A. (2003). The timing of kimberlite magmatism in North America: implications for global kimberlite genesis and diamond exploration. *Lithos*, **71**, 153–184.

Hearn, B. C., Jr. (2003). Upper-Mantle Xenoliths in the Homestead Kimberlite, Central Montana, USA: depleted and Re-Enriched Wyoming Craton Samples. *Proc. 8th Int. Kimberlite Conf.*, Victoria, BC, Canada, June 2003: FLA-0126.

Heest, R. T. and Crough, S. T. (1981). The effect of hot spots on the oceanic age-depth relation. *J. Geophys. Res.*, **86**, 6107–6114.

Heinson, G. S. and Lilley, F. E. M. (1993). An application of thin-sheet electromagnetic modeling to the Tasman Sea. *Phys. Earth Planet. Int.*, **81**, 231–251.

Heinson, G. and Constable, S. C. (1992). The electrical conductivity of oceanic upper mantle. *Geophys. J. Int.* **110**, 159–179.

Heinson, G. S., White, A., Law, L. K., *et al.* (1993). EMRIDGE – The Electromagnetic Investigation of the Juan-De Fuca Ridge. *Marine Geophys. Res.*, **15**, 77–100.

Heintz, M., Debayle, E. and Vauchez, A. (2005). Upper mantle structure of the South American continent and neighboring oceans from surface wave tomography. *Tectonophys.* **406**, 115–139.

Heirtzler, J. R., Dickson, G. O., Herron, E. M., Pitman, W. C., and Le Pichon, X. (1968). Marine magnetic anomalies, geomagnetic field reversals, and motions of the ocean floor and continents. *J. Geophys. Res.*, **73**, 2119–2136.

Heiskanen, W. A. (1931). Isostatic tables for the reduction of gravimetric observations calculated on the basis of Airy's hypothesis. *Bull. Géodesiqué*, **30**, 110–129.

Helmstaedt, H. H. and Schulze, D. J. (1989). Southern African kimberlites and their mantle sample: implications for Archean tectonics and lithosphere evolution. In: Ross, J. (Ed.), *Kimberlites and Related Rocks, Vol.1: Their Compositition, Occurrence, Origin, and Emplacement. Geol. Soc. Aust. Spec. Pub.*, Vol. **14**, pp. 358–368.

Henry, P., Stevenson, R. K., Larbi, Y., and Gariepy, C. (2000). Nd isotopic evidence for early to late Archean (3.4–2.7 Ga) crustal growth in the Western Superior Province (Ontario, Canada). *Tectonophys.*, **322**, 135–151.

Hermance, J. F. (1979). The electrical conductivity of materials containing partial melt: A simple model from Archie's Law, *Geophys. Res. Lett*, **6**, 613–616.

Hermance, J. F. (1995). Electrical conductivity models of the crust and mantle. In: *Global Earth Physics. A Handbook of Physical Constants*. AGU Ref. Shelf 1, 190–205.

Hermance, J. F. and Neumann, G. A. (1990). The Rio Grande Rift: New electromagnetic constraints on the Socorro magma body, *Phys. Earth Planet. Int.*, **66**, 101–117.

Herring, C. (1950). Diffusional viscosity of a polycrystalline solid, *J. Appl. Phys.*, **21**, p. 437.

Herz, N. (1969). Anorthosite belts, continental drift and anorthosite event. *Science*, **164**, 944.

Herzberg, C. (2004). Geodynamic Information in Peridotite Petrology. *J. Petrol.*, **45**: 2507–2530.

Herzberg, C. (1999). Phase equilibrium constraints on the formation of cratonic mantle. In: Y. Fei, C. M. Bertka, and B. O. Mysen (Eds.), *Mantle Petrology, Field Observations and High Pressure Experimentation, A Tribute to Francis R. (Joe) Boyd, Geochem. Soc. Spec. Pub.*, 241–257.

Herzberg, C. (1995). Generation of plume magmas through time: An experimental perspective. *Chem. Geol.*, **126**, 1–16.

Hess, H. H. (1964). Seismic anisotropy of the uppermost mantle under oceans, *Nature*, **203**, 629–631.

Hickman, S. H. (1991). Stress in the lithosphere and the strength of active faults. *Rev. Geophys.*, IUGG Report, 759–775.

Hinze, W. J., and Zietz, I. (1985). The composite magnetic anomaly map of the contermi-nous United States, In: Hinze, W. J., (Ed.) *The Utility of Regional Gravity and Magnetic Anomaly Maps*, edited by 1–24, Soc. Of Expl. Geophys., Tulsa, OK.

Hinze, E., Will, G. and Cemic, L. (1981). Electrical conductivity measurements on synthetic olivines and on olivine, enstatite, and diopside from Dreiser Weiher, Eifel (Germany) under defined thermodynamic activities as a function of temperature and pressure, *Phys. Earth Planet. Int.*, **25**, 245–254.

Hirsch, L. M. (1991). The Fe-FeO buffer at lower mantle pressures and temperatures. *Geophys. Res. Lett.*, **18**, 1309–1312.

Hirsch, L. M. and Shankland, T. J. (1993). Quantitative olivine-defect chemical model; insights on electrical conduction, diffusion, and the role of Fe content, *Geophys. J. Int.*, **114**, 1, 21–35.

Hirsch, L. M., Shankland, T. J., and Duba, A. G. (1993). Electrical conduction and polaron mobility in Fe-bearing olivine. *Geophys. J. Int.*, **114**, 1, 36–44.

Hirschmann, M. M. (2006). Water, melting, and the deep Earth H2O cycle. *Ann. Rev. Earth Planet. Sci.*, **34**: 629–653.

Hirschmann, M. M. (2000). Mantle solidus: experimental constraints and the effects of peridotite composition. *Geochem. Geophys. Geosyst.*, **1**, doi 2000GC000070.

Hirth, G. and Kohlstedt, D. L. (2003). Rheology of the upper mantle and the mantle wedge: a view from the experimentalists. In: J. E. Eiler (Ed.), *Inside the Subduction Factory*. American Geophysical Union, Washington DC, *Geophys. Monogr.*, 138, 83–105.

Hirth, G. and Kohlstedt, D. L. (1996). Water in the oceanic upper mantle – implications for rheology, melt extraction and the evolution of the lithosphere. *Earth Planet. Sci. Lett.*, **144**, 93–108.

Hirth, G. and Kohlstedt, D. L. (1995), Experimental constraints on the dynamics of the partially molten upper mantle: Deformation in the diffusion creep regime, *J. Geophys. Res.*, **100**(B2), 1981–2001.

Hirth, G., Evans, R. L., and Chave, A. D. (2000). Comparison of continental and oceanic mantle electrical conductivity: Is Archean lithosphere dry? *Geochem. Geophys. Geosyst.*, **1**: 10.1029/2000GC000048.

Hirth, G., Teyssier, C., and Dunlap, W. J. (2001). An evaluation of quartzite flow laws based on comparisons between experimentally and naturally deformed rocks. *Int. J. Earth Sci.*, **90**, 77–87.

Hitchings, R. S., Paterson, M. S., and Bitmead, J. (1989). Effects of iron and magnetite additions in olivine pyroxene rheology. *Phys. Earth Planet. Int.*, **55**, 277–291.

Hjelt, S.-E. and Korja, T. (1993). Lithospheric and upper-mantle structures: results of electromagnetic soundings in Europe. *Phys. Earth Planet. Int.* **79**, 137–177.

Hobbs, B. E. (1981). The influence of metamorphic environment upon the metamorphic environment upon the deformation of minerals, *Tectonophysics*, **78**, 335–383.

Hoffman, P. F. (1989). Speculations on Laurentia's first gigayear (2.0 to 1.0 Ga). *Geology* **17**, 135–138.

Hoffman, P. F. (1990). Geological constraints on the origin of the mantle root beneath the Canadian shield. *Phil. Trans. Roy. Soc. Lond.* A **331**, 523–532.

Hoffmann, N., Jodicke, H., and Horejschi L. (2005). Regional distribution of the Lower Carboniferous Culm and Carboniferous limestone facies in the North German Basin – derived from magnetotelluric soundings. *Z. dt Ges Geowiss* **156**(2): 323–339.

Hofmann, A. W., Jochum, K. P., Seufert, M., and White, W. M. (1986). Nb and Pb in oceanic basalts: new constraints on mantle evolution. *Earth Planet. Sci. Lett.* **79**, 33–45.

Hofmeister, A. (1999). Mantle values of thermal conductivity geotherm from phonon lifetimes. *Science*, **283**, 1699–1709.

Hofmeister, A. M. and Criss, R. E. (2005). Earth's heat flux revisited and linked to chemistry. *Tectonophysics* **395**, 159–177.

Holmes, A. (1931). Radioactivity and Earth movements. XVII. *Trans. Geol. Soc. Glasgow*, **18**(3), 559–606.

Holtzman, B. K., Kohlstedt, D. L., Zimmerman, M. E., *et al.* (2003). Melt segregation and strain partitioning: implications for seismic anisotropy and mantle flow. *Science*, **301**, 1227–1230.

Horai, K. (1971). Thermal conductivity of rock-forming minerals, *J. Geophys. Res.*, **76**(5), 1278–1308.

Horai, K. and Susaki, J. (1989). The effect of pressure on the thermal conductivity of silicate rocks up to 12 kbar. *Phys. Earth Planet. Inter.*, **55**, 292–305.

Hoshiba, M. (1993). Separation of scattering attenuation and intrinsic absorption in Japan using the multi lapse time window analysis of full seismogram envelope. *J. Geophys. Res.*, **98**, 15,809–15,824.

Houseman, G. A. and P. Molnar (1997). Gravitational (Rayleigh-Taylor) instability of a layer with non-linear viscosity and convective thinning of continental lithosphere. *Geophysical J. International*, **128**, 125–150.

Houseman, G. A., McKenzie, D. P., and Molnar, P. (1981). Convective instability of a thickened boundary layer and its relevance for the thermal evolution of continental convergent belts, *J. Geophys. Res.*, **86**, 6115–6132.

Huang, X., Xu, Y., and Karato, S. (2005). Water content of the mantle transition zone from the electrical conductivity of wadsleyite and ringwoodite. *Nature*, **434**, 746–749.

Huang, Z., Su, W., Peng, Y., Zheng, Y., and Li, H. (2003). Rayleigh wave tomography of China and adjacent regions. *J. Geophys. Res.* **108** (B2), 2073.

Huebner, J. S., Duba, A., and Wiggins, L. B. (1979). Electrical conductivity of pyroxene which contains trivalent cations: laboratory measurements and the lunar temperature profile. *J. Geophys. Res.* **84**, 4652–4656.

Hughes, D. S. and Cross, J. H. (1951). Elastic wave velocities at high pressures and temperatures. *Geophysics*, **16**, 577–593.

Humphreys, E. D., and K. G. Dueker (1994). Physical state of the western U.S. upper mantle, *J. Geophys. Res.*, **99**, 9635–9650.

Hurley, P. M. and Rand, J. M. (1969). Pre-drift continental nuclei. *Science* **164**, 1229–1242.

Hurt, C. P., Moskowitz, B. M., and S. K. Banerjee (1995). Magnetic properties of rocks and minerals. In: *Rock Physics and Phase Relations. A Handbook of Physical Constants*, AGU Reference Shelf 3, 189–204.

Hyndman, R. D. (1988). Dipping seismic reflectors, electrically conductive zones, and trapped water in the crust over a subducting plate. *J. Geophys. Res.*, **93**: 13,391–13,405.

Hynes, A. (2001). Freeboard revisited: continental growth, crustal thickness change and Earth's thermal efficiency. *Earth Planet. Sci. Lett.*, **185**, 161–172.

Hyvonen, T., Tiira, T., Korja, A., *et al.* (2007). A tomographic crustal velocity model of the central Fennoscandian Shield. *Geophys. J. Int.* **168** (3): 1210–1226.

Ichiki, M., Uyeshima, M., Utada, H., *et al.* (2001). Upper mantle conductivity structure of the back-arc region beneath northeastern China. *Geophys. Res. Lett.*, **28**, 3773–3776.

Ide, J. M. (1937). The velocity of sound in rocks and glasses as a function of temperature. *J. Geol.*, **45**, 689–716.

Inoue, H., Fukao, Y., Tanabe, K., and Ogota, Y. (1990). Whole mantle P-wave travel time tomography, *Phys. Earth planet. Inter.*, **59**, 294–328.

Irifune, T. (1987). An experimental investigation of the pyroxene-garnet transformation in a pyrolite composition and its bearing on the composition of the mantle. *Phys. Earth Planet. Inter.*, **45**, 324–336.

Irifune, T., Sekine, T., Ringwood, A. E., and Hibberson, W. O. (1986). The eclogite-garnetite transformation at high pressure and some geophysical implications. *Earth Planet. Sci. Lett.*, **77**, 245–256.

Irvine, G. J., Pearson, D. G., Kjarsgaard, B. A., *et al.* (2003). A Re-Os isotope and PGE study of kimberlite-derived peridotite xenoliths from Somerset Island and a comparison to the Slave and Kaapvaal cratons. *Lithos*, **71**, 461–488.

Irving, E. (1977). Drift of major continental blocks since the Devonian. *Nature*, **270**, 304–306.

Isaac, D. G., Anderson, O. L., Goto, T., and Suzuki, I. (1989). Elasticity of single-crystal forsterite measured to 1700 K. *J. Geophys. Res.* **94**, 5895–5906.

Isaak, D. G. (1992). High-temperature elasticity of iron-bearing olivines. *J. Geophys. Res.*, **97**, 1871–1885.

Ito, K. and Kennedy, G. C. (1967). Melting and phase relations in a natural peridotite to 40 kilobars. *Am. J. Sci.*, **265**: 519–538.

Ito, K. and Kennedy, G. C. (1971). An experimental study of the basalt – garnet granulite – eclogite transformation. *AGU Geophys. Monogr.*, v. **14**, 303–314.

Ito, E. and Takahashi, E. (1989). Postspinel transformations in the system Mg_2SiO_4-Fe_2SiO_4 and some geophysical implications. *J. Geophys. Res.*, **94**, 637–646.

Ito, E., Harris, D. M., and Anderson, A. T. (1983). Alteration of oceanic-crust and geologic cycling of chlorine and water. *Geochim. Cosmochim. Acta*, **47**, 1613–1624

Iyer, H. M. and Hirahara, K. (Eds.) (1993). *Seismic Tomography: Theory and practice*. London, Chapman and Hall.

Iyer, S. S., Choudhuri, A., Vasconcellos, M., and Jordani, U. U. (1984). Radioactive element distribution in the Archean granulite terrane of Jequie-Bahia, Brazil. *Contrib. Mineral. Petrol.*, **85**, 224–243.

Iyer, H. M., Pakiser, L. C., Stuart, D. J., and Warren, D. H. (1969). Project Early Rise: seismic probing of the upper mantle. *J. Geophys. Res.* **74**, 4409–4441.

Jackson, J. (2002). Strength of the continental lithosphere: time to abandon the jelly sandwich? *GSA Today*, **12**, 4–9.

Jackson, I. (1993). Progress in the experimental study of seismic wave attenuation. *Ann. Rev. Earth Planet. Sci.*, **21**, 375–406.

Jackson, I. (2000). Laboratory measurements of seismic wave dispersion and attenuation: recent progress. In: S. I. Karato *et al.* (Eds.), *The Earth's Deep Interior: Mantle Physics and Tomography from the Atomic to Global Scale*, Geophysical Monograph, American Geophysical Union. Vol. 117, 265–289.

Jackson, J. A. and N. J. White (1989). Normal faulting in the upper continental crust: observations from regions of active extension, *J. Struct. Geol.* **11**, 15–32.

Jackson, I., Paterson, M. S., and Fitz Gerald, J. D. (1992). Seismic wave dispersion and attenuation in Åheim dunite: an experimental study. *Geophys. J. Int.*, **108**, 517–534.

Jackson, I., Rudnick, R. L., O'Reilly, S. Y., and Bezant, C. (1990). Measured and calculated elastic wave velocities for senoliths from the lower crust and upper mantle, *Tectonophysics*, **173**, 207–210.

Jackson, I., Webb, S., Weston, L., and Boness, D. (2005). Frequency dependence of elastic wave speeds at high temperature: a direct experimental demonstration. *Phys. Earth Planet. Inter.*, **148**, 85–96.

Jacob, D. and Foley, S. (1999). Evidence for Archean ocean crust with low high field strength element signature from diamondiferous eclogite xenoliths. *Lithos*, **48**, 317–336.

Jacobsen, S. B. and Dymek, R. F. (1988). Nd and Sr isotope systematics of clastic metasediments from Isua, West Greenland: identification of pre-3.8 Ga differentiated crustal components. *J. Geophys. Res.* **93**, 338–354.

Jacobsen, S. and Wasserburg, G. J. (1979). The mean age of mantle and crustal reservoirs. *J. Geophys. Res.* **84**, 7411–7427.

Jaeger, J. C. (1965). Application of the theory of heat conduction to geothermal measurements. *AGU Monograph.*, v. 8.

Jagoutz, E. (1988). Nd and Sr systematics in an eclogite xenolith from Tanzania: evidence for frozen mineral equilirbia in the continental lithosphere. *Geochim. Cosmochim. Acta*, **52**, 1285–1293.

Jagoutz, E., Carlson, R. W., and Lugmair, G. W. (1980). Equilibrated Nd – unequilibrated Sr isotopes in mantle xenoliths. *Nature*: **286**, 708–710.

James, D. E., Fouch, M. J., VanDecar, J. C., and van der Lee, S. (2001). Tectospheric structure beneath southern Africa, *Geophys. Res. Lett.*, **28**, 2485–2488.

James, D., Boyd, F., Schutt, D., Bell, D., and Carlson, R. (2004). Xenolith constraints in seismic velocities in the upper mantle beneath southern Africa. *Geochem. Geophys. Geosys.* **5**, doi:10.1029/2003GC000551.

Janik, T., Yliniemi, J., Grad, M., *et al.* (2002). Crustal structure across the TESZ along POLONAISE'97 seismic profile P2 in NW Poland. *Tectonophysics*, **360**, 129–152.

Jaupart, C. (1983). Horizontal heat transfer due to radioactivity contrasts: causes and consequences of the linear heat flow relation. *Geophys. J. Roy. Astron. Soc.*, **75**, 411–435.

Jaupart, C. and Mareschal, J.-C. (1999). Thermal structure and thickness of continental roots. *Lithos*, **48**, 93–114.

Jaupart, C. and Mareschal, J.-C. (2004). Constraints on crustal heat production from heat flow data. *Treatise on Geochemistry*, v.3, 65–86, Amsterdam, Elsevier.

Jaupart, C. and Mareschal, J.-C., (2010). *Heat Generation and Transport in the Earth*. Cambridge University Press 464 pp.

Jaupart, C., Mann, J. R., and Simmons, G. (1982). A detailed study of the distribution of heat flow and radioactivity in New Hampshire (USA). *Earth Planet. Sci. Lett.*, **59**, 267–287.

Jaupart, C., Mareschal, J.-C., Guillou-Frottier L., and Davaille, A. (1998). Heat flow and thickness of the lithosphere in the Canadian Shield. *J. Geophys. Res.*, **103**, 15269–15286.

Jeffreys, H. (1926). On the nature of isostasy. *Beitrage zur Geophysik*, **XV**, 153–174.

Jegen, M. and Edwards, R. N. (1998). The Electrical Properties of a 2D Conductive Zone under the Juan de Fuca Ridge. *Geophys. Res. Lett.*, **25**, 3647–3650.

Jessop, A. M. (1990). Developments in solid Earth geophysics. v. 17. *Thermal Geophys*. Elsevier, Amsterdam.

Jessop, A. M., Hobart, M. A. and Sclater, J. G. (1976). The world heat flow data collection – 1975. *Geotherm. Ser.*, **5**, 125 pp. (published by Energy, Mines and Resources, Earth Phys. branch, Ottawa, Canada).

Ji, S., Rondenay, S., Mareschal, M., and Senechal, G. (1996). Obliquity between seismic and electrical anisotropies as a potential indicator of movement sense for ductile shear zones in the upper mantle, *Geology*, **24**, 1033–1036.

Jin, D., Karato, S., and Obata, M. (1998). Mechanisms of shear localization in the continental lithosphere: inference from the deformation microstructures of peridotites from the Ivrea zone, northern Italy. *J. Struct. Geol.* **20**, 195–209.

Jochum, K. P., Hofmann, A. W., Ito, E., Seufert, H. M., and White, W. M. (1983). K, U, and Th in mid-ocean ridge basalt glasses and heat production, K/U and K/Rb in the mantle. *Nature*, **306**, 431–436.

Johnson, C.M. and Beard, B.L. (1993). Evidence from hafnium isotopes for ancient suboceanic mantle beneath the Rio-Grande Rift. *Nature*, **362**, 441–444.

Jones, A. G. (1984). The electrical structure of the lithosphere and asthenosphere beneath the Fennoscandian Shield. *J. Geomag. Geoelectr.*, **35**, 811–827.

Jones, A. G. (1999). Imaging the continental upper mantle using electromagnetic methods. *Lithos*, **48**, 57–80.

Jones, A. G. (1993). Electromagnetic images of modern and ancient subduction zones. *Tectonophysics*, **219**: 29–45.

Jones, A. G. and Craven, J. A. (2004). Area selection for diamond exploration using deep-probing electromagnetic surveying. *Lithos*, **77**, 765–782.

Jones, A. G., Katsube, J., and Schwann, P. (1997). The longest conductivity anomaly in the world explained: sulphides in fold hinges causing very high electrical anisotropy. *J. Geomagn. Geoelectr.*, **49**, 1619–1629.

Jones, A. G., Lezaeta, P., Ferguson, I. J., Chave, A. D., Evans, R. L., Garcia, X., and Spratt, J. (2003). The electrical structure of the Slave craton. *Lithos*, **71**, 505–527.

Jones, A. G., Craven, J. A., McNeice, G. A., *et al.* (1993). The North American Central Plains conductivity anomaly within the Trans –Hudson orogen in northern Saskatchewan. *Geology* **21**, 1027–1030.

Jones, A. G., Snyder, D., Hanmer, S., *et al.* (2002). Magnetotelluric and teleseismic study across the Snowbird Tectonic Zone, Canadian Shield: a Neoarchean mantle suture? *Geophys. Res. Lett.*, **29** (10), doi: 10.1029/2002GL015359.

Jordan, T. H. (1975a). The continental tectosphere, *Rev. Geophys. Space Phys.* **13**, 1–12.

Jordan, T. H. (1975b). Lateral heterogeneity and mantle dynamics. *Nature*, **257**, 745–750.

Jordan, T. H. (1978). Composition and development of the continental tectosphere. *Nature*, **274**, 544–548; doi:10.1038/274544a0.

Jordan, T. H. (1979). Mineralogies, densities and seismic velocities of garnet lherzolites and their geophysical implications. In: Boyd, F. R. and Meyer H. O. A. (Eds.), *The Mantle Sample: Inclusions in Kimberlite and Other Volcanics. Proc. Second Intern. kimberlite Conf.*; v. 2; Am. Geophys. Un., Washington, DC, pp. 1–14.

Jordan, T. H. (1981). Continents as a chemical boundary layer. *Phil. Trans. Roy. Soc. London, Ser. A.* **301**, 359–373.

Jordan, T. H. (1988). Structure and formation of the continental tectosphere, *J. Petrol., Special Lithosphere Issue*, **29**, 11–37.

Jordan, T. H. (1997). Petrological controls on the density and seismic velocities of the cratonic upper mantle. *EOS Trans. AGU*, **78**, No. 46, p. F746.

Judenherc, S., Granet, M., Brun, J.-P., *et al.* (2002). Images of lithospheric heterogeneities in the Armorican segment of the Hercynian Range in France. *Tectonophys.*, **358**, 121–134.

Jung, H., and Karato, S. (2001) Water-induced fabric transitions in olivine. *Science*, **293**, 1460–1463.

Jung, H., Katayama, I., Jiang, Z., Hiraga, T., and Karato, S. (2006) Effect of water and stress on the lattice preferred orientation (LPO) in olivine. *Tectonophysics*, **421**, 1–22.

Jurine, D., Jaupart, C., Brandeis, G., and Tackley, P. J. (2005). Penetration of mantle plumes through depleted lithosphere, *J. Geophys. Res.*, **110,** B10104, doi:10.1029/2005JB003751.

Kaban, M. K. and Schwintzer, P. (2001). Oceanic upper mantle structure from experimental scaling of V_s and density at different depths. *Geoph. J. Int.*, **147**, 199–214.

Kaban, M. K., Schwintzer, P., Artemieva, I. M., and Mooney, W. D. (2003). Density of the continental roots: compositional and thermal effects. *Earth Planet. Sci. Lett.*, **209**, 53–69.

Kalnins, L. M. and Watts, A. B. (2009). Spatial variations in effective elastic thickness in the Western Pacific Ocean and their implications for Mesozoic volcanism. *Earth Planet. Sci. Lett.*, **286**, 89–100.

Kameyama, M., Yuen, D., and Karato, S. (1999) Thermal-mechanical effects of low-temperature plasticity (the Peierls mechanism) on the deformation of a viscoelastic shear zone. *Earth Planet. Sci. Lett.*, **168**, 159–172.

Kaminski, E. (2002). The influence of water on the development of lattice preferred orientation in olivine aggregates, *Geophys. Res. Lett.*, **29**, doi:10.1029/2002GL014710.

Kampfmann, W. and Berckhemer, H. (1985). High temperature experiments on the elastic and anelastic behaviour of magmatic rocks. *Phys. Earth Planet. Int.*, **40**, 223–247.

Kanamori, H. (1970). The velocity and Q of mantle waves. *Phys. Earth Planet. Int.*, **2**, 259–270.

Karato, S.-I. (1986). Does partial melting reduce the creep strength of the upper mantle? *Nature* **319**, 309–310.

Karato, S.-I. (1990). The role of hydrogen in the electrical conductivity of the upper mantle. *Nature*, **357**: 272–273.

Karato, S.-I. (1992). On the Lehmann discontinuity. *Geophys. Res.Lett.* **19**, 2255–2258.

Karato, S.-I. (1993). Importance of anelasticity in the interpretation of seismic tomography. *Geophys. Res. Lett.*, **20**, 1623–1626.

Karato, S.-I. (1995). Effects of Water on Seismic Wave Velocities in the Upper Mantle. *Proc. Japan Acad.*, **71**, Ser. B, 61–67.

Karato, S.-I. (1998). A dislocation model of seismic wave attenuation and micro-creep in the earth: Harold Jeffreys and the rheology of the solid Earth. *Pure Appl. Geophys.* **153**(2–4): 239–256.

Karato, S.-I. (2003a). Mapping Water Content in the Upper Mantle. In: Eiler, J. E. (Ed.), *Inside the Subduction Factory*. Washington DC, American Geophysical Union.

Karato, S.-I. (2003b). *The Dynamic Structure of the Deep Earth: An Interdisciplinary Approach*. Princeton, NJ, Princeton Univ. Press, 241 pp.

Karato, S.-I. (2006). Remote sensing of hydrogen in Earth's mantle. *Rev. Mineral. Geochem.* **62**, 343–375.

Karato, S.-I. (2008). *Deformation of Earth Materials: Introduction to the Rheology of the Solid Earth*. Cambridge, Cambridge University Press, 463 pp.

Karato, S.-I. (2010). Rheology of the deep upper mantle and its implications for the preservation of the continental roots: a review. *Tectonophys.*, **481**, 82–98.

Karato, S.-I. and Jung, H. (1998). Water, partial melting and the origin of the seismic low velocity and high attenuation zone in the upper mantle. *Earth Planet. Sci. Lett.*, **157**, 193–207.

Karato, S.-I. and Jung, H. (2003). Effects of pressure on high-temperature dislocation creep in olivine polycrystals. *Phil. Mag., A.*, **83**, 401–414.

Karato, S.-I. and Karki, B. B. (2001). Origin of lateral variation of seismic wave velocities and density in the deep mantle, *J. Geophys. Res.*, **106**, 21771–21783.

Karato, S.-I. and Rubie, D. C. (1997). Toward experimental study of plastic deformation under deep mantle conditions: a new multianvil sample assembly for deformation experiments under high pressures and temperatures. *J. Geophys. Res.*, **102**, 20111–20122.

Karato, S.-I. and Spetzler, H. A. (1990). Effect of microdynamics in minerals and solid-state mechanisms of seismic wave attenuation and velocity dispersion in the mantle. *Rev. Geophys.*, **28**, 399–421.

Karato, S.-I. and Wu, P. (1993). Rheology of the upper mantle: a synthesis. *Science,* **260**, 771–778.

Karato, S.-I., Paterson, M. S., and Fitz Gerald, J. D. (1986). Rheology of synthetic olivine aggregates: influence of grain-size and water. *J. Geophys. Res.*, **91**, 8151–8176.

Karato, S.-I., Riedel, M., and Yuen, D. (2001) Rheological structure and deformation of subducted slabs in the mantle transition zone: implications for mantle circulation and deep earthquakes. *Physi. Earth Planet. Interiors*, **127**: 83–108.

Karato, S. I., Jung, H., Katayama, I. and Skemer, P. (2008). Geodynamic significance of seismic anisotropy of the upper mantle: new insights from laboratory studies. *Annu. Rev. Earth Planet. Sci.*, **36**: 59–95.

Karki, B. B., Stixrude, L., and Wentzcovitch, R. M. (1995). High-pressure elastic properties of major materials of Earth's mantle from first principles. *Rev. Geophys.*, **39**, 507–534.

Katayama, I. and Karato, S.I. (2008). Low-temperature, high-stress deformation of olivine under water-saturated conditions. *Phys. Earth Planet. Inter.*, **168**, 125–133.

Katayama, I. and Korenaga, J. (2009). Is the African cratonic lithosphere wet or dry? In: Beccaluva, L., Bianchini, G., and Wilson, M. (Eds.), *Volcanism and Evolution of the African lithosphere*. Geological Society of America (GSA) Special Paper.

Katayama, I., Jung, H., and Karato, S. (2004). New type of olivine fabric at modest water content and low stress. *Geology,* **32**, 1045–48.

Katsura, T., Yamada, H., Nishikawa, O., *et al.* (2004). Olivine-wadsleyite transition in the system (Mg, Fe)(2)SiO4. *J. Geophys. Res.*, **109**, Article Number: B02209.

Kaufman, G. B. (1997). Victor Moritz Goldschmidt (1888–1947): a tribute to the founder of modern geochemistry on the fiftieth anniversary of his death. *Chem. Educat.* **2** (5), 1–26. doi:10.1007/s00897970143a.

Kaula, W. M. (1967). Geophysical applications of satellite determinations of the Earth's gravitational field. *Space Sci. Rev.*, **7**, 769–794.

Kawazoe, T., Karato, S., Otsuka, K., Jing, Z., and Mookherjee, M. (2009). Shear deformation of dry polycrystalline olivine under deep upper mantle conditions using a rotational Drickamer apparatus (RDA). *Physics Earth Planetary Interiors*, **174**, 128–137.

Kay, R. W. and Mahlburg Kay, S. (1993). Delamination and delamination magmatism, *Tectonophysics*, **219**, 177–189.

Kelemen, P. B. (1995). Genesis of high Mg# andesites and the continental crust. *Contrib. Mineral. Petrol.* **120**, 1–19.

Kelemen, P. B., Hart, S.R., and Bernstein, S. (1998). Silica enrichment in the continental upper mantle via melt/rock reaction. *Earth Planet. Sci. Lett.*, **164**, 387–406.

Kelemen, P. B. *et al.* (2003). Thermal structure in the mantle wedge beneath arcs. In: J. Eiler (Ed): *Inside the Subduction Factory*. AGU Geophys. Monogr., 293–311.

Kelemen, P. B., Kikawa, E., Miller, D. J., *et al.* (2004). Site 1272. In: Drilling mantle peridotite along the Mid-Atlantic Ridge from 14° to 16° N. *Proc. ODP, Init. Repts.*, **209**: College Station, TX (Ocean Drilling Program), 1–134. doi:10.2973/odp.proc. ir.209.107.2004.

Kellett, R. L., Lilley, F. E. M., and White, A. (1991). A 2-dimensional interpretation of the geomagnetic coast effect of southeast Australia, observed on land and sea-floor. *Tectonophysics*, **192**, 367–382.

Kelley, K. A., T. Plank, T. L. Grove, *et al.* (2006), Mantle melting as a function of water content beneath back-arc basins, *J. Geophys. Res.*, **111**, B09208, doi:10.1029/2005JB003732

Kelly, R. K., Kelemen, P. B., and Jull, M. (2003). Buoyancy of the continental upper mantle, *Geochem. Geophys. Geosyst.*, **4**(2), 1017, doi:10.1029/2002GC000399.

Kelvin, Lord (W. Thomson) (1904). *Baltimore Lectures*, New York, Cambridge University Press.

Kemp, A. I. S. and Hawkesworth, C. J. (2003). Granitic Perspectives on the Generation and Secular Evolution of the Continental Crust. In: Rudnick, R.L. (Ed.) *Treatise on Geochemistry, v. 3, The Crust*, 349–410. Oxford, Elsevier-Pergamon.

Kendall, R., Mitrovica, J. X., and Sabadini, R. (2003). Lithospheric thickness inferred from Australian post-glacial sea-level change: The influence of a ductile crustal zone, *Geophys. Res. Lett.*, **30**(9), 1461, doi:10.1029/ 2003GL017022.

Kennedy, C. S. and Kennedy, G. C. (1976). The equilibrium boundary between graphite and diamond. *J. Geophys. Res.*, **81**, 2467–2470.

Kennett, B. L. N. (2006). On seismological reference models and the perceived nature of heterogeneity. *Phys. Earth Planet. Inter.*, **159**, 129–139.

Kennett, B. L. N. (2003). Seismic structure in the mantle beneath Australia. In: Hillis, R., Müller, D. (Eds.), *The Evolution and Dynamics of the Australian Plate.* Geological Societies of Australia and America, Spec. Publ., **22**, 7–23.

Kennett, B. L. N. (1995). Approximations for surface-wave propagation in laterally varying media. *Geophys. J. Int.* **122**, 470–478.

Kennett, B. L. N. and Engdahl, E. R. (1991). Traveltimes for global earthquake location and estimation. *Geophys. J. Int.* **105**, 429–465.

Kennett, B. L. N., Fishwick, S., and Heintz, M. (2004). Lithospheric structure in the Australian region – a synthesis of surface wave and body wave studies. *Expl. Geophys.* **35** (4), 242–250.

Kennett, B. L. N., Engdahl, E. R., and Buland, R. (1995). Constraints on seismic velocities in the Earth from traveltimes: *Geophys. J. Int.*, **122**, 108–124.

Kennett, B. L. N., Gudmundsson, O., and Tong, C. (1994). The upper mantle S and P velocity structure beneath northern Australia from broad-band observations. *Phys. Earth Planet. Int.*, **86**, 85–98.

Keppler, H., Wiedenbeck, M., and Shcheka, S. S. (2003). Carbon solubility in olivine and the mode of carbon storage in the Earth's mantle. *Nature*, **424**, 414–416.

Kern, H. and Siegesmund, S. (1989). A test of the relationship between seismic velocity and heat production for crustal rocks. *Earth Sci. Planet. Lett.*, **92**, 89–94.

Kern, H. (1993). P-wave and s-wave anisotropy and shear-wave splitting at pressure and temperature in possible mantle rocks and their relation to the rock fabric. *Phys. Earth Planet. Inter.*, **78**, 245–256.

Kern, H. (1978). The effect of high temperature and high confining pressure on compositional wave velocities in quartz-bearing and quartz-free igneous and metamorphic rocks. *Tectonophys.*, **44**, 185–203.

Ketcham, R. A. (1996). Distribution of heat-producing elements in the upper and middle crust of southern and west central Arizona: evidence from the core complexes. *J. Geophys. Res.*, **101**, 13611–13632.

Kind, R; Yuan, X; Saul, J., *et al.* (2002). Seismic images of crust and upper mantle beneath Tibet: evidence for Eurasian plate subduction. *Science*, **298**, 1219–1221.

King, T. A. (1998). Mechanisms of isostatic compensation in areas of lithospheric extension: examples from the Aegean. PhD thesis, University of Leeds, 1998.

King, S. D. and Anderson, D. L. (1995). An alternative mechanism of flood basalt formation. *Earth Planet. Sci. Lett.*, **136**: 269–279.

Kinzler, R., Grove, T. (1992). Primary magmas of midocean ridge basalts. *J. Geophys. Res.*, **97**, 6885–6906, 6907–6926.

Kirby, J. F. (2005), Which wavelet best reproduces the Fourier power spectrum?, *Comput. Geosci.*, **31**(7), 846–864.

Kirby, S. (1995). Intraslab earthquakes and phase changes in subducting lithosphere. *Rev. Geophys.* **33**(suppl.), 287–297.

Kirby, S. H. (1983). Rheology of the lithosphere. *Rev. Geophys.*, **21**, 1458–1487.

Kirby, S. H. (1980). Tectonic stresses in the lithosphere: constraints provided by the experimental deformation of rocks. *J. Geophys. Res.*, **85**, 6353–6363.

Kirby, J. F., and Swain, C. J. (2009). A reassessment of spectral Te estimation in continental interiors: the case of North America. *J. Geophys. Res.* **114**, B08401. doi:10.1029/2009JB006356.

Kirby, J. F., and Swain, C. J. (2006). Mapping the mechanical anisotropy of the lithosphere using a 2D wavelet coherence, and its application to Australia. *Phys. Earth Planet. Inter.*, **158**, 122–138.

Kirschvink, J. L. (1992). Late Proterozoic low-latitude global glaciation: the Snowball Earth. In: Schopf, J. W. and Klein, C. (Eds.), *The Proterozoic Biosphere: A Multidisciplinary Study*, Cambridge, Cambridge University Press, 51–52.

Kissling, E. (1993). Deep structure of the Alps – what do we really know? *Phys. Earth Planet. Inter.*, **79**, 87–112.

Kjarsgaard, B. A. (1996). Searching for diamonds in Canada. *Geological Survey of Canada*, Openfile 3228 LeCheminant, A. N., Dilabio, R. N. W. and Richardson, K. A. (Eds.), 55–60.

Klein, E. M. and Langmuir, C. H. (1987). Global correlations of ocean ridge basalt chemistry with axial depth and crustal thickness. *J. Geophys. Res.* **92**, 8089–8115.

Klemperer, S. L., Hauge, T. A., Hauser, E. C., *et al.* (1986). The Moho in the northern Basin and Range province, Nevada, along the COCORP 40° N seismic reflection transect. *GSA Bull.*, **97**, 603–618.

Knapp, J. H., Steer, D. N., Brown, L. D., *et al.* (1996). Lithosphere-scale seismic image of the Southern Urals from explosion-source reflection profiling. *Science*, **274**, 226–228.

Knoche, R., Webb, S., and Rubie, D. C. (1998). Measurements of acoustic wave velocities at P-T conditions of the Earth's mantle. In: Manghnani, M. and Yagi, T. (Eds.), *Properties of Earth and Planetary Materials at High Pressure and Temperature.* AGU Geophys. Monogr., v. **101**, 119–128.

Kobayashi, R. and Zhao, D. (2004). Rayleigh-wave group velocity distribution in the Antarctic region. *Earth Planet. Sci. Lett.*, **141**, 167–181.

Koga, K., Hauri, E., Hirschmann, M., and Bell, D. R. (2003). Hydrogen concentration analyses using SIMS and FTIR: comparision and calibration for nominally anhydrous minerals. *Geochemistry, Geophysics, and Geosystems*, **4**, doi:10.1029/2002GC000378.

Kogan, M. G. and McNutt, M. (1986). Isostasy in the USSR. Admittance data. In: Fuchs, K. and Froidevaux, C. (Eds.), *Composition, Structure, and Dynamics of the Lithosphere–Asthenosphere System.* AGU. Washington.

Kogan, M. G., Magnitskaya, E. I., and Chemova, N. (1987). Cross-spectral method of the investigation of isostasy. *Izvestia Akad. Nauk SSSR, Fizika Zemli*, **11**, 71–84.

Kohler, T. and Brey, G. P. (1990). Calcium exchange between olivine and clinopyroxene calibrated as a geothermobarometer for natural peridotites from 2–60 kb with applications. *Geochim. Cosmochim. Acta,* **54**, 2375–2388.

Kohlstedt, D. L., Evans, B., and Mackwell, S. J. (1995). Strength of the lithosphere: constraints imposed by laboratory experiments. *J. Geophys. Res.*, **100**, 17587–17602.

Kohlstedt, D. L. and Mackwell, S. J. (1998). Diffusion of hydrogen and intrinsic point defects in olivine. *Z. Phys. Chem.*, **207**, 147–162.

Kohlstedt, D. L. and Zimmerman, M. E. (1996). Rheology of partially molten mantle rocks. *Annu. Rev. Earth Planet. Sci.* **24**, 41–62.

Kola, Superdeep (1998). Scientific Results and Research Experience. Orlov, V. P. and Laverov, N. P. (Eds.), Moscow, Technoneftegaz, 260 pp. (in Russian).

Kola, Superdeep (1984). Studies of the Deep Structure of the Crust by Deep Drilling of the Kola Superdeep Drillhole. Kozlovsky, E. A. (Ed.), Moscow, Nedra, 1984, 490 pp. (in Russian).

Kopylova, M. G., and Caro, G. (2004). Mantle xenoliths from the Southeastern Slave craton: The evidence for a thick cold stratified lithosphere. *J. Petrol.* **45** (5), 1045–1067.

Kopylova, M. G. and McCammon, C. (2003). Composition and the Redox State of the Slave Peridotitic Mantle. *Proc. 8th Intern. Kimberlite Conf.*, Victoria, BC, Canada, June 2003: FLA-0195.

Kopylova, M. G. and Russell, J. K. (2000). Chemical stratification of cratonic lithosphere: Constraints from the northern Slave craton, Canada. *Earth Planet. Sci. Lett.*, **181**: 71–87.

Kopylova, M. G., Lo, J., and Christensen, N. I. (2004). Petrological constraints on seismic properties of the Slave upper mantle (Northern Canada). *Lithos*, **77**, 493–510.

Kopylova, M. G., Russell, J. K., and Cookenboo, H. (1999). Petrology of peridotite and pyroxenite xenoliths from the Jericho kimberlite: implications for the thermal state of the mantle beneath the Slave Craton, Northern Canada. *J. Petrol*, **40**, 79–104.

Korenaga, J. (2006). Archean geodynamics and the thermal evolution of Earth. In: Benn, K., Mareschal, J.-C., and Condie, K. (Eds.), *Archean Geodynamics and Environments,* AGU Geophysical Monograph Series, **164**, 7–32.

Korenaga, T. and Korenaga, J. (2008). Subsidence of normal oceanic lithosphere, apparent thermal expansivity, and seafloor flattening. *Earth Planet. Sci. Lett.*, **268**, 41–51.

Korenaga, J., and Karato, S. (2008). A new analysis of experimental data on olivine rheology, *J. Geophys. Res.*, **113**, B02403, doi:10.1029/2007JB005100.

Korenaga, J. and Jordan, T. (2002). On the state of sublithospheric upper mantle beneath a supercontinent. *Geophys. J. Int.* **149**, 179–189.

Korenaga, J. and Kelemen, P. B. (2000), Major element heterogeneity in the mantle source of the North Atlantic igneous province: *Earth Planet. Sci. Lett.*, **184**, 251–268.

Korja, T. (2007). How is the European lithosphere imaged by magnetotellurics? *Surv. Geophys.*, **28**: 239–272.

Korja, T. (1993). Electrical conductivity distribution of the lithosphere in the central Fennoscandian shield. *Precambrian Res.*, **64**, 85–108.

Korja, T., Engels, M., Zhamaletdinov, A. A., *et al.* and the BEAR Working Group (2002). Crustal conductivity in Fennoscandia – a compilation of a database on crustal conductance in the Fennoscandian Shield. *Earth Planets Space*, **54**: 535–558.

Kosarev, G. L., Petersen, N. V., Vinnik, L. P., and Roecker, S. W. (1993). Receiver functions for the Tien Shan analog broadband network: Contrasts in the evolution of structure across Talasso-Fergano Fault, *J. Geophys. Res.*, **98**, 4437–4448.

Kovalenko, V., Yarmoluk, V., Ionov, D., *et al.* (1990). Mantle evolution in central-Asia and development of tectonic structures of the earth's crust. *Geotectonics*, **24**(4), 283–292.

Kramers, J. D. (2007). Hierarchical Earth accretion and the Hadean Eon. *J. Geological Soc.*, **164**, 3–17

Kranz, R. L. (1983). Microcracks in rocks: a review. *Tectonophysics*, **100**, 449–480.

Krige, L. J. (1939). Borehole temperatures in Transvaal and Orange Free State. *Proc. Roy. Soc.*, **173**, 450–474.

Krogh, E. J. (1988). The garnet-clinopyroxene Fe–Mg geothermometer – a reinterpretation of existing experimental data. *Contrib. Mineral. Petrol.* **99**, 44–48.

Kröner, A. (1991). Tectonic evolution in the Archaean and Proterozoic. *Tectonophys.*, **187**: 393–410.

Kröner, A. (1981). In: *Precambrian Plate Tectonics*, Kröner, A. (Ed.) Elsevier, Amsterdam, 57–90.

Kröner, A., Widley, B. F., Badarch, G., *et al.* (2006). Accretionary growth and crustal formation in the Central Asian Orogenic Belt and comparison with the Arabian-Nubian shield. In: Hatcher, R. D., Carlson, M. P., McBride, J. H., *et al.* (Eds.), *4-D Framework of Continental Crust.* Geological Society of America Memoirs, **200**, 181–209.

Kuge, K. and Fukao, Y. (2005). High-velocity lid of East Antarctica: Evidence of a depleted continental lithosphere, *J. Geophys. Res.*, **110**, B06309, doi:10.1029/2004JB003382.

Kukkonen, I. T. and Joeleht, A. (1996). Geothermal modelling of the lithosphere in the central Baltic Shield and its southern slope. *Tectonophysics*, **255**, 24–45.

Kukkonen, I. T. and Lahtinen, R. (Eds.), (2006). *Finnish reflection experiment FIRE 2001–2005.* Geol. Survey of Finland, Sp. Paper 43, Espoo, 247 pp.

Kukkonen, I. T. and Peltonen, P. (1999). Xenolith-controlled geotherm for the central Fennoscandian Shield: implications for lithosphere-asthenosphere relations. *Tectonophysics*, **304**, 301–315.

Kukkonen, I. T. and Peltoniemi, S. (1998). Relationships between thermal and other petrophysical properties of rocks in Finland. *Phys. Chem. Earth*, **23**, 341–349.

Kukkonen, I. T., Kinnunen, K. A., and Peltonen, P. (2003). Mantle xenoliths and thick lithosphere in the Fennoscandian Shield. *Phys. Chem. Earth*, **28**, 349–360.

Kukkonen, I. T., Golovanova, I. V., Khachay, Yu. V., *et al.* (1997). Low geothermal heat flow of the Urals fold belt – implications of low heat production, fluid circulation or paleoclimate? *Tectonophysics*, **276**, 63–85.

Kumar, P., Yuan, X., Kind, R., and Kosarev, G. (2005). The lithosphere-asthenosphere boundary in the Tien Shan-Karakoram region from S receiver functions: Evidence for continental subduction. *Geophys. Res. Lett.*, **32**, Article Number: L07305.

Kumar, P; Kind, R; Priestley, K, *et al.* (2007). Crustal structure of Iceland and Greenland from receiver function studies. *J. Geophys. Res.*, **112**, Article Number: B03301.

Kumazawa, M. and O. L. Anderson (1969). Elastic moduli, pressure derivatives, and temperature derivatives of single crystal olivine and single crystal forsterite. *J. Geophys. Res.* **74**, 5961–5972.

Kurosawa, M., Yurimoto, H., and Sueno, S. (1997). Patterns in the hydrogen and trace element compositions of mantle olivines. *Physics Chemistry Minerals*, **24**, 385–395.

Kurtz, R. D., Craven, J. A., Niblett, E. R., and Stevens, R. A. (1993). The conductivity of the crust and mantle beneath the Kapuskasing Uplift: electrical anisotropy in the upper mantle. *Geophys. J. Int.*, **113**, 483–498.

Kushiro, I., Syono, Y., and Akimoto, S. (1968). Melting of a peridotite nodule at high pressures and high water pressures. *J. Geophys. Res.*, **73**: 6023–6029.

Kuskov, O. L., Kronrod, V. A., and Annersten, H. (2006). Inferring upper-mantle temperatures from seismic and geochemical constraints: Implications for Kaapvaal craton. *Earth Planet. Sci. Lett.*, **244**, 133–154.

Kusky, T. M. (1993). Collapse of Archean orogens and the generation of late- to postkinematic granitoids. *Geology*, **21**, 925–928.

Kusky, T. M., Li, J-H., and Tucker, R. D. (2001). The Archean Dongwanzi ophiolite complex, North China craton: 2.505-billion-year-old oceanic crust and mantle. *Science*, **292**, 1142–1145.

Kustowski, B., Ekstrom, G., and A. M. Dziewonski (2008a). Anisotropic shear-wave velocity structure of the Earth's mantle: a global model, *J. Geophys. Res.*, **113**, B06306, doi:10.1029/2007JB005169.

Kustowski, B., Ekstrom, G., and Dziewonski, A. M. (2008b). The shear-wave velocity structure in the upper mantle beneath Eurasia, *Geophys. J. Int.*, **174**, 978–992.

Kusznir, N. and Karner, G. (1985). Dependence of the flexural rigidity of the continental lithosphere on rheology and temperature. *Nature*, **316**, 138–142.

Kuznir, N. J. and Park, R. G. (1984). Intraplate lithosphere deformation and the strength of the lithosphere. *Geophys. J. Roy. Astron. Soc.*, **79**, 513–538.

Lacam, A. (1983). Pressure and composition dependence of the electrical conductivity of iron-rich synthetic olivines to 200 kbar. *Phys. Chem. Mineral.*, **9**, 127–132.

Lachenbruch, A. H. (1970). Crustal temperature and heat production: implications of the linear heat-flow relation. *J. Geophys. Res.*, **75**, 3291–3300.

Lachenbruch, A. H. and Morgan, P. (1990). Continental extension, magmatism and elevation; formal relations and rules of thumb. *Tectonophysics*, **174**, 39–62.

Lachenbruch, A. H. and Sass, J. H. (1980). Heat flow and energetics of the San Andreas fault zone, *J. Geophys. Res.*, **85**(B11), 6185–6222.

Lachenbruch, A. H. and Sass, J. H. (1978). Models of an extending lithosphere and heat flow in the basin and range province, *Mem. Geol. Soc. Am.*, **152**, 209–250.

Lachenbruch, A. H. and Sass, J. H. (1977). Heat flow in the United States and the thermal regime of the crust, In Heacock, J. G. (Ed.). *The Earth's Crust: Its Nature and Physical Properties*, Geophys. Monogr. Ser., vol. 20, pp. 626–675, AGU, Washington, DC.

Lachenbruch, A., Sass, J., and Morgan, P. (1994). Thermal regime of the southern Basin and Range Province: 2. Implications of heat flow for regional extension and metamorphic core complexes, *J. Geophys. Res.*, **99**(B11), 22121–22133.

Lakes, R. S. (1987). Negative Poisson's ratio materials, *Science*, **238**, 551.

Lambeck, K., Johnson, P., and Nakada, M. (1990). Holocene glacial rebound and sea level change in NW Europe, *Geophys. J. Int.*, **103**, 451–468.

Lambeck, K. (1990). Glacial rebound, sea-level change and mantle viscosity. *Geophys. J. Roy. Astron. Soc.*, **31**, 1–30.

Lambeck, K., Smither, C., and Johnston, P. (1998). Sea-level change, glacial rebound and mantle viscosity for northern Europe. *Geophys. J. Int.*, **134**, 102–144.

Lambert, I. B. and Wyllie, P. J. (1970). Low-velocity zone of the Earth's mantle; incipient melting caused by water. *Science*, **169**, 764–766.

Lamontagne, M., and Ranalli. G. (1996). Thermal and rheological constraints on the earthquake depth distribution in the Charlevoix, Canada, intraplate seismic zone. *Tectonophysics*, **257**, 55–69.

Landau, L. D. (1933). Über die Bewegung der Elektronen in Kristalgitter. *Phys. Z. Sowjetunion,* **3**: 644–645.

Langel, R. A. and Hinze, W. J. (1998). *The Magnetic Field of the Earth's Lithosphere: The Satellite Perspective.* Cambridge University Press, 429 pp.

Langseth, Jr. M. G., Le Pichon, X., and Ewing, M. (1966). Crustal structure of the mid-ocean ridges. 5. Heat flow through the Atlantic Ocean floor and convection currents, *J. Geophys. Res.* **71**, 5321–5355.

Langston, C. A. (1979). Structure under Mount Rainier, Washington, inferred from teleseismic body waves. *J. Geophys. Res.,* **84**, 4749–4762.

Laske, G. and Masters, G. (1996). Constraints on global phase velocity maps by long-period polarization data. *J. Geophys. Res.* **101**, 16059–16075.

Laske, G. and Masters, G. (1997). A global digital map of sediment thickness, *EOS Trans. AGU*, **78**: F483.

Lavier, L. L. and Steckler, M. S. (1997). The effect of sedimentary cover on the flexural strength of continental lithosphere. *Nature*, **389**, 476–479.

Lawrence, J. F. and Shearer, P. M. (2006). A global study of transition zone thickness using receiver functions. *J Geophys Res.,* **111**, B06307, doi:10.1029/2005JB003973.

Lay, T. and Wallace, T. C. (1995). *Modern Global Seismology.* Academic Press, London.

Le Pichon, X., Henry, P., and Goffe, B. (1997). Uplift of Tibet: From eclogites to granulites – Implications for the Andean Plateau and the Variscan belt, *Tectonophysics*, **273**, 57–76.

Lebedev, S. and Nolet, G. (2003). Upper mantle beneath Southeast Asia from S velocity tomography, *J. Geophys. Res.,* **108**, 2048, doi:10.1029/2000JB000073.

Lebedev, S., Boonen J., and Trampert J. (2009). Seismic structure of Precambrian lithosphere: New constraints from broad-band surface-wave dispersion, *Lithosphere*, **109**, 96–111, doi:10.1016/j.lithos.2008.06.010.

Lebedev, S., Meier, T., and van der Hilst, R. D. (2006). Asthenospheric flow and origin of volcanism in the Baikal Rift area. *Earth Planet. Sci. Lett.,* **249**, 415–424.

Ledo, J. and Jones, A. G. (2005). Upper mantle temperature determined from combining mineral composition, electrical conductivity laboratory studies and magnetotelluric field observations: Application to the intermontane belt, Northern Canadian Cordillera. *Earth Planet Sci. Lett.,* **236**: 258–268.

Ledo, J., Jones, A. G., and Ferguson, I. J. (2002). Electromagnetic images of a strike-slip fault: The Tintina fault – Northern Canadian, *Geophys. Res. Lett.,* **29**(8), 1225, doi:10.1029/2001GL013408.

Lee, C.-T. A. (2003). Compositional variation of density and seismic velocities in natural peridotites at STP conditions: Implications for seismic imaging of compositional heterogeneities in the upper mantle, *J. Geophys. Res.,* **108** (B9), 2441, doi:10.1029/2003JB002413.

Lee, C.-T. A. (2006). Geochemical/petrologic constraints on the origin of cratonic mantle. In: Benn, K., Maraschel, J-C., and Condie, K. C. (Eds.), *Archean Geodynamics and Environments.* American Geophysical Union, Geophysical Monograph Series, **164**.

Lee, H.-Y. and Ganguly, J. (1988). Equilibrium composition of coexisting garnet and orthopyroxene: experimental determinations in the system FeO–MgO–Al2O3–SiO2 and applications. *J. Petrol.* **29**, 93–113.

Lee, D.-C. and Halliday, A. N. (1995). Hafnium–tungsten chronometry and the timing of terrestrial core formation. Nature, **378**, 771–774.

Lee, C.-T. and Rudnick, R. L. (1999). Compositionally stratified cratonic lithosphere: petrology and geochemistry of peridotite xenoliths from the Labait tuff cone, Tanzania, in: J. J. Gurney, S. R. Richardson (Eds.), *Proc. 7th Int. Kimberlite Conference,* Red Roof Designs, Cape Town: 503–521.

Lee, W. H. K. and Uyeda, S. (1965). Review of heat flow data. In: W. H. K. Lee (Ed.), *Terrestrial Heat Flow.* AGU, *Geophys. Monograph,* **8**, 87–190.

Lee, C.-T., Lenardic, A., Cooper, C. M., Niu, F., and Levander, A. (2005). The role of chemical boundary layers in regulating the thickness of continental and oceanic thermal boundary layers. *Earth Planet. Sci. Lett.,* **230**, 379–395.

Lee, C.-T., Yin, Q., Rudnick, R. L., and Jacobsen, S. B. (2001). Preservation of ancient and fertile lithospheric mantle beneath the southwestern United States, *Nature,* **411**, 69–73.

LeFevre, L. V. and Helmberger, D. V. (1989). Upper mantle P-velocity structure of the Canadian Shield. *J. Geophys. Res.,* **94**, 17749–17765.

Lehmann, I. (1962). The travel times of the longitudinal waves of the Logan and Blanca atomic explosions and their velocities in the upper mantle. *Bull. Seism. Soc. Am.,* **52**, 519–526.

Lehmann, I. (1961). S and the structure of the upper mantle. *Geophys. J. Roy. Astron. Soc.,* **4**, 124–138.

Lehmann, I. (1959). Velocities of longitudinal waves in the upper part of the earth's mantle. *Ann. Geophys.,* **15**, 93–113.

Lehtonen, M., O'Brien, H. E., Peltonen, P., Johanson, B. S., and Pakkanen, L. (2004). Layered mantle at the Karelian Craton margin: P-T of mantle xenocrysts and xenoliths from the Kaavi–Kuopio kimberlites, Finland. *Lithos,* **77**, 593–608.

Leibecker, J., Gatzemeier, A.., Hönig, M., Kuras, O., and Soyer, W. (2002). Evidence of electrical anisotropic structures in the lower crust and the upper mantle beneath the Rhenish Shield. *Earth Planet. Sci. Lett.,* **202**, 289–302.

Leloup, P. H., Ricard, Y., Battaglia, J., and Lacassin, R. (1999). Shear heating in continental strike-slip shear zones: Model and field examples, *Geophys. J. Int.,* **136**(1), 19–40.

Lemoine, F. G., Pavlis, N. K., Kenyon, S. C., *et al.* (1998). New high-resolution model developed for Earth's gravitational field. *EOS Trans. AGU.,* **79**, 113–118.

Lenardic, A. (1998). On the partitioning of mantle heat loss below oceans and continents over time and its relationship to the Archaean paradox. *Geophys.J. Int.,* **134**, 706–720.

Lenardic, A. and Moresi, L. N. (1999). Some thoughts on the stability of cratonic lithosphere: effects of buoyancy and viscosity. *Jo. Geophys. Res.,* **104**, 12747–12759.

Lenardic, A. and Moresi, L. (2000). A new class of equilibrium geotherms in the deep thermal lithosphere of continents. *Earth Planet. Sci. Lett.,* **176**, 331–338.

Lenardic, A. and Moresi, L. (2001). Heat flow scaling for mantle convection below a conducting lid: Resolving seemingly inconsistent modeling results regarding continental heat flow. *Geophys. Res. Lett.,* **28**, 1311–1314.

Lenardic, A., Moresi, L. N., and Muhlhaus, H. (2003). Longevity and stability of cratonic lithosphere: Insights from numerical simulations of coupled mantle convection and continental tectonics. *J. Geophys. Res.*, **108**, B6, 2303, doi: 10.1029/2002JB001859.

Lenardic, A., Guillou-Frottier L., Mareschal, J.-C., *et al.* (2000). What the mantle sees: The effects of continents on mantle heat flow. *AGU, Geophys. Monogr.*, **121**, 95–112.

Lerner-Lam, A. L. and Jordan, T. H. (1987). How thick are the continents? *J. Geophys. Res.*, **92**, 14007–14026.

Lévêque, J. J., Debayle, E., and Maupin, V. (1998). Anisotropy in the Indian Ocean upper mantle from Rayleigh- and Love-waveform inversion. *Geophys. J. Int.*, **133**, 529–540.

Lévêque, J.-J., Rivera, L., and Wittlinger, G. (1993). On the use of checker-board test to assess the resolution of tomographic inversions, *Geophys. J. Int.*, **115**, 313–318.

Levin, V. and Park, J. (1998). P-SH conversions in layered media with hexagonal symmetric anisotropy: a cook book. *Pure Appl. Geophys*, **151**, 669–97.

Levy, F., Jaupart, C., Mareschal, J.-C., Bienfait, G., and Limare, A. (2010). Low heat flux and large variations of lithospheric thickness in the Canadian Shield. *J. Geophys. Res.*, **115**, B06404.

Lewis, B. T. R. and Meyer, R. P. I. (1968). A seismic investigation of the upper mantle to the west of Lake Superior. *Bull. Seismol. Soc.Am.* **58**, 565–596.

Li, X.-D. and Romanowicz, B. (1996). Global mantle shear-velocity model using nonlinear asymptotic coupling theory, *J. Geophys.Res.*, **101**, 22,245–22,272.

Li, A., Forsyth, D. W., and Fischer, K. M. (2003a). Shear velocity structure and azimuthal anisotropy beneath eastern North America from Rayleigh wave inversion, *J. Geophys. Res.* **108**, 2362.

Li, L. *et al.* (2006). Deformation of olivine at mantle pressure using the D-DIA. *European J. Mineral.*, **18**: 7–19.

Li, L., Weidner, D. J., Raterron, P., Chen, J., and Vaughan, M. T. (2004). Stress measurements of deforming olivine at high pressure. *Phys. Earth Planet. Inter.*, **143**/144, 357–367.

Li, S., Unsworth, M. J., Booker, J. R., Wei, W., Tan, H., and Jones, A. G. (2003c). Partial melt or aqueous fluids in the Tibetan crust: constraints from INDEPTH magnetotelluric data, *Geophys. J. Int.*, **153**, 289–304.

Li, X., Bock, G., Vafidis, A., *et al.* (2003b). Receiver function study of the Hellenic subduction zone: imaging crustal thickness variations and the oceanic Moho of the descending African lithosphere. *Geophys. J. Int.*, **155**, 733–748.

Li, Z.-X. A., Lee, C.-T. A., Peslier, A. H., Lenardic, A., and Mackwell, S. J. (2008). Water contents in mantle xenoliths from the Colorado Plateau and vicinity: Implications for the mantle rheology and hydration-induced thinning of continental lithosphere. *J. Geophys. Res.*, **113**: 10.1029/2007JB005540.

Lifshitz, I. M. (1963). On the theory of diffusion-viscous flow of polycrystalline bodies, *Sov. Phys. JETP*, **17**, 909–913.

Lilley, F. E. M., Wang, L. J., Chamalaun, F. H., and Ferguson, I. J. (2003). Carpentaria electrical conductivity anomaly, Queensland, as a major structure in the Australian plate. In: Hillis, R. R. and Muller, R. D. (Eds.), *Evolution and Dynamics of the Australian Plate*. Geol. Soc. Austral. Spec. Publ., **22**, 141–156.

Lippitsch, R., Kissling, E., and Ansorge, J. (2003). Upper mantle structure beneath the Alpine orogen from high-resolution teleseismic tomography, *J. Geophys. Res.*, **108**(B8), 2376, doi:10.1029/ 2002JB002016.

Lister, C. R. B. (1972). On the thermal balance of a mid-ocean ridge, *Geophys J. Roy. Astron. Soc.*, **26**, 515–535.

Litasov, K. D., Ohtani, E., Sano, A., Suzuki, A., and Funakoshi, K. (2005). Wet subduction versus cold subduction, *Geophys. Res. Lett.*, **32**, L13312, doi:10.1029/2005GL022921.

Lithgow-Bertelloni C. and Richards, M. A. (1998). The dynamics of Cenozoic and Mesozoic plate motions. *Rev. Geophys.*, **36**, 27–78.

Lithgow-Bertelloni, C. and Silver, P. (1998). Dynamic topography, plate driving forces and the African superswell. *Nature*, **395**, 269–272.

Lizarralde, D., Chave, A., Hirth, G., and Schultz, A. (1995). Northeastern Pacific mantle conductivity profile from long-period magnetotelluric sounding using Hawaii-to-California submarine cable data, *J. Geophys. Res.*, **100**, 17837–17854.

Lockner, D. (1995). Rock failure. *A Handbook of Physical Constants*, AGU Reference Shelf 3, AGU, 127–147.

Loper, D. E. and Stacey, F. D. (1983). The dynamical and thermal structure of deep mantle plumes. *Phys. Earth Planet. Inter.*, **33**, 304–317.

Louden, K. E. (1980). The crustal and lithospheric thickness of the Philippine Sea as compared to the Pacific, *Earth Planet. Sci. Lett.*, **50**, 275–288.

Love, A. E. H. (1944). *A Treatise on the Mathematical Theory of Elasticity*, 4th edn., Dover, Mineola, NY.

Lowrie, W. (1997). *Fundamentals of Geophysics*. Cambridge University Press.

Lowry, A. R. and Smith, R. B. (1994). Flexural rigidity of the Basin and Range–Colorado Plateau–Rocky Mountain transition from coherence analysis of gravity and topography. *J. Geophys. Res.* **99** (B10), 20123–20140.

Lowry, A. R. and Smith, R. B. (1995). Strength and rheology of the western US Cordillera, *J. Geophys. Res.*, **100**, 17 947–17 963.

Ludden, J. and Hynes, A. (2000). The Lithoprobe Abitibi–Grenville transect: two billion years of crust formation and recycling in the Precambrian Shield of Canada. *Can. J. Earth Sci.*, **37**, 456–476.

Ludwig, W. F., Nafe, J. E., and Drake, C. L. (1970). Seismic refraction. In: A. E. Maxwell (Ed.), *The Sea. Ideas and Observations on Progress in the Study of the Seas*, v. 4, 53–84. New York, Wiley-Interscience.

Lundin, E. R. and Dore, A. G. (2005), Fixity of the Iceland "hotspot" on the Mid-Atlantic Ridge: Observational evidence, mechanisms, and implications for Atlantic volcanic margins. In: Foulger, G. R., Natland, J. H., Presnall, D. C., and Anderson, D. L. (Eds.), *Plates, Plumes, and Paradigms:* Geological Society Of America, Special Paper 388, Boulder, CO, USA, 627–652.

Lyngsie, S. B., Thybo, H., and Lang, R. (2007). Rifting and lower crustal reflectivity: A case study of the intracratonic Dniepr-Donets rift zone, Ukraine. *J. Geophys. Res.*, **112**, Issue: B12 Article Number: B12402.

Macario, A., Malinverno, A., and Haxby, W. F. (1995), On the robustness of elastic thickness estimates obtained using the coherence method, *J. Geophys. Res.*, **100**, 15,163–15,172.

MacDonald, G. J. F. (1963). The deep structure of continents. *Rev. Geophys.*, **1**, 587–665.

Macgregor, I. D. (1974). The system MgO-Al2O3-SiO2: Solubility of Al2O3 in enstatite for spinel and garnet peridotite compositions. *Am. Mineral.*, **59**, 110–119.

MacGregor, L. M., Constable, S., and Sinha, M. C. (1998). The RAMESSES experiment – III. Controlled-source electromagnetic sounding of the Reykjanes Ridge at 57 o 45'N. *Geophys. J. Int.*, **135**, 773–789.

Mackenzie, G. D., Thybo, H., and Maguire, P. K. H. (2005). Crustal velocity structure across the Main Ethiopian Rift: results from two-dimensional wide-angle seismic modelling. *Geophysical Journal International*, **162**, 994–1006.

Mackie, R. L., Madden, T. R., and Park, S. K. (1996). A three-dimensional magne-totelluric investigation of the California Basin and Range. *J. Geophys. Res.*, **101**, 16221–16239.

Mackwell, S. J. and D. L. Kohlstedt (1990). Diffusion of hydrogen in olivine: implications for water in the mantle, *J. Geophys. Res.* **95**, 5079–5088.

Mackwell, S. J., Kohlstedt, D. L., and Paterson, M. S. (1985). The role of water in the deformation of olivine single crystals. *J. Geophys. Res.*, **90**: 11319–11333.

Maggi, A., Jackson J. A., McKenzie, D., and Priestley, K. (2000). Earthquake focal depths, effective elastic thickness, and the strength of the continental lithosphere. *Geology*, **28**, 495–498.

Mahoney, J. J. and Coffin, M. R. (Eds.) (1997). *Large Igneous Provinces: Continental, Oceanic, and Planetary Flood Volcanism.* American Geophysical Union, Geophysical Monograph. **100**, Washington DC, USA, 438pp.

Mainprice, D. and Silver, P. G. (1993). Interpretation of SKS-waves using samples from the subcontinental lithosphere, *Phys. Earth Planet. Inter.*, **78**, 257–280.

Mainprice, D., Tommasi, A., Couvy, H., Cordier, P., and Frost, D. J. (2005). Pressure sensitivity of olivine slip systems and seismic anisotropy of Earth's upper mantle. *Nature*, **433**, 731–733.

Makeyeva, L. I., Vinnik, L. P., and Roecker, S. W. (1992). Shear-wave splitting and small-scale convection in the continental upper mantle, *Nature*, **358**, 144–146.

Malamud, B. D. and Turcotte, D. L. (1999). How many plumes are there? *Earth Planet. Sci. Lett.*, **174**, 113–124.

Malkovets, V. G., Taylor, L. A., Griffin, W., *et al.* (2003). Cratonic conditions beneath Arkhangelsk, Russia: garnet peridotites from the Grib Kimberlite. *Proc. 8th Intern. Kimberlite Conf.*, Victoria, BC, Canada, June 2003, FLA–0220.

Mandea, M. and Purucker, M. (2005). Observing, modeling, and interpreting magnetic fields of the solid Earth. *Surv. Geophys.*, **26**, 415–459.

Manley, C. R., Glazner, A. F., and Farmer, G. L. (2000). Timing of volcanism in the Sierra Nevada of California: evidence for Pliocene delamination of the batholitic root? *Geology*, **28**, 81–814.

Mann, P. and Taira, A. (2004). Global tecyonic significance of the Solomon Islands and the Ontong Java Plateau convergent zone. *Tectonophysics*, **389**, 137–190.

Manning, C. E., Mojzsis, S. J., and Harrison, T. M. (2006). Geology, age and origin of Supracrustal rocks at Akilia, West Greenland. *Am. J. Sci.*, **306**, 303–366, doi 10.2475/05.2006.02.

Mantovani, M., Shukowsky, W., Freitas, S. R. C., and Brito Neves, B. B. (2005). Lithosphere mechanical behavior inferred from tidal gravity anomalies: a comparison of Africa and South America. *Earth Planet. Sci. Lett.* 397–412.

Mareschal, J. C. (1983). Uplift and heat flow following the injection of magmas into the lithosphere. *Geophys. J. Roy. Astron. Soc.*, **73**, 109–127.

Mareschal, J. C., C. Jaupart, L. Z. Cheng, *et al.* (1999). Heat flow in the Trans-Hudson Orogen of the Canadian Shield: implications for Proterozoic continental growth, *J. Geophys. Res.*, **104**, 29,007–29,024.

Mareschal, M., Kellett, R. L., Kurtz, R. D., *et al.* (1995). Archaean cratonic roots, mantle shear zones and deep electrical anisotropy. *Nature*, **375**: 134–137.

Markowski, A., Quitte, G., Kleine, T., *et al.* (2007). Hafnium–tungsten chronometry of angrites and the earliest evolution of planetary objects. *Earth Planet. Sci. Lett.*, **262**, 214–229.

Marone, F. and Romanowicz, B. (2007). The depth distribution of azimuthal anisotropy in the continental upper mantle. *Nature*, **447**, 198–203.

Marquering, H. and Snieder, R. (1996). Shear-wave velocity structure beneath Europe, the northeastern Atlantic and western Asia from waveform inversions including surface wave mode coupling. *Geophys. J. Int.*, **124**, 283–304.

Marshak, S., Alkmim, F., and Jordt-Evangelista H. (1992). Proterozoic crustal extension and the generation of dome-and-keel structure in an Archaean granite–greenstone terrane. *Nature*, **357**, 491–493.

Martin, H. (1993). The mechanisms of petrogenesis of the Archean continental crust: comparison with modern processes. *Lithos*, **30**, 373–388.

Martin, C. E. (1991). Osmium isotopic characteristics of mantle-derived rocks. *Geochim. Cosmochim. Acta* **55**, 1421–1434.

Marty, J. C. and Cazenave, A. (1989). Regional variations in subsidence rate of oceanic plates: a global analysis. *Earth Planet. Sci. Lett.*, **94**, 301–315.

Masters, T. G. and Shearer, P. M. (1995). Seismic models of the Earth: Elastic and anelastic. In: *A Handbook of Physical Constants*. AGU Reference shelf 1, AGU, 88–103.

Masters, G., Johnson, S., Laske, G., and Bolton, H. (1996). A shear-velocity model of the mantle. *Phil. Trans. Roy. Soc. Lond.*, **A354**, 1385–1411.

Matsukage, K. N., Nishihara, Y., and Karato, S.-I., (2005). Seismological signature of chemical differentiation of Earth's upper mantle. *J. Geophys. Res.*, **110**, B12305, doi: 10.1029/2004JB003504, 18 pp.

Maupin, V. and Park, J. (2008). Seismic anisotropy: theory and observations. In: Romanowicz, B. and Dziewonski A. M. (Ed.), *Treatise in Geophysics*, Vol. 1, 289–321. Amsterdam, Elsevier.

Mavko, G. M. (1980). Velocity and attenuation in partially molten rocks. *J. Geophys. Res.*, **85**: 5173–5189.

Mavko, G., Mukerji, T., and Dvorkin, J. (1998). *The Rock Physics Handbook: Tools for Seismic Analysis in Porous Media*, Cambridge, UK, Cambridge University Press, 329 pp.

McAdoo, D., Martin, C. F., and Poulouse, S. (1985). Seasat observations of flexure: Evidence for a strong lithosphere. *Tectonophysics*, **116**, 209–222.

McConnell, R. K. (1968). Viscosity of the mantle from relaxation time spectra of isostatic adjustment, *J. Geophys. Res.*, **73**, 7089–7105.

McCulloch, M. T. and Bennett, V. C. (1998). Early differentiation of the Earth: an isotopic perspective. In: I. Jackson (Ed), *The Earth's Mantle: Composition, Structure, and Evolution*. Cambridge University Press, 127–158.

McCulloch, M. T. and Bennett, V. C. (1994). Progressive growth of the Earth's continental crust and depleted mantle: geochemical constraints. *Geochim. Cosmochim. Acta*, **58**(21), 4717–4738.

McCulloch, M. T. and Black, L. P. (1984). Sm–Nd isotopic systematics of Enderby Land granulites and evidence for the redistribution of Sm and Nd during metamorphism. *Earth Planet. Sci. Lett.*, **71**, 46–58.

McDonough, W. F. (1990). Constraints on the composition of the continental lithospheric mantle, *Earth Planet. Sci. Lett.*, **101**, 1–18.

McDonough, W. F. and Sun, S. S. (1995). The composition of the Earth. *Chem. Geol.*, **120**, 223–253.

McGreggor, I. D. and Manton, W. I. (1986). Roberts Victor Eclogites: ancient oceanic crust. *J. Geophys. Res.* **91**, 14063–14079.

McKenzie, D. (2003). Estimating Te in the presence of internal loads. *J. Geophys. Res.*, **108** (B9), 2438, doi:10.1029/2002JB001766.

McKenzie, D. (1989). Some remarks on the movement of small melt fractions in the mantle. *Earth Planet. Sci. Lett.*, **95**, 53.

McKenzie, D. (1985). The extraction of magma from the crust and mantle. *Earth Planet. Sci. Lett.*, **74**, 81.

McKenzie, D. (1978). Some remarks on the development of sedimentary basins, *Earth Planet. Sci. Lett.*, **40**, 25–32.

McKenzie, D. P. (1967). Sea-floor spreading, *J. Geophys. Res.* **72**, 6261–6273.

McKenzie, D. and Bickle, M. J. (1988). The volume of and composition of melt generated by extension of the lithosphere. *J. Petrology*, **29**, 625–679.

McKenzie, D. and Fairhead, D. (1997). Estimates of the effective elastic thickness of the continental lithosphere from Bouguer and free-air anomalies, *J. Geophys. Res.*, **102**, 27,523–27,552.

McKenzie, D., Jackson, J., and Priestley, K. (2005). Thermal structure of oceanic and continental lithosphere. *Earth Planet Sci. Lett.*, **233**, 337–349.

McKenzie, D. P., Roberts, J. M., and Weiss, N. O. (1974). Convection in the earth's mantle: towards a numerical simulation. *J. Fluid Mech.*, **62**, 465–538.

McKenzie, D., Stracke, A., *et al.* (2004). Source enrichment process responsible for isotopic anomalies in oceanic island basalts. *Geochim. Cosmochim. Acta*, 68, 2699–2724.

McLennan, S. M. and Taylor, S. R. (1985). *The Continental Crust: Its Composition and Evolution*. Oxford, Blackwell, 312 pp.

McLennan, S. M. and Taylor, S. R. (1996). Heat flow and the chemical composition of continental crust. *J. Geol.*, **104**, 377–396.

McLennan, S. M., and Taylor, S. R. (1999). Earth's Continental Crust. In: Marshall, C. P., and Fairbridge, R. W. (Eds.), *Encyclopedia of Geochemistry*, Dordrecht, Kluwer Academic Publishers, 712 pp.

McLennan, S. M. and Taylor, S. R. (1982). Geochemical constraints on the growth of the continental crust. *J. Geol.*, **90**, 347–361.

McNutt, M. (1998). Superswells, Rev. *Geophys.* **36**, 211–244.

McNutt, M. K. (1980). Implication of regional gravity for state of stress in the Earth's crust and upper mantle. *J. Geophys. Res.*, **B85**, 6377–6396.

McNutt, M. and Kogan, M. G. (1986). Isostasy in the USSR. 2. Interpretation of admittance data. In: Fuchs, K. and Froidevaux, C. (Eds.), *Composition, Structure, and Dynamics of the Lithosphere-Asthenosphere System*. AGU, Washington, DC.

McNutt, M. and Menard, H. W. (1982). Constraints on the yield strength in the oceanic lithosphere derived from observations of flexure. *Geophys. J. Roy. Astron. Soc.*, **59**, 4663–4678.

McNutt, M. K., Diament, M., and Kogan, M. G. (1988). Variations of elastic plate thickness at continental thrust belts. *J. Geophys. Res.* **93**, 8825–8838.

Mechie, J., Fuchs, K., and Altherr, R. (1994). The relationship between seismic velocity, mineral composition and temperature and pressure in the upper mantle – With an application to the Kenya Rift and its eastern flank, *Tectonophysics*, **236**, 453–464.

Mechie, J., Egorkin, A. V., Solodilov, L., *et al.* (1997). Major features of the mantle velocity structure beneath northern Eurasia from long-range seismic recordings of peaceful nuclear explosions. In: Fuchs, K. (Ed.), *Upper Mantle Heterogeneities From Active and Passive Seismology*. Kluwer Academic Publishers, 33–50.

Medford, L. V., Meloni, A., Lanzerotti, L. J., *et al.* (1981). Geomagnetic induction on a Transatlantic communications cable. *Nature*, **290**, 392–393.

Megnin, C. and Romanowicz, B. (2000). The three-dimensional shear velocity structure of the mantle from the inversion of body, surface and higher-mode waveforms. *Geophys. J. Int.*, **143**, 709–728.

Mei, S. and Kohlstedt, D. L. (2000a). Influence of water on plastic deformation of olivine aggregates, 1. Diffusion creep regime. *J. Geophys. Res.*, **105**: 21457–21469.

Mei, S. and Kohlstedt, D. L. (2000b). Influence of water on plastic deformation of olivine aggregates, 2. Dislocation creep regime. *J. Geophys. Res.*, **105**: 21471–21481.

Meibom, A., Sleep, N. H., Chamberlain, C. P. *et al.* (2002). Re–Os isotopic evidence for long-lived heterogeneity and equilibration processes in the Earth's upper mantle. *Nature*, **419**, 705–708.

Meissner, R. (1999). Terrane accumulation and collapse in central Europe: seismic and rheological constraints. *Tectonophysics*, **305**, 93–107.

Meissner, R. and Strehlau, J. (1982). Limits of stresses in the continental crust and their relation to depth-frequency distributions of shallow earthquakes, *Tectonics*. **1**, 73–89.

Meissner, R., Mooney, W. D., and Artemieva, I. M. (2002). Mantle escape inferred from seismic anisotropy in young continental orogens. *Geophys. J. Int.*, **149**, 1–14.

Menard, G. and Molnar, P. (1988). Collapse of a Hercynian Tibetan plateau into a late Paleozoic European Basin and range province. *Nature*, **334**, 235–237.

Mengel, K. and Kern, H. (1992). Evolution of the petrological and seismic Moho implications for the continental crust-mantle. boundary. *Terra Nova*, **4, 109–116.**

Menzies, M. A. and Cox, K. G. (eds) (1988). Oceanic and Continental Lithosphere: Similarities and Differences. *J. Petrology, Lithosphere*. Special issue, **29.**

Menzies, M. A. and Xu, Y. (1998). Geodynamics of the North China craton. In: Flower, M. *et al.* (Eds.), *Mantle Dynamics and Plate Interactions in East Asia*. AGU Geodynam. Ser., **27**, 155–165.

Menzies, A. H., Westerlund, K., Gurney, J. J., *et al.* (2003). Peridotitic Mantle Xenoliths from Kimberlites on the Ekati Diamond Mine Property, NWT, Canada. *Proc. 8th Int. Kimberlite Conf.*, Victoria, BC, Canada, June 2003: FLA–0305.

Mercier, J.-C. C. (1980). Magnitude of the continental lithospheric stresses inferred from rheomorphic petrology. *J. Geophys. Res.*, **85**, 6293–6303.

Mercier, J-C. C. (1985). Olivine and pyroxenes. In: Wenk, H. R. (Ed.) *Preferred Orientation in Deformed Metals and Rocks: an Introduction to Modern Texture Analysis*. New York: Academic Press, 407–430.

Meyer, J., Hufen, J. H., Siebert, M., and Hahn, A. (1985). On the identification of Magsat anomaly charts as crustal part of internal field, *J. Geophys. Res.*, **90**, 2537–2542.

Michaut, C. and Jaupart, C. (2004). Nonequilibrium temperatures and cooling rates in thick continental lithosphere. *Geophys. Res. Lett.*, **31**, L24602.

Mierdel, K., Keppler, H., Smyth, J. R., and Langenhorst, F. (2007). Water solubility in aluminous orthopyroxene and the origin of Earth's asthenosphere. *Science*, **315**, 364–368.

Milev, A. and Vinnik, L. P. (1991). Deformations in the continental mantle from the data of Global Digital Seismograph network. *Dokl. Akad. Nauk SSSR*, **318**, 1132–1136 (in Russian).

Milne, G. A., Davis, J. L., Mitrovica, J. X., *et al.* (2001). Space-geodetic constraints on glacial isostatic adjustment in Fennoscandia, *Science*, **291**, 2381–2385.

Milne, G. A., Mitrovica, J. X., Scherneck, H.-G., *et al.* (2004). Continuous GPS measurements of postglacial adjustment in Fennoscandia. 2. Modelling results, *J. Geophys. Res.* **109**, DOI 10.1029/2003JB002619.

Minster, J. B. and Anderson, D. L. (1981). A model of dislocation-controlled rheology for the mantle. *Phil. Trans. Roy. Soc. Lond.*, **299**, 319–356.

Mitchell, B. J. (1995). Anelastic structure and evolution of the continental crust and upper mantle from seismic surface wave attenuation. *Rev. Geophys.*, **33**, 441–462.

Mitchell, B. J., Cong, L. L., and Ekstroem, G. (2008). A continent-wide map of 1-Hz Lg coda Q variation across Eurasia and its relation to lithospheric evolution. *J. Geophys. Res.*, **113**(B4), Article Number: B04303.

Mitrovica, J. X. and Peltier, W. R. (1993). The inference of mantle viscosity from an inversion of the Fennoscandian relaxation spectrum. *Geophys. J. Int.*, **114**, 45–62.

Mizukami, T., Simon, W., and Yamamoto, J. (2004). Natural examples of olivine lattice preferred orientation patterns with a flow-normal a-axis maximum. *Nature*, **427**, 432–436.

Mojzsis, S. J., Harrison, T. M., and Pidgeon, R. T. (2001). Oxygen-isotope evidence from ancient zircons for liquid water at the Earth's surface 4,300 Myr ago. *Nature*, **409**, 178–181.

Molnar, P. (1988), A review of geophysical constraints on the deep structure of the Tibetan Plateau, the Himalayas and the Karakorum and their tectonic implications, *Phil. Trans. R. Soc. Lond, Ser. A*, **326**, 33–88.

Molnar, P. and Houseman, G. A. (2004). The effects of buoyant crust on the gravitational instability of thickened mantle lithosphere at zones of intracontinental convergence. *Geophys. J. Int.*, **158**, 1134–1150.

Molnar, P. H. and Tapponier, P. (1975). Cenozoic tectonics of Asia: effects of continental collision, *Science*, **189**, 144–154.

Molnar, P., Houseman, G. A., and Conrad, C. P. (1998). Rayleigh–Taylor instability and convective thinning of mechanically thickened lithosphere: effects of non-linear viscosity decreasing exponentially with depth and of horizontal shortening of the layer. *Geophys. J. Int.*, **133**, 568–584.

MONA LISA Working Group (1997). MONA LISA – Deep seismic investigations of the lithosphere in the southeastern North Sea. *Tectonophysics*, **269**, 1–19.

Montagner, J.-P., (1994). Can seismology tell us anything about convection in the mantle? *Rev. Geophys.*, **32**, 115–137.

Montagner, J.-P. and Anderson, D. L. (1989). Constrained mantle reference model. *Phys. Earth Planet. Inter.*, **58**, 205–227.

Montagner, J.-P. and Guillot, L. (2002). Seismic anisotropy and global geodynamics. In: Karato, S. and Wenk, H.-R. (Eds.), *Plastic Deformation of Minerals and Rocks*. Washington, DC: MSA, 353–385.

Montagner, J.-P. and Tanimoto, T. (1990). Global anisotropy in the upper mantle inferred from the regionalization of phase velocities. *J. Geophys. Res.*, **95**, 4797–819.

Montagner, J.-P. and Tanimoto, T. (1991). Global upper mantle tomography of seismic waves and anisotropies. *J. Geophys. Res.*, **96**, 20337–20351.

Montagner, J.-P., Griot-Pommera, D.-A., and Lavé, J. (2000). How to relate body wave and surface wave anisotropy? *J. Geophys. Res.*, **105**, 19015–19027.

Montelli, R., Nolet, G., Dahlen, F. A., *et al.* (2004). Finite-frequency tomography reveals a variety of plumes in the mantle. *Science*, **303**, 338–343.

Mooney, W. D. and Brocher, T. M. (1987). Coincident seismic reflection/refraction studies of the continental lithosphere: a global review. *Rev. Geophys.*, **25**, 723–742.

Mooney, W. D. and Meissner, R. (1992). Multi-genetic origin of crustal reflectivity: a review of seismic reflection profiling of the continental lower crust and Moho. In: Fountain, D. M. *et al.* (Eds.): *Continental Lower Crust*, Amsterdam, Elsevier, 45–79.

Mooney, W. D., Laske, G., Masters, T. G. (1998). CRUST 5.1: A global crustal model at 5°x5°. *J. Geophys. Res.*, **103**, 727–747.

Mooney, W. D., Prodehl, C., and Pavlenkova, N. I. (2002). Seismic velocity structure of the continental lithosphere from controlled source data. *International Handbook of Earthquake and Engineering Seismology*, v. **81A**, 887–910.

Moorbath, S., Whitehouse, M. J., and Kamber, B. S. (1997). Extreme Nd-isotope heterogeneity in the early Archaean – Fact or fiction? Case histories from northern Canada and West Greenland. *Chem. Geol.*, **135**, 213–231.

Morelli, A. and Danesi, S. (2004). Seismological imaging of the Antarctic continental lithosphere: a review. *Global Planet. Change*, **42**, 155–165.

Morelli, R., Nolet, G., Masters, G., Dahlen, F. A., and Hung, S. H. (2004). Finite-frequency tomography reveals a variety of plumes in the mantle. *Science*, **303**, 338–343.

Morgan, P. (1984). The thermal structure and thermal evolution of the continental lithosphere. *Phys. Chem. Earth,* **15**, 107–193.

Morgan, P. (1985). Crustal radiogenic heat production and the selective survival of ancient continental crust. *J. Geophys. Res.*, **90**, C561–C570.

Morgan, P. (2001). Heat flow in the Earth. *The Earth Science Encyclopedia Online*, Springer.

Morgan, P. and Sass, J. H. (1984). Thermal regime of the continental lithosphere. *J. Geodynamics*, **1**, 143–166.

Morgan, J. P. and Smith, W. H. F. (1992). Flattening of the sea-floor depth age curve as a response to asthenospheric flow. *Nature*, **359**, 524–527.

Morgan, W. J. (1972). Plate motions and deep convection. *Geol. Soc. Am. Memoir,* **132**, 7–22.

Morgan, W. J. (1983). Hotspot tracks and the early rifting of the Atlantic. *Tectonophysics*, **94**, 123–139.

Moroz, Y. F. and Pospeev, A. V. (1995). Deep electrical conductivity of East Siberia and the Far East of Russia. *Tectonophysics*, **245**, 85–92.

Morozov, I. B. and Smithson, S. B. (2000). Coda of long-range arrivals from nuclear explosions. *Bull. Seism. Soc. Am.*, **90**, 929–939.

Morozova, E. A., Morozov, I. B., Smithson, S. B., and Solodilov, L. N. (1999). Heterogeneity of the uppermost mantle beneath Russian Eurasia from the ultra-long-range profile QUARTZ. *J. Geophys. Res.* **104**, 20329–20348.

Mott, N. F. and Gurney, R. W. (1948). *Electronic Processes in Ionic Crystals. Part 2*. Oxford University Press.

Mueller, H. J. and Massonne, H. J. (2001). Experimental high pressure investigation of partial melting in natural rocks and their influence om Vp and Vs. *Phys. Chem. Earth (A)*, **26**, 325–332.

Mukherjee, A. and Das, S. (2002). Anorthosites, granulites and the supercontinent cycle. *Gondwana Res.*, **5**, 147–156.

Muller, R. D., Sdrolias, M., Gaina, C., and Roest, W. R. (2008). Age, spreading rates and spreading symmetry of the world's ocean crust, *Geochem. Geophys. Geosyst.*, **9**, Q04006, doi:10.1029/2007GC001743.

Munoz, G., Heise, W., Paz, C., *et al.* (2005). New magnetotelluric data through the boundary between the Ossa Morena and Centroiberian Zones. *Geol. Acta*, **3**(3), 215–223.

Murase, T. and Fukuyama, H. (1980). Shear wave velocity in partially molten peridotite at high pressures. *Year Book Carnegie Institute*, Washington, DC, **79**, 307–310.

Murase, T. and Kushiro, I. (1979). Compressional wave velocity in partially molten peridotite at high pressures. *Year Book Carnegie Institute*, Washington, DC, **78**, 559–562.

Murase, T., Kushiro, I., and Fujii, T. (1977). Electrical conductivity of partially molten peridotite. *Year Book Carnegie Institute*, Washington, DC, **76**, 416–419.

Murase, T. Kushiro, I., and Fujii, T. (1977). Compressional wave velocity in partially molten peridotite. *Annual Report of the Director Geophysical Laboratory, Carnegie Inst.*, Washington D.C., 414–416.

Musgrave, M. J. P. (1959). Propagation of elastic waves in crystals and anisotropic media. *Rep. Progr. Phys.*, **22**, 74–96.

Nabarro, F. R. N. (1967). Steady state diffusional creep, *Phil. Mag.*, **16**, p. 231.

Nabarro, F. R. N. (1948). Deformation of crystals by the motion of single ions, In: *Strength of Solids*, London, The Physical Society, 75.

Nafe, J. E. and Drake, C. L. (1957). Variation with depth in shallow and deep water marine sediments of porosity, density and the velocities of compressional and shear waves. *Geophysics*, **22**, 523–552.

Nagihara, S., Lister, C. R. B., and Sclater, J. G. (1996). Reheating of old oceanic lithosphere: Deduction from observations. *Earth Planet. Sci. Lett.*, **39**, 91–104.

Nair, S. K., Gao, S. S., Liu, K. H., and Silver, P. G. (2006). Southern African crustal evolution and composition: Constraints from receiver function studies, *J. Geophys. Res.*, **111**, B02304, doi:10.1029/2005JB003802.

Nakajima, J. and Hasegawa, A. (2003). Estimation of thermal structure in the mantle wedge of northeatsren Japan from seismic attenuation data. *Geophys. Res. Lett.*, **30** (14), 1760, doi:10.1029/2003GL017185.

Nakamura, A. and Schmalzried, H. (1983). On the nonstoichiometry and point defects of olivine, *Phys. Chem. Miner.*, **10**,27–37.

Nakanishi, I. (1978). Regional difference in the phase velocity and the quality factor Q of mantle Rayleigh waves. *Science*, **200**, 1379–1381.

Nataf, H. C. (2000). Seismic imaging of mantle plumes. *Ann. Rev. Earth Planet. Sci.*, **28**: 391–417.

Nataf, H.-C. and Ricard, Y. (1996). 3SMAC: An a priori tomographic model of the upper mantle based on geophysical modeling. *Phys. Earth Planet. Inter.*, **95**, p. 101–122.

Nell, J. and Wood, B. J. (1991). High-temperature electrical measurements and thermodynamic properties of Fe3O4-FeCr2O4-MgCr2O4-FeAl2O4 spinels. *Am. Mineralogist*, **76**, 405–426.

Neal, S. L., Mackie, R. L., Larsen, J. C., and Schultz, A. (2000). Variations in the electrical conductivity of the upper mantle beneath North America and the Pacific Ocean. *J. Geophys. Res.* **105**: 8229–8242.

Neal, C. R., Mahoney, J. J., Kroenke, L. W., Duncan, R. A., and Petterson, M. G. (1997). The Ontong Java plateau. In: Mahoney, J. J. and Coffin, M. F. (Eds.), *Large Igneous Provinces: Continental, Oceanic, and Planetary Flood Volcanism*. American Geophysical Union, Geophysical Monograph Series, **100**, 183–216.

Nelson, B. K. and DePaolo, D. J. (1985). Rapid production of continental-crust 1.7 to 1.9 b.y. ago – Nd isotopic evidence from the basement of the North-American mid-continent. *Geol. Society Am. Bulletin*, **96** (6): 746–754.

Nettles, M. and Dziewonski, A. M. (2008): Radially anisotropic shear velocity structure of the upper mantle globally and beneath North America. *J. Geophys. Res.*, **113**, B02303.

Neubert, T., M. Mandea, G. Hulot, *et al.* (2001). Ørsted satellite captures high-precision geomagnetic field data, *Eos Trans. AGU*, **82**(7), 81–88.

Neumann, E. R. and Simon, N. (2009). Ultra-refractory mantle xenoliths from ocean islands: how do they compare to peridotites retrieved from oceanic sub-arc mantle? *Lithos*, **107**, 1–16.

Newman, R. and White, N. (1997). Rheology of the continental lithosphere inferred from sedimentary basins. *Nature*, **385**, 621–624.

Nguuri, T. K., Gore, J., James, D. E., Webb, S. J., and Wright, C. (2001). Crustal structure beneath southern Africa and its implications for the formation and evolution of the Kaapvaal and Zimbabwe cratons. *Geophys. Res. Lett.*, **28**, 2501–2504.

Nicolas, A. (1992). Kinematics in magmatic rocks with special reference to gabbros, *J. Petrol.*, **33**, 891–915.

Nicolas, A. and Christensen, N. I. (1987). Formation of anisotropy in upper mantle peridotite – a review. In: Fuchs, K. and Froidevaux, C. (Eds.), *Composition, Structure and Dynamics of the Lithosphere–Asthenosphere System*. AGU Geodynam. Ser., **16**, 111–123.

Nicolas, A. and Poirier, J. P. (1976). *Crystalline Plasticity and Solid State Flow in Metamorphic Rocks,* New York, Wiley, p. 420.

Nicolas, A., Boudier, F., and Boullier, A. M. (1973). Mechanisms of flow in naturally and experimentally deformed peridotites. *Am. J. Sci.* **273**, 853–76.

Nicolaysen, L. O. (1961). Graphic interpretation of discordant age measurements on metamorphic rocks. *Ann. N. Y. Acad. Sci.*, **91**, 198–206.

Nicolaysen, L. O., Hart, R. J., and Gale, N. H. (1981). The Vredefort radioelement profile extended to supracrustal strata at Carletonville, with implications for continental heat flow. *J. Geophys. Res.*, **86**, 10653–10661.

Nielsen, L. and Thybo, H. (2003). The origin of the teleseismic Pn phase: Crustal scattering of upper mantle whispering gallery phases. *J. Geophys. Res.*, **108**, 2460, 10.1029/2003JB002487.

Nielsen, L., Thybo, H., and Solodilov, L. (1999). Seismic tomographic inversion of Russian PNE data along profile Kraton. *Geoph. Res. Lett.*, **26**, 3413–3416.

Nielsen, L., Thybo, H., Levander, A., and Solodilov, L. (2003). Origin of upper mantle seismic scattering – Evidence from Russian PNE data. *Geophys. J. Int.*, **154**, 196–204.

Nimis, P. and Taylor, W. R. (2000). Single clinopyroxene thermobarometry for garnet peridotites. Part 1. Calibration and testing of a Cr-in-Cpx barometer and an enstatite-in-Cpx thermometer. *Contrib. Mineral. Petrol.* **139**: 541–554.

Nisbet, E. G. and Fowler, C. M. R. (1983). Model for Archean plate tectonics. *Geology*, **11**, 376–379.

Nishimura, C. E. and Forsyth, D. W. (1989). The anisotropic structure of the upper mantle in the Pacific. *Geophys. J.*, **96**, 203–229.

Nishimura, C. E. and Forsyth, D. W. (1988). Rayleigh wave phase velocities in the Pacific with implications for azimuthal anisotropy and lateral heterogeneity. *Geophys. J. Roy. Astron. Soc.*, **94**, 497–501.

Nishitani, T. and Kono, M. (1983). Curie temperature and lattice constant of oxidized titanomagnetite. *Geophys. J. Roy. Astron. Soc.* **74**, 585–600.

Niu, Y. and Batiza, R. (1991). An empirical method for calculating melt compositions produced beneath mid-ocean ridges: applications for axis and off-axis (seamounts) melting. *J. Geophys. Res.*, **96**, 21753–21777.

Nixon, P. H. (1987). *Mantle Xenoliths*. Chichester, Wiley.

Nixon, P. H. and Boyd, F. R. (1973). Petrogenesis of the granular and sheared ultrabasic nodule suite in kimberlites. In: Nixon, P. H. (Ed.), *Lesotho Kimberlites*. Lesotho National Development Corp., 48–56.

Nolasco, R., Tarits, P., Filloux, J. H., and Chave, A. D. (1998). Magnetotelluric Imaging of the Society Islands Hotspot. *J. Geophys. Res.*, **103**(B12), 30,287–30,309.

Nolet, G. (2008). *A Breviary of Seismic Tomography: Imaging the Interior of the Earth and Sun*. Cambridge University Press.

Nolet, G. (1990). Partitioned waveform inversion and two-dimensional structure under the network of autonomously recording seismographs. *J. Geophys. Res.*, **95**, 8499–8512.

Nolet, G. (Ed.), (1987). *Seismic Tomography*. The Netherlands, Reidel, Dordrecht.

Nolet, G. and Zielhuis, A. (1994). Low S-velocities under the Tornquist–Teisseyre zone – Evidence for water injection into the transition zone by subduction. *J. Geophys. Res.*, **99**, 15813–15820.

Nolet, G., Grand, S. P., and Kennett, B. L. N. (1994). Seismic heterogeneity in the upper mantle. *J. Geophys. Res.*, **99**, 23753–23766.

Nover, G. (2005). Electrical properties of crustal and mantle rocks – a review of laboratory measurements and their explanation. *Surv. Geophys.*, **26**, 593–651.

Nur, A. (1971). Effects of stress on velocity anisotropy in rocks with cracks. *J. Geophys. Res.*, **76**, 2022–2034.

Nutman, A. P., McGregor, V. R., Friend, C. R. L., Bennett, V. C., and Kinny, P. D. (1996). The Itsaq Gneiss Complex of southern west Greenland: the world's most extensive record of early crustal evolution (3900–3600 Ma). *Precambrian Res.*, **78**, 1–39.

Nyblade, A. A. (1999). Heat flow and the structure of Precambrian lithosphere. *Lithos*, **48**, 81–91.

Nyblade, A. A. and Pollack, H. N. (1993a). A comparative study of parameterized and full thermal-convection models in the interpretation of heat flow from cratons and mobile belts. *Geophys. J. Int.*, **113**, 747–571.

Nyblade, A. A. and Pollack, H. N. (1993b). A global analysis of heat flow from Precambrian terrains: Implications for the thermal structure of Archean and Proterozoic lithosphere. *J. Geophys. Res.*, **98**, 12207–12218.

Nyblade, A. A., and Pollack, H. N. (1993c). Differences in heat flow between east and southern Africa: implications for regional variability in the thermal structure of the lithosphere. *Tectonophysics*, **219**, 257–272.

O'Brien, H., Lehtonen, M., Spencer, R., and Birnie, A. (2003). Lithospheric mantle in eastern Finland: a 250 km 3D transect. *Proc. 8th Int. Kimberlite Conf.*, Victoria, BC, Canada, June 2003: FLA–0261.

O'Hara, M. J. (1975). Is there an Icelandic mantle plume? *Nature*, **253**, 708–710.

O'Neill, H. S. C. (1981). The transition between spinel lherzolite and garnet lherzolite and its use as a geobarometer. *Contrib. Mineral. Petrol.* **77**, 185–194.

O'Neill, H. S. C. and Palme, H. (1998). Composition of the Silicate Earth: implications for accretion and core formation. In: Jackson, I. (Ed), *The Earth's Mantle: Composition, Structure, and Evolution*. Cambridge University Press, 3–125.

O'Neill, C., Lenardic, A., Moresi, L., Torsvik, T.H., and Lee, C.T.A. (2007). Episodic Precambrian subduction. *Earth Planet. Sci. Lett.*, **262**: 552–562.

O'Neill, C., Lenardic, A., Jellinek, A. M., and Moresi, L. (2009). Influence of supercontinents on deep mantle flow. *Gondwana Research*, **15**, 276–287.

O'Neill, H. S. C. and Wood, B. J. (1979). An empirical study of Fe-Mg partitioning between garnet and olivine and its calibration as a geothermometer. *Contrib. Mineral. Petrol.*, **70**: 59–70.

O'Nions, R. K., Evensen, N. M., and Hamilton, P. J. (1979). Geochemical modelling of mantle differentiation and crustal growth. *J. Geophys. Res.*, **84**, 6091–6101.

O'Reilly, S. Y. and Griffin, W. L. (1985). A xenolith-derived geotherm for southeastern Australia and its geophysical implications. *Tectonophysics*, **111**, 41–63.

O'Reilly, S. Y. and Griffin, W. L. (1996). 4-D lithosphere mapping: a review of the methodology with examples. *Tectonophysics*, **262**, 3–18.

O'Reilly, S. Y. and Griffin, W. L. (2000). Apatite in the mantle: implications for metasomatic processes and high heat production in Phanerozoic mantle. *Lithos*, **53**, 217–232.

O'Reilly, S. Y. and W. L. Griffin (2006). Imaging global chemical and thermal heterogeneity in the subcontinental lithospheric mantle with garnets and xenoliths: geophysical implications. *Tectonophysics*, **416**, 289–309.

O'Reilly, S. Y., Griffin, W. L., Poudjom Djomani, Y. H., and Morgan, P. (2001). Are lithospheres forever? Tracking changes in the subcontinental lithospheric mantle through time, *GSA Today*, **11**, 4–10.

Obayashi, M., Sakurai, T., and Fukao, Y. (1997). Comparison of recent tomographic models. Int. Symp. *New Images of the Earth's Interior Through Long-Term Ocean-Floor Observations*. Suyehiro, K. (Ed.), Earthquake Res. Inst., Univ. of Tokyo, Tokyo. p. 29.

Offerhaus, L. J., Wirth, R., and Dresen, G. (2001). High-temperature creep of polycrystalline albite. In: de Meer, S., *et al.* (Eds.), *Deformation Mechanisms, Rheology and Tectonics*. Utrecht University, Noordwijkerhout, The Netherlands, p. 124.

Ogawa, Y., Jones, A. G., Unsworth, M. J. *et al.* (1996). Deep electrical conductivity structures of the Appalachin Orogen in the southwestern US. *Geophys. Res. Lett.*, **23**, 1597–1600.

Ohuchi, T. and Karato, S. (2010). Plastic deformation of orthopyroxene under the lithospheric conditions. *Earth and Planet. Sci. Lett.*

Ojeda, G. Y., and Whitman, D. (2002). Effect of windowing on lithosphere elastic thickness estimates obtained via the coherence method: Results from northern South America, *J. Geophys. Res.*, **107**(B11), 2275, doi:10.1029/2000JB000114.

Okaya, N., Freeman, R., Kissling, E., and Mueller, St. (1996). A lithospheric cross-section through the Swiss Alps – 1. Thermokinematic modelling of the Neoalpine orogeny. Geophys. *J. Int.*, **125**, 504–518.

Oldenburg, D. W. (1981). Conductivity structure of oceanic upper mantle beneath the Pacific plate. *Geophys. J. Roy. Astron. Soc.*, **65**, 359–394.

Oldenburgh, D. W. (1975). A physical model for the creation of the lithosphere, *Geophys. J. Roy. Astron. Soc.*, **43**, 425–451.

Oldenburg, D. W., Whittall, K. P., and Parker, R. L. (1984). Inversion of ocean bottom magnetotelluric data revisited. *J. Geophys. Res.* **89**, 1829–1833.

Olsen, N., Purucker, M. E., Sabaka, N., *et al.* (2000). Ørsted initial field model, *Geophys. Res. Lett.*, **27**, 3607–3610.

Oreshin, S., L. Vinnik, D. Peregoudov, and S. Roecker (2002). Lithosphere and asthenosphere of the Tien Shan imaged by S receiver functions. *Geophys. Res. Lett.*, **29**(8), 1191, doi:10.1029/2001GL014441.

Osako, M., Ito, E., and Yoneda, A. (2004). Simultaneous measurements of thermal conductivity and thermal diffusivity for garnet and olivine under high pressure. *Phys. Earth Planet. Inter.*, **143**, 311–320.

Osipova, I. L., Hjelt, S.-E., and Vanyan, L. L. (1989). Source field problems in northern parts of the Baltic shield. *Phys. Earth Planet. Inter.*, **53**, 337–342.

Owens, T. J. and Zandt, G. (1997). Implications of crustal property variations for models of Tibetan Plateau evolution. *Nature*, **387**, 37–43.

Oxburgh, E. R. (1980). Heat flow and magma genesis. In: Hargraves, R. B. (Ed.), *Physics of Magmatic Processes*. NJ, Princeton Univ. Press, 161–199.

Pajunpaa, K., Olafsdottir, B., Lahti, I., Korja, T., the BEAR Working Group (2002). Crustal conductivity anomalies in central Sweden and Southwestern Finland. *Geophys. J. Int*, **150**, 695–705.

Panning, M. and Romanowicz, B. (2006). A three-dimensional radially anisotropic model of shear velocity in the whole mantle. *Geophys. J. Int.*, **167**, 361–379.

Panza, G. A., Mueller, S., and Calcagnile, G. (1980). The gross features of the lithosphere-asthenosphere system in Europe from seismic surface waves and body waves. *Pageophysics*, **118**, 1209–1213.

Palme, H. and Nickel, K. G. (1985). Ca/Al ratio and composition of the Earth's upper mantle. *Geochim. Cosmochim. Acta*, **49**, 2123–2132.

Papanastassiou, D. A. and Wasserburg, G. J. (1969). Initial strontium isotopic abundances and resolution of small time differences in formation of planetary objects. *Earth Planet. Sci. Lett.*, **5**, 361–376.

Park, S. K. and Ducea, M. N. (2003). Can in situ measurements of mantle electrical conductivity be used to infer properties of partial melts? *J. Geophys. Res.*, **108**(B5), 2270, doi:10.1029/2002JB001899.

Park, C.-H., Tamaki, K., and Kobayashi, K. (1990). Age-depth correlation of the Philippine Sea back-arc basins and other marginal basins in the world. *Tectonophys.*, **181**, 351–371.

Park, S. K., Hirasuna, B., Jiracek, G. R., and Kinn, C. (1996). Magnetotelluric evidence of lithospheric mantle thinning beneath the southern Sierra Nevada. *J. Geophys. Res.*, **101**, 16241–16255.

Parker, R. L. (1983). The magnetotelluric inverse problem. *Geophys. Surv.*, **6**, 5–25.

Parker, R. L. (1980). The inverse problem of electromagnetic induction – existence and construction of solutions based on incomplete data. *J. Geophys. Res.*, **85**, 4421–4428.

Parker, R. L. (1972). The rapid calculation of potential anomalies. *Geophys. J. Roy. Astron. Soc.*, **31**, 447–455.

Parsons, B. and McKenzie, D. P. (1978). Mantle convection and the thermal structure of the plates. *J. Geophys. Res.*, **83**, 4485–4496.

Parsons, B. and Molnar, P. (1976). The origin of outer topographic rises associated with trenches. *Geophys. J. Roy. Astron. Soc.*, **45**, 707–712.

Parsons, B. and Sclater, J. G. (1977). An analysis of the variation of ocean floor bathymetry and heat flow with age. *J. Geophys Res.*, **82**, 803–827.

Pasyanos, M. (2009). Lithospheric thickness modeled from long period surface wave dispersion. *Tectonophysics*, **481**, 38–50.

Patchett, P. J. and Arndt, N. T. (1986). Nd isotopes and tectonics of 1.9–1.7 Ga crustal genesis. *Earth Planet Sci. Lett.*, **78**, 329–338.

Patchett, P. J. and Samson, S. D. (2003). Ages and Growth of the Continental Crust from Radiogenic Isotopes. *Treatise on Geochemistry*, Volume **3**; 321–348, Elsevier.

Patchett, J. P., Kouvo, O., Hedge, C. E., and Tatsumoto, M. (1981). Evolution of continental crust and mantle heterogeneity: evidence from Hf isotopes. *Contrib. Mineral. Petrol.*, **78**, 279–297.

Patterson, C. C. (1956). Age of meteorites and the Earth. *Geochim. Cosmochim. Acta*, **10**, 230–237.

Pavlenkova, N. I., Pavlenkova, G. A., and Solodilov, L. N. (1996). High seismic velocities in the uppermost mantle of the Siberian craton. *Tectonophysics*, **262**, 51–65.

Pavlov, A. and Kasting, J. F. (2002). Mass-independent fractionation of sulfur isotopes in Archean sediments: strong evidence for an anoxic Archean atmosphere. *Astrobiology*, **2**, 27–41.

Peacock, S. M. (2003). Thermal structure and metamorphic evolution of subducting slabs. In: J. Eiler (Ed.), *Inside the Subduction Factory*. AGU Geophys. Monogr., 7–22.

Peacock, SM. (1990). Fluid processes in subduction zones. *Science*, **248**, 329–337.

Pearson, D. G. (1999). The age of continental roots. *Lithos*, **48**, 171–194.

Pearson, D. G., Canil, D., and Shirey, S. B. (2003). Mantle samples included in volcanic rocks: xenoliths and diamonds. *Treatise on Geochemistry*, v.**2**, 171–275, Elsevier.

Pearson, D. G., Ionov, D., Carlson, R. W., and Shirey, S. B. (1998). Lithospheric evolution in circum-cratonic settings. A Re–Os isotope study of peridotite xenoliths from the Vitim region, Siberia. *Min. Mag.*

Pearson, D. G., Carlson, R. W., Shirey, S. B., Boyd, F. R., and Nixon, P. H. (1995). Stabilisation of Archaean lithospheric mantle; a Re–Os isotope study of peridotite xenoliths from the Kaapvaal Craton. *Earth Planet. Sci. Lett.*, **134**, 341–357.

Pearson, D. G., Boyd, F. R., Hoal, K. E. O., *et al.* (1994). A Re–Os isotopic and petrological study of Namibian peridotites; contrasting petrogenesis and composition of on- and off-craton lithospheric mantle. *Mineral. Mag.*, **58A**, No. L-Z, 703–704.

Pearson, N. J., Griffin, W. L., Doyle, B. J., *et al.* (1999). Xenoliths from kimberlite pipes of the Lac de Gras Area, Slave craton, Canada. In: Gurney, J. J., Gurney, J. L., Pascoe M. D., and Richardson, S. H. (Eds.), *Proc. 7th Intern. Kimberlite Conf.*, v.**2**, 644–658.

Pearson, D. G., Shirey, S. B., Carlson, R. W., *et al.* (1995). Re–Os, Sm–Nd and Rb–Sr isotope evidence for thick Archaean lithospheric mantle beneath the Siberia craton modified by multi-stage metasomatism. *Geochim. Cosmochim. Acta* **59**, 959–977.

Pechmann, J. C., Richins, W. D., and Smith, R. B. (1984). Evidence for a "double Moho" beneath the Wasatch front, Utah, *Eos Trans. AGU*, **65**, 988.

Pedersen, H. A., S. Fishwick, D. B Snyder (2009). A comparison of cratonic roots through consistent analysis of seismic surface waves. *Lithos*, doi:10.1016/j. lithos.2008.09.016.

Peltier, W. R. (1998). Postglacial variations in the level of the sea: Implications for climate dynamics and solid-earth geophysics. *Rev. Geophys.*, **36**, 603–689.

Perchu,c, E. and Thybo, H. (1996). A new model of upper mantle P-wave velocity below the Baltic Shield: indication of partial melt in the 95 to 160 km depth range. *Tectonophysics*, **253**, 227–245.

Percival, D. B. and Walden, A. T. (1993). *Spectral Analysis for Physical Applications, Multitaper and Conventional Univariate Techniques*. New York, Cambridge University Press.

Percival, J. A. and Williams, H. R. (1989). Late Archean Quetico accretionary complex, Superior Province. Canada. *Geology*, **17**, 23–25.

Percival, J. A., Bleeker, W., Cook, E. A., Rivers, T. *et al.* (2004). PanLITHOPROBE Workshop IV: intra-orogen correlations and comparative orogenic anatomy. *Geoscience Canada*, **31**, 23–39.

Pérez-Gussinyé, M. and Watts, A. B. (2005). The long-term strength of Europe and its implications for plate forming processes. *Nature*, **436**, 381–384, doi:10.1038/nature03854.

Pérez-Gussinyé, M., Lowry, A. R., Phipps Morgan, J., and Tassara, A. (2008). Effective elastic thickness variations along the Andean margin and their relationship to subduction geometry. *Geochem. Geophys. Geosyst.*, **9**, Q02003, doi:10.1029/2007GC001786.

Pérez-Gussinyé, M., Lowry, A. R., Watts, A. B., and Velicogna, I. (2004). On the recovery of effective elastic thickness using spectral methods: Examples from synthetic data and from the Fennoscandian Shield. *J. Geophys. Res.*, **109**, B10409, doi:10.1029/2003JB002788.

Pérez-Gussinyé, M., Metois, M., Fernández, M., *et al.* (2009). Effective elastic thickness of Africa and its relationship to other proxies for lithospheric structure and surface tectonics. *Earth Planet. Sci. Lett.*, **287**, 152–167.

Perry, H. K. C., Jaupart, C., Mareschal, J.-C., and Bienfait, G. (2006). Crustal heat production in the Superior Province, Canadian Shield, and in North America inferred from heat flow data. *J. Geophys. Res.*, **111**, B04401, doi:10.1029/2005JB003893.

Peslier, A. H., and Luhr, J. F. (2006). Hydrogen loss from olivines in mantle xenoliths from Simcoe (USA) and Mexico: mafic alkalic magma ascent rates and water budget of the sub-continental lithosphere. *Earth Planet. Sci. Lett.*, **242**, 302–319.

Peslier, A. H., Luhr, J. F., Woodland, A. B., *et al.* (2006). Estimating alkali basalt and kimberlite magma ascent rates using H diffusion profiles in xenolithic mantle olivine. *Goldschmidt Conf.*, 2006, A484.

Peslier, A. H., Reisberg, L., Ludden, J., and Francis, D. (2000). Re-Os constraints on harzburgite and lherzolite formation in the lithospheric mantle: a study of Northern Canadian Cordillera xenoliths. *Geochim. Cosmochim. Acta*, **64**, 3061–3071.

Petersen, N. and U. Bleil (1982). Magnetic properties of rocks, In: G. Angenheister (Ed.), *Landolt-Börnstein: Numerical Data and Functional Relationships in Science and Technology.* Berlin, Springer-Verlag, 366–432.

Petit, C., Ebinger, C. (2000). Flexure and mechanical behavior of cratonic lithosphere: gravity models of the East African and Baikal rifts. *J. Geophys. Res.* **105**, 19, 151–19,162.

Petit, C., Burov, E., and Tiberi, C. (2008). Strength of the lithosphere and strain localization in the Baikal rift. *Earth Planet. Sci. Lett.*, **269**, 523–529.

Petit, C., Koulakov, I., and Deverchere, J. (1998). Velocity structure around the Baikal rift zone from teleseismic and local earthquake traveltimes and geodynamic implications. *Tectonophysics,* **296**, 125–144.

Petitjean, S., Rabinowicz, M., and Gregoire, M., and Chevrot, S. (2006). Differences between Archean and Proterozoic lithospheres: Assessment of the possible role of thermal conductivity. *Geochem., Geophys., Geosyst.*, **7**, Q03021.

Pfiffner, O. A., Frei, W. *et al.* (1988). Deep seismic reflection profiling in the Swiss Alps; explosion seismology results for line NFP 20-EAST. *Geology (Boulder),* **16**(11). 987–990.

Phipps, Morgan J. and Smith, W. H. F. (1992). Flattening of the seafloor depth-age curve as a response to asthenospheric flow? *Nature,* **359**, 524–527.

Pilot, J., Werner, C-D., Haubrich, F., and Baumann, N. (1998). Palaeozoic and Proterozoic zircons from the Mid-Atlantic Ridge. *Nature,* **393**, 676–679.

Pinet, C. and Jaupart, C. (1987). The vertical distribution of radiogenic heat production in the Precambrian crust of Norway and Sweden; geothermal implications. *Geophys. Res. Lett.*, **14**, 260–263.

Pinet, C., C. Jaupart, J. C. Mareschal, *et al.* (1991). Heat flow and structure of the lithosphere in the Eastern Canadian Shield. *J. Geophys. Res.*, **96**, 19,941–19,963.

Piromallo, C. and Morelli, A. (2003). P–wave tomography of the mantle under the Alpine–Mediterranean area. *J. Geophys. Res.*, **108** (B2): art. no. 2065.

Plank, T. and Langmuir, C. H. (1998). The chemical composition of subducting sediment and its consequences for the crust and mantle. *Chem. Geol.*, **145**, 325–394.

Plomerová, J., Babuška, V., Vecsey, L., Kozlovskaya, E., and Raita, T., SSTWG, (2005). Proterozoic–Archean boundary in the upper mantle of eastern Fennoscandia as seen by seismic anisotropy. *J. Geodyn.*, doi:10.1016/j.jog.2005.10.008.

Poirier, J.-P. and Nicolas, A. (1975). Deformation-induced recrystallization due to progressive misorientation of subgrains, with special reference to mantle peridotites. *J. Geol.*, **83**, 707–720.

Poirier, J.-P. and Guillope, M. (1979). Deformation-induced recrystallization of minerals, *Bull. Mineral.*, **102**, 67–74.

Pokhilenko, N. P., Pearson, D. G., Boyd, F. R., and Sobolev, N. V. (1991). Megacrystalline dunites and peridotites: hosts for Siberian diamonds. *Ann. Rep. Dir. Geophys. Lab.*, Carnegie Inst., Washington, 1990–1991: 11–18.

Polet, J. and Anderson, D. L. (1995). Depth extent of cratons as inferred from tomographic studies. *Geology*, **23**, 205–208.

Pollack, H. N. (1986). Cratonization and thermal evolution of the mantle. *Earth Planet. Sci. Lett.*, **80**, 175–182.

Pollack, H. N. and Chapman, D. S. (1977). On the regional variation of heat flow, geotherms and the thickness of the lithosphere. *Tectonophys.*, **38**, 279–296.

Pollack, H. N., Hurter, S. J., and Johnson, J. R. (1993). Heat flow from the Earth's interior: analysis of the global data set. *Rev. Geophys.*, **31**, 267–280.

Polyak, B. G. and Smirnov, Y. A. (1968). Relationship between terrestrial heat flow and the tectonics of continents. *Geotectonics*, **4**, 205–213.

Polyakov, A. I., Muravieva, N. S., and Senin, V. G. (1988). Partial melting of the upper mantle in the Baikal rift (from studies of glasses in lherzolite nodules and megachrysts), *Dokl. Akad. Nauk SSSR*, **300**(1), 208–213 (in Russian).

Ponziani, F., De Franco, R., Minelli, G., Biella, G. *et al.* (1995). Crustal shortening and duplication of the Moho in the Northern Apennines: a view from seismic refraction data. *Tectonophysics*, **252**, 391–418.

Popp, T. and Kern, H. (1993). Thermal dehydration reactions characterized by combined measurements of electrical conductivity and elastic wave velocities. *Earth Planet. Sci. Lett.*, **120**, 43–57.

Poudjom Djomani, Y. H., Fairhead, J. D., and Griffin, W. L. (1999). The flexural rigidity of Fennoscandia: reflection of the tectonothermal age of the lithospheric mantle. *Earth Planet Sci. Lett.*, **174**, 139–154.

Poudjom Djomani, Y. H., O'Reilly, S. Y., Griffin, W. L., and Morgan, P. (2001). The density structure of subcontinental lithosphere through time. *Earth Planet. Sci. Lett.*, **184**, 605–621.

Poudjom Djomani, Y. H., O'Reilly, S. Y., Griffin, W. L., *et al.* (2003). Upper mantle structure beneath eastern Siberia: Evidence from gravity modeling and mantle petrology. *Geochem. Geophys. Geosyst.*, **4**(7), 1066, doi:10.1029/2002GC000420.

Pous, J., Ledo, J., Marcuello, A., and Daignieres, M. (1995a). Electrical resistivity model of the crust and upper mantle from a magnetotelluric survey through the central Pyrenees. *Geophys. J. Int.*, **121**: 750–762.

Pous, J., Munoz, J. A., Ledo, J. J., and Liesa M. (1995b). Partial melting of subducted continental lower crust in the Pyrenees. *J. Geol. Soc. (London)* **152**: 217–220.

Pous, J., Munz, G., Heise, W., Carles Melgarejo, J., and Quesada, C. (2004). Electromagnetic imaging of Variscan crustal structures in SW Iberia: the role of interconnected graphite. *Earth Planet Sci. Lett.*, **217**, 435–450.

Pozgay, S. H., Wiens, D. A., Conder, J. A., Shiobara, H., and Sugioka, H. (2009). Seismic attenuation tomography of the Mariana subduction system: Implications for thermal structure, volatile distribution, and slow spreading dynamics. *Geochem. Geophys. Geosyst.*, **10**, 4, doi:10.1029/2008GC002313.

Pratt, J. H. (1845). *The Mathematical Principles of Mechanical Philosophy and their application to Elementary Mechanics and Architecture, but chiefly to The Theory of Universal Gravitation*. Macmillan, Barclay, and Macmillan, Cambridge, U.K., p. 1, 295 pp.

Praus, O., Pecova, J., Petr, V., Babuska, V., and Plomerova, J. (1990). Magnetotelluric and seismological determination of the lithosphere–asthenosphere transition in Central Europe. *Phys. Earth Planet. Inter.* **60**, 212–228.

Preston, R. F. and Sweeney, R. J. (2003). A comparison of clinopyroxene thermobarometric techniques: applied to Jwaneng, Orapa and Markt Kimberlites. *Proc. 8th Intern. Kimberlite Conf.*, Victoria, BC, Canada, June 2003: FLA-0053.

Priestley, K. and Debayle, E. (2003). Seismic evidence for a moderately thick lithosphere beneath the Siberian Platform. *Geophys. Res. Lett.*, **30**, 1118, doi:10.1029/2002GL015931.

Priestley, K. and McKenzie, D. (2006). The thermal structure of the lithosphere from shear wave velocities. *Earth Planet. Sci. Lett.*, **244**, 285–301.

Priestley, K., McKenzie, D., and Debayle, E. (2006). The state of the upper mantle beneath southern Africa. *Tectonophysics*, **416**, 101–112.

Priestley, K., Cipar, J., Egorkin, A., and Pavlenkova, N. (1994). Upper mantle velocity structure beneath the Siberian platform. *Geophys. J. Int.*, **118**, 369–378.

Prodehl, C., Ritter, J. R. R., Mechie, J., *et al.* (1997). The KRISP 94 lithospheric investigation of southern Kenya – the experiments and their main results. *Tectonophysics*, **278**, 121–147.

Purucker, M. and Whaler, K. (2007). crustal magnetism. In Kono, M. (Ed), *Geomagnetism, Treatise on Geophysics*, vol. 5, Chapter 6, Elsevier, 195–237.

Purucker, M., Langlais, B., Olsen, N., Hulot, G., and Mandea, M. (2002). The southern edge of cratonic North America: Evidence from new satellite magnetometer observations. *Geophys. Res. Lett.*, **29**, no. 15, 10.1029/2001GL013645.

Putnam, G. R. (1922). Condition of the Earth crust and the earlier American gravity observations. *Bull. Geol. Soc. Am.*, **33**, 287–302.

Ramo, O. T. and Calzia, J. P. (1998). Nd isotopic composition of cratonic rocks in the southern Death Valley region: evidence for a substantial Archean source component in Mojavia. *Geology* **26**, 891–894.

Ranalli, G. (2003). How soft is the crust? *Tectonophysics*, **361**, 319–320.

Ranalli, G. (1995). *Rheology of the Earth*, 2nd edn. Chapman and Hall, London.

Ranalli, G. (1994). Nonlinear flexure and equivalent mechanical thickness of the lithosphere. *Tectonophysics*, **240**, 107–114.

Ranalli, G. and Murphy, D. C. (1987). Rheological Stratification of the Lithosphere. *Tectonophysics*, **132**, 281–295.

Raterron, P., Chen, J., Li, L., Weidner, D. J., and Cordier, P. (2007). Pressure-induced slip system transition in forsterite: Single-crystal rheological properties at mantle pressure and temperature. *Am. Mineral.*, **92**: 1436–1445.

Ratschbacher, L., Frisch, W., Neubauer, F., Schmid, S. M., and Neugebauer, J. (1989). Extension in compressional orogenic belts: the eastern Alps. *Geology*, **17**, 404–407.

Rauch, M. and Keppler, H. (2002). Water solubility in orthopyroxene. *Contrib. Mineral., Petrol.* **143**, 525–536.

Rayleigh, Lord (John William Strutt), (1883). Investigation of the character of the equilibrium of an incompressible heavy fluid of variable density. *Proc. London Mathem. Soc.* **14**: 170–177, doi:10.1112/plms/s1 − 14.1.170.

Rea, D. K. and Ruff, L. J. (1996). Composition and mass flux of sediment entering the world's subduction zones: Implications for global sediment budgets, great earthquakes, and volcanism. *Earth Planet. Sci. Lett.*, **140**, 1–12

Read, G. H., Grutter, H. S., Winter, L. D. S., Luckman, N. B., and Gaunt, G. F. M. (2003). Stratigraphic relations, Kimberlite emplacement and lithospheric thermal evolution, Quiricó Basin, Minas Gerais State, Brazil. *Proc. 8th Intern. Kimberlite Conf.*, Victoria, BC, Canada, June 2003, FLA-0304.

Reid, F. J. L., Woodhouse, J. H., and van Heijst, H. J. (2001). Upper mantle attenuation and velocity structure from measurements of differential S phases. *Geophys. J. Int.*, **145**, 615–630.

Ren, J., Jiang, C., Zhang, Z., and Qin, D. (1981). The tectonic evolution of China (in Chinese), in *Collected Papers on Tectonics of China and Adjacent Regions*, edited by J. Huang and C. Li, 138–147, Geology, Beijing.

Revelle, R. and Maxwell, A. E. (1952). Heat flow through the floor of the eastern North Pacific Ocean. *Nature*, **170**, 199–200.

Revenaugh, J. and Jordan, T. (1991). Mantle layering from ScS reverberations. 3. The upper mantle. *J. Geophys. Res.*, **96**, 19781–19810.

Reymer, A. P. S. and Schubert, G. (1984). Phanerozoic addition rates to the continental crust. *Tectonics*, **3**, 63–77.

Ribe, N. M. (1989). Seismic anisotropy and mantle flow. *J. Geophys. Res.* **94**, 4213–4223.

Richards, P. G. (1972). Seismic waves reflected from velocity gradient anomalies within the Earth's upper mantle. *J. Geophys.*, **38**, 517–527.

Richards, M. A., Bunge, H.-P., and Lithgow-Bertelloni C. (2000). Mantle convection and plate motion history: Toward general circulation models. In: Richards, M. A., Gordon, R. G., and van der Hilst, R. D. (Eds.), *The History and Dynamics of Global Plate Motions*. AGU Geophys. Monograph. **121**, 289–308.

Richter, F. M. (1973). Convection and large-scale circulation of mantle. *Geophys. Res. Lett.*, **78**, 8735–8745.

Ringwood, A. E. (1966). Chemical evolution of terrestrial planets. *Geochim. Cosmochim. Acta*, **30**, 41–104.

Ringwood, A. E. (1975). *Composition and Petrology of the Earth's Mantle*. NY, McGraw-Hill, 618 pp.

Ritsema, J. and Allen, R. (2003). The elusive mantle plume. *Earth Planet. Sci. Lett.*, **207**, 1–12.

Ritsema, J, and van Heijst, H. (2000a). New seismic model of the upper mantle beneath Africa. *Geology*, **28**, 63–66.

Ritsema, J. and van Heijst, H. J. (2000b). Seismic imaging of structural heterogeneity in Earth's mantle: evidence for large-scale mantle flow. *Sci. Progr.*, **83**, 243–259.

Ritsema, J. and van Heijst, H. J. (2002). Constraints on the correlation of P- and S-wave velocity heterogeneity in the mantle from P, PP, PPP and PKPab traveltimes. *Geophys. J. Int.*, **149**, 482–489.

Ritsema, J., van Heijst, H. J., and Woodhouse, J. H. (2004). Global transition zone tomography. *J. Geophys. Res.*, **109**, B02302.

Ritsema, J., van Heijst, H. J., and Woodhouse, J. H. (1999). Complex shear wave velocity structure imaged beneath Africa and Iceland. *Science*, **286**, 1925–1928.

Ritter, J. R. R. and Christensen, U. R. (Eds.) (2007). *Mantle Plumes: A Multidisciplinary Approach*. Springer. ISBN 3540680454.

Ritter, J. R. R., Jordan, M., Christensen, U. R., and Achauer, U. (2001). A mantle plume below the Eifel volcanic fields, Germany. *Earth Planet. Sci. Lett.*, **186** (1): 7–14.

Ritz, M. (1984). Inhomogeneous structure of the Senegal lithosphere from deep magneto-telluric soundings. *J. Geophys. Res.* **89**, 11317–11331.

Ritzwoller, M. H. and Lavely, E. M. (1995). Three dimensional seismic models of the earth's mantle. *Rev. Geophys.*, **33**, 1–66.

Ritzwoller, M. H., Shapiro, N. M., Levshin, A. L., and Leahy, G. M. (2001). Crustal and upper mantle structure beneath Antarctica and surrounding oceans. *J. Geophys. Res.*, **106**, 30645–30670.

Roberts, J. J., Tyburczy, J. A. (1991). Frequency dependent electrical properties of polycrystalline olivine compacts. *J. Geophys. Res.*, **96**, 16205–16222.

Roberts, J. J. and Tyburczy, J. A. (1994). Frequency-dependent electrical-properties of minerals and partial-melts. *Surv. Geophys.*, **15**, 239–262.

Roberts, J. J. and Tyburczy, J. A. (1999). Partial-melt electrical conductivity: Influences of melt composition. *J. Geophys. Res.*, **104**, 7055–7065.

Rocha, A. and Acrivos, A. (1973). On the effective thermal conductivity of dulite dispersions: General theory for inclusions of arbitrary shape. *Quart. J. Mech. Appl. Math.*, **26**, 217.

Rodgers, A. and Bhattacharyya, J. (2001). Upper mantle shear and compressional velocity structure of the central US craton: shear wave low-velocity zone and anisotropy. *Geophys. Res. Lett.*, **28**, 383–386.

Röhm, A. H. E., Snieder, R., Goes, S. and Trampert, J. (2000). Thermal structure of continental upper mantle inferred from S-wave velocity and surface heat flow. *Earth Planet. Sci. Lett.*, **181**, 395–407.

Rollinson, H. (1997). Eclogite xenoliths in west African kimberlites as residues from Archaean granitoid crust formation. *Nature*, **389**, 173–177.

Romanowicz, B. (1991). Seismic tomography of the Earth's mantle. *Ann. Rev. Earth Planet. Sci.*, **19**, 77–99.

Romanowicz, B. (1994). Anelastic tomography: a new perspective on upper-mantle thermal structure. *Earth Planet. Sci. Lett.*, **128**, 113–121.

Romanowicz, B. (1995). A global tomographic model of shear attenuation in the upper mantle. *J. Geophys. Res.*, **100**, 12375–12394.

Romanowicz, B. (1998). Attenuation tomography of the earth's mantle: A review of current status. *Pure Appl. Geophys.*, **153**, 257–272.

Romanowicz, B. and Durek, J. J. (2000). Seismological constraints on attenuation in the Earth: review. *Geophysical Monograph*, **117**, 161–178.

Rosen, O. M., Condie, K. C., Natapov, L. M., and Nozhkin, A. D. (1994). Archean and early Proterozoic evolution of the Siberian Craton: a preliminary assessment. In: Condie, K. (Ed.), *Archean Crustal Evolution*. Amsterdam, Elsevier, pp. 411–459.

Rossman, G. R. (2006). Analytical methods for measuring water in nominally anhydrous minerals. *Rev. Mineral. Geochemi.*, **62**, 1–28.

Roth, E. G., Wiens, D. A., and Zhao, D. (2000). An empirical relationship between seismic attenuation and velocity anomalies in the upper mantle. *Geophys. Res. Lett.*, **27**, 601–604.

Roth, J. B., Fouch, M. J., James, D. E., and Carlson, R. W. (2008). Three-dimensional seismic velocity structure of the northwestern United States. *Geophys. Res. Lett.*, **35**, L15304, doi:10.1029/2008GL034669.

Roy, R. F., Blackwell, D. D., and Birch, F. (1968). Heat generation of plutonic rocks and continental heat flow provinces. *Earth Planet. Sci. Lett.*, **5**, 1–12.

Royden, L. H. (1993). The tectonic expression of slab pull at continental convergent boundaries. *Tectonics*, **12**, 303–325.

Royden, L., Horvath, F., Nagymarosy, A., and Stegena, L. (1983). Evolution of the Pannonian Basin system, 2, Subsidence and thermal history. *Tectonics*, **2**, 91–137.

Rudnick, R. L. (1995). Making continental crust. *Nature*, **378**: 571–578.

Rudnick, R. L. and Fountain, D. M. (1995). Nature and composition of the continental crust: a lower crustal perspective. *Rev. Geophys.*, **33**, 267–309.

Rudnick, R. L. and S. Gao (2003). Composition of the Continental Crust. In: *Treatise on Geochemistry*, v. 3, *The Crust*, R. L. Rudnick, (Ed.), 1–64. Elsevier-Pergamon, Oxford.

Rudnick, L. R. and Nyblade, A. A. (1999). The thickness and heat production of Archean lithosphere: Constraints from xenolith thermobarometry and surface heat flow. In: Fei, Y., Bertka, C. M., and Mysen, B. O. (Eds.), *Mantle Petrology: Field Observations and High Pressure Experimentation: A Tribute to Francis R. (Joe) Boyd*. Chem. Soc. Spec. Publ. No. 6, 3–12.

Rudnick, R. L. and Presper, T. (1990). Geochemistry of intermediate to high-pressure granulites. In: Vielzeuf, D. and Vidal, P. (Eds.) *Granulites and crustal evolution*. Kluwer, Netherlands, 523–550.

Rudnick, L. R., McDonough, W. F., and O'Connel, R. J. (1998). Thermal structure, thickness and composition of continental lithosphere. *Chem. Geol.*, **145**, 395–411.

Runcorn, S. K. (1964). Measurements of planetary electric currents. *Nature*, **202**, 10–13.

Rutter, E. H. (1986). On the nomenclature of mode of failure transitions in rocks. *Tectonophys.*, **122**, 381–387.

Rutter, E. H. and Brodie, K. H. (2004). Experimental grain size-sensitive flow of hot-pressed Brazilian quartz aggregates. *J. Struct. Geol.*, **26**, 2011–2023.

Ryaboy, V. Z. and Derlyatko, E. K. (1984). Lateral inhomogeneities of the asthenospheric layer of the upper mantle of North Eurasia according to deep seismic and geoelectric soundings. *Dokl. Akad. Nauk SSSR*, **277**(3), 577–581.

Ryan, C. G., Griffin, W. L., and Pearson, N. J. (1996). Garnet geotherms: pressure-temperature data from Cr-pyrope garnet xenocrysts in volcanic rocks. *J. Geophys. Res.*, **101**, 5,611–5,625.

Rybacki, E. and Dresen, G. (2000). Dislocation and diffusion creep of synthetic anorthite aggregates. *J. Geophys. Res.*, **105**, 26017–26036.

Rybacki, E. and Dresen, G. (2004). Deformation mechanism maps for feldspar rocks. *Tectonophysics*, **382**, 173–187.

Ryberg, T., Fuchs, K., Egorkin, A., and Solodilov, L. (1995). Observation of high-frequency teleseismic Pn on the long-range Quartz profile across Northern Eurasia. *J. Geophys. Res.*, **100**, 18151–18163.

Ryberg, T., Wenzel, F., Mechie, J., *et al.* (1996). Two-dimensional velocity structure beneath Northern Eurasia derived from the super long-range seismic profile Quartz. *Bull. Seism. Soc. Am.*, **86**, 857–867.

Rychert, C. A. and Shearer, P. M. (2007). A Global Lithosphere-Asthenosphere Boundary? *Eos Trans. AGU*, **88**(52), Fall Meet. Suppl., Abstract V33D-02.

Ryder, G. (1992). Chronology of early bombardment in the inner solar system. *Geol. Soc. Am. Abstr. Progm.* **21**, A299.

Sacks, I. S. and Snoke, J. A. (1977). The use of converted phases to infer the depth of the lithosphere-asthenosphere boundary beneath South America. *J. Geophys. Res.*, **82**, 2011–2017.

Safonov, A. S., Bubnov, V. M., Sysoev, B. K., *et al.* (1976). Deep magnetotelluric surveys of the Tungus Syneclise and on the West Siberian Plate. In: Adam, A. (Ed.), *Geoelectric and Geothermal Studies*. Akad. Kiado, Budapest, Hungary, 666–672.

Saleeby, J. and Foster, Z. (2004). Topographic response to mantle lithosphere removal in the southern Sierra Nevada region, California. *Geology*, **32**, 245–248.

Sandiford, M., Van Kranendonk, M. J., and Bodorkos, S. (2004). Conductive incubation and the origin of dome-and-keel structure in Archean granite-greenstone terrains: A model based on the eastern Pilbara Craton, Western Australia. *Tectonics*, **23**, TC1009, doi:10.1029/ 2002TC001452.

Sandoval, S., Kissling, E., Ansorge, J., and the SVEKALAPKO STWG (2004). High-resolution body wave tomography beneath the SVEKALAPKO array: II. Anomalous upper mantle structure beneath central Baltic Shield. *Geophys. J. Int.,* **157**, 200–214.

Sandvol, E., J. Ni, S. Ozalaybey, and J. Schlue (1992). Shear-wave splitting in the Rio Grande Rift. *Geophys. Res. Lett.*, **19**, 2337–2340.

Santos, F. A. M., Soares, A., Nolasco, R., *et al.* (2003). Lithosphere conductivity structure using the CAM-1 (Lisbon-Madeira) submarine cable. *Geophys.J. Int.*, **155**, 591–600.

Santosh, M. and G. Zhao (Eds.), (2009). Supercontinent dynamics. *Gondwana Res.* Special Issue, **15**, 225–470.

Sarker, G. and Abers, G. A. (1998). Deep structures along the boundary of a collisional belt: attenuation tomography of P and S waves in the Greater Caucasus. *Geophys. J. Int.*, **133**, 326–340.

Sarker, G. and Abers, G. A. (1999). Lithospheric temperature estimates from seismic attenuation across range fronts in southern and central Eurasia. *Geology*, **27**, 427–430.

Sass, J. H. and Behrendt, J. C. (1980). Heat flow from the Liberian Precambrian Shield. *J. Geophys. Res.*, **85**, 3159–3162.

Sato, H. and Sacks, I. S. (1989). Anelasticity and thermal structure of the oceanic mantle: Temperature calibration with heat flow data. *J. Geophys. Res.*, **94**, 5705–5715.

Sato, H. and Ida, Y. (1984). Low frequency electrical impedance of partially molten gabbro: the effect of melt geometry on electrical properties. *Tectonophysics,* **107**, 105–134.

Sato, H. I., Sacks, E., Takahashi, E., and Scarfe, C. M. (1988). Geotherms in the Pacific Ocean from laboratory and seismic attenuation studies. *Nature*, **336**, 154–156.

Sato, H., Sacks, I. S., Murase, T., Muncill, G., and Fukuyama, H. (1989). Qp-melting temperature relation in peridotite at high pressure and temperature: Attenuation mechanism and implications for the mechanical properties of the upper mantle. *J. Geophys. Res.*, **94**, 10647–10661.

Savage, B. and Silver, P. G. (2008). Evidence for a compositional boundary within the lithospheric mantle beneath the Kalahari craton from S receiver functions. *Earth Planet. Sci. Lett.*, **272**, 600–609.

Savage, M. K. (1999). Seismic anisotropy and mantle deformation: what have we learned from shear-wave splitting? *Rev. Geophys.*, **37**, 65–106.

Schaefer, BF; Turner, S; Parkinson, I, Rogers, N., and Hawkesworth, C. (2002). Evidence for recycled Archaean oceanic mantle lithosphere in the Azores plume. *Nature*, **420**, 6913, 304–307.

Schatz, J. F. and Simmons, G. (1972). Thermal conductivity of Earth minerals at high temperatures. *J. Geophys. Res.*, **77**, 6966–6983.

Schilling, J.-G., (1991). Fluxes and excess temperatures of mantle plumes inferred from their interaction with migrating mid-ocean ridges. *Nature* **52**, 397–403.

Schilling, F. R., Partzsch, G. M., Brasse, H., *et al.* (1997). Partial melting below the magmatic arc in the central Andes deduced from geoelectromagnetic field experiments and laboratory data. *Phys. Earth Planet. Inter.*, **103**, 17–31

Schlindwein, V. (2006). On the use of teleseismic receiver functions for studying the crustal structure of Iceland. *Geophys. J. Int.*, **164**, 551–568.

Schmeling, H. (1985a). Numerical model of partial melt on elastic, anelastic and electric properties of rocks, Part I. Elasticity and anelasticity. *Phys. Earth Planet. Inter.* **41**, 34–57.

Schmeling, H. (1985b). Numerical models on the influence of partial melt on elastic, anelastic and electrical properties of rocks. Part II: Electrical conductivity. *Phys. Earth Planet. Inter.* **43**, 123–136.

Schmidberger, S. S. and Francis, D. (1999). Nature of the mantle roots beneath the North American Craton: mantle xenolith evidence from Somerset Island kimberlites. *Lithos*, **48**, 195–216.

Schmitz, M. D., Bowring, S. A. (2003). Constraints on the thermal evolution of continental lithosphere from U-Pb accessory mineral thermochronometry of lower crustal xenoliths, southern Africa. *Contrib. Mineral. Petrol.*, **144**, 592–618.

Schock, R. N., Duba, A. G., and Shankland, T. J. (1989). Electrical Conduction in Olivine. *J. Geophys. Res.* **94**, 5829–5839.

Scholl, D. W. and von Huene, R. (2007). Crustal recycling at modern subduction zones applied to the past – Issues of growth and preservation of continental basement crust, mantle geochemistry, and supercontinent reconstruction. In: Hatcher, R. D., Carlson, M. P., McBride, J. H., *et al.* (Eds.), *4-D Framework of Continental Crust*. Geological Society of America Memoirs, **200**, 9–32.

Schreiber, D., Lardeaux, J. M., Martelet, G., *et al.* (2010). 3-D modelling of Alpine Mohos in Southwestern Alps. *Geophys. J. Int.*, **180**, 961–975.

Schroeder, W. (1984). The empirical age-depth relation and depth anomalies in the Pacific ocean basin. *J. Geophys. Res.*, **89**, 9873–9883.

Schubert, G. and Sandwell, D. (1989). Crustal volumes of the continents and of oceanic and continental submarine plateaus. *Earth Planet. Sci. Lett.*, **92**, 234–246.

Schubert, G., Turcotte, D. L., and Olson, P. (2001). *Mantle convection in the Earth and planets*. Cambridge University Press, 940 pp.

Schubert, G, Froideveaux, C., and Yuen, D. A. (1976). Oceanic lithosphere and asthenosphere: Thermal and mechanical structure. *J. Geophys. Res.*, **81**, 3525–3540.

Schultz, A. and Larsen, J. (1990). On the electrical-conductivity of the mid-mantle. 2. Delineation of heterogeneity by application of extremal inverse solutions. *Geophys. J. Int.*, **101**, 565–580.

Schultz, A., Kurtz, R. D., Chave, A. D., and Jones, A. G. (1993). Conductivity discontinuities in the upper mantle beneath a stable craton. *Geophys. Res. Lett.* **20**, 2941–2944.

Schutt, D. L. and C. E. Lesher, (2006). Effects of melt depletion on the density and seismic velocity of garnet and spinel lherzolite. *J. Geophys. Res.*, **111**, B05401, doi:10.1029/2003JB002950.

Schwalenberg, K. and Edwards, R. N. (2004). The effect of seafloor topography on magnetotelluric fields: an analytical formulation confirmed with numerical results. *Geophys. J. Int.*, **159**, 607–621.

Sclater, J. G. and Francheteau, J. (1970). Implications of terrestrial heat flow observations on current tectonic and geochemical models of the crust and upper mantle of the Earth. *Geophys. J. Roy. Astron. Soc.*, **20**, 509–534.

Sclater, J. G., Parsons, B., and C. Jaupart (1981). Oceans and continents: Similarities and differences in the mechanisms of heat loss. *J. Geophys. Res.* **86**, 11335–11552.

Sclater, J. G., C. Jaupart, and D. Galson (1980), The heat flow through oceanic and continental crust and the heat loss of the earth. *Rev. Geophys. Space Phys.*, **18**, 269–311.

Sears, J. W., George, G. M.St. and Winne, J. C. (2005). Continental rift systems and anorogenic magmatism. *Lithos*, **80**, 147–154.

Seipold, U. (1992). Depth dependence of thermal transport properties for typical crustal rocks. *Phys. Earth Plan. Int.*, **69**, 299–303.

Selby, N. D. and Woodhouse, J. H. (2002). The Q structure of the upper mantle: Constraints from Rayleigh wave amplitudes. *J. Geophys. Res.*, **107**(5), 2001JB000257.

Sella, G. F., S. Stein, T. H. Dixon, *et al.* (2007). Observation of glacial isostatic adjustment in "stable" North America with GPS. *Geophys. Res. Lett.*, **34**, L02306, doi:10.1029/2006GL027081.

Semenov, V. Y. and Rodkin, M. (1996). Conductivity structure of the upper mantle in an active subduction zone. *J. Geodynamics*: **21**, 355–364.

Sengör, A. M. C., and B. A. Natal'in (1996). Paleotectonics of Asia: fragments of a synthesis. In: Yin, A. and Harrison, T. M. (Eds.), *The Tectonic Evolution of Asia*. Cambridge, UK, Cambridge University Press, 486–640.

Sengör, A. M. C., Natal'in, B. A., and Burtman, V. S. (1993). Evolution of the Altaid Tectonic Collage and Palaeozoic Crustal Growth in Eurasia. *Nature*, **364**, 299–307.

Sergeyev, K. F., Argentov, V. V., and Bikkenina, S. K. (1983). Seismic model of the crust of the southern Okhotsk region and its geological interpretations. *Pacific Ocean Geol.*, **2**, 3–13 (in Russian).

Shankland, T. J., and Duba, A. (1990). Standard electrical conductivity of isotropic, homogeneous olivine in the temperature range 1200–1500 °C. *Geophys. J. Int.*, **103**, 25–31.

Shankland, T. J., O'Connell, R. J., and Waff, H. S. (1981). Geophysical constraints on partial melt in the upper mantle. *Rev. Geophys. Space Phys.*, **19**, 394–406.

Shapiro, N. M. and Ritzwoller, M. H. (2004). Thermodynamic constraints on seismic inversions. *Geophys. J. Int.*, **157**, 1175–1188.

Shapiro, N. M. and Ritzwoller, M. H. (2002). Monte–Carlo inversion for a global shear velocity model of the crust and upper mantle. *Geophys. J. Int.*, **151**, 1–18.

Shapiro, S. S., Hager, B. H., and Jordan, T. H. (1999). Stability and dynamics of the continental tectosphere. *Lithos,* **48**, 135–152.

Shaw, D. M., Cramer, J. J., Higgins, M. D., and Truscott, M. G. (1986). Composition of the Canadian Precambrian shield and the continental crust of the Earth, In: J. B. Dawson, D. A. Carswell, J. Hall, and K. H. Wedepohl (Eds.), Nature of the Lower Continental Crust, Geological Society of London (UK) Special Pub. 24, 275–282.

Shcheka, S. S., Wiedenbeck, M., Frost, D. J. and Keppler, H. (2006). Carbon solubility in mantle minerals. *Earth Planet. Sci. Lett.*, **245**: 730–742.

Shearer, P. M. (1999). *Introduction to Seismology.* Cambridge University Press, 260 pp.

Shearer, P. (1991). Imaging global body wave phases by stacking long-period seismograms. *J. Geophys. Res.*, **96**, 20,353–20,364.

Sheehan, A. F. and Solomon, S. C. (1992). Differential shear wave attenuation and its lateral variation in the North Atlantic region. *J. Geophys. Res.*, **97**, 15,339–15,350.

Sheriff, R. E. and Geldart, L. P. (1982). *Exploration Seismology*, v.1, 2. Cambridge University Press.

Shih, X. R., Meyer, R. P., and Schneider, J. F. (1991). Seismic anisotropy above a subducting plate. *Geology*, **19**, 807–810.

Shimakawa, Y. and Honkura, Y. (1991). Electrical conductivity structure beneath the Ryukyu trench-arc system and its relation to the subduction of the Phillipine sea plate. *J. Geomagn. Geoelectr.*, **43**, 1–20.

Shirey, S. B. and Walker, R. J. (1998). The Re–Os isotope system in cosmochemistry and high temperature geochemistry. *Ann. Rev. Earth Planet. Sci.* **26**, 423–500.

Shito, A. and Shibutan, T. (2003). Nature of heterogeneity of the upper mantle beneath the northern Philippine Sea as inferred from attenuation and velocity tomography. *Phys. Earth Planet. Inter.*, **140**, 331–341.

Shito, A., Karato, S-I., Matsukage, K. N., and Nishihara, Y. (2006). Towards mapping the three-dimensional distribution of water in the upper mantle from velocity and attenuation tomography. In: *Earth's Deep Water Cycle*. American Geophysical Union. Geophysical Monograph Series **168**, 225–236.

Silver, P. G. (1996). Seismic anisotropy beneath the continents: Probing the depths of Geology, *Ann. Rev. Earth Planet. Sci.*, **24**, 385–432.

Silver, P. G. and Chan, W. W. (1988). Implications for continental structure and evolution from seismic anisotropy. *Nature*, **335**, 34–39.

Silver, P. G. and W. W. Chan (1991). Shear-wave splitting and subcontinental mantle deformation, *J. Geophys. Res.*, **96**, 16,429 –16,454.

Silver, P. G., Gao, S. S., Liu, K. H. and Kaapvaal Seismic Group (2001). Mantle deformation beneath southern Africa. *Geophys. Res. Lett.*, **28**, 2493–2496.

Simmons, N. A., Forte, A. M., and Grand, S. P. (2009). Joint seismic, geodynamic and mineral physical constraints on three-dimensional mantle heterogeneity: Implications for the relative importance of thermal versus compositional heterogeneity. *Geophys. J. Int.*, **177**, 1284–1304.

Simon, R. E., Wright, C., Kgaswane, E. M., and Kwadiba, M. T. O. (2002). The P-wavespeed structure below and around the Kaapvaal craton to depths of 800 km, from traveltimes and waveforms of local and regional earthquakes and mining-induced tremors, *Geophys.J. Int.*, **151**, 132–145.

Simons, F. J., van der Hilst, R. D., and Zuber, M. T. (2003). Spatiospectral localization of isostatic coherence anisotropy in Australia and its relation to seismic anisotropy: Implications for lithospheric deformation, *J. Geophys. Res.*, **108**(B5), 2250, doi:10.1029/2001JB000704.

Simons, F. J., Zielhuis, A., and van der Hilst, R. D. (1999). The deep structure of the Australian continent from surface wave tomography. *Lithos*, **48**, 17–43.

Simons, F. J., Zuber, M. T., and Korenaga, J. (2000), Spatiospectral localization of isostatic coherence anisotropy in Australia and its relation to seismic anisotropy: Implications for lithospheric deformation, *J. Geophys. Res.*, **105**, 19163–19184.

Simons, F. J., van der Hilst, R. D., Montagner, J. P., and Zielhuis, A. (2002). Multimode Rayleigh wave inversion for heterogeneity and azimuthal anisotropy of the Australian upper mantle. *Geophys. J. Int.*, **151**, 738–754.

Simpson, F. (2001). Resistance to mantle flow inferred from the electromagnetic strike of the Australian upper mantle. *Nature*, **412**, 632–635.

Simpson, F. (2002a). A comparison of electromagnetic distortion and resolution of upper mantle conductivities beneath continental Europe and the Mediterranean using islands as windows. *Phys. Earth Planet, Int.* **129**: 117–130.

Simpson, F. (2002b). Intensity and direction of lattice preferred orientation of olivine: are electrical and seismic anisotropies of the Australian mantle reconcilable? *Earth Planet. Sci. Lett.*, **203**, 535–547.

Singh, R. P., Kant, Y., and Vanyan, L. (1995). Deep electrical conductivity structure beneath the southern part of the Indo-Gangetic plains. *Phys. Earth Planet. Int.*, **88**, 273–283.

Sinha, M. C., Navin, D. A., MacGregor, L. M., *et al.* (1997). Evidence for accumulated melt beneath the slow spreading Mid-Atlantic Ridge. *Phil. Trans. Royal Soc., A*, **335**, 233–253.

Sipkin, S. A. and Jordan, T. H. (1975). Lateral heterogeneity of the upper mantle determined from the travel times of ScS. *J. Geophys. Res.*, **80**, 1474–1484.

Sisson, T. W. and Grove, T. L. (1993). Experimental investigations of the role of Hz0 in talc-alkaline differentiation and subduction zone magmatism. *Contrib. Mineral. Petrol.*, **113**, 143–166.

Skemer, P. A., Katayama, I., and Karato, S. (2006). Deformation fabrics of a peridotite from Cima di Gagnone, central Alps, Switzerland: evidence of deformation under water-rich condition at low temperatures. *Contrib. Mineral. Petrol.* **152**: 43–51.

Skinner, E. M., Apter, D. B., Morelli, C., and Smithson, N. K. (2004). Kimberlites of the Man craton, West Africa. *Lithos*, **76**, 233–259.

Sleep, N. H. (2006). Mantle plumes from top to bottom. *Earth Science Rev.*, **77**, 231–271.

Sleep, N. H. (2005). Evolution of the continental lithosphere. *Ann. Rev. Earth Planet. Sci.*, **33**, 369–393.

Sleep, N. H. (2003). Geodynamic implications of xenolith geotherms. *Geochem. Geophys. Geosyst.*, **4**(9), 1079, doi:10.1029/ 2003GC000511.

Sleep, N. (1994). Lithospheric thinning by midplate mantle plumes and the thermal history of hot plume material ponded at sublithospheric depths. *J. Geophys. Res.*, **99**, 9327–9343.

Sleep, N. H. (1990). Hotspots and mantle plumes: some phenomenology. *J. Geophys. Res.*, **95**, 6715–6736.

Smirnov, M. Yu., Pedersen, L. B. (2007). Magnetotelluric measurements across Sorgenfrei-Tornquist-zone in southern Sweden and Denmark. *Geophys. J. Int.,* **176**, 443–456.

Smith, D. (1999). Temperatures and pressures of mineral equilibration in peridotite xenoliths: review, discussion and implications. In: Fei, Y., Bertka, C., and Mysen, B. O. (Eds.), *Mantle Petrology: Field Observations and High Pressure Experimentation*. The Geochemical Society, Houston, vol. 6, 171–188.

Smith, M. F. (1989). Imaging the Earth's aspherical structure with free oscillation frequency and attenuation measurements. Ph.D. thesis, 97 pp., Univ. of Calif., San Diego, La Jolla, USA.

Smith, W. H. F. and Sandwell, D. T. (1997). Global sea floor topography from satellite altimetry and ship depth soundings. *Science,* **277**, 1956–1962.

Smythe, D. K., Dobinson, A., McQuillin, R., *et al*. (1982). Deep structure of the Scottish Caledonides by the MOIST reflection profile. *Nature,* **299**, 338–340.

Snow, C. A. (2006). A reevaluation of tectonic discrimination diagrams and a new probabilistic approach using large geochemical databases. *J. Geophys. Res.*, **111**, B6, paper 6.

Sobolev, N. V. (1974). *Deep Seated Inclusions in Kimberlites and the Problem of the Composition of the Upper Mantle*. Nauka, Moscow (in Russian). English translation by D. A. Brown, AGU, Washington, D.C., 279 pp., 1977.

Sobolev, S. V., Zeyen, H., Stoll, G., *et al*. (1996). Upper mantle temperatures from teleseismic tomography of French Massif Central including effects of composition, mineral reactions, anharmonicity, anelasticity and partial melt. *Earth Planet. Sci. Lett.*, **139**, 147–163.

Soller, D. R., Ray, R. D., and Brown, R. D. (1982). A new global crustal thickness model. *Tectonics*, **1**, 125–149.

Souriau, A. and Granet, M. (1995). A tomographic study of the lithosphere beneath the Pyrenees from local and teleseismic data. *J. Geophys. Res.*, **100**, 18117–18134.

Spakman, W. (1991). Delay-time tomography of the upper mantle below Europe, the Mediterranean, and Asia Minor. *Geophys. J. Int.*, **107**, 309–332.

Spakman, W. and Nolet, G. (1988). Imaging algorithms, accuracy and resolution in delay time tomography, In: Vlaar, N. J. *et al.* (Eds.), *Mathematical Geophysics*, Norwell, Mass, D. Reidel, 155–187.

Spakman, W., van der Lee, S., and van der Hilst, R. D. (1993). Travel-time tomography of the European-Mediterranean mantle down to 1400 km. *Phys. Earth Planet. Inter.*, **79**, 3–74.

Spear, F. S. (1993). *Metamorphic Phase Equilirbia and Pressure–Temperature–Time Paths*. Mineral. Soc. America, 799 pp.

Spector, A. and Grant, F. S. (1970). Statistical models for interpreting aeromagnetic data, *Geophysics*, **35**, 293–302.

Spohn, T. and Schubert, G. (1982). Convective thinning of the lithosphere: a mechanism for the initiation of continental rifting. *J. Geophys. Res.*, **87**, 4669–4681.

Stark, C. and Stewart, J. (1997). Mapping lithospheric strength and loading using wavelet transform admittance and coherence, *Abstract AGU Fall meeting*, San Francisco, CA.

Stein, C. A. (1995). Heat flow of the Earth. In: Ahrens, T. J. (Ed.), *A Handbook of Physical Constants*. Am. Geophys. Union, Reference shelf 1, AGU, 144–158.

Stein, C. A. and Abbott, D. H. (1991). Heat flow constraints on the South Pacific Superswell. *J. Geophys. Res.*, **96**, 16083–16099.

Stein, C. A. and Stein, S. (1992). A model for the global variation in oceanic depth and heat flow with lithospheric age *Nature*, **359**, 123–129.

Stephenson, R. and Beaumont, C. (1980), Small scale convection in the upper mantle and the isostatic response of the Canadian Shield. In: Davies, P. A. and Runcorn, S. K. (Eds.), *Mechanisms of Continental Drift and Plate Tectonics*. San Diego, Calif, Academic, 111–122.

Stewart, J. and Watts, A. B. (1997). Gravity anomalies and spatial variations of flexural rigidity at mountain ranges. *J. Geophys. Res.* **102**, 5327–5353.

Stixrude, L. and Lithgow-Bertelloni, C. (2005) Mineralogy and elasticity of the oceanic upper mantle: Origin of the low-velocity zone. *J. Geophys. Res.*, **110**, B03204, doi:10.1029/2004JB002965.

Stocker, R. L. and Gordon, R. B. (1975). Velocity and internal-friction in partial melts. *J. Geophys. Res.*, **80**, 4828–4836.

Stoddard, P. R. and Abbott, D. (1996). Influence of the tectosphere upon plate motion. *J. Geophys. Res.*, **101**, 5425–5433.

Stolper, E. and Newman, S. (1994). The role of water in the petrogenesis of Mariana trough magmas. *Earth Planet. Sci. Lett.*, **121**, 293–325.

Stratford, W, Thybo, H, Faleide, JI, *et al.* (2009). New Moho Map for onshore southern Norway. *Geophys. J. Int.*, **178**, 1755–1765.

Stüwe, K. (2002). *Geodynamics of the Lithosphere: An Introduction*. Springer, 449 pp.

Su, W.-J. and Dziewonski, A. (1991). Predominance of long-wavelength heterogeneity in the mantle. *Nature*, **352** (6331), 121–126.

Su, W., Woodward, R. L. and Dziewonski, A. M. (1994). Degree 12 model of shear velocity heterogeneity in the mantle. *J. Geophys. Res.*, **99**, 6945–6980.

Sumino, Y. and Anderson, O. L. (1982). Elastic constants in minerals. In: Carmichel, R. S. (Ed), *CRC Handbook of Physical Properties of Rocks*. CRC Press, Boca Raton, Fla., 39–138.

Swain, C. J. and Kirby, J. F. (2006). An effective elastic thickness map of Australia from wavelet transforms of gravity and topography using Forsyth's method. *Geophys. Res. Lett.*, **33**, L02314, doi:10.1029/2005GL025090.

Swanberg, C. A., Chessman, M. D., and Simmons, G. (1974). Heat flow-heat generation studies in Norway. *Tectonophys.*, **23**, 31–48.

Tackley, P. J. (2000). The quest for self-consistent generation of plate tectonics in mantle convection models. In: Richards, M. A., Gordon, R. G., and van der Hilst, R. D. (Eds.), *The History and Dynamics of Global Plate Motions*. AGU Geophys. Monograph, **121**, 47–72.

Tackley, P., Stevenson, D. J., Glatzmaier, G. A., Schubert, G. (1993). Effects of an endo-thermic phase transition at 670 km depth in a spherical model of convection in the Earth's mantle. *Nature*, **361**, 699–704.

Takahashi, E. (1986). Melting of a dry peridotite KLB-1 up to 14 GPA – implications on the origin of peridotitic upper mantle. *J. Geophys. Res.*, **91**, 9367–9382.

Tanimoto, T. (1995). Crustal structure of the Earth. In: Ahrens, T. J. (Ed.), *A Handbook of Physical Constants*. AGU Reference Shelf 1. American Geophysical Union, 214–224.

Tanimoto, T. Anderson, D. L. (1984). Mapping convection in the mantle. *Geophys. Res. Lett.*, **11**, 287–290.

Tao, W. and Jarvis, G. T. (2002). The influence of continental surface area on the assembly time for supercontinents. *Geophys. Res. Lett.*, **29**, no. 12, doi 10.1029/2001GL013712.

Tapponnier, P. and Brace, W. P. (1976). Stress induced microcracks in Westerly granite. *Int. J. Rock Mech. Mining*, **13**, 103–112.

Tapponnier, P., Xu, Z., Roger, F., *et al.* (2001). Oblique stepwide rise and growth of the Tibetan Plateau, *Science*, **294**, 1671–1677.

Tarits, P., Chave, A. D., and Schultz, A. (1993). Comment on "The Electrical Conductivity of the Oceanic Upper Mantle" by G. Heinson and S. Constable. *Geophys. J. Int.*, **114**, 711–716.

Tarits, P., Hautot, S., and Perrier F. (2004) Water in the mantle: results from electrical conductivity beneath the French Alps. *Geophys. Res. Lett.* **31**:L06612.

Tassara, A., Swain C., Hackney, R., and Kirby, J. (2007). Elastic thickness structure of South America estimated using wavelets and satellite-derived gravity data. *Earth Planet. Sci. Lett.*, **253**, 17–36.

Tatsumi, Y., Sakuyama, M., Fukuyama, H., *et al.* (1983). Generation of arc basalt magmas and thermal structure of the mantle wedge in subduction zones. *J. Geophys. Res.*, **88**, 5815–5825.

Taylor, W. R. (1998). An experimental test of some geothermometer and geobarometer formulations for upper mantle peridotites with application to the thermobarometry of fertile lherzolite and garnet websterite. *Neus. Jahrb. Mineral. Abh.* **172**, 381–408.

Taylor, Sir Geoffrey Ingram (1950). The instability of liquid surfaces when accelerated in a direction perpendicular to their planes. *Proc. Roy. Soc. Lond. Series A, Math. Phys. Sci.* **201** (1065): 192–196. doi:10.1098/rspa.1950.0052.

Taylor, S. R. and McLennan, S. M. (1995). The geochemical evolution of the continental crust. *Rev. Geophys.* **33**, 241–265.

Taylor, S. R. and McLennan, S. M. (1985). *The Continental Crust: Its Composition and Evolution*. Oxford, Blackwell, 312 pp.

Tejada, M., Mahoney, J. J., Duncan, R. A., and Hawkins, M. P. (1996). Age and geochemistry of basement and alkali rocks of Malaita and Santa Isabel, Solomon Islands, southern margin of the Ontong Java plateau. *J. Petrol.*, **37**, 361–394.

ten Grotenhuis, S. M., Drury, M. R., Spiers, C. J., and Peach, C. J., (2005). Melt distribution in olivine rocks based on electrical conductivity measurements. *J. Geophys. Res.*, **110**, B12201, doi:10.1029/2004JB003462.

Tezkan, B. (1994). On the detectability of a highly conductive layer in the upper mantle beneath the Black Forest crystalline using magnetotelluric methods. *Geophys. J. Int.* **118**, 185–200.

The MELT Seismic Team. (1998). Imaging the deep seismic structure beneath a mid-ocean ridge: The MELT Experiment, *Science* **280**, 1215–1218.

Thompson, A. B. (1992). Water in the Earth's upper mantle. *Nature*, **358**, 295–302.

Thompson, P. H., Judge, A. S., and Lewis, T. J. (1996). Thermal evolution of the lithsophere in the central Slave Province: Implications for diamond genesis, In: LeCheminant, A. N., Richardson, D. G., DiLabio, R. N. W., and Richardson, K. A. (Eds.), *Searching for Diamonds in Canada*. Geol. Surv. of Canada, Open File 3228, pp. 151–160.

Thomson, D. J. (1982). Spectrum estimation and harmonic analysis, *Proc. IEEE*, **70**, 1055–1096.

Thomson, D. J. and Chave, A. D. (1991). Jackknifed error estimates for spectra, coherences, and transfer functions. In *Advances in Spectrum Analysis and Array Processing*, Haykin, S., (Ed.) vol. 1, Old Tappan, N. J., Prentice-Hall, 58–113.

Thybo, H. (2006). The heterogeneous upper mantle low velocity zone. *Tectonophysics*, **416**, 53–79.

Thybo, H. and Nielsen, C. A. (2009). Magma-compensated crustal thinning in continental rift zones. *Nature*, **457**, 873–876.

Thybo, H. and Perchuc, E. (1997a). The seismic 8° discontinuity and partial melting in continental mantle. *Science*, **275**, 1626–1629.

Thybo, H. and Perchuc, E. (1997b). The transition from cold to hot areas of North America interpreted from Early Rise seismic record sections. In: Fuchs, K. (Ed.), *Upper Mantle Heterogeneities From Active and Passive Seismology*. Kluwer Academic Publishers, 131–138.

Thybo, H., Perchuc, E., and Zhou, S. (2000). Intraplate earthquakes and a seismically defined lateral transition in the upper mantle. *Geophys. Res. Lett*, **27**, 3953–3956.

Tian, Y., Zhao, D., Sun, R., and Jiwen Teng, J. (2009). Seismic imaging of the crust and upper mantle beneath the North China Craton. *Phys. Earth Planet. Inter.*, **172**, 169–182.

Tilmann, F. J., INDEPTH III Seismic Team and Ni, J. (2003). Seismic imaging of the downwelling Indian lithosphere beneath central Tibet. *Science*, **300**, 1424–1427.

Timoshenko, S. (1936). *Theory of Elastic Stability*. McGraw-Hill, 518 pp.

Timoshenko, S. and Woinowsky-Krieger S. (1959). *Theory of Plates and Shells*. McGraw-Hill, 580 pp.

Tittgemeyer, M., Wenzel, F., Fuchs, K., and Wenzel, F. (1996). Wave propagation in a multiple scattering upper mantle – observation and modelling, *Geophys. J. Int.*, **127**, 492–502.

Toh, H. (2003). Asymmetric electrical structures beneath mid-ocean ridges, *J. Geog.* **112**(5), 684–691.

Toksöz, M. N. and Anderson, D. L. (1966). Phase velocities of long-period surface waves and structure of the upper mantle: I. Great-circle Love and Rayleigh wave data. *J. Geophys. Res.*, **71**, 1649–1658.

Tozer, DC. (1981). The mechanical and electrical-properties of earths asthenosphere. *Phys. Earth Planet. Inter.*, **25**, 280–296.

Trampert, J. and Woodhouse, J. H. (1995). Global phase velocity maps of Love and Rayleigh waves between 40 and 150 s. *Geophys. J. Int.*, **122**, 675–690.

Trampert, J. and Woodhouse, J. H. (1996). High resolution global phase velocity distributions. *Geophys. Res. Lett.*, **23**, 21–24.

Trofimov, V. A. (2006). Structural features of the earth's crust and petroleum potential: first results of CMP Deep Seismic Survey along the geotraverse across the Volga–Urals petroliferous province. *Doklady Earth Sci.*, **411**(8), 1178–1183.

Trompert, R. and Hansen, U. (1998). Mantle convection simulations with rheologies that generate plate-like behavior. *Nature*, **395**, 686–689.

Tsumura, N., Matsumoto, S., Horiuchi, S., and Hasegawa, A. (2000). Three-dimensional attenuation structure beneath the northeastern Japan arc estimated from spectra of small earthquakes. *Tectonophys.*, **319**, 241–260.

Tugarinov, A. I. and Bibikova, E. V. (1976). Evolution of the chemical composition of the earth crust. *Geokhimiya*, **8**, 1151–1159.

Turcotte, D. L. and Schubert, G. (1982). *Geodynamics*. J. Wiley and Sons, New York – Chichester – Brisbane – Toronto – Singapore, 347 pp.

Turcotte, D. L. and Schubert, G. (2002). *Geodynamics*. 2nd edn., Cambridge University Press, 456 pp.

Turcotte, D. L., Mcadoo, D. C., and Caldwell, J. G. (1978). Elastic-perfectly plastic analysis of bending of lithosphere at a trench. *Tectonophysics*, **47**, 193–205.

Turner, S., Hawkesworth, C., Liu, J., Rogers, N., Kelley, S., and van Calsteren, P. (1993). Timing of Tibetan uplift constrained by analysis of volcanic rocks. *Nature*, **364**, 50–54.

Turner, S. P., Platt, J. P., George, R. M. M., *et al.* (1999). Magmatism associated with orogenic collapse of the Betic-Alboran domain, SE Spain, *J. Petrol.*, **40**, 1011–1036.

Tyburczy, J. A. and Fisler, D. K. (1995). Electrical properties of minerals and melts. In: A *Handbook of Physical Constants*, AGU Reference Shelf 2, 185–208.

Uffen, R. J. (1952). A method of estimating the melting-point gradient in the earth's mantle. *Trans. Am. Geophys. Union*, **33**, 893–896.

Unsworth, M., Wei, W., Jones, A. G., *et al.* (2004). Crustal and upper mantle structure of northern Tibet imaged with magnetotelluric data. *J. Geophys. Res.*, **109**, B02403.

Utada, H., Koyama, T., Shimizu, H., *et al.* (2003). A semi-global reference model for electrical conductivity in the mid-mantle beneath the north Pacific region. *Geophys. Res. Lett.*, **30**, Article Number: 1194.

Uyeda, S. Kanamori, H. (1979). Back-arc opening and the mode of subduction. *J. Geophys. Res.*, **84**, 1049–1061.

Vacher, P. and Souriau, A. (2001). A three-dimensional model of the Pyrenean deep structure based on gravity modelling, seismic images and petrological constraints. *Geophys. J. Int.*, **145**, 460–470.

Vacher, P., Verhoeven, O. (2007). Modelling the electrical conductivity of iron-rich minerals for planetary applications. *Planet. Space Sci.*, **55**, 455–466.

Valley, J., Cavosie, A. J., Fu, B., Peck, W. H., and Wilde, S. A. (2006). Comment on Heterogeneous "Hadean Hafnium: Evidence of Continental Crust at 4.4 to 4.5 Ga". *Science*, **312**, 1139a.

Van Balen, R. T. and Heeremans, M. (1998). Middle Proterozoic–early Palaeozoic evolution of central Baltoscandian intracratonic basins: evidence for asthenospheric diapirs. *Tectonophysics*, **300**, 131–142.

van der Hilst, R. D. and Engdahl, E. R. (1991). On ISC pP and PP data and their use in delay time tomography of the Caribbean region, *Geophys. J. Int.*, **106**, 169–188.

Van der Hilst, R. and Karason, H. (1999). Compositional heterogeneity in the bottom 1000 km of Earth's mantle: toward a hybrid convection model. *Science*, **283**, 1885–1888.

van der Hilst, R. D., Engdahl, E. R., and Spakman, W. (1993). Tomographic inversion of P and pP data for aspherical mantle structure below the northwest Pacific region, *Geophys. J. Int.*, **115**, 264–302.

van der Hilst, R. D., Widiyantoro, S., and Engdahl, E. R. (1997). Evidence for deep mantle circulation from global tomography. *Nature*, **386**, 578–584.

van der Hilst, R. D., Kennett, B. L. N., Christie, D., and Grant, J. (1994). Project SKIPPY explores the lithosphere and mantle beneath Australia. *EOS Trans. Am. Geophys. Union* **75**, 177–181.

van der Lee, S. (2002). High-resolution estimates of lithospheric thickness from Missouri to Massachusetts, USA. *Earth Planet. Sci. Lett.*, **203**,15–23.

van der Lee, S. and Nolet, G. (1997a). The upper mantle S velocity structure of North America. *J. Geophys. Res.*, **102**, 22815–22838.

van der Lee, S. and Nolet, G. (1997b). Seismic image of the subducted trailing fragments of the Farallon plate. *Nature* **386**, 266–269.

Van der Voo, R. and Meert, J. G. (1991). Late Proterozoic paleomagnetism and tectonic models: a critical appraisal. *Precambrian Res.*, **53**, 149–163.

Van der Voo, R., Spakman, W., and Bijwaard, H. (1999a). Tethyan subducted slabs under India. *Earth Planet. Sci. Lett.*, **171**, 7–20.

Van der Voo, R., Spakman, W., and Bijwaard, H. (1999b). Mesozoic subducted slabs under Siberia, *Nature*, **397**, 246–249.

Van Gerven, L., Deschamps, F., and Van der Hilst, R. D. (2004). Geophysical evidence for chemical variations in the Australian continental mantle. *Geophys. Res. Lett.* **31**, L17607, doi:10.1029/2004GL020307.

Van Schmus, W. R. (1995). Natural radioactivity of the crust and mantle. In: Ahrens, T. J. (Ed.), *A Handbook of Physical constants*. Am. Geophys. Union, Reference shelf 1, AGU, 283–291.

van Wijk, J. W., Govers, R., and Furlong, K. P. (2001). Three-dimensional thermal modeling of the California upper mantle: a slab window vs. stalled slab. *Earth Planet. Sci. Lett.*, **186**, 175–186.

Vanyan, L. L. and Cox, C. S. (1983). Comparison of deep conductivities beneath continents and oceans. *J. Geomag. Geoelectr.*, **35** (11–1): 805–809

Vanyan, L. L., Palshin, N. A., and Repin, I. A. (1995). Deep magnetotelluric sounding using submarine cable Australia–New Zealand, 2: Interpretation. *Fizika Zemli*, **5**, 53–57 (in Russian, also available in English).

Vanyan, L. L., Berdichewski, M. N., Fainberg, E. B., *et al.* (1977). Study of asthenosphere of East European platform by electromagnetic sounding. *Phys. Earth Planet. Inter.*, **14**, P1-P2.

Vasco, D. W., Johnson, L. R., and Marques, O. (2003). Resolution, uncertainty, and whole Earth tomography. *J. Geophys. Res.*, **108**, 2022, ESE9.

Vasseur, G. and Singh, R. N. (1986). Effects of random horizontal variations in radiogenic heat source distribution on its relationship with heat flow. *J. Geophys. Res.*, **91**, 10397–10404.

Vauchez, A. and Nicolas, A. (1991). Mountain building: strike-parallel displacements and mantle anisotropy, *Tectonophysics*, **185**, 183–201.

Veizer, J. and Jansen, S. L. (1979). Basement and sedimentary recycling and continental evolution. *J. Geol.*, **87**, 341–370.

Veizer, J. and Jansen, S. L. (1985). Basement and sedimentary recycling. 2. Time dimension to global tectonics. *J. Geol.*, **93**, 625–643.

Vening-Meinesz F. A. (1931). Une nouvelle method pour la réduction isostatique régionale de l'intensité de la pesanteur. *Bull. Géodésique*, **29**, 33–51.

Vening-Meinesz F. A. (1948). *Gravity Expeditions at Sea 1923–1930*. Vol. IV. Complete results with isostatic reduction, interpretation on the results. Nederlandse Commissie voor Geodesie, Delft, Delft, 233 pp.

Vervoort, J. D. and Blichert-Toft J. (1999). Evolution of the depleted mantle: Hf isotope evidence from juvenile rocks through time. *Geochim. Cosmochim. Acta*, **63**, 533–556.

Vidale, J. E. and Benz, H. M. (1992). Upper-mantle seismic discontinuities and the thermal structure of subduction zones. *Nature*, **356**, 678–683.

Villaseñor, A., Ritzwoller, M., Levshin, A., *et al.* (2001). Shear velocity structure of central Eurasia from inversion of surface wave velocities. *Phys. Earth Planet. Inter.*, **123**, 169–184.

Vinnik, L. P. (1977). Detection of waves converted from P to SV in mantle. *Phys. Earth Planet. Inter.*, **15**, 39–45.

Vinnik, L., V. Farra (2002). Subcratonic low-velocity layer and flood basalts. *Geophys. Res. Lett.*, **29**, doi:10.1029/2001GL014064.

Vinnik, L. P., Chevrot, S., and Montagner, J. P. (1998). Seismic evidence of flow at the base of the upper mantle. *Geophys. Res. Lett.*, **25**, 1995–1998.

Vinnik, L. P., Farra, V., and Romanowicz, B. (1989). Azimuthal anisotropy in the Earth from observations of SKS at Geoscope and NARS broadband stations, *Bull. Seismol. Soc. Am.*, **79**, 1542–1558.

Vinnik, L. P., Foulger, G. R., and Du, Z. (2005). Seismic boundaries in the mantle beneath Iceland: a new constraint on temperature: *Geophys. J. Int.*, **160**, 533–538.

Vinnik, L. P., Green, R. W. E., and Nicolaysen, L. O. (1995). Recent deformations of the deep continental root beneath southern Africa. *Nature*, **375**, 50–52.

Vinnik, L., Kurnik, E., and Farra, V. (2005). Lehmann discontinuity beneath North America: No role for seismic anisotropy. *Geophys. Res. Lett.*, **32**, Article Number: L09306.

Vinnik, L. P., Chevrot, S., Montagner, J. P., and Guyel, F. (1999). Teleseismic travel time residuals in North America and anelasticity of the asthenosphere. *Phys. Earth. Planet. Inter.*, **116**, 93–103.

Vinnik, L. P., Makeyeva, L. I., Milev, A., and Yu Usenko, A. (1992). Global patterns of azimuthal anisotropy and deformationsin the continental mantle. *Geophys. J. Int.*, **111**, 433–447.

Vitorello, I. and Pollack, H. N. (1980). On the variation of continental heat flow with age and the thermal evolution of continents, *J. Geophys. Res.*, **85**, 983–995.

von Bargen, N. and Waff, H. S. (1986). Permeabilities, interfacial areas and curvatures of partially molten systems: results of numerical computations of equilibrium microstructures., *J. Geophys. Res.*, **91**, 9261–9276.

von Herzen, R. P. and Uyeda, S. (1963). Heat flow through the eastern Pacific Ocean floor. *J. Geophys. Res.*, **68**, 4219–4250.

Von Herzen, R. P., Cordery, M. J., Detrick, R. S., and Fang, C. (1989). Heat flow and the thermal origin of hotspot swells: the Hawaiian swell revisited. *J. Geophys. Res.*, **94**, 13783–13799.

Von Herzen, R., Davis, E. E., Fisher, A., Stein, C. A., and Pollack, H. N. (2005). Comments on "Earth's heat flux revised and linked to chemistry" by A. M. Hofmeister and R. E. Criss. *Tectonophysics*, **409**, 193–198, doi:10.1016/j.tecto.2005.08.003.

von Huene, R., and Scholl, D. W. (1991). Observations at convergent margins concerning sediment subduction, subduction erosion, and the growth of continental crust. *Rev. Geophys.*, **29**, 279–316.

Vozoff, K. (Ed.), (1986). *Magnetotelluric Methods*. Soc. Expl. Geophys. Reprint Ser., vol. 5. Tulsa, OK, ISBN 0-931830-36-2.

Vozoff, K. (1991). *The Magnetotelluric Method. Electromagnetic Methods in Applied Geophysics – Applications*. Society of Exploration Geophysicists, Tulsa, OK, pp. 641–712. Chap. 8.

Wagner, T. P. and Grove, T. L. (1998). Melt/harzburgite reaction in the petrogenesis of tholeiitic magma from Kilauea volcano, Hawaii. *Contrib. Miner. Petrol.*, **131**, 1–12.

Walcott, R. I. (1970). Flexure of the lithosphere at Hawaii. *Tectonophysics*, **9**, 435–446.

Walker, R. J. and Morgan, J. W. (1989). Rhenium–osmium isotope systematics of carbonaceous chondrites. *Science*, **243**, 519–522.

Walker, R. J., Carlson, R. W., Shirey, S. B., and Boyd, F. R. (1989). Os, Sr, Nd, and Pb isotope systematics of southern African peridotite xenoliths: implications for the chemical evolution of subcontinental mantle. *Geochim. Cosmochim. Acta* **53**, 1583–1595.

Walter, M. J. (1999). Melting residues of fertile peridotite and the origin of cratonic lithosphere. In: Fei, Y., Bertka, C. M., and Mysen, B. O. (Eds.), *Mantle Petrology: Field Observations and High Pressure Experimentation: A Tribute to Francis R. (Joe) Boyd.* Geochem. Soc. Spec. Publ. No. 6, 225–239.

Walter, M. J. (1998). Melting of garnet peridotite and the orgigin of komatiite and depleted lithosphere. *J. Petrol*, 1998, **39**, 29–60.

Walter, M. J. (2005). Melt extraction and compositional variability in mantle lithosphere. In: Carlson, R. W. (Ed.), *Treatise on Geochemistry,* V.2. Elsevier, Amsterdam, pp. 363–394.

Walte, N. P., Bons, P. D., Passchier, C. W., and Koehn, D. (2003). Disequilibrium melt distribution during static recrystallisation, *Geology*, **31**, 1009–1012.

Wanamaker, B. J. and Duba, A. G. (1993). Electrical conductivity of San Carlos olivine along [100] under oxygen- and pyroxene-buffered conditions and implications for defect equilibria. *J. Geophys. Res.*, **98**, 489–500.

Wang, L. J. (1998). Electrical conductivity structure of the Australian continent. Ph.D. Thesis, Australian National Univ., Canberra.

Wang, L. J. and Lilley, F. E. M. (1999). Inversion of magnetometer array data by thin-sheet modeling. *Geophys. J. Int.*, **137**, 128–138.

Wang, Y. and Mareschal, J.-C., (1999). Elastic thickness of the lithosphere in the central Canadian Shield. *Geoph. Res. Lett.*, **26**, 3033–3036.

Wang, D., Mookherjee, M., Xu, Y., and Karato, S.-I., (2006). The effect of water on the electrical conductivity of olivine. *Nature* **443**, 977–980, doi:10.1038/nature05256.

Wang, J. N., Hobbs, B. E., Ord, A., Shimamoto, T., and Toriumi, M. (1994). Newtonian dislocation creep in quartzites: implications for the rheology of the lower crust. *Science*, **265**, 1203–1205.

Wannamaker, P. E. *et al.* (1989). Magnetotelluric section across the Juan de Fuca subduction system in the EMSLAB project. *J. Geophys. Res.*, **94**, 14111–14125; 14127–14144, 14277–14293.

Wark, D. A. and Watson, E. B. (1998). Grain-scale permeabilities of texturally equilibrated, monomineralic rocks. *Earth Planet. Sci. Lett.*, **164**, 591–605.

Warren, J. M. and Hirth, G. (2006). Grain size sensitive deformation mechanisms in naturally deformed peridotites. *Earth Planet. Sci. Lett.*, **248**, 438–450.

Wasserburg, G. J., MacDonald, G. J., Hoyle, F., and Fowler, W. A. (1964). Relative contributions of uranium, thorium and potassium to heat production in the Earth. *Science*, **143**, 465.

Watanabe, T. and Kurita, K. (1993). The relationship between electrical conductivity and melt fraction in a partially molten simple system: Archie's law behavior. *Phys. Earth Planet. Inter.*, **78**, 9–17.

Watanabe, T. and Kurita, K. (1994). Simultaneous measurements of the compressional-wave velocity and the electrical conductivity in a partially molten material. *J. Phys. Earth*, **42**, 69–87.

Watanabe, T., Langseth, M. G., and Anderson, R. N. (1977). Heat flow in back-arc basins of the western Pacific. In: Talwani, M. and Pitman, III, W. (Eds.) *Island Arcs, Deep Sea Trenches, and Back-Arc Basins*, Washington, DC, American Geophysical Union, 137–162.

Watts, A. B. (2001). *Isostasy and Flexure of the Lithosphere.* Cambridge University Press, 458 pp.

Watts, A. B. and Burov, E. (2003). Lithospheric strength and its relationship to the elastic and seismogenic layer thickness. *Earth Planet. Sci. Lett.*, **213**, 113–131.

Watts, A. B. and Talwani, M. (1974). Gravity anomalies seaward of deep-sea trenches and their tectonic implications. *Geophys. J. Roy. Astron. Soc.* **36**, 57–90.

Watts, A. B., Bodine, J. H., and Steckler, M. S. (1980). Observations of flexure and the state of stress of the oceanic lithosphere. *J. Geophys. Res.*, **85**, 6369–6376.

Weaver, B. L. and Tarney, J. (1984). Empirical approach to estimating the composition of the continental crust. *Nature*, **310**, 575–577.

Wedepohl, K. H. (1995). The composition of the continental crust. *Geochim. Cosmochim. Acta*, **59**, 1217–1232.

Weeraratne, D. S., Forsyth, D. W., Fischer, K. M., and Nyblade, A. A. (2003). Evidence for an upper mantle plume beneath the Tanzanian craton from Rayleigh wave tomography. *J. Geophys. Res.*, **108**, B9, 2427, 2002JB002273R.

Wei, W., *et al.* (2001), Detection of widespread fluids in the Tibetan crust by magnetotelluric studies. *Science*, **292**, 716–718.

Weidelt, P. (1972). The inverse problem of geomagnetic induction. *Z. Geophys.*, **38**, 257–289.

Weijermars, R (1997). Pulsating oblate and prolate three-dimensional strains. *Math. Geology*: **29**, 17–41.

Wendlandt, E., DePaolo, D. J., and Baldridge, W. S. (1993). Nd and Sr isotope chronostratigraphy of Colorado Plateau lithosphere: implications for magmatic and tectonic underplating of the continental crust. *Earth Planet. Sci. Lett.* **116**, 23–43.

Whaler, K. A. and Hautot, S. (2006). The electrical resistivity structure of the crust beneath the northern Main Ethiopian Rift. In: Yirgu, G., Ebinger, C. J. and Maguire, P. K. H. (Eds): *The Afar Volcanic Province within the East African Rift System.* Geological Society, London, Sp. Publ., **259**, 293–305.

White, R. S., McKenzie, D., and O'Nions, R. K. (1992). Oceanic crustal thickness from seismic measurements and rare earth element inversions, *J. Geophys. Res.*, **97**, 19683–19715.

White, R. S., Smith, L. K., Roberts, A. W., *et al.* (2008). Lower-crustal intrusion on the North Atlantic continental margin, *Nature*, **452**, 460–U6.

Whitehouse, M. J., Myers, J. S., and Fedo, C. M. (2009). The Akilia Controversy: field, structural and geochronological evidence questions interpretations of >3.8 Ga life in SW Greenland. *J. Geological Society*; **166**; 335–348; DOI: 10.1144/0016-76492008-070.

Widess, M. B. (1982). Quantifying resolving power of seismic systems. *Geophysics*, **47**, 1160–1173.

Widiyantoro, S. (1997). Studies of seismic tomography on regional and global scale, Ph.D. thesis, Aust. Natl. Univ., Canberra, A.C.T., Australia.

Widom, E. (2002). Ancient mantle in a modern plume. *Nature*, **420**, 281–282.

Widom, E. and Shirey, S. B. (1996). Os isotope systematics in the Azores: implications for mantle plume sources. *Earth Planet. Sci. Lett.* **142**, 451–466.

Wilde, S. A., Valley, J. W., Peck, W. H., and Graham, C. M. (2001). Evidence from detrital zircons for the existence of continental crust and oceans on the earth 4.4 Gyr ago. *Nature* **409**, 175–178.

Williams, D. L. and von Herzen, R. P. (1974). Heat loss from the Earth: new estimate. *Geology*, **2**, 327–328.

Wilshire, H. G. and Jackson, E. D. (1975). Problems in determining local geotherms from mantle-derived ultramafic rocks. *J. Geol.*, **83**, 313–329.

Wilson, M. (1992). Magmatism and continental rifting during the opening of the South Atlantic Ocean: a consequence of Lower Cretaceous superplume activity? *Geol. Soc. London*, Sp. Publ. **68**, 241–255.

Windley, B. F. (1995). *The Evolving Continents.* Chichester, New York, Brisbane, Toronto–Singapore, John Wiley and Sons, 489 pp.

Witt-Eickschen G. and Seck, H. A. (1991). Solubility of Ca and Al in orthopyroxene from spinel peridotite: an improved version of an empirical thermometer. *Contrib. Mineral. Petrol.* **106**, 431–439.

Wittlinger, G., J. Vergne, P. Tapponnier, *et al.* (2004). Teleseismic imaging of subducting lithosphere and Moho offsets beneath western Tibet. *Earth Planet. Sci. Lett.*, **221**, 117–130.

Wolf, D. (1985). Thick plate flexure re-examined. *Geophys. J. Roy. Astron. Soc.*, **80**, 265–273.

Wolf, D. (1987). An upper bound on lithosphere thickness from glacio-isostatic adjustment in Fennoscandia, *J. Geophys.*, **61**, 141–149.

Wolfe, C. J. and Solomon, S. C. (1998). Shear-wave splitting and implications for mantle flow beneath the MELT region of the East Pacific Rise. *Science*, **280**, 1230–1238.

Wood, B. J. (1995). The effect of H2O on the 410-kilometer seismic discontinuity. *Science,* **268**, 74–76.

Woodhouse, J. H. and Dziewonski, A. M. (1984). Mapping the upper mantle: Three-dimensional modeling of Earth structure by inversion of seismic waveforms. *J. Geophys. Res.* **89**, 5953–5986.

Wortel, R. and Spakman, W. (2000). Subduction and slab detachment in the Mediterranean-Carpathian region. *Science*, **290**, 1910–1917.

Wu, X., Ferguson, I. J., and Jones, A. G. (2002). Magnetotelluric response and geoelectric structure of the Great Slave Lake shear zone. *Earth Planet. Sci. Lett.*, **196**, 35–50.

Wyatt, B. A., Mitchell, M., Shee, S. R., *et al.* (2003). The Brockman Creek Kimberlite, East Pilbara, Australia. *Proc. 8 Int. Kimberlite Conf.*, FLA_0180.

Wyllie, P. J. (1995). Experimental petrology of upper-mantle materials, process and products. *J. Geodynam.*, **20**, 429–468.

Xu, Y. S., and Shankland, T. J. (1999). Electrical conductivity of orthopyroxene and its high pressure phases, *Geophys. Res. Lett.*, **26**, 2645–2648.

Xu, Y., Poe, B. T., and Shankland, T. J. (1998). Electrical Conductivity of Olivine, Wadsleyite, and Ringwoodite Under Upper Mantle Conditions. *Science,* **280**, 1415–1418.

Xu, Y., Shankland, T. J., and Duba, A. G. (2000a). Pressure effect on electrical conductivity of mantle olivine. *Phys. Earth Planet. Inter.*, **118**, 149–161.

Xu, Y. S., Shankland, T. J., and Poe, B. T. (2000b). Laboratory-based electrical conductivity of the Earth's mantle. *J. Geophys. Res.*, **105**, 27,865–27,875.

Xu, X., O'Reilly, S. Y., Griffin, W. L., and Zhou, X. (2000). Genesis of young lithospheric mantle in southeastern China; an LAM-ICPMS trace element study. *J. Petrol.*, **41**, 111–148.

Xu, Y. S., Shankland, T. J., Linhardt, S., *et al.* (2004). Thermal diffusivity and conductivity of olivine, wadsleyite and ringwoodite to 20 GPa and 1373 K. *Phys. Earth Planet. Inter.*, **143**, 321–336.

Yale, B. L. and Carpenter, S. J. (1998). Large igneous provinces and giant dyke swarms: proxies for supercontinent cyclicity and mantle convection. *Earth Planet Sci. Lett.*, **163**, 109–122.

Yamano, M., Foucher, J. P., Kinoshita, M., *et al.* (1992). Heat flow and fluid flow regime in the western Nankai accretionary prism. *Earth Planet. Sci. Lett.*, **109**, 451–462.

Yan, B., Graham, E. K., and Furlong, K. P. (1989). Lateral variations in upper mantle thermal structure inferred from three-dimensional seismic inversion models. *Geophys. Res. Lett.*, **16**, 449–453.

Yang, Y. J., Forsyth, D. W., and Weeraratne, D. S. (2007). Seismic attenuation near the East Pacific Rise and the origin of the low-velocity zone. *Earth Planet. Sci. Lett.*, **258**, 260–268.

Yegorova, T. P., Starostenko, V. I., and Kozlenko, V. G. (1998). Large-scale gravity analysis of the inhomogeneities in the European-Mediterranean upper mantle. *Pure Appl. Geophys.* **151**, 549–561.

York, D. (1978). A formula describing both magnetic and isotopic blocking temperatures. *Earth Planet. Sci. Lett.*, **39**, 89–93.

Yoshino, T., Matsuzaki, T., Yamashita, S., and Katsura, T. (2006). Hydrous olivine unable to account for conductivity anomaly at the top of the asthenosphere. *Nature*, **443**, 973–976, doi: 10.1038/nature05223.

Yoshizawa, K. and Kennett, B. L. N. (2004). Multimode surface wave tomography for the Australian region using a three-stage approach incorporating finite frequency effects. *J. Geophys. Res.*, **109**, Article Number: B02310.

Yuan, H. and Romanowicz, B. (2010). Lithospheric layering in the North American craton. *Nature*, **466**, 1063–1168.

Yuen, DA (1982). Normal-modes of the viscoelastic Earth. *Geophysical J. Royal Astron. Soc.*, **69**: 495.

Zandt, G. and Ammon, C. J. (1995). Continental crust composition constrained by measurements of crustal Poisson's ratio. *Nature,* **374**, 152–154; doi:10.1038/374152a0.

Zang, A. (1993). Finite element study on the closure of thermal microcracks in feldspar/quartz rocks. I. Grain boundary cracks. *J. Geophys. Res.*, **113**, 17–31.

Zegers, T. E., de Wit, M. J., Dann, J. and White, S. H. (1998). Vaalbara, Earth's oldest assembled continent? A combined structural, geochronological, and palaeomagnetic test. *Terra Nova*, **10**, 250–259.

Zegers, T. E. and van Keken, P. E. (2001). Middle Archean continent formation by crustal delamination. *Geology*, **29**, 1083–1086.

Zeyen, H., Dererova, J. and Bielik, M. (2002). Determination of the continental lithospheric thermal structure in the Western Carpathians: integrated modelling of surface heat flow, gravity anomalies and topography. *Phys. Earth Plan. Int.*, **134**, 89–104.

Zhang, Y.-S. and Tanimoto, T. (1993). High-resolution global upper mantle structure and plate tectonics. *J. Geophys. Res.*, **98**, 9793–9823.

Zhao, W. J. and Nelson, K. D. (1993). Deep seismic-reflection evidence for continental underthrusting beneath Southern Tibet. *Nature*, **366**, 557–559.

Zhao, Y.-H., Zimmerman, M. E., and Kohlstedt, D. L. (2009). Effect of iron content on the creep behavior of olivine: 1. Anhydrous conditions. *Earth Planet. Sci. Lett.*, **287**, 229–240.

Zhao, M., Langston, C., Nyblade, A., and Owens, T. (1999). Upper mantle velocity structure beneath southern Africa from modelling regional seismic data. *J. Geophys. Res.* **104**, 4783–4794.

Zhao, G., Sun, M., Wilde, S. A., and Li, S. (2004). A Paleo-Mesoproterozoic supercontinent: assembly, growth and breakup. *Earth Sci. Rev.*, **67**, 91–123.

Zhao, D., Lei, J., Inoue, T., Yamada, A., and Gao, S. (2006). Deep structure and origin of the Baikal rift zone. *Earth Planet. Sci. Lett.*, **243**, 681–691.

Zhdanov, M. S., Golubev, N. G., Varentsov, I. M., *et al.* (1986) 2-D model fitting of a geomagnetic anomaly in the Soviet Carpathians. *Ann Geophys.*, **4**(B3), 335–342.

Zhou, H.-W., (1996). A high-resolution P wave model for the top 1200 km of the mantle. *J. Geophys. Res.*, **101**, 27791–27810.

Zhou, S. (1991). A model of thick plate deformation and its application to the isostatic movements due to surface, subsurface, and internal loadings. *Geophys. J. Int.*, **105**, 381–395.

Zhou, Y., Nolet, G., and Dahlen, F. A. (2003). Surface sediment effects on teleseismic P wave amplitude. *J. Geophys. Res.*, **108**, Article Number: 2417.

Zhou, Y., Nolet, G., Dahlen, F. A., and Laske, G. (2006). Global upper-mantle structure from finite-frequency surface-wave tomography, *J. Geophys. Res.*, **111**, B04304, doi:10.1029/2005JB003677.

Zhu, L. and Kanamori, H. (2000). Moho depth variation in southern California from teleseismic receiver functions. *J. Geophys. Res.*, **105**, 2969–2980.

Zielhuis, A. and Nolet, G. (1994). Deep seismic expression of an ancient plate boundary in Europe. *Science,* **265**, 79–81.

Zindler, A. and Jagoutz, E. (1988). Mantle cryptology. *Geochim. Cosmochim. Acta*, **52**, 319–333.

Zindler, A., Staudigel, H., Hart, S. R., Enders, R., and Goldstein, S. (1983). Nd and Sr isotopic study of a mafic layer from the Ronda ultramafic complex. *Nature,* **304**, 226–230.

Zorin, Yu. A. and Vladimirov, B. N. (1988). Thermal state of the lithosphere of the Siberian Platform and the problem of trapps origin. *Izv. AN SSSR, Geological Ser.*, 1988, No.**8**, 130–132 (in Russian).

Zoth, G. and Haenel, R. (1988). Thermal conductivity. In: R. Haenel *et al.* (eds.): *Handbook of Terrestrial Heat-Flow Density Determinations*, Kluwer Acad., 449–463.

Subject index

Printed in the United States
By Bookmasters